C000163226

ISBN 978-0-266-23395-4
PIBN 11029514

## No. 3.—MARCH, 1896.



## No. 4.—MAY, 1896.



## No. 5.—JULY, 1896.



## No. 6.—SEPTEMBER, 1896.

CONTENTS.



No. 7.—NOVEMBER, 1896.

# INDEX.

# BULLETIN

OF THE

# DEPARTMENT OF LABOR.

No. 1—NOVEMBER, 1895.

ISSUED EVERY OTHER MONTH.

EDITED BY

CARROLL D. WRIGHT,
COMMISSIONER.

OREN W. WEAVER,
CHIEF CLERK.

WASHINGTON:
GOVERNMENT PRINTING OFFICE.
1895.

# CONTENTS.

g the last session of the Fifty-third Congress Hon. Lawrence ann, chairman of the Committee on Labor of the House of tatives, introduced a bill (H. R. 8713) providing for the pub- of the Bulletin of the Department of Labor. This bill was to the Committee on Labor January 29, 1895, and February 1 mittee made the following report, which was committed to the ee of the Whole House on the state of the Union:

ommittee on Labor, to whom was referred House bill 8713, have same under consideration, and beg leave to report that the bill that the Commissioner of Labor shall publish a bulletin of the ent of Labor, at intervals not to exceed two months, contain- ent facts as to the condition of labor in this and other coun- ndensations of state and foreign labor reports, facts as to the of employment, and such other facts as may be deemed of the industrial interests of the country.

lowing communication from the Commissioner of Labor, Hon. D. Wright, sets forth potent arguments in favor of the passage ll:

DEPARTMENT OF LABOR,
*Washington, D. C., February 1, 1895.*

Sir: I have the honor to acknowledge the receipt of your letter of yes- ing a copy of bill (H. R. 8713) providing for the publication of the the Department of Labor, with suggestion that you would like my views

I have to say that I have very carefully examined, not only the bill, to me to be fully adequate for the purpose for which it is intended, but for which it provides. The Department of Labor is authorized by its to publish an annual report, and also such special reports as may be either by the Commissioner of Labor or in response to resolutions of of Congress or a request of the President, and it has in the past fully these provisions, sending to Congress annually a report relating to extensive investigation, and also various special reports not requir- such as the annual reports.

Department would have the right to make a special report at regular to do so it would need a larger appropriation than that now bill, therefore, supplies this lack, and further, it would enable bring out a regular bulletin without the necessity of delay in . After very careful consideration of the whole subject I fully the purpose of the bill, especially as foreign gov- precisely what your bill aims to accomplish. The English department of labor, which was established only recently, is now publishing, very successfully and with great acceptance to the industrial interests of the country, a labor gazette. The French department of labor does the same thing, and so, too, does that of New Zealand, and now the Russian government, which has recently established a department of labor, is publishing a gazette. It would seem right, therefore, that the United States, which has been the pioneer of labor departments in the world, should publish a bulletin.

This would have been done before, I presume, had it not been for the suggestion that such bulletins should contain information relative to the lack of labor in different parts of the country. I believe that all now agree that such announcements from an official source would do more harm than good, and therefore the movement has never taken shape; but the publication contemplated by your bill avoids this particular, and to my mind objectionable, feature of a bulletin, and with this objection removed I think it would be greatly for the interest of the industries of this country that such a bulletin should be established.

Should you look for precedents in our own government, you will find them in the Department of Agriculture, the Geological Survey, and the Bureau of Education. All of these offices, while not publishing bulletins at regular intervals, publish them quite frequently, and they are of very great use.

I think our Department is now so constituted that it could bring out at least bimonthly the bulletin contemplated by your bill, and fill its pages with most useful facts relative to the condition of labor in this and other countries—facts which do not naturally and would not generally come within the scope of an annual report. Here would be the great use and great advantage of the bulletin. The annual reports must necessarily be the results of patient and laborious investigation. The bulletins would contain more fragmentary matter, but yet of vital importance. As I read your bill, it is not contemplated that a bulletin should contain theoretical matter or introduce discussions on debatable questions, nor should it become the organ of any propaganda, but its whole function is to be confined to the collection and publication of current but important facts.

The increased expense would be so small that I should not suppose that would stand in the way of the passage of the bill. I am very glad to see that your committee has reported it favorably, and I hope it will secure the favorable action of Congress.

I am, very respectfully,

CARROLL D. WRIGHT,
*Commissioner.*

Hon. LAWRENCE E. MCGANN, M. C.,
*Chairman Committee on Labor, House of Representatives.*

Your committee therefore recommend that the bill be passed.

The bill which Mr. McGann introduced, and which the committee reported favorably, provided for a bulletin at intervals not to exceed two months and not to exceed 100 octavo pages; and to contain current facts as to the condition of labor in this and other countries, condensations of state and foreign labor reports, facts as to conditions of employment, and such other facts as may be deemed of value to the industrial interests of the country. This bill passed the House of Representatives February 26, 1895, and was favorably reported in the Senate, but instead of its passing the Senate as a bill, it was incorporated, in an abbreviated form, as a provision in the act making the appropriations for the Department of Labor, as follows:

The Commissioner of Labor is hereby authorized to prepare and publish a bulletin of the Department of Labor, as to the condition of labor in this and other countries, condensations of state and foreign labor reports, facts as to conditions of employment, and such other facts as may be deemed of value to the industrial interests of the country, and there shall be printed one edition of not exceeding ten thousand copies of each issue of said bulletin for distribution by the Department of Labor.

This amendment was accepted by the [illegible]
it was approved March 2, 1895. It [illegible]

...Bulletin of the Department of Labor

...authorization under the legislative appro... different from that contained in the bill as it ... bill there were limitations as to issue and ... law as it stands contains no limitations nor ... the size of the bulletin or the intervals at which ... the only condition being that not more than 10,000 ...sue of the bulletin shall be printed. Notwithstanding ... authorization, we feel it right and just to ... way, to the terms embodied in the House bill. ...re undertake to limit the size of the bulletin to about ... at present, to issue it every other month. The ... will guide us in the preparation of the bulletin are ... the letter to the chairman of the House Committee ...oted. We need not, therefore, make any restatement

... is to have at least five regular departments of infor... ..., as follows:

...l portion of each issue to be occupied with the results ...tigations conducted by the Department or its agents.

...est of foreign labor reports.

...t of state labor reports.

...production, immediately after their passage, of new ... the interests of the working people whenever such are ... legislatures or Congress; also the reproduction of the ...ts interpreting labor laws or passing upon any subject ...e relations of employer and employee; attention like...d to any other matters pertaining to law which may ...l value to the industrial interests of the country and ... be obtained without expense or trouble from other

...llaneous department, in which brief statements of fact ... interest may find a place.

... special investigations, the results of which are to ...lletin, it may be sometimes that such results will take ... the bulletin. The endeavor, however, will be to pre...r departments, as a rule, as just stated, departing ...hen the importance of the facts to be published war...ture.

...ill not be devoted in any way to controversial mat...tion of theories, nor used in any sense for propagan... undertake to present all the matters in an attractive ...ard way, and while statistical tables will have to be ...ly, the aim will still be to give proper space to read... ...are very many questions constantly coming up on which information can not be secured except by inquiry at original sources. Such questions we hope to be able to consider whenever they arise, and to give the results a place in the bulletin.

We shall not attempt in any way to compete with the press, but in general our aim will be to furnish to the public facts and information relating to industrial affairs which can not readily be secured in any other way. So, merely ephemeral matters will not be given a place in the pages of the bulletin, but those matters which have a more or less permanent value and which will take their place in the industrial history of the country will be treated. Readers of the bulletin, therefore, will not look for accounts of passing events, unless such accounts are necessary for future use. In other words, all those matters which are dealt with fully and comprehensively by the press of the country as the days go by ought not to be and will not be used to fill up the pages of the bulletin. The field for the bulletin is wide enough without making it in any sense a newspaper.

The Department now has three channels of communication with the public. By its organic law it is authorized to make an annual report, and special reports when called upon by Congress or by the President or when considered expedient by the head of the Department, and now this more popular way of disseminating information by means of a regularly published bulletin. The annual reports will, as heretofore, consist of the results of investigations which require a large force and considerable time. They are in a sense scientific productions, and can not legitimately be brought to a popular basis in any broad sense. The special reports authorized by the organic law of the Department are those resulting from more thoroughly individual investigations, those where but one or two persons can economically work upon one subject. The annual reports are the results of inquiries made by the schedule system and where any number of people can be employed. The special reports are studies of conditions where the schedule system can not be so generally applied. The bulletin, as against the annual or the special reports, will contain such matters as can not in the nature of things find a place in the annual or special reports; but it is confidently expected that through the bulletin the Department will be able to bring much of its work closer home to the people.

The editors will take personal supervision of the preparation of the bulletin, and it will be their aim to constantly elevate its standard.

## ND LOCKOUTS IN THE UNITED STATES FROM JANUARY 1, 1881, TO JUNE 30, 1894.

nnual Report of the Commissioner of Labor, entitled ›ckouts, furnished tables covering the details of all outs occurring in the United States for the six years January 1, 1881, and ending with December 31, 1886, ummaries recapitulating the facts shown therein. The Report (soon to be printed) is a volume of about 1,200 g of similar tables and summaries for the strikes and occurred during the seven and one-half years beginning , 1887, and ending with June 30, 1894, being modeled on own in the former report.

›ral tables relating to strikes and lockouts in the Tenth furnish the facts in detail for each strike and lockout days' duration which occurred in the United States , 1887, to June 30, 1894. In addition to the strikes and ng within the above period the report shows the facts for and lockouts which occurred in the latter part of 1886, omitted from the Third Annual Report because of the at that time of the data relating to them. A com l number of disturbances of less than one day's dura l, have been excluded from consideration in these tables. ainly of cases of misunderstanding, in which there was ' cessation of work and no financial loss or assistance this reason full information concerning them could red, and they have not been considered sufficiently classed as strikes.

Annual Report it was found necessary to make the he unit in the tabular presentation, and not the strike nerally each line there represented either a strike or a gle establishment, or a general strike or lockout in two hments; but there were some instances where the facts ated. In the Tenth Annual Report experience and a care have made it possible to make the strike or lock all cases.

the increase or diminution of strikes during the years Third and Tenth Annual Reports on this subject may be determined, the following table, showing the number of strikes in each year from January 1, 1881, to June 30, 1894, is presented:

STRIKES BY YEARS, JANUARY 1, 1881, TO JUNE 30, 1894.

| Year. | Strikes. | Establishments. | Average establishments to a strike | Employees thrown out of employment. |
|---|---|---|---|---|
| 1881 | 471 | 2,928 | 6.2 | 129,521 |
| 1882 | 454 | 2,105 | 4.6 | 154,671 |
| 1883 | 478 | 2,759 | 5.8 | 149,763 |
| 1884 | 443 | 2,367 | 5.3 | 147,054 |
| 1885 | 645 | 2,284 | 3.5 | 242,705 |
| 1886 | 1,432 | 10,053 | 7.0 | 508,044 |
| 1887 | 1,436 | 6,589 | 4.6 | 379,726 |
| 1888 | 906 | 3,506 | 3.9 | 147,704 |
| 1889 | 1,075 | 3,786 | 3.5 | 249,559 |
| 1890 | 1,833 | 9,424 | 5.1 | 351,944 |
| 1891 | 1,718 | 8,117 | 4.7 | 299,064 |
| 1892 | 1,298 | 5,540 | 4.3 | 206,671 |
| 1893 | 1,305 | 4,555 | 3.5 | 265,914 |
| 1894 (6 months) | 896 | 5,154 | 5.8 | 482,066 |
| Total | 14,390 | 69,167 | 4.8 | 3,714,406 |

The figures for the years from 1881 to 1886, inclusive, have been taken from the Third Annual Report. As stated in that report, the figures showing the number of strikes in each of these years are estimates, although they are believed to be approximately correct. For the period covered by the Tenth Annual Report, namely, January 1, 1887, to June 30, 1894, inclusive, the figures showing the number of strikes may be accepted as absolute. The figures showing the number of establishments and the number of employees thrown out of employment by strikes may be accepted as correct for the whole period from 1881 to 1894, inclusive. In using this table it should be borne in mind that the figures for 1894 are for the first six months of that year only, the investigation having been closed June 30, 1894.

By this table it is shown that the average number of establishments to each strike for the thirteen and one-half years was 4.8, the highest average being 7 establishments to each strike in 1886, the lowest average being 3.5 establishments to each strike in 1885, 1889, and 1893. As stated in the Third Annual Report, the strikes for 1880 were reported by Mr. Joseph D. Weeks, special agent of the Tenth Census, according to whose report the number was 610. The number of establishments involved was not reported. Commencing with 1881 the number of establishments involved was 2,928. In 1882 the number dropped to 2,105, while in 1883 it rose to 2,759, or nearly that of 1881. In 1884 and 1885 the number fell rapidly, there being 2,367 in 1884, while in 1885 the number of establishments involved in strikes was smaller than in any previous or succeeding year of the period, namely, 2,284. In 1886 the number rose to 10,053, the greatest number in any of the years considered. In 1887 it dropped to 6,589; in 1888 it dropped still further, to 3,506, and remained nearly stationary in [illegible], while in 1890 the number again rapidly rose to [illegible] great as that for 1886. In the next year, 1891, the [illegible] to 8,117, dropping still further in [illegible]

of 1894 the number was 5,154,
proportionately large number in the last
it would reach in round numbers 10,300, a
that for 1886, in which the largest number
involved in strikes.

establishments involved in strikes during the
of thirteen and one-half years was 69,167. Of this num-
ent had strikes in 1881, 3.04 per cent had strikes in 1882,
had strikes in 1883, 3.42 per cent had strikes in 1884, 3.30
strikes in 1885, 14.53 per cent had strikes in 1886, 9.53
strikes in 1887, 5.07 per cent had strikes in 1888, 5.47 per
kes in 1889, 13.63 per cent had strikes in 1890, 11.74 per
tes in 1891, 8.01 per cent had strikes in 1892, 6.59 per cent
1893, and 7.45 per cent had strikes in the first half of

7 establishments having lockouts during the period of
one-half years 0.15 per cent were in 1881, 0.69 per cent
1.93 per cent were in 1883, 5.83 per cent were in 1884, 3.02
in 1885, 24.87 per cent were in 1886, 21.11 per cent were
per cent were in 1888, 2.18 per cent were in 1889, 5.34 per
1890, 9 per cent were in 1891, 11.80 per cent were in 1892,
vere in 1893, and 6.08 per cent were in the first half of 1894.
ge is highest for both strikes and lockouts in 1886. The
percentages occur in 1890 and 1891 for strikes, and in 1887
lockouts.

seven and one-half years included in the Tenth Annual
s shows the largest number of establishments affected, both
l lockouts, there being 10,060 of the former and 1,193 of the
come New York, with 9,540 establishments involved in
23 in lockouts, and Pennsylvania with 8,219 involved in
0 in lockouts. During the six years immediately preceding
l in this report, the facts for which appeared in the Third
rt, the state in which the greatest number of establish-
ffected by strikes was New York, with 9,247, followed by
,768, and Pennsylvania, with 2,442. The greatest number
ckouts was 1,528, found in New York, followed by 147 in
s and 130 in Pennsylvania, the number in Illinois being
ing the facts for both these periods, in order to secure a
the thirteen and one-half years included in both of the
Department on strikes and lockouts, we find the greatest
establishments affected by strikes to have been in New
followed by Illinois, with 12,828, and Pennsylvania, with
states appear in the same order in lockouts, the number
affected being 2,251 in New York, 1,320 in Illinois,
Pennsylvania.

affected by strikes during the seven and one-
in the Tenth Annual Report were the building

trades, with 20,785 establishments involved; coal and coke, with 5,958; clothing, with 3,041; tobacco, with 2,506; food preparations, with 2,398; stone quarrying and cutting, with 1,993; metals and metallic goods, with 1,834; transportation, with 1,327; printing and publishing, with 608; boots and shoes, with 607; furniture, with 459; wooden goods, with 409, and brick, with 406 establishments. The industries most affected by lockouts were the building trades, with 1,900; stone quarrying and cutting, with 489; clothing, with 431; brewing, with 150; boots and shoes, with 130; metals and metallic goods, with 128, and transportation with 112 establishments involved. For the immediately preceding period of six years, 1881 to 1886, included in the Third Annual Report, the greatest frequency of strikes was found in the building trades, with 6,075 establishments affected; tobacco, with 2,959; mining (practically the same as coal and coke in the Tenth Annual Report), with 2,060; clothing, with 1,728; metals and metallic goods, with 1,570; transportation, with 1,478; food preparations, with 1,419; furniture, with 491; cooperage, with 484; brick, with 478; stone quarrying and cutting, with 468; lumber, with 395; boots and shoes, with 352; wooden goods, with 240, and printing and publishing, with 223 establishments. The lockouts for that period involved 773 establishments in the clothing industry, 531 in the building trades, 226 in the tobacco industry, 155 in boots and shoes, 76 in metals and metallic goods, etc.

A combination of the facts for strikes for the two periods, selecting the 13 industries most largely affected, shows that out of 69,167 establishments involved in strikes during the period from January 1, 1881, to June 30, 1894, 62,038, or 89.69 per cent, were in the following 13 industries: Building trades, 26,860 establishments; coal and coke, 8,018; tobacco, 5,465; clothing, 4,769; food preparations, 3,817; metals and metallic goods, 3,454; transportation, 2,805; stone quarrying and cutting, 2,461; boots and shoes, 959; furniture, 950; brick, 884; printing and publishing, 831, and cooperage, 765.

In the lockouts which occurred during the thirteen and one-half years, six industries bore a very large proportion of the burden, involving 4,914 establishments, or 81 per cent, out of a total of 6,067 establishments. The industries and number of establishments involved in each are as follows: Building trades, 2,431; clothing, 1,204; stone quarrying and cutting, 513; boots and shoes, 285; tobacco, 277, and metals and metallic goods, 204.

The total number of employees involved or thrown out of employment in the whole number of strikes from 1881 to 1886, inclusive, as shown by the Third Annual Report, was 1,323,203. The number as shown by the Tenth Annual Report, for the period from January 1, 1887, to June 30, 1894, was 2,391,203. Adding these numbers together, it is seen that 3,714,406 persons were thrown out of employment by reason of strikes during the period of thirteen and one-half years [illegible] 1, 1881, to June 30, 1894. The number of [illegible] years of this period was shown to have been [illegible].

[illegible]. The number of strikers [illegible] and one-half years was therefore [illegible] there were 103,068 new employees [illegible] 37,483 were brought from other [illegible] the strikes occurred. The per cent of new [illegible] total number of employees before strike— [illegible]. The per cent of the employees brought [illegible] of new employees after strike was 36.38. [illegible] and one-half years, the period involved in [illegible] there were 239,431 new employees after the [w]hich 115,377 were brought from other places. For this [pe]r cent of new employees after strike of the total number [employee]s before strike, 4,300,410, was 5.57, and the per cent of [b]rought from other places of the number of new employees 48.19. Combining the facts for both periods, it is seen [that th]ere, during the thirteen and one-half years, 342,460 new [employees e]ngaged after the strikes, and that 152,860 of that number [were brough]t from other places. The new employees after the strikes [were ... pe]r cent of the total number of employees before the strikes, [wh]ile 44.63 per cent of the new employees after the strikes [were brough]t from other places than those in which the strikes occurred.

[In the Th]ird Annual Report it was shown that during the period [... to] 1886, inclusive, 2,214 establishments were involved in lock[outs, emb]racing 175,270 employees in the establishments before the [lockouts oc]curred, while the number actually involved or locked out [was ...]. There were 13,976 new employees secured at the close [...], 5,682 being brought from other places than those in [which the l]ockouts occurred. For the period of seven and one-half [years cover]ed in the Tenth Annual Report lockouts were ordered in [... establi]shments, having 274,657 employees before the lockouts, of [whom ...]7 were thrown out of employment in consequence thereof. [These establ]ishments secured 27,465 new employees after the lockouts, [of w]hom were brought from other places. Combining these [facts for lo]ckouts for the two periods involved, it is seen that during [the thirteen] and one-half years from January 1, 1881, to June 30, 1894, [lockouts occ]urred in 6,067 establishments in which 449,927 employees [were involve]d. Of this number 366,690, or 81.50 per cent, were thrown [out of emplo]yment by the lockouts. In these establishments there were [... new e]mployees engaged after the lockouts, of whom 21,982 were [brought fro]m other places than those in which the lockouts occurred. [The per cent o]f new employees after the lockouts of the total number [of employees] before lockouts was, therefore, 9.21, and of employees [brought from o]ther places of the number of new employees after lock-[outs ...]

[It should be re]membered in considering the figures relating to the [number of establi]shments, the number of employees, etc., that they [do not represent the] actual number of different individual establishments or different individual employees who were involved in strikes or lockouts in a given industry or in a given year, because in many cases there have been two or more strikes or lockouts in the same establishments in the same year, and in such cases the establishment and the number of employees are duplicated or triplicated, as the case may be, in the totals derived by addition. In the figures showing the number of "employees for whom strike was undertaken" in the tables for strikes there is even more duplication of the kind mentioned. For instance, a sympathetic strike may occur in which the employees strike to enforce the demands of certain employees in another establishment. The number of employees for whom the strike was undertaken, would in that case be the number for whom it was undertaken in that other establishment. The same employees would, therefore, appear in that column in two places in the primary tables, first in connection with the establishment in which they were employed, and second in the establishment in which the sympathetic strike occurred, thus unavoidably being duplicated in tables derived by addition.

As previously stated, a small number of strikes occurring in 1886, 21 in all, which were unavoidably omitted from the Third Annual Report, have been tabulated in the later one. Wherever the facts shown by the two reports are given by years this number and the various facts relating thereto have been added to the figures for 1886 as shown by the Third Annual Report. In the statements previously made, by states and by industries, they have not been so added, but appear in the totals for the period involved in the later report. The number is so small as to make no appreciable difference when comparing the two reports, and to have eliminated them in the later and added them to the Third Annual Report would have involved the reader in many difficulties.

The following table, classifying the employees involved in strikes and lockouts as to sex, combines the facts shown in the Third Annual Report with those shown in the Tenth Annual Report:

SEX OF EMPLOYEES THROWN OUT OF EMPLOYMENT, JANUARY 1, 1881, TO JUNE 30, 1894.

| Year. | Strikes. | | | Lockouts. | | |
|---|---|---|---|---|---|---|
| | Employees thrown out of employment. | Males (per cent). | Females (per cent). | Employees thrown out of employment. | Males (per cent). | Females (per cent). |
| 1881 | 129,521 | 94.08 | 5.92 | 655 | 83.21 | 16.79 |
| 1882 | 154,671 | 92.15 | 7.85 | 4,131 | 93.80 | 6.20 |
| 1883 | 149,763 | 87.66 | 12.34 | 20,512 | 73.58 | 26.42 |
| 1884 | 147,054 | 88.78 | 11.22 | 18,121 | 78.93 | 21.07 |
| 1885 | 242,705 | 87.77 | 12.23 | 15,424 | 83.77 | 16.23 |
| 1886 | 508,044 | 86.17 | 13.83 | 101,980 | [illegible] | [illegible] |
| 1887 | 379,726 | 91.77 | 8.23 | 59,630 | 95.76 | 4.24 |
| 1888 | 147,704 | 91.50 | 8.50 | 15,176 | 79.53 | 20.47 |
| 1889 | 249,559 | 90.48 | 9.52 | 18,731 | 72.91 | [illegible] |
| 1890 | 351,944 | 90.53 | 9.47 | 21,555 | [illegible] | [illegible] |
| 1891 | 299,064 | 94.90 | 5.10 | 31,014 | [illegible] | [illegible] |
| 1892 | 206,671 | 93.57 | 6.43 | [illegible] | [illegible] | [illegible] |
| 1893 | 265,914 | 93.06 | 6.94 | 21,842 | [illegible] | [illegible] |
| 1894 (6 months) | 482,066 | 95.13 | 4.87 | [illegible] | [illegible] | [illegible] |
| Total | 3,714,406 | [illegible] | [illegible] | [illegible] | [illegible] | [illegible] |

[illegible] the six years immediately pre- [illegible] in the Third Annual Report that [illegible] in the country, so far as strikes [illegible] of all the establishments in the coun- [illegible], were found in the same states. [illegible] these facts for the entire period of thir- [illegible] in the Third and Tenth Annual [illegible] for each year for the five states named:

[illegible] ILLINOIS, MASSACHUSETTS, NEW YORK, OHIO, [illegible], JANUARY 1, 1881, TO JUNE 30, 1894.

| [illegible] | | Lockouts. | | |
|---|---|---|---|---|
| [illegible] in the five selected states. | Per cent of establishments in the five selected states. | Total establishments in the United States. | Establishments in the five selected states. | Per cent of establishments in the five selected states. |
| 2,[illegible] | 72.97 | 9 | 4 | 44.44 |
| [illegible] | 71.21 | 42 | 23 | 54.76 |
| [illegible] | 74.16 | 117 | 105 | 89.74 |
| [illegible] | 80.10 | 354 | 306 | 86.44 |
| [illegible] | 69.44 | 183 | 140 | 76.50 |
| [illegible] | 75.35 | 1,509 | 1,403 | 92.98 |
| [illegible],751 | 72.36 | 1,281 | 1,188 | 92.74 |
| 2,[illegible] | 63.57 | 180 | 114 | 63.33 |
| 2,375 | 60.09 | 132 | 65 | 49.24 |
| [illegible] | 74.17 | 324 | 203 | 62.65 |
| [illegible],776 | 71.16 | 546 | 339 | 62.09 |
| [illegible] | 57.76 | 716 | 522 | 72.91 |
| [illegible] | 69.95 | 305 | 190 | 62.30 |
| [illegible] | 72.99 | 369 | 304 | 82.38 |
| [illegible],210 | 71.15 | 6,067 | 4,906 | 80.86 |

[illegible] contained 51 per cent of all the manufacturing [illegible] employed 56 per cent of the capital invested in the [illegible] of the United States, taking the census of 1890 [illegible]ation.

[illegible] strikes and lockouts, by cities, during the seven [illegible] January 1, 1887, to June 30, 1894, is shown in

## SUMMARY OF STRIKES IN THE PRINCIPAL CITIES, JANUARY 1, 1887, TO JUNE 30, 1894.

[In the case of many general strikes extending through different cities it was found impossible to subdivide the facts and credit them to the several cities involved. In such cases the whole strike has generally been tabulated against the city most largely affected.]

| City. | Total strikes. | Establishments. | Employees thrown out of employment. | Wage loss of employees. | Assistance to employees by labor organizations. | Loss of employers. |
|---|---|---|---|---|---|---|
| New York, N. Y | 2,614 | 6,467 | 215,649 | $6,449,385 | $798,817 | $3,545,786 |
| Brooklyn, N. Y | 671 | 1,271 | 31,768 | 914,045 | 145,848 | 582,780 |
| Chicago, Ill | 528 | 8,325 | 282,611 | 8,846,494 | 1,895,788 | 14,444,034 |
| Boston, Mass | 257 | 911 | 25,574 | 800,893 | 173,564 | 589,992 |
| Allegheny and Pittsburg, Pa | 251 | 4,142 | 160,823 | 7,379,785 | 722,706 | 2,599,457 |
| Philadelphia, Pa | 240 | 1,122 | 89,527 | 2,002,219 | 194,277 | 898,568 |
| Saint Louis, Mo | 111 | 1,064 | 19,693 | 848,357 | 96,506 | 572,983 |
| Cincinnati, Ohio | 109 | 580 | 17,577 | 796,806 | 78,[illegible] | 572,272 |
| Milwaukee, Wis | 100 | 1,237 | 29,778 | 1,265,049 | 112,962 | 799,700 |
| Lynn, Mass | 100 | 110 | 4,027 | 147,086 | 9,871 | 88,489 |
| Fall River, Mass | 96 | 156 | 30,232 | 500,264 | 22,429 | 118,319 |
| San Francisco, Cal | 92 | 237 | 7,254 | 480,387 | 96,854 | 415,625 |
| Baltimore, Md | 92 | 280 | 11,192 | 424,149 | 18,604 | 187,558 |
| New Haven, Conn | 82 | 205 | 5,287 | 206,340 | 26,586 | 40,568 |
| Newark, N. J | 69 | 324 | 11,538 | 500,896 | 58,734 | 154,460 |
| Cleveland, Ohio | 64 | 814 | 11,322 | 306,738 | 36,324 | 117,207 |
| Rochester, N. Y | 56 | 237 | 9,314 | 478,702 | 11,781 | 300,621 |
| Indianapolis, Ind | 51 | 309 | 7,851 | 116,429 | 10,353 | 161,102 |
| Haverhill, Mass | 51 | 76 | 5,271 | 97,239 | 6,660 | 78,495 |
| Minneapolis, Minn | 50 | 169 | 7,615 | 167,524 | 18,399 | 189,400 |
| Paterson, N. J | 47 | 117 | 22,326 | 1,019,768 | 26,757 | 555,200 |
| Buffalo, N. Y | 46 | 408 | 14,079 | 459,758 | 19,950 | 818,015 |
| Jersey City, N. J | 46 | 113 | 7,819 | 90,020 | 1,330 | 12,275 |
| Saint Paul, Minn | 45 | 255 | 22,475 | 780,825 | 24,520 | 1,017,795 |
| Troy, N. Y | 42 | 123 | 3,649 | 68,031 | 3,760 | 39,802 |
| Total | 5,909 | 28,662 | 955,250 | 34,988,100 | 4,590,177 | 28,786,446 |

## SUMMARY OF LOCKOUTS IN THE PRINCIPAL CITIES, JANUARY 1, 1887, TO JUNE 30, 1894.

[In the case of many general lockouts extending through different cities it was found impossible to subdivide the facts and credit them to the several cities involved. In such cases the whole lockout has generally been tabulated against the city most largely affected.]

| City. | Total lockouts. | Establishments. | Employees thrown out of employment. | Wage loss of employees. | Assistance to employees by labor organizations. | Loss of employers. |
|---|---|---|---|---|---|---|
| New York, N. Y | 42 | 393 | 19,959 | $587,801 | $83,112 | $370,442 |
| Boston, Mass | 19 | 126 | 3,072 | 212,424 | 40,450 | 97,111 |
| Chicago, Ill | 18 | 1,151 | 48,612 | 3,576,817 | 70,050 | 2,7[illegible],910 |
| Allegheny and Pittsburg, Pa | 15 | 385 | 11,572 | 5,[illegible],764 | [illegible] | 737,[illegible] |
| Philadelphia, Pa | 14 | 82 | 9,262 | 447,958 | 62,565 | 510,575 |
| Cincinnati, Ohio | 13 | 96 | 2,906 | 211,375 | 37,568 | [illegible] |
| San Francisco, Cal | 12 | 42 | 776 | 67,783 | 12,179 | 18,300 |
| Haverhill, Mass | 10 | 57 | 7,436 | 101,[illegible] | 5,990 | 63,490 |
| Saint Paul, Minn | 10 | 55 | 1,056 | 20,780 | 8,393 | 46,150 |
| Brooklyn, N. Y | 9 | 64 | 2,300 | 68,424 | 6,[illegible] | 241,225 |
| Saint Louis, Mo | 8 | 42 | 1,006 | 217,247 | 45,249 | 48,140 |
| Milwaukee, Wis | 7 | 35 | 752 | 946,755 | 12,275 | 596,[illegible] |
| Minneapolis, Minn | 7 | 7 | 1,650 | 28,250 | 2,[illegible] | 28,100 |
| Indianapolis, Ind | 6 | 116 | 1,135 | 66,224 | 900 | 20,[illegible] |
| Rochester, N. Y | 6 | 81 | 18,271 | 462,280 | 3,150 | 296,545 |
| Richmond, Va | 6 | 15 | 117 | 10,508 | 2,274 | 650 |
| Buffalo, N. Y | 5 | 30 | 930 | 72,435 | 395 | 13,670 |
| Seattle, Wash | 5 | 23 | 423 | 19,600 | 2,[illegible] | 4,040 |
| Detroit, Mich | 5 | 11 | 1,264 | 56,201 | 14,842 | 3,500 |
| Springfield, Mass | 5 | 8 | 208 | 1,230 | ........ | 11,758 |
| New Haven, Conn | 5 | 8 | 64 | 3,344 | 340 | 15,735 |
| Baltimore, Md | 5 | 6 | 230 | 7,240 | 3,[illegible] | 7,[illegible] |
| Newark, N. J | 4 | 40 | 2,974 | 130,[illegible] | 13,[illegible] | [illegible] |
| Woburn, Mass | 4 | 19 | 1,577 | 76,[illegible] | 4,[illegible] | [illegible] |
| Albany, N. Y | 3 | 77 | 1,812 | 62,[illegible] | ........ | 12,[illegible] |
| Total | 244 | 2,970 | 149,135 | 12,108,[illegible] | [illegible] | 8,[illegible] |

In the case of both strikes and lockouts the cities [illegible] in which the greatest number of these disturbances [illegible] the period included in the report. [illegible]

[illegible] of disturbances, only 6 cities in

[illegible], it is seen that out of a total of [illegible] country, 5,909, or 56.34 per cent, occurred [illegible] in that table. The number of establishments [illegible] the United States during the period was shown [illegible] number 23,662, or 61.16 per cent, occurred in the 26 [illegible] loss to employees through strikes in the 26 cities was [illegible] $111,993,143 for the entire country, and the loss [illegible] as against $51,888,833. These 26 cities con- [illegible] er cent of all the manufacturing establishments, and [illegible] per cent of the capital invested in the mechanical indus- [illegible] ited States, taking the census of 1890 as the basis of [illegible] of the 26 cities, New York, Brooklyn, Chicago, [illegible], Pittsburg, and Philadelphia, reported 4,561 strikes, [illegible] of all the strikes which occurred in the United States [illegible] riod involved, and 22,248 establishments, or 47.47 per [illegible] ole number of establishments involved.

[illegible] establishments involved in strikes during the six years [illegible] Third Annual Report (1881 to 1886), as was there shown, [illegible], or 82.24 per cent of the whole, were ordered by [illegible] tions, while of the 2,214 establishments in which lock- [illegible],753, or 79.18 per cent, were ordered by combinations of [illegible] e facts for the seven and one-half years included in the [illegible] Report (January 1, 1887, to June 30, 1894) are as follows: [illegible] consideration seven strikes for which no report touch- [illegible] could be secured, 7,295, or 69.60 per cent of the whole [illegible] kes (10,481), were ordered by labor organizations, while [illegible] outs occurring during this period but 81, or 18.33 per [illegible] red by an employers' organization. It will be noticed, in [illegible] statement, that for the former period the establishment [illegible] s of the percentages, while for the latter the strike or [illegible] the basis. This is unavoidable, owing to the difference [illegible] n of the facts for this point in the two reports. It does [illegible] naterially affect the comparableness of the percentages. [illegible] mind, the facts for each year in this respect may be [illegible] the percentage table which follows:

[illegible] LOCKOUTS ORDERED BY ORGANIZATIONS, JANUARY 1, 1881, TO JUNE 30, 1894.

| [illegible] | Strikes (per cent). | Lockouts (per cent). | Year. | Strikes (per cent). | Lockouts (per cent). |
|---|---|---|---|---|---|
| [illegible] | 75.80 | 22.22 | 1888 | 68.14 | 20.00 |
| [illegible] | 78.01 | 26.10 | 1889 | 67.35 | 11.11 |
| [illegible] | [illegible] | 41.03 | 1890 | 71.33 | 11.06 |
| [illegible] | 82.85 | 79.10 | 1891 | 71.84 | 13.04 |
| [illegible] | 79.93 | 71.58 | 1892 | 70.72 | 22.95 |
| [illegible] | 87.23 | 84.89 | 1893 | 69.13 | 21.43 |
| [illegible] | [illegible] | 25.37 | 1894 (6 months) | 63.80 | 14.29 |

Combining the facts for the period involved in the Third Annual Report with those for the period included in the later report, the following table shows by years, in the form of percentages, the proportion of the establishments involved in both strikes and lockouts which were closed in consequence of such disturbance:

ESTABLISHMENTS CLOSED, JANUARY 1, 1881, TO JUNE 30, 1894.

| Year. | Strikes (per cent). | Lockouts (per cent). | Year. | Strikes (per cent). | Lockouts (per cent). |
|---|---|---|---|---|---|
| 1881 | 55.81 | 33.33 | 1889 | 61.89 | 59.09 |
| 1882 | 54.01 | 59.52 | 1890 | 56.25 | 63.89 |
| 1883 | 63.57 | 58.12 | 1891 | 56.66 | 65.93 |
| 1884 | 64.72 | 37.85 | 1892 | 65.60 | 66.90 |
| 1885 | 71.58 | 79.23 | 1893 | 65.64 | 40.98 |
| 1886 | 58.24 | 67.93 | 1894 (6 months) | 60.50 | 14.91 |
| 1887 | 57.55 | 83.84 | | | |
| 1888 | 53.45 | 55.00 | Average | 59.56 | 63 90 |

Referring to the Third Annual Report, it is seen that from 1881 to 1886, inclusive, of the 22,304 establishments subjected to strikes, 13,411, or 60.13 per cent, were temporarily closed, and of the 2,214 establishments in which lockouts occurred, 1,400, or 63.23 per cent, were closed. The duration of stoppage, or the average days closed, for strikes was 23 days and for lockouts 28.4 days. The facts as shown in the Tenth Annual Report for the seven and one-half years from January 1, 1887, to June 30, 1894, are that of 46,863 establishments subjected to strikes 27,787, or 59.29 per cent, were closed, 42 being closed permanently or having strikes still pending June 30, 1894, the remainder being only temporarily closed; while of the 3,853 establishments involved in lockouts 2,477, or 64.29 per cent, were closed, 23 being closed permanently or having lockouts still pending June 30, 1894, the remainder being only temporarily closed. The average days closed on account of strikes, excluding the 42 above mentioned, was 22.3 days, and on account of lockouts, excluding the 23 above mentioned, 35.4 days.

Combining the facts for the two periods, as shown by the preceding figures, it is seen that during the thirteen and one-half years from January 1, 1881, to June 30, 1894, out of a total of 69,167 establishments in which strikes occurred, 41,198, or 59.56 per cent, were closed, while of the 6,067 establishments subjected to lockouts, 3,877, or 63.90 per cent, were closed. The duration of stoppage, or days closed, in the 41,156 establishments which were temporarily closed, was 22.5 days, while in the 3,854 establishments temporarily closed by reason of lockouts the average time closed was 32.8 days.

The duration of strikes or lockouts themselves—that is, the average length of time which elapsed before the establishments resumed operations and were running normally, either by reason of the strikers or employees locked out having returned to work or by their places having been filled by others—applies to all establishments, [illegible] or not, and differs of course from the figures given [illegible] entire stoppage of work, which applies only to [illegible]

[illegible] included in the Third and [illegible] number of establishments which were [illegible] strikes or lockouts, or in which [illegible] pending, have of course been omitted in

[illegible] LOCKOUTS, JANUARY 1, 1881, TO JUNE 30, 1894.

[illegible] from date of strike or lockout to date when employees [illegible] their places were filled by others.]

| [illegible] | Lockouts. | |
|---|---|---|
| [illegible] | [illegible] | Average duration (days). |
| [illegible] | 3 | 22.2 |
| [illegible] | 4[illegible] | 1[illegible].0 |
| [illegible] | 117 | 67.5 |
| [illegible] | [illegible] | 41.4 |
| [illegible] | [illegible] | 27.1 |
| [illegible] | 1,[illegible] | [illegible].1 |
| [illegible] | 1,[illegible] | [illegible].8 |
| [illegible] | 1[illegible] | 74.8 |

| Year. | Strikes. | | Lockouts. | |
|---|---|---|---|---|
| | Establishments. | Average duration (days). | Establishments. | Average duration (days). |
| 1889 | 3,786 | 26.3 | 132 | 57.5 |
| 1890 | 9,424 | 24.2 | 324 | 73.9 |
| 1891 | 8,117 | 34.9 | 546 | 37.8 |
| 1892 | 5,540 | 23.4 | 716 | 72.0 |
| 1893 | 4,555 | 20.6 | 305 | 34.7 |
| 1894 (6 months) | 5,154 | 37.8 | 3[illegible] | 18.7 |
| Total | 69,167 | 25.4 | 6,067 | 47.6 |

[illegible] Third Annual Report, for the years 1881 to 1886, [illegible] whom strikes were instituted 46.52 per cent granted [illegible] their employees; in 13.47 per cent of the establish[illegible] in attaining the objects for which the strikes were [illegible], while failure followed in 39.95 per cent of the [illegible] small number of establishments, constituting 0.06 [illegible] whole number, had strikes still pending December 31, [illegible] outs during those years the firms gained their point [illegible] the establishments; in 8.58 per cent they partially [illegible] per cent failed; in 5.47 per cent of the whole [illegible] ishments involved the lockouts were still pending

[illegible] included in the Tenth Annual Report, out of the [illegible] establishments affected by strikes, viz, 46,863, success [illegible] gained by the employees in 20,397 establishments, [illegible] partial success was gained in 4,775 establishments, or [illegible] failure followed in 21,687 establishments, or 46.28 [illegible] whole number; for 4 establishments, or 0.01 per cent, [illegible] were not reported or the strikes were still pending [illegible] of the 3,853 establishments having lockouts, 1,883, [illegible] the whole number, succeeded in gaining their [illegible] per cent, partially succeeded, and 1,558, or 40.44 [illegible] establishments, or 0.54 per cent of the whole [illegible] were still pending June 30, 1894. The percentages for each of the years included in the two reports are shown as follows:

RESULTS FOR ESTABLISHMENTS, JANUARY 1, 1881, TO JUNE 30, 1894.

| Year. | Per cent of establishments in strikes which— | | | Per cent of establishments in lockouts which— | | |
|---|---|---|---|---|---|---|
| | Succeeded | Succeeded partly. | Failed. | Succeeded | Succeeded partly. | Failed. |
| 1881 | 61.37 | 7.00 | 31.63 | 88.89 | 11.11 | ........ |
| 1882 | 53.59 | 8.17 | 38.24 | 64.29 | .......... | 35.71 |
| 1883 | 58.17 | 16.09 | 25.74 | 56.41 | .......... | 43.59 |
| 1884 | 51.50 | 3.89 | 44.61 | 27.97 | .28 | 71.75 |
| 1885 | 52.80 | 9.50 | 37.70 | 38.25 | 3.28 | 58.47 |
| 1886 | a 34.45 | a 18.82 | a 46.58 | b 19.48 | b 12.06 | b 60.44 |
| 1887 | 45.64 | 7.19 | 47.17 | 34.19 | 1.25 | 64.56 |
| 1888 | 52.22 | 5.48 | 42.30 | 74.44 | 3.89 | 21.67 |
| 1889 | 46.49 | 18.91 | 34.60 | 40.91 | 25.76 | 33.33 |
| 1890 | c 52.64 | c 10.01 | c 37.34 | 65.74 | 5.56 | 28.70 |
| 1891 | 37.87 | 8.29 | 53.84 | 63.92 | 14.29 | 21.79 |
| 1892 | 39.31 | 8.70 | 51.99 | 69.13 | 25.28 | 5.59 |
| 1893 | d 50.82 | d 10.32 | d 38.79 | e 39.02 | e 17.05 | e 37.05 |
| 1894 (6 months) | 23.83 | 15.66 | 60.51 | 21.95 | 1.36 | 76.69 |
| Total | f 44.49 | f 11.25 | f 44.23 | g 40.33 | g 9.58 | g 47.75 |

a Not including 15 establishments in which strikes were still pending December 31, 1886.
b Not including 121 establishments in which lockouts were still pending December 31, 1886.
c Not including 1 establishment not reporting.
d Not including 3 establishments in which strikes were still pending June 30, 1894.
e Not including 21 establishments in which lockouts were still pending June 30, 1894.
f Not including 19 establishments for the reasons stated in notes a, c, and d.
g Not including 142 establishments for the reasons stated in notes b and e.

For the thirteen and one-half years ending June 30, 1894, as shown by this table, out of a total of 69,167 establishments affected by strikes the employees were successful in gaining their demands in 30,772, or 44.49 per cent, and partly successful in 7,779, or 11.25 per cent, while in 30,597 establishments, or 44.23 per cent, they failed; in a very small number of establishments, constituting 0.03 per cent of all the establishments involved, the results of strikes were not obtainable. Of the 6,067 establishments in which lockouts occurred during the same period, the firms gained their point in 2,447 establishments, or 40.33 per cent of the whole number involved; in 581, or 9.58 per cent, they were partly successful, while in 2,897, or 47.75 per cent, they failed; in the remaining 142, or 2.34 per cent of the establishments, the results of the lockouts were not obtainable.

The results of strikes from 1881 to 1886, so far as they concerned employees, as shown in the Third Annual Report, were as follows: The number of persons thrown out of employment, in the 10,375 establishments having successful strikes, was 518,583; in the 3,004 establishments in which strikes were partly successful 143,976 employees were involved, while in the 8,910 establishments in which the strikes were failures 660,396 persons were thrown out of employment. The results of strikes in 15 establishments, involving 248 persons, were not reported. While the establishments in which strikes succeeded constituted 46.52 per cent of the establishments in which strikes occurred, the number of persons thrown out of employment in the successful [illegible] strikes was only 10.88 per cent of the whole [illegible]. The number of establishments in which strikes [illegible] per cent of the whole number, while 49.91 per [illegible] persons thrown out of employment were involved [illegible] number of establishments in which the results of [illegible] constituted 0.06 per cent of the entire num- [illegible] persons thrown out of employment in such estab- [illegible] 0.02 per cent of the entire number of persons thrown [illegible].

[illegible] for the succeeding seven and one-half years, from Jan- [illegible] June 30, 1894, so far as they concerned employees, as [illegible]nth Annual Report, are as follows: In the 20,397 estab- [illegible]g successful strikes 669,992 persons were thrown out [illegible]; in the 4,775 establishments in which strikes were [illegible]l 318,801 employees were involved, while in the 21,687 [illegible] in which strikes failed, 1,400,988 persons were thrown [illegible]ent. The results were not reported in 19 establish- [illegible] 1,422 persons were involved. While the establishments [illegible]s succeeded constituted 43.52 per cent of the establish- [illegible] strikes occurred, the number of persons thrown out of [illegible] the successful strikes constituted 28.02 per cent of the [illegible] of persons involved; the number of establishments [illegible]tly successful strikes was 10.19 per cent of all establish- [illegible]e number of persons involved in such strikes was 13.33 [illegible] whole number. The number of establishments in which [illegible]as 46.28 per cent of the whole number, while 58.59 per [illegible]ole number of persons thrown out of employment were [illegible]ch strikes. In 0.01 per cent of the entire number of [illegible], including 0.06 per cent of the number of persons thrown [illegible]ent, the results of strikes were not reported.

The following table combines the facts for the two reports, showing the results, so far as employees are concerned, for the strikes during an uninterrupted period of thirteen and one-half years, beginning January 1, 1881, and ending June 30, 1894:

RESULTS OF STRIKES FOR EMPLOYEES, JANUARY 1, 1881, TO JUNE 30, 1894.

| Year. | Number thrown out of employment. | | | | Per cent thrown out of employment. | | |
|---|---|---|---|---|---|---|---|
| | In successful strikes. | In partly successful strikes. | In strikes which failed. | In total strikes. | In successful strikes. | In partly successful strikes. | In strikes which failed. |
| 1881 | 55,600 | 17,482 | 56,439 | 129,521 | 42.93 | 13.50 | 43.57 |
| 1882 | 45,746 | 7,112 | 101,813 | 154,671 | 29.58 | 4.60 | 65.82 |
| 1883 | 55,140 | 17,024 | 77,599 | 149,763 | 36.82 | 11.37 | 51.81 |
| 1884 | 52,736 | 5,044 | 89,274 | 147,054 | 35.86 | 3.43 | 60.71 |
| 1885 | 115,375 | 23,855 | 103,475 | 242,705 | 47.54 | 9.83 | 42.63 |
| 1886 | a 196,400 | a 74,167 | a 238,229 | 508,044 | a 38.46 | a 14.60 | a 46.90 |
| 1887 | 127,629 | 26,442 | 225,655 | 379,726 | 33.61 | 6.96 | 59.43 |
| 1888 | 41,106 | 11,130 | 95,468 | 147,704 | 27.83 | 7.54 | 64.63 |
| 1889 | 72,099 | 62,607 | 114,853 | 249,559 | 28.89 | 25.09 | 46.02 |
| 1890 | b 158,787 | b 48,444 | b 144,681 | 351,944 | b 45.12 | b 13.76 | b 41.11 |
| 1891 | 80,766 | 22,885 | 196,413 | 299,064 | 27.01 | 7.65 | 65.34 |
| 1892 | 61,125 | 16,429 | 129,117 | 206,671 | 29.58 | 7.95 | 62.47 |
| 1893 | c 62,018 | c 41,765 | c 160,741 | 265,914 | c 23.32 | c 15.71 | c 60.45 |
| 1894 (6 months) | 65,048 | 88,391 | 328,627 | 482,066 | 13.49 | 18.34 | 68.17 |
| Total | d 1,188,575 | d 462,777 | d 2,061,384 | 3,714,406 | d 32.06 | d 12.46 | d 55.50 |

a Not including 248 engaged in strikes still pending December 31, 1886.
b Not including 32 engaged in strikes not reporting result.
c Not including 1,390 engaged in strikes still pending June 30, 1894.
d Not including 1,670 for the reasons stated in the preceding notes.

The totals as given in this table show that the number of persons thrown out of employment in the 30,772 establishments having successful strikes was 1,188,575. In the 7,779 establishments in which partial success was gained 462,777 employees were involved, while in the 30,597 establishments in which strikes failed 2,061,384 persons were thrown out of employment. The last three columns of the table show for each year, and for the thirteen and one-half years, the per cent of employees in establishments in which the strikes succeeded, partly succeeded, or failed. Taking the total for the period of thirteen and one-half years, it is seen that 32 per cent of the whole number of persons thrown out of employment succeeded in gaining the object for which they struck; 12.46 per cent succeeded partly, while 55.50 per cent, or over half of the whole number, failed entirely in gaining their demands. A small proportion of the whole number, 0.04 per cent, for the various reasons stated in the notes to the table, made no report as to the result.

The Third Annual Report shows that for the years included therein (1881 to 1886) seventeen of the causes for which strikes were undertaken included 90.28 per cent of all the establishments, leaving the remaining 297 causes operative in only 9.72 per cent of establishments in which strikes occurred. Even four leading causes were found to cover 77.16 per cent of the establishments. The following table was given as clearly bringing out these facts:

| [illegible] | | |
|---|---|---|
| [illegible] | 9,439 | 42.32 |
| [illegible] | 4,344 | 19.48 |
| [illegible] | 1,734 | 7.77 |
| [illegible] | 1,[illegible] | 7.[illegible] |
| [illegible] to board with employer | [illegible] | [illegible] |
| [illegible] | [illegible] | 1.[illegible] |
| [illegible] | 238 | 1.07 |
| [illegible] of union men | 215 | .[illegible] |
| [illegible] | 173 | .77 |
| [illegible] | 172 | .77 |
| [illegible], etc. | 162 | .[illegible] |
| [illegible] rules | 145 | .[illegible] |
| [illegible] | 143 | .[illegible] |
| [illegible] | 138 | .[illegible] |
| [illegible] of union [illegible] rules | 132 | .[illegible] |
| [illegible] | 126 | .[illegible] |
| [illegible], firemen, etc. | 124 | .[illegible] |
| [illegible] | 20,136 | [illegible] |
| [illegible] | 2,168 | [illegible] |
| [illegible] | 22,304 | 100.00 |

[illegible] the causes for which strikes were undertaken [illegible] seven and one-half years included in the Tenth [illegible] that the seventeen principal causes included [illegible] the establishments, leaving the remaining 574 [illegible] 18.77 per cent of the establishments subjected to [illegible]. Five of the leading causes included a very [illegible] establishments, the per cent being 61.42 of the [illegible]. The following table, showing the number and [illegible] falling under each of the seventeen principal [illegible] period of seven and one-half years involved in this [illegible] percentages in detail:

[illegible]SES OF STRIKES, JANUARY 1, 1887, TO JUNE 30, 1894.

| Cause or object. | Establishments. | Per cent. |
|---|---|---|
| [illegible] | 12,041 | 25.09 |
| [illegible] | 6,199 | 13.23 |
| [illegible] | 3,830 | 8.17 |
| [illegible] | 3,620 | 7.73 |
| [illegible] of hours | 3,005 | 6.60 |
| [illegible] union men | 1,688 | 3.60 |
| [illegible] | 1,559 | 3.33 |
| [illegible] | 1,314 | 2.80 |
| [illegible] | 844 | 1.80 |
| [illegible] union scale | 783 | 1.67 |
| [illegible] of union | 686 | 1.46 |
| [illegible] to employ none but union men in building trades | 472 | 1.01 |
| [illegible] employees | 468 | 1.00 |
| [illegible] | 383 | .82 |
| [illegible] of hours on Saturday | 378 | .81 |
| [illegible] with employer, and for reduction of hours and [illegible] | 366 | .78 |
| [illegible] | 342 | .73 |
| [illegible] | 38,068 | 81.23 |
| [illegible] | 8,795 | 18.77 |
| [illegible] | 46,863 | 100.00 |

One of the most important features of the tabulation is the statement of the losses of the employees and of the employers by reason of strikes and lockouts. These figures were collected with the greatest possible care, and although in many cases only an estimate could be secured the results as given are believed to be a very close approximation to the exact losses. It is natural to suppose that after the lapse of several years exact figures could not be secured concerning facts of which no record is kept in most instances. The figures here given are for the immediate, and in many instances only temporary, losses of employees and employers. In most businesses there are seasons of entire or partial idleness among its employees, owing to sickness, volnotary lay-offs, running slack time, etc., the working days per year being on an average from 200 to 250 days out of a possible 313. When a strike or lockout occurs in an establishment whose business is of such a character it is often followed by a period of unusual activity, in which the employee and employer both make up the time lost by reason of the temporary cessation of business on account of the strike.

The employer may in some instances be subjected to an ultimate loss by reason of his inability to fill contracts already made, but it may be accepted as a fact that much of the loss in the cases of both employer and employee is only temporary. It was found impossible, however, for the agents of the Department to take these facts into consideration, inasmuch as in many instances a period of six months or even a year must have elapsed before the whole or even a part of such loss was made up. The computation of wage loss has, therefore, been based on the number of employees thrown out of employment, their average wages, and the number of working days which elapsed before they were reemployed or secured work elsewhere. The amounts representing employers' losses are the figures (in most cases, estimates) furnished by the firms themselves, the Department's agents being instructed to consider, as well as they could, their probable correctness. In the summaries by years the figures can not represent absolute accuracy for a given year, because many strikes beginning in one year ended in another; the entire loss and assistance, as well as the other facts included in the tabulation, have been placed in the year in which the strike or lockout began. These differences may, however, counterbalance each other, and the reported results thus be nearly accurate.

Bearing in mind, then, the difficulties in ascertaining the exact losses of employees and employers as a result of strikes and lockouts, reference may be had to the following table showing the amount of loss to employees and to employers and the amount of assistance granted employees by their labor organizations for a period of thirteen and one-half years from January 1, 1881, to June 30, 1894.

| [illegible] | [illegible] Assistance to employees by labor organizations. | Loss of employers. | To date when employees locked out were reemployed or employed elsewhere. | | Loss of employers. |
|---|---|---|---|---|---|
| | | | Wage loss of employees. | Assistance to employees by labor organizations. | |
| [illegible] | [illegible] | $2,919,483 | $16,619 | $3,150 | [illegible] |
| [illegible] | [illegible] | 4,[illegible],[illegible] | 466,345 | 47,[illegible] | 112,[illegible] |
| [illegible] | [illegible] | 4,[illegible],027 | 1,0[illegible],312 | 102,253 | 297,087 |
| [illegible] | [illegible],671 | 3,[illegible],072 | 1,421,410 | 314,027 | 660,847 |
| [illegible] | [illegible],237 | 4,[illegible],[illegible] | 901,173 | 89,488 | 455,477 |
| [illegible] | 1,1[illegible],180 | 12,357,606 | 4,281,058 | 549,452 | 1,940,466 |
| [illegible] | 1,1[illegible],554 | 6,[illegible],496 | 4,[illegible],700 | 146,848 | 2,819,738 |
| [illegible] | 7,[illegible] | 6,809,017 | 1,100,057 | 65,981 | 1,217,199 |
| [illegible] | [illegible],617 | 2,[illegible],753 | 1,379,722 | 115,289 | 307,125 |
| [illegible] | [illegible],265 | 5,125,404 | 957,968 | 77,210 | 464,256 |
| [illegible] | 1,[illegible],557 | 6,177,286 | 888,709 | 50,195 | 616,[illegible] |
| [illegible] | [illegible],574 | 5,145,691 | 2,856,013 | 537,684 | 1,[illegible],080 |
| [illegible] | [illegible],188 | 2,466,195 | 6,650,401 | 264,268 | 1,034,420 |
| [illegible] | [illegible] | 15,557,166 | 457,221 | 31,737 | 596,484 |
| [illegible] | 10,914,406 | 82,590,386 | 26,685,516 | 2,524,298 | 12,235,451 |

[illegible]ployees in the establishments in which strikes occurred, [illegible] thirteen and one-half years, was $163,807,866; the [illegible]s through lockouts for the same period was $26,685,516; [illegible] employees by reason of these two classes of industrial [illegible] $190,493,382. The number of establishments involved [illegible]g this period was 69,167, making an average loss of [illegible]oyees in each establishment in which strikes occurred. [illegible] persons thrown out of employment by reason of strikes [illegible]making an average loss of $44 to each person involved. [illegible] establishments involved in lockouts was 6,067, making [illegible] of $4,398 to employees in each establishment in which [illegible]ed, while the number of employees locked out was [illegible] an average loss of $73 to each person involved. Com[illegible]es for strikes and lockouts, it is seen that the wage loss [illegible] above stated was $190,493,382 and the number of [illegible] involved 75,234, while 4,081,096 persons were thrown [illegible]nt. These figures show an average wage loss of $2,532 [illegible]s in each establishment and an average loss of $47 to [illegible]lved.

[illegible]e given to strikers during the thirteen and one-half [illegible] ascertainable, was $10,914,406; to those involved in [illegible],298, or a total sum of $13,438,704. This sum repre[illegible] per cent of the total wage loss incurred in strikes and [illegible] probably too low. In addition to this sum, which [illegible]stance from labor organizations, much assistance was [illegible] sympathizers, the amount of which the Depart[illegible] ascertaining.

[illegible]ployers through strikes during this thirteen and one-[illegible] to $82,590,386; their losses through lockouts [illegible], making a total loss to the establishments or [illegible] and lockouts during this period of $94,825,837.

## STRIKES AND LOCKOUTS IN GREAT BRITAIN AND IRELAND IN RECENT YEARS.

Since 1888 the statistical and other information concerning labor disturbances in Great Britain and Ireland has been published in the annual reports of the Labor Department of the Board of Trade under the title of Reports by the Chief Labor Correspondent on Strikes and Lockouts, and the information presented herewith has been obtained from those reports. The report for 1888, being the first, is not so comprehensive as those for subsequent years; for this reason, and also to enable a uniform presentation of the various facts, the report for the year 1889 is taken as the starting point, the figures being shown as far as practicable for each year up to and including 1893.

The number of strikes reported for each year is shown in the following statement:

STRIKES, 1889 TO 1893.

| Division. | 1889. | 1890. | 1891. | 1892. | 1893. | Total. |
|---|---|---|---|---|---|---|
| England | 813 | 716 | 667 | 512 | 509 | 3,217 |
| Wales | 53 | 88 | 63 | 52 | 48 | 304 |
| Scotland | 246 | 156 | 125 | 110 | 175 | 812 |
| Ireland | 33 | 68 | 38 | 18 | 36 | 193 |
| Total | 1,145 | 1,028 | 893 | 692 | 768 | 4,526 |

In counting the number of strikes that occurred in 1889, wherever full details were obtained of separate establishments engaged in a general strike, each establishment was considered as one strike. It was not always possible, however, to obtain full details for all the separate establishments affected by a general strike. In those instances a large number of establishments were counted in the annexed table as only one strike. Owing to the difficulty of ascertaining the actual number of establishments affected and of distinguishing between the number of distinct strikes and the number of establishments involved, the system was changed for 1890 and subsequent years so that each strike, whether general or merely local, was counted as one, irrespective of the number of establishments affected. Under these circumstances it can not be inferred that the strike movement in 1890 was not as violent as in 1889, as the above table seems to indicate. A more accurate comparison for the two years may be made by saying that in 1889 there were 3,164 distinct establishments affected by the 1,145 strikes, but the system of enumeration then adopted was not so clear as in 1890, when 4,382 distinct establishments were reported, [illegible]

ferred to in connection with strikes:

KOUTS, 1889 to 1893.

| Number. | Year. | Number. |
|---|---|---|
| 66 | 1892 | 8 |
| 12 | 1893 | 14 |
| 13 | Total | 113 |

ause or object of strikes and their success
re importance and interest than those on
bject. Space will not permit a detailed
causes and objects of strikes as presented
grouping adopted in the following state-
ry with this office:

RIKES BY CAUSES, 1889 TO 1893.

| | Year. | Succeeded. | Succeeded partly. | Failed. | Not reported. | Total. |
|---|---|---|---|---|---|---|
| combined | 1889 | 342 | 290 | 76 | 60 | 768 |
| | 1890 | 208 | 152 | 109 | 45 | 514 |
| | 1891 | 149 | 74 | 68 | 26 | 317 |
| | 1892 | 125 | 59 | 59 | 17 | 260 |
| | 1893 | 116 | 60 | 68 | 12 | 256 |
| ........ | ........ | 940 | 635 | 380 | 160 | 2,115 |
| ame com- | 1889 | 12 | 8 | 20 | 5 | 45 |
| | 1890 | 34 | 14 | 30 | 5 | 83 |
| | 1891 | 52 | 31 | 32 | 14 | 129 |
| | 1892 | 40 | 19 | 50 | 15 | 124 |
| | 1893 | 72 | 45 | 67 | 14 | 198 |
| ........ | ........ | 210 | 117 | 199 | 53 | 579 |
| of prices. | 1889 | 20 | 10 | 4 | 2 | 36 |
| | 1890 | 23 | 4 | 10 | 5 | 42 |
| | 1891 | 17 | 8 | 12 | 1 | 38 |
| | 1892 | 4 | 3 | 3 | 3 | 13 |
| | 1893 | 6 | 3 | 7 | 2 | 18 |
| ........ | ........ | 70 | 28 | 36 | 13 | 147 |
| of hours, rrespond- | 1889 | (a) | (a) | (a) | (a) | (a) |
| | 1890 | 10 | 6 | 5 | 2 | 23 |
| | 1891 | 14 | 6 | 3 | ........ | 23 |
| | 1892 | 7 | 4 | 2 | 1 | 14 |
| | 1893 | 6 | 1 | 2 | 1 | 10 |
| ........ | ........ | b 37 | b 17 | b 12 | b 4 | b 70 |
| , subcon- | 1889 | c 78 | c 40 | c 57 | c 5 | c 180 |
| | 1890 | 57 | 36 | 59 | 12 | 164 |
| | 1891 | 87 | 34 | 60 | 14 | 195 |
| | 1892 | 52 | 21 | 58 | 10 | 141 |
| | 1893 | 39 | 23 | 48 | 5 | 115 |
| ........ | ........ | c 313 | c 154 | c 282 | c 46 | c 795 |

1889 for the cause immediately following.
rably combined with those of the same year for the cause

*immediately preceding.*

RESULTS OF STRIKES BY CAUSES, 1889 TO 1893—Concluded.

| Cause or object. | Year. | Succeeded. | Succeeded partly. | Failed. | Not reported. | Total. |
|---|---|---|---|---|---|---|
| Against employment of nonunion men, and for adoption or enforcement of union rules, etc. | 1889 | 5 | 2 | 17 | 5 | 29 |
| | 1890 | 30 | 4 | 56 | 10 | 100 |
| | 1891 | 24 | 5 | 50 | 5 | 84 |
| | 1892 | 24 | 3 | 27 | 5 | 59 |
| | 1893 | 32 | 9 | 32 | 1 | 74 |
| Total | ....... | 115 | 23 | 182 | 26 | 346 |
| Disputes between classes of work people as to work, wages, etc | 1889 | 9 | 4 | 3 | 2 | 18 |
| | 1890 | 6 | 7 | 12 | ....... | 25 |
| | 1891 | 12 | 7 | 10 | 2 | 31 |
| | 1892 | 7 | 4 | 5 | 2 | 18 |
| | 1893 | 18 | 11 | 14 | 2 | 45 |
| Total | ....... | 52 | 33 | 44 | 8 | 137 |
| Defense of or objection to fellow work people (apart from unionism) | 1889 | 7 | 6 | 12 | 4 | 29 |
| | 1890 | 9 | 1 | 23 | 2 | 35 |
| | 1891 | 10 | 11 | 14 | 1 | 36 |
| | 1892 | 14 | 4 | 19 | 2 | 39 |
| | 1893 | 5 | 6 | 15 | 2 | 28 |
| Total | ....... | 45 | 28 | 83 | 11 | 167 |
| Defense of or objection to superior officials | 1889 | 3 | 1 | 11 | ....... | 15 |
| | 1890 | 2 | 4 | 5 | 2 | 13 |
| | 1891 | 3 | 3 | 6 | 4 | 16 |
| | 1892 | 10 | ....... | 4 | 1 | 15 |
| | 1893 | 8 | ....... | 5 | 1 | 14 |
| Total | ....... | 26 | 8 | 31 | 8 | 73 |
| In sympathy with other strikes and disputes | 1889 | ....... | 7 | 5 | 8 | 20 |
| | 1890 | 4 | 1 | 12 | 2 | 19 |
| | 1891 | ....... | 1 | 4 | 2 | 7 |
| | 1892 | ....... | ....... | 1 | 1 | 2 |
| | 1893 | 1 | 1 | 6 | 2 | 10 |
| Total | ....... | 5 | 10 | 28 | 15 | 58 |
| Cause not known | 1889 | ....... | ....... | 2 | 3 | 5 |
| | 1890 | 1 | 1 | 1 | 7 | 10 |
| | 1891 | 1 | 1 | 4 | 11 | 17 |
| | 1892 | ....... | ....... | ....... | 7 | 7 |
| | 1893 | ....... | ....... | ....... | ....... | ....... |
| Total | ....... | 2 | 2 | 7 | 28 | 39 |
| All causes | 1889 | 476 | 368 | 207 | 94 | 1,145 |
| | 1890 | 384 | 230 | 322 | 92 | 1,028 |
| | 1891 | 369 | 181 | 263 | 80 | 893 |
| | 1892 | 283 | 117 | 228 | 64 | 692 |
| | 1893 | 303 | 159 | 264 | 42 | 768 |
| Total | ....... | 1,815 | 1,055 | 1,284 | 372 | 4,526 |

The vast majority of the labor troubles in the United Kingdom have their origin in disputes as to wages. Chiefly they are differences as to amount of wages, although sometimes they are disputes concerning the principle or mode of payment, or of altered systems of work affecting the amount or mode of payment. Considering the total for five years it appears that over half, or 59.5 per cent, of all the strikes were caused by questions concerning the advance or reduction of wages, and that of the strikes for this object 42.7 per cent were successful, 27.9 per cent partly successful, 21.5 per cent unsuccessful, and for 7.9 per cent the result was not reported. Of the total number of strikes for all purposes that occurred during the five years 40.1 per cent were successful, 23.3 per cent partly successful, 28.4 per cent unsuccessful, and for 8.2 cent the result was not reported.

In connection with the success or failure of strikes it is instructive to consider the number of persons affected. While the number of persons affected is not shown for all of the strikes reported, it is given for a sufficient number to indicate the relative number of persons affected by the disturbances that terminated successfully or otherwise for the workmen, and a summary of the totals for the different years is as follows:

PERSONS AFFECTED BY STRIKES, 1889 TO 1893, BY RESULTS.

[Persons affected means persons thrown out of work, whether actually striking or not.]

| Result. | Year. | Total strikes. | Strikes for which persons affected were reported | |
|---|---|---|---|---|
| | | | Number. | Persons affected. |
| Succeeded | 1889 | 476 | 304 | 93,524 |
| | 1890 | 384 | 275 | 213,867 |
| | 1891 | 369 | 289 | 68,247 |
| | 1892 | 283 | 235 | 48,852 |
| | 1893 | 303 | 271 | 400,141 |
| Total | ........ | 1,815 | 1,374 | 824,631 |
| Succeeded partly | 1889 | 368 | 274 | 177,476 |
| | 1890 | 230 | 188 | 66,029 |
| | 1891 | 181 | 156 | 98,127 |
| | 1892 | 117 | 103 | 113,414 |
| | 1893 | 159 | 148 | 155,249 |
| Total | ........ | 1,055 | 869 | 610,295 |
| Failed | 1889 | 207 | 171 | 40,472 |
| | 1890 | 322 | 254 | 101,902 |
| | 1891 | 263 | 212 | 92,763 |
| | 1892 | 228 | 203 | 70,978 |
| | 1893 | 264 | 233 | 76,430 |
| Total | ........ | 1,284 | 1,073 | 382,545 |
| Not reported | 1889 | 94 | 32 | 10,528 |
| | 1890 | 92 | 21 | 11,183 |
| | 1891 | 80 | 19 | 7,748 |
| | 1892 | 64 | 27 | 3,554 |
| | 1893 | 42 | 13 | 1,709 |
| Total | ........ | 372 | 112 | 34,722 |
| Aggregate | 1889 | 1,145 | 781 | 322,000 |
| | 1890 | 1,028 | 738 | 392,981 |
| | 1891 | 893 | 676 | 266,885 |
| | 1892 | 692 | 568 | 236,798 |
| | 1893 | 768 | 665 | 633,529 |
| Total | ........ | 4,526 | 3,428 | 1,852,193 |

Of the 4,526 strikes that occurred during the five years covered by this statement, particulars concerning the number of persons affected and the results were obtained for 3,428, or 75.7 per cent. These strikes affected 1,852,193 persons. The strikes that terminated successfully affected 44.5 per cent of the total number of persons; those that succeeded partly, 32.9 per cent; unsuccessful, 20.7 per cent, and those for which the result was not reported, 1.9 per cent. The successful and partly successful strikes combined affected 77.4 per cent of the total number of persons. The average number of persons affected by each of the successful or partly successful strikes was 640, by the unsuccessful strikes 357, and by the strikes for which definite information as to the result was not obtained 310.

The time over which industrial stoppages extend, when considered in connection with the number of persons affected, conveys an idea of the magnitude of the disturbances. The statistics on this subject for the different years are presented in the following statement:

DURATION OF STRIKES, 1889 TO 1893.

| Year. | Total strikes. | Strikes for which duration was reported. | | |
|---|---|---|---|---|
| | | Number. | Days of duration. | |
| | | | Number. | Average per strike. |
| 1889 | 1,145 | 840 | 15,100 | 18.0 |
| 1890 | 1,028 | 794 | 13,724 | 17.3 |
| 1891 | 893 | 687 | 16,528 | 24.1 |
| 1892 | 692 | 555 | 17,800 | 32.1 |
| 1893 | 768 | 575 | 16,927 | 29.4 |
| Total | 4,526 | 3,451 | 80,079 | 23.2 |

The number of really large strikes is shown by the following analysis: In 1891 there were 9 strikes, of those reporting the number of persons affected, in which 5,000 persons and upward were involved; 45 in which 1,000 to 5,000 persons were affected, and 622 in which less than 1,000 persons were affected. The Report by the Chief Labor Correspondent on the Strikes and Lockouts of 1892 reports for that year but 8 strikes and lockouts affecting 5,000 persons and upward, 34 affecting 1,000 to 5,000, and 530 affecting less than 1,000. In 1893, 10 strikes and lockouts involved 5,000 persons and upward, 31 from 1,000 to 5,000, and 638 less than 1,000.

The number of persons affected by labor disputes, and the duration of such disputes, though interesting in themselves, become more important when brought into relation with each other. This has been done for both strikes and lockouts in the statements which follow so as to show the average days of time lost by the persons affected.

TIME LOST AND PERSONS AFFECTED BY STRIKES AND LOCKOUTS, 1890 TO 1893.

[Persons affected means persons thrown out of work.]

| Year. | Total strikes and lockouts. | Strikes and lockouts for which both persons affected and lost time were reported. | | | |
|---|---|---|---|---|---|
| | | Number. | Persons affected. | Days of lost time. | |
| | | | | Number. | Average per person affected. |
| 1890 | 1,040 | [illegible] | [illegible] | [illegible] | [illegible] |
| 1891 | 906 | [illegible] | [illegible] | [illegible] | [illegible] |
| 1892 | 700 | [illegible] | [illegible] | [illegible] | [illegible] |
| 1893 | 782 | [illegible] | [illegible] | [illegible] | [illegible] |
| Total | 3,426 | [illegible] | [illegible] | [illegible] | [illegible] |

DURATION OF AND PERSONS AFFECTED BY STRIKES AND LOCKOUTS IN 1893, BY INDUSTRIES.

[Persons affected means persons thrown out of work.]

| Industries. | Total strikes and lock-outs. | Strikes and lockouts for which both persons affected and duration were reported | | | | | |
|---|---|---|---|---|---|---|---|
| | | Number. | Persons affected. | Days of duration. | | Days of lost time. | |
| | | | | Number. | Average per dispute. | Number. | Average per person affected |
| Building and furnishing trades, coach making and coopers | 198 | 155 | 19,976 | 5,882 | 37.9 | 866,971 | 43.4 |
| Clothing (including saddle and harness trade) | 82 | 59 | 10,266 | 1,885 | 31.9 | 204,513 | 19.9 |
| Domestic (a) | 25 | 16 | 5,529 | 501 | 31.3 | 388,569 | 70.3 |
| Labor (b) | 30 | 19 | 1,247 | 134 | 7.0 | 7,646 | 6.1 |
| Metal (including shipbuilding, engineering, etc.) | 136 | 113 | 29,662 | 3,802 | 33.6 | 863,578 | 29.1 |
| Mining and quarrying | 156 | 110 | 501,724 | 3,063 | 27.8 | 27,977,893 | 55.7 |
| Printing, paper, and book trades | 7 | 4 | 286 | 116 | 29.0 | 7,119 | 24.9 |
| Textile trades | 105 | 80 | 44,790 | 1,583 | 19.8 | 422,184 | 9.4 |
| Transport (land and water) | 43 | 30 | 14,489 | 370 | 12.3 | 466,589 | 32.2 |
| Total | 782 | 586 | 627,969 | 17,336 | 29.6 | 31,205,062 | 49.7 |

a Comprises food and drink preparation, tobacco, brush makers, and glass and pottery trades.
b Comprises chemical and gas workers, public cleansing, agricultural, general, unskilled, and female labor.

In the majority of cases the largest disputes in point of numbers were also those for which the duration was the longest. For this reason the average duration per dispute is considerably less than the average number of working days lost per person involved.

With one exception the preceding statements have presented the statistics by totals for years only. In the following summary the strikes and lockouts of the four years from 1890 to 1893 are arranged by general groups of trades. This statement shows the number of disturbances in each group, the number for which the persons affected were reported, and the number of persons affected by such strikes and lockouts.

PERSONS AFFECTED BY STRIKES AND LOCKOUTS, 1890 TO 1893, BY INDUSTRIES.

[Persons affected means persons thrown out of work. It will be noticed that the figures reported below do not agree in every case with the figures given on page 30. The explanation is not known.]

| Industries. | Total strikes and lockouts. | | | | Strikes and lockouts for which persons affected were reported. | | | | | | | | Average persons affected per dispute. |
|---|---|---|---|---|---|---|---|---|---|---|---|---|---|
| | | | | | Number. | | | | Persons affected. | | | | |
| | 1890. | 1891. | 1892. | 1893. | 1890. | 1891. | 1892. | 1893. | 1890. | 1891. | 1892. | 1893. | |
| Building trades | 117 | 149 | 140 | 170 | 83 | 123 | 115 | 152 | 12,558 | 25,229 | 18,175 | 17,738 | 156 |
| Chemical and gas works | 10 | 4 | 3 | 5 | 7 | 4 | 3 | 5 | 1,218 | 118 | 193 | 427 | 103 |
| Cabinetmaking and furniture trades | 18 | 18 | 9 | 20 | 15 | 11 | 7 | 15 | 2,142 | 317 | 312 | 366 | 65 |
| Clothing trades | 78 | 66 | 56 | 80 | 47 | 55 | 49 | 71 | 29,317 | 40,992 | 36,431 | 10,821 | 530 |
| Coach building and coopers | 5 | 9 | 6 | 8 | 3 | 8 | 5 | 6 | 200 | 689 | 477 | 2,495 | 175 |
| Domestic trades | 10 | 11 | 8 | 5 | 5 | 7 | 6 | 5 | 487 | 627 | 428 | 58 | 69 |
| Food, tobacco, and drink preparation | 21 | 18 | 12 | 9 | 18 | 17 | 12 | 9 | 3,704 | 3,271 | 1,510 | 549 | 161 |
| Glass and pottery trades | 11 | 12 | 8 | 11 | 6 | 10 | 7 | 10 | 3,070 | 3,534 | 20,369 | 5,211 | 975 |
| Labor (agricultural and general unskilled) | 29 | 18 | 19 | 25 | 21 | 11 | 12 | 17 | 2,293 | 1,967 | 1,031 | 956 | 102 |
| Leather and rubber trades | 11 | 5 | 5 | 2 | 9 | 3 | 5 | 2 | 498 | 169 | 717 | 30 | 74 |
| Metal trades (including shipbuilding) | 201 | 165 | 129 | 136 | 149 | 123 | 108 | 124 | 81,938 | 60,502 | 39,756 | 30,200 | 422 |
| Mining and quarrying | 101 | 132 | 109 | 156 | 80 | 96 | 96 | 133 | 140,292 | 51,427 | 120,386 | 506,182 | 2,072 |
| Paper, printing, and bookbinding trades | 11 | 20 | 7 | 7 | 8 | 14 | 7 | 7 | 366 | 1,291 | 709 | 381 | 76 |
| Textile trades | 241 | 217 | 187 | 105 | 183 | 164 | 117 | 89 | 42,035 | 44,837 | 102,723 | 45,274 | 425 |
| Transport | 164 | 62 | 41 | 43 | 105 | 42 | 35 | 34 | 72,875 | 22,499 | 12,878 | 15,569 | 630 |
| Theatrical employees | ...... | ..... | 2 | ..... | ..... | ..... | 2 | ..... | ........ | ....... | 700 | ....... | 350 |
| Total | 1,028 | 906 | 700 | 782 | 739 | 688 | 576 | 679 | 392,981 | 267,460 | 256,798 | 636,356 | 617 |

Considering the totals for the four years, the greatest number of disturbances are reported for the textile trades, while those involving the greatest number of employees are in mining and quarrying. The textiles rank second in the number of persons affected, the metal trades second in number of disturbances and third in number of persons affected, the building trades third in number of disturbances and fourth in number of persons affected, while mining and quarrying, the first in the number of persons, is fourth in number of disturbances. The magnitude of the average disturbance in the different trades is also indicated in the above statement by the average number of persons affected, the average being obtained from the totals of the four years. The disturbances in the mining and quarrying industries affected, on an average, the largest number of persons, and were followed in point of magnitude by the glass and pottery trades.

Having presented data as to the number, magnitude, and immediate results of strikes, the statistics next in order are those pertaining to the modes of settling the disputes. The different methods of settling

| | | | Number. | Persons affected. |
|---|---|---|---|---|
| [illegible] | 1891 | [illegible] | [illegible] | [illegible] |
| | 1892 | 341 | [illegible] | 55,7[illegible] |
| | 1893 | [illegible] | 247 | [illegible] |
| [illegible] | 1891 | 9 | 9 | [illegible] |
| | 1892 | 4 | 4 | 76,144 |
| | 1893 | 8 | 8 | [illegible] |
| [illegible] | 1891 | 18 | 15 | 12,207 |
| | 1892 | 19 | 17 | [illegible] |
| | 1893 | 26 | 24 | 12,134 |
| | 1891 | 130 | 103 | 65,724 |
| | 1892 | 115 | 108 | 37,224 |
| | 1893 | 119 | 104 | 63,676 |
| | 1891 | 87 | 67 | 6,149 |
| | 1892 | 79 | 66 | 2,729 |
| | 1893 | 104 | 90 | 4,273 |
| | 1891 | 41 | 36 | 20,249 |
| | 1892 | 23 | 30 | 21,[illegible] |
| | 1893 | 44 | 43 | 114,277 |
| [illegible]...ployed | 1891 | 11 | 9 | 1,927 |
| | 1892 | 13 | 11 | 1,748 |
| | 1893 | 6 | 6 | 803 |
| ...placed | 1891 | 39 | 37 | 30,410 |
| | 1892 | 22 | 20 | 3,891 |
| | 1893 | 22 | 20 | 1,977 |
| of cause of dispute without | 1891 | 12 | 9 | 2,268 |
| | 1892 | 2 | 2 | 112 |
| | 1893 | 3 | 3 | 425 |
| ...to settlement, or establish- | 1891 | 78 | 18 | 7,728 |
| | 1892 | 64 | 27 | 3,554 |
| | 1893 | 49 | 20 | 2,959 |

...ortion of the strikes for each of the three years ...ent were settled by conciliation. Next to con- ...umber of strikes, according to the classification ...e been settled by the submission of the work

...information concerning loss or gain resulting ...complete as could be desired, the information ...n considered in connection with the other sta- ...subject. According to returns received from ...to the cost of strikes in 1893, there were 257 ...affected (a) 139,168 persons whose weekly wages

...between those directly affected and those indirectly ...but it is believed that the former expression refers to ...to others thrown out of employment in consequence of

before the strikes amounted to $863,045, and 43 strikes which indirectly affected 18,714 persons whose weekly wages before the strikes amounted to $124,904. In 123 strikes, affecting 120,127 persons, an estimated fixed capital of $78,522,559 was laid idle, and 80 strikes, affecting 104,811 persons, laid idle property whose estimated ratable value was $3,416,911. In 109 strikes, affecting 135,230 persons, the estimated outlay by employers in stopping and reopening works and in payment of fixed charges and salaries was $1,676,354, and the cost to employers in resisting 6 strikes, affecting 8,487 persons, was $34,980.

Reports from trade unions relating to loss and gain from strikes indicate that in 1893 there were 265 strikes, affecting 239,898 persons, whose weekly wages before the strikes amounted to $1,260,107. Weekly wages both before and after the strikes were reported for 209 strikes, affecting 236,527 persons, whose weekly wages were $1,237,931 before and $1,287,554 after the strikes. In 73 strikes the weekly gain in wages to the 116,249 persons affected was $39,024, and in 21 strikes the weekly reduction to the 2,523 persons affected was $1,557. In 224 strikes, which affected 223,679 persons, the estimated wage loss during the strikes was $8,952,929, the amount expended by trade unions in support of 240 strikes affecting 92,608 persons was $617,457, and the amount expended from other than trade-union funds in support of 37 strikes, affecting 21,171 persons, was $119,701. In 313 strikes 88,940 of the number affected belonged to trade unions.

The statistics for lockouts have, of necessity, been included in some of the preceding statements presenting the data for strikes. In some instances it was practically impossible to obtain a separation of the persons affected by the lockout from those affected by the strike; therefore the statistics for lockouts as a distinct class of labor disturbances are not as complete as may be desired. The following statement gives the number of persons affected by and the duration of the lockouts in the United Kingdom, so far as reported, for the years 1891, 1892, and 1893:

DURATION OF AND PERSONS AFFECTED BY LOCKOUTS, 1891 TO 1893.

[Persons affected means persons thrown out of work.]

| Year. | Total lockouts. | Lockouts for which persons affected were reported. | | | Lockouts for which duration was reported. | | |
|---|---|---|---|---|---|---|---|
| | | Number. | Persons affected. | Average persons affected per lockout. | Number. | Days of duration. | Average days of duration per lockout. |
| 1891 | 13 | 11 | 575 | 52 | 6 | 224 | 37.3 |
| 1892 | 8 | 8 | 120,001 | 15,000 | 6 | 126 | 21.0 |
| 1893 | 14 | 14 | 2,857 | 204 | 21 | 460 | 21.9 |
| Total | 35 | 33 | 123,433 | 3,740 | 33 | 810 | 24.5 |

| | | | | | | |
|---|---|---|---|---|---|---|
| [illegible] wages and other demands | 1891 | 1 | 1 | 2 | | 4 |
| | 1892 | | | | | |
| | 1893 | 1 | 1 | | | 2 |
| [illegible] of wages | 1891 | 2 | | | | 2 |
| | 1892 | | 1 | | | 1 |
| | 1893 | 1 | | | | 1 |
| [illegible] of working arrange- [illegible] of payment, foreign mate- [illegible] | 1891 | 2 | | | | 2 |
| | 1892 | | 1 | 2 | 1 | 5 |
| | 1893 | 2 | 4 | 1 | | 7 |
| [illegible] of union or nonunion [illegible], etc. | 1891 | 4 | | 1 | | 5 |
| | 1892 | 1 | | | | 1 |
| | 1893 | 3 | 1 | | | 4 |
| [illegible] | 1892 | | | 1 | | 1 |
| [illegible] | 1891 | 9 | 1 | 3 | | 13 |
| | 1892 | 1 | 2 | 4 | 1 | 8 |
| | 1893 | 7 | 6 | 1 | | 14 |
| [illegible] | | 17 | 9 | 8 | 1 | 35 |

the 35 lockouts reported for the three years the employers [ga]ed the object for which the lockout was organized, in 8 [th]ey failed, in 9 cases they were partly successful, there being [f]or which the result was not reported.

[Accordi]ng to returns received from employers with regard to the [loc]kouts in 1893 there were 6 lockouts which directly affected [person]s whose weekly wages before the lockouts amounted to [I]n 1 lockout, affecting 637 persons, an estimated capital of [w]as laid idle, and 2 lockouts, affecting 664 persons, laid idle [w]hose estimated ratable value was $39,467. In 3 lockouts, [..]2 persons, the cost to the employers in stopping and reopen- and in payment of fixed charges and salaries, was $1,898.

from trade unions relating to loss and gain from lockouts [th]at in 1893 there were 7 lockouts, affecting 1,763 persons, [week]ly wages before the lockouts amounted to $13,436. Weekly [..] before and after the lockouts were reported for 5 lockouts, [1],437 persons, whose weekly wages were $10,604 before and [aft]er the lockouts. In 1 lockout the weekly gain in wages to [per]sons affected was $788. In 5 lockouts, affecting 1,743 per- [e]stimated wage loss during the lockouts was $48,874, the [ex]pended by trade unions in defense against 5 lockouts, affect- [..], was $15,680, and the amount expended from other [..] funds in defense against 2 lockouts, affecting 364 [..]. In 9 lockouts 1,706 of the number affected belonged

## STRIKES IN FRANCE IN RECENT YEARS.

The report of the French Office du Travail, Statistique des Grèves et des Recours à la Conciliation et à l'Arbitrage Survenus Pendant l'Année, 1894, gives some interesting figures as the result of its annual inquiry into the subject of strikes and lockouts in France. The report shows that during 1894 there were 391 strikes, involving 1,731 establishments and 54,576 strikers, and a loss of work on the part of the strikers and their fellow-employees amounting to 1,062,480 days. In 1893 there were strikes affecting 4,286 establishments and 170,123 strikers. The loss to employees reached 3,174,000 working days.

In 1894, out of the total of 391 strikes, 84, or 21.48 per cent of them, succeeded; 129, or 32.99 per cent, succeeded partly, and 178, or 45.53 per cent, failed entirely. In 1893 the proportions were: 158 strikes, or 24.92 per cent, succeeded; 206, or 32.49 per cent, succeeded partly, and 270, or 42.59 per cent, failed. In 1894, taking into account the number of strikers involved, 23.63 per cent of the strikers succeeded, 45.41 per cent succeeded partly, and 30.96 per cent failed. In 1893 the proportions in regard to number of strikers were 21.27 per cent, 26.36 per cent, and 52.37 per cent, respectively.

Of the strikes reported in 1894, in 295 cases but 1 establishment was involved; in 32 cases from 2 to 5 establishments were involved; in 18 cases from 6 to 10 establishments; in 26 cases from 11 to 25; in 17 cases from 26 to 50; in 2 cases from 51 to 100, and in 1 case 125.

The two following tables summarize the strikes and strikers for 1894, classifying them by industries and by results. The first table shows for each industry the number of strikes and the number of establishments involved, classifying them according as the strikes succeeded, succeeded partly, or failed. The total strikes and establishments are also shown. The second table shows for each industry the number of strikers, classifying them according as they were involved in successful strikes, partly successful strikes, or in strikes that failed. The total strikers and days of work lost are also given. The column in this table headed "Days of work lost" refers here, as well as in the tables which follow, to days lost not only by strikers but by those employees who were thrown out of work by the strike.

STRIKERS IN 1894, BY INDUSTRIES.

| [illegible] | In successful strikes. | In partly successful strikes. | In strikes which failed. | Total strikers. | Days of work lost. |
|---|---|---|---|---|---|
| [illegible] fisheries | 530 | 1,255 | 628 | 2,413 | 23,003 |
| [illegible] | 150 | 2,415 | 200 | 2,765 | 178,964 |
| [illegible] | 427 | 620 | 380 | 1,427 | 13,216 |
| [illegible] | ........ | 143 | 178 | 321 | 1,277 |
| [illegible] | 850 | 781 | 926 | 2,557 | 7,068 |
| [illegible] | 10 | 65 | 86 | 161 | 3,413 |
| [illegible] | 1,567 | 507 | 2,554 | 4,628 | 47,086 |
| [illegible] | 4,044 | 14,549 | 4,868 | 23,461 | 308,225 |
| [illegible] | 130 | 235 | 630 | 995 | 50,524 |
| [illegible] | 518 | 265 | 423 | 1,206 | 26,151 |
| [illegible]ink) | 108 | 396 | 32 | 536 | 10,197 |
| [illegible] | 18 | 1,198 | ........ | 1,216 | 57,112 |
| [illegible] | 583 | 521 | 873 | 1,977 | 36,758 |
| [illegible] | 8 | ........ | 11 | 19 | 301 |
| [illegible]ing, glass and pottery [illegible] | 2,268 | 74 | 1,459 | 3,801 | 266,978 |
| [illegible]rthenware, glass, etc.) | 1,466 | 1,740 | 2,752 | 5,958 | 29,763 |
| [illegible]ling | 220 | 20 | 895 | 1,135 | 2,464 |
| [illegible] | 12,897 | 24,784 | 16,895 | 54,576 | 1,062,480 |

les it appears that the textile industries proper had ber of strikes during the year, 112, or nearly 30 per number—building trades (stone, earthenware, glass, th 55, and metallic goods with 48. Judged by impor- rbances as shown by the number of strikers and days extile industries proper still lead with 23,461 strikers s lost. According to number of strikers, building thenware, glass, etc.) come second with 5,958 strik- leather third with 4,628 strikers. According to days [illegible], stonecutting and polishing, glass and pottery [illegible] mining, which had but 7 strikes, comes third.

The following summaries, by causes, show for each cause the same facts that in the preceding tables were shown for each industry:

STRIKES IN 1894, BY CAUSES.

| Cause or object. | Succeeded. | | Succeeded partly. | | Failed. | | Total. | |
|---|---|---|---|---|---|---|---|---|
| | Strikes. | Establishments. | Strikes. | Establishments. | Strikes. | Establishments. | Strikes. | Establishments. |
| For increase of wages | 37 | 290 | 69 | 604 | 73 | 215 | 179 | 1,109 |
| Against reduction of wages | 18 | 50 | 28 | 28 | 34 | 83 | 80 | 161 |
| For increase of hours of labor | 2 | 2 | ........ | ........ | 1 | 17 | 3 | 19 |
| For reduction of hours of labor, with present or increased wages | 12 | 145 | 6 | 61 | 12 | 105 | 30 | 311 |
| Relating to time and method of payment of wages, etc | 4 | 5 | 2 | 2 | 3 | 34 | 9 | 41 |
| For or against modification of conditions of work | 8 | 49 | 6 | 34 | 19 | 138 | 33 | 221 |
| Against piecework | 4 | 5 | 3 | 5 | 2 | 2 | 9 | 12 |
| For or against modification of shop rules | 2 | 2 | 3 | 3 | 3 | 3 | 8 | 8 |
| For abolition or reduction of fines | 2 | 2 | 4 | 4 | 5 | 5 | 11 | 11 |
| Against discharge of workmen, foremen, or directors, or for their reinstatement | 3 | 3 | 6 | 10 | 19 | 37 | 28 | 50 |
| For discharge of workmen, foremen, or directors | 14 | 16 | 2 | 2 | 34 | 35 | 50 | 53 |
| Against the employment of women | 1 | 1 | ........ | ........ | 4 | 4 | 5 | 5 |
| For discharge of apprentices or limitation in number | ........ | ........ | ........ | ........ | 2 | 2 | 2 | 2 |
| To support the demands of neighboring woodcutters | 1 | 1 | ........ | ........ | ........ | ........ | 1 | 1 |
| Total (a) | 108 | 571 | 129 | 753 | 211 | 680 | 448 | 2,004 |

a A considerable number of strikes were due to two or three causes, and the facts in such cases have been tabulated under each cause. Hence the totals for this table necessarily do not agree with those for the table on the preceding page.

STRIKERS IN 1894, BY CAUSES.

| Cause or object. | In successful strikes. | In partly successful strikes. | In strikes which failed. | Total strikers. | Days of work lost. |
|---|---|---|---|---|---|
| For increase of wages | 7,664 | 16,602 | 6,434 | 30,700 | 601,899 |
| Against reduction of wages | 3,620 | 2,979 | 2,662 | 9,261 | 159,656 |
| For increase of hours of labor | 430 | .......... | 306 | 736 | 1,386 |
| For reduction of hours of labor, with present or increased wages | 1,044 | 895 | 951 | 2,890 | 28,393 |
| Relating to time and method of payment of wages, etc | 198 | 116 | 377 | 691 | 4,623 |
| For or against modification of conditions of work | 1,486 | 316 | 4,540 | 6,342 | 243,734 |
| Against piecework | 324 | 288 | 215 | 827 | 14,045 |
| For or against modification of shop rules | 53 | 128 | 148 | 329 | 1,875 |
| For abolition or reduction of fines | 103 | 507 | 384 | 994 | 6,363 |
| Against discharge of workmen, foremen, or directors, or for their reinstatement | 652 | 2,561 | 2,008 | 5,221 | 346,090 |
| For discharge of workmen, foremen, or directors | 1,601 | 210 | 4,376 | 6,187 | 239,596 |
| Against the employment of women | 30 | .......... | 196 | 226 | 5,978 |
| For discharge of apprentices or limitation in number | .......... | .......... | 32 | 32 | 480 |
| To support the demands of neighboring woodcutters | 30 | .......... | .......... | 30 | 30 |
| Total (a) | 17,345 | 24,083 | 22,585 | b [illegible] | 1,[illegible] |

a A considerable number of strikes were due to two or three causes, and the facts in such cases have been tabulated under each cause. Hence the totals for this table necessarily do not agree with those for the table on the preceding page.

b Figures here apparently should be 63,[illegible]; those given are, however, according to the original.

It will be seen that more than one-half of all the [illegible] by some difference in regard to wages, 179 being [illegible]

| ... | 391 | 54,576 | 19.83 | 1,062,480 |
|---|---|---|---|---|

...with building trades (stone, earthenware, glass, etc.).
...building trades (woodwork).

...perhaps better than any other the relative amount ... the various industries. It will be seen that in all ... 19.83 out of every 1,000 persons employed were en... uring the year. In textile industries proper, where ...0 per cent of all the strikes, are found 22.65 strikers ...s, a slight excess over the average of all industries, ... trades (stone, earthenware, glass, etc.) and metallic ... second and third as regards number of strikes, are ...d 6.51 strikers per 1,000 employees.

...strikers, classified by results and by the duration of ...sented in the table which follows:

...AND STRIKERS, BY DURATION OF STRIKES, 1894.

| Strikes. | | | | Strikers. | | | |
|---|---|---|---|---|---|---|---|
| ... | Succeeded partly. | Failed. | Total. | Succeeded. | Succeeded partly. | Failed. | Total. |
| ... | 72 | 102 | 232 | a 8,501 | 7,722 | 8,270 | 24,713 |
| ... | 24 | 34 | 82 | 1,796 | 4,298 | 2,587 | 8,681 |
| ... | 10 | 20 | 36 | 234 | 2,024 | 3,353 | 5,611 |
| ... | 12 | 18 | 35 | 2,076 | 9,075 | 1,435 | 12,586 |
| ... | 1 | 4 | 6 | 70 | 1,665 | 1,250 | 2,985 |
| ... | 120 | 178 | 391 | 12,897 | 24,784 | 16,895 | 54,576 |

...should be 8,721; those given are, however, according to the original.

It will be seen that most of the strikes, whether successful or not, are of short duration. Nearly 60 per cent in 1894 lasted but a week or less, and only about 10 per cent were of more than a month's duration.

Duration of strikes is presented in another way in the following short table. The strikes are here classified according to number of strikers involved, and for each group the results and days of duration are shown.

DURATION OF STRIKES IN 1894, BY NUMBER OF STRIKERS INVOLVED.

| Strikers involved. | Strikes. | | | | Days of duration. | | | | |
|---|---|---|---|---|---|---|---|---|---|
| | Succeeded. | Succeeded partly. | Failed. | Total. | 1 to 7. | 8 to 15. | 16 to 30. | 31 to 100. | 101 or over. |
| 25 or under | 21 | 27 | 71 | 119 | 75 | 20 | 11 | 12 | 1 |
| 26 to 50 | 16 | 28 | 34 | 78 | 49 | 18 | 7 | 4 | |
| 51 to 100 | 23 | 36 | 35 | 94 | 58 | 16 | 9 | 9 | 2 |
| 101 to 200 | 10 | 21 | 22 | 53 | 26 | 16 | 6 | 4 | 1 |
| 201 to 500 | 10 | 10 | 10 | 30 | 16 | 12 | | 2 | |
| 501 to 1,000 | 2 | 1 | 4 | 7 | 6 | | 1 | | |
| 1,001 to 5,000 | 2 | 5 | 2 | 9 | 2 | | 2 | 3 | 2 |
| 5,001 or over | | 1 | | 1 | | | | 1 | |
| Total | 84 | 129 | 178 | 391 | 232 | 82 | 36 | 35 | 6 |

This table shows that a large number of the strikes are not only of short duration, but that they are small as regards number of strikers involved. In 119 cases 25 strikers or less were involved, and in 291 cases, 100 strikers or less.

Earlier reports of the Office du Travail give the facts for the years 1890 to 1893. A comparison of the figures for the several years shows a considerable variation in the number and importance of labor disturbances during the several years presented.

The first of the following tables gives the facts, so far as reported, as to number of strikes, establishments involved, strikers, and days of work lost for the years 1890 to 1894.

STRIKES, ESTABLISHMENTS, STRIKERS, AND DAYS OF WORK LOST, 1890 TO 1894.

| Year. | Strikes. | Establishments. | Strikers. | Days of work lost. |
|---|---|---|---|---|
| 1890 | 313 | a 813 | b 118,929 | |
| 1891 | 267 | c 402 | d 108,944 | |
| 1892 | 261 | e 466 | f 47,903 | 920,000 |
| 1893 | 634 | 4,286 | 170,123 | 3,174,000 |
| 1894 | 391 | 1,731 | 54,576 | 1,062,480 |
| Total | 1,866 | 7,698 | 500,475 | 5,156,480 |

a In 33 strikes the number of establishments was not reported.
b In 8 strikes the number of strikers was not reported.
c In 24 strikes the number of establishments was not reported.
d In 2 strikes the number of strikers was not reported.
e In 16 strikes the number of establishments was not reported.
f In 8 strikes the number of strikers was not reported.

A second table shows the number and per cent of strikes which succeeded, succeeded partly, and failed in each of the years of the same period, 1890 to 1894.

NUMBER AND PER CENT OF STRIKES BY RESULTS, 1890 TO 1894

| Year. | Succeeded. | | Succeeded partly. | | Failed. | | Not reported. | | Total | |
|---|---|---|---|---|---|---|---|---|---|---|
| | Number. | Per cent. | Number. | Per cent. | Number. | Per cent. | Number. | Per cent. | Number. | Per cent. |
| 1890 | 82 | 26.20 | 64 | 20.45 | 161 | 51.44 | 6 | 1.91 | 313 | 100 |
| 1891 | 91 | 34.08 | 67 | 25.10 | 106 | 39.70 | 3 | 1.12 | 267 | 100 |
| 1892 | 56 | 21.46 | 80 | 30.65 | 118 | 45.21 | 7 | 2.68 | 261 | 100 |
| 1893 | 158 | 24.92 | 206 | 32.49 | 270 | 42.59 | ........ | ........ | 634 | 100 |
| 1894 | 84 | 21.48 | 129 | 32.99 | 178 | 45.53 | ........ | ........ | 391 | 00 |
| Total | 471 | 25.24 | 546 | 29.26 | 833 | 44.64 | 16 | .86 | 1,866 | 100 |

The per cent of successful strikes during the five years reported varies from 21.48 in 1894 to 34.08 in 1891, and the average for the period is 25.24 per cent. The average of those which succeeded partly is 29.26 per cent and of those which failed 44.64 per cent.

The last table deals with strikers in the same way. It shows the number and per cent of strikers who were involved in strikes which succeeded, succeeded partly, and failed.

NUMBER AND PER CENT OF STRIKERS, BY RESULTS OF STRIKES, 1890 TO 1894.

| Year. | In successful strikes. | | In partly successful strikes. | | In strikes that failed. | | Not reported | | Total. | |
|---|---|---|---|---|---|---|---|---|---|---|
| | Number. | Per cent. | Number. | Per cent. | Number. | Per cent. | Number. | Per cent. | Number. | Per cent. |
| 1890 | 13,361 | 11.23 | 28,013 | 23.55 | 76,075 | 63.97 | 1,480 | 1.25 | 118,929 | 100 |
| 1891 | 22,449 | 20.61 | 54,237 | 49.78 | 32,109 | 29.47 | 149 | .14 | 108,944 | 100 |
| 1892 | 9,774 | 20.40 | 23,820 | 49.73 | 14,179 | 29.60 | 130 | .27 | 47,903 | 100 |
| 1893 | 36,186 | 21.27 | 44,836 | 26.36 | 89,101 | 52.37 | ........ | ........ | 170,123 | 100 |
| 1894 | 12,897 | 23.63 | 24,784 | 45.41 | 16,895 | 30 96 | ........ | ........ | 54,576 | 100 |
| Total | 94,667 | 18 92 | 175,690 | 35 10 | 228,359 | 45 63 | 1,759 | .35 | 500,475 | 100 |

## STRIKES IN ITALY IN RECENT YEARS.

The Statistica degli Scioperi avvenuti nell' Industria e nell' Agricoltura durante gli anni 1892 e 1893 furnishes the data for the accompanying statement concerning strikes in Italy in recent years.

The number of these industrial disturbances during each of the years from 1879 to 1893 is shown in the following table, together with the number of strikers involved in strikes reporting as to number involved:

NUMBER OF STRIKES AND STRIKERS, 1879 TO 1893.

| Year | Strikes. | | Strikers. | | Year. | Strikes. | | Strikers. | |
|---|---|---|---|---|---|---|---|---|---|
| | Total. | Reporting number of strikers. | Total. | Average per strike. | | Total. | Reporting number of strikers. | Total. | Average per strike. |
| 1879 | 32 | 28 | 4,011 | 143 | 1887 | 69 | 66 | 25,027 | 368 |
| 1880 | 27 | 26 | 5,900 | 227 | 1888 | 101 | 99 | 28,974 | 293 |
| 1881 | 44 | 39 | 8,273 | 212 | 1889 | 126 | 125 | 23,322 | 187 |
| 1882 | 47 | 45 | 5,854 | 130 | 1890 | 139 | 138 | 38,402 | 289 |
| 1883 | 73 | 67 | 12,900 | 193 | 1891 | 132 | 128 | 34,733 | 271 |
| 1884 | 81 | 81 | 23,967 | 296 | 1892 | 119 | 117 | 30,800 | 263 |
| 1885 | 89 | 86 | 34,166 | 397 | 1893 | 131 | 127 | 32,109 | 253 |
| 1886 | 96 | 96 | 16,951 | 177 | | | | | |

During the last two years included in the table strikes occurred with greater frequency in Lombardy and Piedmont than in any other of the provinces of Italy, they being the centers of industrial activity. In Sicily, however, quite a large number occurred also. These were confined mostly to the sulphur mines, where almost the whole of the workmen struck and where the difficulties were most frequent.

The distribution of the strikes as to the causes for which undertaken for the period from 1878 to 1891, for 1892, and for 1893 is shown in the following table:

CAUSES OF STRIKES, 1878-1891, 1892, AND 1893.

[In each of the years under consideration some of the strikes have been omitted, neither cause of strike nor number of strikers having been reported.]

| Cause or object. | 1878-1891. | | | | 1892. | | | | 1893. | | | |
|---|---|---|---|---|---|---|---|---|---|---|---|---|
| | Strikes. | | Strikers. | | Strikes. | | Strikers. | | Strikes. | | Strikers. | |
| | Number. | Per cent. | Number. | Per cent. | Number. | Per cent. | Number. | Per cent. | Number. | Per cent. | Number. | Per cent. |
| For increase of wages | 522 | 52 | 152,908 | 60 | 39 | 34 | 8,643 | 22 | 51 | 42 | 22,334 | 68 |
| For reduction of hours | 70 | 7 | 10,084 | 4 | 4 | 3½ | 1,750 | 6 | 11 | 9 | 1,529 | 5 |
| Against reduction of wages | 110 | 11 | 23,207 | 9 | 23 | 20 | 7,581 | 35 | [illegible] | [illegible] | [illegible] | [illegible] |
| Against increase of hours | 20 | 2 | 5,046 | 2 | 4 | 3½ | 150 | 2 | [illegible] | [illegible] | [illegible] | [illegible] |
| Other causes | 276 | 28 | 62,542 | 25 | 44 | 38 | 12,971 | 49 | [illegible] | [illegible] | [illegible] | [illegible] |
| Total classified | 998 | 100 | 254,898 | 100 | 114 | 100 | [illegible] | 100 | [illegible] | [illegible] | [illegible] | [illegible] |
| Not classified | 77 | | | | 5 | | | | [illegible] | | | |
| Grand total | 1,075 | | | | 119 | | | | [illegible] | | | |

**ing under each cause.** During the period from **ent of the strikers engaged** in strikes for the pur- **ig their condition,** while in 1892 28 per cent and in all persons involved were engaged in such **strikes.** **, only 11 per cent of the strikers were engaged in** **a decrease of wages for the period from 1878 to** **1892, and only 13 per cent in 1893.**

**follows shows** the results of strikes for the period or 1892, and for 1893, so far as the strikers were con- the strikes as having succeeded, partly succeeded, ct or cause for which they were undertaken.

**SULTS OF STRIKES, 1878-1891, 1892, AND 1893.**

[consideration some of the strikes have been omitted, neither cause of the nor number of strikers having been reported.]

| Succeeded. | | | | Succeeded partly. | | | | Failed. | | | |
|---|---|---|---|---|---|---|---|---|---|---|---|
| Strikes. | | Strikers. | | Strikes. | | Strikers. | | Strikes. | | Strikers. | |
| Num-ber. | Per cent. | Number. | Per cent. | Number. | Per cent. | Number. | Per cent. | Number. | Per cent. | Number. | Per cent. |
| [illegible] | 17 | 43,931 | 29 | 242 | 46 | 74,650 | 49 | 192 | 37 | 34,327 | 22 |
| [illegible] | 28 | 1,078 | 16 | 13 | 33 | 2,050 | 31 | 17 | 44 | 3,514 | 53 |
| [illegible] | 30 | 6,071 | 45 | 18 | 35 | 4,713 | 35 | 18 | 35 | 2,602 | 20 |
| [illegible] | 20 | 2,612 | 36 | 28 | 40 | 2,449 | 34 | 28 | 40 | 4,002 | 40 |
| [illegible] | 25 | 1,500 | 84 | 1 | 25 | 40 | 2 | 2 | 50 | 250 | 14 |
| [illegible] | 46 | 581 | 38 | 4 | 36 | 815 | 54 | 2 | 18 | 123 | 8 |
| [illegible] | 13 | 2,700 | 12 | 47 | 43 | 11,744 | 50 | 49 | 44 | 8,763 | 38 |
| [illegible] | 30 | 2,080 | 48 | 7 | 30 | 1,628 | 22 | 9 | 40 | 2,263 | 30 |
| [illegible] | 23 | 840 | 21 | 10 | 45 | 1,341 | 34 | 7 | 32 | 1,750 | 45 |
| [illegible] | 36 | 2,840 | 45 | 8 | 40 | 2,750 | 49 | 5 | 25 | 356 | 6 |
| ..... | ..... | ........ | ..... | 2 | 50 | 350 | 56 | 2 | 50 | 280 | 44 |
| ..... | ..... | ........ | ..... | 1 | 100 | 300 | 100 | ..... | ..... | ........ | ..... |
| [illegible] | 13 | 9,553 | 15 | 104 | 38 | 27,441 | 44 | 136 | 49 | 25,849 | 41 |
| [illegible] | 16 | 2,396 | 18 | 10 | 23 | 1,764 | 13 | 27 | 61 | 9,409 | 69 |
| [illegible] | 25 | 1,705 | 14 | 13 | 36 | 6,601 | 53 | 14 | 39 | 4,186 | 33 |
| [illegible] | 16 | 62,336 | 24 | 429 | 43 | 119,084 | 47 | 410 | 41 | 73,298 | 29 |
| [illegible] | 21 | 8,636 | 29 | 23 | 29 | 5,832 | 19 | 57 | 50 | 15,710 | 52 |
| [illegible] | 23 | 9,197 | 29 | 46 | 38 | 13,770 | 44 | 41 | 34 | 8,661 | 27 |

total columns, during the years from 1878 to 1891 trikers, or persons involved in strikes, succeeded in or which they struck, 47 per cent succeeded partly, iled. In 1892 29 per cent succeeded, 19 per cent nd 52 per cent failed, while in 1893 29 per cent suc- t succeeded partly, and 27 per cent failed. The results of strikes for any of the causes shown may be found in the same manner by reference to the table.

The classification of strikes, for 1892 and 1893, according to the industries in which strikers were engaged is shown in the following table:

STRIKES, BY INDUSTRIES, 1892 AND 1893.

| Industry. | 1892. | | | 1893. | | |
|---|---|---|---|---|---|---|
| | Strikes. | | | Strikes. | | |
| | Total. | Reporting number of strikers. | Strikers. | Total. | Reporting number of strikers. | Strikers. |
| Weavers, spinners, and carders | 41 | 41 | 7,679 | 44 | 44 | 14,061 |
| Miners and ore diggers | 19 | 19 | 8,280 | 19 | 18 | 3,840 |
| Mechanics | 3 | 3 | 568 | 5 | 5 | 415 |
| Founders | 1 | 1 | 70 | 2 | 2 | 380 |
| Day laborers | 13 | 12 | 2,026 | 9 | 9 | 3,960 |
| Masons and stonecutters | 5 | 5 | 1,940 | 6 | 4 | 380 |
| Kiln and furnace tenders | 6 | 6 | 439 | 2 | 2 | 250 |
| Printers | 5 | 5 | 345 | 1 | 1 | 10 |
| Hat makers | 3 | 3 | 306 | 1 | 1 | 32 |
| Tanners | 1 | 1 | 12 | 6 | 6 | 447 |
| Joiners | 3 | 3 | 500 | ....... | ....... | ....... |
| Omnibus drivers and conductors | 3 | 3 | 2,470 | 5 | 5 | 3,627 |
| Cart drivers | 1 | 1 | 60 | 4 | 3 | 220 |
| Porters and coal carriers | 4 | 4 | 2,610 | 7 | 7 | 1,200 |
| Other industries | 11 | 10 | 3,495 | 20 | 20 | 3,177 |
| Total | 119 | 117 | 30,800 | 131 | 127 | 32,109 |

The textile, mineral, and metallurgic industries, and that of public works, in which most of the common labor is engaged, are more largely represented in strikes because in those industries the workmen are more generally organized.

Immediately following is shown, for a series of years, from 1879 to 1893, the total and average days lost by reason of strikes.

DAYS OF WORK LOST BY REASON OF STRIKES, 1879 TO 1893.

| Year. | Strikes. | | Days of work lost. | | Year. | Strikes. | | Days of work lost. | |
|---|---|---|---|---|---|---|---|---|---|
| | Total. | Reporting number of days lost. | Total. | Average per strike. | | Total. | Reporting number of days lost. | Total. | Average per strike. |
| 1879 | 32 | 28 | 21,896 | 782 | 1887 | [illegible] | [illegible] | [illegible] | [illegible] |
| 1880 | 27 | 26 | 91,890 | 3,535 | 1888 | 101 | 95 | 191,204 | 2,012 |
| 1881 | 44 | 38 | 95,578 | 2,515 | 1889 | 126 | 122 | [illegible] | [illegible] |
| 1882 | 47 | 45 | 25,119 | 558 | 1890 | 130 | 129 | [illegible] | [illegible] |
| 1883 | 73 | 65 | 111,697 | 1,718 | 1891 | 132 | 128 | [illegible] | [illegible] |
| 1884 | 81 | 78 | 149,215 | 1,913 | 1892 | 119 | 114 | [illegible] | [illegible] |
| 1885 | 89 | 83 | 244,[illegible] | 2,980 | 1893 | 131 | 122 | [illegible] | [illegible] |
| 1886 | 96 | 95 | 56,772 | 598 | | | | | |

[illegible] Zusammenstellungen der in den Jahren [illegible] Arbeitseinstellungen im Gewerbebe- [illegible]

[illegible] following are summaries of strikes by [illegible] 1891 and 1892, respectively. They show the [illegible] involved, total employees in such [illegible] striking, and total days of work lost by [illegible] show also the causes or objects for which [illegible], classifying them under three heads, viz: [illegible] that and other demands, against reduction [illegible] The number of strikes which succeeded, suc- [illegible] is also shown:

[illegible]IKES IN 1891, BY INDUSTRIES.

[illegible] of the items in this table will not in all cases produce the totals [illegible] all as given in the report before referred to, the original not being

| Employees. | | | Cause or object. | | | Result. | | |
|---|---|---|---|---|---|---|---|---|
| Total. | Strikers. | Days lost. | For increase of wages or that and other demands. | Against reduction of wages. | All other. | Succeeded. | Succeeded partly. | Failed. |
| [illegible] | 972 | a 21,855 | 7 | ........ | ........ | ........ | 3 | 4 |
| [illegible] | 1,[illegible] | 5,221 | 5 | ........ | 2 | 1 | 3 | 1 |
| 17 | 17 | 17 | ........ | ........ | 1 | ........ | ........ | 1 |
| 107 | 79 | 766 | 2 | ........ | ........ | ........ | 2 | ........ |
| [illegible] | [illegible] | a 75,251 | 3 | ........ | 4 | 2 | ........ | 5 |
| 30 | 30 | 24 | ........ | ........ | 1 | 1 | ........ | ........ |
| 268 | 155 | 155 | 3 | ........ | ........ | 2 | ........ | ........ |
| [illegible] | [illegible] | 4,363 | 7 | 2 | 2 | 2 | 2 | 7 |
| 6 | 6 | 12 | ........ | ........ | 1 | ........ | ........ | 1 |
| [illegible] | 32 | 64 | 1 | ........ | ........ | ........ | ........ | 1 |
| [illegible] | 34 | 656 | ........ | ........ | 2 | 1 | ........ | 1 |
| [illegible] | 5 | 5 | ........ | ........ | 1 | ........ | ........ | 1 |
| 10 | 7 | ........ | 1 | ........ | ........ | ........ | 1 | ........ |
| 691 | 126 | 249 | ........ | 1 | 3 | 1 | ........ | 3 |
| 148 | 148 | 148 | ........ | ........ | 1 | ........ | ........ | 1 |
| [illegible] | 1,[illegible] | 12,911 | 5 | ........ | 7 | 4 | 1 | 7 |
| [illegible] | 104 | 542 | 1 | 2 | ........ | ........ | ........ | - |

a About.

### STRIKES IN 1891, BY INDUSTRIES—Concluded.

[It will be observed that the addition of the items in this table will not in all cases produce the totals given. The figures, however, are all as given in the report before referred to, the original not being accessible.]

| Industry. | Strikes. | Establishments. | Employees. Total. | Employees. Strikers. | Employees. Days lost. | Cause or object. For increase of wages or that and other demands. | Cause or object. Against reduction of wages. | Cause or object. All other. | Result. Succeeded. | Result. Succeeded partly. | Result. Failed. |
|---|---|---|---|---|---|---|---|---|---|---|---|
| Pipe manufacturers | 1 | 1 | 8 | 8 | 64 | ....... | 1 | ...... | 1 | ...... | ...... |
| Brush makers | 1 | 1 | 17 | 3 | 27 | 1 | ...... | ...... | ...... | ...... | 1 |
| Shipping | 1 | 1 | 100 | 100 | 300 | ...... | 1 | ...... | ...... | ...... | 1 |
| Dockers | 1 | 1 | 800 | 30 | 90 | 1 | ...... | ...... | ...... | ...... | 1 |
| Manufacturers of fancy boxes | 1 | 1 | 35 | 35 | 490 | 1 | ...... | ...... | 1 | ...... | ...... |
| Tailors | 1 | 79 | 1,656 | 492 | 3,444 | 1 | ...... | ...... | ...... | 1 | ...... |
| River conservators | 1 | 1 | 72 | 72 | 720 | 1 | ...... | ...... | ...... | 1 | ...... |
| Shoemakers | 8 | 556 | 10,846 | 2,547 | a 87,363 | 4 | 3 | 1 | 1 | 4 | 3 |
| Textile manufacturers | 18 | 18 | 6,585 | 2,929 | 26,529 | 10 | 4 | 4 | 4 | 6 | 8 |
| Joiners and cabinetmakers | 6 | 70 | 401 | 300 | 4,678 | 2 | ...... | 4 | 2 | 2 | 2 |
| Sugar refiners | 1 | 1 | 537 | 130 | 1,170 | ...... | 1 | ...... | ...... | ...... | 1 |
| Total | 104 | b 1,916 | 40,486 | b 14,025 | b 247,076 | 55 | 15 | 34 | 23 | 26 | b 51 |

a About.

b See prefatory note to table.

### STRIKES IN 1892, BY INDUSTRIES

[It will be observed that the addition of the items in this table will not in all cases produce the totals given. The figures, however, are all as given in the report before referred to, the original not being accessible.]

| Industry. | Strikes. | Establishments. | Employees. Total. | Employees. Strikers. | Employees. Days lost. | Cause or object. For increase of wages or that and other demands. | Cause or object. Against reduction of wages. | Cause or object. All other. | Result. Succeeded. | Result. Succeeded partly. | Result. Failed. |
|---|---|---|---|---|---|---|---|---|---|---|---|
| Painters | 1 | 1 | 9 | 9 | 9 | ...... | ...... | 1 | ...... | ...... | 1 |
| Builders | 6 | 34 | 2,241 | 2,049 | 19,401 | 5 | 1 | ...... | 2 | 2 | 2 |
| Binders | 4 | 6 | 749 | 158 | 1,548 | 2 | ...... | 2 | 2 | ...... | 2 |
| Manufacturers of cardboard goods | 1 | 2 | 111 | 111 | 222 | ...... | ...... | 1 | 1 | ...... | ...... |
| Manufacturers of cellular linen clothing and military stocks | 1 | 1 | 14 | 14 | 140 | ...... | ...... | 1 | ...... | ...... | 1 |
| Turners | 4 | 92 | 1,250 | 442 | 13,941 | 3 | ...... | 1 | 2 | 2 | ...... |
| Innkeepers | 1 | 1 | 4 | 4 | 4 | 1 | ...... | ...... | ...... | ...... | 1 |
| Glass and china manufacturers | 5 | 64 | 2,341 | 2,222 | 62,609 | 3 | 1 | 1 | 1 | 2 | 2 |
| Coffee sorters | 1 | 1 | 200 | 200 | 400 | 1 | ...... | ...... | ...... | ...... | 1 |
| Leather manufacturers | 1 | 1 | 180 | 8 | 8 | ...... | ...... | 1 | ...... | 1 | 1 |
| Hackney coachmen | 1 | 1,200 | 1,200 | 1,200 | 2,400 | 1 | ...... | ...... | ...... | 1 | ...... |
| Malt manufacturers | 1 | 1 | 65 | 42 | 42 | ...... | ...... | 1 | ...... | ...... | ...... |
| Metal workers | 15 | 16 | 4,415 | 1,444 | 12,607 | 4 | 4 | 7 | 2 | 5 | 8 |
| Millers | 2 | 2 | 22 | 13 | 25 | 1 | ...... | 1 | ...... | ...... | 2 |
| Pipe makers | 1 | 1 | 18 | 11 | 22 | 1 | ...... | ...... | 1 | ...... | ...... |
| Compositors | 1 | 1 | 4 | 1 | 4 | ...... | ...... | 1 | ...... | ...... | 1 |
| Shoemakers | 5 | 5 | 363 | 238 | 1,756 | ...... | 5 | ...... | 2 | 2 | 1 |
| Trunk makers | 1 | 1 | 19 | 10 | 60 | ...... | ...... | ...... | ...... | ...... | 1 |
| Textile manufacturers | 39 | 48 | 10,515 | 5,420 | 33,114 | 17 | 8 | 14 | 10 | 12 | 17 |
| Joiners and cabinetmakers | 4 | 24 | 206 | 206 | 1,607 | 3 | ...... | 1 | 2 | ...... | 2 |

STRIKES IN 1892, BY INDUSTRIES—Concluded

[It will be observed that the addition of the items in this table will not in all cases produce the totals given The figures, however, are all as given in the report before referred to, the original not being accessible.]

| Industry | Strikes. | Establishments. | Employees. | | | Cause or object | | | Result. | | |
|---|---|---|---|---|---|---|---|---|---|---|---|
| | | | Total. | Strikers. | Days lost. | For increase of wages or that and other demands. | Against reduction of wages. | All other. | Succeeded | Succeeded partly. | Failed. |
| Wheelwrights.... | 1 | 12 | 30 | 20 | 460 | ........ | ........ | 1 | ....... | 1 | ........ |
| Manufacturers of underlinen ...... | 3 | 3 | 509 | 123 | 533 | 2 | 1 | ...... | ....... | 1 | 2 |
| Water company .. | 1 | 1 | 86 | 66 | 66 | 1 | ........ | ...... | ....... | ....... | 1 |
| Match manufacturers .......... | 1 | 1 | 70 | 14 | 14 | ........ | 1 | ...... | 1 | ....... | ........ |
| Total... .... | 101 | 1,519 | 24,621 | 14,025 | 150,992 | *a* 46 | 21 | 34 | 26 | 29 | 46 |

*a* See prefatory note to table.

The following table shows the distribution of strikes in Austria in 1891 and 1892 by districts, giving the number of strikes, establishments involved, employees, and strikers:

STRIKES IN 1891 AND 1892, BY DISTRICTS

[It will be observed that the addition of the items in this table will not in all cases produce the totals given. The figures, however, are all as given in the report before referred to, the original not being accessible.]

| District. | Strikes | | Establishments | | Employees. | | Strikers | |
|---|---|---|---|---|---|---|---|---|
| | 1891. | 1892. | 1891. | 1892. | 1891. | 1892. | 1891. | 1892. |
| Lower Austria........................ | 35 | 28 | 22 | 1,336 | 17,111 | 7,285 | 5,875 | 2,520 |
| Upper Austria.... ................... | 3 | 1 | 3 | 1 | 60 | 4 | 59 | 4 |
| Salzburg .............................. | ........ | 1 | ........ | 4 | ....... | 19 | ........ | 19 |
| Styria ................................ | 2 | 8 | 2 | 3 | 476 | 18 | 474 | 16 |
| Carinthia.............................. | 2 | ........ | 2 | ........ | 80 | ........ | 36 | ........ |
| Carniola .............................. | 2 | 2 | 2 | 2 | 641 | 410 | 124 | 260 |
| Coast lands ........................... | 1 | 3 | 1 | 9 | 800 | 1,270 | 30 | 1,270 |
| Tyrol and Vorarlberg................... | 4 | 1 | 68 | 1 | 440 | 46 | 275 | 44 |
| Bohemia................................ | 27 | 35 | 599 | 127 | 16,852 | 10,740 | 5,023 | 8,004 |
| Moravia................................ | 24 | 24 | 26 | 24 | 4,737 | 4,645 | 1,892 | 1,855 |
| Silesia................................ | ........ | 1 | ........ | 1 | ........ | 66 | ........ | 22 |
| Galicia ........................... ... | 3 | 2 | 90 | 11 | 255 | 118 | 227 | 109 |
| Bukowina............................... | 1 | ........ | 1 | ........ | 34 | ........ | 10 | ........ |
| Dalmatia .............................. | ........ | ........ | ........ | ........ | ........ | ........ | ........ | ........ |
| Total ................................. | 104 | 101 | *a* 1,916 | 1,519 | *a* 40,486 | 24,621 | 14 025 | 14,123 |

*a* See prefatory note to table.

# PRIVATE AND PUBLIC DEBT IN THE UNITED STATES.

BY GEORGE K. HOLMES.

There is an elaborate network of debts and credits associated with production and trade, and growing out of the numerous wants and necessities of men, to satisfy which they in many cases use borrowed or hired wealth. The maufacturer may have a mortgage on his factory and be in debt for materials, the jobber and wholesale merchant are indebted to him, while the retail merchants owe them. The retail merchants have customers who are indebted to them, and these customers are more or less creditors. It is therefore practically impossible to ascertain the true amount of the private debts of the people. The difficulty can be illustrated by the following familiar example: A owes B $10, B owes C $10, and C owes A the same amount; a ten-dollar bill handed by the first to the second, by the second to the third, and by the third to the first will satisfy the three debts, yet in any statistics of private debt under this illustration the total would be $30.

In undertaking to arrive at the amount of private debt it is impossible to offset credits against debts in cases similar to the foregoing. The best that can be done is to ascertain the amounts of the various classes of debts which are offset little, if any, by credits and regard their sum as the minimum amount of debt, somewhere above which is the true amount.

The results of an effort to do this are presented in the accompanying statement of the minimum debt of the United States in 1890. The amounts of the funded and unfunded debt of railroad and street railway companies, and the amount of the funded debt of telephone companies are obtained from the reports of the Eleventh Census of the United States. To the reported debt of railroad companies has been added an estimate of the debt not reported. The totals for the other items in the statement have been taken from similar official or authentic reports (*a*), or are carefully prepared estimates.

---

*a* Bulletins and final reports of the Eleventh Census, Poor's Manual of [illegible], the Manual of American Waterworks, reports of the [illegible] light commissioners, and reports of the Comptroller of [illegible]

| | | |
|---|---|---|
| [illegible] | [illegible] | ........ |
| [illegible] (... to mileage) | [illegible] | ........ |
| [illegible] | 797, [illegible] | ........ |
| [illegible] (... to mileage) | 42, 762, [illegible] | ........ |
| [illegible] | 5, [illegible], 491, 114 | ........ |
| *[illegible] companies.* | | |
| [illegible] | 151, 872, [illegible] | ........ |
| [illegible] | [illegible] | ........ |
| [illegible] | 182, 240, 754 | ........ |
| [illegible] railway companies | 5, 851, 671, [illegible] | ........ |
| *[illegible] corporations.* | | |
| [illegible] (estimated) | 5, 669, 491, 114 | 91.44 |
| [illegible] | 182, 240, 754 | 2.94 |
| [illegible] | 4, 892, 585 | .08 |
| [illegible] | 20, 000, 000 | .32 |
| [illegible] by municipalities ([illegible] estimated) | 89, 127, 489 | 1.44 |
| [illegible] | 75, 000, 000 | 1.21 |
| [illegible] (estimated) | 45, 000, 000 | .73 |
| [illegible], and canal, turnpike, [illegible] (estimated to make round total) | 114, 206, 078 | 1.84 |
| [illegible] | 6, 200, 000, 000 | 100.00 |
| *[illegible] aggregates.* | | |
| *[illegible] occupied by owners.* | | |
| [illegible] from 5,000 to 100,000 population | 292, 611, 974 | 27.96 |
| [illegible] and over | 393, 029, 833 | 37.54 |
| [illegible] population and over | 361, 311, 796 | 34.51 |
| [illegible] | 1, 046, 953, 603 | 100.00 |
| *[illegible] occupied by owners.* | | |
| [illegible] | 1, 085, 995, 960 | 50.92 |
| [illegible] | 1, 046, 953, 603 | 49.08 |
| [illegible] | 2, 132, 949, 563 | 100.00 |
| *On acre tracts.* | | |
| [illegible] | 1, 085, 995, 960 | 49.16 |
| [illegible] tracts | 1, 123, 152, 471 | 50.84 |
| [illegible] | 2, 209, 148, 431 | 100.00 |
| *On lots.* | | |
| [illegible] | 1, 046, 953, 603 | 27.48 |
| [illegible], and all other lots | 2, 763, 577, 951 | 72.52 |
| [illegible] | 3, 810, 531, 554 | 100.00 |
| *On all real estate.* | | |
| [illegible] | 2, 209, 148, 431 | 36.70 |
| [illegible] | 3, 810, 531, 554 | 63.30 |
| [illegible] | 6, 019, 679, 985 | 100.00 |
| [illegible] AND PRIVATE CORPORATIONS. | | |
| [illegible] | 6, 019, 679, 985 | 50.16 |
| [illegible] | 300, 000, 000 | 2.50 |
| [illegible] and chattel mortgages (estimated) | 350, 000, 000 | 2.92 |
| [illegible] | 1, 904, 167, 351 | 15.87 |
| [illegible], not including real estate mortgages | 1, 172, 918, 415 | 9.77 |
| [illegible] | 1, 040, 473, 013 | 8.67 |
| [illegible] (to make round total) | 1, 212, 761, 236 | 10.11 |
| [illegible] | 12, 000, 000, 000 | 100.00 |

MINIMUM DEBT OF THE UNITED STATES, 1890—Concluded.

| Description of debt. | Amount. | Per cent of group total. |
|---|---|---|
| AGGREGATE PRIVATE DEBT. | | |
| Quasi public corporations | $6,200,000,000 | 34.07 |
| Individuals and private corporations | 12,000,000,000 | 65.93 |
| Total | 18,200,000,000 | 100.00 |
| PUBLIC DEBT, LESS SINKING FUND. | | |
| United States | 891,980,104 | 44.00 |
| States | 228,997,389 | 11.30 |
| Counties | 145,048,045 | 7.15 |
| Municipalities | 724,463,060 | 35.74 |
| School districts | 36,701,948 | 1.81 |
| Total | 2,027,170,546 | 100.00 |
| AGGREGATE PRIVATE AND PUBLIC DEBT. | | |
| Private debt | 18,200,000,000 | 89.98 |
| Public debt | 2,027,170,546 | 10.02 |
| Total | 20,227,170,546 | 100.00 |

Some of the classes of corporations enumerated in the foregoing statement, such as transportation companies not otherwise specified, canal, turnpike, and bridge companies, do a cash business, and others a business that is so nearly for cash that there is comparatively little in amount of credits to offset against their debt. The amount of credits of such corporations is undoubtedly much more than balanced by the wages that they owe just before pay day.

Debtors who place mortgages on their real or personal estates are creditors to some extent, how far it is impossible to estimate; but these persons are not regarded as appreciably a creditor class, as they would need to be if their combined debt of $6,669,679,985 was to be reduced much on this account. On the other hand, the borrowers from banks, not including borrowers on real estate security, may be supposed to be creditors to a considerable degree. National banks can not lend on real estate mortgages, and therefore these securities are excluded from the loans of other banks.

The public revenue, too, is derived from persons who are creditors as well as debtors, and a large portion of it, as in the case of crop liens, is not a debt that continues throughout the whole year. Notwithstanding this, it is included in the statement of debt, partly because it is a conspicuous and disagreeable debt burden and partly to account for some of the debt which can not be ascertained in its entirety.

It is believed that the total of the preceding statement expresses the minimum debt of the people of the United States in 1890. Only 12.14 per cent of it is estimated, no part of it is duplicated, and the supposition is that the accepted debt offset by credit is more than equaled by the omitted debt. In addition to showing the amounts, [illegible] gives the percentages that the different amounts are of [illegible] the respective groups of debts. For instance, of the [illegible] statement, in which the various items of debt are con- centages of the total debt of $20,227,170,546:

T OF EACH CLASS OF DEBT OF THE AGGREGATE DEBT, 1890.

| Description of debt. | Per cent. |
|---|---|
| QUASI PUBLIC CORPORATIONS. | |
| partly estimated) | 28.03 |
| nies | .90 |
| , funded debt | .03 |
| (partly estimated) | .10 |
| ies, not owned by municipalities (partly estimated) | .44 |
| nated) | .37 |
| power companies (estimated) | .22 |
| panies, not otherwise specified, and canal, turnpike, bridge, and other ations (estimated) | .56 |
| | 30.65 |
| INDIVIDUALS AND PRIVATE CORPORATIONS. | |
| s | 29.76 |
| th (estimated) | 1.48 |
| the South, and chattel mortgages (estimated) | 1.73 |
| s and overdrafts | 9.41 |
| nd overdrafts, not including real estate mortgages | 5.80 |
| ocal taxes | 5.15 |
| t (estimated) | 6.00 |
| | 59.33 |
| PUBLIC DEBT, LESS SINKING FUND. | |
| | 4.41 |
| | 1.13 |
| | .72 |
| | 3.58 |
| | .18 |
| | 10.02 |
| | 100.00 |
| | 12.14 |
| ned | 87.86 |

ent groups of debts that of individuals and private cor- ds first, forming 59.33 per cent of the total, while quasi tions form 30.65 per cent. The real-estate mortgage debt per cent of the grand total, and is followed by that of anies, 28.03 per cent. Among the items of public debt ited States is first, and forms 4.41 per cent, while the palities is 3.58 per cent of the total.

eat difference between the significance of a debt incurred ownership of capital or the more durable property to be vely and to be retained and used by the debtor and to ble for the payment on his debt, and that of a debt in- purchase of property soon to be consumed unproductively nase of evanescent property. The debt of the quasi public riginally stood for substantially an equal amount of stands for the same at the present time, except in the

ciated with any other purpose, rank second, being 13.96 per cent of the combined debt. The details given in this statement show at a glance the different incentives for debt. By a further condensation of purposes, it appears that real estate purchase and improvements, when not associated with any other purpose, are represented by the following percentages: For farms, 74.22 per cent; for homes, 81.24 per cent; for farms and homes, 77.67 per cent. Real estate purchase and improvements, business, and the purchase of the more durable kinds of personal property are: For farms, 93.68 per cent; for homes, 95.56 per cent; for farms and homes, 94.65 per cent (a). Thus it appears that almost the entire incumbrance on farms and homes occupied by owners was due to the acquiring of capital and the more durable kinds of property.

The crop lien of the South was a necessity that grew out of the conditions in which the farmers found themselves at the close of the civil war. They had their farms and some mules and implements, but beyond that they were poor and could not maintain themselves, to say nothing of paying wages until the harvesting of the next crop, and the ex-slaves, perhaps hardly more than their former masters, were in need of immediate subsistence. In this strait, credit was obtained with the merchants for an advance of supplies until the harvesting of the crop, which, being mostly cotton, but partly tobacco, was as good as cash at the time of harvest. The plantations were next more or less subdivided into holdings to be cultivated by the negroes on shares. Landlords and tenants secured the merchants for advancements by crop liens and by mortgages on farm animals. That system has continued with little abatement until the present time, and the debt that accompanies it is mostly a subsistence debt, but to some extent a debt for capital. The crop liens and chattel mortgage debt of the more recently settled regions partake largely of the latter character.

The purposes of the loans obtained from banks can not be definitely described. It is a matter of common understanding that they are mostly for capital, since banks would not lend to persons, and friends would not indorse for them, if they intended to use the borrowed money so as to weaken their financial responsibility.

The tax debt aims to have for its compensation the maintenance of justice, the promotion of public works, of education, and of undertakings for the general good; and the same may be said of the public debt.

The miscellaneous undescribed debts are those that grow out of trade, production, and services of many varieties. It is impossible to say how far they stand for capital, or for wealth to be preserved or to be consumed.

After the foregoing review of the significance of the various classes of debt, it is apparent that at least about nine-tenths of it was incurred for the acquirement of capital and of the more durable kinds of prop-

a These per cents do not appear in the statement, as they are partly composed of incumbrance taken from some of the "various combinations of purposes, not otherwise specified."

| | | | |
|---|---|---|---|
| [illegible] the 19 cities of 100,000 popu-[illegible] | 393,029,889 | 22,584,500 | 5.75 |
| [illegible] outside of cities and towns of [illegible] | 361,811,796 | 24,179,775 | 6.69 |
| [illegible] | 2,209,148,421 | 162,652,944 | 7.36 |
| [illegible] | 3,810,831,564 | 234,780,848 | 6.16 |
| [illegible] | 6,019,679,985 | 397,442,792 | 6.60 |

on real estate mortgages is given at 6.60 per cent, which ...ed for the loans of all banks. On the crop liens of the average rate is paid, how high it is not known. Numer- ...aive inquiries, many of them answered by merchants ...yers who hold or have held crop liens, point to the con- ...he average rate on these liens must be as high as 40 per ...oing as low as 25 per cent, and often going as high as 75 more.

...eport of the bureau of labor statistics of Illinois it ap- ...e average rate of interest on chattel mortgages in that ...3 per cent in 1887. There is reason to believe that the ...l mortgages farther west and in the South, and the crop the Mississippi River, is higher than this, and the gen- ...0 per cent is adopted for crop liens outside of the South ...l mortgages.

...e estimated "other net private debt," which has been ...12,761,236, does not bear interest, such as the debt owing ..., to lawyers, for labor, and the like, and for the want of ...e of its proportions its amount is arbitrarily assumed to ... of the total of the class to which it belongs, and the ... of interest on the remaining three-fourths to be 7 per

...e rates of interest on the total public debt, the debt of ...tates, and the local public debt are taken from the report ...bt, and taxation, which forms a part of the report of the ...ns of the United States. No attempt has been made to ...of these rates according as the bonds of corporations, ...l as private and quasi public, have been sold above or ...r according as there has been default of payment, except ...ase, for railroad companies.

...nterest-bearing private debt is $16,074,228,836, and the ... paid $1,071,561,924, the average rate being 6.67 per cent, ...erage rate on the debt of the United States is 4.08 per cent, this low rate being partly determined by the debt's freedom from taxation. While the average rate on real estate mortgages is 6.60 per cent, it goes as high as 7.07 per cent on farms occupied by owners and 7.36 per cent on acre tracts.

The material is not statistically or otherwise ascertainable to determine with what ease or difficulty, as the case may be, debtors pay their debts and the interest on them. Numerous voluntary explanations bearing on this point have been made by mortgage debtors and by debtors who own their farms and homes subject to incumbrance. From these explanations it would be impossible to form any definite or tangible conclusion; they are too often doubtful, because remote and involved in political and economic theories.

Whether a ratio between the debt and the wealth possessed by the debtors indicates more than the debt's security to the creditors depends upon the earnings of the borrowed wealth, or, if it has no earnings, upon the income of the debtors.

Subject to these qualifications the accompanying statement is presented, which gives the amount of debt, the wealth, and the percentage that the debt is of the wealth. The figures given in the column headed wealth represent in some cases only the value of the property on which the debt given is a lien; in other cases they represent the value of all of the property in the class to which the debt belongs, although some of the property is not incumbered. These latter are railroad, street railway, and telephone companies, the gas companies in Massachusetts first mentioned, and taxed real estate and untaxed mines.

PER CENT OF DEBT OF WEALTH, 1890.

| Description of wealth. | Debt. | Wealth. | Per cent debt is of wealth. |
|---|---|---|---|
| Railroad companies | $5,669,421,114 | $8,401,508,904 | 67.48 |
| Street railway companies | 182,240,754 | 283,888,519 | 64.19 |
| Telephone companies | 4,982,565 | 72,341,736 | 69.01 |
| Gas companies in Massachusetts (a) | 6,[illegible],329 | b 20,[illegible] | [illegible] |
| Gas companies in Massachusetts owing debt (a) | 6,[illegible] | b 14,475,[illegible] | 47.61 |
| Incumbered farms occupied by owners | 1,085,995,960 | 3,054,923,165 | 35.55 |
| Incumbered homes occupied by owners | 1,046,953,603 | 2,632,374,904 | 39.77 |
| Incumbered farms and homes occupied by owners | 2,132,949,563 | 5,687,298,069 | 37.50 |
| Incumbered homes occupied by owners in the 420 cities and towns of from 8,000 to 100,000 population | 292,611,974 | 739,846,067 | 39.55 |
| Incumbered homes occupied by owners in the 28 cities of 100,000 population and over | 393,029,083 | 934,191,811 | 42.07 |
| Incumbered homes occupied by owners outside of cities and towns of 8,000 population and over | 361,311,796 | 958,337,008 | 37.70 |
| Taxed real estate and untaxed mines | 6,019,679,985 | 36,035,071,490 | 16.71 |
| The United States | 20,227,170,546 | 65,037,091,197 | 31.10 |

a 1891. b Capital stock and bonds.

The percentage that the debt is of the wealth with which it is compared ranges from 16.71 per cent for the taxed real estate and untaxed mines to 69.01 per cent for telephone companies. The percentage in the case of incumbered farms or homes occupied by owners, ranges from 35.55 per cent on farms to 42.07 on homes in the [illegible]

| [illegible] | Amount. |
|---|---|
| [illegible] | $[illegible], [illegible], 170, 548 |
| [illegible] public debt | 1, 166, 101, 303 |
| [illegible] of materials | 4, 211, 239, 271 |
| [illegible] | 2, 480, 107, 464 |
| [illegible] | 44, 277, 514 |
| [illegible] | 587, 230, 082 |
| [illegible] industries | 7, 302, 854, 901 |
| [illegible] | 6, 130, 297, 785 |
| [illegible] | 13, 379, 252, 849 |
| [illegible] wages, farm implements, and machines | 2, 708, 015, 040 |
| [illegible] | 43, 092, 122 |
| [illegible] products on hand | 1, 291, 391, 579 |
| [illegible] productive industries | 23, 456, 559, 176 |
| [illegible] and wages, farm implements, and machines | 15, 982, 267, 689 |
| [illegible] | 2, 282, 216, 529 |
| [illegible] products on hand, raw and manufactured | 3, 058, 593, 441 |
| [illegible] property, shipping, and canals | 701, 755, 712 |
| [illegible] bullion | 1, 158, 774, 948 |
| [illegible] | 1, 204, 235, 951 |
| [illegible] | 18, 691, 434, 190 |
| [illegible] | 21, 395, 091, 197 |
| [illegible] | 2, 819, 902, 791 |

…es based on the number of debtors convey an idea …of debt among them, but when the averages are …population the idea conveyed must be that of the …

The per capita, or social level of debt, is shown in …nent for some of the principal groups of debts:

PER CAPITA DEBT, 1890.

| Description of debt. | Per capita debt. |
|---|---|
| Quasi public corporations: | |
| Railroad and street railway companies | $93 |
| Other quasi public corporations | 6 |
| Total | 99 |
| Individuals and private corporations: | |
| Real estate mortgages on incumbered farms, etc | 34 |
| Other real estate mortgages | 62 |
| Banks, loans and overdrafts, not including real estate mortgages | 49 |
| National, state, and local taxes | 17 |
| Other | 30 |
| Total | 192 |
| Total quasi public and private debt | 291 |
| Public debt | 32 |
| Total private and public debt | 323 |
| On incumbered farms and homes occupied by owners, crop liens, chattel mortgages, taxes, and "other net private debt" | 80 |

The per capita private debt is $291, public debt $32, the total being $323, or $1,594 per family. Few families owe this amount; and the foregoing analysis shows the sources of the debt that contributes to most of the averages. It comes largely from the capital of railroad and other quasi public corporations, from real estate purchases and improvements, and from the loans of banks.

If to the crop liens and chattel mortgages are added the taxes, "other net private debt," and the public debt, the total will include most of the debt to which debtors are most sensitive, although some of it is capital. The total of this group of debts is $4,930,404,795, that is, $79 per capita, or $388 to each family of 4.93 persons in 1890.

The reports of the eleventh census supply some averages of debt computed upon the number of debtors. Each family owning the farm it occupies under incumbrance owes an average incumbrance of $1,224; home, $1,293; farm or home, $1,257; home in the 420 cities and towns of from 8,000 to 100,000 population, $1,363; home in the 28 cities of 100,000 population and over, $2,337; home outside of the cities and towns of 8,000 population and over, $846. The average mortgage on acre tracts made during the ten years from 1880 to 1889, inclusive, was for $1,032; on lots, $1,509; on all real estate, $1,271; on lots in the 27 counties containing the 28 cities of 100,000 population and over, $2,798; mortgages made by quasi public corporations are not included.

To what extent real estate may be mortgaged is a matter of opinion, depending in the aggregate upon the consensus of opinion of lenders as to the degree of risk they will take. The degree of risk varies as between city and country, as between improved and unimproved real estate, and as between one region and another. The real estate mortgage debt has reached $6,019,679,965, and the estimated true value of taxed real estate and untaxed mines with which this [illegible] compared is $36,625,071,499. The taxed real [illegible]

more, the existing mortgage debt is 33.42 per cent of the three-fifths of its value, 27.85 per cent of the limit; if for 5.06 per cent of the limit.

estate values make possible a large mortgage debt, and as ile where real estate values are highly concentrated the of mortgage debt. Among the 2,781 counties covered by nvestigation of mortgages there are 27 that contain the 28 000 population and over, and the mortgage debt on the real se counties is 40.51 per cent of the entire real estate mort- the whole country. In the 338 counties containing the d towns of 8,000 population and over the mortgage debt is t of the total. There are 29 counties in each of which the bt is $25,000,000 and over, and the total mortgage debt on te in these counties is 43.34 per cent of the total for the ry. The 76 counties each having real estate with a mort- $10,000,000 and over, represent 55.20 per cent of the total, ounties each having an existing mortgage debt of $5,000,000 71 per cent of the total.

| | |
|---|---|
| [illegible] | $44,605.76 |
| [illegible] of loans repaid | 272,447.58 |
| [illegible] | 101,993.99 |
| Total | 419,047.33 |
| Disbursements: | |
| [illegible] | $192,947.68 |
| [illegible] | 134,043.95 |
| Expense account | 6,092.64 |
| Other disbursements, including cash on hand | 85,963.06 |
| Total | 419,047.33 |
| Profits: | |
| Net (expenses of management deducted) | $37,947.77 |
| Gross profits | 42,960.12 |
| Number of loans to pay for homes | 59 |
| Number of loans for other purposes | 157 |
| Total number of loans | 216 |
| Loans repaid, number | 1,078 |
| Loans repaid, amount | $964,615.18 |
| Loans outstanding at end of last fiscal year | $732,968.97 |
| Number of shares outstanding at end of last fiscal year | 29,469 |
| Value of shares at maturity | $4,108,850.00 |
| Per cent management expenses are of receipts | 1.97 |
| Per cent borrowers are of shareholders | 22.03 |
| Per cent homeowners of shareholders | 28.37 |
| Number of shareholders at end of last fiscal year, men | 2,251 |
| Number of shareholders at end of last fiscal year, women | 568 |
| Number of shareholders at end of last fiscal year, minors | 169 |
| Total number of shareholders | 2,988 |

**Effects of the Industrial Depression.**—The statistics presented on this subject were obtained directly from the books of 378 leading establishments, representing the principal industries in different parts of the state and giving employment to 48.17 per cent of the total number of employees in all industries, according to the United States census of 1890. As 1892 was a fairly prosperous year it was requested that the number of employees, wages paid, and hours of labor for that year be used in comparison with similar data for each month of the period of depression extending from June, 1893, to August, 1894, inclusive. The number of days entirely shut down and changes in wage rates during the period of depression were also called for. The results are given in detail for each establishment, and summarized in convenient form for the different industries. The extent of the depression is indicated by a tabular statement showing the percentages that the time, number of employees, wages, etc., are of the totals for each establishment when working under the conditions existing in 1892. The summary for all industries shows that the working time during the period of depression was about two-thirds of the full time, and the average number of employees was 84.83 per cent of the average number in 1892, while the average monthly payment in wages had decreased about 25 per cent. A large majority of the industries retained on the pay rolls a large percentage of the ordinary number of employees. The reduction made necessary by the depression was largely in the working time, and this is reflected with the nearest approach to accuracy in the lessening of the payments on account of wages. Of the larger industries woolen goods manufacture

nting on the results of child labor and educational laws in as compared with Connecticut the commissioner states: employed in New York factories 15 children for each 1,000 in Massachusetts 18 per 1,000, and in Connecticut 21 per e calculations are based on the United States census of beneficial results of extending the age limit are treated ppropriate heads of "Strength of body and of character," till and increased comforts," "Would not intensify competi-

OF REDUCED WORKING TIME ON PRODUCTION.—Informa- subject was obtained from about 100 establishments. The lled for the effect of a decrease in working time on piece d on production per employee. The answers are published ach establishment, by industries.

wing statement concerning establishments engaged in the e of hats is illustrative of the data furnished for the different

AKING ESTABLISHMENTS CLOSING AT NOON ON SATURDAY.

answers of each establishment, are discussed in detail. The statistical presentation is preceded by a general treatise on the subject under consideration.

TRADE AND INDUSTRIAL EDUCATION.—This subject is treated in textual form, and covers the methods prevailing in foreign countries and in various institutions in the United States, the data being gathered largely from the Eighth Annual Report of the Commissioner of Labor of the United States.

## INDIANA.

The Fifth Biennial Report of the Department of Statistics of Indiana for the years 1893 and 1894 opens with a summary of the data concerning different industries and a reproduction of the labor laws of the state. The subjects discussed in the report are as follows: Women wage earners of Indianapolis, 108 pages; labor organizations, 57 pages; domestic labor, 57 pages; coal mining statistics, 123 pages; the iron industries, 35 pages; the wood industries, 70 pages; miscellaneous industries, 53 pages; the glass industry, 33 pages; economic, social, and census statistics, 83 pages; cereal crops and farm animals, 5[illegible] pages; railroad statistics, 31 pages.

WOMEN WAGE EARNERS OF INDIANAPOLIS.—The statistics relating to this subject are compiled from the reports of 500 working women in Indianapolis engaged in 20 different industries. The reports were secured by a personal canvass, a representative number being selected for each industry. The questions were designed to obtain detailed information under the following heads:

1. Origin: viz, nativity of girl and of parents, whether city or country reared, and occupation of father.
2. Personal and industrial surroundings.
3. Wages and earnings.
4. Expenses and savings.

All but 31 of the 500 girls involved were born in the United States, and 359 were born in Indiana. Eleven working girls' parents were natives to every 9 girls' parents who were foreign. Eighty-four per cent of the girls were reared in the city. Forty per cent of the girls' fathers were mechanics, 31 per cent laborers, 15 per cent tradesmen, 6 per cent professional men, and 8 per cent in miscellaneous occupations. Nine-tenths of the girls were unmarried, and 86 per cent were living at home.

The facts presented in the following statement have been selected from the tables showing statistics concerning the origin and personal and industrial surroundings of the working girls of Indianapolis, and

| Industry or occupation. | Natives in— | | Occupation of father. | | | | | Conjugal condition. | | | Average age. |
|---|---|---|---|---|---|---|---|---|---|---|---|
| | City. | Country. | Laborer. | Mechanic. | Tradesman. | Professional man. | Miscellaneous. | Single. | Married. | Widowed. | |
| Bindery | 59 | 1 | 23 | 28 | 9 | 7 | 3 | 53 | 3 | 4 | 24.4 |
| Bookkeeping | 9 | 1 | ...... | 5 | 2 | ...... | 3 | 10 | ...... | ...... | 24.7 |
| Candy factory | 10 | ...... | 2 | 5 | ...... | ...... | 2 | 9 | 1 | ...... | 22.6 |
| Card factories | 15 | 5 | 11 | 7 | ...... | ...... | 2 | 20 | ...... | ...... | 19.7 |
| Carpet sewers | 10 | ...... | 1 | 4 | 2 | ...... | 3 | 8 | ...... | 2 | 32.5 |
| Chairs, etc., makers | 6 | 4 | 4 | 3 | 1 | ...... | 2 | 10 | ...... | ...... | 19.8 |
| Cotton mills | 15 | 5 | 10 | 9 | 1 | ...... | ...... | 18 | 1 | 1 | 20.3 |
| Dressmaking | 23 | 8 | 11 | 12 | 4 | 2 | 1 | 26 | 1 | 3 | 28.5 |
| Hair dressing | 8 | 2 | 3 | 6 | ...... | 1 | ...... | 10 | ...... | ...... | 20.9 |
| Laundry | 27 | 12 | 24 | 10 | 4 | 1 | 1 | 36 | 2 | 2 | 28.2 |
| Millinery | 26 | 4 | 6 | 10 | 6 | 4 | 4 | 26 | 3 | 1 | 25.2 |
| Pants, shirts, etc. | 39 | 21 | 20 | 31 | 6 | 3 | ...... | 50 | 4 | 6 | 27.2 |
| Paper box factory | 9 | 1 | 4 | 6 | ...... | ...... | ...... | 10 | ...... | ...... | 19.9 |
| Pork packing house | 9 | 1 | 7 | 2 | 1 | ...... | ...... | 9 | ...... | 1 | 26.0 |
| Saleswomen | 95 | 5 | 17 | 36 | 28 | 7 | 12 | 95 | 2 | 3 | 23.3 |
| Stenographers, etc. | 17 | 3 | 2 | 4 | 6 | 2 | 6 | 18 | 2 | ...... | 22.7 |
| Telephone | 10 | ...... | 1 | 7 | 1 | 1 | ...... | 10 | ...... | ...... | 21.3 |
| Tile works | 15 | 5 | 6 | 12 | 1 | ...... | 1 | 12 | 7 | 1 | 23.4 |
| Tobacco factory | 10 | ...... | 1 | 7 | 2 | ...... | ...... | 9 | 1 | ...... | 22.5 |
| Woolen mills | 10 | ...... | 3 | 6 | 1 | ...... | ...... | 10 | ...... | ...... | 23.7 |
| Total | 421 | 79 | 157 | 200 | 75 | 28 | 40 | 449 | 27 | 24 | a 23.7 |

The averages obtained from some of the other important subdivisions of the inquiry are given in the following summary:

WORKING TIME, EARNINGS, AND EXPENSES OF WOMEN WAGE EARNERS, INDIANAPOLIS.

| Industry or occupation. | Average— | | | | | | | Girls who save money. | |
|---|---|---|---|---|---|---|---|---|---|
| | Age of beginning work. | Hours of work. | | Weeks of vacation. | Unproductive weeks. | Earnings of past year. | Expenses of past year. | Number. | Average savings for year. |
| | | Daily, except Saturday. | Saturday. | | | | | | |
| Bindery | 17.4 | 10.0 | 9.0 | 1.0 | 5.3 | $260 | $253.28 | 7 | $57.60 |
| Bookkeeping | 19.7 | 8.7 | 8.3 | 1.6 | 4.0 | 494 | 448.60 | 4 | 113.25 |
| Candy factory | 15.7 | 10.0 | 8.5 | .8 | 7.0 | 220 | 215.00 | 1 | 50.00 |
| Card factories | 14.9 | 10.0 | 9.5 | .2 | 12.3 | 169 | 166.50 | 1 | 25.00 |
| Carpet sewers | 18.5 | 9.5 | 9.2 | .7 | 7.3 | 260 | 266.50 | 1 | 25.00 |
| Chairs, etc., makers | 17.2 | 10.3 | 8.0 | ...... | 16.2 | 156 | 146.00 | 1 | 100.00 |
| Cotton mills | 14.5 | 10.0 | 9.5 | 1.3 | 8.1 | 201 | 201.00 | ...... | ...... |
| Dressmaking | 18.7 | 9.4 | 10.1 | .6 | 9.1 | 255 | 231.78 | 10 | 69.66 |
| Hair dressing | 15.4 | 9.7 | 13.3 | 1.6 | 4.3 | 248 | 245.60 | 2 | 12.00 |
| Laundry | 16.1 | 10.0 | 8.0 | 1.0 | 5.4 | 306 | 283.00 | 10 | 67.92 |
| Millinery | 17.0 | 9.4 | 13.0 | 1.7 | 14.9 | 427 | 392.87 | 10 | 102.39 |
| Pants, shirts, etc. | 17.6 | 10.2 | 7.6 | .2 | 6.9 | 250 | 230.90 | 13 | 88.15 |
| Paper box factory | 17.3 | 9.5 | 8.3 | ...... | 7.9 | 124 | 124.00 | ...... | ...... |
| Pork packing house | 17.5 | 8.9 | 8.9 | ...... | 11.0 | 190 | 185.00 | 2 | 25.00 |
| Saleswomen | 16.6 | 9.3 | 13.0 | 1.7 | 4.8 | 265 | 246.10 | 25 | 75.60 |
| Stenographers, etc. | 18.3 | 8.3 | 7.9 | 1.6 | 7.1 | 346 | 319.90 | 6 | 87.00 |
| Telephone | 15.8 | 9.0 | 9.0 | 1.8 | 6.3 | 241 | 223.00 | 4 | 45.00 |
| Tile works | 17.2 | 8.0 | 5.0 | .2 | 5.1 | 158 | 155.50 | 1 | 25.00 |
| Tobacco factory | 14.9 | 8.2 | 7.6 | .8 | 4.8 | 200 | 204.50 | 2 | 22.50 |
| Woolen mills | 13.8 | 10.0 | 10.0 | ...... | 20.0 | 227 | 219.50 | 2 | 37.50 |
| Average | a 16.7 | a 9.4 | a 9.1 | a .9 | a 8.3 | a 250 | a 237.97 | 102 | a 51.43 |

a These averages were apparently obtained by adding together the industry averages and dividing the sum by the total industries, 20, and hence take no account of the number of individuals in each grouping. True averages might vary considerably from those here given.

| City. | Housework. | Cook. | Other domestic work. | Age. | Earnings of past year. | Places employed in. | Saved in past year. |
|---|---|---|---|---|---|---|---|
| Indianapolis | 120 | 61 | 53 | 26 | $145.77 | 1.8 | $23.96 |
| Evansville | 99 | 24 | 18 | 26 | 121.27 | 1.6 | 17.19 |
| Fort Wayne | 54 | 7 | 9 | 23 | 118.35 | 1.7 | 15.45 |
| Terre Haute | 66 | 8 | 4 | 26 | 124.40 | 1.9 | 14.45 |
| New Albany | 30 | 11 | 7 | 27 | 117.22 | 1.6 | 21.46 |
| South Bend | 28 | 5 | 3 | 23 | 145.12 | 1.7 | 27.36 |
| Richmond | 28 | 5 | 7 | 26 | 122.25 | 1.5 | 25.67 |
| Lafayette | 27 | 6 | 3 | 24 | 130.94 | 1.5 | 28.61 |
| Total | 467 | 127 | 104 | a 25 | 131.97 | a 1.7 | a 21.76 |
| Per cent | 64 | 20 | 16 | ...... | .......... | .......... | .......... |

a These averages were apparently obtained by adding together the city averages and dividing the sum by the total cities, 8, and hence take no account of the number of individuals in each city. True averages might vary considerably from those here given.

Coal Mining.—The statistics relating to coal mining, as reported by the operators of 71 coal mines in Indiana, representing an invested capital of $1,374,440 and a yearly wage account of $2,473,806, are shown for each mine; also individual reports for 961 miners representing 81 mines. The data were obtained by a personal canvass, "and may be said to show, not approximately, but correctly, the matters which it was designed to call out by the questions." The questions addressed to the miners obtained informaton concerning age, social relations, nativity, hours of work, cost of and price paid for mining coal, daily wages, net earnings, etc. The presentation is a complete showing for the coal mining industries of the state.

Iron, Wood, Glass, and Miscellaneous Industries.—The statistics of iron industries, wood industries, miscellaneous industries, and the glass industry were compiled from returns secured on a personal canvass of 375 establishments, 101 of which were engaged in various iron industries, exclusive of blacksmith and repair shops, 163 in the manufacture of articles in which wood is the exclusive or chief material, 45 in the manufacture of glass, and 66 in miscellaneous industries. The establishments report the employment of 40,253 hands, and of this number individual reports were secured from 2,423, distributed as follows: 577 in the iron, 1,035 in the wood, 134 in the glass, and 677 in the miscellaneous industries. The reports of the proprietors and of the employees, respectively, are published in detail by cities, industries, and occupations.

The proprietors' reports furnish data as to capital, cost of materials, value of products, working time, number of employees, total wages, highest and lowest daily wages, and average wages of boys and of women and girls; also as to strikes and increase or decrease in wages.

VALUE OF PRODUCTS, WAGES, ETC., IN VARIOUS INDUSTRIES.

| Industry. | Establishments. | Buildings, grounds, and machinery. | Cost of materials. | Value of products | Total wages. |
|---|---|---|---|---|---|
| Iron | 101 | $5,830,231 | $9,146,897 | $18,069,340 | $4,174,891 |
| Wood | 163 | 4,615,430 | 9,994,589 | 18,403,267 | 4,900,008 |
| Glass | 45 | 4,987,635 | 1,865,805 | 6,493,518 | 2,950,758 |
| Miscellaneous | 66 | 4,358,993 | 15,816,082 | 21,009,450 | 2,459,808 |
| Total | 375 | 19,792,289 | 36,823,373 | 63,975,575 | 14,485,465 |

AVERAGE DAILY WAGES IN VARIOUS INDUSTRIES.

| Industry. | Employees. | | | Average daily wages. | | | | | |
|---|---|---|---|---|---|---|---|---|---|
| | | | | | | Skilled labor. | | Unskilled labor. | |
| | Men. | Boys. | Women and girls. | Boys. | Women and girls. | Highest. | Lowest. | Highest. | Lowest. |
| Iron | 10,514 | 1,250 | 146 | $0.74 | $1.04 | $3.75 | $1.92 | $1.50 | $1.13 |
| Wood | 11,393 | 1,842 | 310 | .71 | .90 | 2.85 | 1.72 | 1.48 | 1.11 |
| Glass | 5,163 | 1,536 | 195 | .82 | .67 | 3.48 | 2.34 | 1.90 | 1.26 |
| Miscellaneous | 4,035 | 466 | 3,403 | .79 | .88 | 3.20 | 1.82 | 1.45 | .91 |
| Total | 31,105 | 5,094 | 4,054 | ........ | ........ | ........ | ........ | ........ | ........ |

The employees' statements contain data as to age, apprenticeship, number of years engaged in present occupation, working time, highest, lowest, and average wages, social condition, income, expenses, etc., for the different classes of employees in each industry treated.

The principal facts reported by the employees are summarized as follows:

CONDITION, EARNINGS, ETC., OF EMPLOYEES IN VARIOUS INDUSTRIES.

| Industries. | Employees. | | | | | Average. | | | | | | Savings. | |
|---|---|---|---|---|---|---|---|---|---|---|---|---|---|
| | | | | | | Wages. | | | | | | | |
| | Number reporting. | Married. | Single. | Owning homes. | Renting. | Highest. | Lowest. | Average. | Hours per day. | Days per year. | Annual income. | Employees who saved. | Total savings for year. |
| Iron | 577 | 468 | 109 | 223 | 232 | $2.65 | $2.26 | $2.45 | 9.1 | 287 | $610 | 168 | $24,164 |
| Wood | 1,035 | 774 | 261 | 356 | 397 | 2.15 | 1.93 | 2.03 | 9 | 274 | 556 | 199 | 22,621 |
| Glass | 134 | 108 | 26 | 22 | 85 | 4.80 | 4.08 | 4.41 | 8.4 | 212 | 1,022 | 83 | 11,345 |
| Miscellaneous | 677 | 255 | 422 | 76 | 161 | 1.86 | 1.44 | 1.50 | 10 | 250 | 448 | 61 | 8,459 |
| Total | 2,423 | 1,606 | 818 | 677 | 875 | ...... | ...... | ...... | ...... | ...... | ...... | 511 | 66,589 |

ECONOMIC, SOCIAL, AND CENSUS STATISTICS.—Under this caption are presented county, city, and town indebtedness and expenses, also real estate transfers, mortgages, and satisfactions recorded in the several counties of the state. These tables show also the number and condition of inmates of asylums, number of divorces, with the causes of complaint, number of persons naturalized, and number of jail incarcerations. Some of the results of the Eleventh Census of the United States are reproduced.

CEREAL CROPS AND FARM ANIMALS.—The figures and analysis given relating to cereal crops and farm animals constitute a full presentation of the agricultural industries of the state, by county and by state totals.

RAILROADS.—The statistics relating to railroads show in the usual form the totals for 31 roads that were in operation in the state in 1893 and 25 in 1894, some companies not furnishing their reports for 1894 in time to be included.

## MICHIGAN.

The Twelfth Annual Report of the Bureau of Labor and Industrial Statistics of Michigan, for the year ending February 1, 1895, presents the results of investigations into the following subjects: Farm laborers, male, 236 pages; domestic labor, female, 101 pages; statistics from farm proprietors, 109 pages; miscellaneous agricultural statistics, 55 pages; strikes, 21 pages; prisons and prison labor, 4 pages.

FARM AND DOMESTIC LABOR.—The statistics presented under the titles of "Male farm laborers" and "Female domestic labor" are the results obtained from reports made by 5,600 male farm laborers and 2,300 female domestic laborers. The data were collected by the enumerators while engaged in taking the state census. The schedules contained numerous questions as to nationality, age, working time, wages, extras, increase or decrease in wages during given periods, effect of immigration on occupation, etc., as well as questions concerning social conditions. Some of the important results of both investigations are combined in the following summary:

LABORERS ON FARMS AND DOMESTIC SERVANTS.

| Items. | Male farm laborers. | Female domestics. |
|---|---|---|
| Total number considered | 5,600 | 2,300 |
| Americans | 3,219 | 1,431 |
| Germans | 726 | 312 |
| All other nationalities | 1,655 | 557 |
| Average monthly wages | $17.84 | .......... |
| Average weekly wages | .......... | $1.85 |
| Average daily wages | $0.92 | $0.59 |
| Total earnings past year | $1,018,388 | $168,464 |
| Average yearly earnings | $181.85 | $73.24 |
| Amount of money saved past year | $196,891 | $34,528 |
| Average amount for those who saved | $77.67 | $34.80 |
| Number reporting increase in wages past five years | 335 | 324 |
| Number reporting decrease in wages past five years | 3,395 | 675 |
| Number who say times better than five years ago | 146 | 177 |
| Number who say times worse than five years ago | 4,542 | 1,367 |
| Number who say immigration injures their occupation | 3,466 | 814 |
| Average daily wages of foreigners in native land | $0.557 | $0.25 |
| Number of foreigners who say conditions for saving money are better than in native land | 1,089 | 348 |

In some of the returns answers were not given to all the questions. It therefore does not follow that the difference between the number given for any particular item in the above summary and the total number considered represents the number reporting the reverse from what is shown. The report presents the statistics in detail for each laborer, male and female, from whom returns were received.

STATISTICS FROM FARM PROPRIETORS.—These facts were furnished by 935 farmers in Michigan. The effort was made to obtain reports from a reasonable number in each county, that the showing might be general for the state. The inquiries not only covered the question of wages and the condition of wage workers on farms, but also questions pertaining to the staple products of the farms. The average yield and cost of production of leading crops are shown. The number of farmers reporting profit and no profit in stock raising, in dairying, and in poultry raising is also given.

The details shown by the tables are numerous and worthy of careful study, but only a few of the many important results can be stated. The average number of years in which those reporting had been engaged in farming was 25.7. Four hundred and fifty-four employed female help, the average weekly wages for such labor being $1.94. The average monthly wages for males was $18.85. Adding the value of extras, such as fuel, pasture for cow, house rent, etc., made the average daily wages paid male farm laborers for the entire state over $1. The average yield and the average cost of raising per acre, including interest on value of land, is shown for a number of farm products, the results being summarized as follows:

AVERAGE YIELD AND COST PER ACRE OF RAISING CERTAIN FARM PRODUCTS.

| Product. | Yield per acre. | | Cost of raising per acre. | |
|---|---|---|---|---|
| | Farmers reporting. | Average (bushels). | Farmers reporting. | Average. |
| Wheat | 859 | 18.8 | 737 | $9.78 |
| Corn | 849 | 57.4 | 696 | 10.35 |
| Oats | 864 | 35.1 | 692 | 7.74 |
| Barley | 161 | 28.5 | 167 | 7.84 |
| Potatoes | 763 | 107.9 | 558 | 14.84 |
| Beans | 254 | 16.1 | 237 | 8.43 |
| Peas | 239 | 16.4 | 189 | 7.71 |
| Clover seed | 268 | 2.2 | 214 | 4.90 |
| Hay | 881 | a 1.4 | 686 | 5.42 |
| Mint | 13 | b 18.3 | 11 | 15.16 |

a Tons. b Pounds.

About 56 per cent of the farmers reporting are satisfied that there is a profit in dairying, less than 37 per cent that there is a profit in fattening cattle for market, and only 15 per cent that there is a profit in raising horses for sale. Two-thirds of those canvassed say there is a profit in raising poultry for market, and 82 per cent that there is profit in fattening hogs for market. Six hundred and thirty-four of the 935 reporting say there is profit in farming, 162 say there is no profit, and 139 do not answer the question.

These statistics are followed by general remarks from a number of leading farmers in different sections of the state on methods, profits, and the desirability of farming as an industry.

MISCELLANEOUS AGRICULTURAL STATISTICS.—The presentations under the head of miscellaneous agricultural statistics are compilations from the United States census of 1890 and the state census of 1894. They show the size and value of farms with the value and quantity of farm products, by counties and by townships.

STRIKES.—Each strike that occurred in the state during 1894 is described, and is followed by general information concerning some of the large strikes that occurred elsewhere.

PRISONS AND PRISON LABOR.—The number of inmates in the state prison and in the different houses of correction during 1894 is given. The number engaged on contract work, with the average price per day for their work, is shown; also the number engaged on state work.

## MINNESOTA.

The Fourth Biennial Report of the Bureau of Labor of Minnesota is for the years 1893 and 1894. In the introduction to the report the law approved April 19, 1893, changing the name of the office from the Bureau of Labor Statistics to the Bureau of Labor, is quoted, and the general work of the bureau outlined. The contents of the report are as follows: Chattel mortgages and pawnbrokers' loans, 43 pages; agricultural statistics, 66 pages; the apprentice system, 257 pages; mortgage statistics, 164 pages; factory inspection, 125 pages.

CHATTEL MORTGAGES AND PAWNBROKERS' LOANS.—The statistical information presented under this title is the result of an examination of the contracts, leases, mortgages, and other instruments, having the force of chattel mortgages, filed at the city clerk's office of Minneapolis during the year 1893. For the pawnbrokerage business of the city during the same year the data were obtained from the returns made to the chief of police.

The instruments classified as chattel mortgages are divided into two general classes—the first including those executed to secure the cost price of goods purchased and the second those executed to guarantee the repayment of borrowed money. Some of the principal facts concerning the first class are summarized as follows:

INTEREST ON CHATTEL MORTGAGES, MINNEAPOLIS, 1893.

| Interest or credit charge. | Number of instruments. | Goods purchased. | | | | | |
|---|---|---|---|---|---|---|---|
| | | Household goods. | Musical instruments. | Carriages, wagons, live stock, etc. | Merchandise. | Farm machinery. | Miscellaneous. |
| Six per cent per annum | 20 | .......... | 7 | .......... | 2 | .......... | 11 |
| Seven per cent per annum | 100 | 5 | 76 | 2 | 1 | 1 | 15 |
| Eight per cent per annum | 842 | 13 | 763 | 4 | 6 | .......... | 56 |
| Ten per cent per annum | 842 | 781 | 23 | 15 | 2 | .......... | 21 |
| No interest charged | 3,591 | 2,888 | 124 | 20 | 68 | 3 | 488 |
| Five per cent addition | 5,540 | 5,540 | .......... | .......... | .......... | .......... | .......... |
| Total | 10,935 | 9,227 | 993 | 41 | 79 | 4 | 591 |

The number of instruments under the head of "no interest charged" is slightly greater than it actually should be by the inclusion of a few for which the record contained no information as to the interest charged.

Household goods and musical instruments taken together make up about 93 per cent of the sales where chattel mortgages were executed to secure the cost price of the articles purchased. The selling price of the goods purchased by residents of Minneapolis, on the chattel mortgage system, during the year 1893 amounted to $772,537.36. The instruments making the record of these sales show a cash payment at the time of purchase of $110,827.90, leaving a debt of $661,709.46. The average duration of the credit was 5.35 months. These amounts do not include sales for cash or unsecured credit, nor for secured credit to parties residing outside of the city limits.

The chattel mortgages given to secure the repayment of borrowed money are also divided into two classes—those at legal and those at usurious and extortionate rates of interest. The division, however, can only be made approximately. Of chattel mortgages made to secure loans and not known to be extortionate in their interest charges there were 2,171 in 1893, representing an indebtedness of $515,845.06, the average for each mortgage being $237.61, with a duration of 5.36 months. It is believed, however, that some 500 of these loans were at usurious interest, which would reduce the number at strictly legal interest to 1,671, representing a mortgage debt of $495,600.06, the average of the loans being $296.59, and the life of the mortgage 5.75 months.

There were 2,211 usurious loans reported for the year, the face of the mortgage debt amounting to $89,310.02, on which the borrowers probably realized about $80,000 in cash. The borrowers giving these mortgages, so far as could be ascertained, always executed liens for sums about 10 per cent greater than the loans secured by them. The average debt for these loans was $40.49, hence the average loan or money obtained was, approximately, $36. Two-thirds of these loans were secured on household goods. Selecting 95 typical usurious loans, the rate of interest was found, upon inquiry of the borrowers, to range from 41 to 480 per cent per annum. Including the loans classed as legal, but probably usurious, there were, approximately, 2,700 usurious loans in the city during the period covered, representing $110,000, upon which the borrowers obtained less than $100,000 in cash.

There were twenty-five licensed pawnbrokers doing business in Minneapolis in 1893, who paid as license fees $2,458.34. Twenty-three thousand and ninety loans were reported by these brokers, the total amount borrowed being $142,248.12, and the average for each pledge $6.16. There were 5,425 purchases reported by pawnbrokers, the total amount paid therefor being $15,055.19.

The statistics of chattel mortgages and pawnbrokers' loans are presented in detail and accompanied by an extended textual [illegible], in which various loan institutions in the United States and [illegible] countries, established primarily for the relief of the poor, [illegible]

[illegible]—This is the result of an inquiry started [illegible] of 1895 and designed to ascertain something of the [illegible] prosperity, the elements of success, and the causes of failure among the farmers of the state. The data were obtained by agents of the bureau, who secured reports from 1,555 farm owners and 243 farm tenants. In securing these reports counties and townships were selected that were supposed to be representative of the entire state. All the farms in each township selected were visited, and so far as possible returns were secured from each. The following summaries indicate the character of some of the principal branches of the inquiry and the results obtained:

VALUE OF AGRICULTURAL PROPERTY.

| State or country of birth. | Farmers. | Tenants. | Years' farming in Minnesota. | | Value of possessions at beginning. | | Value of present possessions. | |
|---|---|---|---|---|---|---|---|---|
| | | | Farmers. | Tenants. | Farmers. | Tenants. | Farmers. | Tenants. |
| Minnesota | 144 | 39 | 1,725 | 382 | $196,305 | $10,755 | $792,466 | $43,960 |
| United States | 377 | 78 | 6,950 | 740 | 437,707 | 26,440 | 2,891,937 | 119,504 |
| Germany | 317 | 49 | 6,157 | 307 | 266,930 | 15,525 | 1,843,318 | 67,690 |
| Great Britain | 95 | 5 | 2,269 | 63 | 72,100 | 1,050 | 615,234 | 3,906 |
| Scandinavia | 484 | 56 | 6,980 | 518 | 142,486 | 17,375 | 1,630,047 | 55,312 |
| Bohemia | 94 | 7 | 1,576 | 33 | 71,660 | 1,000 | 338,096 | 5,752 |
| British Possessions | 47 | 8 | 1,044 | 25 | 48,700 | 850 | 274,898 | 6,091 |
| Other countries | 17 | 1 | 382 | 8 | 8,610 | ........ | 62,708 | 2,722 |
| Total | 1,555 | 243 | 27,083 | 2,076 | 1,239,498 | 72,995 | 8,449,704 | 304,838 |
| Averages | ........ | ........ | 17 | 9 | 797 | 300 | 5,434 | 1,254 |

INDEBTEDNESS OF FARMERS AND AGRICULTURAL TENANTS.

| State or country of birth. | Amount of indebtedness. | | Net possessions. | |
|---|---|---|---|---|
| | Farmers. | Tenants. | Farmers. | Tenants. |
| Minnesota | $103,237 | $5,560 | $690,229 | $38,300 |
| United States | 311,997 | 17,016 | 2,579,940 | 102,488 |
| Germany | 166,818 | 9,784 | 1,676,500 | 57,906 |
| Great Britain | 53,234 | 600 | 562,010 | 3,306 |
| Scandinavia | 284,922 | 14,255 | 1,345,125 | 41,058 |
| Bohemia | 60,691 | 2,295 | 277,405 | 3,457 |
| British Possessions | 23,965 | 1,500 | 250,923 | 4,591 |
| Other countries | 3,900 | ........ | 58,808 | 2,722 |
| Total | 1,008,754 | 51,010 | 7,440,950 | 253,828 |
| Averages | 649 | 210 | 4,785 | 1,045 |

The information contained in each of the 1,798 reports is shown in detail, including the several items constituting the total value of present possessions given in the above statement. The results are summarized and the averages shown by counties and by nationalities. The following statements are taken from the comments on the figures:

The possession of a sufficient amount of capital at the outset is the most potent single factor in the accumulation of farm wealth.

Of the 1,798 farmers visited, 17, or less than 1 per cent, had, by reason of [illegible] and insufficient capital, dropped back from farm ownership to tenancy, while 235 had risen from tenancy to farm ownership after [illegible] as tenant of four years.

A little less than one-half of the tenants visited had such a small amount of capital that they rented farms for one-half of the produce, the landlord furnishing live stock, farm implements, and seed, or a large proportion thereof.

Thirty-one of the 1,555 owners, at some time in their lives, had lost a farm by mortgage foreclosure, but were able in a short time to retrieve their fortunes and regain their earlier place as farm owners.

The American-born farmer is seen to succeed considerably better than any body of newcomers from Europe.

THE APPRENTICE SYSTEM.—The treatment of the subject of the apprentice system is almost entirely textual. The history of the apprentice system is traced from its origin in the ancient craft or trade guilds of the Middle Ages. The relation between apprenticeship and strikes is treated at considerable length. The statistics of strikes involving the apprentice question, as published for Great Britain, the United States, and the state of New York, are reproduced to show the extent of the disturbances into which it enters as a factor. While in all three reports the apprentice question is shown to have been the source of some trouble, only a very small proportion of the industries have any serious trouble over it.

In order to secure as much information as possible in regard to the relation between apprenticeship and trade unions, and especially to ascertain whether the unions were controlled by the foreign-born population and whether the American boys were discriminated against in securing membership, the bureau obtained from members of trade unions in the state statements showing for each workman his birthplace, where he learned his trade, the years served as apprentice, and kindred information. Returns were received from 1,985 workmen, and of this number 58.54 per cent were born in the United States and 41.46 per cent were foreign born. On the other hand returns from 133,762 males of voting age in the state showed that only 38 per cent were native born. In other words, the percentage of native born workmen in the trade unions, or 58.54 per cent, was 1.5 times as great as the percentage of native born in the voting population, or 38 per cent. There were 1,624 members of the trade unions, or 81.86 per cent, who acquired their trade in the United States, while only 361, or 18.14 per cent, acquired their trade in foreign lands.

The attitude of a number of national and international labor organizations toward apprentices and cheap labor is discussed. The rules and regulations of the several organizations on this subject are quoted, and in those unions where the membership is composed largely of foreign-trained craftsmen facts are presented showing the cause or reason for the same. Where the unions are known to have had strikes in recent years relating to the employment of apprentices, all available facts relating to the dispute are presented. The actions of several associations of employers on the apprentice question are referred to, particular attention being given to the attitude of the National Association of Builders of the United States of America on this subject.

The last 16 pages of the discussion of the apprentice system consist of a summary in which the history of the system and its present status in the industries and trade unions of the United States is given in concise statements.

MORTGAGE STATISTICS.—The different sections of this subject are treated under the following heads: First, real estate mortgage indebtedness; second, mortgage foreclosures; third, redemptions of mortgage foreclosures. Under the first head are shown data relating to the mortgages placed on record, the amount of taxable land as reported by the state auditor, and the general agricultural statistics gathered by the United States census, and comparisons between the same and deductions therefrom. These statistics, as a rule, cover the period from 1880 to 1889, but for eight typical counties the bureau secured and presents statistics of mortgages and taxable land for each year from 1859 to 1893. All of the statistics presented under this general head of mortgages were gathered with the thought that possibly such information would throw new light upon the true relation of mortgage debt to the development and financial prosperity of the average Western community, agricultural or urban. In addition to statistical tables presenting the data of mortgages, agriculture, and taxable property by counties and groups of counties, the report contains graphic tables showing the leading facts for the different branches of the investigation. In the discussion of the figures the increase or decrease of the actual or relative mortgage debt in the different counties is traced, and careful explanation given of the various causes controlling the results shown.

The amount of mortgages placed on record in Minnesota and the acres mortgaged increased relatively, as well as actually, with some irregularity, from 1861 until about 1880. Since that date it has relatively continuously, though irregularly, decreased. In 1893 there was relatively 1 acre of farm land mortgaged for every 2.2 acres thus mortgaged in 1880, and there was $1 of incumbrance on such farms for every $1.80 of such incumbrance in the earlier year. While there had been this relative decrease of farm mortgages there had been a slight increase in the total actual amount of outstanding mortgage debt. But the farm debt of 1893 was, if any larger than that of 1880, increased by an amount so slight that such addition could not have exceeded 1 per cent of the property accumulated by the farmers of Minnesota and added to their former possessions between 1880 and 1893. The statistics relating to mortgage foreclosures are contained in five tables, which give the number, amount, and acreage of foreclosures by counties and groups of counties for each year during the period from 1880 to 1893, and for eight typical counties from 1859 to 1893, with percentages of taxable land sold on foreclosure and of mortgaged acres foreclosed.

The percentage of foreclosures of the mortgages executed and the general movement of foreclosure in city and agricultural property are treated separately. The discussion shows the salient changes in the condition of agriculture and the causes affecting wheat prices and farm prosperity and the foreclosure of farm mortgages in the past thirty-five years in Minnesota.

The following extracts are selected from a list of fifteen conclusions reached after a careful analysis of the figures:

When the foreclosures of one year are compared with the mortgages recorded four years before [four years being the life of the average mortgage], it is found that the foreclosures on farm and acre property in the agricultural counties of the state in 1892 and 1893 were relatively 40 to 50 per cent smaller in number and in acres and amounts involved than in 1884 and 1885.

Between the years 1880 and 1881 and the years 1892 and 1893 the foreclosures on acre property so decreased that relatively only one farm was sold in the latter years by foreclosure where three farms were sold in the earlier, and that one acre of land was foreclosed where two had formerly been, and that the amounts of foreclosure sales had declined so, relatively, that only $1 of such sales is now occurring where in 1880 there were $4 of the same.

The foreclosures of 1892 and 1893 were relatively only one-fifth as numerous as twenty-four years before, in 1869 and 1870. The acres sold were only one-fourth and the amounts involved one-fifth as great in the latter as in the earlier years.

In the history of the state there can be traced two sources of mortgage foreclosure: One arises from the imperfection of the farm owner, that which is due to his lack of experience, his shiftlessness and want of character, or knowledge, or energy; the other is crop failures and varying prices for wheat.

In thirty-five years the rates of interest for farm loans have decreased from the prevailing rate of from 3 to 10 per cent a month in Mower County in 1859 and 1860 to an average of not far from 8 per cent per annum in 1893.

Crop failures by the introduction of diversified farming have ceased to be as great a possible factor for evil as between 1876 and 1881. Wheat prices as a special disturbing factor are becoming of less and less importance with the passage of years.

The data relating to the redemptions of mortgage foreclosures are not considered as complete or perfect. The redemptions for which statistics were secured include only those transactions whereby the original owner recovered possession by means of a legal instrument, placed upon record, usually designated a redemption. Many owners whose lands had been sold under foreclosure proceedings, instead of securing a redemption, obtained a quitclaim deed of the land. These redemptions by quitclaim deeds make up at least one-third of the total redemptions of the state, and in some counties one-half.

The statistics are presented only by groups of counties, and cover the period from 1880 to 1893, and for a group of eight typical counties from 1859 to 1893. From the textual consideration of the subject are taken the following extracts:

In the state as a whole there is an increase in the foreclosures on acres, but a greater one in redemptions. The reverse is the case with lots, and shows that the financial condition of the farmers and owners of acre property has increased more than their debts, while the opposite is the case with the owners of other real estate.

In the sixty-five agricultural counties of the state there were in 1880 and 1881 for every 100 foreclosures on acre property 16 redemptions, while in 1892 and 1893 there were 22.6. In the earlier years there were for every $100 of foreclosures $12.03 of redemptions, while in the latter years there were $16.21.

Comparing all foreclosures and redemptions in the city counties it is found that in 1880 and 1881 there were for every 100 foreclosures 33.3 redemptions, while in 1892 and 1893 there were only 6.4, or only one-sixth as many. In 1880 and 1881 for every $100 of foreclosures on property in city counties there were $20.84 of redemptions, while in 1892 and 1893 there were only $5.53, or barely one-fourth as much.

Making allowance for the redemptions by quitclaim deeds in Minnesota [for which no data were secured] it becomes apparent that from one-fourth to one-third of all farm mortgages foreclosed in the state during the last few years were, or will be, redeemed by the owners of the farms.

FACTORY INSPECTION.—This subject constitutes Part II of the report of the bureau. Guards for dangerous machinery is the first subject treated, the discussion containing 16 illustrations of various machines to which different forms of guards have been attached. The statistics of accidents in the factories and mines of the state show the character of the machine on which the accident happened, or the cause of the same, and the character of the injury. There were 631 accidents reported between April 1, 1893, and December 31, 1894. Forty-three of these were reported by mines and 588 by factories.

The laws of the state regulating the employment of women and children, and various laws bearing on labor and labor organizations, are quoted and amendments recommended.

The condition of guards for switch rails, guard rails, and frogs in 1893 and 1894 on the various railroads in the state is shown.

Between May 1, 1893, and December 31, 1894, the inspectors visited 1,388 different factories and mills in the state. At the time of the first inspection in 1893 these establishments employed 38,866 operatives, of whom 34,436 were males and 4,430 females. The name and address, facts concerning employees and wages, and the various changes in the buildings and machinery ordered by the inspectors are given for each factory inspected. The detail tables are summarized according to the character of the changes ordered and by industries.

## MISSOURI.

The Sixteenth Annual Report of the Bureau of Labor Statistics and Inspection of Missouri is for the year ending November 5, 1894. The first pages of this report contain a discussion of existing conditions and tendencies of the times, and a synopsis of the current work of other labor bureaus. The substance of the report is divided as follows: Earnings of employees in lead mines, 33 pages; statistics of manufactures, 149 pages; factory inspection, 35 pages; [illegible] costs, 125 pages; building and loan associations, 188 pages; [illegible], 19 pages.

EARNINGS OF EMPLOYEES IN LEAD MINES.—Under this subject reports for 1,281 employees in three representative lead mines in different sections of the state, concerning the number of working days, days actually worked, and actual and average earnings are given in detail for each employee. The results are summarized as follows:

TIME AND EARNINGS OF EMPLOYEES OF LEAD MINES, [illegible].

| Items. | Total. | Name of company. | | |
|---|---|---|---|---|
| | | Doe Run. | Center Creek. | Victor. |
| Total number of men employed | 1,281 | 709 | 161 | 411 |
| Number of men required to have done the work if each man had worked each working day in the year | 318+ | [illegible] | 71+ | [illegible]+ |
| Per cent of days worked of working days in period | 79+ | 79.73 | [illegible]+ | [illegible]+ |
| Average daily wages for days worked in period | $1.69 | $1.49 | $1.[illegible] | [illegible] |
| Average daily wages for working days in period | $1.27 | $1.19 | $1.51 | [illegible] |
| Average earnings for each man | $128.41 | [illegible] | [illegible] | [illegible] |
| Average number of days each man worked | 77 − | 90.50+ | 127 | [illegible]+ |
| Average number of working days to each man | 97 − | 112.75+ | [illegible] | [illegible]+ |
| What the average annual earnings would have been if each man had worked every working day in period at the average rate of daily wages for days worked | $497+ | $440.20 | [illegible] | [illegible] |

STATISTICS OF MANUFACTURES.—The statistics of manufactures collected by the bureau are preceded by a reproduction and discussion of the results of the United States census. The bureau secured reports from 757 private firms and 716 corporations. The number of male and female partners and stockholders, aggregate and average values of capital, stock used, wages, goods made, and proportion of business done are shown by industries, the summary for all industries being as follows:

MANUFACTURING INDUSTRIES, 1893.

| | |
|---|---|
| Establishments reporting: | |
| Number of private firms | 757 |
| Number of corporations | 716 |
| Total | 1,473 |
| Number of partners: | |
| Males | 1,[illegible] |
| Females | [illegible] |
| Total | 1,[illegible] |
| Number of stockholders: | |
| Males | [illegible] |
| Females | [illegible] |
| Banks, trustees, etc | [illegible] |
| Total | 7,[illegible] |
| Amount of capital invested | $[illegible] |
| Stock of material used | [illegible] |
| Other supplies | [illegible] |
| Total | [illegible] |

The building and loan associations in the city of Saint Louis are treated separately from those in the state exclusive of the city, and the totals combined. Three hundred and fifty-five active associations are given for the year 1894, reports being received from 314, while 41 known to be in existence failed to make reports. One hundred and eighty-nine of the associations were in the city of Saint Louis and 166 in the state outside of the city. Three hundred and seventy-six associations were reported for the state in 1893; 21 others had been incorporated in 1893, making the total number of associations having a nominal existence 397. If to the number 355, supposed to be active in 1894, be added those in liquidation and chartered in 1894 previous to July 1, the total number will be about the same as 1893. Some of the totals shown for all associations for 1894 are summarized as follows:

BUILDING AND LOAN ASSOCIATIONS, 1894.

| Items. | Saint Louis. | State, exclusive of Saint Louis. | Total. |
|---|---|---|---|
| Assets | $22,303,446.15 | $11,101,149.88 | $33,404,596.03 |
| Liabilities: | | | |
| Value of shares outstanding, including gain | $17,995,099.59 | $9,136,053.10 | $27,131,152.69 |
| Other liabilities, including undivided profits | 4,308,346.56 | 1,965,096.78 | 6,273,443.34 |
| Total | 22,303,446.15 | 11,101,149.88 | 33,404,596.03 |
| Receipts: | | | |
| Cash on hand at close of last fiscal year | $230,535.81 | $225,219.72 | $455,756.53 |
| Cash receipts in last fiscal year, exclusive of loans repaid | 8,728,687.45 | 3,345,618.68 | 12,074,346.13 |
| Loans repaid | 1,866,675.96 | 829,169.11 | 2,695,845.07 |
| Total | 10,825,839.21 | 4,400,007.52 | 15,225,846.73 |
| Disbursements: | | | |
| Loans on mortgage security | $1,929,614.96 | $1,682,241.23 | $3,611,856.18 |
| Withdrawals | 2,787,142.86 | 1,630,897.03 | 4,418,039.89 |
| Borrowed money repaid | 4,329,279.38 | 379,253.81 | 4,708,533.19 |
| Other disbursements, including cash on hand | 1,779,802.02 | 707,615.45 | 2,487,417.47 |
| Total | 10,825,839.21 | 4,400,007.52 | 15,225,846.73 |
| Profits: | | | |
| Expenses | $2,374,584.35 | $776,321.93 | $3,150,906.28 |
| Net profits | 5,307,559.68 | 2,340,237.16 | 7,647,796.84 |
| Gross profits | 7,682,144.03 | 3,116,559.09 | 10,798,703.12 |
| Number of shares issued during the year | 21,687.96 | 39,560.00 | 61,247.95 |
| Number of shares withdrawn during the year | 61,069.33 | 42,289.75 | 105,359.08 |
| Number of shares loaned on during the year | 7,201.51 | 10,129.29 | 17,330.80 |
| Present total number of shares loaned on | 96,162.01 | 46,112.39 | 142,274.40 |
| Present total number of free shares | 122,858.98 | 112,411.67 | 235,270.65 |
| Present total number of all shares | 218,020.99 | 160,524.06 | 378,545.05 |
| Total number of borrowers | 7,287 | 10,474 | 17,761 |
| Total number of nonborrowers | 20,608 | 22,730 | 43,338 |
| Total number of persons who are shareholders | 27,895 | 33,204 | 61,099 |
| Homes secured and paid for | 111 | 1,082 | 1,193 |
| Number of homes partially paid for | 7,236 | 7,004 | 14,240 |

STRIKES.—A brief account is given of two interstate strikes—the strike originating with the employees of Pullman's Palace Car Com[illegible] pany, of Pullman, Illinois, and the coal miners' strike of April 21, [illegible] of several minor strikes in the city of Saint Louis.

The report is accompanied with an industrial map [illegible] tion of all mines and railroads in the state and a [illegible] statistics.

Europe, trouble in hiring first-class workmen, . Reports from employers, representing 2,674 [illegible] wages per hour, are also shown, the by occupations and cities and covering all g trades. The wages paid per hour in these ilwaukee are placed in comparison with the in the several cities of the United States. l average rate per hour for all the trades some of the cities are given as follows:

UR IN BUILDING TRADES IN VARIOUS CITIES, 1893.

| City. | Wages per hour. |
|---|---|
| [illegible] | $0.296 |
| [illegible] | .213 |
| [illegible] | .266 |
| [illegible] | .478 |
| [illegible] | .216 |
| [illegible] | .301 |
| [illegible] | .240 |
| [illegible] | 286 |
| [illegible] | .274 |
| [illegible] | .332 |
| [illegible] | .344 |
| [illegible] | 326 |

ACTURES.—The statistics of manufactures show dustries the number and per cent of employees ges including the per cent receiving less than nount of wages paid in different industries in each year from 1888 to 1893, inclusive, and the per employee in the different industries for 893, inclusive.

The following statement presents the average annual wages in some of the leading industries treated in the summary table:

AVERAGE ANNUAL WAGES PAID IN VARIOUS INDUSTRIES, 1889 TO 1893.

| Industry. | 1889. | 1890. | 1891. | 1892. | 1893. |
|---|---|---|---|---|---|
| Agricultural implements | $427.58 | $558.41 | $315.87 | $543.[illegible] | $649.34 |
| Beef and pork packing | 531.17 | 498.89 | 571.75 | 498.30 | 438.17 |
| Clothing | 538.17 | 271.55 | 519.90 | 500.63 | 378.48 |
| Coffee and spice mills | 785.60 | 723.12 | 500.00 | 597.56 | 922.65 |
| Flour and feed | 656.72 | 609.32 | 709.84 | 657.84 | 470.22 |
| Furniture, not including chairs | 400.69 | 370.21 | 467.67 | 366.64 | 338.84 |
| Iron works, malleable | 350.75 | 546.03 | 409.77 | 394.73 | 405.21 |
| Lumber, laths, and shingles | 384.84 | 534.12 | 582.29 | 348.35 | 341.10 |
| Marble, cut stone | 456.99 | 522.38 | 587.32 | 479.40 | 390.10 |
| Paper and pulp | 404.10 | 404.03 | 368.54 | 412.96 | 406.00 |
| Plumbers' and gas-fitters' supplies | ........ | ........ | ........ | 485.83 | 409.05 |
| Printing, publishing, and bookbinding | 518.09 | 441.14 | 447.09 | 455.97 | 476.82 |
| Railway shops | 509.14 | 526.93 | 560.12 | 496.51 | 536.28 |
| Rolling mills | 562.71 | 562.08 | 475.42 | 784.25 | 668.19 |
| Sash, doors, and blinds | 373.15 | 349.88 | 378.46 | 309.54 | 387.22 |
| Textiles | 314.14 | 289.06 | 253.62 | 330.18 | 276.92 |
| Tobacco | 453.21 | 456.77 | 426.34 | 324.48 | 542.60 |
| Wagon stock | ........ | ........ | ........ | 418.02 | 389.06 |
| Wagons, carriages, etc | 366.35 | 463.09 | 421.44 | 411.30 | 354.50 |
| Windmills, tanks, and pumps | 467.26 | 825.56 | 500.01 | 539.71 | 530.57 |

Comparisons are also made between the total wages and the number of employees in different industries in the city of Milwaukee and those in the state exclusive of the city.

An idea of the magnitude of the different industries in Milwaukee and in the state exclusive of the city may be obtained from the following statement, which shows the total for fifteen selected industries:

EMPLOYEES AND TOTAL WAGES IN FIFTEEN SELECTED INDUSTRIES IN MILWAUKEE AND IN WISCONSIN, 1893.

| Industry. | Milwaukee. | | State, exclusive of Milwaukee. | |
|---|---|---|---|---|
| | Employees. | Wages. | Employees. | Wages. |
| Agricultural implements | 363 | $212,071 | 2,389 | $1,456,980 |
| Beer and malt | 3,678 | 1,926,289 | 654 | 542,661 |
| Boots and shoes | 1,090 | 427,532 | 1,777 | 828,082 |
| Cigars | 400 | 136,651 | 505 | 221,199 |
| Clothing | 1,065 | 547,527 | 1,038 | 239,651 |
| Cut stone, marble | 173 | 81,080 | 309 | 87,768 |
| Flour and feed | 457 | 257,616 | 1,572 | 360,567 |
| Furniture, chairs | 694 | 329,508 | 3,773 | 1,456,163 |
| Iron works, foundries, and machine shops | 2,976 | 1,641,811 | 2,064 | 863,270 |
| Nails, tacks | 459 | 154,109 | 179 | 64,963 |
| Railway repair shops | 306 | 118,640 | 5,787 | 3,139,365 |
| Sash, doors, and blinds | 1,306 | 541,910 | 2,481 | 1,311,264 |
| Tobacco | 312 | 126,643 | ........ | ........ |
| Wagons, carriages, and sleighs | 430 | 147,941 | 3,758 | 909,694 |
| Woolens and worsteds | 1,039 | 306,212 | 1,458 | 590,201 |

The percentages of employees at stated daily rates of wages in factories, in 1893, are shown for the state, and on examining the total representing 102,865 employees, it is found that 48.55 per cent received $1.25 and under $2 per day, while but 1.47 per cent received $3.50 and under

The amount of loss by fire in factories for each industry from 1885 to 1893, inclusive, is also shown.

FACTORY INSPECTION.—The report headed "Synoptical report of and orders issued by inspectors of factories and workshops" is full of interesting detail concerning the different factories inspected. It gives the description and value of each building, with the number of male and female employees. The summary table shows for each industry the value of new factory buildings, also the value of new machinery added, for 1891–92 and 1893–94, respectively. The totals for 1893–94 are also shown by localities, and the orders of the inspectors for repairs or additions are given in full.

## REPORT BY MISS COLLET ON THE STATISTICS OF EMPLOYMENT OF WOMEN AND GIRLS IN ENGLAND AND WALES.

This report of 152 pages, prepared for the labor department of the British Board of Trade, gives statistics bearing on the employment of females in England and Wales, based principally on the following sources of information:

1. Returns made to the labor department in 1894 by cotton, woolen, and worsted manufacturers as to the employment of married women in their mills—specially procured for this report.

2. The statistics of occupations of women and girls at different ages in urban sanitary districts with over 50,000 inhabitants—compiled from the census sheets.

3. The published returns of inquiries recently conducted on the required scale and according to uniform methods—to be found in the census returns of occupations in 1891, and in the board of trade returns of rates of wages in textile trades in 1886.

The report is divided into three parts, dealing, respectively, with census returns of occupations in 1891, labor department returns of the employment of married women in 1894, and board of trade returns of rates of wages in the cotton, woolen, and worsted industries in 1886.

In Part I the census returns of the employment of women and girls in 1891 are compared with those for 1881. In making comparison the increase of population has been taken into account, the numbers of working females being expressed in ratios of the female population over 10 years of age, and these numbers are brought into comparison with the numbers of working males in the same occupations, expressed in similar ratios of the male population.

In Part II statistics from employers in the cotton, woolen, and worsted industries are given, showing the number and proportion of females employed in their mills who in 1894 were married or widowed, and the summarized results are compared with those of the census, as far as possible, with a view to testing their accuracy.

In Part III the broad results of the board of trade rates of wages returns for 1886 are summarized so far as they relate to the employment of women and girls in the cotton, woolen, and worsted industries in England, and an attempt is made to discover whether there is any indication of a relation between the rates of wages and the employment of married women.

The census statistics, presented in Part I of the [illegible] more than four-fifths of the working [illegible] [illegible] in 16 classes or [illegible] such classes. Less than 1 per cent of the working females were employed in each of the remaining classes or groups of occupations, and the total number employed in the 331 classes was only 67 per 1,000 of the total female population over 10 years of age.

The following table, derived from tables in Part I, shows the number of working females per 10,000 females 10 years of age and over engaged in each of the 18 classes of occupations referred to above, in each of which upward of 1 per cent of the total number of working females were employed in either 1891 or 1881, and the number of working females per 10,000 females 10 years of age and over employed in the remaining 331 classes of occupations, in each of which less than 1 per cent of the total number of working females were employed. It also shows the number of working females in 1891, at certain age periods, per 10,000 females at such periods, by occupations, and the decennial increases or decreases in the numbers employed:

WORKING FEMALES IN 1891 AT CERTAIN AGE PERIODS PER 10,000 FEMALES AT SUCH PERIODS.

| Occupation. | 10 and under 15 years. | | | 15 and under 25 years. | | | 25 and under 45 years. | | |
|---|---|---|---|---|---|---|---|---|---|
| | Number. | Compared with 1881. | | Number. | Compared with 1881. | | Number. | Compared with 1881. | |
| | | Increase. | Decrease. | | Increase. | Decrease. | | Increase. | Decrease. |
| Employing more than 1 per cent of females who work: | | | | | | | | | |
| Domestic servants | 665 | ........ | 37 | 2,744 | ........ | 189 | 902 | 57 | ........ |
| Milliners, dressmakers, and staymakers | 108 | 41 | ........ | 732 | 90 | ........ | 331 | ........ | 40 |
| Cotton goods operatives | 305 | 27 | ........ | 555 | ........ | 25 | 258 | ........ | 22 |
| Laundry and bath employees | 11 | 2 | ........ | 130 | 17 | ........ | 164 | ........ | 21 |
| School teachers, professors, and lecturers | 37 | ........ | ........ | 245 | ........ | 5 | 137 | 16 | ........ |
| Charwomen | 2 | 1 | ........ | 24 | 2 | ........ | 102 | 2 | ........ |
| Tailoresses | 32 | 20 | ........ | 148 | 64 | ........ | 66 | 11 | ........ |
| Worsted goods operatives | 87 | 9 | ........ | 117 | ........ | 7 | 45 | ........ | 7 |
| Woolen goods operatives | 32 | 1 | ........ | 106 | ........ | 14 | 51 | ........ | 3 |
| Nurses, midwives, etc | ........ | ........ | ........ | 18 | 14 | ........ | 48 | 25 | ........ |
| Shirtmakers and seamstresses | 8 | ........ | 3 | 45 | ........ | 47 | 39 | ........ | 40 |
| Shoe, boot, patten, and clog makers (a) | 28 | 11 | ........ | 83 | 16 | ........ | 33 | ........ | ........ |
| Drapers and mercers | 8 | 4 | ........ | 91 | 24 | ........ | 35 | 11 | ........ |
| Grocers and chocolate makers and dealers | 6 | 3 | ........ | 31 | 11 | ........ | 38 | 14 | ........ |
| Boarding and lodging house keepers | ........ | ........ | ........ | 4 | ........ | ........ | 38 | 9 | ........ |
| Hotel servants | 5 | ........ | ........ | 93 | 24 | ........ | 35 | 13 | ........ |
| Silk, satin, velvet, and ribbon factory operatives | 23 | ........ | 3 | 46 | ........ | 19 | 23 | ........ | 12 |
| Farm laborers and servants | 8 | ........ | 7 | 30 | ........ | 31 | 18 | ........ | 13 |
| Total | 1,365 | 69 | ........ | 5,242 | ........ | 75 | 2,363 | ........ | ........ |
| Employing under 1 per cent of females who work | 261 | 51 | ........ | 1,094 | 197 | ........ | 597 | 60 | ........ |
| Total | 1,626 | 120 | ........ | 6,336 | 122 | ........ | 2,960 | 60 | ........ |
| Total female population | 1,612,769 | 214,606 | ........ | 2,884,756 | 389,921 | ........ | 4,006,447 | 511,665 | ........ |

a Dealers, who were included in the census returns for 1881, were not included in those for 1891.

WORKING FEMALES IN 1891 AT CERTAIN AGE PERIODS PER 10,000 FEMALES [illegible] SUCH PERIODS—Concluded.

| Occupation. | 45 and under 65 years. | | | 65 years and over. | | | All ages over 10 years. | | |
|---|---|---|---|---|---|---|---|---|---|
| | Number. | Compared with 1881. | | Number. | Compared with 1881. | | Number. | Compared with 1881. | |
| | | Increase. | Decrease. | | Increase. | Decrease. | | Increase. | Decrease. |
| Employing more than 1 per cent of females who work: | | | | | | | | | |
| Domestic servants | 479 | 36 | ...... | 276 | 19 | ...... | 1,209 | ...... | 22 |
| Milliners, dressmakers, and staymakers | 218 | ...... | 47 | 94 | ...... | 16 | 363 | 5 | ...... |
| Cotton goods operatives | 87 | ...... | 14 | 13 | ...... | 5 | 200 | ...... | 13 |
| Laundry and bath employees | 297 | ...... | 44 | 200 | ...... | 43 | 162 | ...... | 15 |
| School teachers, professors, and lecturers | 53 | ...... | 5 | 16 | ...... | 13 | 126 | 3 | ...... |
| Charwomen | 213 | ...... | 2 | 129 | ...... | 23 | 92 | ...... | ...... |
| Tailoresses | 58 | 8 | ...... | 30 | 4 | ...... | 78 | 26 | ...... |
| Worsted goods operatives | 16 | ...... | 4 | 3 | ...... | ...... | 61 | ...... | 3 |
| Woolen goods operatives | 21 | ...... | 4 | 5 | ...... | 4 | 54 | ...... | 5 |
| Nurses, midwives, etc. | 102 | ...... | 4 | 93 | ...... | 31 | 47 | 9 | ...... |
| Shirtmakers and seamstresses | 73 | ...... | 37 | 93 | ...... | 31 | 46 | ...... | a 36 |
| Shoe, boot, patten, and clog makers (b) | 17 | ...... | 5 | 9 | ...... | 3 | 40 | 4 | ...... |
| Drapers and mercers | 19 | 6 | ...... | 9 | 3 | ...... | 40 | 11 | ...... |
| Grocers and chocolate makers and dealers | 73 | 23 | ...... | 63 | 14 | ...... | 40 | 14 | ...... |
| Boarding and lodging house keepers | 102 | 16 | ...... | 81 | 5 | ...... | 30 | 6 | ...... |
| Hotel servants | 10 | 6 | ...... | 2 | 1 | ...... | 39 | 13 | ...... |
| Silk, satin, velvet, and ribbon factory operatives | 21 | ...... | 16 | 13 | ...... | 12 | 28 | ...... | a 13 |
| Farm laborers and servants | 26 | ...... | 21 | 19 | ...... | 25 | 21 | ...... | a 19 |
| Total | 1,885 | ...... | 118 | 1,148 | ...... | 161 | 2,775 | ...... | 26 |
| Employing under 1 per cent of females who work | 612 | 7 | ...... | 450 | ...... | 69 | 667 | 73 | ...... |
| Total | 2,497 | ...... | 111 | 1,598 | ...... | 230 | 3,442 | 37 | ...... |
| Total female population | 2,191,964 | 240,251 | ...... | 768,014 | 112,982 | ...... | 11,461,880 | 1,469,377 | ...... |

a An actual decrease as well as a relative one.
b Dealers, who were included in the census returns for 1881, were not included in those for 1891.

The decrease in the numbers employed in occupations connected with the textile industries is shown at every age period above 15 years.

The increase in the number of laundry and bath employees 15 and under 25 years of age is explained by the statement that laundry work in steam laundries attracts girls and young women more than was the case under the hand system.

The numbers employed as teachers, professors, and lecturers, and as nurses, midwives, etc., show an increase at the most efficient age periods, which, it is said, indicates an advance in the quality of their work.

A large decrease is shown, at every age period, in the [illegible] of shirtmakers and seamstresses. The report says that the [illegible] these occupations would have been still more [illegible]

[illegible] system in the manufacture of shirts and under- [illegible] in the factory system, and the consequent growth of the [illegible] trade, must be traced the great increase in the [illegible] tailoresses.

[illegible] table shows that in 9 of the 18 specified occupations in each of which over 1 per cent of the working females were employed in 1891 or 1881 the employment of females increased relatively to population; these 9 occupations in 1891 employed 812 in every 10,000 females 10 years of age and over, or 90 more than in 1881. The other 9 specified occupations in 1891 employed 1,963 females in every 10,000 of 10 years of age and over, or 126 less than in 1881.

In the remaining occupations, in each of which less than 1 per cent of the working females were employed, 667 females per 10,000 of 10 years of age and over were employed in 1891, or 73 more than in 1881.

A striking fact shown by this table is the decrease in the proportion of females between the ages of 15 and 25 employed in domestic service, and the increase in the number so employed above the age of 25. The decrease is said to be due to a probably diminished supply of young servants; and the consequent improved condition of older servants accounts for the increase in their number. It is also said that as the proportion of children under 10 years of age and the proportion of married to single persons in 1891 were less than in 1881, the need for servants had to some extent diminished.

In order to compare the rate of progress in the employment of females with that of males in certain occupations employing both sexes the following table is given, showing the numbers of working males in 1891, at certain age periods, per 10,000 males at such periods, in the selected occupations, and the increase or decrease in the number employed in each occupation since 1881.

WORKING MALES IN 1891 AT CERTAIN AGE PERIODS PER 10,000 MALES AT SUCH PERIODS, IN OCCUPATIONS LARGELY FOLLOWED BY WOMEN.

| Occupation. | 10 and under 15 years. | | | 15 and under 25 years. | | | 25 and under 45 years. | | |
|---|---|---|---|---|---|---|---|---|---|
| | Number. | Compared with 1881. | | Number. | Compared with 1881. | | Number. | Compared with 1881. | |
| | | Increase. | Decrease. | | Increase. | Decrease. | | Increase. | Decrease. |
| [illegible] | [illegible] | 22 | ........ | 277 | ........ | ........ | 195 | 8 | ........ |
| [illegible] | [illegible] | 10 | ........ | 46 | ........ | 2 | 31 | ........ | ........ |
| [illegible] | [illegible] | 2 | ........ | 75 | 2 | ........ | 59 | ........ | 3 |
| [illegible] | 12 | 1 | ........ | 15 | ........ | 2 | 13 | ........ | 3 |
| [illegible] | [illegible] | 7 | ........ | 130 | 18 | ........ | 128 | ........ | 1 |
| [illegible] | [illegible] | 21 | ........ | 201 | 21 | ........ | 199 | ........ | 22 |
| [illegible] | 11 | 1 | ........ | 94 | 1 | ........ | 64 | ........ | 3 |
| [illegible] | 7 | ........ | ........ | 81 | 7 | ........ | 51 | 8 | ........ |
| [illegible] | [illegible] | ........ | 3 | [illegible] | ........ | 25 | 66 | 13 | ........ |
| [illegible] | [illegible] | ........ | 76 | [illegible] | ........ | 200 | 589 | ........ | 127 |

WORKING MALES IN 1891 AT CERTAIN AGE PERIODS [illegible] PERIODS, IN OCCUPATIONS LARGELY FOLLOWED BY WO[illegible]

| Occupation. | 45 and under 65 years. | | | [illegible] | | |
|---|---|---|---|---|---|---|
| | Number. | Compared with 1881. | | Number. | [illegible] | |
| | | Increase. | Decrease. | | In-crease. | [illegible] |
| Cotton goods operatives | 126 | ........ | 14 | [illegible] | ........ | |
| Worsted goods operatives | 26 | ........ | 4 | 16 | ........ | |
| Woolen goods operatives | 50 | ........ | 9 | [illegible] | ........ | |
| Silk, satin, velvet, and ribbon factory operatives | 20 | ........ | 12 | [illegible] | ........ | |
| Tailors | 141 | ........ | [illegible] | [illegible] | ........ | |
| Shoe, boot, patten, and clog makers | 256 | ........ | 43 | 245 | ........ | |
| Drapers | 45 | 6 | ........ | [illegible] | ........ | ........ |
| Hotel servants | 20 | 3 | ........ | 7 | ........ | ........ |
| School teachers, professors, and lecturers | 37 | 5 | ........ | 13 | ........ | |
| Farm laborers and servants, teamsters, etc | 809 | ........ | 250 | [illegible] | ........ | |

In four occupations employing females and males the former ha[illegible] made distinct advances; in one of these, hotel servants, the numb[illegible] of males also show an advance; in another, drapers, the number of ma[illegible] employees 25 and under 45 years of age show a decrease; in the oth[illegible] two, tailors and shoemakers, boys and youths show an increase. T[illegible] decrease in the last two occupations in the numbers employed at t[illegible] age periods above the 15–25 period is said to be probably due [illegible] changes in the organization of the trades to which these occupatio[illegible] belong, and the decrease in the number of males 25 and under 45 yea[illegible] of age in the shoe trade is partly due to the exclusion of dealers [illegible] the census returns for 1891, who were included in those for 1881. [illegible] decrease in the numbers of both sexes employed as farm laborers a[illegible] servants, teamsters, etc., is shown at every age period.

There was a remarkable increase in the employment of children [illegible] both sexes under the age of 15 years. It is suggested that the increa[illegible] shown by the census of 1891 over the number as reported by the ce[illegible]sus of 1881 may be partly due to concealment of employment of ch[illegible]dren in 1881. As to the employment of female children, the increa[illegible] seems to be attributable to the growth of urban population, su[illegible] increase having occurred in counties containing one or more towns [illegible] over 50,000 inhabitants in which the population has increased at [illegible] higher rate than the urban population generally. This inference is su[illegible]ported by the fact that the increase in the employment of girls und[illegible] 15 years of age was greatest in industries in which the chances [illegible] employment are much greater in towns than in rural districts, [illegible] several branches of which the extended use of machinery and t[illegible] minute subdivision of labor render it easier for children to find wo[illegible] than formerly.

In the table giving the employment of females at certain age peri[illegible] it is shown that in every 10,000 females 10 and under 15 years [illegible] 1,626 were employed, equivalent to 16.26 per cent. The [illegible] for 1891 show that in 34 towns, including London, with [illegible] inhabitants each, the percentage was lower than [illegible]

[illegible] per cent, while in 28 towns of over 50,000 [illegible] was higher, ranging from 17 to 58.1 per cent. [illegible] at which the largest percentage of females is employed [illegible] years. The census of 1891 shows that in England [illegible] per cent of females between these ages were employed in the various occupations. In 30 towns each with a population of upward of 50,000 the percentage was lower than this, ranging from 49.2 to 67.1 per cent, while in 32 towns each having populations of over 50,000 the percentage was higher, ranging from 68.8 to 95.3 per cent.

In discussing the employment of women over 20 years of age, with special reference to married and widowed women, the report says that the age period between 20 and 25 years is that at which the female worker has, perhaps, the most industrial freedom; she is then not only in her prime industrially, but generally has the option of exchanging wage-earning employment for domestic life. In England and Wales 70 per cent of the females at this age period were returned as unmarried; but the large towns showed considerable divergence from this average, the percentages ranging from 41 to 85.

Tables are given showing the percentages of working married and widowed females at different age periods in 19 industrial towns in England in 1891 and 1881, and the inference is drawn that in these towns, most affected by female labor, to which the tables relate, the percentage of working married women is diminishing. It is mentioned as a noteworthy fact that in all these factory centers there is a marked diminution in the proportion of working married women between the ages of 20 and 25 years.

The conditions governing married female labor in the north of England are quite different from those in the south. In the north there has been a large demand for female labor, and married women have been attracted by the high wages obtainable in the textile industries, especially in the cotton trade. The women of the north have not regarded industrial employment as being merely a means of support prior to marriage, but have looked upon it, more than upon domestic management, as their life occupation, and they work with a view to [illegible] greater comfort in living. These causes tend to make [illegible] workers and to develop industrial ambition.

[illegible] of England, where the factory industries are small and [illegible] of female employment is in domestic service, work- [illegible] forward to marriage as a release from wage-earning [illegible] in the upper industrial classes marriage usually gives [illegible]. The girl before marriage rarely aims at becoming a [illegible], and if in later life she finds it necessary to again [illegible] is unable to gain employment except in ordinary [illegible] the lower industrial grades females frequently [illegible] marriage because of the small earnings or irregular employment of their husbands. The effect of [illegible] the quality of female labor is disadvantageous, and the [illegible] labor is of a poor kind.

Part II of the report deals with statistics collected by [illegible] department from manufacturers in 1894, relative to unmarried, married, and widowed females employed by them in cotton, woolen, and worsted mills in Lancashire, Cheshire, Yorkshire, and the west of England. Returns were received from 1,654 manufacturers, of whom 968 were in the cotton industry, 315 in the woolen, 340 in the worsted, and 31 in the mixed woolen and worsted. These returns relate to the employment of 246,825 females, distributed among the industries as follows: Cotton, 176,456; woolen, 20,045; worsted, 46,540; mixed woolen and worsted, 3,784.

The females employed in the cotton and other industries, to whom the labor department statistics relate, are classified as "half-timers," who were 11 and under 13 years of age; as "young persons," who were 13 and under 18 years of age, and as "women 18 years of age and over." This classification is in accordance with the terms of the English factory act, which defines the terms "half-timer," "young person," and "women" in such a manner as to include all females legally employed in factories under these headings.

In the cotton industry 12,536 of the 176,456 females employed, or 7.1 per cent, were half-timers; 45,398, or 25.7 per cent, were young persons; 118,522, or 67.2 per cent, were women 18 years of age and over. Of the 118,522 women, 38,991, or 32.9 per cent, were either wives or widows; the ratio of married and widowed to the total number of females, exclusive of half-timers, was 23.8 per cent, or nearly one-fourth; of the total number of females, including half-timers, 22.1 per cent, or more than one-fifth, were married or widowed; of the 38,991 women who were married or widowed, 4,841, or 12.4 per cent, were widowed.

Great differences exist in the percentages in different localities; for example, in 10 urban sanitary districts each with over 50,000 inhabitants the percentages of women over the age of 18 who were married or widowed ranged from 9.8 to 44.7 per cent.

In the woolen industry the statistics show that of the 20,045 females to whom they relate, 200, or 1 per cent, were half-timers; 3,364, or 16.8 per cent, were young persons, and 16,481, or 82.2 per cent, were women 18 years of age and over. Of the 16,481 women, 4,906, or 29.8 per cent, were either married or widowed. The ratio of married and widowed to the total number of females, exclusive of half-timers, was 24.7 per cent, or about one-fourth.

The relative number of half-timers reported in this industry was so small that the ratio of wives and widows to the total number of working females was nearly the same as their ratio to the total number exclusive of half-timers, being 24.5 per cent. Of the 4,906 [illegible] or widowed, 844, or 17.2 per cent, were reported as

industry 3,944, or 8.5 per cent, of the 46,540 working
f-timers; 13,288, or 28.5 per cent, were young persons,
per cent, were women 18 years of age and over. Of
n, 6,269, or 21.4 per cent, were either married or wid-
of married and widowed to the total number of work-
sive of half-timers, was 14.7 per cent, or slightly more
; the ratio of married and widowed to the total num-
females, inclusive of half-timers, was 13.5 per cent, or
th. Of the 6,269 wives and widows, 1,111, or 17.7 per
vs.

roolen and worsted industry 83, or 2.2 per cent, of the
ratives were half-timers; 792, or 20.9 per cent, were
nd 2,909, or 76.9 per cent, were women 18 years of age
e 2,909 women, 686, or 23.6 per cent, were either married
e ratio of the married and widowed to the total num-
males, exclusive of half-timers, was 18.5 per cent, and
ber, inclusive of half-timers, 18.1 per cent. Of the 686
vs, 108, or 15.7 per cent, were widows.

ber of females in all the industries to which the statis-
46,825. Of this number 16,763, or 6.8 per cent, were
42, or 25.5 per cent, were young persons, and 167,220,
were women 18 years of age and over. Of the 167,220
r 30.4 per cent, were either married or widowed. The
ed and widowed to the total number of working females,
-timers, was 22.1 per cent, and to the total number,
timers, 20.6 per cent. Of the 50,852 wives and widows,
cent, were widows.

statement shows the number of working females, by
erning whom statistics were gathered by the labor
894, classified as "half-timers," those 11 and under 13
oung persons," those 13 and under 18 years of age, and
s of age and over," of whom the number unmarried,
lowed are given. The percentage that each class is of
of working females in each industry is also given.

The following statement presents the average annual wages [illegible] some of the leading industries treated in the summary table:

AVERAGE ANNUAL WAGES PAID IN VARIOUS INDUSTRIES, 1889 TO 1893.

| Industry. | 1889. | 1890. | 1891. | 1892. | 1893. |
|---|---|---|---|---|---|
| Agricultural implements | $427.68 | $558.41 | $315.87 | $543.98 | $640.34 |
| Beef and pork packing | 531.17 | 498.89 | 571.75 | 496.30 | 438.17 |
| Clothing | 538.17 | 271.56 | 519.90 | 500.03 | 376.43 |
| Coffee and spice mills | 785.60 | 723.12 | 500.00 | 597.56 | 922.65 |
| Flour and feed | 656.72 | 609.32 | 709.84 | 657.64 | 470.22 |
| Furniture, not including chairs | 400.69 | 370.21 | 467.67 | 366.64 | 338.84 |
| Iron works, malleable | 350.75 | 546.03 | 409.77 | 394.78 | 405.21 |
| Lumber, laths, and shingles | 334.84 | 524.12 | 582.29 | 348.25 | 341.10 |
| Marble, cut stone | 456.99 | 522.38 | 587.32 | 479.40 | 390.10 |
| Paper and pulp | 404.10 | 404.03 | 368.54 | 412.96 | 406.00 |
| Plumbers' and gas-fitters' supplies | ........ | ........ | ........ | 485.83 | 409.05 |
| Printing, publishing, and bookbinding | 518.09 | 441.14 | 447.09 | 455.97 | 476.82 |
| Railway shops | 509.14 | 525.93 | 560.12 | 496.51 | 536.28 |
| Rolling mills | 552.71 | 502.06 | 475.42 | 784.25 | 668.19 |
| Sash, doors, and blinds | 373.15 | 349.88 | 378.46 | 309.54 | 387.22 |
| Textiles | 314.14 | 269.06 | 253.63 | 230.18 | 276.92 |
| Tobacco | 453.21 | 455.77 | 426.24 | 324.48 | 542.60 |
| Wagon stock | ........ | ........ | ........ | 418.02 | 399.06 |
| Wagons, carriages, etc | 366.35 | 463.09 | 421.44 | 411.30 | 354.50 |
| Windmills, tanks, and pumps | 467.25 | 825.56 | 500.01 | 539.71 | 530.57 |

Comparisons are also made between the total wages and the number of employees in different industries in the city of Milwaukee and those in the state exclusive of the city.

An idea of the magnitude of the different industries in Milwaukee and in the state exclusive of the city may be obtained from the following statement, which shows the total for fifteen selected industries:

EMPLOYEES AND TOTAL WAGES IN FIFTEEN SELECTED INDUSTRIES IN MILWAUKEE AND IN WISCONSIN, 1893.

| Industry. | Milwaukee. | | State, exclusive of Milwaukee. | |
|---|---|---|---|---|
| | Employees. | Wages. | Employees. | Wages. |
| Agricultural implements | 363 | $212,071 | 2,250 | $1,459,980 |
| Beer and malt | 2,678 | 1,926,289 | 654 | 542,001 |
| Boots and shoes | 1,080 | 427,532 | 1,777 | 823,662 |
| Cigars | 400 | 136,651 | 505 | 221,189 |
| Clothing | 1,085 | 547,527 | 1,023 | 330,661 |
| Cut stone, marble | 173 | 81,060 | 309 | 87,766 |
| Flour and feed | 487 | 287,616 | 1,579 | 560,567 |
| Furniture, chairs | 694 | 229,503 | 3,772 | 1,485,163 |
| Iron works, foundries, and machine shops | 2,976 | 1,641,811 | 2,054 | 883,370 |
| Nails, tacks | 489 | 154,100 | 179 | 64,962 |
| Railway repair shops | 206 | 115,640 | 5,787 | 3,129,262 |
| Sash, doors, and blinds | 1,395 | 541,910 | 2,481 | 1,311,354 |
| Tobacco | 313 | 126,843 | ........ | ........ |
| Wagons, carriages, and sleighs | 430 | 147,941 | 2,786 | 900,694 |
| Woolens and worsteds | 1,030 | 306,312 | 1,456 | 630,301 |

The percentages of employees at stated daily rates of wages in factories, in 1893, are shown for the state, and on examining the total representing 102,865 employees, it is found that 48.55 per cent received $1.25 and under $2 per day, while but 1.47 per cent received $3.50 and under $4 per day. The results are summarized so as to permit of a ready comparison of the relative number at each rate in the [illegible] industries.

The following statement shows the number and per cent of females working full time, above referred to, in the industries specified, at and between different weekly wage rates:

FEMALES WORKING FULL TIME IN CERTAIN INDUSTRIES AT AND BETWEEN CERTAIN WEEKLY WAGE RATES, 1886.

| Industry. | Under $2.43. | | $2.43 and under $3.65. | | $3.65 and under $4.87. | | $4.87 and under $6.08. | | $6.08 and over. | | Total. | |
|---|---|---|---|---|---|---|---|---|---|---|---|---|
| | Number. | Per cent. | Number. | Per cent. | Number. | Per cent. | Number. | Per cent. | Number. | Per cent. | Number. | Per cent. |
| Cotton (a) | 7,243 | 10.7 | 30,482 | 44.9 | 21,708 | 32.0 | 8,216 | 12.1 | 192 | 0.3 | 67,843 | 100 |
| Woolen (b) | 1,159 | 10.6 | 6,973 | 64.0 | 2,767 | 25.3 | 10 | .1 | ...... | ...... | 10,909 | 100 |
| Woolen (c) | 796 | 45.3 | 961 | 54.7 | ...... | ...... | ...... | ...... | ...... | ...... | 1,757 | 100 |
| Total | 1,955 | 15.4 | 7,934 | 62.6 | 2,767 | 21.9 | 10 | .1 | ...... | ...... | 12,666 | 100 |
| Worsted and stuff (d) | 6,902 | 36.6 | 11,838 | 62.8 | 115 | .6 | ...... | ...... | ...... | ...... | 18,855 | 100 |
| Grand total | 16,192 | 16.2 | 50,254 | 50.6 | 24,590 | 24.7 | 8,226 | 8.3 | 192 | .2 | 99,364 | 100 |

a In Lancashire and Cheshire.
b In Yorkshire and Lancashire.
c In west of England.
d In Yorkshire.

The board of trade statistics for 1886 are considered in connection with those collected by the labor department in 1894. From the latter it appears that in 1894 the proportion of young persons to women was considerably higher in worsted than in cotton mills, 31.2 per cent of the full timers being young persons in the former case and 27.7 per cent in the latter. Moreover, only 21.4 per cent of the adult females in the worsted mills were married or widowed, as compared with 32.9 per cent in the cotton mills. Supposing somewhat similar conditions to have prevailed in 1886 as in 1894, it would follow that the average age of the cotton operatives was higher than that of the worsted operatives. Making all allowance for such difference in age, it would seem that the average wages were lower in the worsted than in the cotton industry.

The change that has taken place in the woolen manufacture since 1886 makes it most unlikely that the proportion of young persons to adult women employed in 1894 approximated to that prevailing in 1886. The proportion of adult females in woolen mills was abnormally high in 1894, having been 83 per cent of full timers, as compared with 72.3 per cent in the cotton mills. This high proportion of adult females was most probably due to the employment of girls in worsted instead of in woolen mills. Notwithstanding the high proportion of adult females, the percentage of females either married or widowed in the Yorkshire woolen mills was less than the percentage in the cotton mills, having been 28.1 per cent in the former case and 32.9 per cent in the latter. In 1886, before the stream of young workers was diverted from the woolen to the worsted cloth manufacture, the percentage of married females was probably lower still.

Comparing the three industries, it was found that the [illegible] of wages coincide with a higher percentage of adult [illegible] married or widowed. Comparing estimated average wages of young persons in the cotton and worsted mills, the average in the worsted mills was considerably lower than in the cotton mills, although the average age of young persons would be about the same.

An examination of the relation between average wages and the percentage of married women employed in cotton mills in different districts pointed to the conclusion that in the north of England one of the causes of an exceptionally high rate of employment of married women was the high rate of wages that could be earned. In so far as this conclusion is correct, it may be inferred that a falling in wages of working females in the great textile trades would be followed by a diminution in the employment of married women, if the wages of male operatives remained unchanged.

With the relation between wages and the employment of married women in the north of England must be compared the conditions found in the woolen mills in the west of England, where the average weekly wage in 1886 was much lower than the average in Yorkshire and Lancashire, but where the percentage of working females who were either married or widowed was extremely high in 1894.

In conclusion, it is said that the current view that the employment of female labor is rapidly extending, and that women are replacing men to a considerable extent in industrial occupations, is not confirmed. On the whole, the proportion of working females remained practically stationary in the decade 1881–1891, there having been 34.05 working females over 10 years of age per 100 in 1881 and 34.42 per 100 in 1891, the slight increase being attributed to the increased number of females under 25 years of age with definite occupations, and to the increased employment of middle-class women.

The employment of married and elderly women has, on the whole, diminished, as has also the employment of women in casual occupations. There has been an increase in the employment of females under the age of 25 years, which has, however, been concurrent with a [illegible] extension in the employment of young men and boys.

As to the substitution of female for male labor, the census [illegible] show that 83.24 per cent of males over 10 years of age were indu[illegible] employed in 1881 and 83.10 per cent in 1891. In either year there [illegible] less than 17 males in every 100 who could possibly have been add[illegible] the ranks of the employed, whereas there were nearly 66 fema[illegible] every 100 upon which to draw for an increase in wage earners, [illegible] 1891 this available female surplus had only been diminished [illegible] than 1, and it appears to be clearly shown that male labor has [illegible] displaced to any marked extent by the employment of females.

## EMPLOYER AND EMPLOYEE UNDER THE COMMON LAW.

BY VICTOR H. OLMSTED AND STEPHEN D. FESSENDEN.

The relations existing between employers of labor and their employees, and the reciprocal duties, obligations, and rights growing out of these relations, are, in the absence of legislative enactments, governed by the common law in regard to master and servant, the words master and servant being legally synonymous with the words employer and employee.

The common law consists of principles, usages, and rules of action, applicable to the government and security of persons and property, which have grown into use by gradual adoption, without legislative authority, and have received, from time to time, the sanction of the courts of justice.

The great body of the common law of the United States consists of the common law of England, and such statutes thereof as were in force prior to the separation of this country from England, and applicable to circumstances and conditions prevailing here. These laws have been adopted as the basis of our jurisprudence in all the states except Louisiana, and many of the most valued principles of the English common law have been embodied in the constitutions of the United States and the several states.

In many details, however, the common law of the United States now differs widely from that of England by reason of modifications arising from different conditions and established by American adjudications. That branch of the common law governing the relation of master and servant has undergone some changes, although in the main it is the same in this country as in England. It is not the purpose of this article to point out such changes or differences, but to state the principles and rules of the common law now prevailing throughout the United States, except where they have been changed or modified by legislative enactments.

The statement which follows is derived from articles in the American and English Encyclopedia of Law on the subject of "Master and servant" and kindred topics, and from standard legal works treating of the subjects under consideration. The reader should bear in mind that [illegible] principle of the common law, as given in this statement, conflicting with a statute which has not been declared invalid or unconstitutional by the courts, is modified or changed by the statute, [illegible] the statute *instead of the* common law now governs.

MASTER AND SERVANT: DEFINITIONS.—A master [illegible] fined as one who has in his employment one or more [illegible] by contract to serve him either as domestic or common [illegible] who has the superior choice, control, and direction, whose will [illegible] sented not merely in the ultimate result of the work in hand, but [illegible] all its details; one who is the responsible head of a given industry; o[illegible] who not only prescribes the end, but directs, or may at any time dire[illegible] the means and methods of doing the work; one who has the pow[illegible] to discharge; a head or chief; an employer; a director; a governo[illegible]

A servant is one who is employed to render personal service to h[illegible] employer otherwise than in the pursuit of an independent calling, a[illegible] who, in such service, remains entirely under the control and directi[illegible] of the latter.

THE RELATION: ITS CREATION AND EXISTENCE.—The relation [illegible] master and servant is created by contract, either express or implie[illegible] where both parties have the requisite legal qualifications for enteri[illegible] into a valid contract. The relation exists only where the person soug[illegible] to be charged as master employs and controls the other party to t[illegible] contract of service, or expressly or tacitly assents to the rendition [illegible] the particular service by him. The master must have the right [illegible] direct the action of the servant, and to accept or reject his servic[illegible] The relation does not cease so long as the master retains his control [illegible] right of control over the methods and manner of doing the work, [illegible] the agencies by which it is effected. Furthermore, the relation exis[illegible] where the servant is employed, not by the master directly, but by [illegible] employee in charge of a part of the master's business with authori[illegible] to engage assistance therein.

THE CONTRACT OF SERVICE.—A contract of employment is one [illegible] which an employer engages an employee to do something for the ben[illegible] fit of the employer, or of a third person, for a sufficient considerati[illegible] expressed or implied. The authority of a subordinate to employ [illegible] agent or servant includes, in the absence of restrictive words, autho[illegible] ity to make a complete contract, definite as to the amount of wages, [illegible] well as to all other terms.

Ordinarily, when an adult person solicits employment in a particul[illegible] line of work, the solicitation carries with it an implied assertion th[illegible] the one seeking employment is competent to perform the ordina[illegible] duties of the position sought; and it is an implied condition of eve[illegible] contract of service that the employee is competent to discharge t[illegible] duties of his employment.

A servant is presumed to have been hired for such length of [illegible] the parties adopt for the estimation of wages; for example, a hiring [illegible] yearly rate is presumed to be for one year; at a daily rate, for [illegible] a hiring by piecework, for no specified time; but such fact [illegible] in the absence of other evidence, necessarily fix the period [illegible] Where an employee has been hired to work by the week [illegible]

sation, he, prima facie, agrees to give his employer
his rule is not inflexible.

ice running for a longer period of time than one
it be in writing and signed by the party against
ht to be enforced, or by his authorized agent.

express contract of hiring, a person may recover
vices where the same were rendered under such
show that he expected such compensation as a
hat the person for whom they were rendered was
he claimed compensation, or was legally entitled
person performs labor for another, a request and
reasonable worth of such labor are presumed by
erstood that the labor is to be gratuitously per-
ormed under such circumstances as to repel the
mise to pay.

express contract the servant must be furnished
he master during the period covered by its terms.
contract the servant is employed to work by the
year, and nothing is said as to the time of pay-
the wages are due and may be demanded at the
eek, month, or year, as the case may be; but in
l questions relating to the interpretation of con-
strong bearing.

ct to furnish his own services and those of his
s no separate claim can sue for them; and if such
ring, the testimony of the wife in support of his
icient ratification.

onsible for the wages of her husband's employee,
fact that she sometimes pays such wages.

grees to pay his servant what he considers the
be reasonably worth, or, where he agrees to pay
shall be paid to other men in his employ filling
l there is no showing that the master has other
positions, the servant is entitled to recover, in a
his services were actually worth. And where the
gree as to the existence of the contract of service,
wages to be paid, the question of compensation
r.

greed, the wages of an employee must be paid in
s no right to handle, or invest, or in any manner
hether beneficial to the servant or not, but must

An employer may discharge an employee before the term of service stipulated in the contract for good as, for incompetency. The discharge must be as to leave no doubt in the employee's mind of the terminate the relation.

In a majority of the states a contract for service for a specified is considered apportionable, and an employee who has been discha for cause is entitled to compensation for the work he has performed.

Where one has contracted to employ another for a certain peri time, at a specified price for the entire time, and discharges him fully before the expiration thereof, the wrongfully discharged empl is entitled to recover an amount equal to the stipulated wages fo whole period covered by the contract, less the sum earned, or might have been earned in other employment during the period co by the breach. Upon dismissal a servant, under the law, must seek employment, but extraordinary diligence in such seeking is not requ of him. He is only required to use reasonable efforts, and he is bound to seek employment or render service of a different kind or g from that which he was engaged to perform under the violated cont nor to seek employment in a different neighborhood; and if he fai secure employment and works on his own account the value of work can not be deducted from his claim.

Where an employee for a fixed period, at a salary for the period, able at intervals, is wrongfully discharged, he may pursue one of courses—

1. He may sue at once for the breach of contract, in which cas can only recover his damages up to the time of bringing the suit.

2. He may wait until the end of the contract period, and then for the breach.

3. He may treat the contract as existing, and sue at each peric payment for the wages then due.

4. He may treat the contract as rescinded, and sue immediatel the value of his services performed, in which case he can only rec for the time he actually served.

An employee is entitled to recover damages from a person who ciously procures his discharge, provided he proves that the disch resulted in damage to him.

An employer is entitled to maintain an action against anyone knowingly entices away his servant, or wrongfully prevents the ant from performing his duty, or permits the servant to stay with and harbors such servant with the intention of depriving the of his services.

COMBINATIONS AND COERCION OF SERVANTS.—Everyone has right to work or to refuse to work for whom and on what pleases, or to refuse to deal with whom he pleases, and

[illegible] price or without certain conditions.
[illegible] to refuse to work, either singly or in combi- [illegible] terms and conditions satisfactory to themselves, is [illegible] right of employers to refuse to engage the services of [illegible] they deem proper. The master may fix the wages, [illegible] not unlawful, upon which he will employ work- [illegible] right to refuse to employ them upon any other terms. [illegible] employers and employees are entitled to exercise the [illegible] entering into contracts of service, and neither party [illegible] responsible for refusing to enter into such contracts. [illegible]d, however, that employers in separate, independent [illegible] have no right to combine for the purpose of preventing [illegible] have incurred the hostility of one of them, from secur- [illegible] upon any terms, and by the method commonly known [illegible] debarring such workmen from exercising their voca- [illegible] bination being regarded as a criminal conspiracy.
[illegible] hand, a combination of employees having for its pur- [illegible] ishment of an illegal object is unlawful; for instance, [illegible] extort money from an employer by inducing his work- [illegible] and deterring others from entering his service, is [illegible] association which undertakes to coerce workmen to [illegible] thereof or to dictate to employers as to the methods or [illegible] their business shall be conducted, by means of force, [illegible] idation interfering with their traffic or lawful employ- [illegible] rsons is, as to such purposes, an illegal combination. [illegible] ference by employees, or former employees, or persons [illegible] hy with them, with the business of a railroad com- [illegible] s of a receiver, renders the persons interfering liable [illegible] r contempt of court.
[illegible] LIABILITY FOR INJURIES OF EMPLOYEES.—Where [illegible] s an independent contractor to do work for him, and [illegible] es no control over the means or methods by which the [illegible] omplished, he is not answerable for the wrongful acts [illegible] r; and the same rule governs as between a contractor [illegible] tor. Under these circumstances an employer would [illegible] an injury sustained by a workman in the course of his [illegible] which he would have been liable had the work been [illegible] his own direction.
[illegible] is ordinarily liable in damages to his employee who [illegible] ry through the employer's negligence. Such negli- [illegible] in the doing of something by the employer which, [illegible] of ordinary care and prudence, he ought not to have [illegible] omission of any duty or precaution which a prudent, [illegible] ought to have taken.

An important duty on the part of a master is to [illegible] with such appliances, tools, and machinery as are suited [illegible] ment and may be used with safety; and if a master fails [illegible] nary care in the selection or care of such appliances his [illegible] a defect therein will not excuse him from liability for an injury cau thereby; he is responsible for all defects in machinery or appliance which he should have known, but failed through negligence to le of, or which, having learned of, he has failed to remedy.

A railroad company is liable for injuries to its employees occasio by the company's negligence in failing to keep its track or roadbe proper condition; but such company is not bound to furnish an ab lutely safe track or roadbed, its duty only being to use all reasona care in keeping them in safe condition.

A railroad company is likewise liable if it fails to keep its track cl of obstructions and structures dangerously near the same; but s company is not negligent because it erects and maintains structu and contrivances for use in the operation of its road merely for reason that they may be dangerous to employees operating the co pany's trains.

It is negligence for such a company to fail to use safe and appro ate engines; or to have the boilers of its engines properly tested; to furnish suitable freight or passenger cars, and proper and s attachments and appliances to be used in connection therewith; such company can not divest itself of its duty to use due care and d gence with respect to the cars of other companies to be moved a handled by its employees, in seeing that such cars are in safe condit to be so moved and handled, by contracts with such other compa that they shall keep their cars in repair.

It is negligence in such a company to permit its employees to diso its orders, and it is liable for injuries arising from the careless or re less running of its trains, or the starting thereof without notice, or running of its trains at immoderate speed.

Railroad companies, and employers of every description, are ne gent if they fail to protect a servant who is exposed to danger; such a company is not absolutely bound to take all possible precauti against storms, or against washouts, landslides, or other obstructi which may be dangerous to its employees. And if the mill of a m ufacturing corporation is properly constructed for the carrying on of ordinary business, the corporation is not liable to an employee who been injured by a fire, not caused by the negligence of the corporati because it failed to provide means of escape from the fire; nor is corporation liable for an accident resulting in injury to an [illegible] from its failure to fence the ordinary machinery used in the [illegible] employment; if, however, there is a custom in reference to [illegible] of certain safeguards in a given business, so general that [illegible] is presumed to have knowledge of it, he is guilty of [illegible] ving unexpected danger, without notice to his

a rule, required to furnish the best and latest ; only such as is reasonably safe and suitable. ever, are ordinarily bound to adopt new inven- e been proved by satisfactory tests to be safer s.

f an employer to exercise reasonable care in nery, tools, etc., in suitable and safe condition should frequently inspect the machinery, etc., ie system of inspection need not be carried to barrass the operation of his business.

ervant at work in a place of danger without and instruction as the youthfulness, inexpe- ty on the part of the servant reasonably re- gence, and liable to the servant for an injury fact, however, that a master sets a minor e dangerous occupation than that in which he does not, in itself, render the master liable for from, unless under all the circumstances the : was a negligent act; but the master will be ntable in such a case than in the case of an

cturing establishments are charged with the ry care in providing their employees with suit- r can work in reasonable safety, and without within the usual scope of their employment.

yers to make and promulgate such rules and ument of their employees as will, if observed, tection; and employees are bound to obey all e commands of their employers, though such sh and severe.

nployers to have a sufficient number of trust- yees to properly and safely perform the labor in which they are engaged.

re imposed upon an employer by legislative ordinance, designed for the protection of his ce on his part to fail to comply with such re- ble to his employees for injuries arising from ; can be clearly shown that they assumed the

...and repair machinery, tools, and appliances; to ...for the servant to work, the ordinary risks of the ...; to guard against a danger to a servant of which the ...notified, or which he has promised to obviate, or which ...servant did not exist; to make and promulgate proper ...tions for the conduct of the employment in which the ...ed; to employ and retain a sufficient number of com...worthy servants to properly and safely carry on the ...mployee does not assume the risk of injury by reason ...failure of his employer in fulfilling any of the duties ...him, and, as before stated, is not guilty of contributory ...injured by such failure, if he himself was without fault ...of his duty.

...egligence is purely a matter of defense in actions by ...mages resulting from injuries sustained during the ...mployment, and the burden of proving it is upon the ... thereby to avoid liability for such damages.

...OF RISKS BY EMPLOYEES.—Where an employment ...with risks of which those who enter it have, or are ...e, notice, they can not, if they are injured by exposure ...over compensation for the injuries from their employer; ... perform hazardous duties the employee assumes such ...ent to their discharge, and he assumes not only the ...the beginning of his employment, but also such as arise ..., if he had or was bound to have knowledge thereof. ...ever, assume the risk of dangers arising from unsafe ...hods, machinery, or other instrumentalities, unless he ...resumed to have, knowledge or notice thereof, and the ...g that an injured employee had such knowledge or ...efect or obstruction causing the injury is upon the

... assumes all risk of latent defects in appliances or ...s the master was negligent in not discovering the ...xperience, or lack of experience, of the employee is to ... determining whether or not he is chargeable with ...uch defects as are not obvious and of the danger ...

...ssumed by employees is that of the master's method ...s business. If the employee enters upon the service ... of the risk attending the method, he can not hold ...nsible for injuries arising from the use of such method ...ne might have been adopted; but in order to relieve ...liability the method must amount to a custom or mode ... the business, and not consist merely of an instance ...of instances of culpable negligence on the part of the

NEGLIGENCE OF FELLOW-SERVANTS.—The [illegible] law is that he who engages in the employment of [illegible] formance of specified duties and services, for compen[illegible] himself the natural and ordinary risks and perils incident [illegible] formance of such services. The perils arising from the [illegible] and negligence of those who are in the same employment are no e[illegible] tion to this rule, and where a master uses due diligence in the sele[illegible] of competent, trusty servants and furnishes them with suitable [illegible] to perform the services in which he employs them, he is not answe[illegible] to one of them for an injury received in consequence of the care[illegible] ness or negligence of another, while both are engaged in the [illegible] service.

Various attempts have been made by judges and text writers t[illegible] down some rule or formula by which to determine what servants [illegible] common master may be said to be fellow-servants assuming the ri[illegible] each other's negligence. The following are well-known definition[illegible]

Persons are fellow-servants where they are engaged in the same [illegible] mon pursuit under the same general control.

All who serve the same master, work under the same control, d[illegible] authority and compensation from the same common source, and [illegible] engaged in the same general business, though it may be in diff[illegible] grades or departments of it, are fellow-servants who take the ri[illegible] each other's negligence.

The true test of fellow-service is community in that which is the [illegible] of service; which is subjection to control and direction by the same [illegible] mon master in the same common pursuit. If servants are employed [illegible] paid by the same master, and their duties are such as to bring [illegible] into such a relation that the negligence of the one in doing his [illegible] may injure the other in the performance of his, then they are eng[illegible] in the same common pursuit, and being subject to the same control [illegible] are fellow-servants.

All servants in the employ of the same master, subject to the [illegible] general control, paid from a common fund, and engaged in prom[illegible] or accomplishing the same common object, are to be held fellow-ser[illegible] in a common employment.

It is said that these definitions are faulty, and of little practical [illegible] by reason of their being stated so broadly and in such general [illegible] comprehensive terms, nevertheless they give a correct idea as to [illegible] have been determined by many courts to be fellow-servants withi[illegible] rule exempting the master from liability for the negligence of [illegible] them resulting injuriously to another.

The principal limitation contended for on the general rule [illegible] to fellow-servants is that there is such a servant as vice-principal, [illegible] takes the place of the master and is not a fellow-servant with [illegible] beneath him; and there is a variation of this idea to the [illegible] every superior servant is a vice-principal as to those beneath [illegible] doctrine of vice-principal is, however, repudiated by the [illegible] of the states.

[illegible] fellow-servants. The liability of the master for the [illegible] such duties as the law implies from the contract of [illegible] not rest upon the ground of guarantee of their performance, [illegible] the fact of the presence or absence of negligence of the [illegible] their performance.

Whether one is acting as the representative of the master or merely as [illegible] with others employed by the same master, does not depend upon his rank or title, but upon the character of the duties he is performing at the time another servant is injured through his negligence; if at such time the offending servant was in the performance of a duty which the master owed his servants, he was not a fellow-servant with the one injured, but a vice-principal, for the rule is fundamental that a master can not rid himself of a duty he owes to his servants by delegating his authority to another and thus escape responsibility for negligence in the performance of such duty.

If, however, at the time of the injury the negligent servant was not engaged in the performance of duty due from the master to his servants, but was discharging a duty which was due from the servant to the master, he was a fellow-servant to the one injured, engaged in the same [illegible] business, and the master would not be liable for the injuries [illegible] by reason of his negligence.

It is held by the courts of some of the states that, as industrial [illegible] have grown, and, because of the division of labor and the [illegible] operations, have been divided into distinct and separate [illegible], a laborer in one department is not a fellow-servant with a [illegible] in another and separate department of the same establishment.

[illegible] OF FELLOW-SERVANTS.—If an employer knowingly employs or retains an incompetent servant he is liable for an injury [illegible] sustained through the incompetency of the servant [illegible] or retained, provided the injured servant did not know and [illegible] means of knowing the incompetency of his fellow-servant. [illegible], however, liable for injuries to one servant by the negligence of another on the ground of unskillfulness of the latter unless the [illegible] caused by such unskillfulness.

[illegible] not warrant the competency of his servants, but must [illegible] care and diligence in their selection and retention. If [illegible] negligent in selecting a servant, and subsequently [illegible] of the servant's incompetence and still retains him, [illegible] another servant for any injury resulting from said [illegible] incompetence. If the employer had no actual ... incompetence, if it was notorious and of such ... proper care he would have known of it, he will still ...

If a person, knowing the hazards of his employment ... voluntarily continues therein without any promise by the ... any act to render the same less hazardous, the master will ... for an injury he may sustain therein, unless it is caused by the ... act of the master. No servant is entitled to damages resultin ... the incompetence of a fellow-servant when he knew of such in ... tence and did not inform his employer of the same.

When it is alleged that the master has been guilty of selec ... retaining an incompetent servant, the burden of proof of said ... tion is on the plaintiff. Neither incompetency nor unskillfulness ... presumed; they must be proved.

A master who has employed skillful and competent general ... or superintendents is liable for injuries received by inferior se ... through the negligence of those employed by such general ag ... superintendents without due care or inquiry, or retained by thei ... knowledge of their incompetence.

While the servant assumes the ordinary risks, and, as a gener ... such extraordinary risks of his employment as he knowingly a ... untarily encounters, he is not required to exercise the same de ... care as the master in investigating the risks to which he n ... exposed; he has the right to assume that the appliances and mac ... furnished him by the master are safe and suitable for the emplc ... in which he is engaged; and to assume, when engaged in an c ... tion attended with danger and requiring engrossing duties, th ... master will not, without proper warning, subject him to other d ... unknown to him, and from which his occupation necessarily di ... his attention; and he has the right to rely upon the taking ... master of all usual and proper precautions against accident, a ... faithful fulfillment of all the duties devolving upon him.

If an employee is ordered by his master into a situation of ... and obeys, he does not assume the risk unless the danger was s ... ous that no prudent man would have obeyed the order; and the ... will be liable for any injury resulting to him by reason of such d ... ous employment. If, however, he leaves his own place of work f ... more dangerous, in violation of the master's direction, he c ... recover for an injury sustained after such change.

If the servant, upon being ordered to perform duties more dan ... than those embraced in his original employment, undertakes th ... with knowledge of their dangerous character, unwillingly an ... fear of losing his employment, he can not, if injured, recover da ... from the master; nor can he recover such damages where the ...

**results from an unexpected cause during the course of his employment; nor where the injury is sustained in the performance of a service not within the scope of his duty, if his opportunity for observing the danger is equal to that of his employer; and where an employee voluntarily assumes a risk he thereby waives the provisions of a statute made for his protection.**

In Ontario, Canada, a Bureau of Industries was organized under the Commissioner of Agriculture, March 10, 1882, the official head of the Bureau being styled Secretary. Annual and occasional special reports are issued.

In New Zealand a Bureau of Industries was created in 1892. In the following year the designation of the bureau was changed to that of Department of Labor. Its publications consist of annual reports and a monthly journal commenced in March, 1893, under the title Journal of Commerce and Labor, which after the issue of a few numbers was changed to that of Journal of the Department of Labor.

We have been informed unofficially that an office for the collection of labor statistics has recently been established in Spain.

The above statement is believed to include information concerning all bureaus of foreign governments specially created for the collection and publication of statistics relating to labor. It is not a statement, however, of the extent to which foreign governments publish labor statistics, as a great deal of valuable information on this subject is contained in the publications of the central statistical bureaus or other offices of foreign governments.

## BUREAUS OF LABOR STATISTICS IN THE UNITED STATES.

[In some instances there have been changes in the official titles of officers. They are given as they exist at present.]

| Date of act of establishment. | Year of organization. | Locality of the office (post-office). | Title of head of office. | Issue of reports. |
|---|---|---|---|---|
| June 23, 1869 | 1869 | Boston | Chief | Annual. |
| April 12, 1873 | 1872 | Harrisburg | Chief (*a*) | Annual. |
| April 23, 1885 (*b*) | *c* 1885 | Hartford | Commissioner | Annual. |
| May 5, 1877 | 1877 | Columbus | Commissioner | Annual. |
| March 27, 1878 | 1878 | Trenton | Chief | Annual |
| March 29, 1879 | 1879 | Indianapolis | Chief | Biennial (*d*) |
| May 19, 1879 (*e*) | 1879 | Jefferson City | Commissioner | Annual. |
| May 29, 1879 | 1879 | Springfield | Secretary | Biennial. |
| March 3, 1883 | 1883 | San Francisco | Commissioner | Biennial. |
| April 3, 1883 | 1883 | Madison | Commissioner | Biennial. |
| May 4, 1883 | 1883 | Albany | Commissioner | Annual. |
| June 6, 1883 | 1883 | Lansing | Commissioner | Annual. |
| March 27, 1884 | 1884 | Baltimore | Chief | Biennial. |
| April 3, 1884 | 1884 | Des Moines | Commissioner | Biennial. |
| June 27, 1884 (*f*) | *f* 1885 | Washington | Commissioner | Annual and special. (*g*) |
| March 5, 1885 | 1885 | Topeka | Commissioner | Annual. |
| February 28, 1887 | 1887 | Raleigh | Commissioner | Annual |
| March 7, 1887 | 1887 | Augusta | Commissioner | Annual. |
| March 8, 1887 | 1887 | Saint Paul | Commissioner | Biennial. (*h*) |
| March 24, 1887 | 1887 | Denver | Deputy Commissioner (*i*) | Biennial. |
| March 29, 1887 | 1887 | Providence | Commissioner | Annual. |
| March 31, 1887 | 1887 | Lincoln | Deputy Commissioner (*j*) | Biennial. |
| February 22, 1889 | (*k*) | Charleston | Commissioner | Annual. |
| October 1, 1889 | 1889 | Bismarck | Commissioner | Biennial. |
| March 13, 1890 | (*l*) | Salt Lake City | Territorial Statistician. | Annual. |
| March 23, 1891 | 1891 | Nashville | Commissioners | Annual. |
| February 17, 1893 | 1893 | Helena | Commissioner | Annual. |
| March 30, 1893 | 1893 | Concord | Commissioner | Annual. |
| March 19, 1895 | (*l*) | Olympia | (*i*) | (*l*) |

*f* This office was created June 27, 1884, under the title of Bureau of Labor and the Commissioner appointed January 31, 1885. By an act passed June 13, 1888, the office was established as the Department of Labor.

*g* Also, bimonthly bulletins are to be published beginning with November, 1895.

*h* To April 24, 1889, annual.

*i* The secretary of state is ex officio commissioner.

*j* The governor is ex officio commissioner.

*k* First report issued December 1, 1894.

*l* No report yet issued.

# BULLETIN

OF THE

# EPARTMENT OF LABOR.

---

**No. 2—JANUARY, 1896.**

ISSUED EVERY OTHER MONTH.

---

EDITED BY

CARROLL D. WRIGHT,

COMMISSIONER.

OREN W. WEAVER,

CHIEF CLERK.

---

WASHINGTON:

GOVERNMENT PRINTING OFFICE.

*1896.*

# CONTENTS.

# THE POOR COLONIES OF HOLLAND.

BY J. HOWARD GORE, PH. D., COLUMBIAN UNIVERSITY. (a)

The poor colonies here described are not a creation; they are a development. They have not been elaborated out of speculation as to what they ought to be, but forged into their present organic form under the fire of criticism and the shocks of adversity.

General van den Bosch, very soon after the devastating war which was terminated by the battle of Waterloo, saw in Holland thousands of families reduced to helplessness and poverty. He realized that workhouses as well as poorhouses very often feed pauperism; that they systematize it, place their stamp of recognition, if not approval, upon it, and by so doing increase it. Moreover, these provincial or municipal agencies were usually located in large cities into whose overcrowded streets the released or acquitted paupers were cast adrift to again become amenable to the poor law.

The problem, therefore, that presented itself to General van den Bosch was how to help the poor in their life struggle, not merely how to help them to tide over the demands of a single week or month. The first point that received decision was, that in whatever shape his hopes might be ultimately realized the location at least should be in the country. If the land is improved by man, he reasoned, then man can be improved by the land.

Through his influence, and chiefly through his instrumentality, there [illegible] in 1818 the Society of Beneficence (*Maatschappy van Veld-* [illegible] thereafter the society purchased a large tract of

(a) [illegible] States Commissioner-General to the International Expo- [illegible]. The information embodied in this article was collected [illegible] colonies in July of that year.

barren uncultivated heath, which, with additions [illegible] now contains 5,100 acres. It is situated near the [illegible] dom, northeast of the Zuyder Zee, about 5 miles from the [illegible] wyk, and at the junction of the three provinces, Drenthe, [illegible] and Overyssel.

With the alluring motto, "Help the people and improve the considerable enthusiasm was aroused, and in a short time the enrolled a membership paying annually into the treasury of the $22,000. The organization was in such a prosperous condition, was able to do so much for its beneficiaries, that it attracted the tion of the State. The proposition was soon made that the societ charge of the wards of the Government, that is, the beggars, foun and orphans. The conditions offered were so favorable that the accepted them.

It was at once realized that it would be unwise to put those de by judicial acts to be incapable of self-support by the side of who were being encouraged to believe in their ability to become or later not only independent but contributing members of s Consequently the organization, in order to keep these two classes secured land and established two beggar colonies—one in Ove named Ommerschans, and the other in Drenthe, named Veenhui

At the two latter the society continued the generous policy wl had already inaugurated. Thus the beggar colonies became p and a man released at the expiration of his sentence did not res he succeeded in securing a second conviction. It therefore b necessary to add to each sentence such a term of imprisonment a labor as would efficaciously rob the stay at the colonies of the g part of its charm.

Besides acting as an encouragement to begging and profliga generous policy of the society toward the beggar colonists had a effect. There was so much said in police and court circles abo beggar colonies that people failed to discriminate between the the free colony. They thought the Government was meeti expenses, and soon lost interest in the institution which relied maintenance upon their contributions, and so it became a sort of asylum. The free colonists felt that they were being placed on th footing with a convict class, at least in the minds of many, [illegible] the institution which was originally intended to act [illegible] against pauperism by helping sinking families up to a [illegible] ing condition—now acted repressively against this very [illegible]

In this condition of disrepute into which the free colony [illegible] were so slow to take up the lands of the society that it [illegible] to farm them itself, using the weak, unwilling [illegible] colonies. This resulted in great loss to the society. [illegible] ble to break the contract with the Government, and [illegible] needed to administer the conditions [illegible]

[illegible] such as basket making, [illegible] the long winters when the farmer [illegible] It is therefore well when he can [illegible] profitable indoor occupation.
[illegible] vicinity of the colony, are paid [illegible] the installment on the debt incurred [illegible] exceeding 20 cents; 1 cent infirmary fee [illegible] clothing fund; and a reserve for the [illegible] to 10 per cent of the gross earnings. [illegible] thoroughly understood each debit and [illegible].

[illegible] interesting to note that the boys and girls, [illegible] are paid for each merchantable [illegible] makes a good basket he knows exactly [illegible] but should the work be defective, his [illegible] This of course furnishes an important [illegible] endeavors and also tends to hasten the com- [illegible] jobs can be turned out.

[illegible] if the head of the family has given [illegible] and a commendable desire to pay his [illegible] ship, and is called a "vrijboer," or "free [illegible] available he is put on it—a farm of 7.7 [illegible] one, but it is so fertile that it will readily [illegible] This plot of land is either one just vacated [illegible] been in the hands of the society; there- [illegible] provided with such planted crops as would [illegible] tenancy had begun months before. If [illegible] in midwinter, the farmer is furnished with [illegible] rye to sow 2.4 acres, and 33 bushels of [illegible] not gifts; he becomes responsible for their [illegible] which was provided him. His wants are [illegible] that all he receives are gifts. He has just [illegible] paying debts on the installment plan, and [illegible] debt and the slow and tedious process of [illegible]

[illegible] made during the probationary period [illegible] privileges. He has the full enjoyment [illegible] deems best, can work for others when he

...have been spoken of as the colony, ...so important that each has its ...Wilhelminasoord, Willemsoord, ... The principal offices are at the first named, ... the town of Steenwyk.

... in charge of the colony, and who is responsi... is the director—at present Mr. Job van ...ties are clearly defined in the regulations of the ...een distinct heads, but they may be summarized as ...ts the correspondence; executes the orders of the ...ners; looks after the receipts and expenditures; ...account with all the departments and employees; ...intenance, liberty, and safety of the roads, bridges, ...nd open squares; protects the real and personal ...ety; controls the public health; inspects the schools; ...ual interests of public worship; appoints or dis... of the society denominated second class; makes up ...coming year; formulates plans for the furtherance of ...f the colony; keeps a close watch over all the facto... in short, does everything that can possibly be done ...comfort of all concerned.

...such clerical help as is needed, including a book... ...lls the task of keeping several hundred rather com... Then he has directly under him subdirectors, each ...ediate charge of one of the seven districts into ...divided. The subdirector gives to the bookkeeper ... showing the amount of services rendered during ...hom. From these statements the bookkeeper makes ...of his accounts.

...e a still more important function. They, from fre... determine whether each farmer is getting the best ...his land. If not, they give such advice as will enable ...uccessfully. The undivided tracts, or large farms, ...these subdirectors, and, since the laborers are em...rms, it is they who are in the position to make the ...hich, at the end of the two years already mentioned, ... laborer to citizenship or dismisses him from the

...e directors in agricultural matters is more than ...ing. One of the Government agricultural ...located in the colony. Here experiments are ...specifically the best treatment of the soil in ...and the kind of seed best suited to the ...ditions which there exist.

...death before the age of 18

...admitted to the colony are those who ... of greater or less duration, they ... are also people who have been ... most of the time with insufficient ... a helpless old age. It has ... of sustenance in the colony, even under ... wages, to lay by anything for these years ... has therefore erected an "Old Folks' ... weekly toward the support of those ... are urged to aid in the maintenance of ... cases the appeal to the extent of their ... experience the value of aid and the

...

... in July, 1895, there were 1,826 people in the ... as follows: Farmers, 199 families; laborers ... individual laborers, not belonging to the

... year there had been 35 births and 24 deaths in ... men withdrew to accept positions or regular

...s body of men, coming from all parts of the king... them the traditions and habits of their native ... nor can it be an easy task. It has been ... by the society and made easier by calling on the ...o see to it that the rules were observed. In these ...oticed that there is a strong infusion of the golden ... is manifest to throw around the rising generation ...h will lift it out of the helpless state. The rules

...subordination, or insult offered any officer of the ... employees.

... the peace in any manner.

... violation of either of the above is from 10 ... for a period of from one day to three ... ejection from the colony for a repetition. ... in excess of the permissible twenty-four ... by the director.

[illegible] of the society, and one delegate from [illegible] of the colony. These delegates belong [illegible] elected annually, the laborers not having [illegible].

[illegible]franchised in the election of state or local offi- [illegible] the jurisdiction of the society, nor is one [illegible] to the State or province by being a mem- [illegible], the code of the colony is so satisfactory of law and order within its domain that can be [illegible] are left to it.

[illegible]tion comes, What does all this cost?

[illegible]eet, that for 1893, shows that the estimated value [illegible], and the indebtedness $43,380.

[illegible] last reorganization the indebtedness was $56,000, [illegible] rebellion in the United States was attributed the [illegible] debt. The contract which the society had for so [illegible] 1843 for making coffee bags had proved so profit- [illegible] after its withdrawal from this contract, continued [illegible]ate bags in competition with private firms. From [illegible]fee trade with the United States was demoralized, [illegible]pending on this trade suffered loss.

[illegible]eet (1893) shows that the receipts and expenses

RECEIPTS.

| | |
|---|---|
| [illegible] | $5,418.40 |
| [illegible] | 3,931.20 |
| [illegible] products | 3,128.52 |
| [illegible] | 615.02 |
| [illegible]ntices' work | 733.44 |
| [illegible] | 13,826.58 |

EXPENSES.

| | |
|---|---|
| [illegible] | $745.94 |
| [illegible] | 1,381.91 |
| [illegible]inistration | 4,790.32 |
| [illegible]cial instruction | 1,097.32 |
| [illegible]productive work, and losses in the various fac-[illegible] | 8,092.00 |
| [illegible]r and above returns | 538.29 |
| [illegible] | 16,645.78 |

[illegible] year of $2,819.20, or $1.54 for each inhabitant. [illegible]ant is taken care of, and to what extent, the [illegible] two extreme cases are presented:

[illegible]; in watching the happy children [illegible] brothers and sisters working in the basket [illegible] drying fruits and vegetables; in looking [illegible] the churches, the professional schools, [illegible] What more is desired? The society would [illegible] these, although they have reduced their bonded [illegible] the interest on this sum was last year one- [illegible].

[illegible] not through any bad management at the colony nor [illegible] departments did not pay their stipulated quota [illegible] they had at the colony. Their payments fell short [illegible] the amount of the deficit. The society naturally [illegible]e amount, they made their plans and promises [illegible]heir disappointment was shown in the untoward

[illegible]y persons have been aided would be a mere recital [illegible]empt to estimate the amount of good accomplished [illegible]le. As may be seen, the assistance is of the best [illegible] people are helped to help themselves, they are [illegible] faith in mankind is engendered by the faith that [illegible]ividual. The class of persons benefited are in gen- [illegible] opportunity, people unable from some fault or mis- [illegible]start. The society practically says: "We will put [illegible] prove your worth, then if found worthy you shall [illegible] man thus addressed works with confidence that the [illegible]t, and knows that starvation will be kept from his [illegible]eriod of probation.

[illegible]own that the best results are obtained with people [illegible] from the country, while those who have lived in a [illegible]ger or a shorter period, chafe under the restrictions [illegible]how a reluctance to exchange the freedom of a city [illegible] This experience also reflects itself in the donations [illegible]erest, they being, per inhabitant, the minimum in [illegible]otterdam and the maximum in Utrecht and the

[illegible]agined that the colony is in any sense an agricul- [illegible], that good farmers are here taught skill, wisdom, [illegible] sent throughout the kingdom to teach others by [illegible] how to farm. As already intimated, but few, [illegible]ingly leave the colony. And why should they? [illegible] be renters wherever they should go, and in the [illegible]ough for generations to come. Here they are not [illegible]rents, their tenancy is secure, the manifested [illegible]portant *step in the* doing, and their landlord

[illegible] When this societ
[illegible] a truer sense a
[illegible] this fact, and giv
[illegible] General van Swiet
[illegible] perpetuate the memo
[illegible] The corps of
[illegible] ance of the worl
[illegible] Her Maje
[illegible]

[illegible] fact that the be
[illegible] they are not caused to f
[illegible] expelled by the acce
[illegible] joy of having found a
[illegible] free from selfish m
[illegible]

[illegible] acquainted with the socie
[illegible] Mr. van Eeghen, sec
[illegible] Mr. Bleeker, member of th
[illegible] gave me freely of his t

... as readily as they have adopted ... advantage. It is often said that the ... people; that they are only imitators; that ... their industries from China, and that their ... acquired by imitating the methods of ... in a measure, but it is not discredita- ... that attend the development of modern ... wanted, but a power of adaptability ... more useful. The Japanese work- ... has ever seen. His ingenuity is astonish- ... complicated mechanism—a watch or an ... will reproduce it exactly and set it run- ... He can imitate any process and copy any ... and skillfully than any other race in ... which has enabled Japan to make such ... her soon among the great manufacturing

... ago that the ports of Japan were forcibly ... It was only twenty-eight years ago that ... was set up within the limits of that ... imports exceed $115,000,000.

... the general character of the merchandise ... the year 1894: (b)

| | |
|---|---|
| [illegible] | $9,551,961 |
| [illegible] | 7,974,543 |
| [illegible] | 6,662,261 |

... based were collected by Mr. Curtis, personally,

... the basis of 2 silver yen to the dollar.

Copper, brass, and lead ..............................
Books and stationery ..............................
Oil cakes ..............................
Hemp and jute..............................
Other textiles..............................
Silk goods..............................
Wines and liquors ..............................
Glassware ..............................
Clothing ..............................

The chief exports from Japan in 1894 were a

Raw silk..............................
Textile fabrics, mostly silk..............................
Food products, mostly rice..............................
Tea..............................
Coal..............................
Metals, mostly copper..............................
Matches ..............................
Drugs and medicines..............................
Floor matting..............................
Porcelain ..............................
Fish, oil, and vegetable wax..............................
Lacquer ware..............................
Umbrellas ..............................
Straw plaiting..............................
Bamboo and wooden ware..............................
Tobacco ..............................
Fans ..............................
Paper and stationery..............................

It is a curious fact that 10,273.401 [illegible]

[illegible] the source from which the imports of [illegible] according to the magnitude of the trade in

[illegible] COUNTRIES, 1890, 1892, AND 1894.

| [illegible] | 1890. | 1892. | 1894. |
|---|---|---|---|
| [illegible] | $12,309,551 | $10,394,666 | $21,094,987 |
| [illegible] | 6,434,942 | 6,254,705 | 8,755,758 |
| [illegible] | 3,437,398 | 2,994,027 | 5,401,279 |
| [illegible] | 4,456,446 | 2,831,007 | 5,290,224 |
| [illegible] | 2,747,964 | 2,492,861 | 4,499,859 |
| [illegible] | 3,428,478 | 2,167,524 | 3,964,771 |
| [illegible] | 1,984,666 | 1,810,250 | 2,174,024 |

[illegible] is based on the decimal system and corre- [illegible] United States. A rin was originally the same [illegible] 1 sen and 100 sen make 1 yen, which used to [illegible] American dollar, but is now worth about 51 cents. [illegible] in Japan from the United States or Europe [illegible] with gold finds his funds almost doubled [illegible]

[illegible] that is now going on in Japan is quite as [illegible] revolution that occurred there thirty years [illegible] to the rest of the world. Until recently [illegible] in Japan has been in the households, and [illegible] labor is still occupied in the homes of the [illegible] independent of the conditions that govern [illegible]. The weaver has his loom in his own [illegible] and daughters take their turns at it during [illegible] the custom for children to follow the [illegible] finest brocades, the choicest silks, the [illegible], and lacquer work are done under the [illegible] the compensation has heretofore been [illegible] of the piece produced.

...attended the sick and the undertakers ...

... endeavors to control the wages and the ... I was able to ascertain, in Japan ... which has its headquarters in Yokohama ... professional guides and couriers in the Empire. ... and applies at the hotel office or at a ... to the general agency, and the first man ... there are both good and poor, edu... and disagreeable, competent and incom... is not always satisfactory, but a traveler ... wants sooner or later by applying for him. ... down on the list his patron must go without a ... temporarily until his name is reached, ... the man who stands at the head to sell out his ... or divide fees with him until his turn arrives ... This arrangement is often annoying to trav... guide an equal chance, and protects them from ... guides and at the same time gives the unpopular ... employment.

... guides in Japan are much less than in Europe, ...g 2 yen ($1) a day, with 50 sen (25 cents) for ...class traveling expenses. Any guide who cuts ...ttempts to collect more, is disciplined by the ...guild.

... two strikes in Japan. One of these occurred ...ruction gang, who were hired for certain wages ... week, and were required to work seven without ...n. When their protests were unheeded they ... and appealed to the police authorities for the ... which makes six days a week's labor, and pro... of the Government or any corporation or pri...e compelled to work more than six days in a ...mpensation. Sunday is the usual day of rest ...n is not due to law nor to religious scruples, ...ence and, perhaps, out of respect to foreign ... known as the six-day law was passed the ...ple by closing its offices on Sunday, and all ... suit. That law was originally suggested for

| | |
|---|---|
| ............ | $1.44 |
| ............ | 1.20 |
| ............ | 1.22 |
| ............ | .96 |
| ............ | .96 |
| ............ | 2.80 to 7.20 |
| ............ | 2.40 to 4.80 |

than these prices. Middleton & Co., shipping houses in Japan, employ in ber of persons, men and women, who g until 6 o'clock at night, with three respectively, when they eat their rice bring with them and rest for twenty hest wages paid by the Messrs. Mid- quivalent to 21 cents io United States en who are experts in handling tea, cy by natural ability and long years

ung boys and girls who pick over the and other foreign substances. They about twelve hours' work, not includ-

ishment 20 are paid 21 cents (United id 18 cents, 50 are paid 15 cents, 335 ) cents, 5 are paid 9 cents, and 30 are emselves.

he tea "go downs," as they are called, s and manufacturing establishments

---

the basis of 2 silver yen to the dollar.

same labor would be paid from 50 cents to $5 a day. The capacity the factory when fully in operation will be 150 watches a day, a owing to the low price of labor they can be sold with a profit for per cent less than the market price in the United States and Euro

The following statement shows the rates of wages per day paid Japanese artisans and laborers: (*a*)

DAILY RATES OF WAGES, JAPAN.

| Occupation. | Highest. | Lowest. |
|---|---|---|
| Blacksmiths | $0.60 | $0.18 |
| Bricklayers | .88 | .20 |
| Cabinetmakers (furniture) | .53 | .17 |
| Carpenters | .50 | .20 |
| Carpenters and joiners (screen making) | .55 | .17 |
| Compositors | .83 | .10 |
| Coolies or general laborers | .33 | .14 |
| Cotton beaters | .45 | .13 |
| Dyers | .60 | .05 |
| Farm hands (men) | .30 | .16 |
| Farm hands (women) | .28 | .06 |
| Lacquer makers | .58 | .15 |
| Matting makers | .50 | .20 |
| Oil pressers | .34 | .16 |
| Paper hangers | .60 | .20 |
| Paper screen, lantern, etc., makers | .55 | .20 |
| Porcelain makers | .50 | .13 |
| Pressmen, printing | .70 | .11 |
| Roofers | .60 | .20 |
| Sauce and preserve makers | .40 | .10 |
| Silkworm breeders (men) | .50 | .10 |
| Silkworm breeders (women) | .25 | .05 |
| Stonecutters | .69 | .22 |
| Tailors, foreign clothing | 1.00 | .25 |
| Tailors, Japanese clothing | .56 | .15 |
| Tea makers (men) | .80 | .15 |
| Tobacco makers | .50 | .11 |
| Weavers | .40 | .07 |
| Wine and sake makers | .50 | .15 |
| Wood sawyers | .50 | .13 |

The following are the rates of wages paid by the month: (*a*)

MONTHLY RATES OF WAGES, JAPAN.

| Occupation. | Highest. | Lowest |
|---|---|---|
| Confectionery makers and bakers | $12.00 | $1.00 |
| Weavers (men) | 12.00 | 1.00 |
| Weavers (women) | 12.00 | 1.00 |
| Farm hands (men) | 5.00 | 1.00 |
| Farm hands (women) | 3.50 | .49 |
| House servants (men) | 5.00 | .50 |
| House servants (women) | 3.00 | .59 |

*a* Values stated in American gold on the basis of 2 silver yen to the dollar.

...and visited five of the principal ... I saw was a party of young ... who were in a boat sailing down ... was given over to great ceremonies ... it sent to the war, and these ... were continuing their celebration ...

... of tea instead of liquor as a beverage. ... of saloons and are about as numer- ... in New York or Chicago. But a ... an entire family can be bought for 2 sen ... strengthens quite as much as malt or alco- ... is, however, increasing so rapidly in ..., and the Government is making it the ... is a brewery or two in nearly every city ... be bought at almost every tea house.

... was negotiated by Secretary Gresham and ...winter at Washington makes Japan as free for ...ed States, with the exception that they can not ...y a straight reading of the text it would seem ...ited. It provides that foreigners may trade by ...gly or with native partners, and says that they ... houses, manufactories, warehouses, shops, ...," conforming, of course, to the laws and ... apply to them and the natives of the country

...vides that foreigners shall enjoy all rights and ... natives "in whatever relates to residence and ...n of goods and effects, to the succession to per- ...disposition of property;" that they shall not be ... higher taxes, imposts, or other charges than ... enjoy their own religion, bury their dead ..., and shall be exempted from military serv- ... other exactions. No higher duties are to be ... of the United States than upon those of ... and there must be perfect equality in the ... natives in the exportation of merchan- ... is customary in all countries, is withheld

[illegible] assistance from foreign nations,
[illegible] machinery and raw material, particu-
ire our sales will be practically limited
rket for machinery will be limited as to
a great deal within the next few years,
of labor-saving apparatus, but they are
ir own machinery, and in a few years
nations in that respect also. Another
ant fact—is that they will buy only one
We will sell them one set, which they
demands themselves. This will go on
effect, when American patents will be

rorking machinery; and very little shoe-
eople do not wear shoes. The same is
or they do not wear hosiery. I do not
at of the 41,388,313 people who compose
shoes and stockings. Ninety per cent
l, women, children, and men, protecting
wooden and straw sandals. The higher
foot gear, but it is made in a more fin-
little cloth affairs that they call "tabis"
nade of white or blue cotton, and do not
the use of shoes and hosiery is increas-
into it as they have grown into other

ice as much in Japan as it is with us.
ll for $10 and $12 a thousand feet, will
. This is due chiefly to the scarcity of
quired to work it up by their primitive
cting timber off their mountains for 2,500

...upon a fan 3 or 4 feet wide, upon ... so as to leave the chaff in the ... of thrashing is to beat the heads of ... bamboo poles.

... districts in a carriage or jinrikisha one ... preempted by the farmers of the ...pose of drying their grain, which is spread ... mats and raked over every now and then ... that the particles at the bottom may get ... the straw, which is still tied together in ...cks along the roadside during the day and ... to protect it from dampness as well as ... racks are 30 or 40 yards long and 18 feet ...s, and the farmer's wife or one of his daugh... ...vals to inspect the straw to see that it is cur... ...t as valuable as the grain.

... is saved, and it is put to many uses. They ...pes, roofs, matting, the partitions and floors ...ats, baskets, boxes, and a thousand and one ... braid it for fences, too, and the finer, softer ...

... Japan. The grass is wiry and indigesti... ... animals. Some alfalfa is grown, but it ... neighborhood of Kobe, which is one of the ..., the soil seems to be better adapted for ... from that locality.

..., which originated in China and is called ... and thrives upon it, but he is small and

... have never been. ... farmers employ their winter ... implements and repairing ... blacksmith or traveling ... traveling from village to village ... portable forge to assist in repairing

... no reason to complain that the ... the load. Whatever may be the ... the household, although she is not ... in the responsibilities that are ... and wives in America, she is at ... men when there is any hard work ... in the cities or villages or the farming ... and mother working side by side with ... planting, and reaping, and at sunset ... the harvest in a big basket on her back. ... a pair of tiny shafts tugging to haul ... there is always a woman pushing from ... protected, except for a pair of straw sandals, ... cotton leggings like tights extending from ... Sometimes the baby is playing with a few ... Sometimes he is strapped to her shoul... one side to the other with every motion ... may fall off.

... exports of Japan, are raised almost en... and in the mechanical arts she appears ... labor, although she gets little or none of ... fashion many of the choicest pieces of ... and in the decoration of lacquer that ... is equal and often superior to the work of ... other articles of straw, she braids bam... and one other articles that are made ... goes out with her husband in fishing boats ... he brings home; she assists in house... and in various other occupations which

... matters that ... ...s and the busin... ... matters that pertain to the household.

State capital, and the address of the
... assembled at Denver in July,
... report may be grouped as follows:
... manufactures, 8 pages; cost of pro-
... smelter production, 14 pages; wages,
... prices, 129 pages; strikes, 25 pages;
... what they contain, 60 pages; mis-

... of these titles, with the exception of
... smelter production, and strikes, consists,
... compilations from numerous publications on
... subjects.

... AND FACTS RELATING TO SMELTER
... production is shown for the two principal
... of the State, and is accompanied with a
... of the smelting plants and smelter and mine
... produced 24,000,000 ounces of fine silver,
... total production of the United States.

... statement is given of the strikes and labor
... State during 1893 and 1894. The account of
...llowed by the conclusions and recommendations
...pointed to investigate the Chicago strike of

... above referred to that in March, 1893, a
... was made in order to ascertain the number
... State. The returns showed an aggregate of
... and female, out of employment. This num-
... the normal unemployed in the State at that
... the decline in silver, in June, a further
... the sixty days ending with August 31,
... increased to 45,000, and of these 22,500 were
... vicinity where they were employed.

## ILLINOIS.

...rt of the Bureau of Labor Statistics of Illi-
..., devoted to a discussion of the subject of
... comprising 59 pages. The appendix con-

| [illegible] | $16,[illegible]7,000 | $1,485,000 | 9.06 |
|---|---|---|---|
| [illegible] | 30,[illegible],000 | 4,771,050 | 15.64 |
| [illegible] | 46,[illegible]1,000 | 6,256,050 | a 11.21 |

[illegible] those given are, however, according to the original.

[illegible] classes of personal property in Cook [illegible] comparison of their valuations with those of [illegible] counties of the State. The following [illegible] for some of the classes:

[illegible] OF MISCELLANEOUS PERSONAL PROPERTY, 1894.

| [illegible] | Cook County. | | The State, exclusive of Cook County. | | | The State. | | |
|---|---|---|---|---|---|---|---|---|
| [illegible] | Persons to each. | Average value. | Number. | Persons to each. | Average value. | Number. | Persons to each. | Average value. |
| [illegible] | 3,[illegible] | [illegible].60 | 9,967 | 264.32 | $21.54 | 10,364 | a 327.10 | $21.84 |
| [illegible] | 1,[illegible] | [illegible]4.91 | 9,538 | 276.26 | b 125.24 | 10,179 | 375.91 | 190.37 |
| [illegible] | [illegible].91 | 2[illegible].[illegible] | 31,757 | 82.96 | 30.24 | 43,667 | 87.50 | 29.74 |
| [illegible] | [illegible].[illegible] | 3.78 | [illegible],016 | 8.06 | 1.94 | 333,613 | 11.47 | 1.98 |
| [illegible] | 237.81 | 4.47 | 223,315 | 11.80 | c 3.96 | 22[illegible],547 | 16.74 | 3.98 |
| [illegible] | [illegible]7,[illegible].10 | d 22.00 | 1,911 | 1,378.56 | 19.09 | 2,065 | 1,852.95 | 19.31 |

[illegible], the figures here should be 369.20; those given are, how-

[illegible] the figures here should be $125.35; those given are, however,

[illegible] the figures here should be $3.97; those given are, however,

[illegible], the figures here should be 7,739.75; those given are, how-

[illegible] the figures here should be $21.86; those given are, however,

...es by years, of a number of [illegible] of property [illegible] of Chicago that were vacant [illegible] 1892, 1893, and 1894. ...roved in one or two of these years [illegible] ... tables are summarized by years, and [illegible] of the ...wn are as follows:

...F UNIMPROVED AND IMPROVED [illegible]

| | | | |
|---|---|---|---|
| [illegible] | [illegible] | [illegible] | [illegible] |
| [illegible] | [illegible] | [illegible] | [illegible] |
| [illegible] | [illegible] | [illegible] | [illegible] |
| ...round— | | | |
| [illegible] | [illegible] | [illegible] | [illegible] |
| [illegible] | [illegible] | [illegible] | [illegible] |
| [illegible] | [illegible] | [illegible] | [illegible] |
| [illegible] | [illegible] | [illegible] | [illegible] |
| [illegible] | [illegible] | [illegible] | [illegible] |
| [illegible] | [illegible] | [illegible] | [illegible] |
| [illegible] | [illegible] | [illegible] | [illegible] |
| [illegible] | [illegible] | [illegible] | [illegible] |
| [illegible] | [illegible] | [illegible] | [illegible] |
| [illegible] | [illegible] | [illegible] | [illegible] |
| ...d | [illegible] | [illegible] | [illegible] |
| ... and improvements | [illegible] | [illegible] | [illegible] |
| ... and improvements | [illegible] | [illegible] | [illegible] |

...ould be $1,959; those given are, however [illegible]
...ould be $346,657; those given are, however [illegible]

...ation of certain property in the business district of ... years from 1890 to 1894, inclusive, the figures being ... each piece of property, and summarized as follows:

[illegible] the assessed value of unimproved and [illegible] series of tables giving a description, and [illegible], of a number of pieces of property in [illegible] that were vacant in 1892, 1893, and 1894, [illegible] one or two of these years but vacant in [illegible] are summarized by years, and some of the [illegible] follows:

[illegible] AND IMPROVED PROPERTY IN CHICAGO.

| | 1892. | 1893. | 1894. |
|---|---|---|---|
| [illegible] | 128 | 94 | 99 |
| [illegible] | 5,192.7 | 3,495.2 | 3,827.7 |
| [illegible] | 573,307.4 | 379,012.6 | 449,353.6 |
| [illegible] | | | |
| [illegible] | $179.682 | $146.349 | $160.676 |
| [illegible] | $1.037 | $1.382 | $1.966 |
| [illegible] | $932,780 | $511,520 | $615,020 |
| [illegible] | 27 | 61 | 56 |
| [illegible] | 1,693.0 | 3,480.5 | 3,058.0 |
| [illegible] | 210,380.3 | 413,635.1 | 334,294.1 |
| [illegible] | $412,100.00 | $838,860.00 | $729,360.00 |
| [illegible] | 174,790.00 | 861,320.00 | 929,550.00 |
| [illegible] | 586,890.00 | 1,700,180.00 | 1,658,910.00 |
| [illegible] | 243.414 | 241.017 | 238.509 |
| [illegible] | a1.954 | 2.028 | 2.182 |
| [illegible] improvements | b346.893 | 488.487 | 542.482 |
| [illegible] improvements | 2.789 | 4.110 | 4.962 |

[illegible]d be $1.959; those given are, however, according to the original.
[illegible]d be $348.657; those given are, however, according to the original.

[illegible]n of certain property in the business district of [illegible]ars from 1890 to 1894, inclusive, the figures being [illegible]h piece of property, and summarized as follows:

[illegible] VALUATIONS IN CHICAGO, 1890 TO 1894.

| [illegible] | Assessors' valuation. | | | | |
|---|---|---|---|---|---|
| | 1890. | 1891. | 1892. | 1893. | 1894. |
| 65 | $848,650 | $707,600 | $869,400 | $903,200 | $857,450 |
| 299 | 2,396,750 | 2,518,000 | 3,492,800 | 3,566,100 | 2,728,900 |
| 294 | 2,097,400 | 3,228,600 | 4,362,200 | 4,469,300 | 3,504,310 |

[illegible] the above, for the years 1892, 1893, and 1894, [illegible] pieces of property owned by forty-two of the [illegible] in the business district of the city.

[illegible]mendations are made for additional [illegible]

[illegible] closed with a series of tables giving in [illegible] coming under the investigation of the [illegible] 1890, 1891, and 1892, and for the years 1893 [illegible], showing the amount of the consideration [w]arranty deeds, and the assessments on the same the years in which the property was sold; also [showi]ng the same facts by classes of property, accord- [illegible] year from 1870 to 1892, inclusive, and for the [co]nsidered together.

## MAINE.

Report of the Bureau of Industrial and Labor [fo]r the year 1894 treats of the following subjects: [return]s of the effects of the business depression, 51 [illegible], and shops built in 1894, 3 pages; census sta- [reta]il prices, 36 pages; pulp and paper making, 13 [bu]siness, 12 pages; part of the proceedings of the [conven]tion of the National Association of Officials of [St]atistics, 24 pages; the labor laws of Maine, 10 [illegible] inspector of factories, workshops, mines, and

RETURNS OF THE EFFECTS OF THE BUSINESS [illegible] statistics are the result of two investigations, [the ot]her in October, 1894, the reports being collected [illegible] Two hundred and twenty-four reports were the first call, representing nearly all the cotton [illegible] mills, and other important manufacturing estab- [lishments]. While the replies to the second call were not [com]plete, they show an improved condition in some [sli]ght changes in others. The returns are presented [fo]r each establishment and industry the working [e]mployees when running on full time and during [whet]her wages have been reduced since April 1, 1893, [r]eduction. The reports for July are followed by [wh]ich show the changes that occurred in each estab- [illegible] and October.

[do]es not present a statistical summary of the effects [depr]ession on all industries, the Commissioner states [illegible]dly, felt the bad effects of the business depres- [illegible] than other sections of the country." It appears

value of all, leaving 3,434
During the latter period
44 per cent, or nearly one-
ue of the estates by classi-
to $100,000 and over, are
tion. The following sum-

TATES PROBATED IN BALTI-
YEARS.

| r. | 1860, inclusive. Value. | 1888 to 1893, inclusive. Number. | Value. |
| --- | --- | --- | --- |
| 4 | $201,902 | 1,259 | $282,226 |
| 1 | 361,407 | 956 | 656,038 |
| 5 | 1,287,063 | 1,258 | 2,165,694 |
| 8 | 1,596,467 | 792 | 2,776,838 |
| 1 | 2,229,146 | 578 | 4,009,751 |
| 7 | 4,981,125 | 452 | 7,128,912 |
| 7 | 4,475,415 | 199 | 6,996,063 |
| 6 | 3,890,194 | 129 | 9,157,922 |
| 4 | 21,038,650 | 89 | 25,836 140 |
| 8 | 40,001,380 | 5,914 | 59,055,566 |

ng, received from 711 individual work-
branches of industry. The information
rouped by industries, and the totals for
ird in summary form. There are numer-
'idual reports concerning the nativity of
y, conjugal condition, number in family,
chool, number dependent in family, age
d in present occupation and for present
: week, days and weeks unemployed,
y and yearly wages, increase or decrease
ges withheld, income other than wages,
of single and of married men, living
operty, number owning homes, number
unt of dues and of benefits.

made reports 84+ per cent were born
foreign countries. The parents of 72+
merican born, the percentage being the
r. The conjugal condition showed 72+
nmarried, and 2+ per cent widowed. Of
ent reported children in the family and
hundred and thirty-nine families reported
occupations in which the wage-earners
from one to ten years each, and 179 from
number of hours worked weekly varied

[illegible] 24+ per cent of the value of product [illegible] of the total wages paid, and 26 per cent

[illegible] days to the year, the average number of [illegible] for the 299 reporting was 269, or 87.33 [illegible] average proportion of business done was [illegible] of the establishments.

[illegible] for the 166 establishments reporting in [illegible] decrease in capital of 3.20 per cent, in cost [illegible] in value of product 13.85 per cent. The [illegible] decreased 5.46 per cent, while the total [illegible] cent. The average yearly earnings per [illegible] sex or age, decreased from $341 to $325. [illegible] establishments were in operation decreased [illegible] average proportion of business done decreased

[illegible]—A description is given of the additions [illegible] and shops, and a statement made of other [illegible] in industrial establishments in the [illegible] the State.

## OHIO.

[illegible] port of the Bureau of Labor Statistics of [illegible] 1894 treats of the following subjects: [illegible]; "sweating" in Cincinnati, 12 pages; [illegible] 29 pages; child labor in Cincinnati, 7

[illegible] CINCINNATI.

| [illegible] | Price per piece. | Reduction in price during year (per cent). | Weekly earnings of sweater's family. | Persons in sweater's family. | Rooms occupied. | Monthly rent. |
|---|---|---|---|---|---|---|
| 3 | [illegible] | 46.35 | $9.30 | 6.00 | 2.70 | $9.40 |
| [illegible] | .61 | [illegible] | 11.93 | 5.00 | 3.87 | a15.42 |
| | .06 | 21.00 | | | | |
| 4 | .21½ | 29.56 | 11.[illegible] | 3.32 | 2. | 12.62 |
| 6 | [illegible] | 21.00 | 8.17 | 4.52 | 3. | 11.42 |
| | [illegible] | 12.00 | 7. | 2.73 | 2. | 5.50 |
| | .37 | 20.90 | 10. | 2.00 | 2. | 7.16 |
| 2 | 6.[illegible] | 42.90 | 9.07 | 4.25 | 2.50 | 13.37 |
| 1 | 4.06 | 33.00 | 16.[illegible] | 3.29 | 2.[illegible] | 10.13 |

[illegible] average is made includes rent of shop, and this is [illegible] rooms, which varies from $10 to $27 a month.
[illegible], and the 18 pieces averaged are included in the [illegible] only to show difference in price.
[illegible] found were engaged exclusively in finishing: the [illegible]-making shops and represent those in which the work

[illegible]NNATI.—This investigation was lim[illegible]ent-house population engaged in the [illegible]mon day labor, the canvass being con[illegible]d of the city. The statistics gathered [illegible]e shown in detail for each return and [illegible]nditions and earnings of those report[illegible]thers of families reported 165, or 22 per [illegible] per cent, were Americans; 498, or 65 [illegible], or 61 per cent, of the mothers, native [illegible] 13 per cent, of the fathers and 75, or [illegible]ing natives of other foreign countries. [illegible]ied 2,160 rooms; the average monthly

[illegible] 1893, the showing is a very [illegible] that the dullness in all lines of [illegible] year than last." The statisti- [illegible] offices, the number of situations [illegible] and females, respectively. The [illegible] during 1894 are as follows:

[illegible] EMPLOYMENT OFFICES, 1894.

| [illegible] |  | Situations wanted. |  | Positions secured. |  |
|---|---|---|---|---|---|
| Females. |  | Males. | Females. | Males. | Females. |
| 1,[illegible] |  | 2,778 | 3,162 | 267 | 1,144 |
| 2,[illegible] |  | 2,[illegible] | 3,517 | 273 | 1,846 |
| 1,[illegible] |  | 2,673 | 2,226 | 456 | 1,343 |
| [illegible] |  | 2,472 | 1,950 | 367 | 1,350 |
| 2,447 |  | 3,657 | 3,761 | 777 | 1,[illegible] |
| [illegible] |  | 14,[illegible] | 14,616 | 2,140 | 7,[illegible] |

[illegible] reporting in 1892. Ninety-[illegible] information is published had [illegible] of 1,[illegible], at the end of 1893. [illegible] were secured, the Labor [illegible] of 118 other unregistered [illegible] no information was obtain-[illegible] membership amounted to 90,660; [illegible] of which information concerning [illegible].

[illegible] total number of members, amount [illegible] and balance of funds of all unions [illegible] obtained:

| | |
|---|---|
| [illegible] received | 687 |
| [illegible] end of 1893 | 1,270,789 |
| [illegible] beginning of 1893 | $9,258,015 |
| [illegible] | 9,718,259 |
| [illegible] year | 10,932,665 |
| [illegible] | 8,044,655 |

[illegible] the above summary is explained by the [illegible] which did not report the amount of their [illegible] of the year, had an excess of expendi-[illegible].

[illegible] chief items of expenditure were fur-[illegible] shown in the following statement:

[illegible], OF 682 TRADE UNIONS, 1893.

| | Amounts. | Unions. | Members. |
|---|---|---|---|
| [illegible] | $2,496,169 | 378 | 827,840 |
| [illegible] | 3,567,364 | 331 | 1,083,904 |
| [illegible] | 1,161,821 | 228 | 622,908 |
| [illegible] | 136,889 | 99 | 414,989 |
| [illegible] | 571,090 | 89 | 458,678 |
| [illegible] | 458,385 | 387 | 983,834 |
| [illegible] | 555,570 | 391 | 842,202 |
| [illegible] | 299,927 | 405 | 906,618 |
| [illegible] | 1,690,919 | 679 | 1,209,070 |
| [illegible] | 10,928,076 | .......... | .......... |

[illegible] the total number of unions making the [illegible] in the preceding statement, and the

[illegible] 1892 AND 1893.

| [illegible] | 1893. | | | Increase (+) or decrease (—) in expenditures. |
|---|---|---|---|---|
| | Unions. | Members. | Expenditures. | |
| [illegible] | [illegible] | [illegible] | $2,[illegible],141 | + $[illegible] |
| [illegible] | [illegible] | 1,[illegible] | 2,[illegible] | + 1,[illegible] |
| [illegible] | [illegible] | [illegible] | 1,[illegible] | + 102,[illegible] |
| [illegible] | [illegible] | [illegible] | 134,[illegible] | + [illegible] |
| [illegible] | [illegible] | [illegible] | 565,[illegible] | + 51,[illegible] |
| [illegible] | 225 | [illegible] | 440,[illegible] | + 27,[illegible] |
| [illegible] | [illegible] | 1,[illegible] | [illegible] | + 165,[illegible] |
| [illegible] | [illegible] | 1,[illegible] | 1,578,376 | — 47,[illegible] |
| [illegible] | ........ | ........ | a10,494,062 | ........ |

[illegible] and that given in the preceding summary is due [illegible] their expenditures.

[illegible] returns of 534 unions which furnished [illegible] of their expenditures in either 1892 or [illegible] unions did not, in any instance, make [illegible] in both years the table contains an [illegible] to secure a proper basis of compari[illegible], which shows for each of the items [illegible] such unions or societies as made an [illegible] both 1892 and 1893.

[illegible] UNIONS REPORTING FOR BOTH 1892 AND 1893.

| Members. | | Expenditures. | | Increase. |
|---|---|---|---|---|
| 1892. | 1893. | 1892. | 1893. | |
| [illegible] | 706,[illegible] | $1,815,380 | $2,359,484 | $544,104 |
| [illegible] | 973,446 | 2,059,381 | 2,[illegible],487 | 1,209,106 |
| [illegible] | [illegible],677 | 1,0[illegible],148 | 1,138,527 | 105,379 |
| [illegible] | [illegible],521 | 80,352 | 120,[illegible] | 40,[illegible] |
| [illegible] | 428,[illegible] | 514,[illegible] | 564,027 | 49,059 |
| [illegible] | 575,[illegible] | 460,946 | 429,746 | 28,[illegible] |
| [illegible] | [illegible] | 620,581 | 768,951 | 148,370 |
| [illegible] | ........ | 6,[illegible] | 8,[illegible],120 | 2,124,[illegible] |

[illegible] excluded from the report:

[illegible] in a trade which are due, not to [illegible] particular classes of work, but to [illegible] the higher and lower paid classes [illegible].

[illegible] individuals due to promotions, or [illegible] wages. In some classes of undertakings [illegible] the rates of pay of various classes of [illegible]. The rates of pay, therefore, of indi- [illegible] altering. Such internal changes, however, [illegible]eal changes in the rates of wages so long as [illegible] each class of employees remain unaltered [illegible].

[illegible]anges in weekly wages which regularly occur [illegible]ear in certain trades. This change as a rule [illegible]e of hours of labor for the summer and win- [illegible]and merely represents the effect on weekly [illegible]hours, the hourly rate of pay remaining the

[illegible]s of employment which merely provide for [illegible] work.

[illegible]affecting seamen and agricultural laborers, [illegible]anges of wages reported for 1893, and in 706 [illegible] obtained. Of the total number of changes [illegible] decreases. There were 549,977 individuals [illegible]706 changes for which full particulars were [illegible]er 151,140 finished the year with their wages [illegible] beginning, and for the purpose of compar- [illegible] and 1893 may be regarded as having their [illegible] remaining 398,837 employees, 142,364, or [illegible] number affected, gained a net increase of sons affected, the decreases of wages during the over the increases. Nevertheless the net result during the year was a slight rise of wages, the ave increases per head being so much greater than that to overbalance the superiority of numbers.

In presenting the statistics the various subdivisions arranged in seven general groups, the totals being given and each subdivision. The details for each change in wage of labor are also given, the reports being arranged by industries. In the following summaries only the totals for the groups are presented and those cases considered for which par lars were secured.

In many cases the same individual was involved in more change during the year, and to obtain the aggregate number ployees affected should be counted more than once. The ments immediately following show the number of changes ( or decreases in weekly wages) during the year, the number vidual employees affected, and the total and average inc decrease:

NUMBER OF INCREASES OR DECREASES IN WEEKLY WAGES, AND EM AFFECTED, 1893.

| Industries. | Changes. | | | Employees affe[illegible] | | |
|---|---|---|---|---|---|---|
| | Increases. | Decreases. | Total. | Wages increased. | Wages decreased. | [illegible] |
| Building | 265 | 22 | 287 | 40,017 | 4,[illegible] | [illegible] |
| Metal, engineering, and shipbuilding | 35 | 111 | 146 | 6,377 | 95,[illegible] | [illegible] |
| Mining and quarrying | 40 | 40 | 80 | 75,594 | 163,[illegible] | a[illegible] |
| Textiles | 28 | 4 | 32 | 4,196 | 90,[illegible] | [illegible] |
| Clothing | 18 | ........ | 18 | 2,500 | ........ | [illegible] |
| Other trades and occupations | 43 | 18 | 61 | 3,210 | 1,944 | [illegible] |
| Employees of public authorities | 79 | 3 | 82 | 10,121 | [illegible] | [illegible] |
| Total | 508 | 198 | 706 | 142,364 | [illegible] | a[illegible] |

a These employees are included in obtaining the average net increase or de[illegible]

AMOUNT OF INCREASE OR DECREASE IN WEEKLY WAGES [illegible]

| Industries. | Increase. | | | | Decre[illegible] | |
|---|---|---|---|---|---|---|
| | Total. | Average per employee. | Net. Total. | Net. Average per employee. | Total. | [illegible] |
| Building | $30,200.35½ | $0.50½ | $15,936.52 | $0.42 | $1,274.[illegible] | [illegible] |
| Metal, engineering, and shipbuilding | 1,992.82½ | .37 | ........ | ........ | 24,[illegible] | [illegible] |
| Mining and quarrying | 86,282.80 | 1.14 | 79,546.24½ | .35 | 12,[illegible] | [illegible] |
| Textiles | 845.96 | .20 | ........ | ........ | [illegible] | [illegible] |
| Clothing | 1,222.63½ | .28½ | 1,222.63½ | .28½ | ........ | [illegible] |
| Other trades and occupations | 1,281.42 | .30 | ........ | ........ | [illegible] | [illegible] |
| Employees of public authorities | [illegible] | [illegible] | [illegible] | [illegible] | [illegible] | [illegible] |

[illegible] of time.

[illegible] per head calculated on the total [illegible] was 85½ cents, while the average [illegible] in similar way was only 20½ cents. [illegible] per head calculated on the total [illegible] of wages during the year was a [illegible] spread over the total estimated [illegible] to which the returns relate, [illegible] per annum. The number of employees [illegible] about 8 per cent of the total [illegible] the industries covered.

[illegible],91½, shown in the above state- [illegible] much larger amounts of increases and [illegible] individuals, and the same is true of the [illegible] internal changes during the year were [illegible] total weekly increases recorded was [illegible] decreases $229,531.15, the difference between [illegible] as that between the increases and decreases

[illegible] employees (not necessarily separate individuals) [illegible] for which full particulars were reported, [illegible] of the aggregate increases and decreases per [illegible] following statements:

[illegible] BY CHANGES IN WEEKLY WAGES, 1893

[[illegible] more than one change, and hence counted two or more times in this table.]

| [illegible] | Increases. | Decreases. | Total. | Average number to each change. | | |
|---|---|---|---|---|---|---|
| | | | | Increases. | Decreases. | Total. |
| [illegible] | [illegible],217 | 5,071 | 45,288 | 152 | 231 | 158 |
| [illegible] | [illegible] | 160,458 | 190,285 | 881 | 1,437 | 1,308 |
| [illegible] | [illegible] | 549,844 | 1,112,368 | 14,063 | 13,746 | 13,905 |
| [illegible] | 4,[illegible] | 50,801 | 55,137 | 152 | 12,728 | 1,723 |
| [illegible] | 3,[illegible] | ......... | 3,500 | 300 | ......... | 300 |
| [illegible] | 3,460 | 2,194 | 5,654 | 80 | 122 | 98 |
| [illegible] | 10,141 | 36 | 10,177 | 128 | 13 | 124 |
| [illegible] | [illegible] | 767,494 | 1,422,508 | 1,289 | 3,876 | 2,015 |

[illegible] their results:

[illegible]

Louisiana.

| Suc-ceed-ed. | Suc-ceed-ed partly. | Fail-ed. | Not re-port-ed. | To-tal. |
|---|---|---|---|---|
| 1 | ...... | ...... | ...... | 1 |
| ...... | ...... | ...... | ...... | ...... |
| ...... | ...... | ...... | ...... | ...... |
| ...... | ...... | ...... | ...... | ...... |
| ...... | ...... | ...... | ...... | ...... |
| ...... | ...... | ...... | ...... | ...... |
| ...... | ...... | ...... | ...... | ...... |
| ...... | ...... | ...... | ...... | ...... |
| ...... | ...... | 1 | ...... | 1 |
| ...... | ...... | ...... | 1 | 1 |
| ...... | ...... | ...... | 1 | 1 |
| ...... | ...... | 1 | ...... | 1 |
| ...... | ...... | 1 | ...... | 1 |
| ...... | ...... | ...... | ...... | ...... |
| ...... | ...... | ...... | ...... | ...... |
| ...... | ...... | ...... | ...... | ...... |
| ...... | ...... | ...... | ...... | ...... |
| ...... | ...... | ...... | ...... | ...... |
| ...... | ...... | ...... | ...... | ...... |
| ...... | ...... | ...... | ...... | ...... |
| ...... | ...... | ...... | ...... | ...... |
| ...... | ...... | ...... | ...... | ...... |
| ...... | ...... | ...... | ...... | ...... |
| ...... | ...... | ...... | ...... | ...... |
| ...... | ...... | ...... | ...... | ...... |
| ...... | ...... | ...... | ...... | ...... |
| ...... | ...... | ...... | ...... | ...... |
| ...... | ...... | 2 | ...... | 2 |
| ...... | ...... | ...... | ...... | ...... |
| ...... | ...... | ...... | ...... | ...... |
| ...... | ...... | ...... | ...... | ...... |
| ...... | ...... | ...... | 1 | 1 |
| 1 | ...... | ...... | ...... | 1 |
| ...... | ...... | ...... | ...... | ...... |
| ...... | ...... | ...... | ...... | ...... |
| 2 | ...... | 5 | 3 | 10 |

ollowing summary, taken by five-year periods, is interesting as
; the progress of the movement for shorter working days:

DS AND STRIKES FOR REDUCTION OF HOURS, BY FIVE-YEAR PERIODS, 1865 TO 1894.

| rs. | Demands and strikes for— 11 hours per day. Succeeded and succeeded partly. | 11 hours per day. Failed and result not reported. | 10 hours per day. Succeeded and succeeded partly. | 10 hours per day. Failed and result not reported. | 9 hours per day. Succeeded and succeeded partly. | 9 hours per day. Failed and result not reported. | Time not reported. Succeeded and succeeded partly. | Time not reported. Failed and result not reported. |
|---|---|---|---|---|---|---|---|---|
| | 4 | | 9 | 2 | | | | |
| | 6 | 4 | 13 | 9 | 1 | | 1 | 5 |
| | 1 | 1 | 2 | 1 | | | | |
| | 1 | | | | | | | |
| | 3 | 1 | 23 | 3 | | | 1 | 2 |
| | 3 | 1 | 49 | 25 | 4 | 5 | 5 | |
| al | 18 | 7 | 96 | 40 | 5 | 5 | 7 | 7 |

next table shows the strikes, lockouts, and demands by occupa-
or the period of thirty-five years from 1860 to 1894:

STRIKES, LOCKOUTS, AND DEMANDS, BY OCCUPATIONS, 1860-1894.

| ccupation | Demands not resulting in strikes. Succeeded and succeeded partly. | Demands not resulting in strikes. Failed | Demands not resulting in strikes. Not reported. | Strikes. Succeeded and succeeded partly. | Strikes. Failed | Strikes. Not reported. | Lockouts. Succeeded and succeeded partly. | Lockouts. Failed. | Lockouts. Not reported. |
|---|---|---|---|---|---|---|---|---|---|
| | 36 | 2 | 1 | 7 | 13 | 3 | 1 | | |
| employees | 2 | 2 | 1 | 2 | 3 | | | | |
| ters | 4 | 2 | 1 | 2 | 1 | 1 | | | 1 |
| | 5 | 5 | 2 | 8 | 2 | 2 | 1 | | |
| | | | 1 | 6 | | 1 | | | |
| ers | 1 | 1 | | 3 | 2 | 2 | | | |
| nd stove makers | 1 | 2 | 2 | 2 | | | | | |
| ers | 2 | | 3 | 15 | 3 | 1 | | | |
| | 13 | 3 | 5 | 22 | 10 | 3 | | 1 | |
| | | | | 8 | 2 | | | 1 | |
| | 6 | 2 | 2 | 3 | 2 | 1 | | | |
| akers | 4 | 1 | | 21 | 1 | 2 | | 1 | 1 |
| rs and watch case | | | | 11 | 4 | 1 | | | |
| | 1 | | | 4 | | 2 | | | |
| | 10 | | | 22 | 4 | 1 | | | |
| ers | 11 | 2 | | 11 | 6 | 4 | | | |
| thing trades | 1 | | | 1 | | 1 | | | |
| | | 1 | | 2 | 2 | 1 | | | |
| ts | 1 | 6 | 2 | 3 | 2 | | | | |
| ters | 6 | | | 2 | 3 | 1 | | | |
| s | 6 | 2 | | 9 | | 1 | 1 | | |
| ithe and wagon | 6 | 2 | 1 | 9 | 3 | | | | |
| es in weaving establishments | 5 | 3 | 2 | 9 | 7 | 3 | | 2 | |
| tc. | 4 | | | 4 | 2 | | | | |
| es in tobacco | 2 | | | 3 | 3 | | | | |
| | 2 | | | | | | | | |
| | | | | 5 | | 1 | | | |
| | 3 | | 2 | 2 | | | | | |
| neous | 6 | 1 | 4 | | 2 | 1 | | | |
| tal | 138 | 37 | 20 | 196 | 77 | 33 | 3 | 5 | 2 |

report calls attention to the fact that the statistics for the print-
ade are the most *complete, because* the printers have long been

well organized, and have published a special trade journal. This t shows that while there were in all 39 demands that were amic settled, only 21 strikes and lockouts are reported. In the totals for other occupations the reverse condition is found, namely, 165 dema as against 295 strikes and lockouts. This result in the case of printers is probably due to their superior skill and better organizat

In order to better illustrate the effects of labor organization in re ence to disputes between employees and their employers, the follow summary is presented of the number of demands amicably settled, of offensive and defensive strikes—lockouts being classed with latter.

DEMANDS NOT FOLLOWED BY STRIKES AND STRIKES COMPARED.

| Years. | Demands not followed by strikes. | Offensive strikes. | Defe stri |
|---|---|---|---|
| 1860-1864 | 10 | 2 | |
| 1865-1869 | 19 | 24 | |
| 1870-1874 | 30 | 43 | |
| 1875-1879 | 2 | 11 | |
| 1880-1884 | ........ | 2 | |
| 1885-1889 | 44 | 44 | |
| 1890-1894 | 90 | 63 | |

During the first five-year period nearly all the disputes affected the printers, which fact accounts for the great excess of demands strikes. In the two succeeding periods other occupations appear, bu the number of offensive strikes is greatest. During the next two peri the industrial crisis reduced the number of labor disputes to a m mum. In the sixth period of the summary the influence of the org zation of the Swiss reserve fund (Caisse Suisse de Réserve) beca apparent. Notwithstanding the rapidly increasing number of la disputes, the number of demands amicably settled equals that of offensive strikes, while the number of defensive strikes is quite consi able. During the last period the number of demands amicably sett considerably exceeds either that of the offensive or of the defen strikes. This result is due to the thorough organization of the fed tion of labor unions and to the activity of their executive committe

undervalued gold, and silver was attracted to [illegible] favorable ratio of 1:15½. The act of January 18, [illegible] make the fineness of the gold and silver coins [illegible] weight of the gold dollar was fixed at 25.8 grains, [illegible] at 23.22 grains. The fineness was, therefore, changed by [illegible] 0.900 and the ratio to 1:15.988+.

Silver continued to be exported. The act of February [illegible] reduced the weight of the silver coins of a denomination [illegible] which the acts of 1792, 1834, and 1837 had made exactly [illegible] to the weight of the silver dollar, and provided that they [illegible] legal tender to the amount of only $5. Under the acts of 17[illegible] and 1837 they had been full legal tender. By the act of 1853 [illegible] weight of the half dollar was reduced to 192 grains and that [illegible] other fractions of the dollar in proportion. The coinage of the [illegible]tional parts of the dollar was reserved to the Government.

The act of February 12, 1873, provided that the unit of value o[illegible] United States should be the gold dollar of the standard weight of [illegible] grains, and that there should be coined besides the following [illegible] coins: A quarter eagle, or 2½-dollar piece; a 3-dollar piece; [illegible] eagle, or 5-dollar piece; an eagle, or 10-dollar piece, and a double [illegible] or 20-dollar piece—all of a standard weight proportional to that [illegible] dollar piece. These coins were made legal tender in all pay[illegible] their nominal value when not below the standard weight and [illegible] tolerance provided in the act for the single piece, and when [illegible] weight they should be legal tender at a valuation in proportion [illegible] actual weight. The silver coins provided for by the act were [illegible] dollar, a half dollar, or 50-cent piece, a quarter dollar, and [illegible] piece; the weight of the trade dollar to be 420 grains Troy; [illegible] dollar, 12½ grams; the quarter dollar and the dime, respectiv[illegible] half and one-fifth of the weight of the half dollar. The silv[illegible] were made legal tender at their nominal value for any [illegible] exceeding $5 in any one payment. The charge for converting [illegible] gold bullion into coin was fixed at one-fifth of 1 per cent. [illegible] silver bullion were allowed to deposit it at any mint of [illegible] States to be formed into bars or into trade dollars, and [illegible] silver for other coinage was to be received.

Section II of the joint resolution of July 22, 1876, [illegible] trade dollar should not thereafter be legal tender, and [illegible] tary of the Treasury should be authorized to limit the [illegible] same to an amount sufficient to meet the export demand [illegible] act of March 3, 1887, retired the trade dollar and prohibit[illegible] That of September 26, 1890, discontinued the coinage [illegible] and 3-dollar gold pieces.

The act of February 28, 1878, directed the coinage [illegible] of the weight of 412½ grains Troy, of standard [illegible] the act of January 18, 1837, and that such co[illegible]

[illegible] of the Treasury notes issued [illegible] of the act of July 14, 1890, was [illegible] 1896.

[illegible] subsidiary silver coins of the United [illegible] of $10. The minor coins are legal [illegible]

COINS OF THE UNITED STATES.—The [illegible] of the gold, silver, and minor [illegible] have been authorized by Congress, [illegible] their coinage, their original [illegible] in the case of the minor coins), [illegible] authorized by subsequent acts of Con- [illegible] their coinage in certain cases, June 30, 1895. In those cases where [illegible] by act of Congress, the figures [illegible] table represent the total amount coined in the column immediately preceding. [illegible] dollar and Columbian quarter dollar, [illegible] amount coined under the special act by [illegible]:

AUTHORITY FOR COINING, CHANGES IN WEIGHT AND FINENESS, AND AMOUNT COINED, FOR EACH COIN.

| Denomination. | Act authorizing coinage or change in weight or fineness. | Weight (grains). | Fineness. | Act discontinuing coinage. | Total amount coined to June 30, 1895. |
|---|---|---|---|---|---|
| GOLD COINS. | | | | | |
| Double eagle ($20) | March 3, 1849 | 516 | .900 | | $1,225,818,760.00 |
| Eagle ($10) | April 2, 1792 | 270 | .916⅔ | | |
| | June 28, 1834 | 258 | .899225 | | 262,380,780.00 |
| | January 18, 1837 | | .900 | | |
| Half eagle ($5) | April 2, 1792 | 135 | .916⅔ | | |
| | June 28, 1834 | 129 | .899225 | | 217,814,395.00 |
| | January 18, 1837 | | .900 | | |
| Quarter eagle ($2.50) | April 2, 1792 | 67.5 | .916⅔ | | |
| | June 28, 1834 | 64.5 | .899225 | | 28,651,115.00 |
| | January 18, 1837 | | .900 | | |
| Three dollar piece | February 21, 1853 | 77.4 | .900 | September 26, 1890. | 1,619,376.00 |
| One dollar | March 3, 1849 | 25.8 | .900 | September 26, 1890. | 19,499,337.00 |
| SILVER COINS. | | | | | |
| Dollar | April 2, 1792 | 416 | .8924 | | |
| | January 18, 1837 | 412½ | .900 | February 12, 1873 | *a* 431,320,457.00 |
| | February 28, 1878 | | | | |
| Trade dollar (*b*) | February 12, 1873 | 420 | .900 | March 3, 1887 | 35,965,924.00 |
| Half dollar | April 2, 1792 | 208 | .8924 | | |
| | January 18, 1837 | 206¼ | .900 | | |
| | February 21, 1853 | 192 | | | 130,857,278.50 |
| | February 12, 1873 | *c* 192.9 | | | |
| Columbian half dollar | August 5, 1892 | 192.9 | .900 | | *d* 2,501,052.50 |
| Quarter dollar | April 2, 1792 | 104 | .8924 | | |
| | January 18, 1837 | 103⅛ | .900 | | |
| | February 21, 1853 | 96 | | | 49,160,461.25 |
| | February 12, 1873 | *e* 96.45 | | | |
| Columbian quarter dollar | March 3, 1893 | 96.45 | .900 | | *d* 10,005.75 |
| Twenty cent piece | March 3, 1875 | *f* 77.16 | .900 | May 2, 1878 | 271,000.00 |
| Dime | April 2, 1792 | 41.6 | .8924 | | |
| | January 18, 1837 | 41¼ | .900 | | |
| | February 21, 1853 | 38.4 | | | 28,775,218.20 |
| | February 12, 1873 | *g* 38.58 | | | |
| Half dime | April 2, 1792 | 20.8 | .8924 | | |
| | January 18, 1837 | 20⅝ | .900 | | 4,880,219.40 |
| | February 21, 1853 | 19.2 | | February 12, 1873 | |
| Three cent piece | March 3, 1851 | 12⅜ | .750 | | 1,282,087.20 |
| | March 3, 1853 | 11.52 | .900 | February 12, 1873 | |
| MINOR COINS. | | | | | |
| Five cent (nickel) | May 16, 1866 | 77.16 | (*h*) | | 13,884,562.20 |
| Three cent (nickel) | March 3, 1865 | 30 | (*h*) | September 26, 1890 | 941,349.48 |
| Two cent (bronze) | April 22, 1864 | 96 | (*i*) | February 12, 1873 | 912,020.00 |
| Cent (copper) | April 2, 1792 | 264 | | | |
| | January 14, 1793 | 208 | | | 1,562,887.44 |
| | January 26, 1796 (*j*) | 168 | | February 21, 1857 | |
| Cent (nickel) | February 21, 1857 | 72 | (*k*) | April 22, 1864 | 2,007,720.00 |
| Cent (bronze) | April 22, 1864 | 48 | (*i*) | | 7,612,226.12 |
| Half cent (copper) | April 2, 1792 | 132 | | | |
| | January 14, 1793 | 104 | | | 39,926.11 |
| | January 26, 1796 (*j*) | 84 | | February 21, 1857 | |

*a* Amount coined to February 12, 1873, $8,031,238.
*b* Coinage limited to export demand, joint resolution, July 22, 1876.
*c* 12½ grams, or 192.9 grains.
*d* Total amount coined.
*e* 6¼ grams, or 96.45 grains.
*f* 5 grams, or 77.16 grains.
*g* 2½ grams, or 38.58 grains.
*h* Composed of 75 per cent copper and 25 per cent nickel.
*i* Composed of 95 per cent copper and 5 per cent tin and zinc.
*j* By proclamation of the President in conformity with act of March 3, 1795.
*k* Composed of 88 per cent copper and 12 per cent nickel.

The total coinages to June 30, 1895, are as follows: Gold, $1,755,813,763; silver, $685,023,701.90; minor, $26,960,711.35; a grand total of $2,467,798,176.25.

**VALUE OF THE SILVER IN A UNITED STATES SILVER DOLLAR.**— he following table shows the value of the pure silver in a United tates silver dollar, reckoned at the commercial price of silver bullion om $0.50 to $1.2929 (parity) per fine ounce:

ALUE OF PURE SILVER IN A UNITED STATES SILVER DOLLAR ACCORDING TO PRICE OF SILVER BULLION.

| Price of silver per fine ounce. | Value of pure silver in a silver dollar. | Price of silver per fine ounce. | Value of pure silver in a silver dollar. | Price of silver per fine ounce. | Value of pure silver in a silver dollar. |
|---|---|---|---|---|---|
| $0.50 | $0.387 | $0.77 | $0.596 | $1.04 | $0.804 |
| .51 | .394 | .78 | .603 | 1.05 | .812 |
| .52 | .402 | .79 | .611 | 1.06 | .820 |
| .53 | .410 | .80 | .619 | 1.07 | .828 |
| .54 | .418 | .81 | .626 | 1.08 | .835 |
| .55 | .425 | .82 | .634 | 1.09 | .843 |
| .56 | .433 | .83 | .642 | 1.10 | .851 |
| .57 | .441 | .84 | .650 | 1.11 | .859 |
| .58 | .449 | .85 | .657 | 1.12 | .866 |
| .59 | .456 | .86 | .665 | 1.13 | .874 |
| .60 | .464 | .87 | .673 | 1.14 | .882 |
| .61 | .472 | .88 | .681 | 1.15 | .889 |
| .62 | .480 | .89 | .688 | 1.16 | .897 |
| .63 | .487 | .90 | .696 | 1.17 | .905 |
| .64 | .495 | .91 | .704 | 1.18 | .913 |
| .65 | .503 | .92 | .712 | 1.19 | .920 |
| .66 | .510 | .93 | .719 | 1.20 | .928 |
| .67 | .518 | .94 | .727 | 1.21 | .936 |
| .68 | .526 | .95 | .735 | 1.22 | .944 |
| .69 | .534 | .96 | .742 | 1.23 | .951 |
| .70 | .541 | .97 | .750 | 1.24 | .959 |
| .71 | .549 | .98 | .758 | 1.25 | .967 |
| .72 | .557 | .90 | .766 | 1.26 | .975 |
| .73 | .565 | 1.00 | .773 | 1.27 | .982 |
| .74 | .572 | 1.01 | .781 | 1.28 | .990 |
| .75 | .580 | 1.02 | .789 | 1.29 | .998 |
| .76 | .588 | 1.03 | .797 | a 1.2929 | 1.00 |

a Parity.

The following table shows the highest, lowest, and average value of a United States silver dollar, measured by the market price of silver, and the quantity of silver purchasable with a silver dollar at the average London price of silver, for each calendar year from 1873 to 1894:

BULLION VALUE AND PURCHASING POWER IN SILVER OF A UNITED STATES SILVER DOLLAR, 1873 TO 1894.

| Year. | Bullion value of a silver dollar. | | | Grains of pure silver at average price purchasable with a United States silver dollar (a). |
|---|---|---|---|---|
| | Highest | Lowest. | Average. | |
| 1873 | $1.016 | $0.981 | $1.004 | [illegible].77 |
| 1874 | 1.008 | .970 | .988 | 375.[illegible] |
| 1875 | .977 | .941 | .964 | [illegible]5.11 |
| 1876 | .991 | .792 | .894 | 415.27 |
| 1877 | .987 | .902 | .929 | [illegible] |
| 1878 | .936 | .839 | .891 | 416.[illegible] |
| 1879 | .911 | .828 | .868 | 427.[illegible] |
| 1880 | .896 | .875 | .886 | 419.[illegible] |
| 1881 | .896 | .862 | .881 | 421.[illegible] |
| 1882 | .887 | .847 | .878 | 422.[illegible] |
| 1883 | .868 | .847 | .858 | 432.[illegible] |
| 1884 | .871 | .839 | .861 | 431.[illegible] |
| 1885 | .847 | .794 | .823 | 451.[illegible] |
| 1886 | .797 | .712 | .769 | 482.77 |
| 1887 | .799 | .733 | .758 | [illegible] |
| 1888 | .755 | .706 | .727 | 510.[illegible] |
| 1889 | .752 | .746 | .724 | 512.[illegible] |
| 1890 | .926 | .740 | .810 | 458.[illegible] |
| 1891 | .827 | .738 | .764 | 485.[illegible] |
| 1892 | .742 | .642 | .674 | 550.79 |
| 1893 | .655 | .513 | .604 | 615.[illegible] |
| 1894 | .538 | .457 | .491 | 75[illegible].64 |

a 371.25 grains of pure silver are contained in a silver dollar.

**PRODUCTION OF GOLD AND SILVER IN THE UNITED STATES AND IN THE WORLD.**—The following table shows the production of the precious metals in the world for each calendar year from 1873 to 1894. The silver product is given at its commercial value, reckoned at the average market price of silver each year, as well as at its coining value in United States dollars:

PRODUCTION OF GOLD AND SILVER IN THE WORLD FOR EACH CALENDAR YEAR FROM 1873 TO 1894

| Year. | Gold. | | Silver. | | |
|---|---|---|---|---|---|
| | Fine ounces. | Value. | Fine ounces. | Commercial value. | Coining value. |
| 1873 | 4,653,675 | $96,200,000 | 63,267,187 | $82,120,800 | $81,800,000 |
| 1874 | 4,390,031 | 90,750,000 | 55,300,781 | 70,674,400 | 71,500,000 |
| 1875 | 4,716,563 | 97,500,000 | 62,261,719 | 77,578,100 | 80,500,000 |
| 1876 | 5,016,488 | 103,700,000 | 67,753,125 | 78,322,600 | 87,600,000 |
| 1877 | 5,512,196 | 113,947,200 | 62,679,916 | 75,278,600 | 81,040,700 |
| 1878 | 5,761,114 | 119,092,800 | 73,385,451 | 84,540,000 | 94,882,200 |
| 1879 | 5,262,174 | 108,778,800 | 74,383,495 | 83,532,700 | 96,172,600 |
| 1880 | 5,148,880 | 106,436,800 | 74,795,273 | 85,640,600 | 96,705,000 |
| 1881 | 4,983,742 | 103,023,100 | 79,020,872 | 89,925,700 | 102,168,400 |
| 1882 | 4,934,086 | 101,996,600 | 86,472,091 | 98,232,300 | 111,802,300 |
| 1883 | 4,614,588 | 95,392,000 | 89,175,023 | 98,984,300 | 115,297,000 |
| 1884 | 4,921,169 | 101,729,600 | 81,567,801 | 90,785,000 | 105,461,400 |
| 1885 | 5,245,572 | 108,435,600 | 91,609,959 | 97,518,800 | 118,445,200 |
| 1886 | 5,135,679 | 106,163,900 | 93,297,290 | 92,793,500 | 120,626,800 |
| 1887 | 5,116,861 | 105,774,900 | 96,123,586 | 94,031,000 | 124,281,000 |
| 1888 | 5,330,775 | 110,196,900 | 108,827,606 | 102,185,900 | 140,706,400 |
| 1889 | 5,973,790 | 123,489,200 | 120,213,611 | 112,414,100 | 155,427,700 |
| 1890 | 5,749,306 | 118,848,700 | 126,095,062 | 131,937,000 | 163,032,000 |
| 1891 | 6,320,194 | 130,650,000 | 137,170,919 | 135,500,200 | 177,352,300 |
| 1892 | 7,102,180 | 146,815,100 | 153,151,762 | 133,404,400 | 198,014,400 |
| 1893 | 7,608,787 | 157,287,600 | 166,092,047 | 129,551,800 | 214,745,300 |
| 1894 | 8,737,788 | 180,626,100 | 167,752,561 | 106,522,900 | 216,892,200 |
| Total | 122,235,638 | 2,526,834,900 | 2,130,397,137 | 2,151,474,700 | 2,754,452,900 |

The total production of gold and silver in the world since 1493 is shown in the first of the following tables. The second table shows the production of gold and silver from the mines of the United States by periods of years from 1792 to 1844, and annually from 1845 to 1894.

| | |
|---|---|
| .. | 186, |
| .. | 230, |
| .. | 273, |
| .. | 219, |
| .. | 237, |
| .. | 273, |
| .. | 266, |
| .. | 281, |
| .. | 297, |
| .. | 346, |
| .. | 412, |
| .. | 613, |
| .. | 791, |
| .. | 665, |
| .. | 571, |
| .. | 571, |
| .. | 367, |
| .. | 457, |
| .. | 652, |
| .. | 1,760, |
| .. | 6,410, |
| .. | 6,486, |
| .. | 5,949, |
| .. | 6,270, |
| .. | 5,591, |
| .. | 5,543, |
| ... | 4,794, |
| .. | 5,135, |
| ... | 5,116, |
| ... | 5,330, |
| ... | 5,973, |
| ... | 5,749, |
| ... | 6,320, |
| ... | 7,102, |

## PRODUCTION OF GOLD AND SILVER IN THE WORLD SINCE THE DISCOVERY OF AMERICA.

[**Production** for 1493 to 1885 is from a table of averages for certain periods compiled by Dr Adolph **Soetbeer. For** the years 1886 to 1894 the production is the annual estimate of the Bureau of the Mint.]

| Silver. | | | | Percentage of production | | | |
|---|---|---|---|---|---|---|---|
| Annual average for period | | Total for period. | | By weight. | | By value. | |
| Fine ounces. | Coining value. | Fine ounces. | Coining value | Gold. | Silver. | Gold. | Silver. |
| 1,511,050 | $1,954,000 | 42,309,400 | $54,703,000 | 11 | 89 | 66.4 | 33.6 |
| 2,899,930 | 3,749,000 | 69,598,320 | 89,986,000 | 7.4 | 92.6 | 55.9 | 44.1 |
| 10,017,940 | 12,952,000 | 160,287,040 | 207,240,000 | 2.7 | 97.3 | 30.4 | 69.6 |
| 9,628,925 | 12,450,000 | 192,578,500 | 248,900,000 | 2.2 | 97.8 | 26.7 | 73.3 |
| 13,467,635 | 17,413,000 | 269,352,700 | 348,254,000 | 1.7 | 98.3 | 22 | 78 |
| 13,596,235 | 17,579,000 | 271,924,700 | 351,579,000 | 2 | 98 | 24.4 | 75.6 |
| 12,654,240 | 16,361,000 | 253,084,800 | 327,221,000 | 2.1 | 97.9 | 25.2 | 74.8 |
| 11,776,545 | 15,226,000 | 35,530,900 | 304,525,000 | 2.3 | 97.7 | 27.7 | 72.3 |
| 10,834,550 | 14,008,000 | 216,691,000 | 280,166,000 | 2.7 | 97.3 | 30.5 | 69.5 |
| 10,992,085 | 14,212,000 | 219,841,700 | 284,240,000 | 3.1 | 96.9 | 33.5 | 66.5 |
| 11,432,540 | 14,781,000 | 228,650,800 | 295,629,000 | 3.5 | 96.5 | 36.6 | 63.4 |
| 13,863,080 | 17,924,000 | 277,261,600 | 358,480,000 | 4.2 | 95.8 | 41.4 | 58.6 |
| 17,140,612 | 22,162,000 | 342,812,235 | 443,232,000 | 4.4 | 95.6 | 42.5 | 57.5 |
| 20,985,591 | 27,133,000 | 419,711,820 | 542,658,000 | 3.1 | 96.9 | 33.7 | 66.3 |
| 28,261,779 | 36,540,000 | 565,235,580 | 730,810,000 | 2 | 98 | 24.4 | 75.6 |
| 28,746,922 | 37,168,000 | 287,469,225 | 371,677,000 | 1.9 | 98.1 | 24.1 | 75.9 |
| 17,385,755 | 22,479,000 | 173,857,555 | 224,786,000 | 2.1 | 97.9 | 25.3 | 74.7 |
| 14,807,004 | 19,144,000 | 148,070,040 | 191,444,000 | 3 | 97 | 33 | 67 |
| 19,175,867 | 24,793,000 | 191,758,675 | 247,930,000 | 3.3 | 96.7 | 35.2 | 64.8 |
| 25,090,342 | 32,440,000 | 250,903,422 | 324,400,000 | 6.6 | 93.4 | 52.9 | 47.1 |
| 28,488,597 | 36,824,000 | 142,442,986 | 184,169,000 | 18.4 | 81.6 | 78.3 | 21.7 |
| 29,095,428 | 37,618,000 | 145,477,142 | 188,092,000 | 18.2 | 81.8 | 78.1 | 21.9 |
| 35,401,972 | 45,772,000 | 177,009,862 | 228,861,000 | 14.4 | 85.6 | 72.9 | 27.1 |
| 43,051,583 | 55,663,000 | 215,257,914 | 278,313,000 | 12.7 | 87.3 | 70 | 30 |
| 63,317,014 | 81,864,000 | 316,585,069 | 409,322,000 | 8.1 | 91.9 | 58.5 | 41.5 |
| 78,775,602 | 101,851,000 | 393,878,009 | 509,256,000 | 6.6 | 93.4 | 53 | 47 |
| 92,003,944 | 118,955,000 | 460,019,722 | 594,773,000 | 5 | 95 | 45.5 | 54.5 |
| 93,297,290 | 120,626,800 | 93,297,290 | 120,626,800 | 5.2 | 94.8 | 46.8 | 53.2 |
| 96,123,586 | 124,281,000 | 96,123,586 | 124,281,000 | 5 | 95 | 45.9 | 54.1 |
| 108,827,606 | 140,706,400 | 108,827,606 | 140,706,400 | 4.6 | 95.4 | 43.9 | 56.1 |
| 120,213,611 | 155,427,700 | 120,213,611 | 155,427,700 | 4.7 | 95.3 | 44.3 | 55.7 |
| 126,095,062 | 163,032,000 | 126,095,062 | 163,032,000 | 4.3 | 95.7 | 42.1 | 57.9 |
| 137,170,919 | 177,352,300 | 137,170,919 | 177,352,300 | 4.4 | 95.6 | 42.4 | 57.6 |
| 153,151,762 | 198,014,400 | 153,151,762 | 198,014,400 | 4.4 | 95.6 | 42.5 | 57.5 |
| 166,092,047 | 214,745,300 | 166,092,047 | 214,745,300 | 4.4 | 95.6 | 42.4 | 57.6 |
| 167,752,561 | 216,892,200 | 167,752,561 | 216,892,200 | 4.9 | 95.1 | 45.6 | 54.4 |
| ........ | ........ | 7,836,325,160 | 10,131,814,100 | 5 | 95 | 45.6 | 54.4 |

| | | |
|---|---|---|
| | 65,000,000 | |
| 1854 | 60,000,000 | |
| 1855 | 55,000,000 | |
| 1856 | 55,000,000 | |
| 1857 | 55,000,000 | |
| 1858 | 50,000,000 | |
| 1859 | 50,000,000 | |
| 1860 | 46,000,000 | |
| 1861 | 43,000,000 | |
| 1862 | 39,200,000 | |
| 1863 | 40,000,000 | |
| 1864 | 46,100,000 | 1 |
| 1865 | 53,225,000 | 1 |
| 1866 | 53,500,000 | 1 |
| 1867 | 51,725,000 | 1 |
| 1868 | 48,000,000 | 1 |
| 1869 | 49,500,000 | 1 |
| 1870 | 50,000,000 | 1 |
| 1871 | 43,500,000 | 23 |
| 1872 | 36,000,000 | 28 |
| 1873 | 36,000,000 | 35 |
| 1874 | 33,500,000 | 37 |
| 1875 | 33,400,000 | 31, |
| 1876 | 39,900,000 | 28, |
| 1877 | 46,900,000 | 39, |
| 1878 | 51,200,000 | 45, |
| 1879 | 38,900,000 | 40, |
| 1880 | 36,000,000 | 39, |
| 1881 | 34,700,000 | 43, |
| 1882 | 32,500,000 | 46, |
| 1883 | 30,000,000 | 46, |
| 1884 | 30,800,000 | 48, |
| 1885 | 31,800,000 | 51, |
| 1886 | 35,000,000 | 51, |
| 1887 | 33,000,000 | 53, |
| 1888 | 33,175,000 | 59, |
| 1889 | 32,800,000 | 64, |
| 1890 | 32,845,000 | 70, |
| 1891 | 33,175,000 | 75, |
| 1892 | 33,000,000 | 82, |
| 1893 | 35,955,000 | 77, |
| 1894 | 39,500,000 | 64, |
| **Total** | 2,013,336,769 | 1,296, |

**VALUES OF FOREIGN COINS.**—The following table gives the value of foreign coins on January 1, 1896, as estimated by the Director of the Mint, in pursuance of the provisions of section 25 of the act of August 28, 1894, as follows:

> That the value of foreign coins as expressed in the money of account of the United States shall be that of the pure metal of such coin of standard value; and the values of the standard coins in circulation of the various nations of the world shall be estimated quarterly by the Director of the Mint and be proclaimed by the Secretary of the Treasury immediately after the passage of this act and thereafter quarterly on the 1st day of January, April, July, and October in each year.

| | | |
|---|---|---|
| [illegible]ican States: | | |
| Costa Rica | Silver | |
| Guatemala | Silver | |
| Honduras | Silver | |
| Nicaragua | Silver | |
| Salvador | Silver | |
| Chile | Gold and silver | |
| China | Silver | |
| Colombia | Silver | I |
| Cuba | Gold and silver | I |
| Denmark | Gold | C |
| Ecuador | Silver | S |
| Egypt | Gold | P |
| Finland | Gold | M |
| France | Gold and silver | F |
| German Empire | Gold | M |
| Great Britain | Gold | P |
| Greece | Gold and silver | D |
| Haiti | Gold and silver | G |
| India | Silver | R |
| Italy | Gold and silver | Li |
| Japan | Gold and silver (a) | Ye |
| Liberia | Gold | Do |
| Mexico | Silver | Do |
| Netherlands | Gold and silver | Fl |
| Newfoundland | Gold | Do |
| Norway | Gold | Cr |
| Persia | Silver | Kr |
| Peru | Silver | Sol |
| Portugal | Gold | Mi |
| Russia | Silver (c) | Ru |
| Spain | Gold and silver | Pe |
| Sweden | Gold | Cr |
| Switzerland | Gold and silver | Fr |
| Tripoli | Silver | Ma |
| Turkey | Gold | Pi |
| Venezuela | Gold and silver | Bol |

*a* Gold the nominal standard. Silver practically the standard.
*c* Silver the nominal standard. Paper the actual currency, the depreciation [illegible]
he gold standard.

## VALUES OF FOREIGN COINS JANUARY 1, 1896, AS ESTIMATED BY THE DIRECTOR OF THE MINT.

| Value in terms of United States gold dollar. | Coins. |
|---|---|
| $0.965 | Gold, argentine ($4.824) and ½ argentine, silver, peso and divisions. |
| .203 | Gold, former system, 4 florins ($1 929), 8 florins ($3.858), ducat ($2 287) and 4 ducats ($9 149); silver, 1 and 2 florins, gold, present system, 20 crowns ($4.052), 10 crowns ($2.026) |
| .193 | Gold, 10 and 20 francs, silver, 5 francs. |
| .491 | Silver, boliviano and divisions. |
| .546 | Gold, 5, 10, and 20 milreis; silver, ½, 1, and 2 milreis. |
| 1.000 | |
| .491 | Silver, peso and divisions. |
| .912 | Gold, escudo ($1.824), doubloon ($4.561), and condor ($9.123), silver, peso and divisions. |
| .725 | |
| .808 | |
| .760 | |
| .759 | |
| .491 | Gold, condor ($9.647) and double condor; silver, peso. |
| .926 | Gold, doubloon ($5.017); silver, peso. |
| .268 | Gold, 10 and 20 crowns. |
| .491 | Gold, condor ($9.647) and double condor, silver, sucre and divisions |
| 4.943 | Gold, pound (100 piasters), 5, 10, 20, and 50 piasters, silver, 1, 2, 5, 10, and 20 piasters. |
| .193 | Gold, 20 marks ($3.859), 10 marks ($1 93). |
| .193 | Gold, 5, 10, 20, 50, and 100 francs; silver, 5 francs. |
| .238 | Gold, 5, 10, and 20 marks. |
| 4.866½ | Gold, sovereign (pound sterling) and ½ sovereign. |
| .193 | Gold, 5, 10, 20, 50, and 100 drachmas, silver 5 drachmas. |
| .965 | Silver, gourde. |
| .233 | Gold, mohur ($7.105); silver, rupee and divisions. |
| .193 | Gold, 5, 10, 20, 50, and 100 lire; silver 5 lire. |
| .997 | Gold, 1, 2, 5, 10, and 20 yen. |
| .529 | Silver, yen. |
| 1.000 | |
| .533 | Gold, dollar ($0.983), 2½, 5, 10, and 20 dollars, silver, dollar (or peso) and divisions. |
| .402 | Gold, 10 florins; silver, ½, 1, and 2½ florins |
| 2.014 | Gold, 2 dollars ($2.027) |
| .268 | Gold, 10 and 20 crowns. |
| .090 | Gold, ⅕, 1, and 2 tomans ($3.409); silver, ¼, ½, 1, 2, and 5 krans. |
| .491 | Silver, sol and divisions. |
| 1.080 | Gold, 1, 2, 5, and 10 milreis. |
| .772 | Gold, imperial ($7.718) and ½ imperial (*b*) ($3.86). |
| .393 | Silver, ¼, ½, and 1 ruble. |
| .193 | Gold, 25 pesetas; silver, 5 pesetas. |
| .268 | Gold, 10 and 20 crowns. |
| .193 | Gold, 5, 10, 20, 50, and 100 francs, silver, 5 francs. |
| .443 | |
| .044 | Gold, 25, 50, 100, 250, and 500 piasters |
| .193 | Gold, 5, 10, 20, 50, and 100 bolivars, silver, 5 bolivars |

*b* Coined since January 1, 1886. Old half-imperial =$3.986.

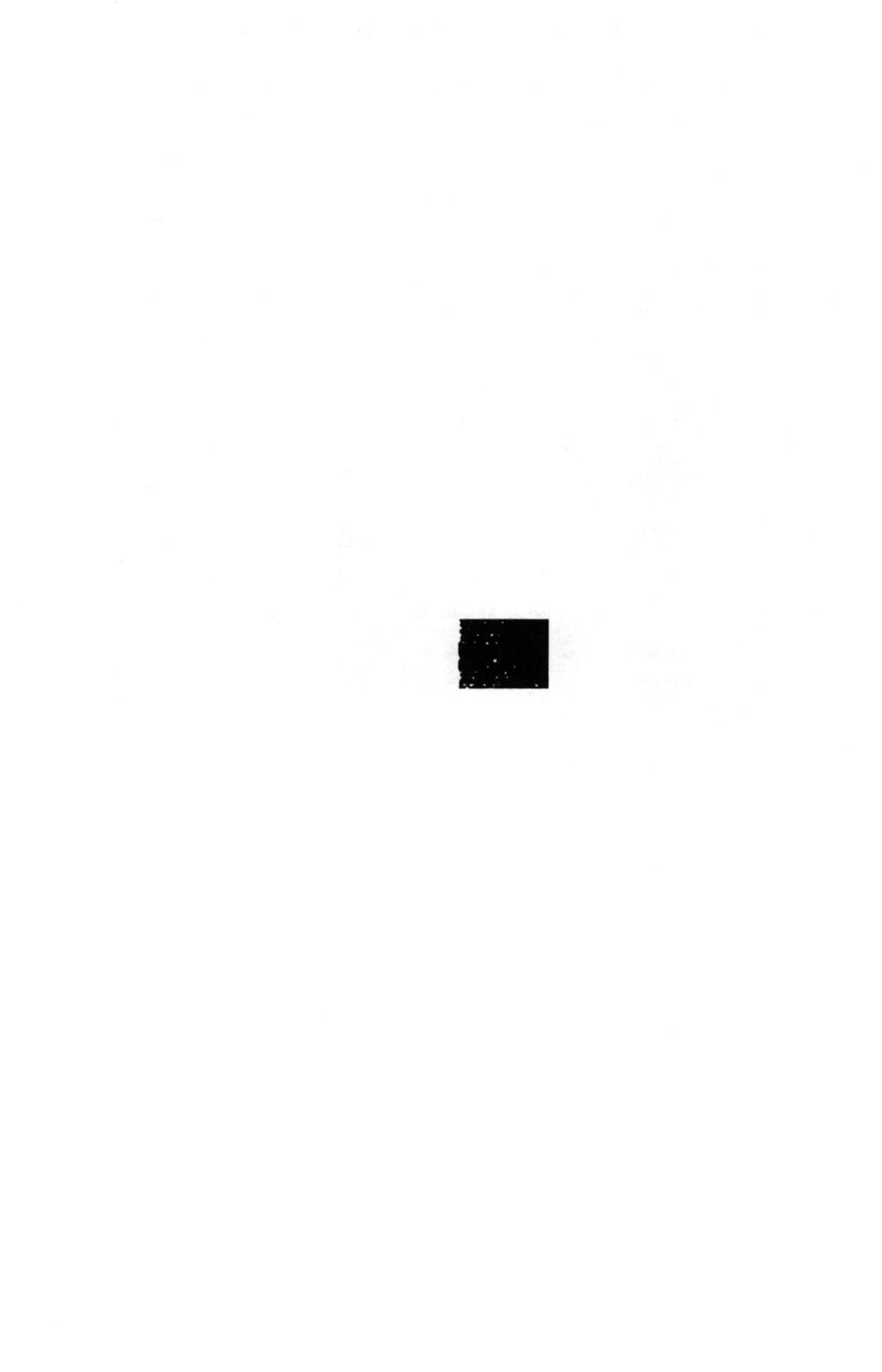

## MONETARY SYSTEMS AND APPROXIMATE STOCKS OF MONEY IN THE AGGREGATE AND PER CAPITA IN THE PRINCIPAL COUNTRIES OF THE WORLD.

paper money in the principal countries of the world has been compiled from the latest information to show approximately the stock of money in the world.]

| Stock of silver | | | Uncovered paper. | Per capita. | | | | |
|---|---|---|---|---|---|---|---|---|
| Full tender. | Limited tender. | Total. | | Gold. | Silver. | Paper | Total. | |
| $548,400,000 | $77,200,000 | $625,600,000 | $416,700,000 | $8.78 | $8.89 | $5.92 | $23.59 | 1 |
| ............... | b115,000,000 | 115,000,000 | c113,400,000 | 14.91 | 2.96 | 2.91 | 20.78 | 2 |
| b430,000,000 | c57,900,000 | 487,900,000 | c32,100,000 | 22.19 | 12.94 | .84 | 35.77 | 3 |
| b105,000,000 | b110,000,000 | 215,000,000 | c60,400,000 | 12.21 | 4.20 | 1.18 | 17.59 | 4 |
| b48,000,000 | b6,900,000 | 54,900,000 | c65,400,000 | 8.73 | 8.71 | 10.38 | 27.82 | 5 |
| bc21,400,000 | bc20,000,000 | 41,400,000 | c191,800,000 | 3.20 | 1.35 | 6.24 | 10.79 | 6 |
| b10,000,000 | 5,000,000 | 15,000,000 | ............... | 4.97 | 5.00 | ........ | 9.97 | 7 |
| b500,000 | b1,000,000 | 1,500,000 | c22,400,000 | .23 | .68 | 10.18 | 11.09 | 8 |
| b126,000,000 | b40,000,000 | 166,000,000 | c83,700,000 | 2.28 | 9.49 | 4.78 | 16.55 | 9 |
| ............... | b24,800,000 | 24,800,000 | c55,100,000 | 7.45 | 4.86 | 10.80 | 23.11 | 10 |
| ............... | c10,600,000 | 10,600,000 | c11,700,000 | 6.65 | 1.83 | 2.02 | 10.50 | 11 |
| ............... | c1,900,000 | 1,900,000 | b3,800,000 | 1.30 | .83 | 1.65 | 3.78 | 12 |
| b80,000,000 | b40,000,000 | 120,000,000 | c204,300,000 | 3.22 | 2.76 | 4.69 | 10.67 | 13 |
| c53,000,000 | c3,200,000 | 56,200,000 | c28,600,000 | 6.21 | 11.96 | 6.08 | 24.25 | 14 |
| ............... | b2,000,000 | 2,000,000 | c3,800,000 | 3.75 | 1.00 | 1.90 | 6.65 | 15 |
| ............... | c4,800,000 | 4,800,000 | c2,100,000 | 1.66 | 1.00 | .43 | 3.10 | 16 |
| ............... | c5,400,000 | 5,400,000 | c5,400,000 | 6.30 | 2.35 | 2.35 | 11.00 | 17 |
| ............... | b48,000,000 | 48,000,000 | c539,000,000 | 3.80 | .38 | 4.28 | 8.46 | 18 |
| b30,000,000 | d10,000,000 | 40,000,000 | ............... | 2.27 | 1.82 | ........ | 4.09 | 19 |
| ............... | b7,000,000 | 7,000,000 | ............... | 24.47 | 1.49 | ........ | 25.96 | 20 |
| ............... | b15,000,000 | 15,000,000 | ............... | 17.65 | 2.20 | ........ | 19.85 | 21 |
| b55,000,000 | ............... | 55,000,000 | b2,000,000 | .41 | 4.54 | ........ | 4.95 | 22 |
| c12,000,000 | ............... | 12,000,000 | c8,000,000 | .09 | 2.14 | 1.43 | 3.66 | 23 |
| b30,000,000 | ............... | 30,000,000 | b550,000,000 | 1.11 | .83 | 15.28 | 17.22 | 24 |
| c68,000,000 | c16,300,000 | 84,300,000 | ............... | 1.95 | 2.05 | ........ | 4.00 | 25 |
| b950,000,000 | ............... | 950,000,000 | b37,000,000 | ........ | 3.21 | .12 | 3.33 | 26 |
| b750,000,000 | ............... | 750,000,000 | ............... | ........ | 2.08 | ........ | 2.08 | 27 |
| b115,000,000 | ............... | 115,000,000 | ............... | ........ | 3.26 | ........ | 3.26 | 28 |
| ............... | b5,000,000 | 5,000,000 | b29,000,000 | 2.92 | 1.04 | 6.04 | 10.00 | 29 |
| b1,500,000 | ............... | 1,500,000 | ............... | 10.00 | .83 | ........ | 10.83 | 30 |
| b2,100,000 | b800,000 | 2,900,000 | c4,200,000 | 3.00 | 2.90 | 4.20 | 10.10 | 31 |
| b3,400,000 | 3,400,000 | 6,800,000 | ............... | .18 | 1.58 | ........ | 1.76 | 32 |
| 3,439,300,000 | 631,200,000 | 4,070,500,000 | 2,469,900,000 | ........ | ........ | ........ | ........ | |

*c* Information furnished through United States representatives.
*d* Haupt.

s in the wealth of the country and in the
he Government.

Eleventh Census it appears that the total
and personal property in the United States
37,091,197; the total value of the products
hich include manufactures, and farm, fishery,
the same year, amounted to $12,148,380,626.
ted during 1890 by the Federal, State, and
orted at $1,040,473,013 and the expenditures
s exceeding the expenditures by $124,518,958,
ss in the transactions of the Federal govern-
service, amounted to $108,936,066, leaving
rising from the financial transactions of the
ents. The total Federal, State, county or
hool district debts of the country, including
stricts in California and the fire districts of
to $2,027,170,546.

cerning the wealth, income, expenses, and
ire country are not available for any year
given in the accompanying tables show the
ferent censuses from 1800 to 1890, inclusive,
e value (a) of real and personal property) and
, and farm, fishery, and mineral products as
ensuses from 1850 to 1890, inclusive, also the
e cash in the Treasury, at the close of the

from 1800 to 1890, inclusive. The receipts [illegible] eral government include premiums and [illegible] expenses into ordinary and extraordinary is [illegible] being to group under the latter title a number of [illegible] may be considered as not forming a part of the [illegible] necessary expenses of the Government. The [illegible] include amounts paid as pensions; for the support of [illegible] improvement of rivers and harbors; the erection and [illegible] arsenals, armories, custom-houses, court-houses, and post-[illegible] ments on interest and as premiums; also expenditures for the [illegible] of the Navy since 1885.

The totals grouped under the respective heads enumerated [illegible] in the following tables:

POPULATION, WEALTH, VALUE OF PRODUCTS OF INDUSTRY, AND [illegible]

| Census year. | Population. | Wealth. | Value of manufactures, and farm, fishery, and mineral products. | [illegible] |
|---|---|---|---|---|
| 1800 | 5, 308, 483 | ........ | ........ | ........ |
| 1810 | 7, 239, 881 | ........ | ........ | ........ |
| 1820 | 9, 633, 822 | ........ | ........ | ........ |
| 1830 | 12, 866, 020 | ........ | ........ | ........ |
| 1840 | 17, 069, 453 | ........ | ........ | ........ |
| 1850 | 23, 191, 876 | $7, 135, 780, 228 | a $1, 0[illegible] | [illegible] |
| 1860 | 31, 443, 321 | 16, 159, 616, 068 | a 1, [illegible] | [illegible] |
| 1870 | 38, 558, 371 | 30, 068, 518, 507 | b 6, [illegible] | [illegible] |
| 1880 | 50, 155, 783 | 43, 642, 000, 000 | [illegible] 7, 874, [illegible] | [illegible] |
| 1890 | 62, 622, 250 | 65, 037, 091, 197 | d 12, 14[illegible] | [illegible] |

a Not including value of farm products.
b Value of farm products includes betterments and addition to stock.
c Certain duplications in statistics of manufactures as published have been [illegible]
d Not including certain manufacturing industries not fully [illegible]

AVERAGE ANNUAL RECEIPTS AND EXPENSES OF THE FEDERAL GO[illegible]

| Decade ending— | Receipts. | Expenses (exclusive of payments [illegible]) | |
|---|---|---|---|
| | | Ordinary. | Extraordinary. |
| 1800 | $6, 776, 891. 80 | $3, 944, 452. 40 | $2, [illegible] |
| 1810 | 13, 056, 864. 78 | 4, 877, 316. 43 | [illegible] |
| 1820 | 21, 086, 958. 73 | 17, 406, 690. 57 | [illegible] |
| 1830 | 21, 923, 071. 60 | 9, 506, 394. 62 | [illegible] |
| 1840 | 30, 461, 109. 62 | 17, 286, 408. 59 | [illegible] |
| 1850 | 27, 117, 363. 77 | 26, 432, 418. 51 | [illegible] |
| 1860 | 60, 237, 254. 13 | 48, 964, 996. 88 | 11, [illegible] |
| 1870 | 304, 100, 060. 83 | 418, 932, 409. 30 | 111, [illegible] |
| 1880 | 312, 476, 778. 23 | 109, 531, 577. 51 | [illegible] |
| 1890 | 371, 204, 582. 21 | 117, 287, 563. 38 | [illegible] |

The values of the products of industry shown for [illegible] not include some elements that enter into the [illegible] years. This fact, combined with the improvement [illegible] and the greater care bestowed on the enumeration [illegible] suses, will not permit of the use of percentages [illegible] or on the total wealth as indicating the exact incre[illegible] given for the other items in the following table[illegible]

various other conditions controlling the receipts, expenses, and debt of the Government, as well as the inflated value of currency in 1870 as compared with gold, and the relatively prosperous or depressed condition of business during the respective census years must also be considered in accepting the percentages given in the following tables:

PERCENTAGES OF INCREASE OF POPULATION, WEALTH, VALUE OF PRODUCTS OF INDUSTRY, AND DEBT.

| Census year. | Population. | Wealth. | Value of manufactures and farm, fishery, and mineral products. | Total national debt, less cash in Treasury. | Census year. | Population. | Wealth. | Value of manufactures and farm, fishery, and mineral products. | Total national debt, less cash in Treasury. |
|---|---|---|---|---|---|---|---|---|---|
| 1800 | ........ | ........ | ........ | ........ | 1850 | 35.87 | ........ | ........ | ........ |
| 1810 | 36.38 | ........ | ........ | ........ | 1860 | 35.58 | 126.46 | 84.51 | ........ |
| 1820 | 33.07 | ........ | ........ | ........ | 1870 | 22.63 | 86.07 | 260.41 | 3,787.59 |
| 1830 | 33.55 | ........ | ........ | ........ | 1880 | 30.08 | 45.14 | 16.52 | a 17.67 |
| 1840 | 32.67 | ........ | ........ | ........ | 1890 | 24.86 | 49.02 | 52.35 | a 53.59 |

a Decrease.

PERCENTAGES OF INCREASE OF RECEIPTS AND EXPENSES OF THE FEDERAL GOVERNMENT.

| Decade ending— | Average annual receipts. | Average annual expenses: Ordinary. | Extraordinary. | Total. | Decade ending— | Average annual receipts. | Average annual expenses: Ordinary. | Extraordinary. | Total. |
|---|---|---|---|---|---|---|---|---|---|
| 1800 | ........ | ........ | ........ | ........ | 1850 | a 10.98 | 53.09 | a 18.58 | 31.94 |
| 1810 | 92.67 | 23.65 | 37.82 | 29.64 | 1860 | 122.14 | 85.25 | 90.46 | 86.20 |
| 1820 | 61.35 | 256.89 | 62.58 | 160.55 | 1870 | 404.84 | 735.58 | 897.43 | 782.00 |
| 1830 | 4.06 | a 45.39 | 2.75 | a 32.34 | 1880 | 2.75 | a 73.85 | 42.70 | a 49.30 |
| 1840 | 38.95 | 81.63 | 8.64 | 51.58 | 1890 | 18.79 | 7.08 | a 5.26 | a .24 |

a Decrease.

These tables, showing increases or decreases from decade to decade, when accompanied by the following tables, which show, per capita, the items given in the tables above, present in convenient summaries the available data pertaining to the growth of the country and the financial operations of the Federal government during the past ninety years:

PER CAPITA WEALTH, VALUE OF PRODUCTS OF INDUSTRY, AND DEBT.

| Census year. | Wealth. | Value of manufactures, and farm, fishery, and mineral products. | Total national debt less cash in Treasury. | Census year. | Wealth. | Value of manufactures and farm, fishery, and mineral products. | Total national debt less cash in Treasury. |
|---|---|---|---|---|---|---|---|
| 1800 | ........ | ........ | ........ | 1850 | $307.68 | $44.37 | ........ |
| 1810 | ........ | ........ | ........ | 1860 | 513.93 | 60.39 | $1.91 |
| 1820 | ........ | ........ | ........ | 1870 | 779.82 | 177.19 | 60.46 |
| 1830 | ........ | ........ | ........ | 1880 | 870.13 | 158.91 | 38.27 |
| 1840 | ........ | ........ | ........ | 1890 | a 1,036.01 | 193.99 | 14.22 |

a Not including the values for Indian Territory.

PER CAPITA RECEIPTS AND EXPENSES OF THE [illegible]

| Decade ending— | Average annual receipts. | Average annual expenses. | | | Decade ending— | Average annual receipts. | [illegible] | |
|---|---|---|---|---|---|---|---|---|
| | | Ordinary. | Extraordinary. | Total. | | | Ordinary. | [illegible] |
| 1800 ..... | $1.28 | $0.74 | $0.55 | $1.29 | 1850 .... | $1.17 | [illegible] | [illegible] |
| 1810 ..... | 1.80 | .67 | .55 | 1.22 | 1860 .... | 1.92 | [illegible] | [illegible] |
| 1820 .... | 2.19 | 1.81 | .67 | 2.48 | 1870 .... | [illegible] | [illegible] | [illegible] |
| 1830 ..... | 1.70 | .74 | .52 | 1.26 | 1880 .... | [illegible] | [illegible] | [illegible] |
| 1840 ..... | 1.78 | 1.01 | .42 | 1.43 | 1890 .... | 5.93 | 1.87 | [illegible] |

While no extended explanation will be made of apparent [illegible] encies in the above tables, attention is called to the tables [illegible] percentages of increase. The large increases shown for 187[illegible] principally to expenses of the civil war and the inflated val[illegible] rency as compared with gold. The largest percentages of [illegible] during the decade ending with 1870 are shown for extraordinary [illegible] and the total debt less cash in the Treasury, the large increase [illegible] these items being due primarily to war expenses. On the oth[illegible] the amount of extraordinary expenses per capita, as shown b[illegible] table, is the smallest per capita item shown for 1870, and is d[illegible] comparatively small total of such expenses.

While, for the reasons given, the totals for wealth and [illegible] products of industry can only be used in a general way for [illegible] the totals for 1860 may be accepted as showing the condition[illegible] ing during the decade immediately preceding the civil war a[illegible] for 1890 as indicating the conditions that prevailed during a de[illegible] the direct effects of the war had disappeared. The two to[illegible] fore indicate normal conditions. Comparing the two years i[illegible] that the total wealth per capita increased from $513.93 to $1,[illegible] 101.59 per cent, and the per capita value of the products of [illegible] increased from $60.39 to $193.99, or 221.23 per cent, while the re[illegible] capita of the Government increased from $1.92 to $5.93, or [illegible] cent, and the total expenses from $1.91 to $4.28, or 124.[illegible] The actual increase per capita in expenses was $2.37, of which [illegible] the ordinary expenses constituted 31 cents, increasing 19.87 pe[illegible] the extraordinary $2.06, increasing 588.57 per cent. One of [illegible] items of expense entering into the average total of ext[illegible] expenses for the decade ending with 1890 is pensions. Exc[illegible] item the yearly average for the decade would be $80,901,[illegible] an extraordinary expense per capita of $1.29, an increase [illegible] 1860 of 268.57 per cent.

While the tables giving the percentages of increase [illegible] total receipts and expenses of the Government have in[illegible] as was necessary to be in keeping with the developme[illegible] in all directions, the tables giving the per capita [illegible] receipts and expenses have not increased abnormal[illegible]

**[illegible] written, wanted for private [illegible] legislative mandate at the will [illegible] the will of the other. Compulsory [illegible] enforced not by suit or action [illegible] where rights are under the [illegible]**

**[illegible] the act of October 21, 1891, entitled [illegible] corporations to give their discharged [illegible] their removal or discharge when dis-[illegible]tional, and that an action founded [illegible] penalty or arbitrary** damages, fixed **[illegible] its mandates, can** not be supported.

---

, 1893, the legislature of Illinois under- facture of clothing, wearing apparel, and ıd to provide for the appointment of State e, and to make an appropriation therefor." actory inspector, appointed under this law, stice of the peace of Cook County against ing section 5 of the statute in question by nale, more than 18 years of age, at work in rs on a certain day in February, 1894. The court of Cook County, on appeal from the ıe peace, and the defendant was convicted e was brought, on writ of error, before the aich tribunal, on March 14, 1895, reversed court and decided that section 5 of the act nale shall be employed in any factory or ours in any one day or forty-eight hours in utional; the court also decided the first ppropriated $20,000 for the salaries of the nstitutional.

hich is published in full in volume 40 of page 454, is followed in the case of Tilt v.

...this, and consequently ...such a power is noth... judgment," and an attempt to ...terms of a constitutional right ...statute is also obnoxious to criti... ...relate to persons or things as a ...only to those who belong to some "law... belonging to a trade union, labor union, etc. ...it is a special, as contradistinguished from a ...union or a nonlabor-union man could be ...without let or hindrance, whenever the ...without reason therefor, while in the case ...union man he could not be discharged if such ...of his being a member of such an organ... ...legislature have undertaken to limit the ...employer as to his right of contract with partic... ...the statute which does this is a special, not a ...is in violation of the State constitution. ...conflict with section 1, article 14, of the Consti... ...States, forbidding that "any State deprive any ...or property without due process of law." ...escape censure by assuming the label of a "police ...none of the elements or attributes which pertain ...for it does not, in terms or by implication, pro... ...the public health, welfare, comfort, or safety; ...would not be allowed, under the guise and pre... ...legislation, to encroach or trample upon any of the ...which the Constitution intended to secure ...abridgment.

---

...Thomas Durkin against the Kingston Coal Company ...damages were recovered in the court of common ...county, Pa., by the plaintiff for injuries received while ...ee in the coal mine of which the defendant company ...defendant Jones was certified foreman, through the ...foreman. The defendants appealed, and the supreme ...vania, on October 7, 1895, reversed the judgment as ...any, holding that the act of 1891, of Pennsylvania, ...cite coal mines (P. L., p. 176) is unconstitutional and ...makes the owner of such mine liable for injuries ...occasioned by the negligence of a mine foreman, ...employed under the requirements of the stat... ...was, however, affirmed as to Jones, the mine fore... ...be liable, independently of, as well as under, the ...an employee due to his want of attention to his

...opinion, delivered by Judge Williams, is published ...Atlantic Reporter, page 237, and so much thereof ...constitutionality of the law in question, and on the ...foreman, is given here:

...constitution of this State, known as the "bill ...all men are possessed of certain inherent and inalienable rights. One of these is ... protect property. The preservation ... every man should be answerable for ... that no man should be required to answer ... of strangers over whom he has no control. A ... such a liability, or that should take the property ... it to another or to the public without making ... for, would violate the bill of rights, and would be ... unconstitutional and void.

It is in furtherance of the right to acquire, ... property that section 17 of the bill of rights prohibits ... laws that shall interfere with or impair the obligation ... The tendency toward class legislation for the protection ... sorts of labor has been so strong, however, that several ... recently been passed that could not be sustained under ... of the bill of rights. Such was the case in Godcharles ... (113 Pa. St., 431 ; 6 Atl., 354); such was the case with some ... visions relating to mechanics' liens, and such is alleged by ... lants to be the case with some of the provisions of the ... under which this action was brought. The title of the ... "An act to provide for the health and safety of persons ... and about the anthracite coal mines of Pennsylvania and ... tection and preservation of property connected therewith." ... the anthracite region into eight districts, and provides for ... ment by the governor of a competent mine inspector in e... who shall have a general oversight of mining operation... district. It creates an examining board for each district, ... to examine candidates, and recommend such as they shall ... fied for the position of mine foreman to the secretary of int... It is made the duty of this officer to issue certificates to tho... therefor and have been recommended by the board of exa...

Article 8, section 1, declares that no person "shall be ... act as mine foreman or assistant mine foreman of any c... colliery" who has not been examined by the board of ex... ommended to the secretary of internal affairs, and provi... officer with a certificate. The employment of a certified m... is made obligatory upon all mine owners and operators, an... do so is punished by a fine of $20 per day, which may be c... the owner, the operator, or the superintendent in charge ... The duties of the mine foreman are prescribed by the ... owner or operator of the mine can not interfere with t... especially to "visit and examine every working place in ... least once every alternate day while the men of such place ... be at work, and direct that each and every working plac... secured by props or timber, and that safety in all respec... by directing that all loose coal or rock shall be pulled dow... and that no person shall be permitted to work in an unsafe ... it be for the purpose of making it secure." (Article 12, ru...

The mine foreman is also required to examine, at least ... day, "all slopes, shafts, main roads, ways, signal appara... and timbering, and see that they are in safe and efficient ... dition." (Rule 13.)

After having thus most effectually taken the manag... mining operations out of his hands, and committed it ... own creation, whose employment is made compulsory, ... statute, in section 8 of article 17, imposes upon the ...

[illegible] incompetency of the men whom he is compelled [illegible]: "That for any injury to person or property [illegible] violation of this act or any failure to comply with [illegible] by any mine foreman, a right of action shall accrue to [illegible] against said owner or operator for any direct damages [illegible] sustained thereby; and in case of loss of life by reason [illegible] neglect or failure aforesaid a right of action shall accrue to the [illegible] lineal heirs of the person whose life shall be lost for like [illegible] of damages for the injury they shall have sustained."

[illegible], regarded as a whole, is an extraordinary piece of legis[illegible] Through it, the lawmakers say to the mine owner: "You can [illegible] trusted to manage your own business. Left to yourself, you will [illegible] care for your own employees. We will determine what [illegible] do. In order to make it certain that our directions are obeyed, [illegible] a mine foreman over your mines, with authority to direct the [illegible] in which your operations shall be conducted, and what precau[illegible] be taken for the safety of your employees. You shall take [illegible] position a man whom we certify to as competent. You shall [illegible] his salary. What he orders done in your mines you shall pay [illegible] notwithstanding our certificate, he turns out to be incompe[illegible] untrustworthy, you shall be responsible for his ignorance or [illegible]." Under the operation of this statute the mine foreman rep[illegible] the Commonwealth. The State insists on his employment by the [illegible], and, in the name of the police power, turns over to him [illegible] determination of all questions relating to the comfort and the secu[illegible] the miners, and invests him with the power to compel compliance [illegible] directions. Incredible as it may seem, obedience on the part [illegible] mine owner does not protect him; but, if the mine foreman fails [illegible] properly what the statute directs him to do, the mine owner is [illegible] to be responsible for all the consequences of the incompetency [illegible] representative of the State. This is a strong case of binding the [illegible]ences of the fault or folly of one man upon the shoulders of [illegible] This is worse than taxation without representation. It is [illegible]ponsibility without blame, and for the fault of another. The [illegible]clusion may be reached by another road.

[illegible] been long settled that a mining boss or foreman is a fellow[illegible] with the other employees of the same master, engaged in a [illegible] business, and that the master is not liable for an injury caused [illegible] negligence of such mining boss. The duty of the mine owner [illegible] employ competent bosses or foremen to direct his operations. [illegible] he does this he discharges the full measure of his duty to his [illegible]es, and he is not liable for an injury arising from the negligence [illegible] foreman. A vice-principal is one to whom an employer delegates [illegible] performance of duties which the law imposes on him, and the [illegible]r is responsible because the duty is his own. As to the acts of [illegible]n, and the manner in which they do their work, the duty of [illegible] employer is to employ persons who are reasonably competent to do [illegible] assigned them, and, if he finds himself mistaken in regard to [illegible] competency, to discharge them when the mistake is discovered. [illegible] is not responsible for the consequences of their negligence as [illegible] they affect each other. Now, the act of 1891 undertakes to reverse [illegible]ed law upon this subject, and declares that the employer shall [illegible] responsible for an injury to an employee resulting from the negli[illegible] fellow-workman. Prior to the act of 1891, the man whose [illegible] caused the injury was alone liable to respond in damages. [illegible] always have property out of which a judgment could be collected, but the plaintiff must, in any case, take ... solvency of the defendant against whom his cause ...

The act of 1891 undertakes to furnish a responsible ... injured person to pursue. Passing over the head of ... at whose hands the injury was received, it fastens on the ... property on which the accident happened, and declares hi... guilty person on whose head the consequences of the acciden... To see the true character of this legislation we must keep th... objection in mind. We must remember that the injury com... is due to the negligence of a fellow-workman, for which th... responsible neither in law nor morals. We must also reme... this fellow-workman has been designated by the State, his dut... and his powers conferred by statute, and his employment ... pulsory, under heavy penalties, by the same statute. Finall... remember that it is the negligence of this fellow-servant, wh... tency the State has certified, and whose employment the Sta... pelled, for which the employer is made liable. The State s... is competent. You must employ him. You shall surrender ... trol the arrangements for the security of your employees. ... says, in effect: "If we impose upon you by certifying to ... tency of an incompetent man, or if the man to whom we c... conduct of your mines neglects his duty you shall pay for o... and for his negligence." We have no doubt that so much, ... section 8 of article 17 of the act of 1891 as imposes liabi... mine owner for the failure of the foreman to comply with the ... of the act which compels his employment and defines hi... unconstitutional and void.

But why should the certified mine foreman be relieved fro... sequences of his negligence? The jury have found that the ... due to his want of attention to his proper duties, and his ... clear, without regard to our mining laws. But the statute re... to examine the roads and ways in use in the mine each day. ... the film of rock separating the upper from the lower workin... 8 feet thick, at best; he knew that the supports for this film ... in line with each other in the upper and lower workings; ... that layers of the rock were falling off, that the thickness o... was reduced under the way on which the accident occurred ... feet, and that, not far away, it had fallen down into the lowe... yet, with all this knowledge he did nothing, so far as we ca... increase the security of the way. Whether his conduct be ... with reference to the statute, or regardless of it, his failure ... he must have known to be necessary was a neglect of dut... should render him liable to his fellow servant who has suffer... We are not prepared to hold the act of 1891 unconstitut... whole. It relates to all anthracite coal mines, and defines ... be regarded as such mines. Coal may be taken out of the ... farm owners for their own use, or it may be taken in such ... tities and for such local purposes as to make the applica... mining laws to the operations so conducted not only ... burdensome to the extent of absolute prohibition. Su... incipient operations are not within the mischief, to reme... mining laws were devised. They are ordinarily cond... poses of exploration, or for family supply, and ... with operations conducted for the supply of the ... of coal mining, like that of insurance or banking, ... legislature. The definition found in the act of ...

licable to them. The ground on which therefore, that the act is local, but that have considered are in violation of the

---

the acts of 1887 of Massachusetts pro-ury is caused to an employee, who is care and diligence at the time, by reason rson in the service of the employer, superintendence, whose sole or principal ice, the employee shall have the same medies against the employer as if the mployee of nor in the service of the ork.

Dane sued the Cochran Chemical Com- an employee, the circumstances of his ollows: The company employed Fred- der a continuing contract to make all d in its works, he to furnish tools and $2.50 per day for his own services, and employed by Johnson in addition to the son agreed to pay the man. He hired, arged the men employed by him, but the k was to be done. Dane was hired by his employment received the injuries for

rsely to Dane by the superior court of ving been brought before the supreme ts on exceptions by the plaintiff, the 9, 1895, overruled the exceptions and superior court, holding that Dane, hav- , was an employee of the carpenter and ce could not recover damages for his

judicial court in this case, delivered by l in full in volume 41 of the Northeastern d on the following reasoning:

the present case is whether the relation ndant, as shown by the evidence, was Could the plaintiff have recovered his y had not been paid by Johnson? Did his own account or as agent for the the defendant, on the evidence, would ecause, if Johnson was a servant of the tiff and the other workmen, then the ants of the defendant, and a master is not liable, at common law, for the injury [illegible] the negligence of his fellow servants; and [illegible] ent contractor, and Dane was his servant, [illegible] not be liable for any injury occasioned by the [illegible] of one of his servants to another of his servants. [illegible] that the only reasonable inference to be drawn from the [illegible] exceptions is that the plaintiff was an employee of Johnson [illegible] the defendant, within the meaning of the statute of 1887, chapter 270, and the amendments thereto.

It does not appear that Johnson was authorized to [illegible] account of the defendant, or that the workmen hired by [illegible] understood that they were to be paid by the defendant, or that the defendant or Johnson so understood. The fact that the defendant retained the right to decide how work should be done on its premises does not of itself make the workmen employed by Johnson employees of the defendant. Apparently Johnson employed whom he pleased, and directed the men employed by him in the performance of their work, whether upon the premises of the defendant or upon other premises where he might be doing work. On the evidence we do not think that the jury could properly find that the relation of employee and employer existed between the parties. If the relation of employer and employee did not exist between the parties, then the action cannot be maintained under the statute of 1887, chapter 270.

A suit for damages was brought by William A. Perry against the Old Colony Railroad Company to recover damages for injuries received while making repairs on a locomotive engine in a roundhouse, through the alleged negligence of one Straw, an engineer claimed to have been in charge of the locomotive, in blowing down the engine.

He recovered in the superior court of Suffolk County, Mass., under chapter 270 of the acts of 1887 of Massachusets, section 1 of which provides that where personal injury is caused to an employee, who is himself in the exercise of due care and diligence at the time, by reason of the negligence of any person in the service of the employer who has the charge or control of any signal, switch, locomotive engine, in train upon a railroad, the employee shall have the same right of compensation and remedies against the employer as if the employee had not been an employee of nor in the service of the employer, nor engaged in its work. The railroad company carried the case, on exceptions, to the supreme judicial court of the State, and that tribunal, on September 14, 1895, sustained the exceptions, holding that the case was not within the purview of the statute above quoted. The decision, delivered by Judge Morton, is published in volume 41 of the Northeastern Reporter, page 289, and so much thereof as bears on the applicability of the statute to the case is given here:

Even if Straw was negligent in blowing down, which we [illegible] decide, we do not think he had charge or control of a locomotive [illegible] a railroad track, within the meaning of the act. The statute [illegible] *said in* Thyng *v.* Fitchburg R. R. Co. (156 Mass., 13; 13 [illegible]

me, instead, where such repairs were made
ily, does not, it seems to us, make any differ-
f great doubt whether the engine was in the
: The testimony tended to show that when
undhouse it was generally assigned to a pit,
d the wheels blocked, and then the engineer
was in charge of the train dispatcher, or, as
engineer had no duties in the roundhouse
little job on his engine to do, he does it,"
hat in the roundhouse he has charge and
blowing down of the engine was in response
and might as well have been done by anyone
s. But even if the engine was in charge or
ot sufficient. In order to make the defend-
ve been upon a railroad track, which we do

W. H. Clune, a local officer of the American
cted, with others, in the United States dis-
rn district of California, of conspiracy to
he United States mails during the time of
n the summer of 1894. Motions for a new
lgment were overruled, and the defendants
y a fine of $1 and to be imprisoned in the
es County, Cal., for the period of eighteen
brought before the Supreme Court of the
error, which court sustained the conviction
15, 1895, as published in volume 16 of the
page 125.
of law in the case was raised in the argu-
ompetency of Congress to impose a heavier
commit a crime than that imposed for the
t Mr. Justice Brewer, in delivering the opin-

ed Statutes United States, the offense of
f the mails is made punishable by a fine of
ection 5440, Revised Statutes United States,
y offense against the United States is pun-
s than $1,000 nor more than $10,000, and by
than two years. Upon this he (the counsel
ded that a conspiracy to commit an offense

can not be punished more severely than the [illegible] when the principal offense itself is, in fact, [illegible] spiracy is merged in it.

The language of the section is plain and not open to [illegible] spiracy to commit an offense is denounced as itself a [illegible] and the punishment thereof is fixed by the statute, and we [illegible] lack of power in Congress to thus deal with a conspiracy. [illegible] may be thought of the wisdom or propriety of a statute making a con-spiracy to do an act punishable more severely than the doing of the act itself, it is a matter to be considered solely by the legislative body. The power exists to separate the conspiracy from the act itself and to affix distinct and independent penalties to each.

With regard to the suggestion that the conspiracy was merged in the completed act, it is enough that we can not, upon the record, hold that the mails were obstructed. All the testimony not being preserved, it may be that the testimony satisfied the jury that there was, in fact, no obstruction of the mails, but only, as charged, a conspiracy to obstruct. If so, the suggestion of a merger falls to the ground.

## DECISIONS UNDER COMMON LAW.

The supreme court of Indiana on October 16, 1895, reversed the judgment of the circuit court of Sullivan County in the case of Margaret C. Tohill *v.* Evansville and Terre Haute Railroad Company, and decided that the plaintiff was not entitled to recover damages from the railroad company for the death of her husband, who was killed while in the performance of his duty as an engineer in the employ of the company in a collision between his train, No. 20, which was run as a "regular," and another train, No. 19, which was being run as an "extra," or inferior train.

The supreme court held that the proximate cause of the collision was not in ordering the running of the extra train and failing to notify the deceased engineer of the regular train of the fact, but was the failure of the deceased engineer's fellow-servants in charge of the extra train to comply with the company's rules, by keeping out of the way of the regular train, and hence that the railroad company was not responsible for the accident.

The substance of the decision, delivered by Judge Hackney, which is published in full in volume 41 of the Northeastern Reporter, page 709, is as follows:

We think it may be conceded to be the law that a railroad company, operating a complicated system of trains, is required to provide for the reasonable safety of the operatives of such trains against collisions; that it would not be a compliance with such requirement to direct one train to run by schedule and another to run over the same track without schedule, in conflict with such schedule train and without notice to the schedule trainmen, by rule or otherwise, or without some limitation upon the extra or nonscheduled train, under which it would so [illegible] to guard against collision with the schedule train. There are [illegible] authorities to the proposition that it is the duty of a railroad [illegible] pany to use ordinary care and prudence in making and [illegible]

~~~~~ taking into consideration the nature of the serv- ~~~~~ does not deny the right of railway companies ~~~~~ in any instance, nor does it mean that every ~~~~~ negligence on the part of the company. From the ~~~~~ business of carrying on an extensive railway system, ~~~~~ expected to run out of schedule time, some from unavoid- ~~~~~ others as extra trains carrying an accumulation of ~~~~~assengers beyond the line of business ordinarily expected ~~~~~y prepared for. As said in the case of Slater *v.* Jewett ~~~~~), "it is at times a necessity to do so, and a necessity so ~~~~~ fall within the occurrences that a railway servant is bound ~~~~~he course of his employment. Even as regards the public ~~~~~rs, a railway manager has a right, when needs press, to ~~~~~ general time-table. All that can then be required from ~~~~~ablic and passengers is that, when he makes a variation, ~~~~~ it with reasonable care and diligence."

~~~~~that a servant could ask or expect. If, therefore, in the ~~~~~ the company made and promulgated such rules as, by their ~~~~~vance, secured reasonable safety to operatives of schedule ~~~~~t collision with extra trains, the company must be held ~~~~~rmed its duty in this respect. The special verdict in this ~~~~~y found that the company maintained rules under which ~~~~~ might be converted into extra trains; that trains of an ~~~~~should clear the right of way for trains of a superior class by ~~~~~track at least five minutes before the arrival of any schedule ~~~~~ast station to which it was safe for such inferior train to ~~~~~ 19 [the extra train] wholly failed to act in accordance with ~~~~~d did not take the side track at Pursell, where No. 20 [the ~~~~~] could have passed in safety, but continued beyond said ~~~~~ere the collision occurred. Thus it appears that obedience ~~~~~by No. 19 [the extra train] would have made the passage ~~~~~ regular train], running by the schedule, entirely safe. A ~~~~~lowed by servants, and when followed securing safety to ~~~~~ a reasonable compliance with the duty owing by the mas- ~~~~~rvants upon this question. As held in Rose *v.* Railway ~~~~~ N. Y., 217), obedience to the regulations of a railway com- ~~~~~d to the running of trains is a matter of executive detail, ~~~~~r the corporation nor any general agent can personally ~~~~~s to which employees must be relied on. If those employees ~~~~~uty by breaking existing regulations, and in consequence ~~~~~ees are injured, no action can be maintained for the injury, ~~~~~ deemed to have been caused by the negligence of a fellow-

~~~~~e seen, the negligence of those in charge of No. 19 [the ~~~~~as the proximate cause of the collision, and the order of ~~~~~patcher was not negligently issued. That the operatives ~~~~~ extra train] were the fellow-servants of the unfortunate ~~~~~ot questioned by the appellee's counsel, but is conceded. ~~~~~ resulting from the negligent act of a fellow-servant cre- ~~~~~ against the master is not only well settled, but is con- ~~~~~ appellee's counsel. It was error, therefore, to deny the ~~~~~ for judgment upon the special verdict. The judg-
~~~~~

The supreme court of Michigan decided on Oct[illegible] case of Shackleton v. Manistee and Northeastern R[illegible] (reported in volume 64 of the Northwestern Reporter, pa[illegible] under the conditions of the case, the railroad company was [illegible] able for the death of an employee who, while in the discharg[illegible] duty, was thrown from a way car and killed, by reason of the a[illegible] of a hand railing, which had been removed from the car. The c[illegible] stances of the case are set forth in the opinion of the court, de[illegible] by Judge Montgomery, which is as follows:

The plaintiff, as administratrix, sues to recover damages for th[illegible] of her husband, alleged to have been caused by the defendant's[illegible] gence. The deceased was a conductor on a freight train of the [illegible] ant company, and the injuries resulting in his death were cau[illegible] his being thrown off the rear end of the way car to the track a[illegible] train passing over him. The car from which he was thrown ha[illegible] rails provided on either side of the steps, the rear hand rail ext[illegible] to the brake, so that a sudden lurch of the car would not result in [illegible] ing one attempting to alight from the car. On the occasion i[illegible] tion, however, this rear hand rail had been removed, and the tes[illegible] offered by the plaintiff tends to show that while deceased was st[illegible] down from the car he was thrown off by reason of this defect. [illegible] is no room for attributing any negligence to the defendant, unle[illegible] for the absence of the hand rail at this time. A few days bef[illegible] accident deceased and his trainmen, when using the way car in qu[illegible] allowed it to run against a car loaded with logs, which extend[illegible] the ends of the car on which they were loaded, so that, coming [illegible] tact with the hand rail, they bent it nearly against the side of t[illegible] Deceased thereupon called the attention of a workman in the [illegible] ant's repair shops to the hand rail, and asked him to take it [illegible] repair it. The workman replied "All right," and took it off. De[illegible] then started off with the car, without reporting the defect to the [illegible] intendent, as required by the rules of the company, and without [illegible] any objection to using the defective car to those in authority, or [illegible] one connected with the defendant in any way. While on his tr[illegible] day the assistant superintendent saw the car and said to the de[illegible] "You want to see that that is fixed. Get it fixed." Deceased [illegible] called the attention of the workman in the repair shop to it, [illegible] again promised to fix it, but neglected to do so. Deceased, ho[illegible] continued in the use of the car without protest, until he was k[illegible] the manner above described.

The plaintiff's counsel recognize the general rule that the s[illegible] who engages in the use of, or continues in the use of, defective mac[illegible] or appliances, assumes the risk incident to the employment—bu[illegible] to bring this case within the exception to the rule which obtain[illegible] the servant has been induced to continue the use of the defectiv[illegible] ances by reason of the master's promise to repair. The prese[illegible] not within any such exception to the rule. No one represe[illegible] master had induced deceased to continue in the use of the [illegible] then condition. The employee in the car-repair shop g[illegible] deceased no such directions. On the contrary, he wa[illegible] instructions received from the deceased. The statement o[illegible] superintendent, so far from being authority to contin[illegible] car, was more in the nature of a rebuke for using it [illegible]

[illegible], and one which has rarely been judicially [illegible] in the case of William Mattison v. The [illegible] Southern Railway Company, before the [illegible] County, Ohio.

[illegible] of a discharged employee, who had been [illegible] employer, to recover pecuniary damages for [illegible] him by reason of the blacklisting.

[illegible] case were substantially as follows: Mattison had [illegible] railway company as a conductor, at wages of [illegible]. He declared in his petition that, having been [illegible] representative of other workmen, he made objection to [illegible] by the defendant company and "by all other [illegible] the United States," designated as "blacklist [illegible] having made such objection, he was discharged [illegible] the company, "without cause or provocation," and [illegible], conspiring with other railroad companies in [illegible] him from securing employment in his chosen avoca- [illegible] blacklist rules to be enforced against him, thus [illegible] obtaining such employment. He was compelled [illegible]where, and secured employment as a policeman, but as [illegible]een able to earn $720 per year, instead of $1,440 per [illegible]d received as a railroad conductor.

[illegible] demurred to Mattison's complaint, and the question [illegible]her the blacklist, resulting in injury to an innocent [illegible]oyee, is a wrong for which such employee can obtain

[illegible] September 25, 1895, sustained the right of Mattison [illegible] for damages. The report of the case, furnished the [illegible] Labor by the official stenographer of the court, shows the opinion to be as follows:

[illegible] right to employment is equally sacred with the right [illegible] employ him; it is not only a serious right, affecting [illegible] you may say that it is his life. The laboring man's [illegible] only thing that stands between him and starvation, [illegible] than starvation—pauperism—and it is for the pub[illegible] the public good that the right of a man to his own [illegible] honest work which he may seek, should not be [illegible].

This, of course, does not meddle at all with the [illegible] a man, to judge himself who he will have to work [illegible] no difference whether he refuses to let a man work [illegible] is incompetent or because he dislikes him. He has a right to [illegible] his own employees, but, as is frequently said, one man's right ends where another man's commences, and the right of the employer to discharge ends with his own employment, and he must not trench upon the right of the employee to seek other employment by which he may support himself and his family, and it is for the public interest that the largest liberty to seek employment should be before every man, whatever may be his employment or whatever may be his business, trade, or occupation. It is also a matter of public interest to encourage men in becoming proficient in their employment. It is, of course, a matter of public policy that a railroad company should have the right to employ such men as it sees fit and to judge for itself of the competency of its employees. There is no doubt about that. It is, however, for the public interest that a man who is skilled and who has become proficient in his employment should be able to find employment, if not with one railroad, with another railroad, or some other railroad—at least that the field should be open to him, that he should have that right; and while a railroad company may discharge its men and not employ them themselves, they trench upon the rights of the employees whenever they, by one deed or another, seek to prevent their employees from getting employment of other railroad companies, or combine or conspire in any way to prevent it, as is charged in this petition, and the matters alleged in the petition are, on demurrer, to be taken as confessed.

Of course there may be an injury that is not a legal injury resulting from a company discharging one of its employees, and so long as they simply discharge him their right to make the discharge should not be questioned; but if they make a combination, as is charged in this petition, with other companies that they shall not employ him, then it seems to me they go beyond their legal right.

The matters alleged here are sufficient to constitute a cause of action against the defendant, and the demurrer will therefore be overruled.

[illegible] provide by law for a board of labor, con[illegible]
[illegible]tion, which shall fairly represent the interests of
[illegible]bor. The board shall perform duties and receive
[illegible] prescribed by law.
[illegible]gislature shall prohibit:
[illegible]loyment of women or of children under the age of 14
[illegible]ound mines.
[illegible]ontracting of convict labor.
[illegible]bor of convicts outside prison grounds, except on pub-
[illegible] the direct control of the State.
[illegible]olitical and commercial control of employees.
[illegible]change of black lists by railroad companies or other
[illegible]ociations, or persons is prohibited.
[illegible]ght of action to recover damages for injuries resulting
[illegible]ver be abrogated, and the amount recoverable shall
[illegible] any statutory limitation.
[illegible] hours shall constitute a day's work on all works or
[illegible]rried on or aided by the State, county, or municipal
[illegible] the legislature shall pass laws to provide for the
[illegible] of employees in factories, smelters, and mines.
[illegible]gislature, by appropriate legislation, shall provide for
[illegible] of the provisions of this article.

## NOTE REGARDING BUREAUS OF STATISTICS OF LABOR.

The following minor emendations are offered to the table relating to dates of establishment, etc., of bureaus of labor statistics printed on pages 110 and 111 of Bulletin No. 1:

MARYLAND.—February 25, 1892, a new organic act relating to this bureau was passed in which, in addition to ordinary labor statistics, provision was made for gathering statistics of agriculture, mining, transportation by railroad and other means, and of shipping and commerce. Also that reports should be made annually instead of biennially as heretofore. The new bureau is known officially as the Bureau of Industrial Statistics.

CONNECTICUT.—During the existence of the original bureau (July 12, 1873, to July 23, 1875) two annual reports were made, instead of one as stated in Bulletin No. 1, one in May, 1874, and one in May, 1875.

# BULLETIN

OF THE

# DEPARTMENT OF LABOR.

---

No. 3—MARCH, 1896.

ISSUED EVERY OTHER MONTH.

---

EDITED BY

CARROLL D. WRIGHT,
COMMISSIONER.

OREN W. WEAVER,
CHIEF CLERK.

---

WASHINGTON:
GOVERNMENT PRINTING OFFICE.
*1896.*

## CONTENTS.

# EDITORIAL NOTE.

' three years a statement purporting to give ost of production, or the proportion of labor s been going the rounds of the press. This een in the following form:

the national labor statistician, has figured out wages per year paid in the United States is roduct of each laborer is valued at $1,888. 2.2 per cent, while the man who does the work is allowed a paltry 17.8 per cent. In spite of nd high wages, the fact remains that the pro- ' his labor paid to the American workingman t paid to any other workingman in any civil- r on the globe.

varies in its statement, both in percentages tial features, but usually conforms very closely which has been taken literally from one of the peared. The prominence given to this state- n the Bulletin. Ordinarily it is not our pur- urrent items, but the figures quoted and the upon the authority of the Commissioner of xception. The figures themselves are in the more particularly to the census of 1880 than f data. An analysis of the figures and the ows the fallacy of the conclusion drawn from

paid in the manufacturing and mechanical States, as shown by the census of 1880, be nber of employees to whom the wages were paid, the quotient will be 347, thus determining the average wages paid to the employees in the manufacturing and mechanical industries of the country as $347. Dividing the aggregate value of all the products of manufacturing and mechanical industries by the number of employees engaged therein, the quotient is 1,965, showing that the average product per employee was $1,965. Now, $347 is 17.7 per cent of the gross value of the per capita product, as stated, leaving a balance, of course, of 82.3 per cent, or $1,618, which the originator of the statement quoted above assumes goes to the employer. The 82.3 per cent of the total product, or $1,618 per capita, covers all expenses of production, cost of materials, miscellaneous items, profit, deterioration, interest—everything, in fact, which can be counted as cost of production other than wages. Taking the Eleventh Census—that for 1890—it is found that the value of the gross product per capita for the number of employees engaged in manufacturing and mechanical industries was $2,204, and the average annual wages per employee, computed for 1890 as already computed for 1880, was $445. The writer of the statement quoted above would assume, for the Eleventh Census, that while $445 was paid to labor, $1,759 went to the employer. As a matter of fact, of the total product per capita, 20.18 per cent went to labor, 55.08 per cent for materials, and 24.74 per cent to miscellaneous expenses, salaries, interest, profits, etc.

All statements like that quoted above are fallacious in their application. While the figures in themselves are in the main fairly correct, and the percentages so, the balance, or 82.2 per cent, does not go to the employer, but, as shown, largely for raw materials; and of the amount paid for raw materials the bulk goes to labor for their production. That the statement emanates from the Commissioner of Labor is an assumption without any authority. From what sources the comparison with workingmen of other countries is secured is not known, but the concluding statement in the quoted article is undoubtedly as fallacious as the one which gives to the employer 82.2 per cent of the value of the product.

C. D. W.

# CHAPTER I.

## INTRODUCTION.

The growth of the large industry and the creation of special industrial centers constitute two of the most marked industrial changes of recent years. They are the last steps in the evolution of the factory system from the régime of individual handicraft production. This has necessitated the aggregation in one center of large numbers of workingmen, who, with their families, are dependent upon a single industry, and this in turn has given rise to conditions and problems peculiar to such places. The present study deals with the results of an investigation into the conditions of labor and industry in those special industrial centers where a large number of workingmen have been brought together in one place, all dependent upon a single large establishment, and under such conditions as to constitute more or less self-contained communities.

The town of Essen, Germany, the seat of the great iron and steel works of Friedrich Krupp, is the best and most widely known example of this class of industrial communities in Europe. Essen has therefore been taken as the type of communities to be investigated, and the effort has been made to consider all the industrial centers of importance closely approaching it in character.

Inevitably in such centers there have developed systems of special institutions that give to each a special life and character of its own. It has been thought a matter of importance to determine as far as possible what changes have here been introduced into the organization of industry, and the results, beneficial or otherwise, to which they have given rise.

The most important of these results is the enormous development of common interests. Greater interdependence of interests, both between the workingmen themselves and between the workingmen and their

many years as possible. There is thus afforded [illegible] statistical comparison of the conditions of the same [illegible] the general and physical conditions are the same through[illegible] under comparison.

The second feature to which attention is directed is th[illegible] bility of employment. The inability of the workingman [illegible] find employment is one of the greatest hardships that he h[illegible] If the growth of the large industry has the tendency t[illegible] employment as well as of production, there can be no d[illegible] development has rendered a vast service, in this resp[illegible] improvement of the general condition of wage earners.

It is advisable at this point to state the reasons for th[illegible] each particular case of the centers that have been [illegible] Especial prominence has been given to the coal-mining [illegible] reasons that are obvious. Mining occupies a unique po[illegible] industrial world. The location of the mines in the open f[illegible] the creation of special industrial centers in most cases [illegible] Again, the fact that mining has been, and in all pro[illegible] continue to be, carried on during successive generations [illegible] place, differentiates it widely from manufacturing, strict[illegible] The result of this is that there tends to grow up aroun[illegible] a class of workingmen among whom the pursuit of mini[illegible] hereditary. A condition of affairs is created where worki[illegible] the employ of the mining companies as boys, succeeding t[illegible] remain until incapacitated for work through sickness or [illegible] are succeeded in turn by their children. The miners are [illegible] exclusively recruited from among the surrounding populat[illegible] almost be deemed to have acquired prescriptive rights in t[illegible] as far as the right of employment is concerned.

To represent this industry, therefore, the two important [illegible] panies of Anzin and Blanzy, the one in northern and [illegible] southeastern France, have been chosen as the subjects of [illegible] and III, respectively. The reasons dictating their particu[illegible] were that they were the largest mining companies of [illegible] among the largest in Europe, employing together over [illegible] that they have had a continuous existence, the one over [illegible] and the other over fifty years; and, more important than [illegible] acter of the workingmen's institutions that have deve[illegible] renders them especially valuable places for investigation.

It is scarcely necessary to state the reasons for the [illegible] Essen, Germany, to which Chapter IV is devoted. [illegible] of the great iron works of Friedrich Krupp, giving [illegible] 25,000 workingmen, with institutions there organized [illegible] the workingmen that have earned a world-wide [illegible] is this establishment founded that it presents al[illegible] of stability as those offered by mining compan[illegible] a purely self-contained community. It is the most successful example of an absolutely cooperative enterprise, and as effort to put into practice many of the essen- socialistic régime. Whatever the opinion of the regarding the principles involved, the mechanism by were enabled to acquire the ownership of the plant ... without entailing the slightest pecuniary sacri- its owner, their former employer, the method conduct of the business on principles of strict mutual- adopted for preventing the ownership of the into the hands of outsiders, are worthy of the especially as the scheme has now been in successful ... years.

...e mining company of Mariemont and Bascoup, in ... a part of the final chapter (VI) is devoted, is useful ... opportunity afforded of presenting the results of two ...ing workingmen's institutions, viz, that of the coun- ... and arbitration, organized by the company, and that ...old age and invalidity pension funds, which is not ...e company, but pertains generally to all coal-mining ...

... chapter also gives an account of a number of other ..., which, while not of sufficient importance to warrant ...ns, should yet be accorded brief mention.

...nvestigation pursued has been that of selecting par- ... making in each case a detailed study of the condi- ...ey exist there. The advantages of this monographic ...t. In this way only is it possible to investigate in ...us elements going to influence the conditions of life ...ling the workingmen; and the subject of this study ...at the method of organization of industry, existing ...ted, has upon the whole life of the individual.

...of workingmen's institutions to be of value requires ... concrete examples of such institutions, their consti- ...on, and results of operation during a series of years. ...atter of considerable value if exact statistical com- ...made of conditions in these special communities with ...rdinary conditions of production. It is manifest, ... conditions can not be established with sufficient ... the application of the strict statistical method ... not, however, prevent the student, after making a ... entering into the life of the workingmen in ...bed, from making a comparison with conditions ... elsewhere. Such a comparison, if the subtle ... are considered, will be of more value than ... by the mere juxtaposition of figures.

## CHAPTER II.

### COAL MINING COMPANY OF ANZIN, FRAN

No industry offers better opportunities for the study of the of labor and industry in the special industrial communi nature comprehended within the scope of the present repor of the mining of coal. The industry is at once one of the n tant in America or Europe; one that must be carried on scale, and one the conditions surrounding the conduct of wh the aggregation of a great many workingmen and their fami cial industrial communities. Two of the most important n panics of Europe have, therefore, been selected for a deta that of Anzin in northern and Blanzy in southeastern ] study of these centers affords a knowledge, not only of the of labor in two particular places where the community of i the entire population is as marked as in any on the conti the conditions generally under which coal is mined.

In France the mining of coal is carried on under concessio by the central Government, and is under the direct control of mining engineers. In 1893, the last year for which offi are obtainable, there were 298 concessions under which oper actually prosecuted. It must be understood, however, tha company often possesses a number of concessions of terri following table will show the general importance of the industry in France during recent years:

PRODUCTION OF COAL AND EMPLOYEES OF COAL MINES IN FRANCE

| Year. | Production (tons of 2204.6 lbs.). | Total employees. | Year. | Production (tons of 2204.6 lbs.). | Total employees. | Year. | Product (tons 2204.6 l |
|---|---|---|---|---|---|---|---|
| 1870 | 13,330,000 | 82,700 | 1878 | 16,961,000 | 106,400 | [illegible] | [illegible] |
| 1871 | 13,259,000 | 83,600 | 1879 | 17,111,000 | 102,500 | 1887 | [illegible] |
| 1872 | 15,803,000 | 91,900 | 1880 | 19,362,000 | 107,200 | [illegible] | [illegible] |
| 1873 | 17,479,000 | 105,500 | 1881 | 19,776,000 | 106,400 | [illegible] | [illegible] |
| 1874 | 16,908,000 | 106,300 | 1882 | 20,604,000 | 106,300 | [illegible] | [illegible] |
| 1875 | 16,957,000 | 108,700 | 1883 | 21,334,000 | 113,000 | 1891 | [illegible] |
| 1876 | 17,101,000 | 110,800 | 1884 | 20,024,000 | 109,400 | [illegible] | [illegible] |
| 1877 | 16,805,000 | 106,900 | 1885 | 19,511,000 | 101,600 | [illegible] | [illegible] |

Though coal was mined in 1893 in 41 departments, ments produced nine-tenths, and of these the two adjoining of the Nord and Pas-de-Calais, in the north of France, siderably over one-half of the total quantity mined. The [illegible] situated in the former of these two [illegible] of Valenciennes. The company is [illegible] company in France. Alone it produced [illegible] coal mined in its department, and one- [illegible] for the whole country. The first search for coal [illegible] in 1716 by prospectors, who afterwards [illegible] of Anzin. Coal was discovered in 1720. The [illegible] company of Anzin, however, was not effected [illegible]. Since that date its existence has been con- [illegible] table, giving its annual production since 1757, [illegible] importance. The figures for the years prior to [illegible] or estimates based on various data.

THE COAL MINING COMPANY OF ANZIN, 1757 TO 1893.

| Year. | Production (tons of 2204.6 lbs.). | Year. | Production (tons of 2204.6 lbs.). | Year. | Production (tons of 2204.6 lbs.). | Year. | Production (tons of 2204.6 lbs.). |
|---|---|---|---|---|---|---|---|
| 1803. | [illegible], 443 | 1826. | 376, 986 | 1849. | 614, 906 | 1872. | 2, 196, 436 |
| 1804. | [illegible], 332 | 1827. | 400, 668 | 1850. | 669, 999 | 1873. | 2, 191, 500 |
| 1805. | 225, 812 | 1828. | 406, 593 | 1851. | 648, 062 | 1874. | 1, 922, 037 |
| 1806. | [illegible] | 1829. | 410, 632 | 1852. | 706, 633 | 1875. | 2, 058, 566 |
| 1807. | 194, 138 | 1830. | 508, 708 | 1853. | 803, 812 | 1876. | 2, 063, 931 |
| 1808. | [illegible], 106 | 1831. | 460, 964 | 1854. | 856, 295 | 1877. | 2, 042, 035 |
| 1809. | [illegible] | 1832. | 472, 959 | 1855. | 947, 936 | 1878. | 1, 979, 454 |
| 1810. | 279, 965 | 1833. | 541, 504 | 1856. | 920, 574 | 1879. | 1, 980, 934 |
| 1811. | [illegible], 815 | 1834. | 573, 239 | 1857. | 919, 187 | 1880. | 2, 314, 006 |
| 1812. | [illegible] | 1835. | 591, 836 | 1858. | 950, 889 | 1881. | 2, 264, 955 |
| 1813. | [illegible], 840 | 1836. | 623, 546 | 1859. | 907, 543 | 1882. | 2, 215, 611 |
| 1814. | [illegible] | 1837. | 651, 511 | 1860. | 930, 700 | 1883. | 2, 210, 702 |
| 1815. | 247, 404 | 1838. | 659, 644 | 1861. | 956, 610 | 1884. | 1, 720, 306 |
| 1816. | 250, 044 | 1839. | 707, 748 | 1862. | 993, 950 | 1885. | 2, 070, 442 |
| 1817. | 236, 856 | 1840. | 623, 312 | 1863. | 1, 053, 334 | 1886. | 2, 337, 439 |
| 1818. | 234, 482 | 1841. | 643, 623 | 1864. | 1, 067, 017 | 1887. | 2, 504, 412 |
| 1819. | [illegible], 947 | 1842. | 721, 030 | 1865. | 1, 225, 425 | 1888. | 2, 595, 581 |
| 1820. | 230, 189 | 1843. | 642, 280 | 1866. | 1, 348, 812 | 1889. | 2, 857, 663 |
| 1821. | [illegible], 788 | 1844. | 597, 953 | 1867. | 1, 441, 002 | 1890. | 3, 121, 552 |
| 1822. | 240, 489 | 1845. | 714, 755 | 1868. | 1, 617, 621 | 1891. | 2, 933, 724 |
| 1823. | 318, 570 | 1846. | 803, 804 | 1869. | 1, 606, 075 | 1892. | 2, 818, 529 |
| 1824. | 327, 327 | 1847. | 774, 896 | 1870. | 1, 633, 818 | 1893. | 2, 975, 691 |
| 1825. | [illegible], 457 | 1848. | 618, 502 | 1871. | 1, 715, 878 | | |

engaged in the operations of [illegible]
services of the company:

EMPLOYEES OF THE COAL MINING COMPANY OF [illegible] AND OTHER INDUSTRIES, [illegible]

| Year. | Mining proper. | | | Other. | Total. | Year. | Mining proper. | | | [illegible] |
|---|---|---|---|---|---|---|---|---|---|---|
| | Above ground. | Below ground. | Total. | | | | Above ground. | Below ground. | [illegible] | |
| 1870.. | 968 | 8,165 | 9,122 | ........ | ........ | 1883. | 1,7[illegible] | [illegible] | [illegible] | [illegible] |
| 1871.. | 1,102 | 8,481 | 9,583 | ........ | ........ | 1884. | 2,[illegible] | [illegible] | [illegible] | [illegible] |
| 1872.. | 1,293 | 9,290 | 10,583 | ........ | ........ | 1885. | 1,[illegible] | [illegible] | [illegible] | [illegible] |
| 1873.. | 1,564 | 9,953 | 11,517 | ........ | ........ | 1886. | 2,[illegible] | [illegible] | [illegible] | [illegible] |
| 1874.. | 1,687 | 10,300 | 11,987 | ........ | ........ | 1887. | 1,[illegible] | [illegible] | [illegible] | [illegible] |
| 1875.. | 1,637 | 10,649 | 12,286 | ........ | ........ | 1888. | 1,[illegible] | [illegible] | [illegible] | [illegible] |
| 1876.. | 1,800 | 10,900 | 12,700 | ........ | ........ | 1889. | 1,[illegible] | [illegible] | [illegible] | [illegible] |
| 1877.. | 1,807 | 11,074 | 12,881 | ........ | ........ | 1890. | 1,[illegible] | [illegible] | [illegible] | [illegible] |
| 1878.. | 1,830 | 11,085 | 12,915 | ........ | ........ | 1891. | 1,881 | [illegible] | [illegible] | [illegible] |
| 1879.. | 1,988 | 11,013 | 001 | 2,208 | 15,204 | 1892. | 1,[illegible] | [illegible] | [illegible] | [illegible] |
| 1880.. | 1,956 | 10,814 | 770 | 2,558 | 15,328 | 1893. | 1,620 | [illegible] | [illegible] | [illegible] |
| 1881.. | 1,873 | 10,978 | 851 | 2,502 | 15,353 | 1894. | 1,977 | [illegible] | [illegible] | [illegible] |
| 1882.. | 1,796 | 10,407 | 12,203 | 2,291 | 14,494 | | | | | |

The following table gives for the years 1883 to 1892, incl[illegible]
number of employees according to the general division of ser[illegible]

EMPLOYEES OF THE COAL MINING COMPANY OF ANZIN ENGAGED IN CO[illegible] BY NATURE OF WORK PERFORMED, 1883 TO 1892.

[The totals in this table do not agree with those in the preceding table, as they [illegible] of affairs for a particular date rather than an average for the year.]

| Year. | Actual work of mining. | | Maintenance and repairs. | | Transportation. | | Filling in exhausted veins. | | Overseeing. | | Total below ground. | | Total above ground. | |
|---|---|---|---|---|---|---|---|---|---|---|---|---|---|---|
| | Number. | Per cent. | Number. | Per cent. | Number. | Per cent. | Number. | Per cent. | Number. | Per cent. | Number. | Per cent. | Number. | Per cent. |
| 1883.. | 5,475 | 47.2 | 1,493 | 12.9 | 1,261 | 10.9 | 1,361 | 11.8 | 271 | 2.3 | 9,861 | 85.1 | 1,[illegible] | [illegible] |
| 1884.. | 5,284 | 53.2 | 876 | 8.8 | 1,162 | 11.7 | 1,023 | 10.3 | 241 | 2.5 | 8,586 | [illegible] | 1,[illegible] | [illegible] |
| 1885.. | 5,786 | 60.0 | 510 | 5.3 | 995 | 10.3 | 581 | 6.0 | 230 | 2.4 | 8,103 | 84.0 | 1,[illegible] | [illegible] |
| 1886.. | 5,848 | 61.1 | 470 | 4.9 | 1,012 | 10.6 | 482 | 5.1 | 232 | 2.4 | 8,044 | 84.1 | 1,[illegible] | [illegible] |
| 1887.. | 6,118 | 63.1 | 458 | 4.7 | 998 | 10.3 | 396 | 4.1 | 226 | 2.3 | 8,196 | 84.5 | 1,[illegible] | [illegible] |
| 1888.. | 6,344 | 65.0 | 396 | 4.0 | 993 | 10.2 | 307 | 3.1 | 212 | 2.3 | 8,262 | 84.5 | 1,[illegible] | [illegible] |
| 1889.. | 6,517 | 65.7 | 374 | 3.8 | 1,036 | 10.4 | 296 | 2.9 | 236 | 2.3 | 8,478 | 86.1 | 1,[illegible] | [illegible] |
| 1890.. | 6,974 | 65.4 | 444 | 4.2 | 1,115 | 10.5 | 324 | 3.0 | 240 | 2.2 | 9,097 | 85.3 | 1,[illegible] | [illegible] |
| 1891.. | 7,004 | 65.5 | 407 | 3.8 | 1,146 | 10.7 | 294 | 2.7 | 243 | 2.3 | a9,094 | 85.0 | a1,[illegible] | [illegible] |
| 1892.. | 6,793 | 64.3 | 416 | 4.0 | 1,193 | 11.3 | 320 | 3.0 | 243 | 2.3 | 8,985 | 84.6 | 1,[illegible] | [illegible] |

a The addition of the total employees below and above ground does not produce the gr[illegible] shown. The figures are given, however, as reported to the Department.

## THE GENERAL CONDITIONS OF LABOR.

The aggregation of over 12,000 employees in one locality, [illegible] their families, are all dependent upon a single industry, and [illegible] practically the same conditions, furnishes an excellent opport[illegible] the study in detail of the influences surrounding an importan[illegible] workingmen engaged in one of the principal industries of the [illegible] This description will naturally fall into two parts: First, [illegible] conditions of labor; second, the numerous workingmen's in[illegible] such as pension funds, cooperative stores, mutual aid so[illegible] which Anzin has been remarkable during the past half [illegible]

[illegible] age limit was raised to 12, [illegible] 2, 1892, to 13 years. The minimum [illegible] above ground is likewise 13, except that [illegible] be employed provided that they have a [illegible] amount of education and one showing [illegible]. The new recruits are at first employed in [illegible], such as washing and sorting coal, and [illegible] the work below ground as the need for extra [illegible] that moment, if their conduct is satisfactory, [illegible] is assured. It will be seen further on that [illegible] is one of the characteristics of the conduct of [illegible]. Anzin is a remarkable example of this. At [illegible] father, if he desires to do so, can retire from [illegible], from a pension fund maintained by the [illegible] and the miners. The period of active labor [illegible] thus be estimated to be, on an average, 40 years. [illegible] have been introduced to show the ages of work- [illegible] employment of the company for the years 1888 to [illegible] actual ages of all employees in February, 1892. [illegible] because the computation had already been [illegible] and nothing would be gained by making a cal- [illegible].

[illegible] OF EMPLOYEES OF EACH SPECIFIED AGE ENTERING [illegible] COAL MINING COMPANY OF ANZIN, 1888 TO 1893.

| [illegible] | Ages. | | | | | | | | Total. |
|---|---|---|---|---|---|---|---|---|---|
| [illegible] | [illegible] years or under 16. | 16 years or under 20. | 20 years or under 25. | 25 years or under 30. | 30 years or under 35. | 35 years or over. | Unknown. | |
| [illegible] | 37 | 82 | 89 | 43 | 25 | 45 | 60 | 720 |
| [illegible] | 62 | 200 | 136 | 128 | 90 | 71 | ........ | 1,182 |
| [illegible] | 70 | 183 | 140 | 84 | 46 | 36 | ........ | 1,037 |
| [illegible] | 15 | 48 | 300 | 27 | 14 | 9 | ........ | 757 |
| [illegible] | 11 | 19 | 217 | 5 | 5 | 1 | ........ | 646 |
| [illegible] | 30 | 15 | 294 | 4 | 3 | 1 | ........ | 743 |
| [illegible] | 235 | 561 | 1,126 | 291 | 183 | 163 | 60 | 5,085 |
| [illegible] | [illegible] | 11.03 | 22.14 | 5.72 | 3.60 | 3.21 | 1.18 | 100 |

AGE OF EMPLOYEES OF THE COAL [illegible]

| Age. | Employees. Below ground. | Above ground. | Total. | Age. | Below ground. | Above ground. | Total. |
|---|---|---|---|---|---|---|---|
| 12 ...... | ........ | 29 | 29 | 30 ...... | [illegible] | 7 | [illegible] |
| 13 ...... | 151 | 147 | 298 | 31 ...... | [illegible] | [illegible] | [illegible] |
| 14 ...... | 378 | 2 | 380 | 32 ...... | 261 | 15 | [illegible] |
| 15 ...... | 445 | 1 | 446 | 33 ...... | 277 | [illegible] | [illegible] |
| 16 ...... | 393 | 5 | 398 | 34 ...... | 316 | [illegible] | [illegible] |
| 17 ...... | 450 | 3 | 453 | 35 ...... | [illegible] | [illegible] | [illegible] |
| 18 ...... | 358 | 2 | 360 | 36 ...... | 190 | 7 | [illegible] |
| 19 ...... | 384 | 10 | 394 | 37 ...... | 185 | [illegible] | [illegible] |
| 20 ...... | 412 | 9 | 421 | 38 ...... | 167 | 13 | [illegible] |
| 21 ...... | 365 | 4 | 369 | 39 ...... | 167 | [illegible] | [illegible] |
| 22 ...... | 385 | 9 | 394 | 40 ...... | 211 | 13 | [illegible] |
| 23 ...... | 348 | 1 | 349 | 41 ...... | 162 | 17 | [illegible] |
| 24 ...... | 322 | 8 | 330 | 42 ...... | 143 | 14 | [illegible] |
| 25 ...... | 340 | 6 | 346 | 43 ...... | 162 | 12 | [illegible] |
| 26 ...... | 337 | 10 | 347 | 44 ...... | 181 | 20 | [illegible] |
| 27 ...... | 296 | 16 | 312 | 45 ...... | 162 | 8 | [illegible] |
| 28 ...... | 294 | 6 | 300 | 46 ...... | 169 | 12 | [illegible] |
| 29 ...... | 276 | 14 | 290 | 47 ...... | 125 | 15 | [illegible] |

Total. [illegible]

It will be noticed from the first of these two tables that a lar[illegible] tion of the employees of the company enter its service as [illegible] 14 years of age. The apparently large number entering betwe[illegible] of 20 and 25 years is caused by their return from military d[illegible] great majority of them had already been in the employ of the

The two tables following make a comparison of the ages o[illegible] men at Anzin with those of all coal miners in France. T[illegible] November 2, 1892, regulating the hours of labor of women an[illegible] divides the workingmen into three classes: (1) Children, 13 o[illegible] years; (2) minors, 16 or under 18 years; and (3) adults, 18 ye[illegible] or over. The first table compares Anzin for the year 1892 w[illegible] for 1893, according to this grouping. The second is a com[illegible] ages in greater detail. The conditions at Anzin, it will be s[illegible] little from those of France generally.

NUMBER AND PER CENT OF EMPLOYEES, BY AGE PERIODS, AT AN[illegible] ALL COAL MINES OF FRANCE.

| Age periods. | Employees below ground. Anzin, 1892. Number. | Per cent. | France, 1893. Number. | Per cent. | Employees above [illegible] Anzin, 1892. Number. | Per cent. | [illegible] |
|---|---|---|---|---|---|---|---|
| 13 or under 16 years. | 974 | 9.37 | 4,412 | 4.71 | a179 | [illegible] | [illegible] |
| 16 or under 18 years. | 843 | 8.11 | 5,507 | 5.88 | 8 | [illegible] | [illegible] |
| 18 years or over ... | 8,574 | 82.52 | 83,766 | 89.41 | 474 | 71.71 | [illegible] |
| Total ........ | 10,391 | 100.00 | 93,685 | 100.00 | a661 | 100.00 | [illegible] |

a Includes 29 twelve years of age.

led, in general descend into the mines
l return to the surface at 1 or 2 o'clock
r ten hours below ground. If the time
ng the shaft and a half hour's rest for
is from eight to nine hours' actual labor.
bor is suspended, except in certain cases
g table shows the average number of
the number of hours devoted to actual
each mine employee of the company of
of information should be considered in
or of coal miners, as the time consumed
greatly with miners in different mines.
yees working above and below ground

OYEES OF THE COAL MINING COMPANY OF
AND HOURS OF ACTUAL LABOR, 1891.

nsumed in going to and returning from work.]

| ground. | | Employees above ground. | | | |
|---|---|---|---|---|---|
| t actual labor. | | On duty. | | At actual labor. | |
| mber. | Per cent. | Number. | Per cent. | Number. | Per cent. |
| 107 | 1.21 | ........ | ........ | 46 | 3.10 |
| 75 | .85 | ........ | ........ | ........ | ........ |
| [illegible] | 2.33 | ........ | ........ | 22 | 1.52 |
| 45 | .51 | ........ | ........ | ........ | ........ |
| [illegible] | 10.85 | 1 | .07 | 30 | 2.07 |
| 443 | 5.03 | ........ | ........ | ........ | ........ |
| 681 | 7.73 | ........ | ........ | ........ | ........ |
| 648 | 7.36 | ........ | ........ | ........ | ........ |
| [illegible] | 23.08 | ........ | ........ | 80 | 5.52 |
| 1,002 | 11.39 | ........ | ........ | ........ | ........ |
| [illegible] | 11.27 | ........ | ........ | 70 | 4.83 |
| 479 | 5.44 | ........ | ........ | 2 | .14 |
| 454 | 5.15 | 75 | 5.17 | 164 | 11.31 |
| [illegible] | 2.99 | ........ | ........ | 24 | 1.65 |
| 212 | 2.41 | 280 | 19.31 | 777 | 53.59 |
| 57 | .65 | 10 | .69 | ........ | ........ |
| 102 | 1.16 | 97 | 6.69 | 179 | 12.34 |
| ........ | ........ | ........ | ........ | 20 | 1.38 |
| 40 | .45 | 152 | 10.48 | 21 | 1.45 |
| | | 20 | 1.38 | ........ | ........ |
| 12 | .14 | 814 | 56.14 | 16 | 1.10 |
| ........ | ........ | ........ | ........ | ........ | ........ |
| ........ | ........ | 1 | .07 | ........ | ........ |
| [illegible] | 100.00 | 1,450 | 100.00 | 1,450 | 100.00 |

A comparison of average hours of [illegible] year 1891 with the average hours in 1890 [illegible] basin in which Anzin is situated (Nord and [illegible]) the coal mines of France, is made in the following [illegible].

From the comparison here given it is seen that the hours of labor are in general somewhat longer at Anzin than either in its special coal basin or in France. The difference is more pronounced in the case of hours on duty than in that of hours of actual labor.

AVERAGE HOURS OF LABOR PER DAY AT COAL MINES AT ANZIN, IN THE DEPARTMENTS OF NORD AND PAS-DE-CALAIS, AND IN ALL FRANCE.

[The figures for the departments of the Nord and Pas-de-Calais and for France were obtained from Une Notice sur le Nombre, les Salaires et la Durée du Travail des Ouvriers des Mines, en 18[illegible], contained in the annual volume Statistique de l'Industrie Minérale et des Appareils à Vapeur en France et en Algérie pour l'année 1889.]

| Locality. | Average hours per day on duty. | | | Average hours per day [illegible] | | |
|---|---|---|---|---|---|---|
| | Employees below ground. | Employees above ground. | All employees. | Employees below ground. | Employees above ground. | All employees. |
| Anzin, 1891 | 9.52 | 11.47 | [illegible] | [illegible] | [illegible] | [illegible] |
| Nord and Pas-de-Calais, 1890 | 9.40 | 10.50 | [illegible] | [illegible] | [illegible] | [illegible] |
| France, 1890 | 9.45 | 10.46 | 10.00 | [illegible] | [illegible] | [illegible] |

There is probably no one feature of the modern system of [illegible] tion of industry more productive of injurious results to the [illegible] men than the periodic interruptions to which they are subjected [illegible] present conditions in their ability to obtain work. Next to that of the amount of their wages, the question of the regularity of their [illegible] ment is the one in which workingmen are most interested. The [illegible] of a reasonable certainty of continuous employment means not only a curtailing of their earning capacity, but their demoralization [illegible]. The constancy of employment is, then, a prime element in [illegible] the condition of any particular class of workingmen.

The conditions at Anzin in this respect could scarcely be [illegible] upon. During the entire year the intensity of work is equal. The [illegible] that follows shows that in the twenty-four years from 1870 to 18[illegible] [illegible] sive, the mines were operated almost every possible working [illegible] the tables relating to the number of years the workingmen [illegible] continuously employed that are given further on be considered [illegible] nection with this one, it is evident that Anzin includes a [illegible] stable body of workingmen, to whom the evils of lack of [illegible] are almost unknown.

orkingmen of a particular locality or
wages must, under present conditions,
determining their economic well-being.
rarely paid. The wages of employees
of work performed, and a settlement is
paid so much per car of coal mined.
e company's engineer and accepted by
a certain distance along the vein to be
l not transport his own coal nor main-
ne work of the carman was absolutely
or. This gave rise to serious difficulties.
at his coal was not carried away with
n complained that a sufficient quantity
im fully employed. This has now been
nt practice is for the miner to have his
his family work with him to aid in the
necessary, his son can aid him in the
accumulates, he himself can assist in
dvantage of the system, moreover, is
ing the mine serve an apprenticeship
their welfare. Under these conditions
s can not be given. In the table that
daily and yearly wages of all mine
the three great classes of occupations—
other than miners, and laborers above
e three classes combined, for the years

ploying workingmen of widely-varying
ually divergent rates of wages, average
ly meaningless, this is not true of coal
of the work performed falls into a few
ich require about the same degree of

skill, and in which the wages paid differ but slightly. For practical purposes, therefore, these tables give a sufficiently accurate idea of the wages of miners, as well as the variation of wages during the period covered.

AVERAGE DAILY AND YEARLY WAGES OF EMPLOYEES OF THE COAL MINING COMPANY OF ANZIN, 1870 TO 1893.

| Year. | Average daily wages. | | | | Average yearly wages. | | | |
|---|---|---|---|---|---|---|---|---|
| | Employees below ground. | | Employees above ground. | All employees. | Employees below ground. | | Employees above ground. | All employees. |
| | Miners. | Others. | | | Miners. | Others. | | |
| 1870 | $0.71 | $0.58 | $0.45 | $0.57 | $218.19 | $171.39 | $140.10 | $171.96 |
| 1871 | .75 | .61 | .45 | .58 | 228.33 | 181.78 | 151.34 | 174.27 |
| 1872 | .80 | .64 | .49 | .62 | 250.70 | 204.44 | 174.71 | 201.97 |
| 1873 | .92 | .72 | .56 | .70 | 288.43 | 229.15 | 198.94 | 227.08 |
| 1874 | .90 | .71 | .55 | .69 | 275.52 | 218.04 | 185.37 | 214.82 |
| 1875 | .92 | .72 | .56 | .70 | 288.15 | 222.14 | 198.94 | 219.99 |
| 1876 | .92 | .73 | .57 | .71 | 278.27 | 220.90 | 198.03 | 219.58 |
| 1877 | .82 | .70 | .56 | .66 | 240.08 | 202.33 | 186.30 | 195.19 |
| 1878 | .81 | .67 | .55 | .65 | 224.27 | 185.11 | 177.81 | 185.39 |
| 1879 | .79 | .65 | .54 | .64 | 219.94 | 181.89 | 174.16 | 180.88 |
| 1880 | .82 | .67 | .55 | .66 | 251.45 | 206.44 | 189.22 | 203.23 |
| 1881 | .82 | .67 | .61 | .66 | 258.04 | 207.86 | 208.47 | 208.65 |
| 1882 | .85 | .69 | .60 | .68 | 261.90 | 211.34 | 208.54 | 211.14 |
| 1883 | .87 | .73 | .66 | .71 | 265.20 | 218.08 | 227.87 | 220.18 |
| 1884 | .85 | .73 | .70 | .72 | 207.01 | 179.92 | 216.72 | 184.89 |
| 1885 | .85 | .76 | .60 | .73 | 208.66 | 196.96 | 181.61 | 194.46 |
| 1886 | .84 | .76 | .61 | .73 | 233.62 | 214.75 | 191.49 | 211.65 |
| 1887 | .85 | .76 | .58 | .73 | 238.39 | 218.94 | 178.47 | 214.56 |
| 1888 | .85 | .77 | .60 | .74 | 242.87 | 229.78 | 190.06 | 215.91 |
| 1889 | .89 | .80 | .61 | .77 | 252.68 | 232.08 | 195.99 | 226.56 |
| 1890 | 1.02 | .90 | .66 | .86 | 287.30 | 260.96 | 210.84 | 253.49 |
| 1891 | 1.06 | .94 | .68 | .90 | 294.56 | 268.19 | 218.57 | 260.85 |
| 1892 | 1.06 | .94 | .67 | .89 | 287.94 | 260.32 | 213.21 | 253.21 |
| 1893 | 1.05 | .93 | .67 | .89 | 291.46 | 264.76 | 213.25 | 256.87 |

It is scarcely necessary to call attention to the value of a record such as that contained in the table just given. While it can not be used for exact comparison with wages paid elsewhere, inasmuch as it relates to the earnings of several classes of workingmen combined, it shows clearly the relative variations in wages of coal-mine employees at Anzin. The showing is a very gratifying one as regards the increasing economic welfare of the laborers. From an average daily wage of 2.96 francs (57 cents) the rate for all employees increased rather slowly during the first decade, being but 3.42 francs (66 cents) in 1880. In the next ten years, however, wages had increased materially, being 4.48 francs (86 cents) in 1890, while during the next three years a still further increase to 4.61 francs (89 cents) was recorded. The significance of this increase is still more apparent if annual wages be contrasted. In 1870 the average for all employees was 890.96 francs ($171.96). In 1880 the average was 1,053 francs ($203.23), in 1890 had increased to 1,313.44 francs ($253.49), and in 1893 was 1,330.94 francs ($256.87). These figures represent an increase of a little over 49 per cent during the period.

The only possible element that could enter into this showing to vitiate the deduction that a real increase in wages had resulted, not only for all employees combined, but for each of the different classes of workmen entering into the calculation, is that the number of employees in higher-paid classes of work figure to a greater extent in later than

proximately similar proportional distribution of the different classes of work. There is every reason that all classes of workingmen have profited in wages.

however, may serve as an index of the absolute economic condition of the laborers, they need to terpreted in the light of, two important considera- unt of supplementary advantages enjoyed by the cheap housing, medical attendance, free fuel, etc.; f necessary or usual articles of consumption that hase, or, to state it in another way, the relation f wages and the prices of commodities.

st point—that of supplemental advantages—a will be given later on. For the present purpose ert a table showing, for each of the eleven years total and average wages, the total and average l advantages, and the proportional addition to mentary advantages represent. From this table ere has been a fairly constant tendency for the plemental advantages to increase, and that this ghtly increasing proportion of the average amount t the present time, speaking in round numbers, it upplemental advantages enjoyed by workingmen least a 10 per cent addition to their wages.

ES SUPPLEMENTARY TO WAGES OF EMPLOYEES OF THE INING COMPANY OF ANZIN, 1883 TO 1893.

nt of wages and advantages supplementary to wages per employee were Company of Anzin from data which are not known; hence they differ on the basis of the total number of employees given elsewhere in the

| ntages ple- ary to ges. | Total wages and supple- mentary advantages. | Wages per em- ployee. | Supple- mentary advantages per employee. | Total wages and supple- mentary advantages per employee. | Per cent of supple- mentary advantages of wages. |
|---|---|---|---|---|---|
| [illegible].99 | $3,196,729.99 | $220.18 | $17.34 | $237.52 | 7.98 |
| [illegible].52 | 2,482,800.52 | 184.89 | 19.55 | 204.44 | 9.90 |
| [illegible].65 | 2,347,[illegible].65 | 194.46 | 20.40 | 214.86 | 10.12 |
| [illegible].43 | 2,525,994.43 | 211.05 | 19.35 | 230.40 | 9.55 |
| [illegible].97 | 2,675,212.97 | 214.56 | 20.55 | 235.11 | 9.99 |
| [illegible].89 | 2,[illegible],096.89 | 215.91 | 22.05 | 237.96 | 10.70 |
| [illegible].47 | 2,[illegible],943.47 | 226.55 | 22.58 | 249.13 | 10.52 |
| [illegible] | 3,[illegible],[illegible]21.34 | 253.49 | 22.53 | 276.02 | 9.56 |
| [illegible] | 3,427,[illegible].52 | 260.55 | 24.63 | 285.18 | 10.08 |
| [illegible] | 3,317,892.71 | 253.21 | 25.61 | 278.82 | 10.74 |
| [illegible] | 3,375,520.22 | 256.87 | 26.01 | 282.88 | 10.84 |

an be seen, without reproducing the figures which in the table itself, that the inclusion of the addi- plementary advantages accentuates yet more the increase in earnings shown in the [illegible] should be remembered, moreover, that [illegible] advantages as here stated indicates only their [illegible] is probable that if they had been furnished by the [illegible] their cost would have been considerably greater. [illegible] these supplementary advantages is therefore without doubt in excess of that shown in the table.

Concerning the second point, it is of interest to note the study, with accompanying tables, made by M. Georges Michel, of the Économiste Français, on this particular question for the miners of [illegible] included in his book entitled Histoire d'un Centre Ouvrier. ([illegible] cessions d'Anzin). The most important of his tables—the one [illegible] he has brought into correlation the average prices of [illegible] the budget of a typical family, composed of father, mother, and [illegible] children, of which the eldest has commenced work, for each decade [illegible] 1820 to 1887—is here reproduced. It should be borne in mind that this table is but a calculation based on such data as were obtainable. M. Michel first made the calculation for the period 1880 to 1887, in which it is reasonable to suppose that substantial accuracy was secured. Then, using this as a basis, and taking into account, not only the variation in the prices of commodities and of average earnings, but of changes in the habits of the workingmen as well, he was able to calculate the budgets for the preceding decades. The officials of the company, moreover, examined the figures of M. Michel, and after making a [illegible] corrections, expressed themselves as satisfied that they [illegible] a substantial approximation to the true condition of affairs.

COST OF LIVING OF A TYPICAL FAMILY OF SIX AT ANZIN, 1820 TO 1887.

| Objects of expenditure. | Unit. | 1820 to 1830. | | | 1830 to 1840. | | | [illegible] | | | [illegible] | | |
|---|---|---|---|---|---|---|---|---|---|---|---|---|---|
| | | Quantity. | Price. | Value. | Quantity. | Price. | Value. | Quantity. | Price. | Value. | Quantity. | [illegible] | [illegible] |
| Bread | Pound. | 2,337 | $0.026 | $60.76 | 2,337 | $0.028 | $65.44 | 2,337 | $0.02[illegible] | $67.7[illegible] | [illegible] | [illegible] | [illegible] |
| Meat | Pound. | 66 | .061 | 4.03 | 99 | .070 | 6.93 | [illegible] | [illegible] | [illegible] | [illegible] | [illegible] | [illegible] |
| Milk | Qaurt. | 106 | .024 | 2.54 | 137 | .024 | 3.29 | [illegible] | .024 | [illegible] | [illegible] | [illegible] | [illegible] |
| Butter | Pound. | 22 | .131 | 2.88 | 33 | .136 | 4.49 | 40 | .15[illegible] | [illegible] | [illegible] | [illegible] | [illegible] |
| Eggs | Dozen. | 100 | .115 | .96 | 150 | .12[illegible] | 1.5[illegible] | 150 | .137 | 1.[illegible] | [illegible] | [illegible] | [illegible] |
| Fruits and vegetables purchased. | ...... | ...... | ...... | 3.86 | ...... | ...... | 6.69 | ...... | ...... | 9.[illegible] | ...... | [illegible] | [illegible] |
| Beer | Gallon. | 53 | .044 | 2.33 | 132 | .044 | 5.81 | 15[illegible] | .044 | 7.[illegible] | [illegible] | [illegible] | [illegible] |
| Alcoholic drinks | ...... | ...... | ...... | .97 | ...... | ...... | .97 | ...... | ...... | .9[illegible] | ...... | [illegible] | [illegible] |
| Groceries: | | | | | | | | | | | | | |
| Oil | Quart. | 11 | .164 | 1.80 | 16 | .194 | 3.10 | 16 | [illegible] | [illegible] | [illegible] | [illegible] | [illegible] |
| Coffee | Pound. | 18 | .219 | 3.94 | 22 | .218 | 4.80 | [illegible] | .17[illegible] | [illegible] | [illegible] | [illegible] | [illegible] |
| Sugar | Pound. | 11 | .197 | 2.17 | 15 | .185 | 2.78 | [illegible] | [illegible] | [illegible] | [illegible] | [illegible] | [illegible] |
| Soap | Pound. | 99 | .054 | 5.35 | 99 | .056 | 5.54 | 110 | [illegible] | [illegible] | [illegible] | [illegible] | [illegible] |
| Miscellaneous | ...... | ...... | ...... | 1.93 | ...... | ...... | 1.[illegible] | ...... | ...... | [illegible] | [illegible] | [illegible] | [illegible] |
| Haberdashery | ...... | ...... | ...... | 1.93 | ...... | ...... | 2.[illegible] | ...... | ...... | [illegible] | [illegible] | [illegible] | [illegible] |
| Table utensils | ...... | ...... | ...... | .97 | ...... | ...... | .97 | ...... | ...... | [illegible] | [illegible] | [illegible] | [illegible] |
| Furniture | ...... | ...... | ...... | 1.54 | ...... | ...... | 1.[illegible] | ...... | ...... | [illegible] | [illegible] | [illegible] | [illegible] |
| Clothing | ...... | ...... | ...... | 13.51 | ...... | ...... | 17.7[illegible] | ...... | ...... | [illegible] | [illegible] | [illegible] | [illegible] |
| Hats | ...... | ...... | ...... | 1.54 | ...... | ...... | 2.[illegible] | ...... | ...... | [illegible] | [illegible] | [illegible] | [illegible] |
| Shoes | ...... | ...... | ...... | 2.72 | ...... | ...... | [illegible] | ...... | ...... | [illegible] | [illegible] | [illegible] | [illegible] |
| Rent and cultivation of garden. | ...... | ...... | ...... | 11.58 | ...... | ...... | 13.51 | ...... | ...... | [illegible] | [illegible] | [illegible] | [illegible] |
| Various, saloon, savings, etc. | ...... | ...... | ...... | 1.[illegible] | ...... | ...... | [illegible] | ...... | ...... | [illegible] | [illegible] | [illegible] | [illegible] |
| Total expenditures. | ...... | ...... | ...... | [illegible] | ...... | ...... | 161.[illegible] | ...... | ...... | [illegible] | [illegible] | [illegible] | [illegible] |

| | | | | | | | | |
|---|---|---|---|---|---|---|---|---|
| [illegible] | [illegible] | 3.[illegible] | 260 | .1[illegible] | 3.79 | 270 | .174 | 3.[illegible] |
| ........ | ........ | 11.[illegible] | ........ | ........ | 12.51 | ........ | ........ | 11.58 |
| [illegible] | .0[illegible] | 15.81 | 317 | .077 | 24.41 | 317 | .080 | 2[illegible] |
| ........ | ........ | 1.54 | ........ | ........ | 1.[illegible] | ........ | ........ | 1.9[illegible] |
| [illegible] | .250 | 4.73 | 21 | .265 | 5.57 | 21 | .274 | 5.75 |
| [illegible] | .1[illegible] | 6.[illegible] | 3[illegible] | .2[illegible] | 9.41 | 39 | .241 | 9.[illegible] |
| 31 | .1[illegible] | 3.75 | 31 | .131 | 4.96 | 40 | .149 | 5.96 |
| [illegible] | .0[illegible] | 5.61 | 1[illegible] | .0[illegible] | 5.6[illegible] | 122 | .0[illegible] | 5.15 |
| ........ | ........ | 4.[illegible] | ........ | ........ | 5.79 | ........ | ........ | 5.40 |
| ........ | ........ | 5.7[illegible] | ........ | ........ | 6.76 | ........ | ........ | 6.76 |
| ........ | ........ | 2.[illegible] | ........ | ........ | 2.90 | ........ | ........ | 2.90 |
| ........ | ........ | 3.[illegible] | ........ | ........ | 3.[illegible] | ........ | ........ | 3.[illegible] |
| ........ | ........ | 30.[illegible] | ........ | ........ | 42.4[illegible] | ........ | ........ | 44.39 |
| ........ | ........ | 4.[illegible] | ........ | ........ | 5.79 | ........ | ........ | 5.79 |
| ........ | ........ | [illegible] | ........ | ........ | 11.58 | ........ | ........ | 11.58 |
| ........ | ........ | 15.44 | ........ | ........ | 16.41 | ........ | ........ | 16.41 |
| ........ | ........ | 24.4[illegible] | ........ | ........ | 3[illegible].78 | ........ | ........ | 38.60 |
| ........ | ........ | 2[illegible].31 | ........ | ........ | 337.56 | ........ | ........ | 348.21 |

rest in budgets of this character is the information ng the manner in which the workingman spends his rom this, however, it is possible, through reference to ng the prices of commodities which go to make up the consumption of the workman, to determine whether ages shown in previous tables is counteracted or not in the cost of articles he is accustomed to purchase. the author, in part, on this table are as follows: bles it can be seen that the greatest variations occur

Workingmen in the north of France can now eat meat , while fifty years ago meat could not always be had

ing. Although the price of materials for clothing has later years, the expenditure demanded by this item of budget has sensibly increased. The same observa- ll classes of society. Formerly tastes were more sim- less changing. To-day the workingman feels called imself and family in a more elegant manner, and to ire, the frequent changes of fashion.

expenses. This class, in which are included the amusements and at the drinking places, has unfortu- creased. We are of the opinion that it now represents family expenditures. It is the drinking houses

## WORKINGMEN'S INSTITUTIONS.

The most important part of an investigation of [illegible] communities must necessarily be that of the study of the [illegible] in their mutual relations. Such a grouping of men and women with identical interests gives rise to opportunities for the organization of institutions for mutual and collective action that do not exist elsewhere.

These opportunities can be taken advantage of in three ways: either through the creation by the employer of institutions for the benefit of his employees, through the organization by the employees themselves of institutions for their mutual benefit, or through the cooperation of the employer with his employees for the purpose of securing improved conditions. It is a matter of fundamental importance which of these [illegible] policies is in each case pursued. There has been a distinct evolution of sentiment in regard to the question here involved. From the original position that they had no obligation toward their employees other than that of the payment of wages, employers, in many cases, went to the opposite extreme. Though they created admirable institutions for the benefit of their employees, they treated the latter as wards, and retained in their own hands an arbitrary management of their new creations. Since then the workingman has more and more demonstrated his ability to look after his own interests, and, conscious of his own capacities, he has chafed under any species of tutelage. The employer has in many cases responded to this new sentiment, and as far as possible has given over the management of his social institutions into the hands of those for whose benefit they are intended. At the same time the employees are encouraged to organize independent institutions of their own.

This development of self-help and self-reliance by no means resulted in a curtailment of the province or the variety of workingmen's institutions. On the contrary, the development of the association idea has rendered possible the exercise of mutual action in fields that it was impossible for employer-managed institutions to enter.

The workingmen's institutions at Anzin are a notable example of institutions organized on a liberal basis. Whether regarded from the standpoint of the generous sacrifices made by the company for their maintenance, or from that of its liberal attitude regarding the participation of employees in their management, the social institutions of Anzin are the most remarkable of any in France. A study of the constitutions and workings of these institutions will therefore be [illegible] great detail. These institutions are, however, by no means the [illegible] possession of Anzin. Though the same ensemble of institutions [illegible] in no other place in France, each one is but the representative [illegible] of institutions found at a great many of the other important [illegible] centers of the country. The study here made will, therefore, [illegible] the more importance, as it will represent a study, not of [illegible] have an isolated existence, but of those that have received [illegible] application and approval.

OF EMPLOYEES.

... employees by the large industrial ... practices is, in France, almost univer... ... in his report on workingmen's ... tion at Paris, 1889, "there is not in ... astry who has not made efforts to ... s." To secure this end four com... t into practice by employers:

... s in which their workingmen are ... t.

... es that are afterwards rented to the ...

... s and their sale to employees.

... ey or provision of land to employees ... houses of their own.

... hough the first is rarely practiced. ... have been directed in all of the last ... s that it rents at low rates; it has ... yees; it has advanced money to and ... employees that they might acquire ... ir individual tastes and needs. The ... ner than a single device are evident. ... are not similar. Some are satisfied ... willing to make the sacrifices neces... others eagerly embrace the oppor... and yet others prize highly the ... g houses upon plans selected by ...

... the Working People, issued by the ... workingmen's houses in France has ... ouses at Anzin are among those of ... There is thus no necessity of giv... r than a bare statement of what has ...

... company for the employees were ... it has bought or constructed, near ... at it rents to its employees. The ... 3 was 2,582.

... em of constructing houses in solid ... great disadvantage of this system ... however, soon led to its abandon ... houses were built. Regarding these ... M. Picot, in his report above cited, ... has been selected. * * * A cel... ting room and kitchen, a first floor ... of 200 meters (2,152.8 square feet);

such is the house that rents for from [illegible] per month, 42 to 72 francs ($8.11 to $13.90) per year. [illegible] 2,800 francs ($540.40). If the interest on this is [illegible], 112 francs ($21.62), and maintenance and taxes at [illegible] the rent ought to be 156 francs ($30.11). The company [illegible], 72 francs ($13.90), or a loss of 84 francs ($16.21) from a [illegible], which is equivalent to a loss, on 2,628 houses (the number at the end of 1888), of 220,752 francs ($42,605.14) a year. In other words, the company obtains a net gain of 28 francs ($5.40) from a rent of 72 francs ($13.90), or 1 per cent on the capital invested. In no other place have we found similar figures or efforts on so large a scale."

In 1867, in order to encourage saving among its workingmen, the company commenced the construction of isolated houses with gardens, which it sold to its employees at the cost of construction and the land. Those first erected were valued at from 2,200 to 2,700 francs ([illegible] to $521.10) each, but those erected later were of a better model and cost from 2,700 to 3,550 francs ($521.10 to $685.15). Payment for the houses was made in installments until the entire amount was paid. No interest of any kind was charged. Under this arrangement 93 houses had been erected up to 1893, at a total cost of 275,207 francs ($53,114.95). To supplement its former work, the company decided, in 1869, to commence the advancing of money to the most worthy employees who wished to buy or build houses for themselves. Here the same facility for reimbursement by partial payments was offered. The advances were also without interest. In 1888 the company had advanced a total sum of 1,446,604 francs ($279,194.57), of which all but 101,140.09 francs ($19,520.04) had been repaid, and a total of 741 houses had been constructed or otherwise acquired by its employees.

If to the 2,628 houses rented by the company there be added the 93 houses built by it and sold to the workingmen, and the 741 houses acquired through advances made by the company to the employees, there results a total of 3,462 houses that had been provided through the efforts of the company in 1888, and the number remained practically unchanged in 1893. In 1888 the company estimated that it had lost rent to the amount of 84 francs ($16.21) per year on each of [illegible] houses, or a total of 220,752 francs ($42,605.14); interest on houses sold and not paid for, 3 per cent on 67,558.12 francs ($13,038.72), or [illegible] francs ($391.16); interest on sums still due on advances made to build, 1,022.20 francs ($197.28); making the total cost of its effort for the housing of its employees 223,800.95 francs ($43,193.58).

## OLD-AGE PENSIONS.

The continuous existence, during a long period of years, of [illegible] industrial establishment employing thousands of men [illegible] relations of responsibility on its part for the welfare of its [illegible] that do not exist where an industry is carried on on a [illegible] *smaller scale.* In a company such as that of Anzin [illegible]

[illegible] them for work. Sons have suc- [illegible] succeeded by their sons. It is diffi- [illegible] appreciate the importance that the European [illegible] this question of the provisions against old [illegible] within the last ten years by Germany, [illegible] of the State have been used to secure this [illegible] importance of the problem and its acuteness at the [illegible] France the efforts made by the company of Anzin [illegible] employees are the most important made by a [illegible].

[illegible] the company followed the practice of according to [illegible] had grown old in its service, and had become unable [illegible] for the remainder of their lives, that constituted a [illegible] the resources of the company. The employees par- [illegible] in the regulation of these pensions, nor was any [illegible] wages during previous years retained to aid in the [illegible] insurance fund. There were serious objections to [illegible] workingmen did not like the feeling that they were [illegible] bounty of the company; and the pension being [illegible] financial prosperity of the company, they did not feel [illegible]ty that an independent insurance scheme would have [illegible]nizing this, on January 1, 1887, the company inaugu- [illegible]stem of old-age pensions. By this system it frankly [illegible]men into partnership and provided for the constitution, [illegible] sacrifices, of an insurance fund that should be wholly [illegible] the company's funds or management. Though the [illegible] sacrifices equal in amount to those under the old sys- [illegible]n was no longer a bounty but a right to which the [illegible]ould acquire a title by years of voluntary sacrifices. [illegible]he more remarkable, for in 1894 the Government, as we [illegible]ed in toto its principles in framing its law regulating [illegible] miners generally throughout France. The regulations [illegible] the company concerning the granting of pensions might [illegible] have formed a model after which the French law was

[illegible] provisions of these regulations may be summarized in [illegible]ragraphs:

[illegible] January 1, 1887, the company agreed to deposit in the [illegible] for Old-Age Pensions,(a) in the name of each working-

[illegible] Bank for Old-Age Pensions is a State institution created in 1850 [illegible] by the law of July 20 and decree of December 28. Its opera- [illegible] by the Government and controlled by a commission organized [illegible] of Commerce, Industry, Posts, and Telegraphs. Its object is [illegible] by small annual payments the right to a life pension, [illegible] ($[illegible]1.60) as a maximum, at the age of 50 years, or later, as [illegible]. The especial idea of Parliament was to offer to the ordi- [illegible] to insure *for himself* through a small regular deduc-

man who would make an equal payment, [illegible] the wages of the workingman. The two deposits [illegible] an individual account book, which remains the property of the workingman. For workingmen employed below ground the payments of the company commence from the time of their entering upon work. For those employed above ground the payments of the company commence when the workingmen are at least 18 years of age and have been in the employ of the company during three years. The payments of the company cease when the workingman has reached the age of 50 years. The latter, however, can defer the enjoyment of his pension, if he so desires, by continuing his personal payments. Through these payments the workingman is enabled to acquire the right to an annuity from the National Bank for Old-Age Pensions, on reaching the age of 50 years, for the remainder of his life. In case of permanent disability before reaching that age, he enters into the immediate enjoyment of a pension proportionate to his age and the amount of deposits to his credit.

In addition to these provisions whereby the company agreed to contribute toward the acquisition of pensions by workingmen an amount equal to their own payments, the company further provided for the increase of these pensions as a reward for long and faithful service. When a workingman has fulfilled the double condition of being at least 35 years of age and has been ten years without interruption in the employ of the company, a special account is opened with him for the succeeding years of his connection with the company, or until he has reached the age of 55 years, or has been retired on his pension. The total can in no case exceed fifteen years. For each of these years a special supplement to the pension, when due, of 3 francs (58 cents) for workingmen below and 1½ francs (29 cents) for those above ground will be added. The total supplementary pension, except in cases of severe injuries or infirmities contracted during work, is not paid unless the workingman remains with the company until he is 50 years of age. The latter, also, can not enter upon the enjoyment of his supplementary pension until he ceases to work for the company. For workingmen employed as overseers below ground the supplemental pension is raised

tion from his wages a provision for his old age. To this end the bank receives deposits of the smallest amounts which are increased by the accumulation of interest. With the exception of an amount sufficient to meet the daily payments, all the funds are invested in bonds of the French Government or other obligations guaranteed by the State. All interest thus earned is placed to the credit of the depositors. No deduction is made for the expenses of administration. In case of absolute incapacity to work, as the result of permanent infirmities, the depositor enters immediately into the enjoyment of a pension calculated according to his age and the amount of the payments he has made. Insurance through this institution is purely optional. No engagement, moreover, is entered into by the depositor as regards the amount or frequency of his payments. He can interrupt, diminish, or increase [illegible] as he desires. A separate account is kept with each depositor. The [illegible] has been made of this bank is its utilization by large industrial [illegible] aid societies to provide for the insurance of their employees [illegible]

pplemental pension is doubled.
le in all cases for the payment of pensions to
. Transitory provisions make special arrange-
already in the employment of the company but
iem from acquiring pensions according to the
in the new regulation.
and at the same time admirable provision of the
reby the service of the old-age pensions proper
m the control of the company. There is thus
re principle. Each workingman knows exactly
ards his ultimate right to a pension, and feels
pension is in no way dependent upon his remain-
e company. Independence is thus not sacrificed

regulations was purely optional. Ninety-five
gmen, however, recognized the great advantages
signified their approval. The following table
the efforts of the company for pensioning its
ie years immediately preceding the adoption of
he years succeeding, including 1893. The table
e ages of all pensioners of the company on March

ITURES FOR PENSIONS OF THE COAL MINING COMPANY OF ANZIN, 1883 TO 1893.

| d employees. | | Pensions to widows of employees. | | | Amount paid by company to National Bank for Old-Age Pensions. | Total amount expended by company for pensions. |
|---|---|---|---|---|---|---|
| al ons. | Average pension. | Pensioners. | Total pensions. | Average pension. | | |
| 0.41 | $35.80 | 651 | $13,426.80 | $20.62 | ............ | $38,917.21 |
| 1.50 | 35.81 | 621 | 13,417.29 | 21.61 | ............ | 41,848.79 |
| 9.10 | 36.18 | 633 | 13,687.98 | 21.62 | ............ | 53,417.08 |
| 3.51 | 36.90 | 664 | 14,468.02 | 21.79 | ............ | 56,201.53 |
| 9.56 | 37.10 | 684 | 14,714.28 | 21.51 | $5,899.82 | 63,953.66 |
| 3.68 | 38.24 | 686 | 14,857.83 | 21.66 | 20,926.55 | 83,707.46 |
| 0.82 | 38.38 | 725 | 15,654.84 | 21.59 | 26,023.73 | 90,769.39 |
| 1.30 | 38.27 | 740 | 15,961.05 | 21.57 | 31,436.81 | 96,970.06 |
| 6.14 | 38.97 | 787 | 16,616.28 | 21.11 | 55,938.73 | 123,301.15 |
| 2.52 | 39.07 | 804 | 17,072.34 | 21.23 | 55,736.86 | 126,261.72 |
| 0.06 | 39.69 | 827 | 17,400.31 | 21.04 | 56,775.39 | 128,915.76 |

SIONED BY THE COAL MINING COMPANY OF ANZIN, MARCH 15, 1894.

| ge. | Number. | Age. | Number. | Age. | Number. | Age. | Number. | Age. | Number. |
|---|---|---|---|---|---|---|---|---|---|
| 47 | 2 | 56 | 58 | 65 | 56 | 74 | 20 | 83 | 5 |
| 48 | 7 | 57 | 69 | 66 | 44 | 75 | 13 | 84 | 3 |
| 49 | 3 | 58 | 86 | 67 | 60 | 76 | 11 | 85 | 1 |
| 50 | 12 | 59 | 73 | 68 | 47 | 77 | 12 | 86 | 1 |
| 51 | 36 | 60 | 70 | 69 | 44 | 78 | 8 | 87 | 3 |
| 52 | 27 | 61 | 54 | 70 | 41 | 79 | 12 | 88 | 1 |
| 53 | 41 | 62 | 65 | 71 | 25 | 80 | 5 | 89 | 1 |
| 54 | 58 | 63 | 66 | 72 | 22 | 81 | 7 | | |
| 55 | 59 | 64 | 73 | 73 | 27 | 82 | 4 | Total. | 1,363 |

In explanation of the first [illegible] the average amount of the pensions [illegible] granted by the company as reward for [illegible] this, since 1887 the workingmen have been [illegible] the National Bank for Old-Age Pensions. [illegible] company paid for this purpose the sum of 294,[illegible] in connection with this the workingmen have to [illegible] the total amount paid into the national bank being, [illegible] francs ($113,550.78).

The year 1893 practically closes the record of voluntary [illegible] efforts on the part of mine owners to pension their old [illegible] widows.

Mention has been made of a general law concerning old-age [illegible] for mine employees. This law was passed June 29, 1894, and by [illegible] insurance of all mine employees was made obligatory.

According to it each mine operator was required to pay [illegible] National Bank for Old-Age Pensions, or into a special institution [illegible] by the operator for his own employees or in connection with [illegible] operators for the mutual insurance of their employees, the [illegible] of which had received the authorization of the Government, [illegible] of each workingman the wages of whom did not exceed 2,400 [illegible] ($463.20) a year, a sum equal to 4 per cent of the latter's wages, [illegible] of which was to be deducted from the wages of the workingman [illegible] the other half be borne directly by the operators. These [illegible] then were devoted to the ultimate acquisition of an old-age [illegible] according to the regular rules of the National Bank for Old-Age Pensions.

For the company of Anzin and its employees, it will be [illegible] the law made necessary but few changes. The company was [illegible] making such payments to the amount of 3 per cent of its [illegible] wages, in addition to the supplemental pensions for length of [illegible]. The general effect of the law was to make obligatory upon all [illegible] owners the adoption of a system that had been practiced at [illegible] since 1887.

The obligation to maintain aid societies for the aid of [illegible] injured workingmen was likewise imposed upon mine operators [illegible] law, but this provision should be considered in connection [illegible] account of the mutual aid societies of Anzin that follows.

### MUTUAL AID SOCIETIES.

A temporary relief organization is a necessary [illegible] system of old-age and invalidity insurance. The latter [illegible] for the time when employees, through old age or disability, [illegible] longer to earn wages. The former provides for temporary [illegible] or misfortune. Experience has demonstrated that it is [illegible] keep these two services distinct from each other. The [illegible]

...has developed in almost every commune and village ... mutual aid societies (sociétés de secours ... organization and operation afford a study of the most ... that has been developed in France for improving ...e laboring classes. They are to France what the ...s are to America. Though ministering to different ...ilar to our building associations in that they are ...aged for the most part by the workingmen them- ... the workingmen together for purposes of mutual ...ir methods of operation and control are simple, and ...ministration are reduced to a minimum. M. Lafitte, ... matters relating to mutual aid societies, thus tersely ...s (a): "Mutual aid societies have for their essential ... to their members when sick medical attendance and ...s; to pay to them a daily sum of money during their ... to them a small pension after they have reached a ... defray the expense of a suitable burial on their ... this each member pays into the funds of the society ...n, as dues, usually divided into monthly or weekly

...ave had an existence in France for over fifty years. ...ment has encouraged their organization through the ...al laws, and a general control is exercised over their ... a special bureau created in 1852 under the Depart- ...r. Annual reports are made by the societies to this ...u issues an annual report on their operations dur- ... important part that these societies play in the life ... of France is shown in the following statistics of ...ount of business for 1892, the latest year obtainable:

| | |
|---|---|
| ... | 9,662 |
| ... | 248 |
| ... | 1,503,397 |
| ... | 31,112 |
| ... | $6,052,520.53 |
| ... | 138,846.90 |
| ... | 37,816,056.58 |
| ... | 2,383,582.23 |

... whole question of mutual aid societies in France ...ing one. Here only a brief statement of their gen- ... importance has been given in order that the purpose ... exist at Anzin may be understood.

---

...ationnelle des Sociétés de Secours Mutuels, Paris, 1892.

Seven mutual aid societies [illegible] the company. The constitutions [illegible] identical.

Briefly stated, each constitution [illegible] workingmen into a society by which, for [illegible] centimes ($9\frac{6}{10}$ cents) a month, the members [illegible] sickness or accident to free medical [illegible] benefit of 1 franc ($19\frac{3}{10}$ cents) during [illegible] sickness is not more than one year in duration, [illegible] relief, such as burial expenses, etc. To accomplish [illegible] receipts of the society from dues have to be largely [illegible] other sources. The company, therefore, turns over [illegible] all fines collected by it for the infraction of any of [illegible] practice that removes the criticism often made [illegible] selfish purposes, and also makes to it liberal gifts [illegible] the society are not sufficient to meet its obligations.

In the following tables the combined operations of the [illegible] ties at Anzin during the eleven years from 1888 to 1898 [illegible] From the first table it will be seen that the number of [illegible] are members of the societies is now over 8,000, or 94 per [illegible] eligible for membership. The increasing percentage [illegible] of those eligible for membership who have become [illegible] increasing appreciation of the benefits that the societies [illegible]

In the table of receipts and expenditures two points are [illegible] special attention. The first is the very small percentage [illegible] expense of administration is of the total expenditures of [illegible] the average for the period covered being less than 2 per [illegible] second is the policy of creating as rapidly as possible a [illegible] fund, so that the societies may always have on hand an [illegible] fund and at the same time profit from interest on their invest[illegible] this connection there should be noted a feature that [illegible] workingmen's institutions whether organized at Anzin or [illegible] France, and that is, that a system of mutual assessment [illegible] sickness or death has never found the slightest favor with [illegible] ingmen. They desire to know in all cases the exact amount [illegible] obligations. In other words, they prefer the system of [illegible] There are no details of the table that are worthy of special [illegible] The growth of receipts and expenditures has been normal, [illegible] to the growth in membership.

| | Employees eligible but not members. | Employees eligible for membership. | Per cent of active members of eligible employees. | Days of sickness. | | Expenditure per active member. |
|---|---|---|---|---|---|---|
| | | | | Total. | Per active member. | |
| | 5,206 | 7,780 | 33 | 26,751 | 10.4 | $1.78 |
| | 4,259 | 7,429 | 43 | 28,097 | 8.9 | 1.55 |
| | 2,630 | 7,300 | 64 | 40,932 | 8.8 | 2.00 |
| | a 1,627 | a 7,573 | 78 | 60,623 | 10.2 | 2.58 |
| | 1,352 | 7,648 | 82 | 63,590 | 10.1 | 2.57 |
| | 1,126 | 7,676 | 85 | 74,162 | 11.3 | 2.74 |
| | 933 | 7,835 | 88 | 75,498 | 10.9 | 2.74 |
| | 1,044 | 8,649 | 88 | 97,202 | 12.8 | 3.21 |
| | 1,076 | 8,128 | 88 | 89,019 | 11.6 | 2.73 |
| | a 590 | a 8,800 | 93 | 95,851 | 11.7 | 2.99 |
| | 541 | 8,926 | 94 | 114,728 | 13.7 | 3.71 |

and employees eligible but not members does not produce the
The figures are given, however, as reported to the Department.

## OF MUTUAL AID SOCIETIES AT ANZIN, 1883 TO 1893.

| | 1883. | 1884. | 1885. | 1886. | 1887. |
|---|---|---|---|---|---|
| ....... | $1,930.22 | $2,275.83 | $2,918.62 | $6,537.09 | $8,338.72 |
| ....... | 68.32 | 354.35 | 683.41 | 732.63 | 709.40 |
| ....... | 4,796.44 | 4,975.16 | 8,792.16 | 12,102.21 | 12,699.74 |
| ieties. | ........... | 1.16 | 1,700.95 | 2,527.21 | 2,241.57 |
| ....... | ........... | 19.30 | 1,042.20 | 1,166.68 | 1,209.76 |
| ....... | 7.29 | 27.84 | 146.49 | 247.67 | 128.25 |
| ....... | 35.73 | 34.91 | 119.90 | 93.35 | 189.14 |
| ....... | 14.73 | 140.14 | 494.69 | 313.33 | 183.79 |
| ....... | 6,852.73 | 7,828.69 | 15,898.42 | 23,720.17 | 25,880.37 |
| ....... | 59.25 | 86.72 | 136.32 | 238.93 | 259.10 |
| ....... | 4,392.59 | 4,560.98 | 7,025.30 | 11,033.23 | 12,109.74 |
| ....... | 85.69 | 123.33 | 183.35 | 217.90 | 186.63 |
| nced. | 34.74 | 73.34 | 1,871.13 | 3,139.43 | 3,142.85 |
| ....... | 4.63 | 65.70 | 145.23 | 751.96 | 480.23 |
| ....... | 2,275.83 | 2,918.62 | 6,537.09 | 8,338.72 | 9,701.82 |
| ....... | 6,852.73 | 7,828.69 | 15,898.42 | 23,720.17 | 25,880.37 |

| 8. | 1889. | 1890. | 1891. | 1892. | 1893. |
|---|---|---|---|---|---|
| .82 | $9,455.45 | $10,807.56 | $8,903.26 | $11,152.59 | $13,004.34 |
| .14 | 720.28 | 681.00 | 833.95 | 819.96 | 774.99 |
| .96 | 13,748.74 | 15,243.39 | 16,793.80 | 18,117.10 | 17,856.94 |
| .52 | 2,195.65 | 2,439.78 | 3,103.62 | 2,553.99 | 6,795.87 |
| .92 | 2,528.69 | 3,396.80 | 1,779.46 | 4,272.85 | 4,517.38 |
| .97 | 128.92 | 173.00 | 91.24 | 103.83 | 46.90 |
| .81 | 271.27 | 236.77 | 264.86 | 341.07 | 290.66 |
| .62 | 649.22 | 350.76 | 1,375.80 | 136.98 | 984.83 |
| .76 | 29,698.22 | 33,329.06 | 33,145.99 | 37,498.37 | 44,271.91 |
| .81 | 352.19 | 385.62 | 510.86 | 520.45 | 643.58 |
| .21 | 13,267.50 | 17,940.64 | 15,427.55 | 17,451.74 | 21,144.21 |
| .73 | 254.10 | 334.08 | 454.52 | 429.04 | 825.07 |
| .47 | 4,548.26 | 5,220.17 | 3,976.67 | 5,666.88 | 6,729.94 |
| .09 | 468.61 | 545.29 | 1,623.78 | 425.92 | 1,769.02 |
| .45 | 10,807.56 | 8,903.26 | 11,152.59 | 13,004.34 | 13,160.09 |
| .76 | 29,698.22 | 33,329.06 | 33,145.99 | 37,498.37 | 44,271.91 |

The enactment of the law of June 29, [illegible] changes in the organization of these societies. [illegible] organization of aid societies for the relief of employees in case of sick-ness and accidents obligatory upon all mine operators. While it pro-vided that wherever such societies were already in existence it would not be necessary to organize new societies, it required that they should be reorganized to such an extent as to comply with the provisions of the law. The chief innovation required was that concerning the sources of receipts. While heretofore the company had voluntarily contributed to the resources of the societies through the turning over to them of the product realized from fines, and through gifts from time to time, it was now rendered obligatory upon it to contribute toward the main-tenance of the fund to an equal extent with the workingmen. This was in accordance with the provision of the law, which provided that the receipts of such societies should come from the following five sources: (1) A deduction from the wages of each employee, the amount of which is determined by the administration of the society and which can not exceed 2 per cent of his wages; (2) an equal payment by the mine operator; (3) subsidies granted by the State; (4) gifts and legacies, and (5) the product of fines for the infraction of certain regulations relating to the conduct of work below ground.

The law further made certain general provisions regarding the nature of the relief to be granted, for the government of the society through the mutual participation of the members and of representatives of the mine owners, etc. These sections, however, left a great deal of liberty to the individual societies and necessitated but few changes in societies already in existence.

## COOPERATIVE DISTRIBUTIVE SOCIETY.

The organization of cooperative stores in the mining centers of France has, in a number of instances, achieved notable success. At Anzin a society was organized as early as 1865, and it is thus one of the first cooperative distributive societies created in France. Its creation was the direct result of the influence of the success of the Rochdale Pioneers in England. Operations commenced in 1865, but the society, properly speaking, had a legal existence only after December [illegible], 18[illegible], when it took advantage of the law of July 24–29, 1867, to become a legal corporation.

The organization of the society is that of a joint stock company, but on such a basis that all speculative interest in the stock is eliminated. Each member of the society is required to own one, and can not own more than two shares of stock. The value of the shares of [illegible] is 50 francs ($9.65) each. The society is absolutely independent of [illegible] control by the company of Anzin. Membership is strictly [illegible] employees of the company, and only members can trade at the [illegible] The object of the society has always been to buy merchandise [illegible]

are purchased directly from the producers
of the middleman, a considerable profit is
each year for distribution among the mem-
it of necessary running expenses the profits
e payment of a 5 per cent dividend on the
purchasers in proportion to the value of their
ies.

has developed rapidly. The two following
years since organization, the membership of
of capital, the amount of dividends paid, the
amount of profits earned, and the proportion
he value of all sales:

DIVIDENDS OF THE COOPERATIVE DISTRIBUTIVE
IETY OF ANZIN, 1866 TO 1894.

| al. | Dividends paid. | Year. | Members. | Capital. | Dividends paid. |
|---|---|---|---|---|---|
| .05 | $96.04 | 1881 | 2,544 | $24,549.60 | $1,144.45 |
| .25 | 148.43 | 1882 | 2,625 | 25,331.25 | 1,206.16 |
| .00 | 300.02 | 1883 | 2,373 | 22,899.45 | 1,249.92 |
| .05 | 654.75 | 1884 | 2,683 | 25,890.95 | 1,223.91 |
| .65 | 695.28 | 1885 | 2,928 | 28,255.20 | 1,303.94 |
| .50 | 786.48 | 1886 | 3,021 | 29,152.65 | 1,396.01 |
| .35 | 835.73 | 1887 | 3,043 | 29,364.95 | 1,412.52 |
| .95 | 791.50 | 1888 | 3,123 | 30,136.95 | 1,441.57 |
| .90 | 935.15 | 1889 | 3,227 | 31,140.55 | 1,482.47 |
| .35 | 962.08 | 1890 | 3,319 | 32,028.35 | 1,535.72 |
| .60 | 987.82 | 1891 | 3,386 | 32,674.90 | 1,574.28 |
| .55 | 978.62 | 1892 | 3,497 | 33,746.05 | 1,605.87 |
| .19 | 978.86 | 1893 | 3,629 | 35,019.85 | 1,684.19 |
| .70 | 1,016.28 | 1894 | 3,760 | 36,284.00 | 1,758.96 |
| .90 | 1,077.13 | | | | |

...quired to be granted in such cases, viz, medical ...food for the sick, such as bouillon and wine, and

...is made between the different services, and espe... ...orkingmen employed below and those employed ...ds the amount of assistance granted. A special ...exercised over the miners proper and the other ...below ground. In the case of the former the ...sistance is extended to their wives, children, and ...n them at the time. To workingmen employed ... attendance is not extended to the other members ... medical supplies are issued for their use. Medi... ...dical supplies are also granted to all working... ...roll, provided their individual pensions do not ...($193) per annum. Regarding the granting of ...npany has issued the following regulations and

...es of severe wounds, such as fracture of the ...rus, accidents to the head affecting the brain, ...s the result of an accident, serious burns from ... of fire damp, or wounds causing the loss of one ...ll be granted per fortnight to married working... ...; to unmarried workingmen, other than putters, ...l to putters 10 francs ($1.93).

...se of fracture of the clavicle without internal ... of the forearm, mutilation of the fingers or toes, ...s endangering the sight, to married workingmen, ...unmarried workingmen, other than putters, 10 ...putters, 6 francs ($1.16).

...e of slight injuries of any kind, to married work... ...1.93); to unmarried workingmen, other than ...cents), and to putters, 3 francs (58 cents).

...ion of the health service the company has in its ...ians, each of whom, with the exception of one, ...o diseases of the eye, has his particular district. ...a room, where, during certain hours, he receives ...e to him. The other sick are cared for in their ...e and carriage is furnished each physician with ...its.

...tailed by the maintenance of this medical and ...ive table on page 255.

### SAVINGS BANK.

But a few words will be required concerning the [illegible] the company of savings among its employees. Previous to the passage of the act of 1881 creating a national postal savings bank, the company of Anzin, in common with other industrial companies, stimulated in every way the spirit of saving among its employees. To this end the company created, in 1869, a savings bank in which its employees were encouraged to deposit their savings and upon which they received interest at the rate of 5 per cent for amounts under 2,000 francs ([illegible]) and 4 per cent for amounts over that sum. The national savings bank created in 1881 offered all necessary inducements and guaranties of security. The company therefore reduced the rate of interest paid to 3 per cent, the same as that paid by the Government, and ceased to encourage deposits in its own bank. Thus the number of depositors, that had reached in 1877 a total of 1,431, with deposits of [illegible] francs ($374,422.27), has now sunk to less than one-third that number and the amount of deposits has been correspondingly decreased.

### EDUCATION.

The company first began to occupy itself with the provision of school facilities for the children of its employees in 1873. Previous to [illegible], the year in which the law providing for free public instruction was passed, the efforts of the company represented a considerable expense. It erected numerous infant and primary schools and, in connection with the latter, workshops for manual training. At the present time, however, it possesses actually but one school, and its total expenditure for schooling in 1888 was 31,875.45 francs ($6,151.96), divided as follows:

| | |
|---|---|
| Subsidies to teachers and infant schools | [illegible] |
| Fuel for schools | [illegible] |
| Prizes in the form of books and savings-bank deposits | [illegible] |
| Maintenance of its own school | [illegible] |
| Total | [illegible] |

In addition it maintains a special advanced school conducted by its own engineers for the purpose of educating skilled workingmen, the pupils of which are taken from among the best scholars of the primary schools. The company also pays the tuition and board of [illegible] young workingmen at the school for boss miners at Douai ([illegible] Maitres-mineurs de Douai).

Four churches have also been erected and are now owned by the company. All are consecrated to the Catholic faith.

### MISCELLANEOUS AID TO WORKINGMEN.

In addition to the various ways for aiding workingmen [illegible] been enumerated, the company contributes to their support [illegible] of ways that can not well be classified. Of these, the [illegible]

[illegible] of 50 centimes (9⅗ cents) per day [illegible] the latter's period of military service, [illegible] (1⅗ cents) per day for each child they [illegible] orphans until they are of an age to commence [illegible] workingmen generally and their families in [illegible]

[illegible] the company for these purposes, as well as for [illegible] medical service for the eleven years, 1883 to [illegible] given in the following table:

[illegible] COAL MINING COMPANY OF ANZIN FOR MEDICAL [illegible] SERVICE, AID, ETC., 1883 TO 1893.

| [illegible] | [illegible] | Employees in need. | Wives and children of employees doing military service. | First working suits, grants to mothers on first communion of their children, etc. | Total expenditures (aid to employees and medical services). | Value of fuel given to employees and pensioners. |
|---|---|---|---|---|---|---|
| [illegible] | [illegible] | $1,[illegible] | $1,[illegible]3.47 | $1,537.87 | $56,270.85 | $87,764.32 |
| [illegible] | 12,122.64 | 1,[illegible] | 1,[illegible].61 | 1,502.44 | 52,343.39 | 66,923.60 |
| [illegible] | [illegible] | 4,[illegible] | 1,[illegible].11 | 1,042.[illegible] | 53,476.52 | 59,845.48 |
| [illegible] | 12,[illegible] | [illegible] | 1,758.[illegible] | 1,243.[illegible] | 52,762.80 | 63,679.00 |
| [illegible] | 16,[illegible].12 | 3,200.76 | 1,441.27 | 1,287.43 | 52,139.86 | 69,763.71 |
| [illegible] | 18,[illegible].97 | 2,[illegible] | 1,[illegible]0.02 | 1,206.72 | 52,497.17 | 69,212.09 |
| [illegible] | 19,[illegible] | 4,[illegible].18 | 1,270.95 | 1,288.70 | 55,350.38 | 71,991.70 |
| [illegible] | 9,072.24 | 5,[illegible].46 | 1,186.28 | 1,314.[illegible] | 61,006.25 | 76,544.96 |
| [illegible] | [illegible] | 4,[illegible].56 | 1,[illegible]0.65 | 1,376.40 | 61,011.05 | 81,[illegible].26 |
| [illegible] | 9,748.05 | 2,270.74 | 2,[illegible].61 | 1,210.42 | 63,615.14 | 84,386.89 |
| [illegible] | 19,[illegible].47 | 3,907.96 | 1,188.97 | 1,386.36 | 73,306.40 | 81,249.91 |

OF THE EXPENDITURES OF THE COAL MINING ANZIN FOR THE BENEFIT OF ITS EMPLOYEES.

en made in considerable detail in the foregoing pages s institutions in which the workingmen of Anzin par- ing the general question of workingmen's institutions know what is the total expense that the maintenance s entails upon the company, what pecuniary advan- of these efforts confers upon the workingmen, and latter bears to the amount they receive in the way of ulation of the total expenditures of the company for mployees has therefore been made in such a way as facts for the eleven years, 1883 to 1893.

[illegible] is another one similarly constructed, giving the [illegible] a number of the most important coal mining com- [illegible] of Anzin. The data for this table were taken [illegible] L'Organisation du Travail dans les Mines et Par- [illegible] Houillères, by Charles Ledoux, engineer in chief [illegible] relate to the year 1888.

From the second table it is evident [illegible] means holds a unique position in respect [illegible] ingmen's institutions. The existence of [illegible] described for Anzin is almost universal [illegible] companies of France. Wherever material for [illegible] at Anzin with those existing in other mining [illegible] obtained, the original statement, that in choosing [illegible] study a typical center had been selected, seems to be [illegible]

EXPENDITURES OF THE COAL MINING COMPANY OF [illegible] EMPLOYEES, BY OBJECTS OF [illegible]

| Year. | Expenditures for— | | | | | | Total expenditures. | [illegible] |
|---|---|---|---|---|---|---|---|---|
| | Pensions. | Housing. | Schools. | Medical service. | Free fuel. | Direct pecuniary and other aid. | | |
| 1883 | $38,917.21 | $45,159.70 | [illegible] | [illegible] | [illegible] | [illegible] | [illegible] | [illegible] |
| 1884 | 41,848.79 | 44,763.66 | 7,169.10 | [illegible] | [illegible] | [illegible] | [illegible] | [illegible] |
| 1885 | 53,417.08 | 44,349.28 | 4,750.[illegible] | [illegible] | [illegible] | [illegible] | [illegible] | [illegible] |
| 1886 | 56,201.52 | 42,324.60 | 5,255.30 | [illegible] | [illegible] | [illegible] | [illegible] | [illegible] |
| 1887 | 64,953.66 | 42,878.19 | 5,194.55 | 28,716.[illegible] | [illegible].71 | [illegible] | [illegible] | [illegible] |
| 1888 | 83,707.45 | 42,193.66 | 5,223.51 | [illegible] | [illegible] | [illegible] | [illegible] | [illegible] |
| 1889 | 90,769.40 | 43,125.42 | 5,160.57 | [illegible].21 | 71,[illegible] | [illegible].17 | [illegible] | [illegible] |
| 1890 | 96,970.05 | 42,401.23 | 5,121.47 | 27,[illegible].79 | 76,544.[illegible] | [illegible] | [illegible] | [illegible] |
| 1891 | 123,301.15 | 43,284.97 | 5,342.25 | 28,643.[illegible] | 81,[illegible] | [illegible] | [illegible] | [illegible] |
| 1892 | 126,261.71 | 42,221.30 | 5,452.11 | 30,209.12 | 84,[illegible] | [illegible] | [illegible] | [illegible] |
| 1893 | 128,915.76 | 42,179.65 | 5,447.67 | 34,737.[illegible] | 81,[illegible].81 | 37,[illegible].41 | [illegible] | [illegible] |
| Total | 904,263.79 | 475,861.86 | 62,377.22 | 290,621.83 | 813,516.[illegible] | 342,[illegible] | a[illegible] | [illegible] |
| Average for the 11 years. | 82,205.80 | 43,261.99 | 5,670.66 | 26,420.17 | 73,[illegible] | 31,166.46 | [illegible] | [illegible] |

a This total is $12,616.99 more than the sum of the items. The explanation [illegible] figures are given as reported to the Department.

b See preceding note.

EXPENDITURES OF VARIOUS COAL MINING COMPANIES IN FRANCE FOR BENEFIT OF EMPLOYEES, BY OBJECTS OF EXPENDITURE, 1888.

| Name of company. | Expenditures for— | | | | | | Total expenditures. | [illegible] |
|---|---|---|---|---|---|---|---|---|
| | Pensions. | Housing. | Schools. | Medical service. | Free fuel. | Direct pecuniary and other aid. | | |
| Anzin | $83,707.45 | $43,193.66 | $5,223.51 | $23,946.22 | $69,312.0[illegible] | $[illegible],550.94 | [illegible] | [illegible] |
| Douchy | 8,051.60 | 12,886.94 | 227.16 | 2,688.77 | 9,919.10 | 7,017.[illegible] | [illegible] | [illegible] |
| Liévin | 181.03 | 35,291.63 | 5,268.28 | 5,519.55 | 11,904.[illegible] | 7,7[illegible] | [illegible] | [illegible] |
| Bessèges | 8,370.17 | .......... | 7,561.91 | 7,847.96 | 20,059.77 | 22,[illegible] | [illegible] | [illegible] |
| Blanzy | 10,251.46 | 30,176.03 | a33,701.11 | 7,776.97 | 61,716.[illegible] | 50,554.[illegible] | [illegible] | [illegible] |
| Courrières | 3,849.76 | 21,616.00 | a5,450.38 | 8,479.40 | 24,704.[illegible] | 7,[illegible].57 | 71,[illegible] | [illegible] |
| Béthune | .......... | .......... | .......... | .......... | .......... | .......... | [illegible] | [illegible] |
| Nœux | .......... | .......... | .......... | .......... | .......... | .......... | 114,[illegible] | [illegible] |
| Lens (b) | .......... | .......... | .......... | .......... | .......... | .......... | 1[illegible] | [illegible] |
| Montrambert | .......... | .......... | .......... | .......... | .......... | .......... | 40,146.[illegible] | [illegible] |
| Roche-la-Molière et Firminy | .......... | .......... | .......... | .......... | .......... | .......... | [illegible] | [illegible] |

a This includes contributions for religious purposes.

b The figures given are for 1886.

From the first of these tables it is possible to follow [illegible] way the progress of institutions and work undertaken [illegible] of Anzin for the benefit of its employees. Taken in [illegible]

...the amount increased every year, with ...attained at 134.77 francs ($26.01) in 1893. ...further examined in order to determine the par... ...expenditures that are responsible for this increase, it ...the item of pensions accounts for the entire augmen... ...have increased from 201,643.57 francs ($38,917.21) in ...than that sum, or 667,957.30 francs ($128,915.76) ...it is but illustrative of the direction toward which ...has the attention of employers and employees alike ...has been turned during the past decade for the ...the condition of the working classes. The one great ...ing to better the condition of these classes is to ...ne way or other provision shall be made for working... ...ave become old and incapacitated for labor. At the ...nearly a consensus of opinion has been reached that ...accomplished through an insurance system to be main... ...e mutual efforts of employers and their employees.

...n of this table has been introduced in order to show, ..., the relative value of these supplemental advantages ... the amount the workingmen receive in the way of ...eaking. It is a matter of considerable importance in ...whole question of workingmen's institutions, to deter... ...h their development the tendency is for the working... ...arger or smaller portion of their reward in this indirect ...hrough the payment of cash wages. The information ...direct upon this point. Though the absolute value of ...y advantages per member has increased from 89.83 ... 1883 to 134.77 francs ($26.01) in 1893, this represents ...ater percentage of wages in later than in earlier years. ...he percentage of supplementary advantages of wages, ...ave been abnormally low. In 1884, however, it was ... 1 per cent less than it was ten years later, in 1893, ...age of supplementary advantages of wages was 10.84.

## CONCLUSION.

...now been gained from which it is possible to make a ...f the life of miners and other mine employees at ... years of age the future workingman attends a ...by the joint efforts of the State and the mining com... ...When 13 he enters the employ of the company, and ...ge surface work, such as sorting or washing coal. ...few years he is drafted below ground and commences ...a miner. Meanwhile he has become a member ...society, from which he is entitled to receive certain ...sickness or accident. If at all industrious he can

they have been employed, and the num-
tablishment and for what reason. Infor-
been secured for the mining company of

ave presented all the facts necessary for
e personnel at Anzin. In the main the
call for but little comment. In them is
y of employment which, under the condi-
uld seem almost impossible to exceed.
hen one considers that Anzin is in the
ing region of France and in close prox-
ium, and it would be easy for a miner to
of employment.
page 231, giving the ages of employees
r the company of Anzin, it will be seen
y is almost exclusively recruited by the
15 years of age. During the six years
nt of the total number of new working-
at age, and if there be omitted those
between the ages of 20 and 25, as the
r terms of military service, the statement
hirds of the employees of the company
tween the ages of 13 and 14 years.
llow show in detail the ages and length
t is important to know their ages, for, as
s, practically all of those who have been
o youthful that their longer employment

ature of a summary, with the element of
ws that 12.17 per cent of all employees
loyed 30 years, 27.63 per cent 20 years,
over. The percentages of all employees
pectively. If, in calculating these per-
30 years of age be omitted, it will be seen
of 94.58 have been employed 10 years,
yed 20 years, and 28.14 per cent 30 years

33 shows a calculation of the stability of
dates, in order to determine if there has
this respect in recent years. An estab-
ding to the number of its employees is
g the number of employees that have
ime, an element which should be taken
g to make any comparisons.
s of the reasons for which employees left
The number leaving voluntarily bears but

an insignificant relation to the ... larger number leaving resulted ... sity of entering the military service ...

EMPLOYEES OF THE COAL MINING COMPANY ... BY AGES AND YEARS OF SERVICE ...

| Age. | Years of service. | | | | | | | | | | |
|---|---|---|---|---|---|---|---|---|---|---|---|
| | Under 1. | 1 or under 2. | 2 or under 5. | 5 or under 10. | 10 or under 15. | 15 or under 20. | 20 or under 25. | 25 or under 30. | [illegible] | [illegible] | [illegible] |
| 12 | ...... | ...... | ...... | ...... | ...... | ...... | ...... | ...... | ...... | ...... | ...... |
| 13 | 141 | 8 | 2 | ...... | ...... | ...... | ...... | ...... | ...... | ...... | ...... |
| 14 | 126 | 224 | 16 | ...... | ...... | ...... | ...... | ...... | ...... | ...... | ...... |
| 15 | 55 | 132 | 256 | ...... | ...... | ...... | ...... | ...... | ...... | ...... | ...... |
| 16 | 43 | 39 | 310 | 1 | ...... | ...... | ...... | ...... | ...... | ...... | ...... |
| 17 | 50 | 26 | 222 | 140 | ...... | ...... | ...... | ...... | ...... | ...... | ...... |
| 18 | 24 | 22 | 66 | 246 | ...... | ...... | ...... | ...... | ...... | ...... | ...... |
| 19 | 23 | 23 | 29 | 300 | ...... | ...... | ...... | ...... | ...... | ...... | ...... |
| 20 | 19 | 19 | 22 | 351 | 1 | ...... | ...... | ...... | ...... | ...... | ...... |
| 21 | 9 | 2 | 16 | 331 | 7 | ...... | ...... | ...... | ...... | ...... | ...... |
| 22 | 9 | 2 | 6 | 90 | 378 | ...... | ...... | ...... | ...... | ...... | ...... |
| 23 | 4 | 1 | 2 | 44 | 297 | ...... | ...... | ...... | ...... | ...... | ...... |
| 24 | 5 | 1 | 6 | 28 | 282 | ...... | ...... | ...... | ...... | ...... | ...... |
| 25 | 2 | 1 | 2 | 20 | 310 | 3 | ...... | ...... | ...... | ...... | ...... |
| 26 | 5 | 5 | 7 | 27 | 227 | 56 | ...... | ...... | ...... | ...... | ...... |
| 27 | 3 | 3 | ...... | 29 | 60 | 182 | ...... | ...... | ...... | ...... | ...... |
| 28 | 3 | 5 | 47 | 230 | ...... | ...... | ...... | ...... | ...... | ...... | ...... |
| 29 | 3 | ...... | 1 | 12 | 38 | 264 | ...... | ...... | ...... | ...... | ...... |
| 30 | ...... | ...... | 1 | 19 | 24 | 284 | 2 | ...... | ...... | ...... | ...... |
| 31 | 4 | ...... | 4 | 13 | 35 | 226 | 29 | ...... | ...... | ...... | ...... |
| 32 | 2 | 1 | 1 | 14 | 23 | 97 | 113 | ...... | ...... | ...... | ...... |
| 33 | 2 | ...... | 1 | 12 | 36 | 63 | 128 | ...... | ...... | ...... | ...... |
| 34 | 4 | 1 | ...... | 13 | 20 | 39 | 141 | ...... | ...... | ...... | ...... |
| 35 | 1 | ...... | 1 | 6 | 12 | 29 | 142 | 13 | ...... | ...... | ...... |
| 36 | 1 | ...... | 2 | 5 | 13 | 34 | 92 | 53 | ...... | ...... | ...... |
| 37 | ...... | ...... | 1 | 9 | 12 | 21 | 61 | 80 | ...... | ...... | ...... |
| 38 | ...... | 2 | 11 | 13 | 32 | 44 | 55 | ...... | ...... | ...... | ...... |
| 39 | ...... | ...... | ...... | 7 | 15 | 18 | 32 | 115 | ...... | ...... | ...... |
| 40 | 1 | ...... | 9 | 10 | 32 | 37 | 83 | 39 | 1 | ...... | ...... |
| 41 | ...... | ...... | ...... | 3 | 9 | 17 | 28 | 55 | 55 | ...... | ...... |
| 42 | ...... | ...... | 1 | 4 | 5 | 16 | 28 | 18 | 76 | ...... | ...... |
| 43 | ...... | ...... | 1 | 1 | 9 | 18 | 19 | 21 | 85 | ...... | ...... |
| 44 | ...... | ...... | 1 | 7 | 11 | 26 | 8 | 20 | 108 | ...... | ...... |
| 45 | ...... | ...... | 2 | 3 | 11 | 23 | 7 | 14 | 114 | 9 | ...... |
| 46 | ...... | ...... | ...... | ...... | 10 | 16 | 22 | 10 | 94 | 15 | ...... |
| 47 | ...... | ...... | ...... | 1 | 5 | 13 | 13 | 6 | 81 | 66 | ...... |
| 48 | ...... | ...... | ...... | 8 | 1 | 17 | 15 | 9 | 33 | 79 | ...... |
| 49 | ...... | ...... | ...... | 2 | 7 | 16 | 18 | 7 | 20 | 71 | ...... |
| 50 | ...... | ...... | ...... | 1 | 4 | 12 | 8 | 8 | 21 | 67 | |
| 51 | ...... | ...... | ...... | 2 | 4 | 5 | 6 | 13 | 16 | 38 | [illegible] |
| 52 | ...... | ...... | ...... | ...... | 4 | 14 | 9 | 5 | 9 | 18 | [illegible] |
| 53 | ...... | ...... | ...... | ...... | 4 | 2 | 6 | 6 | 5 | 10 | [illegible] |
| 54 | ...... | ...... | ...... | ...... | 3 | 7 | 10 | 5 | 5 | 7 | [illegible] |
| 55 | ...... | ...... | ...... | ...... | 1 | 3 | 8 | ...... | 3 | 3 | |
| 56 | ...... | ...... | ...... | ...... | 3 | 2 | 4 | 1 | 4 | ...... | |
| 57 | ...... | ...... | ...... | ...... | 2 | 1 | 2 | 2 | 2 | ...... | |
| 58 | ...... | ...... | ...... | ...... | 1 | 4 | ...... | ...... | ...... | 1 | |
| 59 | ...... | ...... | ...... | 1 | ...... | 1 | ...... | 1 | 1 | ...... | ...... |
| 60 | ...... | ...... | ...... | ...... | ...... | 2 | 3 | ...... | ...... | ...... | ...... |
| 61 | ...... | ...... | ...... | ...... | ...... | ...... | ...... | ...... | ...... | ...... | ...... |
| 62 | ...... | ...... | ...... | ...... | 1 | 2 | 1 | ...... | 1 | ...... | ...... |
| Total. | 552 | 519 | 1,069 | 2,096 | 1,845 | 1,528 | 1,107 | 560 | 694 | 378 | [illegible] |

EES OF THE COAL MINING COMPANY OF ANZIN AT WORK ABOVF GROUND, BY AGES AND YEARS OF SERVICE, FEBRUARY, 1892.

| Years of service. | | | | | | | | | | | | |
|---|---|---|---|---|---|---|---|---|---|---|---|---|
| nder 1. | 1 or under 2. | 2 or under 5. | 5 or under 10. | 10 or under 15. | 15 or under 20. | 20 or under 25. | 25 or under 30. | 30 or under 35. | 35 or under 40. | 40 or under 45. | 45 or under 50. | Total. |
| 29 | ... | ... | ... | ... | ... | ... | ... | ... | ... | ... | ... | 29 |
| 138 | 8 | 1 | ... | ... | ... | ... | ... | ... | ... | ... | ... | 147 |
| 2 | ... | ... | ... | ... | ... | ... | ... | ... | ... | ... | ... | 2 |
| ... | ... | 1 | ... | ... | ... | ... | ... | ... | ... | ... | ... | 1 |
| 2 | ... | 3 | ... | ... | ... | ... | ... | ... | ... | ... | ... | 5 |
| ... | ... | 2 | 1 | ... | ... | ... | ... | ... | ... | ... | ... | 3 |
| ... | ... | 1 | 1 | ... | ... | ... | ... | ... | ... | ... | ... | 2 |
| ... | 2 | 1 | 7 | ... | ... | ... | ... | ... | ... | ... | ... | 10 |
| 2 | 1 | 2 | 4 | ... | ... | ... | ... | ... | ... | ... | ... | 9 |
| ... | ... | ... | 4 | ... | ... | ... | ... | ... | ... | ... | ... | 4 |
| ... | ... | 1 | 4 | 4 | ... | ... | ... | ... | ... | ... | ... | 9 |
| ... | ... | ... | ... | 1 | ... | ... | ... | ... | ... | ... | ... | 1 |
| ... | 1 | 1 | 2 | 4 | ... | ... | ... | ... | ... | ... | ... | 8 |
| 1 | ... | 2 | ... | 3 | ... | ... | ... | ... | ... | ... | ... | 6 |
| 1 | ... | ... | 3 | 6 | ... | ... | ... | ... | ... | ... | ... | 10 |
| ... | 1 | ... | ... | 5 | 10 | ... | ... | ... | ... | ... | ... | 16 |
| 2 | ... | ... | 2 | 1 | 1 | ... | ... | ... | ... | ... | ... | 6 |
| ... | 1 | 1 | 2 | 2 | 8 | ... | ... | ... | ... | ... | ... | 14 |
| ... | ... | 1 | 1 | 1 | 4 | ... | ... | ... | ... | ... | ... | 7 |
| ... | ... | 1 | 1 | 3 | 5 | ... | ... | ... | ... | ... | ... | 10 |
| ... | 1 | ... | 6 | 1 | 5 | 2 | ... | ... | ... | ... | ... | 15 |
| ... | ... | 1 | 1 | 7 | 5 | 5 | ... | ... | ... | ... | ... | 19 |
| ... | ... | 5 | 6 | 3 | 3 | 7 | ... | ... | ... | ... | ... | 24 |
| 1 | ... | 1 | 2 | 2 | 3 | 1 | 2 | ... | ... | ... | ... | 12 |
| ... | ... | ... | 1 | ... | 2 | 2 | 2 | ... | ... | ... | ... | 7 |
| ... | ... | ... | 2 | 3 | 1 | 3 | 4 | ... | ... | ... | ... | 13 |
| 1 | ... | ... | 2 | 1 | ... | 4 | 4 | ... | ... | ... | ... | 12 |
| ... | ... | ... | 1 | 2 | 2 | 2 | 2 | 1 | ... | ... | ... | 10 |
| ... | 1 | 1 | 1 | 1 | 2 | 1 | 4 | 1 | ... | ... | ... | 12 |
| ... | 2 | ... | 2 | 3 | 3 | ... | 5 | 2 | ... | ... | ... | 17 |
| ... | ... | ... | ... | 5 | 1 | 2 | 3 | 3 | ... | ... | ... | 14 |
| ... | ... | ... | 1 | 1 | 3 | 1 | 1 | 5 | ... | ... | ... | 12 |
| ... | ... | ... | 5 | 3 | 3 | 4 | 1 | 4 | ... | ... | ... | 20 |
| ... | ... | ... | ... | ... | 2 | ... | 1 | 5 | ... | ... | ... | 8 |
| ... | ... | ... | 2 | 1 | ... | 2 | ... | 7 | ... | ... | ... | 12 |
| ... | ... | ... | 1 | 1 | 1 | 5 | 1 | 3 | 3 | ... | ... | 15 |
| ... | ... | ... | ... | 2 | 2 | 2 | 1 | 2 | 8 | ... | ... | 17 |
| ... | ... | ... | 1 | ... | 4 | 5 | ... | 1 | 9 | ... | ... | 20 |
| ... | ... | ... | 1 | 2 | 6 | 5 | 1 | 1 | 6 | ... | ... | 22 |
| ... | ... | ... | 2 | ... | 2 | 2 | ... | 2 | 5 | 1 | ... | 14 |
| ... | ... | ... | ... | ... | 2 | 3 | 2 | 1 | 5 | 3 | ... | 16 |
| ... | ... | ... | ... | 1 | 2 | 3 | 1 | 1 | 1 | 2 | ... | 11 |
| ... | ... | ... | 1 | 1 | ... | 1 | ... | 2 | ... | 1 | ... | 6 |
| ... | ... | ... | 1 | 2 | ... | 3 | 1 | ... | 1 | 1 | ... | 9 |
| ... | ... | ... | 1 | ... | 2 | ... | 2 | ... | 1 | ... | 2 | 8 |
| ... | ... | ... | ... | 1 | ... | ... | ... | ... | ... | ... | 2 | 3 |
| ... | ... | ... | ... | ... | ... | 1 | 1 | ... | ... | ... | 2 | 4 |
| ... | ... | ... | ... | 2 | 1 | ... | ... | 1 | 1 | ... | ... | 5 |
| ... | ... | ... | ... | 1 | ... | ... | 1 | ... | ... | ... | ... | 2 |
| ... | ... | ... | ... | 1 | 1 | ... | ... | ... | ... | ... | ... | 2 |
| ... | ... | ... | ... | ... | ... | ... | 1 | ... | ... | ... | ... | 1 |
| 79 | 18 | 26 | 72 | 77 | 86 | 66 | 41 | 42 | 40 | 8 | 6 | 661 |

| | | | | | | | | | |
|---|---|---|---|---|---|---|---|---|---|
| ...... | ...... | ...... | ...... | ...... | ...... | ...... | ...... | ...... | ...... |
| 3 | ...... | ...... | ...... | ...... | ...... | ...... | ...... | ...... | ...... |
| 16 | ...... | ...... | ...... | ...... | ...... | ...... | ...... | ...... | ...... |
| 259 | ...... | ...... | ...... | ...... | ...... | ...... | ...... | ...... | ...... |
| 313 | 1 | ...... | ...... | ...... | ...... | ...... | ...... | ...... | ...... |
| 234 | 141 | ...... | ...... | ...... | ...... | ...... | ...... | ...... | ...... |
| 67 | 247 | ...... | ...... | ...... | ...... | ...... | ...... | ...... | ...... |
| 40 | 307 | ...... | ...... | ...... | ...... | ...... | ...... | ...... | ...... |
| 24 | 355 | 1 | ...... | ...... | ...... | ...... | ...... | ...... | ...... |
| 16 | 335 | 7 | ...... | ...... | ...... | ...... | ...... | ...... | ...... |
| 7 | 94 | 282 | ...... | ...... | ...... | ...... | ...... | ...... | ...... |
| 2 | 44 | 298 | ...... | ...... | ...... | ...... | ...... | ...... | ...... |
| 7 | 30 | 286 | ...... | ...... | ...... | ...... | ...... | ...... | ...... |
| 5 | 20 | 313 | 8 | ...... | ...... | ...... | ...... | ...... | ...... |
| 7 | 30 | 243 | 56 | ...... | ...... | ...... | ...... | ...... | ...... |
| ...... | 29 | 74 | 202 | ...... | ...... | ...... | ...... | ...... | ...... |
| 47 | 241 | 1 | 1 | ...... | ...... | ...... | ...... | ...... | ...... |
| 2 | 14 | 28 | 242 | ...... | ...... | ...... | ...... | ...... | ...... |
| 2 | 20 | 25 | 238 | 2 | ...... | ...... | ...... | ...... | ...... |
| 5 | 14 | 28 | 228 | 29 | ...... | ...... | ...... | ...... | ...... |
| 1 | 20 | 24 | 102 | 115 | ...... | ...... | ...... | ...... | ...... |
| 2 | 13 | 43 | 68 | 128 | ...... | ...... | ...... | ...... | ...... |
| 5 | 19 | 23 | 42 | 148 | ...... | ...... | ...... | ...... | ...... |
| 2 | 8 | 14 | 32 | 143 | 14 | ...... | ...... | ...... | ...... |
| 2 | 6 | 13 | 26 | 94 | 55 | ...... | ...... | ...... | ...... |
| 1 | 11 | 16 | 22 | 64 | 84 | ...... | ...... | ...... | ...... |
| 11 | 14 | 33 | 44 | 89 | 4 | ...... | ...... | ...... | ...... |
| ...... | 8 | 17 | 20 | 34 | 117 | 1 | ...... | ...... | ...... |
| 10 | 11 | 33 | 39 | 83 | 43 | 2 | ...... | ...... | ...... |
| ...... | 5 | 12 | 20 | 23 | 60 | 57 | ...... | ...... | ...... |
| 1 | 4 | 10 | 17 | 25 | 21 | 79 | ...... | ...... | ...... |
| 1 | 2 | 10 | 19 | 20 | 22 | 100 | ...... | ...... | ...... |
| 1 | 12 | 14 | 29 | 12 | 21 | 112 | ...... | ...... | ...... |
| 2 | 3 | 11 | 24 | 7 | 15 | 119 | 9 | ...... | ...... |
| ...... | 2 | 11 | 16 | 24 | 10 | 101 | 16 | ...... | ...... |
| ...... | 2 | 6 | 14 | 18 | 7 | 34 | 69 | ...... | ...... |
| ...... | 3 | 3 | 19 | 17 | 10 | 35 | 78 | ...... | ...... |
| | 3 | 7 | 20 | 23 | 7 | 21 | 80 | | |

[illegible] OF THE COAL MINING COMPANY OF [illegible] SEPTEMBER 6, 18[illegible], DECEMBER 31, 18[illegible], AND [illegible]

| [illegible] | Number: [illegible] | Number: February, 18[illegible]. | Per cent: September 6, 18[illegible]. | Per cent: December 31, 18[illegible]. | Per cent: February, 18[illegible]. |
| --- | --- | --- | --- | --- | --- |
| [illegible] | [illegible] | 131 | ............ | [illegible] | [illegible] |
| [illegible] | [illegible] | [illegible] | ............ | [illegible] | 4.[illegible] |
| [illegible] | 1,[illegible] | 1,[illegible] | ............ | [illegible] | [illegible] |
| [illegible] | [illegible] | [illegible] | ............ | 17.1[illegible] | [illegible] |
| [illegible] | [illegible] | 1,[illegible] | ............ | 17.[illegible] | 17.[illegible] |
| [illegible] | [illegible] | [illegible] | ............ | 1[illegible].74 | 14.[illegible] |
| [illegible] | 7,[illegible] | 7,970 | 6[illegible] | [illegible] | 7[illegible].17 |
| [illegible] | 1,[illegible] | 1,173 | [illegible].37 | [illegible] | 1[illegible].61 |
| [illegible] | [illegible] | 541 | 7.[illegible] | [illegible] | 4.[illegible] |
| [illegible] | [illegible] | 735 | 6.94 | 6.33 | [illegible] |
| [illegible] | [illegible] | [illegible] | [illegible] | [illegible] | [illegible].74 |
| [illegible] | 131 | [illegible] | 1.70 | 1.[illegible] | 1.40 |
| [illegible] | [illegible] | 97 | .[illegible] | .[illegible] | .[illegible] |
| [illegible] | [illegible] | 11,[illegible] | 100.00 | 100.00 | 100.00 |

[illegible] OF THE COAL MINING COMPANY OF ANZIN, [illegible] FOR LEAVING, 18[illegible] TO 18[illegible].

| [illegible] | Total. | Dismissed. | To enter military service. | Pensioned. | Died. | Total. | Per cent of employees leaving voluntarily and dismissed of total employees. |
| --- | --- | --- | --- | --- | --- | --- | --- |
| [illegible] | [illegible] | [illegible] | 34 | [illegible] | 21 | 372 | 2.4 |
| [illegible] | [illegible] | [illegible] | [illegible] | [illegible] | 37 | 637 | 2.3 |
| [illegible] | [illegible] | 120 | 304 | [illegible] | 74 | 750 | 2.3 |
| [illegible] | [illegible] | 75 | 277 | 1[illegible] | 56 | 78[illegible] | 2.0 |
| [illegible] | [illegible] | 72 | [illegible] | 117 | 79 | [illegible] | 1.3 |
| [illegible] | [illegible] | 405 | 1,[illegible] | 467 | [illegible] | 3,244 | 2.0 |

In conclusion, attention should be drawn to [illegible] a study such as the one just made affords information of [illegible] usual value. The material has been presented in such a way as to furnish an opportunity for a statistical comparison of present with former conditions of a body of men, the general and physical conditions of whose labor have remained practically identical. The evidence afforded by such a comparison is irresistible that there has been a steady betterment of the condition of the coal miners of Anzin in almost every element that enters into their life. The age at which they commence work has been advanced, and they consequently enjoy a longer period of schooling. Their hours of labor have been steadily reduced. Average wages have constantly advanced, while there is every reason to believe that the amount of commodities that they will purchase has increased in like or greater ratio. The single matter of housing shows an enormous increase in comfort. A contrasting of the types of houses erected by the company at different periods shows a striking advance by each period over the preceding one. From a dweller in a barrack apartment the miner has become the occupier of an individual cottage with garden attached, and in many cases the owner of his own home. The uncertainties of a possible lack of employment or the cares and anxieties of sickness and approaching old age have been lessened. At the same time the workingman enters more into public life. He comes more into contact with his fellowmen through the exercise of his political rights, and through his participation in the management of the mutual benefit, cooperative, and recreative societies of which he is a member.

The comparisons which have been made throughout the report with conditions elsewhere indicate that the experience of Anzin has been repeated in the other great coal mining centers of France.

owns in the State asking for informa-
[illegible] and the expense incurred
[illegible] 1894. The amounts reported
State and the different towns. They
[illegible] by individuals or private boards

distributing aid and of recording the
[illegible] of data, especially for the earlier
what incomplete. These and various
sability of certain classes of relief, the
rs assisted and the amounts expended,
different towns, are explained in the
ollow present the totals given for the
and for the State.

| Items. | 1894. | 1895. |
|---|---|---|
| | $704,523.13 | $696,723.29 |
| | 28,442.84 | 40,608.82 |
| | 48,285.14 | 58,010.94 |
| | 9,352.91 | 185,286.06 |
| | 790,605.02 | 980,685.61 |

**MANUFACTURES.**—Under this title individual reports
shments, grouped by industries, are published and an
made of the returns for each industry. The totals
or all industries are as follows:

| | |
|---|---|
| rting | 1,000 |
| 93 | 115,139 |
| 94 | 98,617 |
| 95 | 112,002 |
| ncing wages since July 1, 1894 | 33 |
| cing wages since July 1, 1894 | 28 |
| cing former rates | 106 |

ny establishments reported in 1895 that were not in
2. These establishments reported 1,509 employees.
number from the total for 1895, the actual decrease
percentage of decrease 4.04.

rates of wages reported were only those that were
ter. There were 2,624 employees affected by advances
rage of the percentages of increase being 8.52. The
d in establishments reducing wages numbered 1,287,
he percentages of decrease being 8.53. There were
affected by the restoration of wages to a former rate,
he percentages of increase being 8.5. The average
labor in the 1,000 establishments for the year ending
considering the days closed, were 58.07; deducting
on of days closed, the average was 54.46.

LOCKOUTS.—An historical statement is made for each
that occurred during the year.

TING.—The treatment of this subject is confined to a
which includes a description of various institutions
ng in Connecticut and elsewhere.

AND MEDIATION AND ARBITRATION.—The laws relat-
ted at the January session of the legislature, 1895, are
short account given of the action of the board of
bitration in assisting in a compromise of a strike of

## IOWA.

*Sixth Biennial Report of the Bureau of Labor Statistics for the State of Iowa, 1894–95.* W. E. O'Bleness, Commissioner. Printed by Order of the General Assembly. 199 pp.

The report presents individual tabulations of returns from [illegible] working men and women engaged in different industries throughout the State. The questions for which the answers were presented were designed to show the actual condition of the laboring classes. No totals for the State or conclusions are presented, the individual reports only being given as a fair and unbiased showing of actual conditions. In addition to the individual tabulations, quotations are given from answers made by workmen to questions concerning the desirability of labor organizations, foreign immigration, the character of work that is preferable (piece or time), and what action would be of the greatest benefit to the wage earners of the country.

## MONTANA.

*Second Annual Report of the Bureau of Agriculture, Labor, and Industry of Montana for the year ended November 30, 1894.* James H. Mills, Commissioner; A. C. Schneider, Chief Clerk. v, 191 pp.

In the extended introductory, reference is made to various subjects, such as methods of work of the bureau, disturbed industrial conditions, free public employment offices, and protection of human life by means of fire escapes. The following subjects are treated in the report proper: Wages and cost of living, 32 pages; investment, wages, and production, 16 pages; precious and semiprecious metals, 8 pages; agriculture and stock growing, 40 pages; miscellaneous, 49 pages.

WAGES AND COST OF LIVING.—The schedule used in collecting the statistics from wage earners contained 48 questions, designed to [illegible] all material facts of public interest relating to the economic and social condition of the workmen. All employees receiving $2,000 or more per annum were excluded, and the presentation limited to those coming clearly under the denomination of "wage earners." The results are summarized for 80 occupations and for the principal labor-employing counties. The tables show the percentage of the total number [illegible] affirmatively or otherwise to the different questions, with the [illegible] and aggregate wages, expenses, etc.

The opinions and suggestions of wage earners given in reply to the query, "What legislation, if any, would, in your opinion, [illegible] general welfare, particularly of wage earners following [illegible] are reproduced in full.

| [illegible] | Wages per day. | Occupation. | Wages per day. |
|---|---|---|---|
| Employees in mines, smelters, and quartz mills—concluded. | | Employees in mines, smelters, and quartz mills—concluded. | |
| [illegible] | [illegible] | Furnacemen's helpers | $2.60 |
| [illegible] | 4.01 | Laborers | 2.73 |
| [illegible] | 2.30 | Machinists | 4.02 |
| [illegible] | 2.60 | Machinists' helpers | 2.85 |
| [illegible] | 2.19 | Millmen | 2.77 |
| [illegible] | 3.00 | Refiners | 3.00 |
| [illegible] | 4.00 | Roasters | 2.98 |
| [illegible] | 2.50 | Skimmers | 4.25 |
| [illegible] | 3.25 | Trammers | 2.67 |
| [illegible] | 3.05 | Weighers | 2.44 |
| [illegible] | 3.00 | Vannermen | 3.00 |
| [illegible] | 2.25 | | |

[illegible] by a presentation of the average rates [illegible] boarding houses, average prices of articles of [illegible] and expenses for representative families in the State.

[illegible] organizations of the State is accompanied with [illegible].

[illegible] and Production.—Reports from 146 manu[illegible] establishments engaged in various industries con[illegible] investment, product, employees, wages, [illegible] grouped and published in detail. As far as [illegible] wages for the various classes of labor in the dif[illegible] 1893 and 1894 are placed in comparison.

[illegible] Precious Metals.—The statistical presenta[illegible] consists of extended quotations from the reports of [illegible]ssay office at Helena, Mont., and the Director of the [illegible]

[illegible] Stock Growing.—Comparative figures are [illegible] the number and wages of employees of stock [illegible] acreage of land owned and fenced, with char[illegible] of the different agricultural products. The [illegible] classes of farm animals that perished during [illegible] killed by wolves are shown, and details con[illegible] 1893 and 1894.

as J. Dowling, Commissioner. 675 pp.

l in this report are as follows: Part I, labor organ- Part II, prison-made goods, 18 pages; Part III, boycotts, 77 pages; Appendix, 126 pages. The list of the bureaus of labor in the United States of the tenth annual convention of the National ls of Bureaus of Labor Statistics.

RGANIZATIONS.—The bureau sent to each trade State a letter of inquiry containing the following ganization; number of members at time of organ- embers at present time; rate of wages previous to wages at present time; hours of labor per day ion; hours of labor per day at present time. Is used in your trade or calling? Has the use of he number employed in your trade or calling; and s the use of machinery decreased the number le or calling; and what per cent? In your opinion ges been prevented by the fact of the existence Have the general conditions in your trade or d owing to the existence of your organization? n rendered any aid, financially or otherwise, to its past year? How much? The answers of the to these questions are printed in detail.

ved from 695 organizations. The following state- ts of the summarization of the detail tables which relative to wages and hours of labor:

OF LABOR PRIOR TO ORGANIZATION AND IN 1894, BY INDUSTRIES.

| | Wages, number of organizations reporting— | | | | Hours of labor, number of organizations reporting— | | | |
|---|---|---|---|---|---|---|---|---|
| | Increase. | Decrease. | No change | Total. | Increase. | Decrease. | No change | Total. |
| | 136 | 18 | 51 | 205 | ....... | 177 | 30 | 207 |
| | 31 | 3 | 7 | 41 | ....... | 40 | 9 | 49 |
| | 30 | 2 | 6 | 38 | 1 | 27 | 12 | 40 |
| ployees .... | 3 | ....... | 1 | 4 | ....... | 1 | 4 | 5 |
| | 8 | 1 | 10 | 19 | 2 | 12 | 5 | 19 |
| | 1 | 1 | 3 | 5 | ....... | 3 | 3 | 6 |
| | 4 | ....... | 2 | 6 | ....... | 3 | 3 | 6 |
| | 1 | 1 | 1 | 3 | ....... | 2 | 2 | 4 |
| nd ........ | 4 | 1 | 2 | 7 | ....... | 5 | 2 | 7 |
| | 25 | 18 | 21 | 64 | ....... | 20 | 45 | 65 |
| | 6 | 3 | 1 | 10 | ....... | 1 | 12 | 13 |
| nd mineral ........ | 16 | 1 | 1 | 18 | 1 | 12 | 5 | 18 |
| | 2 | 2 | 4 | 8 | ....... | 1 | 5 | 6 |
| | 2 | 2 | 2 | 6 | ....... | ....... | 6 | 6 |
| [illegible]....... | 10 | ....... | 3 | 13 | ....... | 3 | 10 | 13 |

WAGES AND HOURS OF LABOR PRIOR TO ORGANIZATION AND IN 1894, BY INDUSTRIES—Concluded.

| Industries. | Wages, number of organizations reporting— | | | | Hours of labor, number of organizations reporting— | | | |
|---|---|---|---|---|---|---|---|---|
| | Increase. | Decrease. | No change. | Total. | Increase. | Decrease. | No change. | Total. |
| Printing, binding, engraving, stereotyping, and publishers' supplies | 21 | 2 | 11 | 34 | ...... | 17 | 20 | 37 |
| Railroad employees (steam) | 57 | 3 | 25 | 85 | 2 | 27 | 51 | 80 |
| Railroad employees (street surface) | 1 | ...... | 1 | 2 | ...... | 3 | ...... | 3 |
| Stone workers | 13 | 1 | 6 | 20 | ...... | 19 | 3 | 22 |
| Street paving | 3 | ...... | 6 | 9 | ...... | 3 | 5 | 8 |
| Textiles | 8 | 1 | 1 | 10 | ...... | 8 | 2 | 10 |
| Theatrical | 3 | ...... | ...... | 3 | ...... | 2 | 2 | 4 |
| Wood workers | 11 | ...... | 1 | 12 | ...... | 10 | 2 | 12 |
| Miscellaneous | 6 | 2 | 8 | 16 | ...... | 9 | 8 | 17 |
| Total | 402 | 62 | 174 | 638 | 6 | 404 | 247 | 657 |

There were 49 divisions of working time reported by 656 organizations. Eight hours constituted a day's work in 42 branches of trade, and the eight-hour day was enjoyed by 48,411 members of 169 organizations. The number is nearly one-third of the 155,843 members reported. The daily hours of work and the number of members observing the indicated working time is shown for each organization reported.

Four hundred and seventy-four organizations, with a membership of 121,957, report $511,817.59 as having been expended in benefits during the year, of which amount $106,801.69 was to assist those out of work, $60,207.98 to assist the sick, $93,437.92 in cases of death, $89,[illegible] to support strikes, $10,676.74 donated to other labor organizations, and $151,543.22 not classified.

Out of 695 organizations, 371 report that improved machinery is used, 285 report that it is not, and 39 failed to answer the question. Sixty-three organizations report that the introduction of machinery has increased the working force, while 208 state that it has resulted in a reduction of the number of employees, and 47 failed to answer the question.

Five hundred and forty-four organizations reported that the existence of the organization had prevented a reduction in wages, and 96 reported that it had not, while 22 failed to answer the question, and 33 reported that there had been no attempt at reduction of wages. Six hundred and twenty-two organizations reported that the general conditions of labor in their trades had been improved by the existence of the union, 49 that the union had not improved general conditions, while 24 failed to answer the question.

There were 667 organizations that reported their membership [illegible] 46,455 at the date of organization, and at the time of reporting in [illegible] 691 organizations reported their membership at 155,843.

Extended quotations are made from remarks contained in the [illegible] of organizations relative to desired legislation, immigration, and [illegible]laneous subjects affecting labor.

| [illegible] | [illegible] | [illegible] | [illegible] | [illegible] | [illegible] |
|---|---|---|---|---|---|
| | [illegible] | 212 | 1,680 | 89,462 | ........ |
| | [illegible] | 1,714 | 2,086 | 175,368 | 6,280 |
| | [illegible] | 755 | 1,977 | 54,389 | 8,241 |
| | [illegible] | 446 | 1,027 | 24,082 | 2,371 |
| | [illegible] | 959 | 1,374 | 33,708 | 4,208 |
| | 178 | 512 | 6,228 | 96,894 | 5,220 |
| | [illegible] | 985 | 4,480 | 63,590 | 2,581 |
| | [illegible] | 770 | 2,326 | 36,826 | 1,987 |
| | [illegible] | 285 | 2,638 | 27,545 | 2,374 |
| [illegible] | 1,[illegible] | 8,767 | 36,587 | 587,788 | 34,236 |

## [illegible] STRIKES, LOCKOUTS, AND BOYCOTTS, BY YEARS, [illegible] TO [illegible].

| [illegible] | Cost to labor organizations. | Estimated gain in wages. | Number engaged and who received increase of wages where wages were involved. | Loss to employers. |
|---|---|---|---|---|
| | $172,[illegible].73 | $[illegible],152.86 | 10,661 | $416,426.00 |
| | [illegible],897.25 | 2,[illegible],616.90 | 31,196 | 2,696,404.00 |
| | [illegible].75 | [illegible],[illegible].55 | 11,512 | 1,166,768.20 |
| | [illegible],897.05 | 410,[illegible].66 | 3,[illegible] | 290,720.00 |
| | [illegible].26 | [illegible],519.76 | 10,624 | 533,368.95 |
| | 146,[illegible].65 | 3,[illegible].16 | 42,907 | 549,974.42 |
| | 204,[illegible].85 | 787,622.65 | 23,194 | 374,946.50 |
| | 241,[illegible].60 | [illegible],151.06 | 7,977 | 366,215.90 |
| | [illegible],[illegible].34 | [illegible],164.17 | 7,867 | 102,680.10 |

## NORTH CAROLINA.

*Eighth Annual Report of the Bureau of Labor Statistics of North Carolina for the year 1894.* B. R. Lacy, Commissioner. 394 pp.

The report treats of the following subjects: Statistics of and letters concerning cotton, woolen, and other factories, 86 pages; agricultural statistics and views of farmers, 100 pages; reports from and views of mechanics, 61 pages; the fishery industry, 9 pages; statistics of employees and wages of railroads, 5 pages; organized labor, 17 pages; miscellaneous, 26 pages.

STATISTICS OF AND LETTERS CONCERNING COTTON, WOOLEN, AND OTHER FACTORIES.—These statistics were obtained by means of a circular letter. Reports were received from a number of factories engaged in various industries in different sections of the State. These reports relate to the character of goods manufactured, number of spindles and looms, days in operation, hours constituting a day's work, advisability of reducing working time, average daily wages, etc. The data are presented in detail for each establishment, arranged by counties. The facts are summarized by counties for some industries, but no general average for the State is attempted.

AGRICULTURAL STATISTICS AND VIEWS OF FARMERS.—The information given under this title was obtained from the best and most influential farmers in the State. The wages and other compensation of farm laborers and their condition morally, socially, and financially are shown. The individual returns are presented in full and the averages given by counties and for the State. The averages for the State show that the working day for the year is about nine hours, and the average wages per month for laborers, $9, with extras for married men. The average for women was $5 and for children $3 per month. About 66⅔ per cent of the farmers report a decrease in wages, and the remainder say there has been no change. Numerous letters from farmers and others in different sections of the State expressing views on various phases of agricultural pursuits follow the statistics.

REPORTS FROM AND VIEWS OF MECHANICS.—This presentation covers information concerning the condition of trade, wages, methods of payment, effect of labor-saving machinery on wages, apprenticeship, and age at which children should engage in the different trades, cost of living, and social and moral conditions. The individual reports are published and summarized by trades. The statistics are accompanied by letters from a number of mechanics expressing views as to the legislation needed for the elevation of the labor classes.

STATISTICS OF EMPLOYEES AND WAGES OF RAILROADS.—These statistics show the number of the different classes of employees and the average daily wages for each class for each railroad of

of persons who have had experience in ss of the State, giving reasons for the he past two or three years. Attention sed under the direction of the bureau ng the Russian thistle. The subjects ouped as follows: Agricultural statisisus statistics, 131 pages; cost of profarmers, 159 pages; cost of producing ages; industrial statistics, 11 pages.

—The average yield per acre, and in different crops and other farm products, nd State totals. Comparisons are made ge, and average yield per acre of the otals, from 1888 to 1893, inclusive, and ven, with the acreage under cultivation for some of the products enumerated

VARIOUS AGRICULTURAL PRODUCTS, 1893.

| 1893. | | | | Acres sown, 1894. |
|---|---|---|---|---|
| wn. | Acres harvested. | | Total product (bushels). | |
| Average yield per acre (bushels). | Number. | Average yield per acre (bushels). | | |
| 10.51 | 2,902,301 | 10.96 | 31,732,169 | 3,087,643 |
| 28.89 | 483,844 | 21.70 | 10,496,451 | 546,369 |
| 17.57 | 218,256 | 18.30 | 3,983,236 | 258,252 |
| 5.75 | 58,398 | 6.19 | 330,214 | 110 365 |
| 8.96 | 38,296 | 9.62 | 367,976 | 52,099 |
| a 12.80 | ............ | .......... | 345,734 | 38,696 |
| 62.31 | ............ | .......... | 1,299,090 | 19,627 |

7; those given are, however, according to the original.

STATISTICS.—The number of persons assessed value of live stock and of all unty and State totals. Assessed valu-

[illegible] quotations from numerous letters from [illegible] industries and from men engaged in the [illegible] pertain to the treatment of employees, [illegible] Tabulations are also given for a number [illegible] engaged in various industries, presenting infor- [illegible], cost of board, wages, time of payment, [illegible], hours of work, and pay for time lost by [illegible] The reports tabulated for the building trades [illegible] from 1890 to 1894 and answers to numerous [illegible] wages, hours of labor, apprenticeship, etc. The [illegible] information as to conditions prevailing in [illegible] throughout the State.

[illegible] short history is given of the most notable strikes that [illegible] during the year. The total estimated loss in [illegible] and lockouts in Pennsylvania from 1881 to 1894 is [illegible]. The industry, locality, number of persons [illegible] beginning and ending, and other facts are shown for [illegible] lockouts that occurred in the State during 1894.

[illegible]RE OF PIG IRON.—A detailed description and a his- [illegible]are given of the pig-iron industry of the State. The [illegible]nical analyses are shown for the different kinds of ore, [illegible] is compared with the production of other States, and [illegible]tus of the industry and the methods and cost of manufac- [illegible]d.

[illegible] OF MANUFACTURES.—Facts are given relative to the [illegible]sons employed, wages paid, and value of product for 412 [illegible] plants for which returns were received for 1894 and [illegible] given for 1892 and 1893. Reports were not secured [illegible] in the State, but from a sufficient number, it was [illegible]rm correct general deductions. The results are sum- [illegible]

[illegible] AND VALUE OF PRODUCT OF 412 MANUFACTURING ESTABLISHMENTS, 1892, 1893, AND 1894.

| [illegible] | 1892. | 1893. | 1894. | Per cent of decrease. 1892 to 1893. | Per cent of decrease. 1893 to 1894. |
|---|---|---|---|---|---|
| [illegible] | [illegible] | 132,[illegible]3 | 116,310 | 11.38 | 12.32 |
| [illegible] | [illegible] | [illegible],740 | $48,368,006 | 16.46 | 20.39 |
| [illegible] | [illegible],751 | $[illegible],819,[illegible] | $191,492,115 | 17.28 | 19.17 |

Various reasons are given for the [illegible]
The following statement is presented [illegible]
by industries:

EMPLOYEES OF 412 MANUFACTURING ESTABLISHMENTS, [illegible] AND 1894.

| Industry. | [illegible] | [illegible] |
|---|---|---|
| Iron | [illegible] | [illegible] |
| Carpets | [illegible] | [illegible] |
| Hosiery | [illegible] | [illegible] |
| Woolen | [illegible] | [illegible] |
| Cotton | [illegible] | [illegible] |
| Glass | [illegible] | [illegible] |
| Miscellaneous | [illegible] | [illegible] |
| Total | [illegible] | [illegible] |

MINE ACCIDENTS.—The statistics relating to [illegible]
were obtained from the reports of the mine inspectors, and [illegible]
follows:

ACCIDENTS IN COAL MINES, [illegible] TO [illegible].

| | [illegible] | [illegible] | [illegible] | [illegible] |
|---|---|---|---|---|
| Anthracite coal: | | | | |
| Product per employee, tons | 342 | [illegible] | [illegible] | [illegible] |
| Fatal accidents | [illegible] | [illegible] | [illegible] | [illegible] |
| Employees to each fatal accident | 312 | 311 | [illegible] | [illegible] |
| Employees to each nonfatal accident | [illegible] | 116 | [illegible] | [illegible] |
| Tons mined to each fatal accident | 101,490 | [illegible] | [illegible] | [illegible] |
| Tons mined to each nonfatal accident | [illegible] | [illegible] | [illegible] | [illegible] |
| Bituminous coal: | | | | |
| Product per employee, tons | 565 | [illegible] | [illegible] | [illegible] |
| Fatal accidents | 105 | 146 | [illegible] | [illegible] |
| Employees to each fatal accident | 561 | [illegible] | [illegible] | [illegible] |
| Employees to each nonfatal accident | [illegible] | 177 | [illegible] | [illegible] |
| Tons mined to each fatal accident | [illegible] | [illegible] | [illegible] | [illegible] |
| Tons mined to each nonfatal accident | 114,803 | [illegible] | [illegible] | [illegible] |

The principal provisions of the different laws that have [illegible]
mining and mine inspection in the State are quoted. The [illegible]
enforcing the various provisions and the effect the [illegible]
had in preserving the health and lives of those engaged in [illegible]
discussed, numerous quotations being made from the reports [illegible]
different inspectors.

## RHODE ISLAND.

*Eighth Annual Report of the Commissioner of Industrial Statistics to the General Assembly at its January Session, 1895.* Henry E. [illegible] Commissioner. viii, 327 pp.

This report contains the result of an investigation [illegible]
manufactures of the State, the condition of skilled labor [illegible]
of industry, and the retail prices of food and fuel. [illegible]
jects were chosen in 1893 as the basis of a [illegible]
which should annually cover certain specific [illegible]
allotted to each subject in the report for 1894 is as [illegible]
returns, textile industries, 224 pages; retail prices, [illegible]
of manufactures, textile industries, 55 pages.

nerous facts concerning their civil and social conditions, re summarized by towns and for the State. The totals e items shown for all three branches of the industry are

| | |
|---|---|
| ns | 2,299 |
| | 1,559 |
| | 690 |
| | 50 |
| | 827 |
| | 1,472 |
| in family | 15 |
| in family | 2 |
| homes | 245 |
| n incumbrance | 88 |
| nements (39 also own homes) | 1,373 |
| ges | $6.00 |
| ges | .40 |
| g an increase in wages during the year | 32 |
| g a decrease in wages during the year | 1,367 |
| yed during a portion of the year | 1,692 |

ICES.—The average retail prices of different articles of are shown by cities, towns, counties, and for the State. for the State is as follows:

RETAIL PRICES OF FOOD AND FUEL FOR THE STATE, 1894.

| Articles. | 1894. | | | |
|---|---|---|---|---|
| | January. | April. | July. | October. |
| | $0.438 | $0.532 | $0.304 | $0.243 |
| | .661 | .640 | .669 | .655 |
| l, per pound | .071 | .068 | .070 | .066 |
| pound | .136 | .131 | .138 | .132 |
| | .074 | .074 | .072 | .066 |
| | .335 | .289 | .264 | .304 |
| | .021 | .026 | .020 | .019 |
| | .159 | .160 | .152 | .155 |
| e), per ton | 6.450 | 6.280 | 5.710 | 5.650 |
| | .072 | .073 | .072 | .072 |
| l | .026 | .026 | .026 | .026 |
| ound | .053 | .035 | .048 | .048 |
| per pound | .076 | .074 | .074 | .073 |
| nd | .134 | .140 | .154 | .125 |
| | .338 | .188 | .224 | .289 |
| pound | .094 | .095 | .092 | .090 |
| , per pound | .080 | .082 | .082 | .079 |
| arrel | 4.830 | 4.630 | 4.510 | 4.180 |
| pound | .125 | .120 | .142 | .131 |
| lon | .100 | .099 | .098 | .097 |
| | .116 | .107 | .107 | .110 |
| ound | .132 | .128 | .124 | .127 |
| | .059 | .053 | .053 | .059 |
| | .517 | .512 | .507 | .507 |
| | .106 | .105 | .123 | .097 |
| | .049 | .048 | .050 | .049 |
| | .043 | .041 | .050 | .042 |
| | .130 | .129 | .132 | .130 |
| d | .117 | .105 | .106 | .110 |
| | .250 | .236 | .260 | .222 |
| | .105 | .101 | .101 | .099 |
| | .031 | .031 | .031 | .030 |
| | .017 | .017 | .016 | .016 |
| nd | .052 | .052 | .053 | .052 |
| | .052 | .051 | .051 | .052 |
| ound | .453 | .444 | .440 | .440 |
| | .072 | .069 | .072 | .070 |
| | .245 | .245 | .246 | .241 |
| ft), per cord | 7.620 | 7.400 | 7.690 | 7.540 |

STATISTICS OF MANUFACTURING [illegible] reau secured reports from 121 [illegible] of their establishments during 1893 [illegible] were for the cotton industry, 44 [illegible] works, dyeworks, and bleacheries, [illegible] for silk and silk goods. The statistics are given [illegible] try. A summary for the 121 establishments [illegible] statement:

STATISTICS OF 121 MANUFACTURING ESTABLISHMENTS [illegible]

| Items. | 1892. | 1893. | [illegible] |
|---|---|---|---|
| Establishments | 121 | [illegible] | [illegible] |
| Private firms | 56 | [illegible] | [illegible] |
| Partners: | | | |
| Male | 98 | [illegible] | [illegible] |
| Special | 7 | [illegible] | [illegible] |
| Total | 105 | [illegible] | [illegible] |
| Corporations | 65 | [illegible] | [illegible] |
| Stockholders: | | | |
| Male | 762 | [illegible] | [illegible] |
| Female | 347 | [illegible] | [illegible] |
| Banks, trustees, etc | 34 | 104 | [illegible] |
| Total | 1,139 | [illegible] | [illegible] |
| Total partners and stockholders | 1,236 | 1,377 | [illegible] |
| Capital invested | $37,578,111 | [illegible] | [illegible] |
| Employees: | | | |
| Greatest number | 30,302 | [illegible] | [illegible] |
| Smallest | 26,570 | [illegible] | [illegible] |
| Average | 28,704 | [illegible] | [illegible] |
| Total wages | $10,466,032 | [illegible] | [illegible] |
| Average annual wages | $364.62 | [illegible] | [illegible] |
| Average days in operation | 262.65 | [illegible] | [illegible] |
| Cost of materials used | $27,496,995 | $21,130,270 | [illegible] |
| Value of goods made and work done | $48,406,877 | $37,404,845 | [illegible] |

## TENNESSEE.

*Fourth Annual Report of the Bureau of Labor, Statistics, and [illegible] the Forty-ninth General Assembly of the State of Tennessee.* [illegible] 1895. John E. Lloyd, Commissioner. 200 pp.

The contents of the report are grouped as follows: Introduction, [illegible] pages; statistics of mines and mine inspection, 56 pages; [illegible] 86 pages.

INTRODUCTION.—A short sketch is given of the financial [illegible] trial depression, which is followed by articles on the [illegible] capital and labor, arbitration, and the importance of [illegible] recommendations concerning additional legislation [illegible] work of the bureau. Sketches are also given of the [illegible] the coal miners' great strike of April 21, 1894.

STATISTICS OF MINES AND MINE INSPECTION.—[illegible] as mined in Tennessee during the year 1894, 2,471,[illegible]

; inlet and outlet, and remarks concerning
nine at time of inspection and additions
ıspector. The results are shown for two
year.

o received injuries resulting fatally are
tion of the mine, also the testimony given
ıer's jury in each case. The report also
sis of the coal for each mine in the State.
;his head are grouped articles on different
posits of the State are treated with con-
al was first discovered in Lewis County,
of 1893, and the deposits are now shown
article is accompanied with a tabular
ıhosphate beds. In an article on "man-
rst manganese mined in the United States
837, but comparatively nothing has been
though manganese of a high grade exists
r every county in east Tennessee. The
und in the State are described in an arti-
mes and locations of the different cotton
are shown; also a synopsis of the statis-
tgages for the State, as published by the
ıd States. An appendix gives the report
United States Strike Commission.

ST VIRGINIA.

*of Labor of the State of West Virginia,*
lenstricker, Commissioner. 211 pp.

ort of the commissioner of labor of West
ending June 30, 1894. The introductory,
the bureau, and gives a copy of the law
, is followed by a discussion of the "func-
tistics," and the volume is closed with a
l industrial depression." The statistics

presented are grouped as follows: Laborers' statistics, 23 pages; railroad statistics, 4 pages; coal-mine operators' statistics, 11 pages; manufacturing and mechanical industries, 33 pages; agricultural statistics, 13 pages; coal, coke, and oil statistics, 31 pages; building and loan associations, 30 pages; farms, homes, and mortgages, 19 pages.

LABORERS' STATISTICS.—The individual reports of 189 laborers engaged in various occupations in different sections of the State are given in full. The information was collected by correspondence, and consists of replies to questions concerning nativity, residence, name of employer, occupation, hours of labor, earnings of self and family, cost of living, character of employment of wife and children, education of children, deductions from wages, apprenticeship, increase or decrease in wages, and cost of living, savings, debts, etc.

RAILROAD STATISTICS.—Statistics are given in detail for each of 25 railroads, showing for the State the miles of road, the average number of employees during the year, the total number at the time of making the report, the number of each class of employees, and the average monthly and total wages for the year.

MANUFACTURING AND MECHANICAL INDUSTRIES.—Reports are published in full for each of 77 establishments showing answers to questions concerning the value of buildings, land, and machinery, cost of materials, value of product, and details concerning employees and wages. In addition the report of the Eleventh Census of the United States relating to manufacturing and mechanical industries is reproduced.

AGRICULTURAL STATISTICS.—The bureau secured reports from a number of farmers in different sections of the State which are published in detail, by counties, and contain information pertaining to the size, the entire value of farms, and the value of the portion used in farming operations, value of personal property and of farm products, expenses of farming, yield, and value of different crops, etc.

COAL, COKE, AND OIL STATISTICS.—The statistics concerning the production for each of these industries in the State are shown in detail by totals for districts and counties, with percentages of increase or decrease. It is stated that the State ranks fourth in the coal-producing States of the country. The product increased from 672,000 short tons in 1873 to 10,708,578 short tons in 1893. The coke product increased from 138,755 short tons in 1880 to 1,062,076 short tons in 1893, and the petroleum from 120,000 barrels in 1876 to 8,445,412 barrels in 1893. A list of the names and addresses of the coal operators of the State is given, with statistics concerning investment, thickness of seam, days mines were worked during the year, output, number of miners and other employees, and total wages paid different classes.

BUILDING AND LOAN ASSOCIATIONS.—The reports for 56 associations in the State are shown in detail, and convenient summaries made of the statistics for a number of representative associations.

## [illegible] THE ANNUAL STATISTICS OF MANUFAC-TURES IN MASSACHUSETTS.

*s of Manufactures, 1894.* Ninth Report. xvi, 229 e Bureau of Statistics of Labor, Horace G. Wadlin,

s of an introduction, 3 pages; tables presenting il, 165 pages; the analysis, 83 pages. There are d to an industrial chronology of the State, which nd city the principal events affecting the industrial g 1894.

ot shown for all the manufacturing and mechanical te, the report being confined to a comparison of e establishments reporting for the different years. le for 4,093 establishments for 1893 and 1894, for for the five years from 1890 to 1894, inclusive, and ts for the ten years from 1885 to 1894, inclusive. ted in this synopsis have been selected principally 1893 and 1894, to which the major portion of the

ived from 4,486 establishments for 1894; of this npared with reports for 1893. These reports are ified industries, and reflect the industrial condi- e State during the two years.

hments were conducted during 1894 by 3,183 pri- rporations, which were managed by 43,337 individ- were partners and 38,281 stockholders. Of the nt were males, 2.49 per cent females, and 1.88 per f the stockholders 56.45 per cent were males, 32.55 d 11 per cent banks, trustees, etc. Considering ckholders in the aggregate, 61.02 per cent were t were females, and 9.93 per cent banks, trustees,

crease in capital invested, wages paid, stock used, d work done in 1894 as compared with 1893 are tements which follow for each of the 9 leading ies, and for the remaining 66, of the 75 referred together.

STATISTICS OF MANUFACTURES [illegible]

| Industries. | Year. | Establishments. | Capital invested. | [illegible] | [illegible] | [illegible] |
|---|---|---|---|---|---|---|
| Boots and shoes | 1893 | [illegible] | [illegible] | [illegible] | [illegible] | [illegible] |
| | 1894 | [illegible] | [illegible] | [illegible] | [illegible] | [illegible] |
| Carpetings | 1893 | 11 | 7,[illegible] | [illegible] | [illegible] | [illegible] |
| | 1894 | 11 | 7,[illegible] | [illegible] | [illegible] | [illegible] |
| Cotton goods | 1893 | 148 | 115,11[illegible] | [illegible] | [illegible] | [illegible] |
| | 1894 | 148 | 114,[illegible] | [illegible] | [illegible] | [illegible] |
| Leather | 1893 | 141 | 7,[illegible] | [illegible] | [illegible] | [illegible] |
| | 1894 | 141 | [illegible] | [illegible] | [illegible] | [illegible] |
| Machines and machinery | 1893 | [illegible] | [illegible],447,[illegible] | [illegible] | [illegible] | [illegible] |
| | 1894 | [illegible] | [illegible] | [illegible] | [illegible] | [illegible] |
| Metals and metallic goods | 1893 | 327 | [illegible] | [illegible] | [illegible] | [illegible] |
| | 1894 | 327 | [illegible] | [illegible] | [illegible] | [illegible] |
| Paper and paper goods | 1893 | 98 | [illegible] | [illegible] | [illegible] | [illegible] |
| | 1894 | 98 | [illegible] | [illegible] | [illegible] | [illegible] |
| Woolen goods | 1893 | 115 | [illegible] | [illegible] | [illegible] | [illegible] |
| | 1894 | 115 | [illegible] | [illegible] | [illegible] | [illegible] |
| Worsted goods | 1893 | 21 | [illegible] | [illegible] | [illegible] | [illegible] |
| | 1894 | 21 | 15,[illegible] | [illegible] | [illegible] | [illegible] |
| Other industries (66) | 1893 | 2,272 | 160,150,654 | [illegible] | [illegible] | [illegible] |
| | 1894 | 2,272 | 147,903,412 | [illegible] | [illegible] | [illegible] |
| Total | 1893 | 4,066 | 431,121,145 | [illegible] | [illegible] | [illegible] |
| | 1894 | 4,063 | 417,647,[illegible] | 111,[illegible] | [illegible] | [illegible] |

DECREASE IN MANUFACTURES IN 1894 AS COMPARED WITH 1893 IN 75 INDUSTRIES.

| Industries. | Decrease in— | | | | | | | |
|---|---|---|---|---|---|---|---|---|
| | Capital. | | Wages. | | Stock used. | | Goods made and work done. | |
| | Amount. | Per cent. | Amount. | Per cent. | Amount. | Per cent. | Amount. | Per cent. |
| Boots and shoes | a $41,069 | a 0.16 | $[illegible],348 | 1.[illegible] | $1,[illegible],118 | [illegible] | [illegible] | [illegible] |
| Carpetings | 715,999 | 8.96 | 259,669 | 16.34 | 1,1[illegible] | [illegible] | [illegible] | [illegible] |
| Cotton goods | 1,096,872 | .95 | 2,683,[illegible] | 10.[illegible] | 4,[illegible] | [illegible] | [illegible] | [illegible] |
| Leather | a 404,406 | a 5.09 | 74,682 | 2.67 | [illegible],211 | [illegible] | [illegible] | [illegible] |
| Machines and machinery | 33,446 | .11 | 1,336,408 | 15.52 | 1,120,412 | [illegible] | [illegible] | [illegible] |
| Metals and metallic goods | 556,761 | 2.79 | 37,175 | .56 | 1,[illegible] | [illegible] | [illegible] | [illegible] |
| Paper and paper goods | a 384,152 | a 1.57 | 121,295 | 2.97 | 1,344,[illegible] | [illegible] | [illegible] | [illegible] |
| Woolen goods | 1,139,544 | 4.53 | 859,285 | 14.96 | [illegible] | [illegible] | [illegible] | [illegible] |
| Worsted goods | a 1,486,728 | a 10.82 | 430,713 | 12.[illegible] | 1,[illegible] | [illegible] | [illegible] | [illegible] |
| Other industries (66) | 12,247,242 | 7.65 | 5,194,884 | 11.63 | 16,778,124 | 11.44 | [illegible] | [illegible] |
| Total | 13,473,509 | 3.13 | 11,392,852 | 9.30 | [illegible] | [illegible] | [illegible] | [illegible] |

a Increase.

The term "capital invested" used in compiling these [illegible] not mean merely cash capital or capital stock, but includes [illegible] of capital devoted to production, such as notes, bills [illegible] value of land, machinery, and stock on hand or in process of [illegible] ture. Inasmuch as some of the elements included as capital [illegible] from year to year, it follows that apparently wide fluctuations [illegible] amount of capital invested will sometimes appear in [illegible] reduction in capital does not, of course, imply retrogression [illegible]

Four of the 9 leading industries show an increase and [illegible] in the amount of capital invested, the decrease for [illegible] amounting to 3.13 per cent. A decrease is shown for [illegible] and value of goods made and work done in each of [illegible] *for the total of* the 75 industries.

| Year. | Establishments. | Number of employees. | | | Average wages per year. |
|---|---|---|---|---|---|
| | | Average. | Smallest. | Greatest. | |
| 1893..... | 696 | 41,253 | 31,506 | 49,744 | $496.32 |
| 1894..... | 698 | 40,363 | 33,164 | 48,151 | 491.45 |
| 1893..... | 11 | 4,335 | 1,917 | 5,081 | 388.56 |
| 1894..... | 11 | 3,744 | 1,963 | 4,667 | 355.06 |
| 1893..... | 148 | 71,568 | 69,720 | 76,711 | 342.20 |
| 1894..... | 148 | 69,236 | 55,164 | 76,094 | 329.43 |
| 1893..... | 141 | 5,698 | 3,806 | 7,289 | 468.65 |
| 1894..... | 141 | 5,738 | 4,531 | 7,129 | 475.27 |
| 1893..... | 232 | 15,808 | 11,773 | 19,108 | 544.94 |
| 1894..... | 232 | 13,561 | 10,686 | 16,140 | 535.81 |
| 1893..... | 227 | 13,087 | 10,299 | 15,172 | 507.20 |
| 1894..... | 227 | 11,754 | 9,560 | 13,636 | 560.79 |
| 1893..... | 96 | 9,924 | 8,244 | 11,012 | 411.42 |
| 1894..... | 96 | 9,665 | 8,148 | 10,787 | 409.89 |
| 1893..... | 115 | 15,520 | 11,215 | 17,969 | 376.31 |
| 1894..... | 115 | 14,261 | 10,163 | 16,708 | 343.75 |
| 1893..... | 21 | 9,404 | 6,979 | 10,790 | a 254.28 |
| 1894..... | 21 | 9,222 | 5,590 | 11,275 | 314.89 |
| 1893..... | 2,272 | 94,257 | 66,734 | 115,917 | b 437.36 |
| 1894..... | 2,272 | 86,245 | 68,432 | 105,586 | 457.28 |
| 1893..... | 4,096 | 280,966 | 214,261 | 322,763 | 436.13 |
| 1894..... | 4,098 | 268,386 | 206,432 | 310,167 | 421.81 |

$354.60; those given are, however, according to the original.
$473.36; those given are, however, according to the original.

stries given in the above statement shows a erage, greatest, and smallest number of per- average annual wages. The decrease in the o $14.32, or 3.28 per cent.

the 75 industries, the per cent of males and er employed at each specified weekly rate of wing statement:

MALES OF THE WHOLE NUMBER EMPLOYED AT WEEKLY WAGES, 1893 AND 1894.

From the above statement it [illegible] in each wage class increases after the [illegible] passed, while the proportion of females [illegible]

In the following statement, which [illegible] total number of males, the total number of [illegible] ber of employees of both sexes are each [illegible] per cent, and the number of employees in each [illegible] parts of this aggregate.

PER CENT OF THE TOTAL MALES AND FEMALES [illegible] WEEKLY WAGES, 1893 AND 1894.

| Weekly wages. | 1893. | | | [illegible] | | |
|---|---|---|---|---|---|---|
| | Males. | Females. | Total. | Males. | Females. | [illegible] |
| Under $5 | 7.50 | 24.41 | 12.20 | 4.36 | [illegible] | [illegible] |
| $5 or under $6 | 4.79 | 17.89 | 9.21 | 5.71 | [illegible] | [illegible] |
| $6 or under $7 | 7.56 | 12.87 | 11.54 | [illegible] | [illegible] | [illegible] |
| $7 or under $8 | 9.00 | 13.33 | 10.46 | [illegible] | [illegible] | [illegible] |
| $8 or under $9 | 8.69 | 9.82 | 9.67 | [illegible] | [illegible] | [illegible] |
| $9 or under $10 | 12.87 | 6.71 | 10.79 | 12.25 | [illegible] | [illegible] |
| $10 or under $12 | 14.67 | 5.15 | 11.46 | 14.65 | [illegible] | [illegible] |
| $12 or under $15 | 17.10 | 2.43 | 12.15 | 16.08 | [illegible] | [illegible] |
| $15 or under $20 | 12.36 | .81 | 9.13 | 11.91 | .74 | [illegible] |
| $20 or over | 4.47 | .08 | 2.90 | 4.18 | .16 | [illegible] |
| Total | 100.00 | 100.00 | 100.00 | 100.00 | 100.00 | [illegible] |

The following comparative statement shows the average proportion of business done and the average number of days in operation for the 9 selected industries and for the 66 other industries considered together, in 1893 and 1894. The proportional amount of business done was computed by considering the maximum production—that is to say, the greatest amount of goods that can be turned out with the present facilities—as representing 100 per cent.

AVERAGE PROPORTION OF BUSINESS DONE AND AVERAGE DAYS IN [illegible] IN 75 INDUSTRIES, 1893 AND 1894.

| Industries. | Number of establishments. | Average proportion of business done. | | Average days in operation. | |
|---|---|---|---|---|---|
| | | 1893. | 1894. | [illegible] | [illegible] |
| Boots and shoes | 638 | 59.19 | [illegible] | [illegible] | [illegible] |
| Carpetings | 11 | 69.27 | [illegible] | [illegible] | [illegible] |
| Cotton goods | 148 | 85.97 | [illegible] | [illegible] | [illegible] |
| Leather | 141 | 61.90 | [illegible] | [illegible] | [illegible] |
| Machines and machinery | 332 | 60.12 | [illegible] | [illegible] | [illegible] |
| Metals and metallic goods | 327 | 59.96 | [illegible] | [illegible] | [illegible] |
| Paper and paper goods | 98 | 74.72 | [illegible] | [illegible] | [illegible] |
| Woolen goods | 115 | 75.42 | [illegible] | [illegible] | [illegible] |
| Worsted goods | 21 | 77.95 | [illegible] | [illegible] | [illegible] |
| Other industries (66) | 2,272 | 59.49 | [illegible] | [illegible] | [illegible] |
| Total | 4,093 | 61.49 | 59.06 | [illegible] | [illegible] |

In the 4,093 establishments making returns in each year [illegible] tion of business done in 1893 is represented by 61.49 [illegible] 1894 by 59.06 per cent. Fifteen industries reported a [illegible] of business done in 1894.

The average number of days in operation during [illegible]

sion of the proceeds of each industry, one
:t is paid to the labor force in the form of
itutes a fund from which are paid freights,
and stock, rents, commissions, salaries, etc.,
ıan those for stock and wages. The remain-
the employer.

.ND PROFIT AND EXPENSES IN NINE SPECIFIED INDUSTRIES, 1894.

| [illegible] | Wages. | Profit and expense fund. | Industry product. | | Per cent of industry product. | |
|---|---|---|---|---|---|---|
| | | | Per $1,000 of capital. | Average per employee. | Paid in wages. | Devoted to profit and expenses. |
| ,496 | $20,068,906 | $13,861,480 | $1,299.23 | $930.67 | 59.16 | 40.84 |
| ,131 | 1,329,350 | 809,781 | 293.95 | 571.35 | 62.14 | 37.86 |
| ,915 | 21,863,642 | 11,228,270 | 290.25 | 484.97 | 66.07 | 33.93 |
| ,156 | 2,722,360 | 1,843,795 | 547.22 | 797.16 | 59.62 | 40.38 |
| ,418 | 7,276,856 | 6,532,557 | 454.06 | 1,016.82 | 52.69 | 47.31 |
| ,044 | 6,591,565 | 3,585,489 | 525.47 | 865.84 | 64.77 | 35.23 |
| ,720 | 3,961,597 | 4,686,123 | 347.56 | 894.75 | 45.81 | 54.19 |
| ,067 | 4,887,984 | 4,377,063 | 384.53 | 649.68 | 52.76 | 47.24 |
| ,532 | 2,906,940 | 1,855,892 | 312.62 | 516.14 | 61.01 | 38.99 |

e statement shows the value of goods made
l by 857 identical establishments in each
usive:

MADE AND WORK DONE, 1885 TO 1894.

**a Decrease.**

e; the effect that the enactment of a prohibitory would have in respect of social conditions, agricultural, industrial and commercial interests, of the revenue, municipalities, provinces, and of the Dominion, and ity of efficient enforcement; all other information ...tion of prohibition."

comprises one volume of 1,003 pages. Numerous ...ined by the commission in the provinces of Ontario, ...hwest Territories, British Columbia, and in the evidence is contained in five volumes, having a

...ted by the liquor traffic in Canada are so varied, ...sive, and the data available so limited, that the ...le to do more than refer to the most prominent of ...supply such information in regard to them as they

...ars from 1889 to 1893 there were manufactured in average each year, 4,538,000 gallons of whisky ...as of beer and ale. Estimating the value of the ...nd of the beer and ale at 30 cents per gallon, and ...l refuse products sold at $800,000, the total value ...t of the establishments manufacturing spirits and ...

...the extent of the interests affected by the industr... ...ted of the values of the products of other ...age, fuel, certain farm crops, transportation, ... The estimates of the various amounts paid annually by the distilleries and [illegible] marized as follows:

| | |
|---|---|
| Raw materials, the products of the farm | [illegible] |
| Wages | [illegible] |
| Fuel | [illegible] |
| Transportation | [illegible] |
| Casks, bottles, cases, etc | [illegible] |
| Capsules, corks, etc | [illegible] |
| Printing, advertising, show cards, etc | [illegible] |
| Repairs, blacksmiths' work, etc | [illegible] |
| Insurance | [illegible] |
| Taxes, gas, water supplies, etc | [illegible] |
| Ice | [illegible] |
| Sundries | [illegible] |
| Total | [illegible] |

There was paid annually $1,038,671 for imported [illegible] $4,001,235 as the sum paid for Canadian products, wages, [illegible]

There are, based on an average for five years, [illegible] spirits, malt liquors, and wines imported annually into the [illegible] which are valued at $1,736,897.

Taking an average of the total amount of spirits, wine, [illegible] liquors entered for consumption for the five years ending [illegible] 1893, it was found to be 21,676,749 gallons per annum. [illegible] population for the same period was 4,834,876, making the [illegible] consumption 4.48 gallons. The valuation of the annual con[illegible] was placed at $15,030,064. Taking an average of the qua[illegible] wine, spirits, and malt liquors entered for consumption in the [illegible] ending 1893, but excluding cider and native wines, and takin[illegible] age of the retail prices, the calculation shows the sum of [illegible] to be paid for liquor by the consumers.

The total annual Government revenue derived from the tra[illegible] on the reports for five years, is given at $7,101,557.22.

How much of the crime, poverty, and insanity of the [illegible] be attributed to the use of intoxicating liquors could not be [illegible] determined.

The average number of convictions per year to each 1,000 [illegible] tion for different offenses is shown in the following statement [illegible]

CONVICTIONS PER 1,000 OF POPULATION, 1881 TO 1893.

| Periods. | For drunkenness. | For offenses against liquor laws. | [illegible] |
|---|---|---|---|
| Five years ending 1885 | 2.49 | 0.[illegible] | [illegible] |
| Five years ending 1890 | 2.72 | .[illegible] | [illegible] |
| Three years ending 1893 | 2.46 | .[illegible] | [illegible] |

In summarizing, it is stated that the statistics show [illegible]

1. An increase in the number of insane.

*2. A decrease in the number of commitments [illegible] and of those remaining therein.*

[illegible] the number of those arrested for offenses in the [illegible] towns, more particularly in those arrested for

[illegible] in the number of convictions for offenses of all kinds, [illegible] five years ended 1890 with the five years ended 1885, [illegible] in the convictions per 1,000 of the population in the [illegible]ded 1893 as compared with those for the five years [illegible]d a steady reduction in the yearly ratios from 1889 to

for the earlier years for which the statistics are given [illegible]o be less accurate than those for the later ones.

[illegible]e statistics of convictions for drunkenness for the whole [illegible]ill be found that the average for the five years ended [illegible] per 1,000 of the population. In the five years ended [illegible]e greater portion of which the Scott act(a) was in force [illegible]nber of counties in Ontario, the average was 2.72 per [illegible]opulation. In the three years ended 1893, the average [illegible]o 2.46. The highest ratios were in the years 1889 and [illegible]ere the years immediately following the abandonment of throughout the counties in Ontario. In them the ratio [illegible]rom that point there was a gradual reduction until, in reached 2.35 per 1,000.

[illegible]tics of the committals to, and those remaining in, the of the Dominion show a large decrease in the period [illegible]nd 1893.

[illegible] impracticable to make a summarization of the legisla- the liquor traffic or of the results of such legislation. [illegible]detail presentation concerning this, as well as the other [illegible] inquiry, should be consulted to obtain a correct idea of [illegible]scussed.

*of the Bureau of Industries for the Province of Ontario,* James, Secretary. Published by Ontario Department [illegible]re. xvi, 339 pp.

[illegible]s for the year 1894, and presents statistics on the follow- [illegible]Weather and the crops, 58 pages; live stock, the dairy, [illegible], 42 pages; values, rents, and farm wages, 39 pages; [illegible]tment companies, 28 pages; chattel mortgages, 4 pages; [illegible]stics, 174 pages.

[illegible]ND THE CROPS.—Tables giving temperature, sunshine, [illegible]ion, as observed at various well-distributed points [illegible]e province, furnish an interesting exhibit of weather the years 1893 and 1894, also the average for thirteen 1894) for temperature and precipitation, and for twelve [illegible]) for sunshine.

[illegible]3,038,974 acres of land assessed in the rural area of the [illegible]g 1894, of which 12,292,610 acres were cleared, there

marsh, or waste land. [illegible] the year and 8,315,153 in crops. [illegible] year sown in crops during the [illegible] 7,655,848. The acreage and yield of [illegible] the following statement:

ACREAGE AND YIELD OF [illegible]

| Products. | [illegible] | [illegible] |
|---|---|---|
| Fall wheat | [illegible] | [illegible] |
| Spring wheat | [illegible] | [illegible] |
| Barley | [illegible] | [illegible] |
| Oats | [illegible] | [illegible] |
| Rye | [illegible] | [illegible] |
| Pease | [illegible] | [illegible] |
| Corn, for husking | [illegible] | [illegible] |
| Corn, for silo and fodder | [illegible] | [illegible] |
| Buckwheat | [illegible] | [illegible] |
| Beans | [illegible] | [illegible] |
| Potatoes | [illegible] | [illegible] |
| Mangel-wurzels | [illegible] | [illegible] |
| Carrots | [illegible] | [illegible] |
| Turnips | [illegible] | [illegible] |
| Hay and clover | [illegible] | [illegible] |

a Tons.

LIVE STOCK, THE DAIRY, AND THE APIARY.—[illegible] the statistics of 1894, given under this title, are shown in [illegible] summarized statement:

Horses, number ....

Hogs, number ....

Horned cattle, number ....

Sheep, number ....

Wool clip:

Number of fleeces ....

Pounds of wool ....

Pounds of wool per fleece ....

Poultry, number of fowls ....

Cheese factories:

Number reporting .... [illegible]

Milk used, pounds .... [illegible]

Cheese made, pounds .... [illegible]

Gross value of cheese .... [illegible]

Creameries:

Number reporting ....

Butter made, pounds ....

Value of butter ....

Apiary outfit:

Hives of bees, number .... [illegible]

Value of outfit .... [illegible]

VALUES, RENTS, AND FARM WAGES.—The [illegible] property for 1894 are summarized in the following [illegible]

Farm land .... [illegible]

Buildings .... [illegible]

Implements .... [illegible]

Live stock on hand .... [illegible]

| | [illegible] | [illegible] |
|---|---|---|
| [illegible] | [illegible] | [illegible] |
| [illegible] | [illegible] | [illegible],047,711 |
| [illegible] | [illegible] | 50,[illegible] |
| [illegible] | 84,916,[illegible] | [illegible] |
| [illegible] | 134,[illegible],4[illegible] | 137,541,741 |
| [illegible] | 118,040,915 | 120,[illegible],818 |
| [illegible] | 16,161,573 | 17,311,[illegible] |
| [illegible] | 134,[illegible],4[illegible] | 137,541,741 |

[illegible]—During the year ending December 31, 1894, [illegible] mortgages, representing $11,220,205, on record [illegible]. Of this number 11,687, represent-[illegible] farmers. In 1893 the chattel mortgages [illegible] represented $9,333,385, of which 10,684, repre-[illegible] farmers.

[illegible]STICS.—The details presented for the municipal [illegible]ince for the year 1893 are summarized in the fol-

| | |
|---|---|
| [illegible] | 1,910,059 |
| [illegible] | $825,530,052.00 |
| [illegible] | 12,522,660.00 |
| [illegible] | 6.56 |
| [illegible] | 15.17 |
| [illegible] | 48,083,243.00 |
| [illegible] | a 28.17 |
| [illegible] | 6,796,422.00 |
| [illegible] | 2,508,691.00 |

[illegible] be $25.17; those given are, however, according

*Die Arbeitseinstellungen im Gewerbebetriebe* [illegible] "Statistischen Monatschrift," 1894 [illegible] *Gewerbebetriebe in Österreich Während* [illegible] geben vom Statistischen Departement im [illegible] 31, 128 pp.

The Austrian Government has been collecting [illegible] each year since 1891. Those for the years 1891 and 1892 [illegible] by the Government, but not for general distribution. [illegible] 1893 was published in the form of a supplement to the monthly [illegible] tical bulletin, Statistische Monatschrift. The last report, [illegible] the first that appeared in the form of a special report of the [illegible] statistics of the Imperial Ministry of Commerce.

The statistics for 1891 and 1892 appeared in Bulletin No. [illegible] Department of Labor, in an article on strikes in Austria. [illegible] was prepared from data obtained from Volume XI of the foreign [illegible] of the British Royal Commission on Labor. The statistics [illegible] in the present article are obtained from the above-mentioned [illegible] reports of the Austrian Government.

The strike statistics in these reports do not cover agricultural [illegible] estry, or mining industries. These will be separately treated [illegible] report soon to be published by the Austrian Minister of Agriculture [illegible]

The statistics for the two years here presented were collected [illegible] ing to such different methods, that it will be necessary to show [illegible] information in separate tables for each year. The report for [illegible] embraces but two general tables, one showing strikes according [illegible] localities, and the other by industries affected. The essential [illegible] of the second are reproduced in the following table:

STRIKES IN 1893, BY INDUSTRIES.

| Industries. | Strikes. | Establishments. | Employees. | | | Cause or object. | | | Results [illegible] | |
|---|---|---|---|---|---|---|---|---|---|---|
| | | | Total. | Strikers. | Days lost. | For increase of wages or that and other demands | Against reduction of wages. | All other. | Succeeded. | [illegible] |
| Building | 10 | 249 | 12,406 | 9,892 | 209,155 | 7 | ........ | 3 | ........ | [illegible] |
| Brewing | 7 | 16 | 1,225 | 222 | 3,067 | 7 | ........ | ........ | 1 | [illegible] |
| Stonecutting | 1 | 135 | 700 | 700 | 52,500 | 1 | ........ | ........ | ........ | [illegible] |
| Dyeing, bleaching, and finishing | 7 | 18 | 1,291 | 1,183 | 19,109 | 5 | ........ | 2 | 4 | [illegible] |
| Board sawing | 1 | 1 | 44 | 19 | 38 | 1 | ........ | ........ | ........ | [illegible] |
| Printing | 3 | 3 | 48 | 31 | 88 | ........ | ........ | 3 | ........ | [illegible] |
| Paper-box making | 1 | 1 | 85 | 70 | 350 | 1 | ........ | ........ | ........ | [illegible] |
| Cement | 1 | 1 | 30 | 27 | 54 | ........ | ........ | 1 | ........ | [illegible] |
| Piano | 1 | 1 | 55 | 55 | 985 | 1 | ........ | ........ | ........ | [illegible] |
| Wood turning | 10 | 37 | 570 | 370 | 10,153 | 7 | ........ | 3 | [illegible] | [illegible] |
| Ribbon printing | 1 | 1 | 196 | 150 | 1,350 | 1 | ........ | ........ | [illegible] | [illegible] |
| Gas and water works | 1 | 1 | 33 | 33 | 66 | 1 | ........ | ........ | [illegible] | [illegible] |
| Glass and ceramics | 9 | 229 | 3,617 | 2,051 | 32,580 | 2 | 6 | 1 | [illegible] | [illegible] |
| Rubber goods | 2 | 2 | 1,063 | 343 | 3,[illegible] | 2 | ........ | ........ | [illegible] | [illegible] |
| Hats | 3 | 3 | 574 | 348 | 534 | ........ | 3 | [illegible] | [illegible] | [illegible] |
| Cartographing and lithographing | 1 | 1 | 30 | 23 | 368 | ........ | ........ | 1 | [illegible] | [illegible] |

| | | | | | |
|---|---|---|---|---|---|
| 9 | ........ | 2 | 2 | 3 | 6 |
| 1 | ........ | ...... | ...... | 1 | ...... |
| 1 | 1 | ...... | 1 | ...... | 1 |
| ...... | ........ | 1 | ...... | 1 | ...... |
| 1 | ........ | ...... | ...... | 1 | ...... |
| 101 | 20 | 51 | 33 | 55 | 84 |

details concerning strikes, with
s are given in more comprehen-
rmation concerning each strike,
; to localities and industries, the
ing to the months in which they
o their duration, and the results
n these tables is also much more
ng them than in the report for
ne same information as that for
concerning strikes in 1894:

INDUSTRIES.

| lish- ts. | Total employees. | Strikers. | Strikers reemployed. | New employees after strikes. |
|---|---|---|---|---|
| 130 | 7,717 | 6,415 | 6,235 | 104 |
| 38 | 4,606 | 2,752 | 2,523 | 165 |
| 7 | 579 | 194 | 103 | 45 |
| 666 | 13,818 | 9,793 | 9,579 | 104 |
| 19 | 765 | 641 | 421 | 107 |
| 46 | 10,467 | 6,317 | 5,624 | 529 |
| 145 | 422 | 194 | 194 | ........ |
| 22 | 837 | 668 | 511 | 114 |
| 1 | 25 | 24 | 23 | ........ |
| 97 | 1,021 | 299 | 283 | 14 |
| 3 | 1,612 | 1,268 | 468 | 932 |
| 858 | 18,921 | 14,975 | 14,397 | 34 |
| 5 | 152 | 85 | 60 | 24 |
| 1 | 166 | 104 | ........ | 100 |
| 2 | 509 | 249 | 50 | 118 |
| 3 | 98 | 97 | 97 | ........ |
| 466 | 60,718 | 44,075 | 40,567 | 2,[illegible] |

ON AND RESULTS OF STRIKES IN 1894, BY INDUSTRIES.

| Duration of strikes. | | | | | Results of strikes | | | | |
|---|---|---|---|---|---|---|---|---|---|
| 10 days and under | 11 to 20 days. | 21 to 30 days | 31 to 40 days. | Over 40 days. | Succeeded. | | Succeeded partly. | | N b |
| | | | | | Number. | Per cent. | Number. | Per cent. | |
| 17 | 2 | 1 | ...... | 2 | 6 | 27.27 | 10 | 45.46 | |
| 15 | 3 | 1 | 3 | 1 | 6 | 26.09 | 6 | 26.09 | |
| 4 | 3 | ...... | ...... | ...... | ...... | ...... | ...... | ...... | |
| 16 | 3 | 1 | 1 | 2 | 8 | 34.78 | 6 | 26.09 | |
| 5 | 2 | ...... | ...... | 2 | 3 | 33.33 | 1 | 11.11 | |
| 25 | 6 | 1 | ...... | 2 | 2 | 5.88 | 10 | 29.41 | |
| ...... | 1 | ...... | ...... | ...... | ...... | ...... | 1 | 100.00 | .. |
| 6 | 1 | ...... | 1 | 1 | 4 | 44.45 | 3 | 33.33 | |
| 1 | ...... | ...... | ...... | ...... | ...... | ...... | ...... | ...... | |
| 7 | ...... | ...... | ...... | ...... | 2 | 28.57 | 2 | 28.57 | |
| 2 | ...... | ...... | ...... | ...... | ...... | ...... | ...... | ...... | |
| 9 | 1 | 1 | ...... | ...... | 4 | 36.36 | 2 | 18.18 | |
| 3 | 2 | ...... | ...... | ...... | 3 | 60.00 | ...... | ...... | |
| 1 | ...... | ...... | ...... | ...... | ...... | ...... | ...... | ...... | |
| 2 | ...... | ...... | ...... | ...... | ...... | ...... | 1 | 50.00 | |
| 2 | ...... | ...... | ...... | ...... | 1 | 50.00 | 1 | 50.00 | .. |
| 115 | 24 | 5 | 5 | 10 | 39 | 24.53 | 43 | 27.04 | |

CAUSES OF STRIKES IN 1894, BY INDUSTRIES.

| Against reduction of wages. | For increase of wages. | For regular payments | For reduction of | For discharge of foremen, | Against discharge of employees | For reinstatement of discharged em- | Against obnoxious rules. | For Labor Day, May 1. | C c |
|---|---|---|---|---|---|---|---|---|---|

[illegible] each group figures is shown in the following [illegible]:

| [illegible] | Per cent of strikes. | Per cent of days lost. |
|---|---|---|
| [illegible] | 28.98 | 23.14 |
| [illegible] goods | 27.21 | 49.86 |
| [illegible] | 14.65 | 5.48 |
| [illegible] | 14.23 | 8.06 |
| [illegible] | 14.93 | 13.46 |

ation of strikes was, in general, very short. Out of a total ikes, 115 lasted less than 11 days. The longest strike lasted while the average duration was 11.68 days.

ig the causes of strikes, the Austrian bureau has adopted a method of presentation. As strikes may, and usually do, m a variety of causes, it has been thought preferable to use instead of the strike as the unit. The table, therefore, shows er of times that each cause figured as the incentive to a hus there is shown a total of 318 causes for 159 strikes.

be seen that the demands for an increase of wages and for a of hours are by far the most frequent causes of strikes. Of of 318 causes, 88, or 27.67 per cent, were due to the former, 13.52 per cent, to the latter cause.

reau has, however, also made a calculation of the causes of cording to the more usual method of using the strike as the ch a presentation, together with the percentage of strikes ch cause, is given in the following statement:

ER CENT OF STRIKES DUE TO EACH SPECIFIED CAUSE, 1894.

| Cause. | Strikes. | |
|---|---|---|
| | Number. | Per cent. |
| [illegible] of wages | 11 | 6.92 |
| [illegible] of wages in connection with various other demands | 7 | 4.40 |
| [illegible] wages | 21 | 13.21 |
| [illegible] wages and reduction of hours | 9 | 5.66 |
| [illegible] wages and reduction of hours in connection with other demands | 24 | 15.09 |
| [illegible] wages in connection with other demands, but not including [illegible] hours | 34 | 21.38 |
| [illegible] | 2 | 1.26 |
| [illegible] in connection with other demands | 1 | 0.63 |
| [illegible] of hours | 3 | 1.89 |
| [illegible] of hours in connection with other demands, but not including [illegible] | 7 | 4.40 |
| [illegible] foremen or superintendents | 6 | 3.77 |
| [illegible] foremen or superintendents in connection with other demands | 3 | 1.89 |
| [illegible] discharged employees | 13 | 8.18 |
| [illegible] discharged employees in connection with other demands | 5 | 3.14 |
| [illegible] | 1 | 0.63 |
| [illegible] | 12 | 7.55 |
| [illegible] | 159 | 100.00 |

It is thus seen that 21, or [illegible] single demand for higher wages, [illegible] higher wages and shorter hours [illegible] demands, and 34, or 21.38 per cent, [illegible] connection with other demands, not including [illegible]

Regarding the results of strikes, the first table [illegible] 39, or 24.53 per cent of all strikes, were [illegible] cent were partly successful, and 77, or 48.43 per cent, [illegible]

The most important information, however, that can be [illegible] cerning the results of strikes is that where they are [illegible] to the causes for which strikes were undertaken. It is [illegible] obtain this for the first time for the year 1894. This is done in th[illegible] lowing table:

RESULTS OF STRIKES, BY CAUSES, 1894.

| Cause. | Succeeded. | | | Succeeded partly. | | | Failed. | | | [illegible] | |
|---|---|---|---|---|---|---|---|---|---|---|---|
| | Strikes. | Establishments | Strikers. | Strikes. | Establishments | Strikers. | Strikes. | Establishments | Strikers. | [illegible] | [illegible] |
| Against reduction of wages | 10 | 25 | 1,525 | ...... | ...... | ...... | 6 | 9 | [illegible] | [illegible] | [illegible] |
| For increase of wages | 20 | 74 | 3,122 | 31 | 314 | 7,510 | 37 | [illegible] | 12,779 | [illegible] | [illegible] |
| For regular payments | 3 | 3 | 162 | ...... | ...... | ...... | ...... | ...... | ...... | 3 | [illegible] |
| For reduction of hours | 15 | 41 | 1,788 | 3 | 210 | 467 | 25 | [illegible] | [illegible] | [illegible] | [illegible] |
| For discharge of foremen, etc | 2 | 2 | 98 | ...... | ...... | ...... | 14 | 14 | [illegible] | [illegible] | [illegible] |
| Against discharge of employees | 6 | 6 | 479 | ...... | ...... | ...... | 17 | 18 | [illegible] | [illegible] | [illegible] |
| For reinstatement of discharged employees | 3 | 3 | 154 | 1 | 1 | 68 | 24 | 24 | [illegible] | [illegible] | [illegible] |
| Against obnoxious rules | 1 | 9 | 295 | ...... | ...... | ...... | ...... | ...... | ...... | [illegible] | [illegible] |
| For Labor Day, May 1 | 6 | 10 | 4,498 | 1 | 11 | 488 | 17 | [illegible] | [illegible] | [illegible] | [illegible] |
| Other causes | 29 | 40 | 7,089 | 8 | 83 | 1,564 | 87 | [illegible] | [illegible] | [illegible] | [illegible] |
| Total (b) | 95 | 213 | 19,210 | 44 | 619 | 10,083 | 179 | 7,437 | [illegible] | [illegible] | [illegible] |

a These figures do not represent the totals as shown by the other columns; they are, [illegible] as reported.

b A considerable number of strikes were due to two or three causes, and the [illegible] have been tabulated under each cause. Hence the totals for this table necessarily [illegible] those for some of the preceding tables.

As already explained, the systems of presentations of [illegible] and 1894 differ so materially that a comparison of one [illegible] is difficult. However, from figures for the four years [illegible] report, a reasonably accurate idea may be gained of the [illegible] in Austria during that period, namely, 1891 to 1894, [illegible]

ng, and the number of working days lost. 891 and 1894, it is found that in every case year than for the first year.

ean material contained in the reports, shows of strikes according to the principal causes

S, BY CAUSES AND RESULTS, 1891 TO 1894.

| | 1891. | | 1892. | | 1893. | | 1894. | |
|---|---|---|---|---|---|---|---|---|
| | Number. | Per cent. | Number. | Per cent. | Number. | Per cent. | Number. | Per cent. |
| ... | 26 | 25.00 | 19 | 18.81 | 38 | 22.09 | 21 | 13.21 |
| ... | 16 | 15.39 | 19 | 18.81 | 20 | 11.63 | 11 | 6.92 |
| th ... | 28 | 26.92 | 32 | 31.69 | 63 | 36.63 | 67 | 42.13 |
| on of ... | 7 | 6.73 | 9 | 8.91 | 5 | 2.91 | 10 | 6.29 |
| ... | 7 | 6.73 | 15 | 14.85 | 21 | 12.21 | 19 | 11.95 |
| ... | 20 | 19.23 | 7 | 6.93 | 25 | 14.53 | 31 | 19.50 |
| ... | 19 | 18.63 | 26 | 25.74 | 33 | 19.18 | 39 | 24.53 |
| ... | 29 | 28.43 | 29 | 28.71 | 55 | 31.98 | 43 | 27.04 |
| ... | 54 | 52.94 | 46 | 45.55 | 84 | 48.84 | 77 | 48.43 |
| ... | a 104 | 100.00 | 101 | 100.00 | 172 | 100.00 | 159 | 100.00 |

ported, hence they are omitted under "Strikes by results."

at the percentage of strikes resulting from ges alone has decreased materially during that of strikes resulting from demands for ection with reduction of hours and other ased. At the same time, the percentage of reased, while there was a corresponding ailures.]

*Fourth Annual Report of the Department* [illegible]
W. P. Reeves, Minister of [illegible]

This report, which is for the year ending [illegible] condition of the labor market; assistance rendered by the [illegible] in procuring employment; the establishment of labor [illegible] farms; the effect of certain features of the factory [illegible] and shop-assistant's acts of 1894; labor disturbances; [illegible] ing with the unemployed; reports of factory inspectors and [illegible] employees in various industries.

During the year the department assisted 2,007 married and 1,023 single men in finding employment. The total number of men assisted in this manner since the organization of the department is as follows:

MEN ASSISTED IN PROCURING EMPLOYMENT.

| Year. | Number. | [illegible] |
|---|---|---|
| June 1, 1891, to March 31, 1892 | [illegible] | [illegible] |
| April 1, 1892, to March 31, 1893 | [illegible] | [illegible] |
| April 1, 1893, to March 31, 1894 | [illegible] | [illegible] |
| April 1, 1894, to March 31, 1895 | [illegible] | [illegible] |
| Total | [illegible] | [illegible] |

Of the men assisted during the year ending March 31, 1895, there were 894 sent to private employment and 2,136 to Government [illegible] The nonemployment of 3,004 of the number was due to [illegible] trade and similar causes, while sickness was given as the [illegible] 26 cases. Of the persons dependent on those assisted, [illegible] wives, 330 parents and others, and 6,546 children.

The provision of the factory inspection law of New Zealand [illegible] makes it compulsory that the written permit of the inspector [illegible] spicuously fastened to the wall of the room in which overtime [illegible] worked, is proving a great safeguard. The requirement of the [illegible] law that all goods given out as piecework to be done in a [illegible] dwelling, or in any place not registered as a factory, shall have [illegible] to it a printed label describing the place where the work was [illegible] stating that it is an unregistered workshop, which label [illegible] removed before the goods are finally sold, has had good [illegible] venting owners of factories giving out material to be made [illegible] whose dwellings are unfit to be used as workshops for the [illegible] of clothing. It has probably not prevented poor women [illegible] ing work, because where any two persons (such as mother [illegible] or two friends) choose to work together they can register [illegible] and their workshop be under preper inspection.

The number of persons working under the factories [illegible] of whom 22,324 were men and 7,555 women. This was [illegible] 4,028 over the year ending March 31, 1894, the difference [illegible]

2 vols.: xxvii, 726, 916 pp.

ndertaken as a purely private enterprise by awarding to it the Rossi prize in 1890 and ral and Political Sciences, unable to pub- earnestly recommended that the Comité Ministère de l'Instruction Publique, issue Documents Inédits sur l'Histoire de France. therefore now appears as a public docu-

essentially documentary. In general the undertaken much the same work as that Mr. Thorold Rogers in his History of Agri- Britain. Of the 1,669 pages embraced within only are devoted to introductory remarks naining 1,148 pages consisting of quotations ese latter pages all but 33 consist of a mere notations of prices, values, and rents. Each e whence derived, the locality, the date, the quantity expressed in the measure now in the old measure quoted and the corre- the modern measure, and finally, the price easure in use at the present time. All the article are given in one place, according to for each article a series of quotations in g the entire period from 1201 to 1800. These the value of different kinds of agricultural es at Paris and elsewhere in France, (3) the , (4) the rents of houses, and (5) the prices and of bread.

es the attempt is made to calculate the aver- ducing power of land at different periods prices of the more important cereals, both f France and for the whole country gen- g the average price of wheat for each year be obtained during the period 1201 to 1800, nd oats during the period 1601 to 1800, and ds 1201 to 1800, are of such general impor- reproduction. As regards the single article contains a chart prepared by M. Levasseur,

based on material contained in [illegible]
showing graphically the course of [illegible]

AVERAGE PRICE PER BUSHEL OF [illegible]

| Year. | Price. | Year. | Price. | Year. |
|---|---|---|---|---|
| 1201 | $0.208 | 1326 | $0.498 | 1409 |
| 1202 | .332 | 1327 | .489 | 1410 |
| 1203 | .162 | 1328 | .307 | 1411 |
| 1211 | .231 | 1329 | .450 | 1412 |
| 1220 | .277 | 1331 | .380 | 1413 |
| 1222 | .324 | 1332 | .453 | 1414 |
| 1224 | .171 | 1333 | .584 | 1415 |
| 1226 | .524 | 1334 | .871 | 1416 |
| 1228 | .303 | 1335 | .127 | 1417 |
| 1229 | .174 | 1337 | .236 | 1418 |
| 1233 | .204 | 1338 | .154 | 1419 |
| 1237 | .595 | 1339 | .407 | 1420 |
| 1238 | .317 | 1340 | .285 | 1421 |
| 1239 | .473 | 1341 | .535 | 1422 |
| 1241 | .171 | 1342 | .224 | 1423 |
| 1247 | .441 | 1343 | .569 | 1424 |
| 1249 | .372 | 1344 | .606 | 1425 |
| 1250 | .743 | 1345 | .293 | 1426 |
| 1251 | .218 | 1346 | .748 | 1427 |
| 1253 | .394 | 1347 | .540 | 1428 |
| 1255 | .280 | 1348 | .756 | 1429 |
| 1256 | .583 | 1349 | .565 | 1430 |
| 1258 | .396 | 1350 | 2.049 | 1431 |
| 1259 | .318 | 1351 | 1.183 | 1432 |
| 1260 | .206 | 1353 | .318 | 1433 |
| 1261 | .564 | 1354 | .619 | 1434 |
| 1263 | .224 | 1355 | .330 | 1435 |
| 1264 | .148 | 1356 | .511 | 1436 |
| 1265 | .373 | 1357 | .212 | 1437 |
| 1268 | .112 | 1358 | .958 | 1438 |
| 1269 | .596 | 1359 | .915 | 1439 |
| 1271 | .628 | 1360 | .529 | 1440 |
| 1272 | .602 | 1361 | 1.059 | 1441 |
| 1273 | .136 | 1362 | .326 | 1442 |
| 1276 | .152 | 1363 | .900 | 1443 |
| 1277 | .182 | 1364 | .628 | 1444 |
| 1278 | .197 | 1365 | .516 | 1445 |
| 1281 | .333 | 1366 | .644 | 1446 |
| 1282 | .531 | 1367 | .488 | 1447 |
| 1284 | .080 | 1368 | .318 | 1448 |
| 1285 | .762 | 1369 | .867 | 1449 |
| 1287 | .240 | 1370 | .660 | 1450 |
| 1288 | .169 | 1371 | 1.450 | 1451 |
| 1289 | .250 | 1372 | .310 | 1452 |
| 1290 | .484 | 1373 | .536 | 1453 |
| 1291 | .642 | 1374 | 1.048 | 1454 |
| 1293 | .662 | 1375 | .345 | 1455 |
| 1294 | .516 | 1376 | .316 | 1457 |
| 1295 | .645 | 1378 | .374 | 1458 |
| 1296 | .380 | 1379 | .224 | 1459 |
| 1297 | .302 | 1380 | .192 | 1460 |
| 1298 | .484 | 1381 | .162 | 1461 |
| 1299 | .884 | 1382 | .272 | 1462 |
| 1300 | 1.020 | 1384 | .258 | 1463 |
| 1301 | .350 | 1385 | .340 | 1464 |
| 1302 | .380 | 1386 | .227 | 1465 |
| 1303 | .373 | 1387 | .169 | 1466 |
| 1304 | 1.192 | 1388 | .214 | 1467 |
| 1305 | .466 | 1389 | .364 | 1468 |
| 1307 | .569 | 1390 | .380 | 1469 |
| 1309 | .367 | 1391 | .441 | 1470 |
| 1310 | .216 | 1392 | .478 | 1471 |
| 1311 | .687 | 1393 | .596 | 1472 |
| 1312 | .571 | 1394 | .498 | 1473 |
| 1313 | 1.711 | 1395 | .150 | 1474 |
| 1314 | .363 | 1396 | .336 | 1475 |
| 1315 | .840 | 1397 | .289 | 1476 |
| 1316 | 1.472 | 1398 | .309 | 1477 |
| 1317 | .288 | 1399 | .340 | 1478 |
| 1318 | .534 | 1400 | .171 | 1479 |
| 1319 | .547 | 1401 | .330 | 1480 |
| 1320 | .813 | 1402 | .224 | 1481 |
| 1321 | .351 | 1403 | .316 | 1483 |
| 1322 | .779 | 1404 | .186 | 1484 |
| 1323 | .406 | 1405 | .414 | 1485 |
| 1324 | .728 | 1406 | .354 | 1486 |
| 1325 | [illegible] | 1408 | .198 | 1487 |

| Oats. |
| --- |
| $0.211 |
| a.563 |
| .240 |
| .236 |
| .311 |
| ....... |
| .380 |
| .348 |
| .341 |
| .350 |
| .436 |
| .172 |
| ....... |
| .210 |
| .308 |
| .128 |
| .301 |
| .237 |
| .341 |
| .231 |
| .285 |
| ....... |
| .378 |
| ....... |
| .305 |
| .151 |
| .340 |
| .283 |
| .261 |
| .190 |
| .182 |
| .141 |
| .150 |
| .[illegible] |
| 1.[illegible] |
| .365 |
| ....... |
| .483 |
| .233 |

AVERAGE PRICE PER BUSHEL [illegible]

| Year. | Prices. | | |
|---|---|---|---|
| | Rye. | Barley. | Oats. |
| [illegible] | $0.437 | | $0.[illegible] |
| 1720 | .[illegible] | $0.456 | .[illegible] |
| 1725 | .[illegible] | .575 | .[illegible] |
| [illegible] | .424 | .[illegible] | .[illegible] |
| 1727 | | | .[illegible] |
| 1730 | .[illegible] | .[illegible] | .[illegible] |
| 1735 | .[illegible] | .178 | .[illegible] |
| [illegible] | | | .145 |
| [illegible] | .348 | | |
| [illegible] | .518 | .[illegible] | .[illegible] |
| 1740 | .[illegible] | .[illegible] | .[illegible] |
| 1741 | | .475 | .[illegible] |
| 1745 | .[illegible] | .[illegible] | .[illegible] |
| 1746 | | | .[illegible] |
| 1747 | .375 | .517 | .[illegible] |
| 1748 | .379 | | .[illegible] |
| 1750 | .441 | .[illegible] | .[illegible] |
| 1751 | .538 | .412 | .[illegible] |
| 1752 | .765 | .[illegible] | .700 |
| 1755 | .[illegible] | .[illegible] | .[illegible] |
| 1756 | .[illegible] | | |
| 1758 | 1.058 | | .[illegible] |
| 1760 | .[illegible] | .[illegible] | .[illegible] |
| 1761 | .353 | .[illegible] | .[illegible] |
| 1762 | .581 | .449 | .[illegible] |
| 1763 | .[illegible] | .[illegible] | .[illegible] |
| 1764 | .784 | .[illegible] | .[illegible] |
| 1765 | .[illegible] | .303 | .210 |

AVERAGE PRICE PER BUSHEL OF WHEAT, RYE, BARLEY, AND [illegible]
1201 TO 1890, BY PERIODS.

| Period. | Prices. | | | | Period. | Prices. | |
|---|---|---|---|---|---|---|---|
| | Wheat. | Rye. | Barley. | Oats. | | Wheat. | [illegible] |
| 1201-1225 | $0.258 | $0.129 | $0.088 | $0.104 | 1526-1550 | $0.[illegible] | [illegible] |
| 1226-1250 | .280 | .256 | .109 | .[illegible] | 1551-1575 | .[illegible] | [illegible] |
| 1251-1275 | .394 | .340 | .131 | .[illegible] | 1576-1600 | 1.[illegible] | [illegible] |
| 1276-1300 | .436 | .417 | .237 | .090 | 1601-1625 | .[illegible] | [illegible] |
| 1301-1325 | .589 | .408 | .272 | .156 | 1626-1650 | 1.[illegible] | [illegible] |
| 1326-1350 | .456 | .340 | .272 | .[illegible] | 1651-1675 | 1.[illegible] | [illegible] |
| 1351-1375 | .612 | .340 | .224 | .181 | 1676-1700 | .[illegible] | [illegible] |
| 1376-1400 | .317 | .190 | .136 | .[illegible] | 1701-1725 | 1.[illegible] | [illegible] |
| 1401-1425 | .490 | .238 | .204 | .[illegible] | 1726-1750 | .[illegible] | [illegible] |
| 1426-1450 | .456 | .313 | .214 | .[illegible] | 1751-1775 | .[illegible] | [illegible] |
| 1451-1475 | .221 | .156 | .105 | .071 | 1776-1800 | 1.[illegible] | [illegible] |
| 1476-1500 | .272 | .204 | .110 | .[illegible] | [illegible] | 1.[illegible] | [illegible] |
| 1501-1525 | .272 | .224 | .104 | .109 | | | |

*Conseil Supérieur du Travail, Ministère de l'Agri*[illegible] *et des Travaux Publics.* 1re Session, 1892: Applic[illegible] 4, 6, et 7 de la Loi du 13 décembre, 1889. 2e Session, [illegible] 1re Partie: Minimum de Salaire. 2e Partie: Durée [illegible] les Briqueteries. 3e Partie: Règlements d'Atelier[illegible] tistiques du Travail.

By royal order of April 7, 1892, the King of Belgi[illegible] the Department of Agriculture, Industry, and Publi[illegible] council of labor, to consist of 16 representatives [illegible] labor, 16 representatives of the laborers themselve[illegible] selected on account of their special [illegible] social questions, or 48 members in all. [illegible] industrial education, factory rules, in factories and workshops; the accidents, etc., in a word, all matters labor and capital, and, finally, to organization of a statistical service formation concerning labor.

as the council up to the present time ated above.

investigation concerning the modifi- in articles 4, 6, and 7 of the law of e hours of labor and conditions of children in industrial establishments. or, created by the law of August 16, industrial centers of Belgium, were is question; and on the basis of the perior council of labor prepared vari- commendations upon which its mem- of the different councils of industry liberations and recommendations of contents of the volume. No attempt gathered in a statistical form.

cussion had by the superior council making it obligatory upon the Gov- or public works a provision requiring ployees wages not inferior to a mini- Government; in other words, to fix a engaged on work for the Government. sted, expression being given at the Government ought to encourage the ng of minimum wages was declared izations, and it was highly desirable nized and therefore in a position to he intervention of the State.

nsideration of the regulation of the nd tile works.

he results of an investigation of the ught to intervene in the way of regu- es, the imposition of fines for their uiries covering this subject was first industry and labor, the answers to cussions by the council. The council he proposed law, the nature of which ught to be covered by a set of shop blicity, methods of enforcement, etc. 's devoted to a consideration of the

yzed. The second, or chronological index, gives a decrees of each country according to the order of [illegible], with reference to the pages where they are

inces its intention to continue the work here begun f periodical bulletins reproducing new legislation

*ers aux États-Unis.* Par M. Isidore Finance. Ex-des Délégués Ouvriers, Exposition Internationale Ministère du Commerce, de l'Industrie, des Postes s. 214 pp.

portion of the official report of a delegation of sited America on the occasion of the World's Fair inder the auspices of the French Government, to is of labor in the United States. It is devoted and description of those labor organizations in the i are of a national character. Two brief chapters concerning the general history of the attempts of i organizations, and the other on the general labor nited States, especially as it relates to the right of ndividual national labor organization is then taken ite chapter, and an account given of its organization, ites and places of its annual conventions, and other s history. The histories of 57 organizations, com- of the Knights of Labor and of the American , are given in this way. Two concluding chapters the principles and work of building trades' coun- ngman's budget (that of an employee of the build- York City earning $3.50 per day), and brief com- ral features of the condition of organized labor in

to be based on original sources. It should be said author, is the chief of the division in the Office du particularly to labor legislation and labor organ-

1892. In these eleven volumes, representing ... to the present time, is presented the most ... relating to the question of accidents to ... of workingmen against sickness, accidents, ... in any language.

*... Française des Habitations à Bon Marché.* M. ...rin, Secrétaire Général. 1890–1895, 6 vols.

...des Habitations à Bon Marché was founded at ... 1889, as the direct result of the Congrès Interna... à Bon Marché held in connection with the Inter... of Paris of that year. Its object is to encourage the ...viduals, manufacturers, or local societies, of sanitary ... for workingmen, or the improvement of existing ...pecially to diffuse information concerning the best ... workingmen to become the owners of their own ...

...de of action is to place at the disposition of indi... , plans, models of constitutions, forms of official ... loan, sale, etc., and to act as a technical consulting ...dvice, when sought, will be given gratuitously con... for the organization of societies or for the pro...en's houses. All direct work by it in the way of ... purchase of ground, or the erection of houses is ... by its constitution.

...t work is, therefore, the publication of a bulletin ... regularly four times a year since 1890. The greater ...nts of this bulletin is given up to detailed descrip...ons of societies or individuals for the provision of ...es. These accounts embrace not only a general ...xtent of the work, but statements in detail of the ... the buildings, accompanied by elaborate architec... of the constitutions of the societies, or the condi...hey are occupied or can be acquired, and analyses ...ts obtained. In this way the society not only pre...ncerning what has been accomplished in the past, ..., models of constitutions, etc., which may serve as ... guide to others wishing to undertake operations

...statistics, by C. P. Leland.
... the United States since 1840, by Carroll

... die Fortschritte der Statistik in Oesterreich seit
the progress of statistics in Austria since 1891], by
...

...nung eines internationalen Sterblichkeitsmasses
) [Concerning the calculation of an international
of mortality)], by Joseph Körösi.
ionale Classificierung der Berufsarten [An interna-
... of occupations], by Joseph Körösi.
...grammes et cartogrammes sur la femme en France,
... dans le "Woman Building," à la "World Fair"
and charts relating to woman in France, exhibited
... Women's Building of the World's Fair], by Victor

...ifications of occupations in the United States, by

...ata for the study of the assimilation of races and
... United States, by Richmond Mayo-Smith.
...hical distribution of the population of the United
...annett.
... and mode of statistics of marriage and divorce, by

...e des professions: Rapport sur les observations faites
...teurs de services statistiques, à propos du projet de
... professions présenté à leur examen par l'Institut
...Statistique (session de Vienne, 1891) [Nomenclature
...port on the observations made by different directors
...us in regard to the proposition for a nomenclature
...sented to them for examination by the International
...e, Vienna meeting, 1891], by Dr. Jacques Bertillon.
...menclature des professions [Proposed nomenclature
... Dr. Jacques Bertillon.
...y report on anthropometry in the United States, by
...artwell.
... the theory of anthropometry, by Franz Boas.
...ication to individual school children of the means
...pological measurements by the generalizing method,
...Porter.
...ric statistics of Amherst College, by Edward Hitch-

...metrical study of the effects of gymnastic training
... by Claës J. Enebuske.

42 of the Northeastern Reporter, page 40.
brought by John A. Newbauer and others against
nd others for the foreclosure of a mechanic's lien.
in favor of the plaintiffs by the circuit court of
the defendants appealed to the supreme court, bas-
he ground, among others, that the circuit court erred
demurrer to the complaint. In passing upon the
tute Chief Justice Howard, delivering the opinion of

e demurrer to the complaint it is first contended that
law of this State is invalid, as repugnant to section 1,
onstitution of the United States, which provides that
ive any person of life, liberty, or property without due
his contention is based upon the provisions of section
's lien law (sec. 7257, Rev. Stat., 1894; sec. 1690,
hich provides that any person wishing to acquire
y property shall file in the recorder's office, "at any
days after performing such labor or furnishing such
f his intention to hold such lien. This notice—the
for in the statute—is insufficient, say counsel, to
cess of law referred to by the Federal Constitution
a lien upon the citizen's property. Under the law
contend, anyone may perform labor or furnish
nstruction of a building for a landowner, without
ledge or consent, and then secure a lien upon the
notice filed after the work is done or materials
the property owner should have notice at or
work or the supplying of the materials, so that
vent the doing of such work or the furnishing
keep his property free of the lien.

It has often been held that every statute [illegible] made enters into and forms a part of such [illegible] in the contract for the erection of the dwelling [illegible] erty, are therefore chargeable with knowledge [illegible] the provisions of our mechanic's lien law then in [illegible] of the agreement entered into, the contractors were [illegible] rials necessary for the construction of the building [illegible] that such materials were to be furnished; and the [illegible] contract was made was further notice that the bui[illegible] upon which it was to be erected would be liable to [illegible] of the materials so furnished. The only uncertaint[illegible] those who should furnish the material would claim [illegible] That uncertainty is provided for in the statute, whi[illegible] notice of intention to hold the lien be filed in the [illegible] sixty days. The owner has, consequently, ample me[illegible] and is not liable to a lien without notice, nor to have [illegible] without due process of law.

It is intimated that the law hampers the freedo[illegible] property owner; that he may desire to pay the cont[illegible] or to pay him by an exchange of other property [illegible] the buildings; and that it may be an inconvenience [illegible] tractor to bid higher for the work, if payment is [illegible] sixty days after the work is done. These, however, [illegible] that should be addressed to the legislature and [illegible] Besides, it is to be remembered that without the rig[illegible] property laborers and material men would in ma[illegible] security for their toil or the materials furnished by [illegible] is worthy of his hire, and the seller of goods ought to [illegible] As the law stands, all parties are secured in their right[illegible] seeing that laborers and material men are paid, or b[illegible] sixty days from the contractor sufficient to make su[illegible] danger of having to pay twice for his building; whil[illegible] the man whose labor or materials have gone into the [illegible] to the building itself, and to the ground upon whic[illegible] security. The property owner enjoys the benefit o[illegible] this material, and it is but just that he should be c[illegible] sixty days, with the responsibility of seeing that th[illegible]

---

CONSTITUTIONALITY OF PROPOSED LAW REQUI[illegible] MENT OF WAGES.—By resolution of the house of [illegible] the legislature (general court) of Massachusetts, [illegible] court of that State was required to give its opinion [illegible] important question of law: "Is it within the [illegible] the legislature to extend the application of the [illegible] the weekly payment of wages by corporations, [illegible] and partnerships, as provided in the bill entitled [illegible] the weekly payment of wages,' now pending before [illegible]

The opinion of the court, given in reply to [illegible] May 6, 1895, was to the effect that under [illegible] article 4, of the State constitution, which [illegible] authority shall be given to the general [illegible]

[illegible] the general court by the constitu- [illegible] more comprehensive than that found [illegible] other States. The constitution of [illegible] 1, art. 4) provides as follows: "And [illegible] hereby given and granted to the [illegible] to make, ordain, and establish all [illegible] orders, laws, statutes, and ordi- [illegible] either with penalties or without, [illegible] contrary to this constitution, as they [illegible] welfare of this Commonwealth, and [illegible] thereof, and of the subjects of the [illegible] and defense of the government [illegible] constitution of Massachusetts any- [illegible] the freedom or liberty of contract, as [illegible] the press. The constitution declares [illegible] equal, and have certain natural, essen- [illegible] among which may be reckoned the right [illegible] lives and liberties; that of acquiring, [illegible] property; in fine, that of seeking and obtain- [illegible]" and it is also declared that "no sub- [illegible] despoiled, or deprived of his property, [illegible] of the protection of the law, exiled, [illegible] estate, but by the judgment of his [illegible]" (Declaration of Rights, articles 1, 12.) [illegible] from Magna Charta, and in substance [illegible] fourteenth amendment of the Constitu- [illegible] as follows: "No State shall make or [illegible] abridge the privileges or immunities of [illegible] nor shall any State deprive any person [illegible] without due process of law, nor deny to [illegible] the equal protection of the laws."

ealth, and for the government and ordering ects of the same," are not for us to weigh, except ary to determine whether the legislation pro- ontrary to the constitution. The legislation on t Britain and in other foreign countries which ution limiting the powers of the legislature is inent to the present inquiry; but, considering n in England concerning servants or laborers and the statutes which in modern times have oreign countries and many of the States of this employment of laborers in factories, we can not hat the legislation proposed is so plainly not e that the general court may not judge it to be e of the Commonwealth. We know of no rea- stitution of the Commonwealth or of the United be a distinction made in respect to such legis- tions and persons engaged in manufacturing, kind of business. The existing statutes on the nufacturing corporations, we do not regard as essarily in amendment of their charters. They rations described, whether there is any power ure to amend their charters or not, and they do n passed for the purpose of restricting the cor- rporations.

o define the limits of the power of the general to control the right of its inhabitants to make can not say that a statute requiring manufac- of their employees weekly is not one which the onstitutional power to pass, if it deems it expedi- not examined in detail the provisions of the bill ed whether the bill may not need amendment to ; but the question submitted, we think, should mative.

---

T OF WAGES.—The Cumberland Glass Manu- John F. Perry were convicted before the court mberland County, N. J., of unlawful payments n of the first section of an act approved March ct to secure to workmen the payment of wages plement to the Revision of the Statutes of New

before the supreme court of the State, which 7, 1895, that if a workman agrees with his r his work in part in merchandise, the merchan- not constitute a ground of set-off; it is a pay- nution of the claim for work; also, that such a

bargain is in violation of the first section of
The court, however, retained the case for future
power of the legislature to prevent a workman
the character of the compensation to be given him

The decision of the supreme court, delivered by Chief
ley, as published in volume 33 of the Atlantic Reporter,
follows:

The defendants were convicted before the Cumberland
sions upon an indictment charging them with being
manufacture of glass, and with unlawfully paying to one
a workman in the employ of the corporate defendant, the
in store goods and merchandise, as and for the wages
while in the employ of said corporation. At the trial
that the workman above named, at the time of his engage
into the following agreement, to wit: "Bridgeton, N. J.,
In consideration of the Cumberland Glass Manufactur
furnishing me with groceries, merchandise, and money,
to work for them at glass blowing for the blast of 1890
should I fail to do so, I hereby waive any plea in defense
ing goods and money under false pretenses." The work
was done and the goods furnished under that contract.

The act alleged to have been violated was the statute
act to secure to workmen the payment of wages in lawful
in 188[illegible]. The first section of this law makes it unlawful
manufacturer, iron master, foundry man, collier, factory m
cranberry grower, or his agent or company, their agents
pay the wages of workmen or employees by them empl
store goods, merchandise, printed, written, verbal orders
any kind." By the fourth section it is provided as follow
any glass manufacturer, iron master, foundry man, collier,
employer or company offending against the provisions of
same shall be a misdemeanor, and punishable by a fine of
ten dollars, or more than one hundred dollars for each and
or imprisonment not to exceed the term of thirty days, at
of the court; but nothing in this act shall apply to or affe
individual giving orders as aforesaid on a store in the busi
whereof he has no interest, directly or indirectly, or to th
debt due from such workman to any glass manufacturer
foundry man, collier, factory man, employer or company w
debt is voluntarily contracted by the employee or to th
any debt due from such workman to any glass manufacture
foundry man, collier, factory man, employer, or company

By an act approved March 13, 1888 (P. L., p. 174),
of the original was amended so as to eliminate from
restrictive clause just recited; and it was the validity
ment that forms the topic of the discussion in the
This argument proceeded on the assumption that
ferred upon the defendants the right to set off
furnished to the employee, and that, if that provision
defendants were guiltless. But the court is of
discussion is irrelevant to the case before us.
exceptive clause in section 4 of the original act
debts due from the employee to the employer,

...of the case it is proper to say that we ... the conviction of the defendant Perry ... only connection with the transaction is that ... a stockholder of the glass company. On ... he is not responsible for the violation of ... It is section 4 that denounces the punish- ... it is the act of the employer himself, ... is made the punishable misdemeanor. ... the judgment must be reversed.

... of the case so far as it is exhibited in the briefs ... another problem that must be resolved before ... decide upon the alleged criminality of the defend- ... whether the legislature, in enacting the law of ... its authority. It is obvious that the general effect ... prevent a workman who is entirely sui juris from ... character of the compensation to be given him for ... thus arising is one of great importance, touch- ... the essential rights of the citizen and the extent ...thority, and therefore should not be settled except ...sideration. The result is that the case will be ...nsel can send in briefs on the point thus reserved.

---

...ILITY—RAILROAD COMPANIES.—In an action by ...ainst the Northern Pacific Railroad Company to ... injuries received while in the performance of his ... of said company, through the negligence of fellow- ... States circuit court, district of Minnesota, fifth ... October 31, 1895, that under section 1, chapter 13, ... of Minnesota, Mitchell was entitled to recover

... court, delivered by Judge Nelson, as published in ...leral Reporter, page 15, is as follows:

...this case was submitted to a referee to report ...conclusions of law; and, upon confirmation thereof ...to be entered accordingly. The referee reported ...laintiff, on the 25th day of February, 1893, was ...ner for defendant at Staples, Minn., and while so ...senger coach on a side track, another coach was ... at a dangerous and unusual rate of speed by a ...isting of a locomotive engineer, fireman, foreman,

ı a more favorable footing as to this than
ıem from the bar before held to arise from
efective conditions. It is not a defense, but
for consideration, among others, in order to
absence of contributory negligence, which
before, but is not to be made out against an
of his knowledge.

---

ACT OF TEXAS.—Chapter 24 of the acts of
llow-servants," provided, in section 2, "that
ed in the common service of such railway
s so engaged, are working together at the
common purpose, of same grade, neither of
ed by such corporations, with any superin-
their fellow-employees, are fellow-servants
that nothing herein contained shall be so
oyees of such corporation, in the service of
vants with other employees of such corpora-
department or service of such corporation.
within the provisions of this section shall
vants."

ed to was repealed by chapter 91 of the acts
n quoted was practically reenacted and its
lude, in addition to the employees of any
ployees of the "receiver, manager, or per-

was considered by the court of civil appeals

...either Kinchelow or the switch- ... together at the same time and place, to ... by their ordinary every day labor ... Kinchelow, who was in charge of the ... was the "hostler," whose duty it ... and carry them to the roundhouse, ... the roundhouse and deliver them ... was a car repairer, working in a separate ... in another department of the service, ... with appellee. At the time the collision ... inflicted appellee was on a car, under ... to assist in repairing damages caused by ... statement of this fact demonstrates that he was ... and place to a common purpose with either ...

## ...S UNDER COMMON LAW.

...—In the case of Burke et al. v. Anderson, ... court of appeals, seventh circuit, on October ... judgment of the United States circuit court for ... Wisconsin, by which $4,000 damages were ...son for personal injuries caused by an explo- ...the following circumstances: Matthew C. Burke ...ed in making a roadbed for a railroad, and his ..., had sole charge of the work for him as general ...ndent. The work was carried on by blasting dynamite and other explosives and afterwards ...s, John Burke having personal charge of the ... common laborer, unfamiliar with the use of ...y John Burke and set to work digging with a ... the blasting had been done the day before, ...owledge of possible danger. Anderson was ... caused by striking with his pick a piece of ...om the blast, which was found to have been ...

...Matthew C. Burke had created the risk due ...sives for his own purposes, and was bound not ... care and every available precaution against ...men, but to give them warning of the risk, ...

have worked safely in the place to which he
pany is undisputed that he had engaged in
before, had no experience in or knowledge of
given thus employed, and had no information
was incurred by digging in this ground. He
the superintendent to enter and work there,
to rely, upon the implied assurance of the
reasonably safe; that there was no other
was obvious and necessary." The master
servants to work, and if his acts create special
chargeable with the positive duty to exercise
available precaution against possible injury to
; but if danger impends notwithstanding
is further obligated to give due information
those in his service who are ignorant of its
them to incur the risk.

ment of the plaintiff and the directions for his
and conceded that the superintendent repre-
principal. In the same relation he is charge-
the danger in using explosives, and with the
and notify them of risk. If the plaintiff was
which compliance with the order involved, or
it, the risk thus created can not be held to
the service in which he was engaged, and
assumed by him in his employment. The
behalf of the principal defendant, and the
as well, rest upon the claim that the opera-
on labor, and not the work of a superintendent
performance by the superintendent was in
servant, and the master was not liable for any
the exercise of ordinary care in selecting his
connection it is argued that the use and care of
personal duty of the master.

risk was created by the master or for his pur-
ate finding by the jury of negligence on the
the performance, causing the injury; and,
was ignorant of the risk, and had not assumed
exempts the master from liability arising out
-servants is based upon the assumption by
risks of his employment, in which the negli-
included, but it has no application to risks
by him in entering upon the service, and
for this extraordinary risk interposed by the

---

Y—MINING COMPANIES.—In the case of
Company v. Ingraham, the United States
circuit, on September 16, 1895, affirmed
court of the United States for the western

[illegible] it to an officer or [illegible] the place of the master; [illegible] is the negligence of the [illegible] negligence may recover from [illegible] the relation the injured servant [illegible] whose negligence resulted in inflict-

[illegible] therefore, and the issue as it was pre- [illegible] whether the "pit boss" was a fellow- [illegible] immaterial, and the court might well [illegible] all instructions relating to that question. [illegible] the pit boss, but the negligence of the mas- [illegible] reasonably safe place to work that was com- [illegible] on the master to exercise reasonable care [illegible] the plaintiff a reasonably safe place in which [illegible] relieved from responsibility for failing to [illegible] proper, instead of performing it him- [illegible] to a servant who neglected the duty. [illegible] cases is the negligence of the master. [illegible] the mule played in the accident * * * [illegible] of the rule which the defendant attempts [illegible] should have anticipated that this mule [illegible] to the room in the mine where the plain- [illegible] that, while there, the mule would come in contact [illegible] supported the roof of the mine, and knock them [illegible] insecurely set, and that as a result of all this [illegible] be injured, and that, anticipating all [illegible] the defendant's service. The case does not [illegible] what is a primary, proximate, or remote cause. [illegible] the accident, whether remote or proximate, were [illegible] negligence, which the plaintiff was not [illegible]

---

[illegible]IABILITY—RAILROAD COMPANIES.—In an action by [illegible]inst the Oregon Short Line and Utah Northern Rail- [illegible] recover damages for the death of her husband, James [illegible] an engineer in the employ of said company and who [illegible] in Montana while in the performance of his [illegible] a verdict, whereupon the railway company [illegible] basing the motion on the refusal of the court to [illegible] in a verdict for the defendant on the ground [illegible] not liable for the death of Frost, the engineer, as [illegible] the negligence of a fellow-servant, and on the [illegible] to the jury to the effect that the employee [illegible] the collision occurred was the representative [illegible] his acts and negligence were the acts and [illegible]

[illegible] was denied by the United States circuit [illegible] S. D., on September 24, 1895. The cir- [illegible] and the decision of the court rendered by

This is usually done by the train dis- termed a "time-table." This is the act If a time-table is changed temporarily, this dispatcher. He acts in both cases in the of the company or of its road. A railway not perform its whole duty to its employees time-table, either general or temporary. It care, under all the circumstances, to bring of all persons who are charged by it with its railway track. The notice of a temporary as necessary as the notice of the general time- danger to be apprehended from the establishment when a general one has been in use than from general time-table in the first place.

performed is one which it was the duty of the master, to execute, can it, in any way, transfer and exonerate itself from liability in case this in its performance? I think, under estab- it can not.

the list of duties required of a master toward the duty of a railroad company to establish notice thereof to those engaged in managing and ehend we must class that duty under the head master to provide a suitable place for his servant the case, the duty of giving notice to those es upon the railroad company, and those who ted with this office personally represent it. A e the duty of providing a safe place in which pon to work, so as to escape responsibility, if r care in providing such place.

tted that the establishing of a temporary time- railroad company, and the duty of giving notice al or temporary, devolves upon it. How can it one case more than another this duty of the mas- to a fellow-servant of those who are operating ved from liability? The duty of giving notice tables is the duty of the master, and the master ty to another without being responsible for his rule the telegraph operator Stuerer at Dillon epresenting the company in the duty assigned the temporary change of the time-table, or in intrusted to him to deliver to the conductor of ge in the time-table. In doing this duty he was those operating the road, but a personal repre- any, for whose negligence the company was

---

ITY—RAILROAD COMPANIES.—The supreme decided, on October 7, 1895, in the case of Elkins d Company, that a railroad company is respon- by one of its brakemen, through a defect in the while acting as one of a crew sent to a shipper's

...the case of Hermann v. Little-
...the Pacific Reporter, page 443) the
...ed, on October 9, 1895, that a contract
...devote his whole time and services to
...not broken by doing a little work for
...night, such work not resulting in dam-
...when a person performing labor at an
...e continued in the same employment
... without a new agreement, it is pre-
...inal contract were continued, and the
... evidence in a suit by the employee to
... the expiration of the time covered

...d in the opinion delivered by Judge

... for work and services performed by
... conducting and carrying on the busi-
... went for plaintiff, and this appeal is
...and from the order denying the motion

...ntered into a written contract, by the
... to give his services "as a draftsman
...he necessary and reasonable working
... the term of three years." Hermann
...e time and services to the interest of
...an and assistant architect, to use at all
...tion in and for the true and best inter
...f he were a partner with him." This
...e of time, and plaintiff continued to
... defendant, and performed the same
... before such expiration. By his answer,
... had made a substantial default in the
...ontract upon his part to be performed,
...urt, by its finding of fact, declared, in
...ed the contract as agreed upon, and
...ime of his employment he performed
...defendant, from which employment he
... of $178; and it thereupon deducted
...due to plaintiff, and ordered judgment

... committed an error in admitting the
...ence. We think there is no weight in
...is clearly admissible, as showing the
...h plaintiff performed the labor. It is
... performing labor at an agreed price
... the same employment after the expi-
... agreement, it is presumed by the law,
... contrary, that the terms of the origi-
... the fact that the present action is one
... way deprives the plaintiff of the right
...ence.
...erformance of the contract by plaintiff
...ce. We see nothing in the record dis-

in this respect could hardly be de

# BULLETIN

OF THE

# ?ARTMENT OF LABOR.

No. 4—MAY, 1896.

ISSUED EVERY OTHER MONTH.


EDITED BY

CARROLL D. WRIGHT,
COMMISSIONER.

OREN W. WEAVER,
CHIEF CLERK.



WASHINGTON:
GOVERNMENT PRINTING OFFICE.
1896

# CONTENTS.

carried on. Its prominence and age, it being
nining company of France, and one whose his-
th the history of coal mining in that country,
ons for giving it the preference over other com-
ation, all the conditions influencing the welfare
e been considered in the greatest possible de-
sary to enter into the same amount of detail in
center now to be described, for the conditions
similar to those existing at Anzin. The chief
on of the Blanzy company lies in the opportu-
y of other forms of workingmen's institutions,
applements the description of Anzin and empha-
lescription there given of conditions and insti-
exceptional coal mine, but of all of the most
companies of France.

cial interest attaches to the experiences of the
he company at first attempted the system of
everything was done directly by the company.

The far from satisfactory results of its efforts led to a complete change and the adoption of a more liberal system. The testimony of a company of such importance as that of Blanzy on this vital point of the policy to be pursued in the organization of workingmen's institutions is deserving of serious attention. The company describes, in a brochure published by it in 1894, on its workingmen's institutions, the results of its efforts under the two policies pursued. The following is a condensed translation:

From its very foundation the Coal Mining Company of Blanzy has appreciated the peculiar duties that it owes toward its employees. It has always been among those who thought that the duties of an employer toward his employees did not cease with the payment of their wages, but that there were social and moral duties as well that should be regarded. The company, therefore, as early as 1834, commenced the creation of institutions for the welfare of its employees that have since developed until their maintenance requires an expenditure of over 1,000,000 francs ($193,000) annually. These "company institutions," however, while rendering important service, have not given results proportionate to the sacrifices made by the company. Benefits obtained without efforts are rarely appreciated. Favors soon get to be considered as rights, and selfish interests are thought to have dictated their granting. Worse still, the workingman ceases to depend upon himself; he loses the habit of foresight and economy; individual initiative is lost; self-dignity is lessened. These results, which are the consequences of a patronage too highly developed, commenced to be felt a few years ago by the company of Blanzy.

At the same time, by a natural reaction, the spirit of association was awakened among the employees. Cooperative stores, mutual aid and kindred societies developed rapidly throughout the country. The new movement denoted a certain spirit and the presence of aspirations which the company of Blanzy could not afford to disregard. It appreciated the situation, and, while preserving its institutions that had a real raison d'être, resolved to take advantage of the new spirit. It took as its principle that the employer should make use of associations of employees wherever possible. With this system the company is no longer solely responsible for the welfare of its employees. Associations, also, when well directed, contribute powerfully to the maintenance of social peace.

We can not but felicitate ourselves upon the results of our new policy. The interest of the workingmen has exceeded our expectations. Associations are increasing in number and importance, and the day can be anticipated when they will replace all the institutions now controlled by the company, or, at least, so modify them that the workingmen will share with the company in their management.

The coal basin of Blanzy is situated in the central portion of the department of Saône-et-Loire, and next to the coal basin in the [illegible] the most important coal deposit of France. Authority was [illegible] to mine coal here in 1769, under the title of Concession of [illegible] In 1832 the concession was divided into two parts, named [illegible] Concession of Blanzy and Concession of Creusot. Dur[illegible] coal was mined by various companies, among which a [illegible] ization and reorganization was taking place. The [illegible]

[illegible]-1833, but has since undergone [illegible] capital of the company is now [illegible] the property covered by its conces- [illegible] acres). Though the company , the place where coal was first mined, s named, the real seat of its operations,

pany's town in the sense of being spe- s so entirely dominated by and depend- possesses all the characteristics of a re the commencement of mining opera- l on the maps as a farm. In 1834, when d its operations, the commune of Mont- nts. In 1856 this number had increased ncorporated as a separate village. In was 3,337; in 1866, 5,548; in 1872, 8,287; nt population, according to the latest the meantime Montceau has become, e company, a substantially built town. ginning no resources, no property, and g a good many years the company sup- nd improved the most important roads queducts. It laid out, most of them on quares of the new town. It built and expense several very pretty churches. f erecting a town hall, it contributed building fund and donated the ground supported the whole expense of build- e opinion of the state inspectors one of of between 300,000 ($57,900) and 400,000 me it paid the salaries of the officials of ided the necessary buildings for their forts made by the company to provide nhabitants will be described in another

any itself is best shown by the amount and the total number of its employees. rom 1833 to 1850 had never exceeded ) tons in 1860, 400,000 in 1870, 700,000 in in 1892, and 1,105,317 in 1893.

more important colliery companies of represents but a part of the activities of anufacture of coke have been installed, bricks of coal dust are operated, and a y the company, though its management

---

equals 2,204.6 pounds.

[illegible] gives for the years 1872–73 to 1891–92:

[illegible] EMPLOYEES OF THE COAL MINING COMPANY OF BLANZY, 1872-73 TO 1891-92.

| [illegible] | | Year. | Average daily wages of— | | |
|---|---|---|---|---|---|
| [illegible] | All employees. | | Employees at pits, below and above ground. | Other employees. | All employees. |
| [illegible] | $0.58 | 1882-83...... | $0.72 | $0.62 | $0.68 |
| [illegible] | .60 | 1883-84...... | .74 | .63 | .68 |
| [illegible] | .58 | 1884-85...... | .72 | .62 | .68 |
| [illegible] | .60 | 1885-86...... | .73 | .63 | .68 |
| [illegible] | .61 | 1886-87...... | .76 | .63 | .70 |
| [illegible] | .62 | 1887-88...... | .78 | .64 | .71 |
| [illegible] | .61 | 1888-89...... | .78 | .64 | .71 |
| [illegible] | .63 | 1889-90...... | .79 | .67 | .75 |
| [illegible] | .64 | 1890-91...... | .80 | .69 | .77 |
| [illegible] | .66 | 1891-92...... | .80 | .71 | .79 |

### [illegible]KINGMEN'S INSTITUTIONS.

rtaken by the company of Blanzy, and by the t Montceau follows along the same general lines the latter place, the main efforts of the company ovision of houses and of facilities for working- owners, the organization of pension funds, the etc.; and those of the employees, the organiza- ciety, in which, however, the company largely ocieties. There are, of course, important points cy of the company of Blanzy in encouraging en particularly liberal and has met with remark-

The company, again, has done a very useful ovide employment for miners' wives and daugh- at purpose workshops, in connection with its of articles of wearing apparel, and a completely mill for the older women. On the other hand, ensions, whereby the workingmen are not per- any way to the maintenance of the fund, is one commend itself as that practiced at Anzin.

rganized by the workingmen themselves, the nce, which the workingmen have organized for business affairs, serves a useful purpose, while, rovince of distributive cooperation seems to be

THE HOUSING OF [illegible]

The situation of the mines of Blanzy, located, [illegible] agricultural community, of necessity demanded [illegible] the company should be directed toward the provision their employees. The work of the Blanzy company [illegible] deserves especial attention, both on account of the ext tions and because of the liberal policy that it has purs house ownership.

M. Simonin, a mining engineer, writing concernin institutions at Montceau, after a visit in 1867, says:

The question of lodgings for its employees has almos ning been an object of solicitude to the Blanzy Coal M In 1834 it established dwellings for workingmen; but covered all the drawbacks that the old system of barra as regards tranquillity, hygiene, and morals, especiall relieve each other by turns, it adopted for the future th arate houses, each surrounded by a garden. Each ho tenements, having each two rooms on the ground floor, a and a pigsty attached. In front of the house there i the back a garden, the whole well fenced in. The ya street, which is broad and well kept. * * * Th an area of from 400 to 500 square meters (4,306 to 5, The workingman raises vegetables for his table and his home. A garden is absolutely necessary to the co connected by his birthplace and by the nature of his w cultural population. * * * The village looks chee houses, with their roofs of red tile, produce a pictures

Since the time that M. Simonin wrote the above im all in the nature of improvements, have been made. tage described by that author has now been complete new and more graceful model was adopted in 1867. It and in some cases four rooms. The gardens have also be average area being now about 700 square meters (7, Instead of grouping the houses all together in one been erected in separate groups, and in many cases the country as isolated dwellings, but in close proximit

The following table, prepared by the company for t the condition of affairs in 1894 as regards the direct pr by the company:

NUMBER AND COST OF HOUSES ERECTED FOR EMPLOYEES BY COMPANY OF BLANZY, 1894.

| Groups of houses. | Dates of erection. | Houses. | Tenements. | Cost. | |
|---|---|---|---|---|---|
| | | | | Land. | [illegible] |
| Alouettes | 1844–1859 | 90 | [illegible] | [illegible] | [illegible] |
| Magny | 1859–1876 | 96 | [illegible] | [illegible] | [illegible] |
| Bel-Air | [illegible]–1876 | 74 | [illegible] | [illegible] | [illegible] |
| Bois-du-Verne | [illegible] | 101 | [illegible] | [illegible] | [illegible] |
| Other houses | | 120 | [illegible] | [illegible] | [illegible] |

atceau, so aptly described by M. Simonin,
[illegible]. Well-kept carriage roads run
rters; trees and plants of all kinds have
ir foliage makes a pleasant contrast with
here a cottage somewhat larger than the
miner, gang-leader or checker, or some
breaks the monotony. In the center the
esque style give to the various groups a
small towns of Burgundy.

y the company to obtain a commercial
dings. While an ordinary workingman's
s ($2.32) per month in the neighboring
rged by the company of Blanzy is from
3). Thirty-nine cottages are let at 2.50
(58 cents), 534 at 4.50 francs (87 cents),
t 6 francs ($1.16), 4 at 9 francs ($1.74),
pecial arrangements.

ses by the company, such as has been
r, the least interesting part of its efforts
ses. Much greater interest attaches to
d for encouraging the workingmen to
n homes. To do this, the first step taken
e several tracts of ground adjoining its
nto lots of from 20 to 25 ares (21,528 to
r having laid out the streets of a village.
osition of its miners at their cost price,
enth as much as they would have to pay
. Any miner could ask for the conces-
was granted an advance of 1,000 francs
at the same time made to him, the whole
installments.

appreciated the advantages offered to
g figures: In 1888 the total number of
concessions of ground was 316, and the
s 77.7393 hectares (192 acres), at an aver-
507.42) per hectare. To enable them to
ed to 303 of their number 235,492 francs
30 francs ($150) each. In 1893, five years
400 for the total number of workingmen
d to 88.9936 hectares (220 acres) for the
age value of 2,721.95 francs ($525.34) per
$46,709.67) had been advanced to 387
about 625 francs ($120.63) each.

was provided and money advanced to
ld was adopted in 1857. To supplement
ns dated May 2, 1874, offered to advance
workingmen who were either the owners

of land elsewhere or for any reason desired to that offered by the company. Including these these later regulations the total number of working advances in 1893 was 540, and the total amount of francs ($90,669.86).

The system of grants of land and advances of money regulations that have just been described developed, ho serious drawbacks. Many of the workingmen, when t to build, were in possession of totally inadequate mean pose, and the advances of the company did not suffice building expenses. The workingmen were, nevertheless, availing themselves of the advantages offered that they elsewhere relatively large sums, sometimes at high rat and, as a result, they would be kept a long time under tl heavy debt. The houses, likewise, were occasionally of a able character. The greatest hardship, however, was those cases where death intervened before either the con creditors had been repaid their advances. The prope quence, would be sold or otherwise disposed of under u cumstances; lawyers' fees would often have to be paid, lies would be left in great distress. It was desirabl devise some means by which these disadvantages could As a result important modifications were made in 1893 tions of 1874. Henceforth the amount of the advan workingmen with a view to building was not limited t ($193), but, if need be, could equal the total value of th built, without, however, exceeding the sum of 2,500 fr The maximum delay for the repayment of the advance w time extended to fifteen years. The essential innovation that by which the prospective builder was required to co or tontine life insurance policy to the amount of his adv

The premiums paid on this policy, therefore, take th annual installments formerly paid to the company. T workingman pays, or rather agrees that there shall be his wages, during the time indicated in his policy, first, his insurance premiums, and, second, interest at 3 per c that has been advanced to him. At the date fixed upon, of the workingman, if his death takes place before the ex term, the insurance money serves to repay the amount a complete title to the property is secured.

A most important feature of the new regulations is administration of the whole system of advances Prudence, a society that the laborers and officials organized to take charge of their business description is given in another place. The work build addresses himself to La Prudence [illegible] in favor of the society, and it advances ary for building. On this, as has already been stated, st at the rate of 3 per cent. In order to stimulate the [illegible] the acquiring of homes the company pays to it s ([illegible]) during fifteen years for every house built by [illegible] these auspices.

of these encouragements and aid by the company has of remarkably successful results. A number of impor- [illegible] have grown up that have an appearance om that of a regular workingmen's village. In 1889 workingmen who were owners of the houses that they the surrounding ground. In 1892 this number had 6, or over 25 per cent of the total number of working- ads of families. But the new regulations have suc- more remarkable degree. In the year succeeding the lication, April 15, 1893, 113 workingmen took out insur- l commenced the acquisition of homes, representing a rrowed of 266,456 francs ($51,426.01). The popularity m is very natural if one takes into consideration that orkingman can become the owner of a house of the ,500 francs ($482.50) in fifteen years while making an of little if any in excess of an ordinary rent. The ove concerning house ownership represent, moreover, ngmen who have become house owners by profiting ges offered by the company. As to those who have ir own savings or have acquired homes in any other t represented in the total as given, though their num- o be an important one.

### OLD-AGE PENSIONS.

atter of housing, that of the care of their old employees st continue to be the subject of the greatest solicitude e large employers of labor in Europe. The company e honor to be the first mining company to establish at a fund for the pensioning of its workingmen in their ation dates from January 25, 1854. Until 1882, how- s of the fund were limited to employees working below stituted during this period, provision was made for 180 to 240 francs ($34.74 to $46.32) a year to unmar- 240 to 300 francs ($46.32 to $57.90) to married work- y had reached a certain age and had worked a certain for the company, half of the pension to be continued to

[illegible]. There is almost a consensus of [illegible] insurance is that whereby the work- [illegible] to an equal extent if possible with [illegible] of an insurance fund. They are at [illegible]d to participate in the management of the fund; [illegible]e that when acquired the recipient accepts with a [illegible] not a bounty but rather a return for his [illegible]l economy.

[illegible]s question much depends on the character of the locality, but it is difficult to believe that working- [illegible] large an extent become house owners are not in a [illegible]e actively in the constitution and management of [illegible] is done by their brother miners in the north. It [illegible]fficult to understand, as these principles have been [illegible] in the insurance fund for the higher grades of [illegible]pany.

[illegible]ny organized a second pension fund for the benefit [illegible]sicians, and office employees. In this case, how- [illegible] not entirely borne by the company. The receipts [illegible] of an annual payment by each member of a sum [illegible] of the total amount he has received in wages, a [illegible]pany of an equivalent amount, and miscellaneous [illegible]rest on funds invested, fines, etc. A member may [illegible] pension after he is 55 years of age and has been [illegible]s years. The value of the pension equals one-half [illegible] salary that the pensioner has earned during the [illegible] employment, and is increased by one-fifteenth for [illegible] of service if he remains in the employ of the com- [illegible]enty-five years before retiring, but this pension [illegible]-thirds of the average annual salary based upon [illegible]f service. The minimum and maximum pensions [illegible].60) and 9,000 francs ($1,737), respectively, per an- [illegible]th the pension to half the amount is continued to

[illegible]le gives the results of the operation of this fund [illegible] 1884–85 to 1893–94:

[illegible]S OF THE PENSION FUND FOR HIGHER EMPLOYEES AT BLANZY, 1884-85 TO 1893-94.

| Total receipts. | Total expenditures for pensions. | Surplus. | Capital at end of year. | Pensioners. | Average pension. |
|---|---|---|---|---|---|
| [illegible] | $616.57 | $3,631.18 | $26,129.82 | 7 | $88.08 |
| [illegible] | 820.49 | 3,541.22 | 29,670.84 | 8 | 102.56 |
| [illegible] | 2,466.37 | 2,162.03 | 31,832.87 | 13 | 189.72 |
| [illegible] | 2,874.35 | 2,917.93 | 34,750.80 | 14 | 205.31 |
| [illegible] | 2,673.89 | 4,240.59 | 38,991.39 | 16 | 167.12 |
| [illegible] | 3,978.42 | 3,154.03 | 42,145.42 | 18 | 221.02 |
| [illegible] | 4,114.36 | 7,492.39 | 49,637.81 | 19 | 216.55 |
| [illegible] | 4,477.91 | 7,898.12 | 57,535.93 | 19 | 235.68 |
| [illegible] | 4,073.67 | 4,894.65 | 62,430.58 | 18 | 226.32 |
| [illegible] | 3,904.74 | 5,436.13 | 67,866.72 | 19 | 205.51 |

[illegible] of the second class of a sum equal to 2½ [illegible] wages. The subsidies of the company [illegible] paid in as dues by the second class. In s subsidy, the company obligates itself to maintain, at its infant schools for the children and primary schools in h workshops, where all the children of employees will be itously; to provide the necessary ground and buildings repair a hospital for workingmen injured by accidents, cases for those suffering from serious illness, a pharoratory and pharmaceutic stock, a sufficient number of ooms for physicians, and also the payment of the salaans, pharmacists, nurses, etc.

eat category of relief granted by the society is that to sick. In general all members have a right to receive dance and medicines gratuitously after they have been nonth. The wives and children of members who have two months are likewise entitled to similar aid.

class of assistance afforded by the society is that of cash mbers during the time they are incapacitated for work as ickness or accident. The constitution contains elaborate d schedules, showing when such benefits are due and the ording to which their amounts are determined. These shown in the following table:

ENEFITS FOR SICKNESS AND ACCIDENT PAID TO MEMBERS OF THE MUTUAL AID SOCIETY AT BLANZY.

| Beneficiaries. | Daily benefit for— | |
|---|---|---|
| | Injury or sickness resulting from work. | Ordinary sickness lasting more than seven days. |
| [illegible] or widower, 17 years of age or over | \$0.24 to \$0.34 | \$0.14 to \$0.19 |
| [illegible] living with his wife | .29 to .34 | .19 to .24 |
| [illegible] under 17 years of age, as supplement to [illegible] | .06 to .07 | .05 to .06 |
| 17 years of age or over employed by company | .14 to .19 | .12 to .14 |
| [illegible] of age employed by company | .14 | .10 to .12 |

ticed that a distinction is made between sickness or accif from the prosecution of the work and that contracted nditions, the benefits being greater in the former than in . In exceptional cases, where the injured or sick works unusually large number of persons dependent upon him e amount of the benefit as fixed in the regular schedules y increased. In no case, however, can the pecuniary aid married man, for himself, his wife, and children, exceed [illegible]) in case of injury from an accident, or 1.75 francs [illegible] sickness. No benefit is paid for an injury or sick[illegible] drunkenness or other misconduct.

[illegible] occasioned directly by the work, continues more

t them a monthly pension not to exceed ing table:

AGE PAID TO MEMBERS OF THE MUTUAL ETY AT BLANZY.

| | Years of service. | Minimum age. | Maximum aid (per month). |
|---|---|---|---|
| | 15 | 45 | $2.90 |
| | 15 | 45 | 1.93 |
| | 20 | 50 | 3.86 |
| | 20 | 50 | 2.51 |
| | 25 | 55 | 4.83 |
| | 25 | 55 | 3.09 |

ewise made for the pensioning of the men pensioned according to these pro-

the conditions under which relief is cases under which such aid is granted pecial conditions under which relief is o so would have necessitated the trans- enso. Only the general classes of aid

is administered by a council of 23 mem- ger, the engineer in chief, and the gen- are ex officio president, and first and vely. The other members consist of a from among the office force, and 18 other eted from among the engineers, chief om among the master miners, foremen, rs, and yard masters; six from among

e as many if necessary. From the main wards
apel, which allows the sick patients to receive
olations of the church. Halls for consultation,
macy, and dispensary—everything has been
rranged. There is a garden and terrace for
It seems as though one had sought to take from
unpleasant associations which it arouses in the
, but which are less repugnant to him when he
ch kindness.

incent de Paul have charge of the new hospital,
one. Mention should also here be made of a
pened that is intended both as a hospital and
ngmen. Several months before his death M.
eral manager of the company, commenced the
ding at his own expense. After his death his
donated it to the company, with the proviso
r the care of the old workingmen who, through
alone in the world. Its objects are exactly
e Soldiers' Homes in this country. No expense
ke it a comfortable and attractive home. It
ersons. The company has accepted the gift,
support all the expenses connected with its

SCHOOLS.

ompanies have exceeded that of Blanzy in the
ovision of educational facilities for the children
e company at the present time supports 21
infant, 8 boys', and 7 girls' primary schools.
re in attendance a total of 6,292 scholars—1,830
girls' primary, and 2,265 in the boys' primary
s number 148, of whom 61 are Marist Brothers
Vincent de Paul and of Saint Joseph de Cluny.
every year, and scholarships of 450 and 225
are awarded to those making the best record
hem to continue their studies in advanced

establishment of these schools and of the
, an account of which follows, amounted to

[illegible] time to time until now it amounts to [illegible] into shares of 50 francs ($9.65) each. [illegible], three-fourths of whom at least are work- [illegible] into two classes of honorary and ordinary [illegible] pledge themselves not to receive a dividend 'hey have a right to their share in the reserve any number of shares, and have in the general f votes that their shares confer upon them.

ders can not possess more than 20 shares each. e society is intrusted to a general manager ssembly of shareholders, assisted by a board a similar way.

ntention of the founders was that the society siness manager of its members, while at the ns for the investment of savings. This bring- men in business relations has, however, been sults. The demonstration of the advantages such that the members have been led to use eation of various other institutions for their here have been organized within the society partments.

at of the business agency, for which the soci- d. In the beginning, business was transacted he society, all of whom were employees of the y degrees the work spread, and to-day La Pru- from every calling, and deals with the public of members is transacted gratuitously. Out- a fixed schedule of charges. This department d by the members and by the public as well.

is that of the bank. This department has portant part, at least from the financial point of the society.

contrasting the operations of the bank during will serve to convey an idea of its importance work that it performs:

CASH STATEMENT.

| | |
|---|---|
| [illegible] | $19,504.44 |
| [illegible] | 15,171.80 |
| | 724,373.52 |
| | 728,706.17 |
| [illegible] | 1,453,079.69 |
| [illegible] | 1,127,986.53 |
| | 325,0[illegible]3.16 |

PAPER DISCOUNTED.

| | |
|---|---|
| Value of paper discounted, 1892 | [illegible] |
| Value of paper discounted, 1891 | [illegible] |
| Increase | [illegible] |

NEGOTIATION OF PROMISSORY NOTES.

| | |
|---|---|
| On hand January 1, 1892, 79 promissory notes, amounting to | [illegible] |
| On hand January 1, 1893, 142 promissory notes, amounting to | [illegible] |
| Loaned on 3,636 promissory notes | [illegible] |
| Received on 3,573 promissory notes canceled | [illegible] |
| Total of operations during year 1892 | [illegible] |
| Total of operations during year 1891 | [illegible] |
| Increase | [illegible] |

COLLECTION OF BILLS.

| | |
|---|---|
| On hand January 1, 1892, 1,198 bills for collection, amounting to | [illegible] |
| On hand January 1, 1893, 1,108 bills for collection, amounting to | [illegible] |
| Received for collection during year, 19,896 bills, amounting to | [illegible] |
| Collected during year, 19,986 bills, amounting to | [illegible] |
| Total of operations during year 1892 | [illegible] |
| Total of operations during year 1891 | [illegible] |
| Increase | [illegible] |

DEPOSITORS' ACCOUNTS.

| | |
|---|---|
| Deposits on hand January 1, 1892 | [illegible] |
| Deposits on hand January 1, 1893 | [illegible] |
| Deposited during year | [illegible] |
| Withdrawn during year | [illegible] |
| Total of operations during year 1892 | [illegible] |
| Total of operations during year 1891 | [illegible] |
| Increase | [illegible] |

The third branch of La Prudence is the savings bank department. Two distinct savings banks have been created. The first, called La [illegible]lire (The Money Box), is a schools' savings bank. Depositors must be under 21 years of age and either attend the company's schools or be employed in the company's works. Deposits bear interest at 4 per cent, compounded annually. As soon, however, as an account amounts to 200 francs ($38.60) the society has the option of converting it into a savings certificate in the ordinary shareholders' savings bank. The latter bank, called La Fourmi (The Ant), after a similar organization [illegible] Paris, is similar in its operations to other savings banks. The [illegible] figures concerning the operations of the two banks during [illegible] 1891 and 1892 illustrate the importance of their operations:

| | |
|---|---|
| Cash on hand January 1, 1892 | [illegible] |
| Cash on hand January 1, 1893 | [illegible] |
| Receipts during year 1892 | [illegible] |
| Paid out during year 1892 | [illegible] |
| Total of operations during year 1892 | [illegible] |
| Total of operations during year 1891 | [illegible] |
| Increase | [illegible] |

very possible way. It has provided for it a
r story of which is fitted up as a club room
s with it a running account, borrowing from
iterest, thus insuring that the society shall
lative money. It pays a small subsidy for
ature, consultations, representation in the
makes it a medium for the transaction of
.

nent, however, is that whereby it has been
ng the system lately inaugurated by the com-
to workingmen with which to build homes.
ie, as well as the relation of La Prudence to
. under the section devoted to workingmen's
lready become one of the most important
ciety.

tatement of the assets and liabilities of the
and 1893 shows the growth in importance
years:

ILITIES OF LA PRUDENCE, 1889 AND 1893.

| | 1889. | 1893. |
|---|---|---|
| s. | | |
| | $168.89 | $17,243.46 |
| | | 2,065.06 |
| | 6,631.91 | 37,829.97 |
| | 2,288.32 | 10,697.50 |
| | 5,856.33 | 66,520.12 |
| | 2,004.60 | 21,021.28 |
| ild homes | | 29,555.14 |
| | 16,950.05 | 184,952.53 |
| ies. | | |
| le for building | | 17,389.30 |
| | 7,729.65 | 26,219.15 |
| | 7,243.06 | 64,468.44 |
| | 1,033.59 | 2,700.36 |
| | 943.75 | 74,155.28 |
| | 16,950.05 | 184,952.53 |

existing conditions. The conclusions there reached apply in almost every particular to Blanzy as well. In every respect the workingman's conditions seem to have improved, and while the tendency is for the employer to make greater sacrifices for his employees, the latter are allowed a constantly increasing participation in the management of institutions intended for their benefit. The most important series of tables given in the conclusion for Anzin was that showing the stability of employment. A similar set of tables is here given for Blanzy. An analysis of them along the same lines shows almost identical results. These can be easily seen from an inspection of the tables themselves, as they are very simple in construction.

EMPLOYEES OF THE COAL MINING COMPANY OF BLANZY AT WORK BELOW GROUND, BY AGES AND YEARS OF SERVICE, DECEMBER, 1892.

| Ages. | Years of service. | | | | | | | | | | | | Total. |
|---|---|---|---|---|---|---|---|---|---|---|---|---|---|
| | Under 1. | 1 or under 2. | 2 or under 5. | 5 or under 10. | 10 or under 15. | 15 or under 20. | 20 or under 25. | 25 or under 30. | 30 or under 35. | 35 or under 40. | 40 or under 45. | 45 or over. | |
| 13 y'rs or under 18. | 61 | 90 | 484 | ...... | ...... | ...... | ...... | ...... | ...... | ...... | ...... | ...... | [illegible] |
| 18 y'rs or under 23. | 55 | 198 | 295 | 386 | ...... | ...... | ...... | ...... | ...... | ...... | ...... | ...... | 934 |
| 23 y'rs or under 28. | 63 | 191 | 172 | 179 | 214 | ...... | ...... | ...... | ...... | ...... | ...... | ...... | 819 |
| 28 y'rs or under 33. | 37 | 147 | 151 | 143 | 105 | 113 | 25 | 5 | ...... | ...... | ...... | ...... | [illegible] |
| 33 y'rs or under 38. | 33 | 78 | 63 | 86 | 81 | 97 | 49 | ...... | ...... | ...... | ...... | ...... | 487 |
| 38 y'rs or under 43. | 14 | 51 | 26 | 41 | 57 | 49 | 65 | 37 | 4 | ...... | ...... | ...... | 344 |
| 43 y'rs or under 48. | 15 | 11 | 14 | 22 | 26 | 43 | 35 | 49 | 23 | ...... | ...... | ...... | [illegible] |
| 48 y'rs or under 53. | ...... | ...... | 15 | 20 | 21 | 28 | 32 | 23 | 24 | 19 | ...... | ...... | 182 |
| 53 y'rs or under 58. | ...... | ...... | ...... | ...... | 17 | 15 | 23 | 14 | 24 | 15 | 17 | ...... | 125 |
| 58 y'rs or under 63. | ...... | ...... | ...... | ...... | 14 | 13 | 21 | 25 | 17 | 11 | 15 | ...... | 116 |
| 63 y'rs or under 68. | ...... | ...... | ...... | ...... | ...... | ...... | 12 | 14 | 9 | 12 | 7 | 2 | 56 |
| 68 y'rs or over.... | ...... | ...... | ...... | ...... | ...... | 7 | 8 | 7 | 8 | 7 | 3 | ...... | 40 |
| Total....... | 278 | 775 | 1,220 | 877 | 535 | 365 | 270 | 174 | 109 | 64 | 42 | 2 | 4,715 |

EMPLOYEES OF THE COAL MINING COMPANY OF BLANZY AT WORK ABOVE GROUND, BY AGES AND YEARS OF SERVICE, DECEMBER, 1892.

| Ages. | Years of service. | | | | | | | | | | | | Total. |
|---|---|---|---|---|---|---|---|---|---|---|---|---|---|
| | Under 1. | 1 or under 2. | 2 or under 5. | 5 or under 10. | 10 or under 15. | 15 or under 20. | 20 or under 25. | 25 or under 30. | 30 or under 35. | 35 or under 40. | 40 or under 45. | 45 or over. | |
| 13 y'rs or under 18. | 215 | 297 | 222 | ...... | ...... | ...... | ...... | ...... | ...... | ...... | ...... | ...... | 734 |
| 18 y'rs or under 23. | 65 | 89 | 121 | 89 | ...... | ...... | ...... | ...... | ...... | ...... | ...... | ...... | 364 |
| 23 y'rs or under 28. | 86 | 104 | 84 | 101 | 60 | 12 | ...... | ...... | ...... | ...... | ...... | ...... | 447 |
| 28 y'rs or under 33. | 92 | 123 | 72 | 52 | 36 | 28 | ...... | ...... | ...... | ...... | ...... | ...... | 403 |
| 33 y'rs or under 38. | 77 | 97 | 40 | 28 | 32 | 24 | 4 | ...... | ...... | ...... | ...... | ...... | [illegible] |
| 38 y'rs or under 43. | 37 | 35 | 39 | 24 | 16 | 15 | 11 | 11 | ...... | ...... | ...... | ...... | [illegible] |
| 43 y'rs or under 48. | 23 | 34 | 19 | 73 | 36 | 39 | 8 | 19 | 17 | 4 | ...... | ...... | [illegible] |
| 48 y'rs or under 53. | ...... | ...... | ...... | 12 | 27 | 32 | 18 | 15 | 9 | 13 | ...... | ...... | 126 |
| 53 y'rs or under 58. | ...... | ...... | ...... | ...... | ...... | 9 | 7 | 5 | 3 | ...... | ...... | ...... | [illegible] |
| 58 y'rs or under 63. | ...... | ...... | ...... | ...... | 4 | 5 | 3 | 12 | 9 | 5 | 3 | ...... | 41 |
| 63 y'rs or under 68. | ...... | ...... | ...... | ...... | 12 | 9 | 15 | 13 | 3 | ...... | ...... | ...... | 52 |
| 68 y'rs or over.... | ...... | ...... | ...... | ...... | 5 | 7 | 2 | ...... | 7 | ...... | ...... | ...... | 21 |
| Total....... | 595 | 769 | 597 | 374 | 228 | 180 | 68 | 75 | 48 | 22 | 3 | ...... | [illegible] |

[illegible] (204.3 bushels) of coal, representing [illegible], free of cost annually. Other work- [illegible] of buying coal at reduced rates. During [illegible] distributed to underground workingmen [illegible] bushels), worth 296,005 francs ($57,128.97). [illegible] 8,918 fuel grants, worth 12 francs ($2.32) pay 3 francs (58 cents), representing, therefore, ny of 80,262 francs ($15,490.57). In addition, [illegible] pensioned and superannuated workingmen, the vicarage, and a great many poor with coal year 1891–92 the total value of the coal given 716.20 francs ($76,566.23). To but mention any defrays the expense of a band organized has built and kept in repair several churches; ing establishments; it conducts a small circu- cted a large flouring mill, in order to cheapen mployees, and it contributes materially to the ocieties for sport and recreation that the work- among themselves. Among the latter may be , the gymnastic club, the fencing club, and ocial clubs.

L MINING COMPANY OF BLANZY FOR THE BENEFIT [illegible] OF EXPENDITURE, 1887-88, 1891-92, AND 1892-93.

| [illegible]iture. | 1887-88. | 1891-92. | 1892-93. |
|---|---|---|---|
| .......................... | $36,562.90 | $47,907.80 | $42,063.68 |
| .......................... | a 27,931.86 | 42,423.95 | 44,925.56 |
| .......................... | b 19,251.46 | 21,715.83 | 24,368.71 |
| .......................... | 2,096.28 | 6,988.05 | 4,778.07 |
| .......................... | 33,701.11 | 42,059.00 | 45,229.10 |
| .......................... | 7,776.96 | 11,121.63 | 9,257.46 |
| .......................... | 61,716.86 | 76,566.23 | 94,955.20 |
| ocieties (direct aid, etc) | 26,889.98 | 48,318.68 | 48,225.73 |
| .......................... | 215,927.41 | 297,191.17 | 313,824.11 |

[illegible] agree with those given for the calendar year 1888, page 349. [illegible] figures in both cases are as furnished by the company.
[illegible] those given on page 344. The explanation is not known. The [illegible] by the company.

## CONCLUSION.

[illegible]nt of the conditions at Anzin a general sum- the influences surrounding the life of the work- the *standpoint* of a comparison of past with

# THE SWEATING SYSTEM.

BY HENRY WHITE, GENERAL SECRETARY, UNITED GARMENT WORKERS OF AMERICA.

The sweating system, which makes of the home a workshop, even in the crowded tenement, and drafts the members of the family into service, presents a problem so serious as to command the attention of reformers and statesmen in all nations having the modern industrial system fully developed.

The term "sweating system" has a general meaning, but is specifically used to describe a condition of labor in which a maximum amount of work in a given time is performed for a minimum wage, and in which the ordinary rules of health and comfort are disregarded. It is inseparably associated with contract work, and it is intensified by subcontracting in shops conducted in homes. Such conditions prevail to a distressing degree in localities having a large, herded foreign population, and among people known for excessive industry and thrift—virtues otherwise considered indispensable to prosperity and happiness. Recently arrived foreign working people crowded into the big cities are the most helpless, and, in order to barely live, are willing to submit to almost incredible exactions. It is thus that this form of labor soon outcompetes and displaces all other forms and becomes the standard for the particular industry in which it is introduced.

The use of machinery in the making of garments has not figured largely in displacing labor and reducing the standard of skill formerly required, but rather the subdivision of labor, especially in the cheaper grades of work. This has made the garment-making industry an easy refuge for immigrants, and enables them to work in small shops which in many other industries would be inadequate to compete with larger and better-equipped factories. This class of workers, therefore, become wholly dependent upon the knowledge that they have acquired of one small part of the trade, and are incapable of advancement through individual effort.

High rents contribute their share of responsibility for the sweating system. If rents were cheap there would be a distinct advantage in working in separate rooms or shops. This is invariably the result, for in localities where rents are lower, the shops are larger and the evils not so acute. The saving in rent is an important item. To combine a kitchen, bedroom, and workshop, to utilize a garret or a loft over a

ow it, if stripped of some of its obnoxious ... the small shop of the master and journey-... with the household, before the advent of ... small towns and villages of eastern and ... still continues, particularly in the tailor-... heightened by the fact that nearly all of ... workers, and many of the clothing and ... tives of eastern Europe, and it is thus that ... of industry, the sweating system, became so ... engrafted on our manufacturing system. ... primitive workshop, the sweat shop ... rmer the employer was also the dealer, and ... not exist. As the shopkeeper or master ... on his immediate neighbors, the keen com-... enterprise was not a factor, and the pres-... ployment was unknown. Neither were the ... The workers labored more leisurely, their ..., and their wants were but few.

... ap labor afforded by the sweat shop were in ... those manufacturers who evidently eased ... plea of necessity, and apparently it was ... work was done as long as a certain price ... weater who begged for the work.

... ys alert to introduce the newest productive ... irst making an extensive use of the labor-... n. Immediately after the close of the civil ... ollowed, coming largely from the countries ... urope. An excessive demand also at that ... ue to the revival in trade, and the limited ... ed the making of garments in homes and by ... ll probability, the sweating system received ... lesale manufacture and selling of clothing ... ngland, Ireland, France, and Germany fol-... l with it came the sweat shop to the large ... oundings for its development.

... eady-made over custom work is not so large ... ough considerable custom work is made in ... elp is employed. This distinction is impor-... ome on some article of manufacture is not ... The combination of living apartment and factory and the employment of outsiders therein [illegible] the detrimental features which in time become a menace to the community. Much has been done by modern medical science and sanitation to prevent the spread of disease, but, paradoxical as it may seem, the sweat shop as a source of disease in our great centers of population has developed at the same time.

In order to fully comprehend this subject, it is necessary to know the extent and status of the industries in which the sweating evil prevails. Although the sweating system exists in a number of occupations, it is the garment-making industry (comprising men's clothing, ladies' cloak and suit, undergarment, and shirt-making branches) that has given it its real significance. The manufacture of clothing and cloaks at wholesale is the most concentrated of all the garment-making branches, and is confined mainly to the following large cities in the order of their rank, viz: New York and vicinity (including Brooklyn and Newark), Chicago, Philadelphia, Rochester, Baltimore, Boston, Cincinnati, Syracuse, Cleveland, St. Louis, Utica, and Milwaukee. One hundred thousand people, in round numbers, are engaged in this industry in these cities, of whom fully 40,000 are in the vicinity of New York City. By including the shirt and undergarment branches, there are at least 60,000 persons employed at garment making in New York and vicinity, and about 70 per cent of these work in small shops and on contract work. In the ladies' cloak and suit trade the seasons of work are short, and work is usually rushed. The frequent interruptions caused by strikes in this branch have induced many of the wholesale firms to conduct large shops of their own. This is considered to be an improvement.

The clothing and cloak cutters and trimmers number about 8,000, and are credited with being the most intelligent and skilled workmen in the trade. They are employed directly by the firms, usually on the premises, and their condition is in marked contrast to that of the other branches. The hours of labor are nine per day, with the exception of a half holiday on Saturdays during five months. In Chicago, however, the hours of labor are eight per day. The standard wages are from $15 to $24 per week, but the usual rate is $20 per week. This is an indication of the difference between direct employment and indirect contract work. When the latter is undertaken, the middleman as contractor becomes a factor and his profits must be taken from wages. Of course, with subcontracting wages must be reduced still [illegible] worker suffers at every appearance of the contractor or [illegible]

Conditions of labor as degrading, perhaps, as those of [illegible] exist in many industries, but no class of laborers is so [illegible] uated, owing to the difficulty of introducing reforms in [illegible] small shops abounding in the dark corners of the great [illegible] lessness of the victims, and the ignorant tenacity with [illegible] to their tasks.

There are many model shops in which garments are [illegible] the latest appliances for the comfort and health of [illegible]

such work is carried on in large, healthy
ing of competition through this means and
it for petty contractors to successfully com-
set system so disadvantageous.
made possible the sweating evil, and he is
little capital and not much general knowledge
to become a contractor, almost any ordinary
field and compete with the others on even
is in such keen competition between the con-
ity to obtain work that prices are reduced to

do not the wholesale manufacturers conduct
t as the cutting of the clothing is done? The
s frankly acknowledge that that would cost
the contractor makes a profit, and, besides,
n the trouble and expense of supervision are
ns that the wholesale merchant under the con-
esponsibility for the conditions under which
reover, large shops would become more amen-
laws, and it would then be impossible to im-
o near the very life line.
ere, however, for the contractors in turn sub-
making and finishing or "felling" to others.
one by women at their homes, and very often
ler such an arrangement it is easily seen that,
ces and wages must continue to fall, and the
ned until the limit of human endurance is
eems, has been touched through the task sys-
e coat-making branch by which the contractor
ge in a sort of cooperation, under which the
is employees to solicit work from the ware-
s refused by another, provided they (a set of
ive persons) are willing to do a certain task
o much wages, even though it takes two or
ecified "day's work." This "set," of course,
in a day as it chooses, the only limit being
more objectionable plan be devised to obtain
m a human being, whether man, woman, or
h is insufficient to maintain such a high pres-
truly the very height of the sweating system.
victims are grasping at a chance to preserve
t any cost. Piecework, which is the rule in
shirt-making trades, has been brought, also,
the contract method of work. The above

description I have endeavored to give temperately and ac based upon careful observation. In fact, it would be difficult gerate the unfavorable conditions of this system of labor and it effects.

It is usually supposed that only cheap and common clothing in sweat shops and in homes, but such is not the fact, becau clothing is generally manufactured on a large scale so that t can be systematically divided into many divisions—in some many as 12—in order that it may be turned out very quickly; well-made garments require the long, continuous, and careful w the workers at home or in small shops can give.

Factory legislation is now generally recognized as being of value and in accord with public policy. The factory acts of have been largely copied in this country with marked succ there is a decided tendency in all manufacturing States t greater efficiency in the factory inspection service, and the laws applied to workshops are being made more stringent. Th made in such regulations, particularly in regard to minors an tion of working hours, can best be realized when compared factory act of England, introduced in 1833, which prohibited under 11 years of age from working longer than nine hours per obliged them to attend school two hours a day. What a wretchedness does this tell! The act was passed only after t intense opposition notwithstanding the fearful disclosures m Parliamentary committee.

The limitations set for such protective legislation are con present to the enforcement of sanitary rules and rules for safety, the employment of minors, and the limitation of the labor for women and children, and even for men in [illegible] do not touch the contract system, wages, etc.; [illegible] gap can only be filled by the working people [illegible] own endeavors. Factory laws promote the cause of [illegible] ing children of school age from replacing adults in the [illegible] providing rules framed in the interest of cleanliness [illegible] this must naturally have a wholesome tendency to [illegible] independence, and self-help, and to make a higher [illegible] possible. Some manufacturers have pursued the [illegible] opposing all factory restrictions which eventually [illegible] efit to the fair manufacturer and the industry in [illegible]

The possibility of contagion spreading [illegible] coming from filthy shops has been widely [illegible] physicians have acknowledged this to be an [illegible] beginning of 1894 in Chicago a smallpox [illegible] confined mainly to the clothing districts. [illegible] three different tenement houses were [illegible] to be infected, and the health [illegible] on their list.

...to health in addition to the other impor... led to radical special legislation aiming at the ... sweat shop. New York, Massachusetts, Illinois, have thus grappled with the problem, and the results ...dopted by these States are watched with close inter... ...kers of the other States, and similar legislation is ...d in Ohio. The laws of Massachusetts, Illinois, and ...modeled after the following section of the New York ...3:

...rtment in any tenement or dwelling house shall be ...e immediate members of the family living therein, ...e of coats, vests, trousers, knee pants, overalls, cloaks, ...iders, jerseys, blouses, waists, waistbands, under... ...furs, fur trimmings, fur garments, shirts, purses, ... flowers, cigarettes, or cigars. No person, firm or ...hire or employ any person to work in any room or ... rear building or building in the rear of a tenement ... at making in whole or in part any of the articles ...section, without first obtaining a written permit from ...tor, his assistant, or one of his deputies, stating the ... of persons allowed to be employed therein. Such ...be granted until an inspection of such premises is ...ry inspector, his assistant, or one of his deputies, and ...y the factory inspector at any time the health of the ...hose so employed may require it. It shall be framed ...nspicuous place in the room or in one of the rooms .... Every person, firm, company or corporation, con... ...manufacture of any of the articles mentioned in this ...out the incomplete material from which they or any ...made, or to be wholly or partially finished, shall keep ... of the names and addresses of all persons to whom ...en to be made, or with whom they may have con... ...ame. Such register shall be produced for inspection ...f shall be furnished on demand made by the factory ...stant, or one of his deputies. No person shall know... ...e for sale any of the articles mentioned in this section ...in any dwelling house, tenement house, or building ...enement or dwelling house, without the permit re... ...tion, and any officer appointed to enforce the provi... ...so shall find any of such articles made in violation ...reof, shall conspicuously affix to such article a label ...ds "tenement made" printed in small pica capital ... less than two inches in length; and such officer ... owning or alleged to own such articles that he ... person shall remove or deface any tag or label so ...ticle mentioned in this section is found by the ... assistant, or any of his deputies, to be made ...healthy conditions, he shall affix thereto the label ..., and shall immediately notify the local board ... shall be to disinfect the same and thereupon

...ibit not only the manufacture of garments ...by the immediate members of the family\ ...but seek to interfere with the sale of such goods by making it necessary to have a label attached and by forbidding their sale until properly disinfected and the label removed.

The Massachusetts law is similar in its provisions, but permits goods made in violation of the law to be sold upon the following conditions:

> Whoever knowingly sells or exposes for sale any ready-made coats, vests, trousers or overcoats which have been made in a tenement house used as a workshop, as specified in section forty-four of this act, shall have affixed to each of said garments a tag or label not less than two inches in length and one inch in width, upon which shall be legibly printed or written the words "tenement made" and the name of the State and the town or city where said garment or garments were made.
>
> No person shall sell or expose for sale any of said garments without a tag or label as aforesaid affixed thereto, nor sell or expose for sale any of said garments with a false or fraudulent tag or label, nor willfully remove, alter or destroy any such tag or label upon any of said garments when exposed for sale.

The special laws of Illinois dealing with the sweat shops are similar to those of New York and Massachusetts with the exception of the tag or label provision. The factory inspectors, however, are given the power to order garments destroyed when found infectious or containing vermin, and to prevent the employment in any dwelling room of any person not a member of the family living therein. These distinctions are of considerable importance, as in New York and Massachusetts the inspector is first obliged to make complaint to the board of health before goods can be so destroyed, and persons can only be prohibited from being employed in dwelling rooms through process of law.

Pennsylvania last year adopted a special sweat-shop law worded similarly to the New York law, with the exception that it omits the tag or label provision.

The age limitations for the employment of children in these four States are: New York, 14 years; Massachusetts, 13 years; Illinois, 14 years; Pennsylvania, 13 years.

The hours of labor in New York are limited to 60 per week for persons under 18 years of age and women under 21 years of age. No person under 18 years of age and no woman under 21 can be employed before 6 a. m. or after 9 p. m.

In Massachusetts the law limits the hours of work to 58 a week for minors under 18 years and women. No child under 14 years of age can be employed before 6 a. m. or after 7 p. m. In Illinois the law of [illegible] provided that no female could be employed more than 8 hours in any day, but the supreme court, in November, 1894, declared this provision unconstitutional. In Pennsylvania minors must not be employed [illegible] one day longer than 12 hours nor in any one week more than [illegible]

Ohio, Maryland, and Missouri have not enacted any special [illegible] legislation, although the sweating evil flourishes in their [illegible] general factory laws in Ohio are above the average [illegible] which children may be employed in manufacturing [illegible] years, with the provision that children of more than [illegible] employment during the time they are not school. The number of inspectors employed a chief inspector and a clerk. The Missouri provision for the age limitation of child labor, shall not be required to clean machinery, etc. Mary- ot employ a single inspector; it simply makes violations of lent factory laws subject to prosecution by individuals and partments.

ules at their best can deal only with one phase of the sub- not lessen the tasks of the sweating classes. Neither can ficials, in addition to other duties, hope to maintain super- the evasive sweater. It is only by those who have made itions, particularly in the tenement districts, a study that s can be properly enforced and the spread of disease pre- e evil is so extensive and so difficult to reach that the ordi- inspectors, whose duties are not alone to investigate the l, are plainly unable to cope with the abuses. Legislation he employment of minors, restricting the hours of labor of children, and specifying the amount of air space required son employed all tends to make the sweat shops as such le. All of the manufacturing States have legislated in this t there is a most remarkable laxity in enforcing these laws for their enforcement. A large, populous State has a force spectors not as large as the force required to police a city st the duties of each inspector are as important and more ainly than those of a policeman.

tory inspectors' work is still regarded as experimental, large sults are expected, and public criticism is based upon the nade on the whole vast factory system of the State. To te all the workshop evils, very careful and close inspection Besides, the prosecutions for violations are made very , owing to the difficulty of obtaining sufficient evidence. es of a sweat shop, usually through intimidation and igno- en giving testimony endeavor to shield the employer, even rant infringement of the law is apparent. The ease with ater can change his abode, thus necessitating a new inspec- ner great obstacle. As the tenement shop is the cheapest manufacturing, it is the fittest under the sharp competition it, and consequently the tendency is irresistibly that regulated clothing workshop, being at a disadvantage, the combination of dwelling and workshop, which name of the one nor is suited to the other. Thus are called upon to stem this tide—not alone to To judge what really has been accomplished in deal with New York, Massachusetts, Illinois, States that have made a serious attempt shop evil.

The State of New York [illegible]
women), a chief, and an [illegible]
State. Of these factories [illegible]
York City, and the subfactory [illegible]
City has 14 inspectors attached (7 [illegible]

Massachusetts is better provided [illegible]
for 26,923 factories.

Illinois has 11 inspectors (including 4 women [illegible]
factories in the State.

In 1895 Pennsylvania had 20 deputy [illegible]
for 39,339 factories.

From the forthcoming Tenth Annual Report [illegible]
ors of New York is taken the following list of [illegible]
City for violations of the law regulating the [illegible]
in tenement workshops during the year 1895:

PROSECUTIONS FOR VIOLATIONS OF THE LAW REGULATING THE [illegible] OF CLOTHING IN TENEMENT WORKSHOPS, NEW YORK CITY, [illegible]

| Offense. | [illegible] |
|---|---|
| Working in tenement houses | [illegible] |
| Employing children under 14 years of age | [illegible] |
| Having unclean water-closets | [illegible] |
| Having insufficient water-closets | [illegible] |
| Employing children without certificates | [illegible] |
| Working in rear buildings without permit | [illegible] |
| Employing illiterate children | [illegible] |
| Overcrowding | [illegible] |
| Obscene writing in water-closets | [illegible] |
| Total | [illegible] |

The report for the State for 1894 shows that 10,426 [illegible]
issued requiring changes to be made in or about the [illegible]
which the following represent the most important:

| | |
|---|---|
| Factories ordered to stop overworking minors | [illegible] |
| Children under 14 years of age ordered discharged | [illegible] |
| Illiterate children under 16 years of age ordered discharged | [illegible] |
| Elevators and hoistways ordered guarded | [illegible] |
| Fire escapes ordered erected | [illegible] |
| Machines ordered protected | [illegible] |
| Separate toilet rooms for women ordered | [illegible] |
| Factories ordered renovated | [illegible] |
| Running water for workrooms ordered provided | [illegible] |
| Buildings condemned as unsafe | [illegible] |
| Ordered to cease making clothing in sweat shops | [illegible] |
| Overcrowding ordered stopped | [illegible] |
| Better ventilation ordered | [illegible] |

Since the enactment of the law prohibiting [illegible]
other than by members of the family, in 189[illegible]
(one year and a half), there were erected [illegible]
formerly occupied by tenement [illegible]

new factory buildings were built
clothing trade under the new conditions.
and have legal space for 15,477 work
tenements, formerly used for both working
cleared entirely of workers not members of
and these tenements are now used for
There were also 85 tenement buildings, which
esidents and remodeled into shop buildings. These
d the condition of 17,147 persons who manufacture
York City. This was during the "hard times," when
a standstill, and the number of the persons given
eir conditions improved is very low. The figures were
al count. But this relates only to their sanitary wel-
hing to do with the serious question of their ill-paid

pection department of Massachusetts reported for
on of 5,069 manufacturing, mechanical, and mercan-
ts in which 13,892 children were employed (302 of
en 13 and 14 years of age), and 232,317 adult males
t females, making a total of 403,331. Orders were
rkshops.

nia inspection department reports the following work
ber 1, 1893, to November 30, 1894:

| | |
|---|---|
| outside work | 11 |
| | 4,234 |
| inspections have been made | 175,791 |
| here inspections have been made | 84,945 |
| and 16 years of age employed where inspections have | 22,397 |
| ars of age found employed and discharged | 21 |
| stablishments that have been inspected | 260,736 |
| | 648 |
| shops where inspections have been made | 2,914 |
| | 2,516 |
| | 1,480 |

for the year 1894 shows by statistical tables the
factories and workshops, employing 97,600 men,
children, a total of 130,065 employees. In 1893
ted 2,362 factories and workshops, employing 52,480
and 6,456 children, a total of 76,224 employees.
of 1,078 factories and workshops inspected,
women, and 1,674 children more than in
inspected in 1894, 1,437 were sweat shops,
women, and 721 children, an increase over
2,304 women, and 121 children.
indicate the number of inspections made,
been inspected monthly and others semi-

monthly. On the other hand, no account is made in the tables of the many places visited but not found working at any time during the year.

The above summaries, taken from the factory inspectors' annual reports, show much salutary work being done. Each one of the cases enumerated is tabulated and separately described in voluminous books indicating painstaking methods in the compiling of facts and the enforcement of the laws. Each prosecution in addition serves as an object lesson to other violators who take for granted that such laws exist only to be ignored. It is evident that if factory inspection was extended and the force of inspectors increased commensurate with the vast importance of the work, many of the detrimental features of the workshops would be removed, children of school age would be replaced by adults in the factories, reasonable hours of labor would prevail, and the health of the workers and of the public would be protected and further reforms through the employees themselves would thus be rendered easier.

The lack of uniformity in the factory laws of the different States interferes very seriously with the efficiency of the laws enacted; and as an inducement is given for sweaters to remove to States where more leniency exists, sweat shops are thus spreading to localities where they were never known. As a means of securing uniform legislation throughout the country, the inspectors of thirteen States and two provinces of Canada have formed the International Association of Factory Inspectors, which has held nine annual conventions. In matters of factory legislation their opinions based upon expert knowledge are of [illegible]sequence.

At the meeting of this association in Chicago, September, [illegible], the following resolution in reference to the restrictions of the hours of labor was adopted:

Recognizing the inequality of the existing laws regulating the employment of women and minors in the different States and Territories, and with a view of bringing into effect more uniformity in the [illegible] which would be just and profitable to all engaged in [illegible] suits; first, by placing the employers in the different States on [illegible] basis of competition so far as hours of labor are [illegible] affording to the employed the same protection from the [illegible] follow the overworking of women and children, wherever [illegible] therefore we recommend the adoption by the several [illegible] ulating the hours of labor of women and minors to 48 per [illegible]

The State of Illinois subsequently adopted the [illegible] substance. An appeal was made to the supreme [illegible] ber, 1894, the part pertaining to women was declared [illegible] on the grounds that adult women were not wards of [illegible] law gave special privileges to one sex as against [illegible] interference with the right of women to work [illegible] they choose or to contract for or dispose [illegible] naturally handicaps further legislation [illegible]

ing done, and the protection of the health of
y worthy of greater consideration. As an
irst proposition, it is only necessary to refer
nses of 1880 and 1890 to show that the State
ly doubled its manufacturing resources and
trade, notwithstanding the stringent factory
New York has been a pioneer. In 1880 the
ries, in 1890 65,840, and an increase of the
,246,575 to $1,130,161,195. The same relative
er States having stricter factory legislation.

he prosperity of a community depends upon
nd consumptive power of the people, and that
particular class of toilers is a disease of the
e cured by such remedies as are available.
was enacting measures for the alleviation of
eat shop, and was seeking its suppression as
ratives themselves, whose poverty seemed to
e and mentality, became aroused, and after a
s abolished the worst features of the sweat-
s akin to a revolution in its suddenness and
York, Brooklyn, and Newark in September,
in Boston and Baltimore. Similar disputes,
curred in Philadelphia, Rochester, Chicago,

organization among the tailors existed and
ut all of a spasmodic nature. Some of these
as victories, but the unions were unable to
gained and suffered a relapse. But the
dormant, waiting favorable opportunities.

York City, the foundation for a successful
gh the formation of a national union compris-
ustry, known as the United Garment Workers
g cutters, who had a long trade-union experi-
orably circumstanced, identified themselves
ment, took an active interest, and thus gave
ization.

was begun for the abolition of the sweating
immediate visible results. The industrial
894 set in, and during that period the tailors
tion bordering on pauperism. Special relief

works were started in the large [illegible]tion. The task system sets no [illegible] so increased that the amount of [illegible] further deprived others of work. When [illegible] in August and September, 1894, the tailors had [illegible] how without the task. The unions, which acted as relief [illegible] during the depression, issued a manifesto for the overthrow of the task system and ordered a general strike. About 16,000 coat makers in [illegible] shops in New York, Brooklyn, and Newark responded. The competing contractors, who were used by the manufacturers as implements to increase the daily tasks, formed an association, but granted the demands and signed individual agreements with the unions after the third week. The terms granted provided for weekly work on a basis of [illegible] per day; a minimum rate of wages of from $9 to $15 per week, according to the branch of work; no overtime to be permitted, and the employment of members of the union. So fearful were the now emancipated operatives that the task system would be again returned, that every contractor was obliged to furnish a real-estate bond as security that all the terms of the agreement would be lived up to. The legal standing of the agreements and bonds obtained are now under consideration by the higher courts of New York through several test cases brought against employers for violation of agreement.

The improvements made through these strikes can not be [illegible] estimated by the great material gain. Hope and ambition have taken the place of the characteristic supineness of the clothing operative. Since the first strike the agreements were renewed, with additions, through another struggle the following year, and a few months ago the organized contractors caused a lockout which was successfully resisted. The other branches of the tailoring trade, although [illegible] working under the piecework system, accomplished [illegible] results. It is estimated that 40,000 tailors, about 70 per [illegible] total number affected by the sweating system in the [illegible] are working in shops conducted under similar conditions, and in [illegible] the work day is but nine hours. The hours of labor have [illegible] shortened by from two to five hours per day, and this has [illegible] noticeable effect of prolonging the working seasons and giving [illegible] employment. A number of small contractors were obliged [illegible] their shops, owing to the refusal of the unions to make [illegible] them. It is remarkable that the wholesale manufacturing [illegible] not suffered through the increased cost of production [illegible] uniformity of the increase, and the manufacturers have [illegible] themselves favorable to the change, which has removed [illegible] odium from the trade and raised the method of [illegible] higher plane. There are still many small shops [illegible] teeming tenements in which the sweating system [illegible] most difficult to reach, but the improved condition[illegible]

in the trade has usually been large. There
; a year ago, but a relapse has taken place
the manufacturers to make encroachments
wages, hours of labor, etc., established and
This caused a very dissatisfied and restless
facturers in the different cities formed asso-
ons. This led to the recent large general
go, and Cincinnati which began at the end
d involved about 11,000 persons, including
e, both cutters and tailors, of which number
0 in Baltimore, and 300 in Cincinnati. The
e important clothing cities acted concertedly
bborn and prolonged. The manufacturers
ige, owing to the extreme dullness in the

ith the cutters in each city. In Baltimore
inly in support of the cutters, who demanded
minimum rate of wages. The strike was
s. In Cincinnati the cutters' union ordered
firm and the other manufacturers resented
rs. The unions thereupon declared for the
r seven weeks the cutters returned to work
nfacturers rescinded the resolution not to
on.

urers' Association precipitated the conflict
declaring for the "merit system" instead of
, and the nine-hour workday in place of the
ree years. The tailors also made issue with
lly successful, and in shops employing about
en-hour workday and minimum rate of wages
em. Ten firms, employing about 100 cutters
eded the terms of the union, and at the end
iors returned to work under the conditions
.

contests which involved so many persons
employee, this fact signifies only a check

to the many gains that have been made. The sweating system is being grappled with in all earnestness, and all facts plainly show that this detrimental system of labor is steadily being suppressed, both in this and in other countries. The reforms introduced in the tailor shops of this country helped to stir up the tailors and seamstresses of Germany, who, after immense strikes in the large cities in February last, succeeded in effecting a compromise through a board of conciliation, by which the manufacturers agreed to pay an increase of 12½ per cent, and to fix a list of minimum rates under which each of the various articles of clothing should not be given out either to contractors or workers. The contractors agreed to pay the workers the full amount of the increase. Manufacturers agreed not to deal with contractors who violated the above conditions. Wages were to be paid weekly. The workers' demand for the erection of special workshops was withdrawn. These concessions were granted to 34,000 persons.

Another factor used in the warfare waged against the sweating system is the influence of the public as purchasers. Quite a number of large manufacturers have been obliged to withdraw work sent to sweating contractors, through the systematic appeals made by unions of the trade upon members of other unions and sympathizers to withhold patronage from dealers handling or keeping such goods on sale. Usually a retail clothier would cease dealing with an objectionable manufacturer rather than incur the opposition of patrons. In line with this method the union label has been of service. It is designed to enable sympathizers to distinguish and give preference to goods guaranteed to be made under union, fair, and sanitary conditions. A number of large manufacturers have adopted this label, which has been actively agitated for during the past three years.

The substantial reform work being done in the clothing industry both by legislation and the trade unions is doing much to correct the evils resulting from the laissez-faire policy which regarded all such interferences with free competition, so called, as pernicious and despotic. To-day the principle of factory legislation is seldom disputed, but for nearly a century, ever since the advent of the factory system in England, there has been a most vigorous and bitter contest waged between the advocates and the opponents of factory laws. Some of the ablest economists and legislators were arrayed against what was called the pernicious interference with supply and demand, even in the face of the degrading conditions existing in the workshops and the mines. But surely and steadily the humanizing influences [illegible] Shaftsbury gained the ascendency, and now such protective [illegible] as well as trade-union regulations are generally conceded [illegible]ficial, as being conducive to more equity in the dealings [illegible] and therefore in accord with public policy.

Social theories and policies, as in physics, depend upon [illegible] results for their value, and our perfected method of [illegible] *does much to* enable us to form more correct conclusions [illegible]

[illegible] so closely those of England [illegible] are similar. In dealing with the [illegible] system, a comparison made with the [illegible]fore of value, and is applicable also [illegible] a large scale.

[illegible] a select committee of the House of [illegible]estigation into the sweating system [illegible] exhaustive report published in 1890 [illegible]e extent of the evil. Most of the [illegible]mittee have since been enacted into [illegible]aken from that report:

[illegible] advantage of the necessities of the [illegible] workers.
[illegible] such workers live.
[illegible]ced to the state of things disclosed.

[illegible]ur inquiry, and ample evidence hav-[illegible]y matter comprised within its scope, [illegible]e can not assign an exact meaning [illegible] that name are shown in the pages [illegible]y low rate of wages; (2) excessive [illegible]tate of the houses in which the work

[illegible]erated. The earnings of the lowest [illegible]nt to sustain existence. The hours [illegible]ves of the workers periods of almost [illegible] the last degree.
[illegible]which the work is conducted are not [illegible]persons employed, but are dangerous [illegible]e of the trades concerned in making [illegible]pread by the sale of garments made [illegible] suffering from smallpox and other

[illegible]on evidence of the truth of which we [illegible]nd to express our admiration of the [illegible]ndure their lot, of the absence of any [illegible]tion, and of the almost unbounded [illegible]ther in endeavoring by gifts of food [illegible]ny distress for the time being greater

[illegible] remembered that the observations [illegible]ly, in the main, to unskilled or only [illegible]oroughly skilled workers can almost

[illegible] causes of and the remedies for the [illegible] labor which go under the name of [illegible]olved in a labyrinth of difficulties. [illegible]uction of subcontractors, or middle-[illegible] Undoubtedly, it appears to us that [illegible]oral obligations which attach to cap-[illegible]supply articles and know nothing of [illegible]hom such articles are made, leaving [illegible]ecting the workers, and giving [illegible] by way of compensation a portion of the profit. [illegible] that the middleman is the consequence, not the cause of the evil; the instrument, not the hand which gives motion to the instrument which does the mischief.

Machinery, by increasing the subdivision of labor, and consequently affording great opportunities for the introduction of unskilled labor, is also charged with being the cause of sweating. The answer to this charge seems to be, that in some of the largest clothing and other factories in which labor is admitted to be carried on under favorable conditions to the workers, machinery and subdivision of labor to the greatest possible extent, are found in every department of the factory.

With more truth it may be said that the inefficiency of many of the lower class of workers, early marriages and the tendency of the residuum of the population in large towns to form a helpless community, together with a low standard of life and the excessive supply of unskilled labor, are the chief factors in producing sweating. Moreover, a large supply of cheap female labor is available in consequence of the fact that married women working at unskilled labor in their homes, in the intervals of attendance on their domestic duties and not wholly supporting themselves, can afford to work at what would be starvation wages to unmarried women. Such being the conditions of the labor market, abundant materials exist to supply an unscrupulous employer with his wretched dependent workers.

The most important question is whether any remedy can be found for this unhappy state of a portion of the laboring class. With respect to the low wages and excessive hours of labor, we think that good may be effected by cooperative societies and combination amongst the workers. We are aware that home workers form a great obstacle in the way of combination, inasmuch as they can not readily be brought to combine for the purpose of raising wages. To remove this obstacle we have been urged to recommend the prohibition by legislation of working at home; but we think such a measure would be arbitrary and oppressive.

We now proceed to make recommendations in respect of the evils which appear to us, under existing circumstances, to require immediate Parliamentary interference.

Under the factory law work places for the purposes of sanitation [illegible] divided into three classes—(1) factories; (2) workshops; (3) domestic workshops.

We are of opinion that all work places included under the above descriptions should be required to be kept in a cleanly state, to be lime-washed or washed throughout at stated times, to be kept free [illegible] noxious effluvia, and not to be overcrowded; in other words [illegible] treated for sanitary purposes as factories are treated under the factory law.

We are also of the opinion that as respects administration, an [illegible]quate number of inspectors should be appointed to enforce [illegible] observance of the law. It has been suggested that the inspect[illegible] be assisted by workmen having practical knowledge of [illegible] inspected, and paid only the wages of artisans, but we de[illegible] the disadvantages arising from the division of responsibility [illegible] outweigh any advantages to be derived from their technical [illegible]

We think that inspectors should have power to enter all [illegible] within their jurisdiction at reasonable times without a wa[illegible]

We consider that the establishment of county councils [illegible] every county a body capable of being trusted with the su[illegible] *of the inspection* by sanitary authorities and of making [illegible] efficient.

that different Departments of the Gov- with matters relating to the labor ques- are appointed by and under the control of trade supplies the public, through its rmation as to the conditions of labor and ses, and requires for that purpose the aid inspectors. The local government board hich, for the purpose of promoting hygi- mmunication with the factory inspectors. advisable to bring the officers employed tas into closer relations with each other ontrol of one department, or otherwise f administration.

ble evidence attributing, to the disuse of incompleteness of the education of the gested are, on the one hand, a renewal of d, on the other, the promotion of a larger . We think that the encouragement of ses of artisans is more likely to prove an nce to the old system of apprenticeship. t is incumbent on all Departments of the l and other public bodies to take care they are satisfied that the workmen by orked out are paid proper wages. We nly in the interest of the workman, but l insure to the public a corresponding f the work. This recommendation may contractor to show the scale of wages , supposing such scale to be satisfactory, e workmen, or otherwise making known e paid.

nt expressing our earnest hope that the been brought to our notice will induce tion to the conditions under which the h goods is conducted, and that the public om traders who are known to conduct hich regards neither the welfare of the e work produced.

mary of some of the provisions of the fac- 878 to 1895, enacted by Parliament, which ops:

laundry must be kept clean, well venti- l overcrowding, and at a reasonable tem- ubic feet of space (400 in overtime) must

In all factories and steam laundries the duty of seeing that these provisions are carried out belongs to the factory inspector, in workshops and hand laundries, to the local sanitary authority.

The working hours are variously limited for children, young persons, and women. A "child" means a person between 11 and 14 years of age; a "young person" means a person between 14 and 18 years of age; a "woman" means a woman of 18 years of age and upward. The period of employment for young persons and women in factories and workshops is limited to the hours between 6 a. m. and 6 p. m., or 7 a. m. and 7 p. m., or 8 a. m. and 8 p. m. In textile factories two hours must be allowed for meals (one of them before 3 p. m.), and work must not be carried on for more than two and one-half hours without an interval of one-half hour for meals. In nontextile factories and workshops one and one-half hours must be allowed for meals (one of them before 3 p. m.), and work must not be carried on for more than five hours without an interval of one-half hour for meals. In textile factories when work begins on Saturdays at 6 a. m. manufacturing processes must cease at 1 p. m. if not less than one hour is allowed for meals. If less than one hour is allowed for meals, manufacturing processes must cease at 12.30 p. m. When work begins at 7 a. m. manufacturing processes must cease at 1.30 p. m. In nontextile factories and workshops the hours of employment on Saturdays may be between 6 a. m. and 2 p. m., or 7 a. m. and 3 p. m., or 8 a. m. and 4 p. m. In every case an interval of not less than one-half hour must be allowed for meals.

Children employed in factories and workshops may only work half-time, that is, either in the morning or afternoon or on alternate days.

Special provisions with regard to employment on Saturday and Sunday are made for young persons and women of the Jewish religion.

Employment outside a factory or workshop, in the business of that factory or workshop, before or after working on the same day inside, is forbidden for children. It is also forbidden for youn persons and women who are employed inside both before and after the dinner hour. Work given out, or allowed to be taken out, is treated as employment on that day.

The occupier of a factory or workshop or laundry may not, to his knowledge, employ a woman within four weeks after she has given birth to a child.

Notice must be sent to the inspector within one month of the time when work is begun in any factory or workshop.

All occupiers of existing workshops must, before the expiration of twelve months from January 1, 1896 (unless they have already done so), send their names and addresses, and particulars of the work carried on in such workshops, to the inspector.

On any evening when it is intended that women shall work overtime notice must be sent to the inspector before 8 p. m.

An abstract of the factory and workshop acts, with the names and addresses of the inspector and surgeon of the district, the hours of employment and times of meals, also a notice stating he total cubic space and the number of persons who may be empl yed, must be affixed in every factory, workshop, and laundry in such a position as to be easily read. Occupiers of "domestic workshops" are not required to send or affix notices.

The occupier of every factory and workshop, and every [illegible] employed by such occupier, shall, if the trade is included [illegible] of the home secretary, keep lists of the names and add[illegible] *persons* employed as outworkers, such lists to be open [illegible]

not being so changeable, the enforce-
ıs rendered easier. In dealing with the
e conditions are apparent. In the very
pless, dependent population, and much
rough immigration and the tendency of
e to the large cities. Englishmen are
to other countries.

## MASSACHUSETTS.

*Twenty-fifth Annual Report of the Bureau of Statistics o* 1895. Horace G. Wadlin, Chief. xvii, 33

This report treats of the following subjects: Comp tain occupations of graduates of colleges for women, 4 tribution of wealth, 254 pages; labor chronology, 33 p

COMPENSATION IN CERTAIN OCCUPATIONS OF GRA LEGES FOR WOMEN.—This investigation was not confi of colleges, and is more accurately described as "The certain occupations of women who have received special training." The presentation is based upon conducted by a committee of the Association of Colleg

There were 451 schedules received from women en from employers of women. These returns were distrib Massachusetts, 59; Minnesota, 55; Connecticut, 44; R California, 61; New York, 90; Indiana, 39; Illinois, 14; ing 153 from various other States.

Of the schedules received from employees, 437 conta concerning occupation, residence, and conjugal con totals are summarized as follows:

RESIDENCE AND CONJUGAL CONDITION OF WOMEN WHO HA LEGE OR OTHER SPECIAL TRAINING.

| Conjugal condition. | Resid |
|---|---|
| | At home. |
| Single | 209 |
| Married | 24 |
| Widowed | 11 |
| Total | 244 |

Seventy-eight of the women from whom schedule failed to state their age. The others are classified periods, and 313 were 20 but under 40 years of age.

The following statement gives the totals of the ansv concerning occupation and means of support:

OCCUPATION AND MEANS OF SUPPORT OF WOMEN WHO HAVE OR OTHER SPECIAL TRAINING.

| Question. | Answer. | |
|---|---|---|
| | Yes. | No. |
| Have you any remunerative occupation besides your main work? | [illegible] | [illegible] |

[illegible] and under $200; 2, $200 and under $300, [illegible] of $300 per month. Forty-eight failed to [illegible] concerning compensation.

[illegible] fifty reports stated that men received more pay [th]e same grade of work, 95 reported the same pay for while 5 reported that men received less pay than [fa]iled to answer the question.

[O]ne of the reasons for paying women less than men [state]ments that a man is called upon to support others [w]hile, as a rule, women in industry do not aid in the . Of the 379 women who answered the question on or 41.42 per cent, aided in the support of others. [gi]ven is that women do not remain continuously in one the 333 who answered the question on this subject but 1 employment, their average term of service [yea]r and eight months, 88 had been in 2 employments, [i]n 5, and 1 in 8 employments since beginning work.

[469]7 males and 3,097 females, a total of 7,794 persons employers who made returns. Ninety employers [que]stion, "Are the services of men and women equally [desirable?]" Of this number 46 answered yes, 29 no, and 7 8 stated that for some work they were as desirable [othe]r they were not.

[Of the]s, 29 indicate that the fact of supply and demand, or e reason for the difference in compensation of the [co]nsider physical and mental differences, or differences to be the real reason. In 17 replies no other reason [off]ered.

[Facts s]ummarized in the above statements are shown in the [tabulat]ions, so that the conditions prevailing in each indus[try are co]mbined, and are followed by condensed text statements both employees and employers.

[DISTRIBU]TION OF WEALTH.—Of the 254 pages devoted to this [are] in statistical tables. The information presented is part of a general inquiry into wealth distribution, covers the initial stage of a projected investigation . The statistics were obtained from the records of the the State and cover four periods of three years each, 1859 to 1861, 1879 to 1881, and 1889 to 1891.

[The cou]rts of the State administer substantially all the estates [of thos]e possessing property worth taking account of. The [estates of which] no trace can be obtained in these courts are almost

ble to determine the total value of all [illegible], and to that extent the value of the statistics is limited. The following statement indicates exactly how far this limitation affects the [illegible]:

PROBATES FILED WITH AND WITHOUT INVENTORIES, BY SELECTED PERIODS.

| Period. | Inventory filed. | | Inventory not filed. | | Total. |
|---|---|---|---|---|---|
| | Number. | Per cent. | Number. | Per cent. | |
| 1829 to 1831 (3 years) | [illegible] | [illegible] | [illegible] | [illegible] | [illegible] |
| 1859 to 1861 (3 years) | [illegible] | [illegible] | [illegible] | [illegible] | [illegible] |
| 1879 to 1881 (3 years) | 11,142 | [illegible] | [illegible] | [illegible] | [illegible] |
| 1889 to 1891 (3 years) | [illegible] | [illegible] | [illegible] | [illegible] | [illegible] |
| Total | 36,370 | [illegible] | [illegible] | [illegible] | [illegible] |

For the State as a whole, and for each county except Dukes, it appears that in recent years the number of probates registered without inventories has considerably increased, and in the comparison of the values for different periods this should be borne in mind. It is probably true that the estates unaccompanied by inventories are, as a rule, larger than those for which inventories are filed. In the following statements only inventoried probates are referred to, and the total for the State by periods considered. The statistics are shown in the report by counties for the different years.

The total number and value of probates, classified according to the sex of the deceased, are shown in the following statement:

NUMBER AND VALUE OF INVENTORIED ESTATES PROBATED DURING EACH SELECTED PERIOD, BY SEX OF DECEASED.

| Period. | Males. | | | Females. | | | Total. | | |
|---|---|---|---|---|---|---|---|---|---|
| | Number. | Value. | | Number. | Value. | | Number. | Value. | |
| | | Total. | Average. | | Total. | Average. | | Total. | Average. |
| 1829 to 1831 | 3,102 | $13,500,099 | $4,352 | 596 | $994,008 | $1,668 | [illegible] | [illegible] | [illegible] |
| 1859 to 1861 | 5,103 | 45,847,981 | 8,985 | 1,819 | 7,408,613 | 4,073 | [illegible] | [illegible] | [illegible] |
| 1879 to 1881 | 7,030 | 114,747,943 | 16,323 | 4,112 | 22,626,316 | 5,503 | 11,142 | [illegible] | [illegible] |
| 1889 to 1891 | 8,349 | 114,032,780 | 13,658 | 6,259 | 41,526,008 | [illegible] | [illegible] | [illegible] | [illegible] |
| Total | 23,584 | 288,128,803 | 12,217 | 12,786 | 72,555,145 | 5,675 | 36,370 | [illegible] | [illegible] |

Out of the 36,370 probates represented by inventories, [illegible] testate and 22,060 were intestate. In the first period considered, [illegible] per cent of the probates were intestate; the percentages in the other three periods being, respectively, as follows: 64.07, 57.86, [illegible]. The larger number of the estates were distributed without [illegible] by will—that is, to heirs in accordance with the provisions of [illegible] utes—but the tendency to dispose of estates by will appears [illegible].

Of the total number of estates considered, 35,304, or [illegible] were solvent, and 1,066, or 2.93 per cent, were insolvent. [illegible] year period ending with 1831, 8.14 per cent of the probates [illegible] vent, the percentages for the other periods being [illegible] respectively.

[illegible] REAL AND PERSONAL ESTATES [illegible] BY SELECTED PERIODS.

| | Real estate. | | | Personal estate. | | |
|---|---|---|---|---|---|---|
| | Number. | Value. | | Number. | Value. | |
| | | Total. | Average. | | Total. | Average. |
| [illegible] | [illegible] | [illegible] | [illegible] | 3,634 | [illegible] | [illegible] |
| [illegible] | [illegible] | [illegible],756,[illegible] | 4,[illegible] | 6,665 | 31,502,026 | 4,736 |
| [illegible] | [illegible] | [illegible],272,[illegible] | 6,443 | 10,690 | [illegible],100,277 | 8,789 |
| [illegible] | [illegible] | [illegible],190,940 | 6,873 | 12,394 | 95,367,842 | 7,159 |
| [illegible] | 22,[illegible] | 128,379,644 | 5,843 | 34,233 | 227,494,204 | 6,645 |

represented in all of the probates considered is owing statement by classes. Within the classes given applicable to each class, estates of approxilue being averaged together. In order that the may be clearly seen, the number of estates within , with the aggregate value which these estates

[illegible] OF INVENTORIED ESTATES PROBATED DURING FOUR SELECTED PERIODS, BY CLASSES.

| Males. | | | Females. | | | Total. | | |
|---|---|---|---|---|---|---|---|---|
| Number. | Value. | | Number. | Value. | | Number. | Value. | |
| | Total. | Average. | | Total. | Average. | | Total. | Average. |
| [illegible] | [illegible],183 | $221 | 2,[illegible] | [illegible],193 | $247 | 6,955 | $1,605,375 | $231 |
| [illegible] | 1,[illegible] | 782 | 1,[illegible] | 1,415,[illegible] | 721 | 4,612 | 3,353,949 | 727 |
| [illegible] | [illegible] | 2,498 | 5,471 | 12,739,018 | 2,328 | 14,886 | [illegible],162,451 | 2,429 |
| [illegible] | [illegible] | 6,978 | 1,304 | 8,998,485 | 6,901 | 4,482 | 31,174,[illegible] | 6,956 |
| [illegible] | [illegible],145 | 15,617 | 899 | 13,453,[illegible] | 14,965 | 3,185 | 49,156,077 | [illegible] |
| [illegible] | 37,[illegible],090 | 24,642 | [illegible] | 9,930,463 | 33,892 | 1,100 | 37,886,463 | 442 |
| [illegible] | [illegible] | [illegible],611 | 121 | 8,731,739 | 72,163 | 600 | 42,075,620 | 126 |
| [illegible] | 34,[illegible] | 137,185 | 54 | 7,171,567 | 132,807 | 303 | 41,318,226 | [illegible],364 |
| [illegible] | 31,[illegible] | [illegible],206 | 12 | 3,024,924 | 252,077 | 100 | 24,[illegible],227 | [illegible],392 |
| [illegible] | 14,[illegible],010 | 343,884 | 10 | 3,321,683 | 332,168 | 53 | 18,103,[illegible] | 341,673 |
| [illegible] | 11,[illegible],481 | 460,[illegible] | 2 | 841,173 | 420,587 | 27 | 12,107,654 | 448,432 |
| [illegible] | 62,[illegible] | 956,079 | 3 | 2,271,276 | 757,092 | 67 | 63,396,329 | 946,214 |
| [illegible] | [illegible] | 12,217 | 12,786 | 72,555,145 | 5,675 | 36,370 | 360,683,948 | 9,917 |

e tables similar to the above for the different years, . For the three-year period ending with 1831 the inventoried probates was $3,919, but 1,431, or not 8 estates considered, were valued at less than $500, of greater value than $500,000. Comparing these ee-year period ending with 1891, in which 14,608 age value of $10,649 were considered, it appears

The average value of the real and personal estates of males and females, respectively, is shown for the classes given in the above statement by counties for each year and period covered by the report. The following statement reproduces the average values given for real and personal estates, respectively, by periods of years:

AVERAGE VALUE OF INVENTORIED REAL AND PERSONAL ESTATES PROBATED DURING EACH SELECTED PERIOD, BY CLASSES.

| Class. | Real estate. | | | | Personal estate. | | | |
|---|---|---|---|---|---|---|---|---|
| | 1829 to 1831. | 1859 to 1861. | 1879 to 1881. | 1889 to 1891. | 1829 to 1831. | 1859 to 1861. | 1879 to 1881. | 1889 to 1891. |
| Under $500 | $220 | $238 | $233 | $249 | $182 | $202 | $210 | $206 |
| $500 but under $1,000 | 725 | 721 | 713 | 717 | 706 | 717 | 718 | 713 |
| $1,000 but under $5,000 | 2,317 | 2,320 | 2,307 | 2,394 | 2,066 | 2,197 | 2,275 | 2,285 |
| $5,000 but under $10,000 | 6,808 | 6,727 | 6,801 | 6,796 | 6,824 | 7,039 | 6,864 | 7,071 |
| $10,000 but under $25,000 | 15,176 | 15,132 | 14,980 | 14,856 | 15,249 | 14,942 | 15,583 | 15,366 |
| $25,000 but under $50,000 | 33,859 | 33,731 | 34,045 | 33,680 | 34,736 | 35,196 | 34,639 | 35,218 |
| $50,000 but under $100,000 | 66,944 | 66,812 | 71,948 | 69,913 | 64,412 | 67,210 | 69,790 | 70,785 |
| $100,000 but under $200,000 | 179,000 | 139,754 | 132,828 | 141,873 | 127,087 | 140,895 | 142,340 | 137,549 |
| $200,000 but under $300,000 | 257,411 | 237,310 | 246,897 | 231,825 | 230,371 | 228,614 | 246,225 | 251,389 |
| $300,000 but under $400,000 | ........ | 341,779 | 348,690 | 339,503 | ........ | 326,368 | 339,600 | 344,549 |
| $400,000 but under $500,000 | ........ | ........ | 422,900 | 439,142 | ........ | 449,316 | 437,330 | 437,344 |
| $500,000 and over | ........ | ........ | 845,939 | 758,063 | 595,194 | 866,225 | 1,141,034 | 713,294 |
| Total | 3,128 | 4,807 | 6,443 | 6,573 | 2,051 | 4,726 | 8,783 | 7,152 |

In the following statement real and personal property have been combined and similar averages shown, together with the number of probates in each class:

NUMBER AND AVERAGE VALUE OF INVENTORIED ESTATES PROBATED DURING SELECTED PERIODS, BY CLASSES.

| Class. | 1829 to 1831. | | 1859 to 1861. | | 1879 to 1881. | | 1889 to 1891. | |
|---|---|---|---|---|---|---|---|---|
| | Number. | Average value. | Number. | Average value. | Number. | Average value. | Number. | Average value. |
| Under $500 | 1,431 | $186 | 1,485 | $233 | 1,822 | $245 | 2,217 | $247 |
| $500 but under $1,000 | 463 | 732 | 960 | 726 | 1,451 | 722 | 1,738 | 721 |
| $1,000 but under $5,000 | 1,274 | 2,372 | 2,827 | 2,403 | 4,568 | 2,458 | 6,197 | 2,432 |
| $5,000 but under $10,000 | 295 | 6,799 | 797 | 6,909 | 1,421 | 6,988 | 1,969 | 6,974 |
| $10,000 but under $25,000 | 157 | 15,455 | 507 | 15,361 | 1,023 | 15,443 | 1,498 | 15,449 |
| $25,000 but under $50,000 | 42 | 35,170 | 168 | 34,880 | 410 | 33,986 | 480 | 34,613 |
| $50,000 but under $100,000 | 25 | 73,166 | 92 | 69,448 | 218 | 70,497 | 265 | 69,789 |
| $100,000 but under $200,000 | 6 | 134,244 | 52 | 128,908 | 111 | 139,698 | 134 | 136,039 |
| $200,000 but under $300,000 | ........ | ........ | 18 | 237,513 | 37 | 239,803 | 45 | 248,650 |
| $300,000 but under $400,000 | 2 | 326,032 | 7 | 340,995 | 22 | 346,316 | 22 | 339,735 |
| $400,000 but under $500,000 | 1 | 415,371 | 3 | 474,523 | 10 | 438,342 | 13 | 452,715 |
| $500,000 and over | 2 | 633,909 | 6 | 848,109 | 29 | 1,144,758 | 30 | 794,722 |
| Total | 3,698 | 3,910 | 6,922 | 7,094 | 11,142 | 12,329 | 14,608 | 16,640 |

If $50,000 is arbitrarily assumed as the dividing line between large and small estates, the following statement shows the number, value, and average value of probates above and below the line, respectively, for the different periods:

ESTATES PROBATED AND INVENTORIED ABOVE AND BELOW $50,000 IN VALUE, BY SELECTED PERIODS.

| Period and classification. | Number. | | Value. | |
|---|---|---|---|---|
| | Total. | Per cent. | Total | Average |
| **1829 to 1831.** | | | | |
| Under $50,000 | 3,662 | 99 03 | $9,536,245 | $2,604 |
| $50,000 and over | 36 | 0 97 | 4,957,862 | 137,718 |
| Total | 3,698 | 100 00 | 14,494,107 | 3,919 |
| **1859 to 1861.** | | | | |
| Under $50,000 | 6,744 | 97.43 | 26,989,881 | 4,002 |
| $50,000 and over | 178 | 2.57 | 26,266,913 | 147,567 |
| Total | 6,922 | 100.00 | 53,256,794 | 7,694 |
| **1879 to 1881.** | | | | |
| Under $50,000 | 10,715 | 96 17 | 52,432,701 | 4 893 |
| $50,000 and over | 427 | 3.83 | 84,941,558 | 198,926 |
| Total | 11,142 | 100.00 | 137,374,259 | 12,329 |
| **1889 to 1891.** | | | | |
| Under $50,000 | 14,099 | 96.52 | 70,379,372 | 4,992 |
| $50,000 and over | 509 | 3.48 | 85,179,416 | 167,347 |
| Total | 14,608 | 100.00 | 155,558,788 | 10,649 |

The total amount represented by the estates above the $50,000 line in the three-year period ending with 1831 was only slightly in excess of one-half the total amount represented by the estates below the line.

In the three-year period ending with 1891, the value of the estates above the line was 54.76 per cent and the value of those below the line 45.24 per cent of the total value of all the estates considered.

The average holding in the estates below the line nearly doubled in the sixty years, while the number of persons who died worth less than $50,000 in the last period was nearly four times as great as in the first. The deceased owners of these estates represented one person in every 476 of the population in the period centering in 1890 and one in every 500 in the period centering in 1830. The average holding in the estates above the line has only exhibited a moderate increase, while the number of persons who died worth $50,000, or over, during the period centering in 1890, was more than 14 times as great as during the period centering in 1830. There was but one such inventoried estate probated in every 50,867 of the population in 1830, as against one in every 13,170 in 1890.

In order to enable a comparison of the probate statistics with those showing death, tables are introduced which give the total number of deaths by sex for the different counties for each year covered by the three periods from 1859 to 1891. The totals for the State are summarized as follows:

TOTAL DEATHS IN THE STATE, BY SELECTED PERIODS.

| Period. | Deaths. |
|---|---|
| 1859 to 1861 (3 years) | 68,129 |
| 1879 to 1881 (3 years) | 102,551 |
| 1889 to 1891 (3 years) | 130,490 |
| Total | 301,170 |

Out of 68,129 deaths in the State during the three years 1859 to 1861 there were registered 6,744 inventoried estates below $50,000 in value, and 178 estates above. Out of 130,490 deaths during the three years 1889 to 1891 there were registered 14,099 inventoried estates below $50,000 in value, and 509 estates above.

LABOR CHRONOLOGY.—Under this title the resolutions and other actions of the labor organizations throughout the State on various subjects are grouped by dates.

## MISSOURI.

*Seventeenth Annual Report of the Bureau of Labor Statistics of the State of Missouri, being for the year ending November 1, 1895.* Lee Meriwether, Commissioner. 354, v pp.

The following are the subjects treated in this report: Truck stores and "checks," 87 pages; machine versus hand labor, 5 pages; franchises and taxation, 4 pages; surplus products, 3 pages; Plasterers' Union, Kansas City, 18 pages; factory inspection, 48 pages; statistics of manufactures, 169 pages. An appendix of 12 pages contains a synopsis of Missouri laws relating to labor.

The presentation concerning truck stores and "checks," machine versus hand labor, franchises and taxation, and the Plasterers' Union of Kansas City, consists almost entirely of a textual discussion. The truck-store system is explained and remedies suggested. A brief account is given of an investigation made by the bureau in September, 1895, respecting the system as conducted in Missouri and the efforts made to have legal proceedings instituted against those who were violating the laws on this subject. The opinions of the courts in four leading cases, construing State laws relating to checks and truck stores, are given in full.

The displacement of labor by reason of the introduction of type-setting machines is treated. The evidence taken by the commissioner and his decision on investigating certain charges against the Plasterers' Union of Kansas City made by nonunion men, and as to the condition of the cigar business in that city, are given.

[illegible] statistics showing the surplus agricul- [illegible] commodities shipped from each county during [illegible] form of a map, in which the quantities of the [illegible] in the respective counties. The totals [illegible] principal products are given in the following statement:

SURPLUS PRODUCTS, 1894.

| Product. | Quantity. | Counties marketing. |
|---|---|---|
| [illegible] bushels.. | 1,406,048 | 71 |
| [illegible] pounds.. | 2,810,880 | 94 |
| [illegible] do.... | 33,461,156 | 12 |
| [illegible] head.. | 894,838 | 107 |
| [illegible] bushels.. | 10,960,732 | 91 |
| [illegible] dozen.. | 23,765,835 | 107 |
| [illegible] pounds.. | 20,136,336 | 17 |
| [illegible] barrels.. | 2,676,277 | 90 |
| [illegible] pounds.. | 6,755,177 | 91 |
| [illegible] head.. | 2,596,077 | 107 |
| [illegible] feet.. | 39,042,000 | 23 |
| [illegible] do.... | 296,130,430 | 97 |
| [illegible] pounds.. | 44,160,662 | 107 |
| [illegible] bushels.. | 12,203,502 | 93 |
| [illegible] pounds.. | 2,503,680 | 96 |

[illegible]TORY INSPECTION.—Statistics under this title are shown for 8 [illegible] throughout the State, in which 600 establishments employing [illegible] men, 10,517 women, 1,431 boys over 12 and under 18 years of [illegible] 590 girls over 14 and under 18, 10 boys under 12, and 88 girls [illegible] 14 were inspected. There were 554 establishments in operation [illegible] time, 45 less than full time, and 1 closed entirely; 252 with full [illegible] 179 with less than full force, and 169 part of the time with full [illegible] of employees. The statistics are shown in detail for each [illegible] establishment inspected.

[illegible] STATISTICS OF MANUFACTURES.—Schedules were sent to 3,000 [illegible] and of this number 864 made returns sufficiently com- [illegible] be presented in the report. The statistics are shown in detail [illegible] industry. The following statement gives the totals for St. [illegible] Kansas City, the State exclusive of these cities, and the total:

STATISTICS OF MANUFACTURES, 1895.

| Items. | St. Louis. | Kansas City. | State, exclusive of the two cities. | Total for State. |
|---|---|---|---|---|
| [illegible] | 283 | 40 | 59 | 382 |
| [illegible] | 297 | 69 | 116 | 482 |
| [illegible] | 580 | 109 | 175 | 864 |
| [illegible] | $64,539,206 | $4,845,100 | $6,233,170 | $75,617,476 |
| [illegible] and [illegible] | $28,034,990 | $2,882,296 | $1,857,322 | $32,774,608 |
| [illegible] | $10,978,386 | $1,285,408 | $1,582,894 | $13,846,688 |
| [illegible] | $46,173,357 | $10,217,655 | $5,079,013 | $61,470,025 |
| [illegible] | $5,486,690 | $1,085,099 | $472,560 | $7,044,349 |
| [illegible] | $89,727,625 | $14,617,585 | $9,503,853 | $113,849,063 |
| [illegible] | 3,370 | 417 | 449 | 4,236 |
| [illegible] | $4,983,449 | $493,835 | $517,667 | $5,994,951 |
| [illegible] | 34,821 | 4,144 | 4,041 | 43,006 |
| [illegible] | $14,810,075 | $1,726,437 | $1,317,516 | $17,854,028 |
| [illegible] | 54,506 | 5,412 | 7,094 | 67,012 |

[illegible] statements concerning wages that may be con- [illegible] the pay rolls of 16 factories throughout the

State were carefully examined, and the amount received by each individual employee noted. In this manner there was obtained the actual amount received in 15 different industries, by 4,401 employees, 3,098 of whom were men and 1,303 women. The average daily income is shown for the different occupations in each industry. Comparison is also made of the daily incomes in certain industries as shown by the reports of 1890 and 1895. The daily income during 1895 is further indicated by assigning the employees in the different occupations of the various industries to classes according to the amount of income from 50 cents to $5. The average daily income and the estimated net yearly wages of each employee in the different occupations in each industry is given with the average wages received for each working day, amount received for days worked, number of days worked, and number of working days in period considered.

## UTAH.

*First Triennial Report of the Bureau of Statistics of Utah, for the year ended December 31, 1894, with Census, 1895.* Joseph P. Bache, Statistician; H. W. Griffith, Chief Clerk. 54 pp.

The bureau issuing this report was organized under an act of the legislature of the Territory, approved March 10, 1892, which designates the statistical year as 1895, and each triennial year thereafter, defines the duties of the Territorial statistician, indicates the character of the statistics to be gathered, and provides for the publication of the reports for each statistical year.

The statistics are presented in detail for cities, towns, and counties. The totals are shown in the following summaries:

POPULATION, 1895.

| | |
|---|---|
| Males | 126,803 |
| Females | 120,521 |
| Native born | 194,825 |
| Foreign born | 52,499 |
| White | 245,985 |
| Colored | 571 |
| Chinese | 768 |
| Total | 247,324 |

INDUSTRIAL AND COMMERCIAL STATISTICS, 1894.

| | |
|---|---|
| Industrial: | |
| Establishments | 880 |
| Horsepower employed | 11,280 |
| Laborers | 5,054 |
| Wages | $2,027,118 |
| Capital invested | $5,476,246 |
| Value of plant | $5,986,215 |
| Cost of materials | $2,640,038 |
| Value of products | $6,678,118 |
| Commercial: | |
| Stores | 1,974 |
| Capital invested | $14,551,345 |
| Sales | $32,865,611 |
| Employees | 5,023 |
| Wages | $2,814,314 |

MINING STATISTICS, 1894.

| | |
|---|---|
| **Gold and silver:** | |
| Mines patented | 271 |
| Mines unpatented | 275 |
| Average number of employees | 2,534 |
| Wages | $2,809,817 |
| Output, tons | 251,924 |
| Value of output | $4,289,606 |
| Cost of plant | $4,592,187 |
| Cost of development work | $7,991,186 |
| **Coal:** | |
| Mines patented | 4 |
| Mines unpatented | 8 |
| Employees | 139 |
| Wages | $59,775 |
| Output, tons | 62,101 |
| Cost of plant | $46,708 |
| Cost of development work | $43,600 |

FARM STATISTICS.

| | |
|---|---|
| Free of incumbrance | 17,684 |
| Mortgaged | 2,132 |
| Total | 19,816 |
| Amount of mortgages | $1,971,352 |
| Expended for buildings, 1894 | $721,229 |
| **Acreage, 1894:** | |
| Under cultivation | 467,162 |
| Irrigated | 417,455 |
| Pasturage, fenced | 294,725 |
| Improved | 806,650 |
| Unimproved | 979,182 |
| Hired laborers | 5,960 |
| Wages paid laborers | $1,015,366 |
| **Expenses:** | |
| Repairs | $226,879 |
| Fertilizers | $110,621 |
| Interest and taxes | $610,820 |
| Sundries | $303,145 |
| **Live stock:** | |
| Milch cows | 60,595 |
| Other cattle | 238,974 |
| Horses | 99,895 |
| Swine over 6 months old | 47,703 |
| Mules | 1,308 |
| Asses | 835 |
| Goats | 2,966 |
| **Sheep, 1894:** | |
| Number | 2,422,802 |
| Value | $3,686,934 |
| Wool, pounds | 12,119,763 |
| Value | $864,260 |
| Value per pound | $0.07⅛ |

PRINCIPAL FARM PRODUCTS, 1894.

| Name. | Acreage. | Yield. | | |
|---|---|---|---|---|
| | | Bushels. | Value. | Average per acre (bushels). |
| Wheat | 144,717 | 3,113,073 | $1,440,096 | 21.5 |
| Corn | 13,893 | 260,697 | 151,433 | 18.8 |
| Oats | 49,334 | 1,387,710 | 470,658 | 28.1 |
| Barley | 8,754 | 271,866 | 100,207 | 31.0 |
| Rye | 3,791 | 42,352 | 20,094 | 11.2 |
| Lucerne | 163,544 | *a* 462,459 | 1,851,639 | *a* 2.8 |
| Hay | 89,255 | *a* 133,646 | 604,399 | *a* 1.4 |
| Potatoes | 13,526 | 1,649,239 | 522,855 | 121.9 |

*a* Tons.

## SECOND ANNUAL REPORT ON THE BUILDING A
## CIATIONS OF CALIFORNIA.

*Second Annual Report on the Building and Loan Associ of California.* By the Board of Commissioners of Loan Associations. May 31, 1895. xiv, 387 pp.

This report is for the year ending May 31, 1895, an ments in detail from the 144 association's doing busi Forty-three pages are devoted to an analysis of the st cussion of exceptional methods prevailing in some appendix of 31 pages presents the laws of the State government, control, and existence of corporations.

The number and the proportional amount of busi of the three classes into which the associations are d the following statement:

NUMBER AND PERCENTAGE OF BUSINESS OF DIFFERENT CL AND LOAN ASSOCIATIONS.

| Class of association. |
| --- |
| Local |
| National |
| Cooperative banks |
| Total |

Of the 135 local associations reported, 134 are on t the terminating plan.

The aggregate resources and liabilities and the a and disbursements of the different classes of associat the following table:

RESOURCES AND LIABILITIES, RECEIPTS AND DISBURSEMENT LOAN ASSOCIATIONS.

| Items. | Locals. | Nationals. | Coo[illegible] |
| --- | --- | --- | --- |
| RESOURCES. | | | |
| Loans | $17,[illegible] | $1,4[illegible] | [illegible] |
| Arrearages | [illegible] | [illegible] | [illegible] |

[illegible] AND DISBURSEMENTS OF BUILDING AND [illegible] ASSOCIATIONS—Concluded.

| | Locals. | Nationals. | Cooperative banks. | Total. |
|---|---|---|---|---|
| | [illegible] | $6,127,[illegible] | $646,902.89 | $14,361,[illegible] |
| | [illegible] | [illegible] | 112,894.81 | 4,743,[illegible] |
| | 87,[illegible].22 | [illegible] | ---- | 70,137.84 |
| | [illegible] | 67,090.91 | 200,000.00 | 1,[illegible],137.22 |
| | 494,491.90 | 36,[illegible] | 2,[illegible] | [illegible].25 |
| | 345,[illegible] | 79,184.92 | 164,745.90 | [illegible].42 |
| | 18,[illegible] | 1,849,818.44 | 1,[illegible].12 | 21,[illegible].01 |
| | 129,716.44 | 16,[illegible].44 | 24,412.46 | 170,755.34 |
| | 2,[illegible] | 294,202.45 | [illegible].77 | 3,[illegible] |
| | 45,[illegible] | 47,981.15 | 67,[illegible].15 | 179,[illegible] |
| | 364,[illegible] | 72,904.19 | 45,864.12 | 482,[illegible] |
| | 1,[illegible] | 72,848.17 | 57,621.05 | 1,350,[illegible].84 |
| | 24,[illegible].78 | 7,[illegible].51 | 2,083.[illegible] | 35,847.51 |
| | 4,[illegible] | 1,762.80 | 2,952.05 | 9,665.[illegible] |
| | 2,942,619.45 | 175,610.50 | [illegible],944.94 | 3,448,074.[illegible] |
| | 1,042,544.61 | 71,534.79 | 40,861.30 | 1,154,940.70 |
| | 136,[illegible].61 | 10,[illegible].06 | 1,800,817.[illegible] | 1,946,[illegible].22 |
| | 182,417.65 | 72,190.97 | 31,166.[illegible] | 286,775.15 |
| | 9,128,648.98 | 944,263.02 | 2,716,128.65 | 12,789,040.00 |
| | 1,[illegible],087.41 | 65,[illegible].84 | 91,516.01 | 1,676,[illegible].46 |
| | 2,[illegible],912.69 | 366,350.25 | 315,833.92 | 3,236,097.86 |
| | 112,[illegible].50 | 6,[illegible].98 | 12,734.01 | 131,[illegible].44 |
| | 2,672,[illegible] | 271,348.57 | 291,112.11 | 3,234,775.51 |
| | 771,619.35 | 44,911.73 | 24,166.65 | 840,697.73 |
| | 132,[illegible].40 | 37,[illegible].92 | 14,356.00 | 184,[illegible].32 |
| | 191,175.56 | 10,745.62 | 16,[illegible] | 218,115.80 |
| | 41,[illegible].15 | 25,482.61 | 24,509.92 | 91,901.69 |
| | 106,977.81 | 5,228.78 | 1,786,130.37 | 1,892,336.96 |
| | 797,[illegible].61 | 87,194.72 | 111,784.08 | 996,136.36 |
| | 272,[illegible].62 | 54,778.45 | 27,789.80 | 356,434.87 |
| | 9,128,648.98 | 944,263.02 | 2,716,128.65 | 12,789,040.00 |

[illegible] presents miscellaneous statistical information [illegible] statements, arranged under appropriate [illegible]d with proper explanation and analysis:

[illegible] OF BUILDING AND LOAN ASSOCIATIONS.

| | Locals. | Nationals. | Cooperative banks. | Total. |
|---|---|---|---|---|
| | $364,200,000 | $222,500,000 | $100,000,000 | $686,700,000 |
| | 1,605,600 | 2,325,000 | 1,000,000 | 4,930,600 |
| | 267,068[illegible] | 83,086[illegible] | 48,778 | 398,922[illegible] |
| | 37,517[illegible] | 31,815[illegible] | 8,887 | 78,220 |
| | 304,575[illegible] | 114,902 | 57,665 | 477,142[illegible] |
| | 54,249[illegible] | 31,518[illegible] | 22,024 | 107,791[illegible] |
| | 250,227[illegible] | 83,383[illegible] | 35,641 | 369,251[illegible] |
| | 89,142[illegible] | 25,444[illegible] | 12,669 | 127,257[illegible] |
| | 161,182[illegible] | 57,939 | 22,972 | 242,094[illegible] |
| | $14,069,[illegible].80 | $1,298,857.30 | $910,665.54 | $16,299,416.64 |
| | 20,426 | 7,160 | 2,585 | 30,184 |
| | 7,222 | 1,643 | 526 | 9,391 |

[illegible] ASSOCIATIO[illegible]

| [illegible] | [illegible] | Cooperative banks. |
|---|---|---|
| | 405 | 125 |
| [illegible] | [illegible].74 | $228,003.73 |
| [illegible] | [illegible].51 | 87,830.19 |
| [illegible] | [illegible].25 | 315,833.92 |
| [illegible] | [illegible].45 | 906,143.44 |
| [illegible] | 17 | 3 |
| [illegible] | $47,386.17 | $5,966.79 |
| [illegible] | 1,811,171.25 | 1,210,781.00 |
| [illegible] | 1,679,128.60 | 753,030.00 |
| [illegible] | 215,403.41 | 159,389.81 |
| [illegible] | 3,705,703.26 | 2,123,200.81 |
| [illegible] | 139,613.63 | 50,287.96 |
| [illegible] | 271,348.57 | 291,113.11 |
| [illegible] | 44,911.73 | 24,166.65 |
| [illegible] | 316,260.30 | 315,279.76 |
| [illegible] | 1,127,926.22 | 846,902.89 |
| [illegible] | 308,682.94 | 113,804.81 |
| [illegible] | 1,436,609.16 | 960,707.70 |
| [illegible] | 1,298,857.30 | 910,665.54 |

## ANNUAL REPORT ON THE COOPERATIVE SAVINGS AND LOAN ASSOCIATIONS OF NEW YORK.

*Annual Report of the Superintendent of Banks Relative to Cooperative Savings and Loan Associations for the year 1894.* Transmitted to the Legislature March 1, 1895. Charles M. Preston, Superintendent. 978 pp.

This report opens with 20 pages devoted to a summarization of the statistics, discussion of exceptional methods prevailing in some associations, and the names and addresses of associations that came into existence and of those that were closed during the year. Nine hundred and fifty-eight pages are contained in the appendix and index. The returns for each of the 393 associations reported are presented in detail in the appendix, and the returns for 368 associations are summarized by county totals. The returns for "lot associations" (25) and for the foreign association doing business in the State are shown separately. The laws of the State governing the organization and supervision of associations are given in full.

The following statements give the aggregate assets and liabilities, receipts and disbursements, and the earnings account of the 368 associations the reports of which have been tabulated:

ASSETS.

| | |
|---|---|
| Loans on bond and mortgage (face value) | $39,584,944 |
| Loans on other securities | 1,085,326 |
| Real estate | 1,636,995 |
| Cash on hand and in bank | 1,799,967 |
| Furniture and fixtures | 75,728 |
| Installments due and unpaid | 572,872 |
| Other assets | 267,426 |
| Add for cents | 477 |
| Total | 45,023,735 |

LIABILITIES.

| | |
|---|---|
| Due shareholders, due installments paid | $33,491,406 |
| Due shareholders, installments paid in advance | 2,582,569 |
| Due shareholders, earnings credited | 4,451,835 |
| Borrowed money | 475,583 |
| Balance to be paid out on loans made | 666,557 |
| Undivided earnings | 3,017,499 |
| Other liabilities | 337,726 |
| Add for cents | 560 |
| Total | 45,023,735 |

RECEIPTS (FROM ALL SOURCES).

| | |
|---|---|
| Cash on hand January 1, 1894 | [illegible] |
| Subscriptions on shares, installment | [illegible] |
| Subscriptions on shares, single payment | 1,417,710 |
| Money borrowed | [illegible] |
| Mortgages redeemed (in whole or in part) | [illegible] |
| Other loans redeemed | [illegible] |
| Premiums received | [illegible] |
| Interest received | 1,956,912 |
| Fines received | [illegible] |
| Initiation, entrance, or membership fees | [illegible] |
| Other receipts | [illegible] |
| Add for cents | [illegible] |
| Total | [illegible] |

DISBURSEMENTS.

| | |
|---|---|
| Loaned on mortgage | [illegible] |
| Loaned on other securities | [illegible] |
| Paid on withdrawals, dues | [illegible] |
| Paid on withdrawals, dividends | [illegible] |
| Salaries, advertising, printing, and rent | [illegible] |
| Other disbursements | [illegible] |
| Cash on hand | [illegible] |
| Add for cents | [illegible] |
| Total | [illegible] |

EARNINGS ACCOUNT.

DR.

| | |
|---|---|
| Interest | [illegible] |
| Premium | [illegible] |
| Fines | [illegible] |
| Transfer fees | [illegible] |
| Passbooks and initiation, membership, or share fee | 58,897.32 |
| Other earnings | [illegible] |
| Total | [illegible] |

CR.

| | |
|---|---|
| Expenses | [illegible] |
| Earnings over expenses | [illegible] |
| Total | [illegible] |

The following is a statement of miscellaneous items of interest compiled from the tabulated reports of the 368 associations:

| | |
|---|---|
| Foreclosures for the year | [illegible] |
| Shares issued during the year | [illegible] |
| Shares in force at the close of the year | [illegible] |
| Shares pledged or borrowed on | [illegible] |
| Borrowing members | [illegible] |
| Nonborrowing members | [illegible] |
| Female members | [illegible] |
| Shares held by female members | [illegible] |
| Loans now secured by mortgages in the State | [illegible] |
| Total expenses | [illegible] |

## DECENNIAL CENSUS OF KANSAS FOR 1895.

*Report of the Kansas State Board of Agriculture for the month ending December 31, 1895.* Part I.—State decennial census, 1895. Part II.—Farm, crop, and live-stock statistics. F. D. Coburn, Secretary. 136 pp.

In Part I, which embraces statistics of population for 1895, the returns are presented by counties and minor civil divisions. The aggregate population by counties for prior census years is also shown. The following statement gives the totals for the State for 1885 and 1895:

POPULATION, 1885 AND 1895.

| Items. | 1885. | 1895. |
|---|---|---|
| Males | 679,300 | 693,928 |
| Females | 589,230 | 640,806 |
| Total | 1,268,530 | 1,334,734 |
| Native born: | | |
| Males | 601,946 | 621,185 |
| Females | 533,909 | 585,147 |
| Total | 1,135,855 | 1,206,332 |
| Foreign born: | | |
| Males | 77,354 | 72,744 |
| Females | 55,321 | 55,658 |
| Total | 132,675 | 128,402 |
| White: | | |
| Males | 654,825 | 644,804 |
| Females | 565,498 | 640,945 |
| Total | 1,220,323 | 1,285,749 |
| Colored: | | |
| Males | 24,379 | 24,411 |
| Females | 23,655 | 24,300 |
| Total | 48,034 | 48,711 |
| Chinese and Indians: | | |
| Males | 96 | 148 |
| Females | 77 | 126 |
| Total | 173 | 274 |
| Number of families | 251,661 | 279,816 |
| Average number of persons to family | 5.04 | 4.77 |
| Children of school age, 5 to 20 years, inclusive: | | |
| Males | 244,440 | 252,264 |
| Females | 232,437 | 247,059 |
| Total | 476,877 | 499,323 |
| Males of military age | 273,628 | 280,693 |
| Males of voting age, 21 years and over | 336,371 | 359,603 |
| Persons 21 years of age and over | 600,183 | 668,568 |
| Persons engaged in agriculture | 202,279 | 185,394 |
| Persons engaged in professional and personal services | 49,746 | 63,694 |
| Persons engaged in trade and transportation | 32,549 | 42,574 |
| Persons engaged in manufacturing and mechanical industries | 43,421 | 43,719 |
| Persons engaged in mining | 3,699 | 7,195 |

Part II relates to farm, crop, and live-stock statistics.

## RECENT FOREIGN STATISTICAL PUBLICATIONS.

*General Report on the Wages of the Manual Labor Classes in the United Kingdom, with Tables of the Average Rates of Wages and Hours of Labor of Persons Employed in Several of the Principal Trades in 1886 and 1891.* 1893. xlviii, 481 pp. (Published by the Labor Department of the Board of Trade.)

This report is the final one of a series on this subject prepared by the labor department of the British Board of Trade, which are based on a census of wages taken in conformity with a resolution of the House of Commons of March 2, 1886. The data were obtained by means of schedules sent by mail to employers calling for a statement as to "the weekly rates of wages actually earned in a particular week in October, 1886, by the numbers actually at work in that week, divided according to the varied classification of their occupations, and distinguishing in all cases the wages paid to men, to lads and boys, to women, and to girls, respectively;" half-timers were also designated, and a distinction made between wages earned by "piece" and wages earned by "time" work. The "numbers and rates of wages for each occupation" were also required. A distinct set of questions called for the total wages paid during 1885, with particulars as to the maximum amount paid as wages in one week with the number employed for that week; also the minimum amount paid as wages in one week with the number employed for that week.

From these reports the actual earnings for 1886 were estimated by multiplying the wages for a given week by 52 and making any deductions that might seem expedient, the totals being verified by the totals reported for 1885. The results obtained from the two processes are summarized in the following statement, which shows the general average annual earnings of men, women, and children:

AVERAGE ANNUAL WAGES IN VARIOUS MANUFACTURING INDUSTRIES, OBTAINED BY DIFFERENT METHODS, COMPARED.

| Industry. | Total wages for 1885 divided by number employed Oct. 1, 1886. | Average rate for a week in 1886 multiplied by 52. |
|---|---|---|
| Pig iron | $256.23 | $301.48 |
| Engineering, etc | 288.30½ | 277.14½ |
| Iron and steel shipbuilding | 370.34 | 328.73 |
| Tin plate | 268.87½ | 283.47½ |
| Brass work and metal wares | 266.88½ | 262.91½ |
| Sawmills | 256.96 | 271.86½ |
| Wood shipbuilding | 277.69½ | 261.82 |
| Cooperage | 276.41½ | 292.06½ |
| Coach and carriage building | 255.73½ | 273.90½ |

| | | |
|---|---|---|
| ........................ | [illegible].92 | [illegible].79½ |
| ........................ | [illegible].90½ | 211.[illegible] |
| ........................ | [illegible].74 | 375.45 |

*a* Not ascertained.

pondence between the high annual average fifty-two weeks and the high percentage of parison, however, is not likely to hold in dis- e rates of wages of men are very different. it, which shows the average weekly wages of, men, lads and boys, women, and girls, are arranged according to the annual wages wo weeks:

D PERCENTAGE OF MEN, LADS AND BOYS, WOMEN, RIOUS MANUFACTURING INDUSTRIES.

| Men. | | Lads and boys. | | Women. | | Girls. | |
|---|---|---|---|---|---|---|---|
| ages. | Per cent. | Wages. | Per cent. | Wages. | Per cent. | Wages. | Per cent. |
| 7.09½ | 43.6 | $1.70½ | 49.1 | $2.57½ | 2.0 | $1.50 | 5.3 |
| 5.90 | 59.2 | 2.03 | 16.3 | 3.04 | 17.7 | 1.34 | 6.8 |
| 5.55½ | 71.0 | 2.19 | 22.2 | 2.27 | 5.8 | 1.84½ | 1.0 |
| 8.19 | 46.5 | 2.09 | 37.6 | 2.86 | 7.0 | 1.36 | 8.9 |
| 4.95 | 98.7 | 2.39½ | .7 | 2.29 | .6 | ........ | ........ |
| 6.59½ | 65.0 | 1.56 | 35.0 | ........ | ........ | ........ | ........ |
| 5.90 | 81.7 | 2.15 | 18.3 | ........ | ........ | ........ | ........ |
| 5.59½ | 88.7 | 2.39½ | 8.5 | 2.13 | 2.8 | ........ | ........ |
| 6.45 | 75.4 | 1.62 | 24.6 | ........ | ........ | ........ | ........ |
| 6.26½ | 76.9 | 2.21 | 23.1 | ........ | ........ | ........ | ........ |
| 7.20 | 65.0 | 2.05 | 29.6 | 3.14½ | 3.9 | 1.50 | 1.5 |
| 8.13 | 52.2 | 2.73½ | 24.1 | 2.51½ | 17.5 | 1.68½ | 6.2 |
| 8.12½ | 82.3 | 2.55½ | 17.1 | 3.22½ | .4 | 1.70½ | .2 |
| 7.40 | 68.1 | 1.88½ | 31.9 | ........ | ........ | ........ | ........ |
| 5.90 | 94.0 | 2.37 | 6.0 | ........ | ........ | ........ | ........ |
| 5.96 | 95.2 | 2.59½ | 4.8 | ........ | ........ | ........ | ........ |
| 7.11½ | 80.8 | 2.96 | 19.2 | ........ | ........ | ........ | ........ |
| 9.02½ | 74.3 | 2.03 | 22.9 | 2.96 | .8 | 1.58 | 2.0 |

The percentage of men employed at stated weekly wages in different industries is shown in the following statement:

PERCENTAGE OF MEN EMPLOYED IN VARIOUS MANUFACTURING INDUSTRIES, BY STATED WEEKLY WAGES.

| Industry. | Percentage of men earning— | | | | | | Total. |
|---|---|---|---|---|---|---|---|
| | Under $4.86½. | $4.86½ or under $6.08½. | $6.08½ or under $7.30. | $7.30 or under $8.51½. | $8.51½ or under $9.73½. | $9.73½ or over. | |
| Printing and engraving, small establishments | 4.8 | 15.7 | 30.6 | 20.3 | 11.2 | 8.4 | 100.0 |
| Boot and shoe factories | 11.9 | 49.0 | 22.9 | 10.9 | 3.3 | 2.0 | 100.0 |
| Brick and tile | 27.8 | 39.6 | 18.4 | 11.5 | 1.6 | 1.1 | 100.0 |
| Printing and engraving, large establishments | 3.1 | 8.7 | 16.4 | 27.5 | 23.9 | 20.4 | 100.0 |
| Distilleries | 52.5 | 30.4 | 8.0 | 4.8 | 3.1 | 1.2 | 100.0 |
| Wood shipbuilding | 11.5 | 18.5 | 14.5 | 43.6 | 9.9 | 2.0 | 100.0 |
| Sawmills | 26.0 | 31.4 | 19.3 | 15.8 | 4.9 | 2.6 | 100.0 |
| Chemical manure | 33.4 | 27.7 | 25.7 | 8.6 | 2.9 | 1.7 | 100.0 |
| Coach and carriage building | 15.0 | 26.3 | 23.9 | 22.6 | 6.3 | 5.9 | 100.0 |
| Engineering, etc | 29.5 | 16.9 | 28.9 | 15.2 | 5.9 | 3.6 | 100.0 |
| Brass work and metal wares | 8.2 | 18.4 | 23.6 | 28.4 | 10.8 | 10.9 | 100.0 |
| Tin plate | 16.1 | 13.3 | 7.1 | 8.8 | 21.5 | 33.2 | 100.0 |
| Railway-carriage and wagon building | 24.0 | 24.5 | 22.6 | 26.0 | 1.6 | 1.3 | 100.0 |
| Cooperage | 6.4 | 14.1 | 23.2 | 27.2 | 19.0 | 10.1 | 100.0 |
| Breweries | 26.2 | 37.0 | 18.9 | 8.2 | 4.4 | 5.3 | 100.0 |
| Pig iron | 33.5 | 23.1 | 22.5 | 15.0 | 2.1 | 3.8 | 100.0 |
| Iron and steel shipbuilding | 18.9 | 14.2 | 20.5 | 21.9 | 11.4 | 13.1 | 100.0 |
| Newspapers | .4 | 8.0 | 13.5 | 18.0 | 21.8 | 37.4 | 100.0 |

Comparing different districts it is found that higher rates of wages prevail in Great Britain than in Ireland, and higher in England than in Scotland. The following statement is a comparison between the average annual wages in England, Scotland, and Ireland, and the United Kingdom in certain industries for which the necessary particulars were secured:

AVERAGE ANNUAL WAGES IN VARIOUS MANUFACTURING INDUSTRIES IN ENGLAND, SCOTLAND, IRELAND, AND THE UNITED KINGDOM, COMPARED.

[The averages in this table were obtained by multiplying the average for a week in 1886 by 52.]

| Industry. | England. | Scotland. | Ireland. (a) | United Kingdom. |
|---|---|---|---|---|
| Engineering, etc | $285.66½ | $264.98 | $[illegible].37 | $277.14½ |
| Brass work and metal wares | 290.04½ | 266.44 | 248.68 | 282.01½ |
| Sawmills | 281.04 | 255.25 | 241.62 | 271.66 |
| Coach and carriage building | 279.33½ | 256.90 | 257.44 | 273.78 |
| Breweries | 300.90½ | 206.92½ | 245.49 | 295.64 |
| Distilleries | 313.64½ | 254.27½ | 229.70 | 253.94 |
| Chemical manure | 285.17½ | 243.33½ | 228.72½ | 272.06½ |
| Printing and engraving: | | | | |
| Large establishments | 269.84½ | 228.72½ | 214.65½ | 255.78 |
| Small establishments | 214.37 | 204.15 | 179.35 | 211.29 |

a Most of the returns used came from Dublin, Belfast, and other large towns.

Particulars of rates of wages, etc., are also given for indoor workers in the tailoring, dressmaking, millinery, mantle making, and linen-underclothing industries. The total amount of wages returned for the year 1885, and the numbers employed, included in many cases persons working at their own homes and only partially employed by the firms who made the returns. The details for 1885 were further complicated in the

S OF INDOOR WORKERS IN TAILORING, DRESSMAKING, AND SIMILAR INDUSTRIES.

re obtained by multiplying the average for a week in 1886 by 52.]

| | Men. | Lads and boys. | Women. | Girls. | Total. |
|---|---|---|---|---|---|
| ........ | $399.54 | $67.40 | $192.95½ | $62.29 | $339.68 |
| ........ | ........ | ........ | 170.08½ | 29.68½ | 136.26 |
| ........ | ........ | ........ | 167.41 | 28.47 | 94.65½ |
| ........ | ........ | ........ | 182.49½ | 46.96 | 165.70½ |
| ........ | ........ | ........ | 160.11 | 47.93½ | 143.32 |

s who, in addition to money wages, were allowed full or partial board

vages shown for tailoring are subject to consider- t time, and are also affected by the number of d in cutting out garments that are made up not ployed on the premises, but by large numbers of e average for the trade would include outworkers rs, but this can not be arrived at from the data of girls, the average for dressmaking and mil- e averages for the other industries, owing to the ndustries a large proportion of the young girls ges, as will be seen by the following statement, ly to those who were not provided with lodging noney wages:

AND GIRLS EMPLOYED AT STATED WEEKLY WAGES DRESSMAKING, AND SIMILAR INDUSTRIES.

| Women. | | | | | Girls. | | | |
|---|---|---|---|---|---|---|---|---|
| [illegible] | $3.65 or under $4.86⅔. | $4.86⅔ or under $6.08⅓. | $6.08⅓ or over. | Total. | Unpaid appren- tices. | $0.48⅔ or under $1.21⅔. | $1.21⅔ or under $1.70⅓. | Total. |
| 4 | 37.5 | 19.5 | ........ | 100.0 | ........ | 26.7 | 73.3 | 100.0 |
| 5 | 25.2 | 2.6 | 1.3 | 100.0 | 42.5 | 30.8 | 26.7 | 100.0 |
| 9 | 29.2 | 7.6 | 2.8 | 100.0 | 47.0 | 36.8 | 16.2 | 100.0 |
| 7 | 35.3 | 2.7 | 1.3 | 100.0 | 14.3 | 52.4 | 33.3 | 100.0 |
| 9 | 50.0 | .5 | ........ | 100.0 | ........ | *a* 92.1 | 7.9 | 100.0 |
| 4 | 34.4 | 3.9 | 1.0 | 100.0 | 33.3 | 43.3 | 23.4 | 100.0 |

of 76.2 per cent of full-timers, paid 73 cents to $1.21⅔ per week, and 61 cents to 85 cents.

n provided with board or lodging in the dress- g, and millinery industries is only about 12 per er employed in all three industries, and their gs are considerably higher than those of the eive either board or lodging. The explanation of this seeming anomaly is that those who are given board and lodging, or partial board, are skilled workers, viz, fitters, heads of tables, cutters, etc.

The following statement shows the number, percentage, and average annual wages of the women who are not allowed board or lodging, allowed full board and lodging, and allowed partial board, respectively:

NUMBER, PERCENTAGE, AND AVERAGE ANNUAL WAGES OF WOMEN EMPLOYED IN THE DRESSMAKING, MILLINERY, AND MANTLE MAKING INDUSTRIES, BY METHOD OF PAYMENT.

| Industry. | Without board or lodging. | | | With full board and lodging. | | | With partial board. | | | Total. | |
|---|---|---|---|---|---|---|---|---|---|---|---|
| | Number. | Per cent. | Average wages. | Number. | Per cent. | Average wages. | Number. | Per cent. | Average wages. | Number. | Per cent. |
| Dressmaking | 616 | 87.6 | $170.08½ | 43 | 6.1 | $238.46 | 44 | 6.3 | $368.14½ | 703 | 100.0 |
| Millinery | 106 | 75.2 | 167.41 | 32 | 22.7 | 232.13 | 3 | 2.1 | 314.12½ | 141 | 100.0 |
| Mantle making | 300 | 94.9 | 182.49½ | 7 | 2.2 | 285.42 | 9 | 2.9 | 411.95 | 316 | 100.0 |
| Total | 1,022 | 88.1 | 173.49 | 82 | 7.1 | 239.92 | 56 | 4.8 | 293.69½ | 1,160 | 100.0 |

The data concerning wages and employees on railways, in shipping, domestic service, the building trades, and various public employments, such as the army, navy, and hospitals, were obtained in a somewhat different manner than that for the other industries. Of this group only the statistics concerning the building trades will be referred to in this synopsis. The information concerning the building trades was obtained from employers, the National Association of Master Builders, and trade union reports, as well as other sources, and covers the years 1886 and 1891.

The first statement presented deals with the total number of employees reported for all occupations coming under this general group, and shows the percentage of men employed at stated weekly rates of wages.

PERCENTAGE OF MEN EMPLOYED IN BUILDING TRADES, BY WEEKLY WAGES, 1886 AND 1891.

| Weekly wages. | Per cent of total. | | | |
|---|---|---|---|---|
| | January. | | July. | |
| | 1886. | 1891. | 1886. | 1891. |
| Under $4.86½ | 22.1 | 14.8 | 6.3 | 6.9 |
| $4.86½ or under $6.08¼ | 12.9 | 22.3 | 23.2 | 27.8 |
| $6.08¼ or under $7.30 | 23.7 | 13.5 | 14.6 | 10.1 |
| $7.30 or under $8.51½ | 33.1 | 32.0 | 30.2 | 19.8 |
| $8.51½ or under $9.73¼ | 6.1 | 14.0 | 21.0 | 28.2 |
| Above $9.73¼ | 2.1 | 3.4 | 4.7 | 7.6 |
| Total | 100.0 | 100.0 | 100.0 | 100.0 |

The average weekly rate for the year would not be the mean of the rates given for January and July, but a figure arrived at on the basis of taking about 37 weeks of the summer rate, and 15 weeks of the winter rate, and dividing the result by 52. That would yield the average rate for a normal week, the result being subject, as in the case of all

### [illegible]RAGE WEEKLY WAGES IN BUILDING TRADES, IN JANUARY [illegible], FOR THE YEARS 1886 AND 1891.

| | January. | | | | July. | | | |
|---|---|---|---|---|---|---|---|---|
| | 1886. | 1891. | | 1886. | | 1891. | |
| [illegible] | Average weekly wages. | Employees. | Average weekly wages. | Employees. | Average weekly wages. | Employees. | Average weekly wages. |
| [illegible] | $[illegible] | [illegible] | $7.50 | 451 | $7.95 | 610 | $[illegible] |
| [illegible] | 7.[illegible] | 426 | 7.84 | 237 | 8.64 | 888 | 8.[illegible] |
| [illegible] | [illegible] | 1,4[illegible] | 8.31 | 1,139 | 8.31 | 1,516 | 8.57 |
| [illegible] | [illegible] | [illegible] | 7.0[illegible] | 466 | 7.72 | 813 | 8.17 |
| [illegible] | 4.[illegible] | 1,740 | 4.[illegible] | 1,401 | 5.25 | 2,845 | 5.53 |

ers shown in the above statement as having been re larger than the numbers for January, there is no ate increase, the rise in a year like 1891 being more year of depression like 1886. There is an increase s for each occupation presented. This increase has when the standard weekly hours of labor have been ough the reduction in hours of work has not been ncrease in rates of wages.

nployees and total wages paid per week in 1886 and he following statement. The figures given in this all occupations included in the returns received for

### [illegible]ES IN ALL BUILDING TRADES OCCUPATIONS, 1886 AND 1891.

| | 1886. | 1891. |
|---|---|---|
| [illegible] | 5,[illegible]68 | 8,743 |
| [illegible] | 3,596 | [illegible],576 |
| [illegible] | 4,538 | [illegible],068 |
| [illegible] | $33,783.24 | $57,025.64 |
| [illegible] | 17,839.34 | 28,191.63 |
| [illegible] | 27,923.97 | 50,197.94 |

ummarized in the preceding statements is presented port, so that similar data can be obtained for each

"In conclusion, the statistics for all classes of labor in different industries are summarized. With the exception of the statistics for railways, mercantile marine, building, and other trades referred to as having been treated separately, this summary includes all data gathered at or in connection with the census of wages of 1886. The totals are representative of perhaps three-fourths of the manual laboring classes of the United Kingdom.

"The general effect of this summary is to show an average rate of wages per head of 24s. 7d. ($5.98) per week, equal to £64 ($311.45½) per annum, if the weekly rate were multiplied by 52. Questions of course arise upon such a statement as to regularity of employment, overtime, and the like; * * * but considering that the year 1886, to which the census primarily relates, was a year of depression, and the great number of trades dealt with where employment is not irregular, I should not myself consider that any great mistake would be made in assuming average earnings for men for the average of the last few years to be not far short of 24s. 7d. ($5.98) per week. The point may also be raised that the rates are for a year a considerable way back; but there has been enough experience in the department to show that rates of wages change rather slowly, and as the tendency has, on balance, been upward since 1886, there is the more reason to believe that even if an average rate, as shown above, was rather above the average earnings of 1886 in particular, it would not be above the average of 1886-1892." Similar averages for women, lads and boys, and girls show the weekly wages to have been $3.08, $2.17, and $1.54, respectively.

The following statement, which includes all industries, as indicated above, shows the percentage of men, women, lads and boys, and girls, respectively, employed at stated weekly wages:

PERCENTAGE OF MEN, WOMEN, LADS AND BOYS, AND GIRLS EMPLOYED IN ALL INDUSTRIES, BY STATED WEEKLY WAGES.

| Weekly wages. | Men. | Women. | Lads and boys. | Girls. |
|---|---|---|---|---|
| [illegible] | | | 11.9 | 27.2 |
| [illegible] | 0.1 | [illegible] | 49.7 | [illegible] |
| [illegible] | 2.4 | 50.0 | 32.5 | 8.9 |
| [illegible] | 21.5 | 18.6 | 5.8 | 1.0 |
| [illegible] | 33.6 | 5.4 | .1 | |
| [illegible] | 24.2 | .1 | | |
| [illegible] | 11.6 | | | |
| [illegible] | 4.2 | | | |
| [illegible] | 2.4 | | | |
| Total | 100.0 | 100.0 | 100.0 | 100.0 |

REPORTED FOR ALL INDUSTRIES, BY OCTOBER 1, [illegible].

| | Employees. | |
|---|---|---|
| | Number. | Per cent. |
| ........................ | [illegible] | [illegible] |
| ........................ | [illegible] | [illegible] |
| ........................ | [illegible] | [illegible] |
| ........................ | [illegible] | [illegible] |
| ........................ | 62,841 | 7.7 |
| ........................ | 816,106 | 100.0 |

l employees, obtained by dividing those paid in Government works, October 1, 1886, as shown in the

ient under $194.66 may be consid- h female or child labor is mainly

nen in the industries referred to as given as follows: Railways (1891), 5.25½; seamen (mercantile marine), etty officers and seamen), $316.32½ 's and men), $233.59 (a); domestic (a); employees in lunatic asylums, and infirmaries, $296.85½ (a).

*twerk veranstaltet im Sommer 1895.* tatistischen Amt. 579 pp.

ference to the condition of skilled rman Government for the use of r officials. According to the intro- German artisans a desire that the on a firmer basis than the present vith the view of obtaining greater f apprentices. In response to this ore taking any stand in the matter, condition of the skilled trades at se instituted an investigation call- ng skilled artisans, and concerning ction with the organization of the

od and lodging where necessary.

A selection was made of 37 localities in Prussia, Bavaria, Saxony, Würtemberg, Baden, Hesse, and Lübeck, having an aggregate population in 1890 of 2,292,525 persons and an area of 18,700 square kilometers (7,220 square miles), care being taken to select such districts as would be typical of the industrial conditions of the Empire generally.

In each of the districts a census enumeration was made of all persons coming under the following three categories: First, all occupations which may be regarded as purely skilled; second, those in which there is a doubt as to whether the labor may be regarded as skilled or factory work; third, those in which skilled work is done at the workingman's home, but where he is in the employ of others. The work was done by means of blank schedules distributed by the communal and police authorities and collected by them within five days. In this manner 64,899 blanks were distributed, of which 61,257 were returned in such shape that they could be used. These give information concerning 134,712 persons. The enumeration was made in the summer of 1895.

The report consists of an introduction explaining the object of and methods pursued in the investigation and an analysis of the results obtained, and a series of statistical tables showing by enumeration districts and for each skilled trade, respectively, the number and sex of employers and independent master workmen, foremen, journeymen, apprentices, and unskilled laborers covered by the investigation, the number of trades learned, and length of apprenticeship of all male employers and independent master workmen, etc.

The results of this investigation of most general interest, as far as those engaged in undoubtedly skilled trades are concerned, can be summarized in two tables. The first of these shows the distribution of these workingmen according to whether they are employers or independent master workmen, journeymen, apprentices, or other employees for each trade separately.

NUMBER OF PROPRIETORS IN THE SKILLED TRADES AND THEIR EMPLOYEES.

| Skilled trades. | Proprietors. | | | Foremen. | Journeymen. | | | Apprentices. | | | Other employees. | Total persons occupied. |
|---|---|---|---|---|---|---|---|---|---|---|---|---|
| | Employing labor. | Working alone. | Total. | | At trades specified. | At other trades. | Total. | At trades specified. | At other trades. | Total. | | |
| Barbers and hairdressers | 613 | 591 | 1,204 | 3 | 451 | 2 | 453 | 583 | ..... | 583 | 11 | 2,254 |
| Wigmakers and hairdressers | 55 | 17 | 72 | ..... | 67 | ..... | 67 | 52 | ..... | 52 | 4 | 195 |
| Bakers | 2,958 | 1,559 | 4,517 | 117 | 2,706 | 25 | 2,731 | 1,775 | 13 | 1,788 | 482 | [illegible] |
| Surgical instrument and bandage makers | 77 | 39 | 116 | 4 | 96 | 1 | 97 | 117 | ..... | 117 | 9 | 343 |
| Coopers | 329 | 566 | 895 | 6 | 329 | 2 | 331 | 211 | 2 | 213 | 24 | [illegible] |
| Brewers | 264 | 73 | 337 | 61 | 380 | ..... | 380 | 126 | ..... | 126 | 178 | [illegible] |
| Well diggers | 30 | 29 | 59 | ..... | 48 | 2 | 50 | 9 | ..... | 9 | 29 | [illegible] |
| Bookbinders and paper-box makers | 200 | 160 | 360 | 5 | 203 | 3 | 206 | 177 | 2 | 179 | 54 | [illegible] |
| Brushmakers | 72 | 98 | 170 | 4 | 86 | ..... | 86 | 46 | ..... | 46 | 7 | [illegible] |
| Confectioners | 219 | 65 | 284 | 9 | 264 | ..... | 264 | 246 | ..... | 246 | 78 | [illegible] |
| Roofers | 320 | 361 | 681 | 12 | [illegible] | 6 | 605 | 166 | 1 | 167 | [illegible] | [illegible] |
| Wire drawers | ..... | 1 | 1 | ..... | ..... | ..... | ..... | ..... | ..... | ..... | ..... | [illegible] |

| | | | | | | | |
|---|---|---|---|---|---|---|---|
| [illegible] | 4 | 73 | [illegible] | 1 | 33 | [illegible] | 185 |
| [illegible] | [illegible] | [illegible] | 48 | 3 | 48 | [illegible] | 170 |
| [illegible] | 7 | [illegible] | 50 | | 50 | [illegible] | 534 |
| [illegible] | [illegible] | [illegible] | 21 | 7 | 28 | [illegible] | 127 |
| [illegible] | [illegible] | [illegible] | 101 | | 101 | 4 | 731 |
| | | | 1 | | 1 | | 3 |
| [illegible] | 6 | [illegible] | 40 | 1 | 41 | [illegible] | [illegible] |
| 76 | | 14 | 17 | | 17 | [illegible] | 73 |
| 11 | 1 | 12 | 8 | 1 | 9 | [illegible] | 71 |
| [illegible] | | [illegible] | 15 | | 15 | 12 | 300 |
| | | 6 | | | | 1 | 80 |
| [illegible] | 10 | [illegible] | 500 | | 500 | 22 | 1,950 |
| 448 | | 448 | 221 | | 221 | 16 | 2,088 |
| 75 | 6 | 81 | 32 | | 32 | 23 | 326 |
| [illegible] | 32 | 165 | 110 | 8 | 118 | 13 | 444 |
| 1,[illegible] | 11 | 1,[illegible] | 1,014 | 5 | 1,019 | 126 | 4,851 |
| 7,771 | 783 | 8,[illegible] | 1,536 | 164 | 1,690 | 2,697 | 15,794 |
| 1,[illegible] | 17 | 1,535 | 1,345 | 1 | 1,346 | 390 | 6,984 |
| 1,[illegible] | 120 | 1,347 | 248 | 3 | 251 | 275 | 3,307 |
| [illegible] | 4 | 85 | 15 | | 15 | 4 | 164 |
| 28 | | 28 | 5 | | 5 | 2 | 79 |
| 6 | | 6 | 6 | | 6 | 1 | 25 |
| [illegible] | | 229 | 4 | | 4 | | 241 |
| 15 | | 15 | 5 | | 5 | 21 | 71 |
| 472 | 10 | 482 | 426 | 1 | 427 | 10 | 2,031 |
| 156 | | 156 | 35 | | 35 | 2 | 288 |
| 404 | | 404 | 576 | | 576 | 11 | 2,709 |
| 593 | | 593 | 1,214 | | 1,214 | 40 | 2,660 |
| 215 | | 215 | 412 | 1 | 414 | 3 | 833 |
| 25 | | 25 | 26 | | 26 | | 82 |
| 34 | | 34 | 30 | | 30 | 1 | 82 |
| 31 | | 31 | 13 | | 13 | 10 | 72 |
| | | | | | | | 2 |
| 84 | | 84 | 57 | | 57 | | 152 |
| | | | | | | | 6 |
| 6 | | 6 | 1 | | 1 | | 12 |
| 1,191 | | 1,191 | 1,060 | | 1,060 | 42 | 4,984 |
| 8 | | 8 | 10 | | 10 | | 30 |
| 11 | | 11 | 9 | | 9 | 1 | 40 |
| 2 | | 2 | | | | | 3 |
| 357 | 3 | 360 | 206 | | 206 | 13 | 1,224 |
| 10 | | 10 | 1 | | 1 | 3 | 19 |
| 54 | | 54 | 18 | | 18 | 4 | 230 |
| 1 | | 1 | | | | | 3 |
| 22 | | 22 | 24 | | 24 | 4 | [illegible] |

NUMBER OF PROPRIETORS IN THE SKILLED TRADES AND T
Concluded.

| Skilled trades. | Proprietors. | | | Foremen. | Journeymen. | | | Appre | |
|---|---|---|---|---|---|---|---|---|---|
| | Employing labor. | Working alone. | Total. | | At trades specified. | At other trades. | Total. | At trades specified. | At other trades |
| Tailors | 2,168 | 3,462 | 5,630 | 55 | 2,734 | 1 | 2,735 | 1,726 | |
| Chimney sweeps | 103 | 22 | 125 | 6 | 102 | 1 | 103 | 42 | |
| Joiners (not specialists) | 1,704 | 2,013 | 3,717 | 92 | 2,386 | | 2,386 | 1,448 | |
| House joiners | 333 | 217 | 550 | 9 | 612 | 2 | 614 | 304 | |
| Wooden toy and ornament makers | 1 | 2 | 3 | | 2 | | 2 | | |
| Trunk makers | 11 | 4 | 15 | 1 | 19 | | 19 | 4 | |
| Cabinet makers | 8 | 4 | 12 | | 18 | | 18 | 10 | |
| Cabinet makers, fancy inlaid and scroll work | 2 | 1 | 3 | | 9 | | 9 | 3 | |
| Pattern makers | 7 | 8 | 15 | | 8 | | 8 | 2 | |
| Furniture makers | 238 | 168 | 406 | 6 | 457 | 6 | 463 | 237 | |
| Floor carpenters | 1 | | 1 | | 1 | | 1 | 1 | |
| Coffin makers | 15 | 5 | 20 | | 27 | | 27 | 9 | |
| Chair makers | 15 | 22 | 37 | | 27 | | 27 | 9 | |
| Loom makers | 1 | | 1 | | 2 | | 2 | | |
| Wooden tool makers | 4 | | 4 | | 5 | | 5 | | |
| Shoe and slipper makers | 2,696 | 6,710 | 9,406 | 31 | 2,787 | 3 | 2,790 | 1,754 | |
| Soap and candle makers | 28 | 25 | 53 | 3 | 32 | | 32 | 5 | |
| Rope makers | 88 | 97 | 185 | 5 | 110 | | 110 | 34 | |
| Sieve makers | 11 | 19 | 30 | | 16 | | 16 | 1 | |
| Spur and screw makers and filers | 34 | 23 | 57 | | 47 | 1 | 48 | 36 | |
| Umbrella and parasol makers | 15 | 12 | 27 | 2 | 18 | | 18 | 8 | |
| Toy and wood-work finishers | 2 | 5 | 7 | 1 | 6 | | 6 | 4 | |
| Stone cutters | 176 | 107 | 283 | 26 | 865 | 4 | 869 | 182 | |
| Stone setters | 57 | 37 | 94 | 5 | 207 | | 207 | 31 | |
| Knitters and embroiderers | 72 | 151 | 223 | 1 | 85 | | 85 | 17 | |
| Stucco workers | 67 | 46 | 113 | | 223 | 4 | 227 | 82 | |
| Paper hangers, decorators | 155 | 122 | 277 | 7 | 176 | 11 | 187 | 158 | |
| Potters | 172 | 137 | 309 | 4 | 311 | | 311 | 148 | |
| Cloth makers | 82 | 111 | 193 | | 77 | | 77 | 29 | |
| Watch and clock makers | 231 | 352 | 583 | 3 | 186 | 2 | 188 | 184 | |
| Gilders | 22 | 8 | 30 | | 30 | 2 | 32 | 20 | |
| Finishers of rough wooden ware | 101 | 314 | 415 | 2 | 93 | | 93 | 47 | |
| Wagon makers | 773 | 1,222 | 1,995 | 8 | 640 | 19 | 659 | 474 | |
| Weavers | 883 | 3,800 | 4,683 | 2 | 736 | 2 | 738 | 161 | |
| Carpenters | 586 | 988 | 1,574 | 48 | 2,870 | 713 | 3,583 | 693 | |
| Total | 27,257 | 33,942 | 61,199 | 1,024 | 40,189 | 1,854 | 42,043 | 21,366 | |

Of the total number of persons engaged in the skilled
in this table, 46 per cent were proprietors of establishm
were foremen, 32 per cent were journeymen, 16 per c
tices, and 6 per cent were employed at unskilled labor
etors, 44½ per cent employed labor and 55½ per cent w
As regards sex, the enumeration shows that out o
in the above table, 1,607 were females. Of th
[illegible] who worked alone, that is, who did not
[illegible] were tailoresses, 144 weavers, 19 bakers, 17
[illegible] knitters and embroiderers, 9 butchers, 8

maker, and 1 stonecutter.
are employers of labor. They rep-
, though the greater part fall
them there were 231, bakers 206,
hoe and slipper makers 43, black-
t specialists) 24, and barbers and
04 foremen, but 5 were females.
3 female journeymen, 405 female
loyed at occupations other than

f technical training undergone by
ngaged in purely skilled trades.

E PROPRIETORS IN SKILLED TRADES.

| ed an apprenticeship of years pecified. | | | | | Number who have served apprentice-ships in— | |
|---|---|---|---|---|---|---|
| Over 4, but not over 5. | Over 5, but not over 6. | Over 6. | Not re-port-ed. | Total. | Their present trades. | Other trades. |
| 21 | ...... | 1 | 11 | 1,111 | 1,106 | [illegible] |
| 8 | ...... | ...... | 1 | 72 | 72 | ...... |
| 35 | 9 | 10 | 39 | 4,024 | 3,983 | 41 |
| 4 | 3 | 2 | 3 | 110 | 94 | 16 |
| 65 | 14 | 4 | 8 | 808 | 806 | 2 |
| 1 | 3 | 2 | 1 | 281 | 278 | 3 |
| ...... | 1 | 1 | ...... | 49 | 46 | 3 |
| 30 | 2 | ...... | 3 | 339 | 339 | ...... |
| 12 | 4 | 1 | ...... | 153 | 148 | 5 |
| 6 | 2 | ...... | 4 | 262 | 259 | 3 |
| 12 | 4 | 8 | 5 | 642 | 640 | 2 |
| ...... | ...... | ...... | ...... | 1 | 1 | ...... |
| 28 | 3 | 1 | 2 | 379 | 376 | 3 |
| 14 | ...... | 2 | 2 | 105 | 102 | 3 |
| 5 | ...... | ...... | 1 | 28 | 28 | ...... |
| ...... | ...... | ...... | ...... | 4 | 3 | 1 |
| 2 | ...... | 1 | 1 | 98 | 96 | 2 |
| 5 | ...... | ...... | 3 | 62 | 62 | ...... |
| 2 | 1 | ...... | ...... | 41 | 28 | 13 |
| 5 | 1 | ...... | 1 | 53 | 50 | 3 |
| 4 | 2 | 1 | 4 | 187 | 187 | ...... |
| 3 | 2 | ...... | 1 | 44 | 43 | 1 |
| 14 | 5 | ...... | 1 | 376 | 374 | 2 |
| ...... | ...... | ...... | ...... | 1 | ...... | 1 |
| 38 | 7 | 5 | 1 | 138 | 138 | ...... |
| 8 | 1 | 1 | ...... | 36 | 35 | 1 |
| 4 | ...... | ...... | ...... | 44 | 43 | 1 |
| 6 | 3 | ...... | 1 | 118 | 116 | 2 |
| 2 | ...... | ...... | ...... | 21 | 21 | ...... |
| 45 | 11 | 2 | 5 | 828 | 823 | 5 |
| 37 | 12 | 9 | 10 | 1,257 | 1,248 | 9 |
| 16 | 1 | ...... | ...... | 182 | 180 | 2 |
| 11 | 4 | ...... | 2 | 141 | 141 | ...... |
| 86 | 34 | 11 | 26 | 1,710 | 1,692 | 18 |
| 62 | 19 | 19 | 19 | 2,610 | 2,598 | 12 |
| 44 | 3 | 7 | 24 | 3,169 | [illegible] | [illegible] |

DURATION OF APPRENTICESHIP (

| Skilled trades. | Male proprietors. | Number who [illegible] | | |
|---|---|---|---|---|
| | | 1 or under. | Over 1, but not over 2. | Ov 2, bu no ove 3. |
| Millers | 1,361 | 29 | 128 | 6 |
| Millwrights | 56 | ...... | 2 | |
| Musical-instrument makers | 40 | ...... | 1 | |
| Pin and needle makers | 12 | 1 | 1 | |
| Nailsmiths | 208 | 19 | 36 | |
| Fringe makers | 36 | ...... | ...... | |
| Saddlers and harness makers | 1,086 | 3 | 22 | 7 |
| Ship carpenters | 92 | ...... | ...... | |
| Grinders (knives, scissors, tools) | 1,713 | 12 | 37 | 2 |
| Iron workers (not specialists) | 780 | 3 | 28 | 5 |
| House and locksmiths | 187 | ...... | 2 | 1 |
| Safe makers | 7 | ...... | 1 | |
| Range makers | 13 | ...... | 2 | |
| Ornamental iron workers | 16 | ...... | ...... | |
| Iron workers on vehicles | 2 | ...... | ...... | .... |
| Iron workers on machinery | 41 | ...... | 2 | |
| Iron turners | 6 | ...... | ...... | |
| Tool makers | 5 | ...... | ...... | |
| Blacksmiths (not specialists) | 2,511 | 7 | 136 | 1,7 |
| Axle and spring makers | 12 | ...... | ...... | |
| Anchor, anvil, and ship smiths | 19 | ...... | 1 | |
| Shapers (façonschmiede) | 1 | ...... | ...... | |
| Horseshoers and gunsmiths | 570 | 2 | 36 | 4 |
| Chainsmiths | 5 | ...... | ...... | |
| Cutlers | 140 | 3 | 3 | |
| Scythe and sickle smiths | 2 | ...... | ...... | |
| Toolsmiths | 56 | 1 | 7 | |
| Tailors | 5,074 | 42 | 453 | 3,1 |
| Chimney sweeps | 121 | ...... | 2 | |
| Joiners (not specialists) | 3,692 | 19 | 290 | 2,2 |
| House joiners | 546 | 3 | 32 | 3 |
| Wooden toy and ornament makers | 3 | ...... | ...... | |
| Trunk makers | 12 | ...... | ...... | |
| Cabinetmakers | 12 | ...... | ...... | |
| Cabinetmakers, fancy inlaid and scroll work | 3 | ...... | ...... | |
| Pattern makers | 15 | ...... | ...... | |
| Furniture makers | 401 | ...... | 22 | 2 |
| Floor carpenters | 1 | ...... | ...... | .... |
| Coffin makers | 19 | ...... | ...... | |
| Chair makers | 37 | ...... | 6 | |
| Loom makers | 1 | ...... | ...... | |
| Wooden tool makers | 4 | ...... | ...... | |
| Shoe and slipper makers | 9,346 | 68 | 920 | 5,[illegible] |
| Soap and candle makers | 51 | 4 | 4 | |
| Rope makers | 121 | ...... | 8 | 1 |
| Sieve makers | 26 | 1 | 6 | |
| Spur and screw makers and rifiers | 87 | ...... | 1 | |
| Umbrella and parasol makers | 36 | 3 | 8 | |
| Toy and wood work finishers | 7 | 1 | 2 | .... |
| [illegible] | 277 | 7 | 88 | [illegible] |

| [illegible] | [illegible] | [illegible] | [illegible] not over 2. | [illegible] not over 3. | [illegible] not over 4. | [illegible] not over 5. | [illegible] not over 6. | Over 6. | [illegible] reported. | Total. | Their present trades. | Other trades. |
|---|---|---|---|---|---|---|---|---|---|---|---|---|
| [illegible] | [illegible] | [illegible] | [illegible] | 67 | 6 | 2 | 1 | ...... | 1 | [illegible] | [illegible] | 1 |
| [illegible] | [illegible] | [illegible] | [illegible] | 97 | 17 | ...... | ...... | ...... | 2 | [illegible] | [illegible] | 4 |
| [illegible] | [illegible] | [illegible] | [illegible] | [illegible] | 17 | 2 | 1 | 1 | 1 | 110 | [illegible] | 5 |
| [illegible] | [illegible] | ...... | 4 | [illegible] | [illegible] | 37 | 1 | ...... | ...... | 270 | [illegible] | 1 |
| [illegible] | [illegible] | ...... | 7 | [illegible] | [illegible] | 50 | 14 | 2 | [illegible] | 261 | [illegible] | 1 |
| [illegible] | [illegible] | 1 | [illegible] | [illegible] | [illegible] | 14 | 2 | ...... | ...... | [illegible] | [illegible] | 3 |
| Watch and clock makers | [illegible] | 7 | [illegible] | [illegible] | [illegible] | 80 | 10 | 4 | 2 | 544 | [illegible] | 2 |
| [illegible] | [illegible] | ...... | 1 | 8 | 12 | 5 | 1 | ...... | 1 | [illegible] | [illegible] | ...... |
| Finishers of rough wooden ware | [illegible] | [illegible] | [illegible] | 114 | 7 | 6 | 2 | ...... | 7 | 261 | [illegible] | 8 |
| [illegible] | 1,[illegible] | [illegible] | [illegible] | 1,[illegible] | 315 | [illegible] | [illegible] | 3 | [illegible] | 1,961 | 1,[illegible] | 12 |
| [illegible] | 4,[illegible] | [illegible] | 1,[illegible] | 2,[illegible] | 173 | [illegible] | [illegible] | [illegible] | [illegible] | 4,[illegible] | 4,[illegible] | 17 |
| [illegible] | 1,572 | [illegible] | [illegible] | 1,[illegible] | [illegible] | 16 | 7 | 10 | 24 | 1,515 | 1,[illegible] | 13 |
| Total | [illegible] | 1,[illegible] | 7,[illegible] | [illegible] | 11,[illegible] | 2,244 | [illegible] | [illegible] | [illegible] | 57,[illegible] | 57,274 | [illegible] |

It is interesting to note that of the 59,592 male proprietors enumerated [illegible] per cent served apprenticeships. Of those who served apprenticeships, 3.4 per cent served one year or under, 12.5 per cent over one year but not more than two years, 58.8 per cent over two but not more than three years, 19.5 per cent over three but not more than four years, 3.9 per cent over four but not more than five years, 0.7 per cent over five but not more than six years, and 0.4 per cent over six years. In the case of [illegible] per cent the time was not reported.

[illegible] *der Kommission für Arbeiterstatistik:* Hefte I, III.—Erhebung über die Arbeitszeit in Bäckereien und Konditoreien (I Theil, [illegible]; II Theil, 1896). Hefte II, V, VII.—Erhebung über Arbeitszeit, [illegible] und Lehrlings-Verhältnisse im Handelsgewerbe (I Theil, [illegible]; II Theil, 1894; III Theil, 1894). Hefte IV, VIII.—[illegible] über die Arbeitszeit in Getreidemühlen (I Theil, 1894; II Theil, [illegible]). Hefte VI, IX.—Erhebung über die Arbeits- und [illegible] Verhältnisse der Kellner und Kellnerinnen (I Theil, 1894; II Theil, [illegible]).

[illegible] investigations were conducted by the commission on [illegible] of the German Empire. The duties of the commission, [illegible] official decree, are (1) to undertake the collection and [illegible] statistics of labor, and to give an opinion on the results [illegible] by the imperial council or the chancellor of the [illegible] recommendations to the chancellor regarding the [illegible] this object.

[illegible] duties, the commission has thus far issued the

The first and third reports (Vols. I, III) are the results of an investigation concerning the hours of labor in bakeries and confectioneries, conducted by means of schedules of inquiry. The information collected relates chiefly to hours of labor for a regular day's work, and at extra time, in all bakeries and confectioneries; conditions of apprenticeship, contract of employment, tuition, length of service, education, and hours of labor of apprentices; housing conditions of employees; application of machinery in the bakery industries. The second part of the inquiry relates to an analysis of replies received from master-bakers' and journeymen's associations in reference to the introduction of the twelve-hour working day, the regulation of the hours of labor of apprentices and children, and Sunday rest. It also contains opinions in reference to the influence of the work of bakers upon their health.

The second, fifth, and seventh reports (Vols. II, V, VII) of the commission contain the results of an investigation covering inquiries regarding the hours of labor, notice given in case of discontinuance of employment, and conditions of apprenticeship of employees in the mercantile industries. The first part contains a series of statistical tables showing the number of employers and employees in each locality enumerated, the sex and standing of the latter (whether apprentices or journeymen), hours of opening and of closing the establishments, actual working hours at regular and at extra time for each sex, length required of notice of discontinuance of employment, and the number of employees boarding or lodging with their employers, or living in dwellings rented of the latter. The inquiry covers 8,235 establishments in 389 localities, and employing 23,725 persons. The second and third parts of this inquiry contain a presentation and discussion of replies to inquiries sent to mercantile associations throughout Germany regarding the extent to which the existing hours of labor may be regarded as detrimental to the physical and intellectual condition of employees; in what manner, under existing conditions, the time of labor may be reduced without being harmful to the business interests and to the public; the reasons for or against the enactment of a law fixing a minimum period of notice to be required for the severance of the relations of employers and their employees; and, finally, the advisability and practicability of adopting a uniform hour for closing establishments.

The fourth and eighth reports (Vols. IV, VIII) relate to similar inquiries regarding the hours of labor in flour mills. Statistics cover chiefly the hours of labor and Sunday rest in steam, water [illegible] wind mills.

The sixth and ninth reports (Vols. VI, IX) contain the [illegible] inquiries regarding the labor and wage conditions of waiters [illegible] resses, the influence of the present working day upon [illegible] tion, and domestic life of employees, the regulation of [illegible]

of a census enumeration taken by the con-
with the cooperation of the Mine Oper-
, on December 16, 1893. It relates to the
of the officials and employees at the
within the jurisdiction of the central mining
) at Dortmund. The territory covered by
provinces of Westphalia and Hanover, the
, the districts of Osnabrück and Aurich, the
ein-Wittgenstein and Wittgenstein-Berleburg,
and Neunkirchen, and in Rhenish Prussia, the
Duisburg, Essen, and parts of Dusseldorf and

are found deposits of coal, iron, zinc, lead, sul-
hief mining industry is that of coal, over one-half
Germany coming from this section.

meration of December 16, 1893, the total number
the mines, salt works, and mineral baths of the
s 158,368. These were distributed as follows:
nt, in 164 coal mines; 2,147, or 1.36 per cent, in
.14 per cent, in 5 salt works; 62, or 0.04 per cent,
neral baths at Oeynhausen.

rsons employed in other than coal-mining indus-
insignificant, about 1½ per cent of the whole, the
aluable as regards coal-mine workers. Concern-
numeration shows that of the total employees of
ed, only 27 were females. These were engaged at
. There were no women engaged in any capac-
nd salt works.

nprising this report consist mainly of statistical
total of 1,217 pages being devoted to text. The
greatest detail, each establishment being sepa-
view of the more important facts is presented

In the establishments considered the ages of [illegible] vary from 14 to over 60 years, as shown in the [illegible]:

AGE OF OFFICIALS AND [illegible].

| Age (years). | Officials. | Engine men and firemen. | Actual mine workers. | Day laborers. | Other [illegible]. | Total. |
|---|---|---|---|---|---|---|
| 14 | 11 | [illegible] | [illegible] | [illegible] | [illegible] | [illegible] |
| 15 | 16 | [illegible] | 1,470 | [illegible] | [illegible] | [illegible] |
| 16 | 12 | 97 | [illegible] | [illegible] | [illegible] | [illegible] |
| 17 | [illegible] | 100 | [illegible] | [illegible] | [illegible] | [illegible] |
| 18 | 13 | 88 | [illegible] | [illegible] | [illegible] | [illegible] |
| 19 | [illegible] | 77 | [illegible] | [illegible] | [illegible] | [illegible] |
| 20 | 11 | 51 | [illegible] | [illegible] | [illegible] | [illegible] |
| 21 | 17 | [illegible] | [illegible] | [illegible] | [illegible] | [illegible] |
| 22 | 27 | 48 | [illegible] | [illegible] | [illegible] | [illegible] |
| 23 | [illegible] | 78 | [illegible] | [illegible] | [illegible] | [illegible] |
| 24 | 56 | 118 | [illegible] | [illegible] | [illegible] | [illegible] |
| 25 | [illegible] | 109 | [illegible] | [illegible] | [illegible] | [illegible] |
| 26 | 80 | 137 | [illegible] | [illegible] | [illegible] | [illegible] |
| 27 | 111 | 141 | [illegible] | [illegible] | [illegible] | [illegible] |
| 28 | 131 | 131 | 4,747 | [illegible] | [illegible] | [illegible] |
| 29 | 141 | 134 | 4,761 | [illegible] | [illegible] | [illegible] |
| 30 | 140 | 151 | [illegible] | [illegible] | [illegible] | [illegible] |
| 31 | 133 | 136 | [illegible] | [illegible] | [illegible] | [illegible] |
| 32 | 144 | 140 | [illegible] | [illegible] | [illegible] | [illegible] |
| 33 | 163 | 160 | [illegible] | [illegible] | [illegible] | [illegible] |
| 34 | 170 | 149 | 3,727 | [illegible] | [illegible] | [illegible] |
| 35 | 187 | 154 | 3,410 | [illegible] | [illegible] | [illegible] |
| 36 | 177 | 155 | [illegible] | [illegible] | [illegible] | [illegible] |
| 37 | 164 | 154 | [illegible] | [illegible] | [illegible] | [illegible] |
| 38 | 141 | 139 | [illegible] | [illegible] | [illegible] | [illegible] |
| 39 | 165 | 153 | [illegible] | [illegible] | [illegible] | [illegible] |
| 40 | 165 | 161 | [illegible] | [illegible] | [illegible] | [illegible] |
| 41 | 156 | 139 | [illegible] | [illegible] | [illegible] | [illegible] |
| 42 | 165 | 147 | [illegible] | [illegible] | [illegible] | [illegible] |
| 43 | 158 | 168 | [illegible] | [illegible] | [illegible] | [illegible] |
| 44 | 160 | 153 | [illegible] | [illegible] | [illegible] | [illegible] |
| 45 | 157 | 133 | 1,873 | [illegible] | [illegible] | [illegible] |
| 46 to 47 | 242 | 276 | [illegible] | [illegible] | [illegible] | [illegible] |
| 48 to 50 | 332 | 345 | [illegible] | [illegible] | [illegible] | [illegible] |
| 51 to 55 | 569 | 479 | [illegible] | [illegible] | [illegible] | [illegible] |
| 56 to 60 | 286 | 281 | [illegible] | [illegible] | [illegible] | [illegible] |
| Over 60 | 188 | 163 | 701 | [illegible] | [illegible] | [illegible] |
| Total | 4,976 | 5,566 | [illegible] | [illegible] | [illegible] | [illegible] |

The age of 19 years, which immediately precedes entry upon military service, is best represented, there being a much smaller number of employees of each age from 20 to 23 years, which are the ages of military service. After the period of military service, the age of 27 years is best represented. From that age onward there is an almost steady decrease in number for each succeeding year, especially in the case of mine workers.

The number of married employees is comparatively large when it is considered that over one-half of all employees were under 30 years of age. Of the 158,368 persons reported 91,648, or 57.87 per [illegible] married; 2,466, or 1.56 per cent, were widowed and divorced, [illegible] or but 40.57 per cent, were single.

Stability of occupations.—This subject receives [illegible] than any other, and is illustrated by tables showing [illegible] officials and employees who follow the occupations of [illegible] length of service in the same occupation of all officials [illegible] and the number remaining in the same establishm[illegible] *entire period of service.*

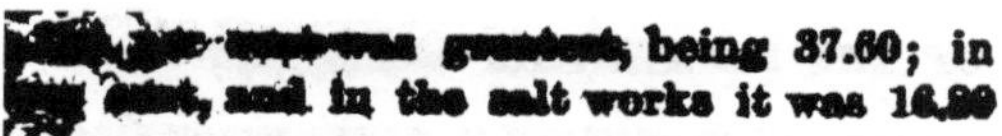

... was greatest, being 37.60; in ... and in the salt works it was 16.89

... in sections where the industries were ..., for instance, the percentage of coal mine ... were also miners was, in South Dortmund, ...; and in Werden, 55.69, while in Gelsen- ... and Herne, districts in which coal mining is ..., the percentages were, respectively, 28.36,

... shows the length of service of officials and ... at occupations, and the number of these who ... establishments during the entire period specified

...CIALS AND EMPLOYEES AT THE SAME OCCUPATIONS ... THE SAME ESTABLISHMENTS.

| ... | Actual mine workers. | | Day laborers. | | Other occupations. | | Total. | |
|---|---|---|---|---|---|---|---|---|
| ... | At same occupation. | In same establishment. | At same occupation. | In same establishment. | At same occupation. | In same establishment. | At same occupation. | In same establishment. |
| [illegible] | [illegible] | [illegible] | 2,970 | 2,970 | 2,172 | 2,172 | 29,206 | 20,206 |
| [illegible] | [illegible] | 6,671 | 1,279 | 1,085 | 841 | 689 | 11,584 | 8,853 |
| [illegible] | [illegible] | [illegible] | [illegible] | 875 | 618 | 473 | 12,890 | 8,911 |
| [illegible] | [illegible] | 5,676 | 761 | 562 | 493 | 334 | 11,470 | 6,853 |
| [illegible] | [illegible] | [illegible] | [illegible] | 330 | 245 | 224 | 8,713 | 4,549 |
| [illegible] | [illegible] | 2,147 | 350 | 234 | 206 | 115 | 6,211 | 2,628 |
| [illegible] | [illegible] | [illegible] | 247 | [illegible] | 170 | 91 | 4,985 | 1,787 |
| 97 | [illegible] | 1,168 | [illegible] | 160 | 136 | 66 | 4,712 | 1,491 |
| [illegible] | [illegible] | 1,141 | [illegible] | 135 | 133 | 67 | 4,805 | 1,459 |
| 67 | [illegible] | 1,347 | 325 | 184 | 186 | 105 | 6,408 | 1,774 |
| [illegible] | [illegible] | [illegible] | 279 | 171 | 152 | 70 | 5,293 | 1,503 |
| 75 | [illegible] | 940 | 275 | 155 | 125 | 62 | 5,310 | 1,276 |
| 75 | [illegible] | [illegible] | [illegible] | 142 | 109 | 53 | 5,030 | 1,149 |
| [illegible] | [illegible] | 541 | 182 | 84 | 70 | 25 | 3,871 | 734 |
| [illegible] | [illegible] | [illegible] | 175 | 83 | 71 | 25 | 3,614 | 581 |
| [illegible] | [illegible] | 301 | 133 | 74 | 51 | 19 | 2,706 | 462 |
| [illegible] | [illegible] | 281 | 150 | 78 | 56 | 25 | 2,734 | 514 |
| [illegible] | [illegible] | [illegible] | 168 | 89 | 63 | 28 | 3,247 | 628 |
| [illegible] | [illegible] | 343 | 178 | 92 | 48 | 20 | 2,872 | 538 |
| 91 | [illegible] | 504 | 272 | 123 | 67 | 20 | 4,438 | 795 |
| [illegible] | [illegible] | [illegible] | 168 | 80 | 56 | 17 | 2,985 | 538 |
| [illegible] | [illegible] | [illegible] | 131 | 60 | 49 | 17 | 2,746 | 464 |
| 41 | [illegible] | 275 | 130 | 51 | 26 | 9 | 2,860 | 419 |
| [illegible] | [illegible] | 173 | 95 | 30 | 28 | 9 | 2,224 | 263 |
| [illegible] | 1,848 | 163 | 114 | 29 | 30 | 9 | 2,253 | 251 |
| 10 | 1,484 | 152 | 120 | 48 | 33 | 7 | 1,870 | 249 |
| [illegible] | 1,430 | 168 | 93 | 18 | 26 | 10 | 1,776 | 242 |
| [illegible] | 1,310 | 139 | 76 | 16 | 16 | 7 | 1,639 | 200 |
| [illegible] | 1,108 | 102 | 65 | 18 | 17 | 5 | 1,393 | 172 |
| 13 | 1,156 | 92 | 108 | 16 | 16 | 5 | 1,487 | 142 |
| 8 | [illegible] | 62 | 54 | 15 | 3 | 2 | 769 | 106 |
| 9 | [illegible] | 46 | 52 | 10 | 9 | ........ | 727 | 78 |
| 7 | [illegible] | 50 | 51 | 15 | 7 | 3 | 722 | 88 |
| 7 | 387 | 40 | 40 | 6 | 6 | 1 | 529 | 67 |
| 10 | 381 | 41 | 57 | 10 | 9 | ........ | 589 | 74 |
| 8 | 376 | 35 | 48 | 7 | 10 | ........ | 568 | 62 |
| [illegible] | [illegible] | 28 | 42 | 4 | 5 | ........ | 409 | 59 |
| 8 | [illegible] | 34 | 35 | 8 | 3 | ........ | 396 | 55 |
| 6 | [illegible] | 15 | 34 | 4 | 3 | 2 | 313 | 35 |
| 6 | [illegible] | 7 | 30 | 7 | 5 | 1 | 300 | 26 |
| 1 | 78 | 4 | 19 | 2 | 3 | ........ | 148 | 10 |
| 3 | 57 | 4 | 9 | 1 | 2 | ........ | 107 | 8 |
| 3 | 66 | 2 | 14 | 1 | 1 | 1 | 101 | 11 |

YEARS OF SERVICE OF OFFICIALS AND EMPLOYEES [illegible] AND IN THE SAME ESTABLISHMENTS—Concluded.

| Years of service. | Officials. | | Engineers and firemen. | | Actual mine workers. | | Day laborers. | | Other occupations. | | Total. | |
|---|---|---|---|---|---|---|---|---|---|---|---|---|
| | At same occupation. | In same establishment. | At same occupation. | In same establishment. | At same occupation. | In same establishment. | At same occupation. | In same establishment. | At same occupation. | In same establishment. | At same occupation. | In same establishment. |
| 44 | 13 | 2 | 6 | 1 | 35 | ........ | 10 | 2 | 1 | ........ | 65 | 5 |
| 45 | 22 | 2 | 7 | 1 | 28 | 2 | 17 | 6 | 2 | ........ | 76 | 11 |
| 46 | 12 | 1 | 8 | 2 | 16 | ........ | 7 | 1 | ........ | ........ | 43 | 4 |
| 47 | 14 | 2 | 4 | ........ | 12 | ........ | 4 | ........ | 1 | ........ | 35 | 2 |
| 48 | 9 | 1 | 2 | ........ | 7 | ........ | 4 | ........ | ........ | ........ | 22 | 1 |
| 49 | 5 | 1 | 3 | ........ | 8 | ........ | 1 | 1 | 1 | ........ | 18 | 2 |
| 50 | 9 | ........ | 3 | ........ | 7 | ........ | 4 | 1 | 1 | ........ | 24 | 1 |
| 51 | 6 | 1 | ........ | ........ | 4 | ........ | 2 | 2 | ........ | ........ | 12 | 3 |
| 52 | 5 | ........ | ........ | ........ | ........ | ........ | 2 | ........ | ........ | ........ | 7 | ........ |
| 53 | 6 | ........ | ........ | ........ | 1 | ........ | 1 | ........ | ........ | ........ | 8 | ........ |
| 54 | 2 | 1 | ........ | ........ | 1 | ........ | ........ | ........ | 1 | ........ | 4 | 1 |
| 55 | 1 | ........ | ........ | ........ | 1 | ........ | 1 | ........ | 1 | ........ | 4 | ........ |
| 56 | 2 | 1 | ........ | ........ | ........ | ........ | ........ | ........ | ........ | ........ | 2 | 1 |
| 57 | 1 | ........ | ........ | ........ | ........ | ........ | ........ | ........ | ........ | ........ | 1 | ........ |
| 58 | ........ | ........ | ........ | ........ | 1 | ........ | ........ | ........ | ........ | ........ | 1 | ........ |
| 59 | 1 | ........ | ........ | ........ | ........ | ........ | 1 | ........ | ........ | ........ | 2 | ........ |
| Total. | 4,976 | 1,362 | 5,586 | 3,113 | 129,602 | 52,810 | 11,731 | 8,348 | 6,468 | 4,788 | [illegible] | 70,341 |

As the industries in this section are largely in a stage of rapid development, a greater number of employees are found to have been engaged in their present occupations for one year or under than for any other period of service. There is an almost steady decrease for each succeeding period. The only decided exception is found in the number employed seven, eight, and nine years. The abnormal decrease shown here is due to military service, for if the persons mostly entered upon their occupations at the minimum age, 14 years, they would have attained the ages of 21, 22, and 23 years, respectively, which ages, as has been shown, correspond to the period of military service.

Of all the employees and officials, 70,341, or 44.42 per cent, remained in the service of the same establishments during the entire time of employment at their present occupations.

The enumeration showed that out of the total number of employees and officials 154,517, or 97.57 per cent, had completed their elementary education. This percentage would undoubtedly have been greater were it not for the large influx of foreigners, which is always present where industries are rapidly developing. In some other sections of this mining district, where the industries are older, the illiterates comprise less than one-half of 1 per cent.

| | | | | | | | |
|---|---|---|---|---|---|---|---|
| ........ | 36 | ........ | 7 | 17 | ........ | ........ | 1,467 |
| ........ | ........ | ........ | 19 | 168 | ........ | ........ | 623 |
| ........ | 2 | ........ | 2 | 19 | ........ | ........ | 57 |
| [illegible] | 87 | 1,069 | 26 | 204 | 429 | 1,756 | 2,167 |
| [illegible] | 60 | ........ | ........ | ........ | ........ | ........ | 181 |
| 3 | 4 | ........ | ........ | 4 | ........ | ........ | 37 |
| 9 | 1 | ........ | ........ | 1 | ........ | ........ | 7 |
| [illegible] | 48 | 96 | ........ | 5 | 29 | 146 | 225 |
| 20 | 2 | ........ | ........ | ........ | ........ | ........ | 47 |
| ........ | ........ | ........ | ........ | 3 | ........ | ........ | 24 |
| 2 | ........ | ........ | ........ | ........ | ........ | ........ | 1 |
| 21 | 3 | 14 | ........ | 3 | 11 | 31 | 62 |
| 23,914 | 2,548 | 74,156 | 975 | 32,021 | 34,753 | 144,454 | 158,368 |

…ts the comparatively large proportion of house or 10.24 per cent. These, however, did not all s, 14.17 per cent of their number living in rented f the total officials and employees, 8.79 per cent es, 1.61 per cent were tenants of their employers, other rented houses, 0.61 lived in lodging houses, parents, and 20.22 per cent boarded with stran- and eighty-four of the house owners were single each family occupied three rooms.

The distance from residence to place of [illegible] in the accompanying table:

DISTANCE BETWEEN RESIDENCE AND PLACE OF WORK OF OFFICIALS AND EMPLOYEES.

| Distance of residence from place of work. | Coal mines. | | Ore mines. | | Salt works. | | Mineral baths. | | Total. | |
|---|---|---|---|---|---|---|---|---|---|---|
| | Persons. | Per cent. | Persons. | Per cent. | Persons. | Per cent. | Persons. | Per cent. | Persons. | Per cent. |
| 2.5 miles and under | 133,79[illegible] | [illegible] | 1,[illegible] | 84.37 | [illegible] | [illegible] | [illegible] | [illegible] | [illegible] | 85.51 |
| Over 2.5 and not exceeding 4.7 miles | 18,969 | 12.16 | [illegible] | [illegible] | 13 | 5.7[illegible] | 3 | 4.84 | [illegible] | 12.41 |
| Over 4.7 and not exceeding 6.8 miles | 2,641 | 1.70 | [illegible] | [illegible] | 2 | [illegible] | ...... | ...... | 2,[illegible] | 1.73 |
| Over 6.8 and not exceeding 9.3 miles | 332 | 0.21 | 5 | [illegible] | ...... | ...... | ...... | ...... | [illegible] | 0.21 |
| Over 9.3 miles | 212 | 0.14 | 4 | 0.19 | 3 | 1.[illegible] | ...... | ...... | [illegible] | 0.14 |
| Total | 155,934 | 100.00 | 2,147 | 100.00 | [illegible] | 100.00 | [illegible] | 100.00 | [illegible] | 100.00 |
| Persons living over 4.7 miles: | | | | | | | | | | |
| In their own houses | [illegible] | 19.79 | 14 | 12.75 | 1 | 20.00 | ...... | ...... | [illegible] | [illegible] |
| In rented dwellings | 1,205 | 37.8[illegible] | 65 | [illegible] | 4 | 80.00 | ...... | ...... | [illegible] | [illegible] |
| Lodging with parents | 697 | 21.8[illegible] | [illegible] | 18.61 | ...... | ...... | ...... | ...... | [illegible] | [illegible] |
| Lodging with strangers | 653 | 20.50 | 13 | 12.74 | ...... | ...... | ...... | ...... | [illegible] | [illegible] |
| Total living over 4.7 miles | 3,185 | 100.00 | [illegible] | 100.00 | 5 | 100.00 | ...... | ...... | [illegible] | 100.00 |
| Average distance of residence from place of work (miles) | 1.5 | | 2.2 | | 1.4 | | 1.4 | | 1.5 | |

It appears from the preceding table that as a rule the employees live within walking distance of their places of work, 85.51 per cent having not over 4 kilometers (2.5 miles) to go. The average distance of all was 2.4 kilometers (1.5 miles). Of those who lived so far distant as to be compelled to use conveyances, namely, 7.5 kilometers (4.7 miles) and over, only 19.59 lived in their own homes. Nearly twice the number, 38.40 per cent, lived in rented dwellings, 21.78 per cent boarded with their parents, and 20.23 per cent boarded with strangers.

The next table shows the number of persons depending upon officials and employees for their support.

PERSONS DEPENDENT UPON OFFICIALS AND EMPLOYEES.

| Industries. | Wives. | Children. Under 14 years of age. | | Children. Over 14 years of age. | | Children. Total. | Children. Wholly dependent. | Parents and grand-parents. | Brothers and sisters. | Total dependents. | [illegible] | [illegible] |
|---|---|---|---|---|---|---|---|---|---|---|---|---|
| | | Sons. | Daughters. | Sons. | Daughters. | | | | | | | |
| Coal mining | 89,955 | 114,101 | 109,051 | 35,290 | 32,047 | 290,48[illegible] | [illegible] | [illegible] | 6,[illegible] | [illegible] | [illegible] | [illegible] |
| Ore mining | 1,466 | 1,849 | 1,[illegible] | 875 | 784 | 5,377 | 4,064 | 545 | 11[illegible] | 7,[illegible] | [illegible] | [illegible] |
| Salt works | 180 | 218 | 204 | 107 | [illegible] | [illegible] | [illegible] | [illegible] | [illegible] | [illegible] | [illegible] | [illegible] |
| Mineral baths | 47 | 65 | 72 | [illegible] | [illegible] | 184 | 164 | [illegible] | 1 | [illegible] | [illegible] | [illegible] |
| Total | 91,648 | 116,223 | 111,196 | 36,300 | 32,959 | 296,672 | 242,[illegible] | 25,[illegible] | 6,[illegible] | [illegible] | [illegible] | [illegible] |

The average number of persons depending upon an [illegible] ployee was 2.66. Of the dependents, 21.79 per cent were [illegible] per cent children, 6.03 per cent parents and grandparents, [illegible] cent brothers and sisters. Over three-fourths of the [illegible] were under 14 years of age.

## ANNUAL STATISTICAL ABSTRACTS.

In the first issue of the Bulletin a brief note was given indicating the Governments of the world which had created statistical bureaus, the sole or special object of which was to investigate and report on the conditions under which labor is prosecuted. In all cases the creation of these bureaus is of recent date. The nations of Europe and America have not waited, however, for the establishment of these special bureaus before entering upon the collection and presentation of statistical information concerning labor and social conditions, and the reports of such bureaus are by no means the sole source of statistical information concerning these questions. The various departments of the different Governments, especially those departments concerned with industry, commerce, or public works, have sooner or later found it necessary to organize more or less complete statistical services, and within the field of their investigations it has often been advisable to include the collection of information concerning matters purely social. Scattered through the various reports of the independent departments or bureaus of all Governments there is, therefore, information of a strictly statistical character of immediate and important bearing upon labor conditions.

The great number and variety of these publications and the greater or less difficulty in obtaining access to them renders it almost impossible to gain even a knowledge of their existence, much less a familiarity with their nature and contents. Recognizing this, practically every nation of importance has undertaken the publication of an annual statistical abstract in which are presented in summary form the most important statistical tables appearing in all of the various official reports of the Government. These abstracts, therefore, do not in themselves represent original investigations, but only brief summaries of the more important inquiries made by various bureaus. In the great majority of cases where information is sought, a reference to these compilations is sufficient.

The following, made from the latest accessible abstracts, is given in order to show for at least the more important nations the periods during which such abstracts have been issued and in a general way the principal subjects concerning which information is given.

[illegible] no effort is made to enter into particulars. Only [illegible] headings are employed. Thus, for instance, the single [illegible] Statistics" is made to embrace all the information [illegible] upon agricultural matters.

## BELGIUM.

*Annuaire Statistique de la Belgique* [Ministère de l'Intérieur et de l'Instruction Publique].

Published since 1870.

CONTENTS.

Area and population:
- Territorial extent, division of property.
- Population, ages, nationality, education, occupations, etc.
- Houses.
- Families.
- Births, deaths, marriages, divorces.
- Immigration and emigration.

Political statistics (elections).

Education.

Publications, libraries, and schools of art and music.

Savings and benevolent institutions:
- Savings banks.
- People's banks (cooperative societies).
- Pawnshops.
- Mutual aid societies.
- Employment bureaus.
- Workingmen's houses.
- Miners' relief and pension funds.

Religious statistics.

Statistics of insane and idiotic.

Medical profession.

Administration of justice:
- Court records.
- Crime.

Administration of justice—Concluded.
- Jails, reformatories, etc.

Army.

Government finance (receipts, expenditures, debt, etc.).

Production:
- Agriculture and forestry.
- Mines and mining (includes statistics of number of employees, wages, accidents, etc.).
- Manufactures (includes statistics of number of employees, wages, accidents, boards of arbitration, etc.).

Trade and commerce:
- Foreign commerce (imports, exports, customs receipts).
- Navigation.
- Fisheries.

Wealth.

Insurance.

Financial institutions (banks, currency).

Transportation:
- Railways (includes statistics of accidents).
- Tramways.
- Waterways.

Postal, telegraph, and telephone services.

The authorities are indicated by footnotes.

## HOLLAND.

*Jaarcijfers* [uitgegeven door de Centrale Commissie voor de Statistiek] *Annuaire Statistique des Pays-Bas* [Publié par la Commission Central de Statistique].

Published since 1881.

CONTENTS.

Area.

Meteorology.

Population:
- Houses.
- Families.
- Births, deaths, marriages, divorces.
- *Emigration.*

Defective classes (blind, deaf and dumb, etc.).

Education.

Economic statistics:
- Occupations.
- Prices of commodities.
- Public charity.

tures, taxation, debt).
Army and navy.

# FRANCE.

la *France* [Office du Travail, Ministère du dustrie, des Postes et des Télégraphes].

CONTENTS.

| | |
|---|---|
| ...reau de la ...al Statisti- | Production: |
| | Agriculture and forestry. |
| | Mines and manufactures. |
| ...from which | Manufactures by the Government (to- |
| ...s taken. | bacco and matches). |
| | Labor: |
| ...ographical | Labor exchanges. |
| ...tion. | Employment bureaus. |
| ...njugal con- | Wages and hours of work. |
| | Strikes. |
| | Labor unions. |
| ...tion. | Accidents and factory inspection. |
| | Cooperation. |
| | Transportation: |
| | WaterWays. |
| ...us. | Public roads. |
| | RailWays. |
| | TramWays. |
| | Postal, telegraph, and telephone services. |
| | Domestic and foreign commerce. |
| ...ent institu- | Navigation and fisheries. |
| | Government finance: |
| | Receipts and expenditures. |
| ...ings banks. | Direct and indirect taxes. |
| | Public debt. |
| | Financial institutions, banks, currency, |
| | real estate values, etc. |
| ...masters, etc. | Election statistics. |
| | NeWspapers and periodicals. |
| of art and | Military and naval statistics. |
| | Algeria, Tunis, and the colonies. |

## GREAT BRITAIN.

***Statistical Abstract for the United Kingdom*** ... **Statistical Department, Board of ...**

**Published** since 1853.

CONTENTS.

Government receipts, expenditures, debts.
Imports and exports.
Finance:
  Bullion and specie.
  Coinage.
  Accumulative Government stocks.
Transshipments.
Prices of products, imports, and exports.
Production:
  Agriculture.
  Fisheries.
  Textiles.
  Mines.
Shipping.
Transportation:
  Railways.
  Tramways.
Joint-stock companies.
Banking:
  Savings bank...
  Bank of Engl...
  Clearing hous...
Provident institut...
  Building soci...
  Industrial and...
  Life insurance...
Postal service.
Population.
Births, deaths, an...
Emigration and in...
Army and police...
Education.
Administration of...
  Crime.
  Bankruptcy.
  Patents and t...
Wrecks and lives l...

## GERMANY.

***Statistisches Jahrbuch für das Deutsche Reich*** [Kaiser... Amt].

**Published** since 1880.

CONTENTS.

Area and population:
  Territorial extent of the German States.
  Population by age, sex, place of birth, and religion.
  Marriages, births, and deaths.
  Emigration.
Production:
  Agriculture and forestry.
  Mines and mining.
  Manufactures.
Trade and transportation:
Consumption and ...
  Wholesale pri...
Government finan... tures, debts, etc...
Administration of...
  Court records.
  Crime.
  Bankruptcy.
Workingmen's ins...
  Sick-benefit in...
  Accident insur...
  Invalid and ol...

## AUSTRIA.

*Oesterreichisches statistisches Handbuch für die im Reichsrathe vertretenen Königreiche und Länder* [k. k. statistische Central-Commission].

Published since 1882.

CONTENTS.

Area and population:
- Territorial extent, division of property.
- Population, ages, places of birth, domicile, sex, religion, etc.
- Marriages, births, and deaths.
- Immigration and emigration.

Production:
- Agriculture and forestry (includes prices of agricultural products and wages).
- Mines and mining (includes number of employees, accidents, and benevolent institutions).
- Manufactures (includes workingmen's accident and sick-benefit insurance).

Trade and transportation:
- Fisheries.
- Foreign commerce.
- Waterways.
- Railways (includes statistics of accidents).

Postal, telegraph, and telephone services.

Education.

Religion.

Health service and benevolent institutions:
- Physicians, surgeons, and apothecaries.
- Hospitals
- Asylums for the insane, blind, deaf and dumb, and foundlings.
- Causes of death.
- Benevolent institutions.

Societies.

Business corporations.

Administration of justice:
- Court records.
- Crime.
- Prison service.

Financial institutions and Government finance:
- Banks.
- Currency.
- Government receipts, expenditures, and debt.

Stock quotations.

Army and navy.

Newspapers and other periodicals.

Damages by fire and hail.

## SWITZERLAND.

*Statistisches Jahrbuch der Schweiz* [Statistisches Bureau des eidg. Departements des Innern]: *Annuaire Statistique de la Suisse* [Bureau de Statistique du Département Fédéral de l'Intérieur].

Published since 1891.

CONTENTS.

Area and population:
- Territorial extent.
- Population.
- Marriages, births, divorces, deaths, etc.
- Emigration.

Production:
- Agriculture.
- Forestry.
- Pisciculture.
- Mining.
- Manufactures.

Postal, telegraph, and telephone services.

Transportation:
- Railways.
- Tramways.
- Steam navigation.

Commerce, domestic and foreign (includes income from customs duties, etc.)

Insurance (life, accident, marine, and fire).

Banks.

Prices of food products

Hygiene, sanitary police, and assistance:
- Asylums for epileptics, insane, alcoholism, etc.
- Hospitals.
- Diseases.

CONTENTS—[illegible]

Accidents.
Education.
Government finance:
  Receipts, expenditures, etc.
  Coinage and currency.

[illegible]
Army.
[illegible]
Meteorology.

## ITALY.

*Annuario Statistico Italiano* [Ministero di Agricoltura, Industria e Commercio, Direzione Generale Della Statistica].

Published since 1878.

CONTENTS.

Meteorology.
Area and population:
  Territorial extent.
  Population.
  Marriages, births, deaths.
  Emigration.
Public health:
  Physicians, surgeons, and apothecaries.
  Diseases.
  Causes of deaths.
Education.
Libraries.
Administration of justice:
  Civil cases.
  Criminal cases.
  Prison service.
Publications:
  Periodicals.
  Other publications.
Mutual aid and cooperative associations:
  Mutual benefit societies.
  Cooperative banks, stores, manufactories, etc.
Public assistance and benevolent institutions.
Army and navy.
Foreign commerce.
Navigation.
Duties on certain imports and exports.
Mortgage debts.
Production:
  Agriculture.
  Manufactures.
  Mines.
  Fisheries.
  Prices of certain food products.
  Workingmen's wages in certain industries.
Postal, telegraph, and telephone service.
Political statistics (elections).
Money and credit:
  Coinage.
  Paper currency.
  Government bonds and exchange.
  Banks and credit institutions.
Transportation:
  Public roads.
  Tramways.
  Railways.
Government finance:
  Commercial receipts, expenditures, debts, etc.
  Provincial receipts, expenditures, debts, etc.
  National receipts, expenditures, debts, etc.
Colonial possessions and protectorates.

The authorities are given in connection with each chapter.

Coinage and currency.
Transportation (railways).

ies, etc. | Postal, telegraph, and telephone services.

Government finance (receipts, expenditures, public debt, etc).
Banks and savings institutions.
Real estate:

of assemblies. | Peasant proprietorship.

Mortgage debts.

nd life). | Publishing houses, libraries, reading rooms, etc.

statistics of | Churches, convents, etc.

Military statistics.

es statistics of | Education.

cidents in the | Criminal statistics.

Meteorological data.

## NORWAY.

Kongeriget Norge [Udgivet af Det Statistiske Cen-
n]: *Annuaire Statistique de la Norvége.*

'9·

CONTENTS.

Production:
Agriculture and forestry.

, conjugal con- | Manufactures (includes statistics of wages).

l deaths. | Fisheries.

Mining.
Commerce and navigation (imports, exports, customs duties).
Transportation:

s. | Public roads.

Railways.
Postal and telegraph services.
Banking institutions.
Fire and marine insurance.
Finance (local and national Government receipts, expenditures, debts, etc.).

## SWEDEN.

*Sveriges Officiela Statistik i Sammandrag* [illegible] af Kungl. Statistiska Centralbyrån.

Published since 1860.

CONTENTS.

Area and population:
- Territorial divisions.
- Population by sex, age, conjugal condition, religious persuasion, etc.
- Marriages, births, and deaths.
- Immigration and emigration.
- Occupations by sex.
- Blind, deaf and dumb, and insane.

Education.

Administration of justice:
- Court records.
- Crimes and misdemeanors.
- Prison service.

Public health.

Statistics of pauperism and public charities.

Production:
- Agriculture.

Production—Concluded.
- Forestry.
- Manufactures.
- Mining.

Commerce and navigation:
- Tariff rates.
- Foreign commerce.
- Prices.
- Pilotage, life-saving and lighthouse service.

Transportation (railways).

Postal, telegraph, and telephone services.

Government finance (receipts, expenditures, debts, etc.).

Banks.

Election statistics.

Wages.

## DENMARK.

*Danmarks Statistik; Statistisk Aarbog* [Udgivet af Statens Statistiske Bureau]: *Statistique du Danemark; Annuaire Statistique* [Publié par le Bureau de Statistique de l'État].

Publication commenced in 1896.

CONTENTS.

Sources of information.

Money, weights, and measures.

Area and population:
- Territorial divisions.
- Population by sex, age, conjugal condition, place of birth, religion, etc.
- Births, marriages, and deaths.
- Emigration.
- Occupations according to sex.

Agricultural statistics (including prices of cereals).

Manufactures.

Imports and exports.

Production of brandy, beer, artificial butter, and beet sugar.

Production of principal articles of consumption.

*Navigation.*

*Fisheries.*

Shipwrecks.

Transportation (railways).

Postal and telegraph services.

Finance:
- Banks.
- Joint-stock companies.
- Mortgage statistics.
- Pawnshops.
- Insurance (fire, marine, [illegible]).

Administration of justice.

Education.

Publications.

Subsidies for old-age pensions.

Mutual aid societies.

Recruitment of army.

Election statistics.

Finances of the communes.

Finances of the State.

Meteorology.

## DECISIONS OF COURTS AFFECTING LABOR.

[This subject, begun in Bulletin No. 2, will be continued in successive issues, dealing with the decisions as they occur. All material parts of the decisions are reproduced in the words of the courts, indicated, when short, by quotation marks, and when long by being printed solid. In order to save space immaterial matter, needed simply by way of explanation, is given in the words of the editorial reviser.]

### DECISIONS UNDER STATUTORY LAW.

Constitutionality of Act Providing for Attorneys' Fees in Suits for Wages.—The constitutionality of the act of June 1, 1889, of Illinois, providing for taxing attorneys' fees as costs in actions brought by servants for wages which they have previously demanded in writing, was drawn in question in the case of Vogel et al. *v.* Pekoc (Northeastern Reporter, vol. 42, p. 386), and its validity was sustained by the supreme court of the State on June 15, 1895, the opinion of the court being delivered by Chief Justice Craig.

A petition for rehearing was filed which emphasized the previous contention that the act in question is a partial and special statute, working deprivation of property without due process of law, and therefore unconstitutional, and reliance was placed on decisions in a number of cases by the Illinois supreme court adverse to the constitutionality of certain legislative enactments as sustaining the position taken. On the rehearing, October 28, 1895, the court again sustained the validity of the law, Judge Magruder dissenting, and in their opinion said:

Those cases do not, however, control the present case, or decide the question here involved. Without discussing separately the facts of the cases relied upon, it may be said generally, that in each of those cases the principal and controlling question was the right of miners of coal (no less than their employers) to make contracts regulating the time and manner of the payment of wages and the method of computing such wages, and in each case cited a law restricting in some manner this important right of contract was held invalid.

The statute here in question interferes with no one's right to contract. It embraces a well defined class of cases and persons, not singled out, and [illegible], wholly without reason and arbitrarily; but upon grounds which may, we think, properly serve as a basis for valid legislative action. Those to whom the wages of labor are due, and who, after [illegible] of a sum no greater than that subsequently recovered, [illegible] to establish, and do establish, their rights as [illegible] judgment of court, are within the provisions of the act; [illegible] that this classification is so arbitrary and unreasonable, or so partial and unequal, as to be beyond legislative discretion and power. Indeed, if this law were to be held unconstitutional for the reasons assigned, then many other acts long in force in this State, hitherto deemed to be salutary, and against which no constitutional objection has been heard, would certainly fall with it. Why, for instance, should the seller of materials for a building have by law a lien for their price, not only upon the specific things sold, but upon the whole structure, with the land it stands on, while the seller of a horse, a piano, or a cornsheller is denied any lien even on the specific thing sold? Why should he whose labor constructs a house be secured by a lien on his product, while he who raises a crop must look only to the personal responsibility of his hirer? Surely, it could be said the lien law makes classes of beneficiaries quite as arbitrary in character as that marked out to receive benefit by the act under discussion.

Again, why should the wages of a defendant, who is head of a family, to an amount not exceeding $50, be exempt from garnishment (Hurds St., p. 783, sec. 14), while sums due other defendants are protected by no such exemption? And why, again, it might be asked, should heads of families earning wages be made the subject of advantageous provisions not applied to all other wage earners if not to all other persons? The general exemption law also makes heads of families a distinct class, who may claim as exempt $300 worth more property than others are allowed, while a further section (id., p. 727, sec. 4) declares that where a judgment is for the wages of a laborer or servant, and noted by the court as such, no personal property whatever shall be free from levy, whatever the estate or condition of the debtor. An analogous case for this purpose is found in the provisions of the general assignment law that "all claims for the wages of any laborer or servant which have been earned within three months next preceding the making of the assignment," etc., "be preferred and first paid to the exclusion of all other demands" (id., p. 166, sec. 6). It is difficult to see how any of these statutes, and many similar ones which might be named, could be sustained if the strict rule of constitutional validity, so strenuously urged in this case, were applied to them.

In dissenting from the foregoing opinion Judge Magruder cited the case of Coal Co. v. Rosser (41 N. E., 263), in which the supreme court of Ohio had occasion to consider and condemn a statute similar to the one sustained by his colleagues. He said that the opinion in that case expressed what seemed to him to be the correct view of the subject, and made the following quotation therefrom as sufficiently indicating the reasons for his dissent:

Upon what principle can a rule of law rest which permits one party, or class of people, to invoke the action of our tribunals of justice at will, while the other party, or another class of citizens, does so at the peril of being mulct in an attorney's fee, if an honest but unsuccessful defense should be interposed? A statute that imposes this restriction upon one citizen, or class of citizens, only denies to him or them [illegible] protection of the law. It is true that no provision of the [illegible] of 1851 declares in direct and express terms that this may [illegible] but nevertheless it violates the fundamental principles upon [illegible] Government rests, as they are enunciated and declared by [illegible] ment in the bill of rights. The first section of the [illegible] that the right to acquire, possess, and protect property [illegible] and the next section declares, among other things, that "[illegible] *instituted for* the equal protection and benefit" of every [illegible] tribunals are provided for the equal protection of ... right to retain property already in possession is as ... right to recover it when dispossessed. The right to defend ... action to recover money is as necessary as the right to defend ... to recover specific real or personal property. An adverse ... the case deprives the defeated party of property. ... general assembly has power to enact the statute in question, ... enact one providing that lawyers, doctors, grocers, or any ... of citizens might make out their accounts, demand in writing ... within a short time, which, if not complied with, would ... plaintiff to an attorney fee in addition to his claim, if he ... the amount demanded. We do not think the general assem... power to discriminate between persons or classes respecting ... invoke the arbitrament of the courts in the adjustment of ... tive rights. The legislative power to compel an unsuccess... an action—generally the defendant—to pay an attorney fee ... ant has received the attention of a number of courts of last ... ell as laws which impose as a penalty double damages or ... r penalty for some wrongful or negligent act injurious to ... Where the penalty has been imposed for some tortious or ... t, the statute has generally, though not always, been sus... on the contrary, where no wrongful or negligent conduct was ... the defeated party, any attempt to charge him with a pen... prevailed. [At this point in the decision a number of cases ... sustaining the foregoing propositions.] Various phases ... ect have received attention in the foregoing cases, as well as ... ers to which we do not deem it necessary to refer. The gen... cy of these authorities is toward the result which we have ... t whether they do or do not support our conclusions, we are ... at the fundamental principles of government declared by ... rights clearly and unequivocally prohibit legislation of the ... that involved in this case.

---

...TIONS—LIABILITY OF STOCKHOLDERS FOR WAGES.— ...c8 of volume 3, Howell's Annotated Statutes of Michigan, ... at stockholders of manufacturing corporations "shall be ... liable for all labor performed for such corporations, which ... y may be enforced against any stockholder by action ... this statute, at any time after an execution shall be ... satisfied in whole or in part against the corporation," etc. ... statute suit was brought before a justice of the peace by ... against Peter Wintermute, a stockholder in the Chees... Manufacturing Company, to recover for personal work ... by Kamp for the company before Wintermute ... therein. The justice gave judgment for Kamp, ... removed to the circuit court of Muskegon County by

...—7

writ of certiorari and there the judgment [illegible]. The case was then carried to the supreme [illegible], which affirmed the judgment of the circuit court, holding [illegible] question does not make a stockholder liable for labor [illegible] he became a stockholder.

The opinion of the supreme court was delivered by Judge Long, and is published in volume 65 of the Northwestern Reporter, page 570. In the course of the opinion Judge Long said:

It is conceded in this court that the defendant was not a stockholder in the corporation at the time the labor was performed; but, as shown on the trial before the justice, he was a stockholder at the time suit was commenced. The claim is that those who are stockholders in a corporation at the time of the commencement of an action against the corporation upon the labor claims are liable upon such demands, although not stockholders at the time corporate liability accrued; that such stockholders impliedly assume all the obligations which rested upon former holders as members of the company, and are liable to the same extent as the former holders.

Whether statutory liability attaches to a stockholder in respect of debts contracted before he became a member of the corporation, is a question turning upon the words of the statute. While the persons who were stockholders at the time the debt was contracted may be held liable, we think the liability is confined to such stockholders, and not to those who thereafter purchase. The stockholders mentioned in this section of the statute must be construed to mean those who were such at the time the liability attaches to the corporation. The statutes of many of the States prescribe when such liability is to attach. Our statute fixes no such time by any express words, but, by necessary implication, was not intended to apply to one who became a purchaser after liability was incurred.

The general doctrine seems to be that a stockholder does not avoid a statutory liability to creditors who were such at the time he transfers his stock; and there is good reason for holding the rule being thus settled that the transferee does not take the stock subject to such statutory liability.

---

Employers' Liability—Railroad Companies.—In the case of Caron v. Boston and Albany Railroad Company, the superior court of Hampden County, Mass., gave judgment for damages [illegible] railroad company in favor of Marie Caron for the death of [illegible], an employee of said company, resulting from injuries [illegible] lision caused by the alleged negligence of some person who [illegible] or control of a certain train in shifting it over upon [illegible] Caron was at work.

The railroad company carried the case, on exceptions, [illegible] judicial court of Massachusetts, and the exceptions [illegible] that tribunal on November 26, 1895. The opinion of [illegible] in vol. 42 of the Northeastern Reporter, p. 112) was [illegible] Morton.

...chapter 270 of the acts of 1887 of Massachu... as follows: "Where, after the passage ... is caused to an employee, who is himself in ... and diligence at the time, (1) by reason of any ... of the ways, works, or machinery connected with ... of the employer, which arose from or had not ... unremedied owing to the negligence of the employer ... in the service of the employer and intrusted by him ... of seeing that the ways, works, or machinery were in ...n, or * * * (3) by reason of the negligence of any ... service of the employer who has the charge or control of ... locomotive, engine, or train upon a railroad, the ... the injury results in death, the legal representative ... shall have the same right of compensation and reme... employer as if the employee had not been an employee ... service of the employer, nor engaged in its work."

... of the opinion sustaining the exceptions of the railroad ... Morton said:

...t contends that the plaintiff's intestate was not in the ... care. There was nothing to show that he had any ... knowledge that the cars which caused the collision were ... track, or that he could see them; and, for aught that ... engaged in the discharge of his duty when injured. ... be inferred from the absence of negligence as well as ... acts of diligence. We think that there was evidence ... the jury in finding that he was in the exercise of due ...

...t contends further that the plaintiff's intestate assumed ... we do not think it fairly can be held that he assumed ... from cars which were sent in, as there was evidence ... that the colliding cars were moving at the rate of ten ... an hour, and with such force as to throw off the track ... train which Collins [the conductor] and his men were ... to break the draw bars of others. Such a manner of ...ness would be unreasonable, and not within any risk ...tiff's intestate assumed.

...t also contends that the cars which were switched onto ... Caron was working did not constitute a train at the ...ident. It is not easy to define what, under all circum... constitute a train, within the meaning of the statute. ... with one or more cars attached to it, with or without ... freight, in motion upon a railroad from one point to ... of power furnished by the locomotive, would ...stitute a train. So it would if the steam was shut off ...tive and the train was moving by its own momentum. ..." as used in the railroad act (Pub. St., c. 112) generally ... motion. Usually the power would be furnished by a ... whether a number of cars coupled together and in ... one connected whole, do or do not constitute a ...pend, we think, upon whether a locomotive engine is ... the time, and they are moved by the power thus ... to accident for which St. 1887, c. 270, was designed to furnish a remedy would be the same in kind, though perhaps not so great in degree, whether the motive power was furnished by a locomotive attached to the cars or in some other manner. And it seems to us that a number of cars, coupled together as these were, forming one connected whole and moving from one point to another upon a railroad, in the ordinary course of its traffic, under an impetus imparted to them by a locomotive, which, shortly before the accident, had been detached, constitute a train within the meaning of the statute.

The next and more difficult question is whether either of the two brakemen or Mozier, the foreman of the switching gang, was in "the charge or control" of the train when the accident occurred. The words "the charge or control" do not seem to have received a final construction anywhere. The implication of our own decisions, so far as they can be said to have given rise to one, is that they are to be regarded, not, perhaps, as synonymous, but as explanatory of each other, and as used together for the purpose of describing more fully one and the same thing; and we think that this is the better construction. If "control" is one thing and "charge" is another, then, inasmuch as to some extent every brakeman upon a train would have "control" of it, every employee injured by an accident resulting from the carelessness of a brakeman would have a right of action against the corporation which employed him, and the defense of common employment as to brakemen would be done away with, even though the brakeman might be acting under an immediate superior. The statute is to be fairly construed, and, while it removes the defense of common employment in some cases, it does not extinguish it altogether; and we do not think that the legislature intended that it should be abolished in all cases where injuries were sustained by the carelessness of a brakeman. We think, therefore, that by the words, "any person * * * who has the charge or control," is meant a person who, for the time being at least, has immediate authority to direct the movements and management of the train as a whole and of the men engaged upon it. It is not necessary that the person in charge or control should be actually upon the train itself. On the other hand, the mere fact that a laborer or brakeman is put in such a position that for the moment he physically controls and directs its movements under the eye of his superior does not constitute him a person in "the charge or control" of the train, though there may be circumstances under which he would have such charge or control. It is possible that more than one person may have "the charge or cootrol" of a train at the same time. The case of the engineer is expressly provided for, and it is not likely that any negligence of his which would affect the train would not be negligence in the management of the locomotive engine of which at the time he had "the charge or control."

Applying these principles to the case before us, we do not think that either O'Brien or Desloury had "the charge or control" of the train as it went onto the side track, and after the engine and caboose had been detached, but that they were fellow-servants of the plaintiff's [illegible]. They certainly had not before that, and after it they were [illegible] as before, under the supervision and direction of Stickles, [illegible] who was on the ground or in the caboose, and had [illegible] given up the direction or control. The duty of each was to [illegible] the brakes on the cars at his own end of the train, and to [illegible] quickly, in conjunction with the other, when it had [illegible] switch. They did not have the charge, and, except in [illegible] was, and was, as we think, not meant, they did not [illegible]. We think, also, whether, quoad this train, Mozier, the [illegible]

[illegible] to show that after he had given the [illegible] anything further to do with the train. But, even [illegible] change or control," we see no evidence of negligence on

count was for a defect in the ways, works, and machinery, alleged, "consisted in an improper and inefficient method rs to be switched upon the track where said Caron was at me while said track was in use by another train, and in cars to be so switched without any lights or other signals persons while on said track." We are of opinion that, ethod adopted was a dangerous one, the plaintiff's inteseld to have taken the risk of it. There is nothing to ad not been customary to switch cars onto tracks already out warnings or signals save such as would naturally be other by those engaged in the work, during the whole plaintiff's intestate had been working in the yard, which, testified, was two years and four months. It does not ny change had been made in the mode of doing the businake it more dangerous after he entered the defendant's law is well settled that under such circumstances the es the risk of such dangers as ordinarily are connected ce in which he is engaged. He enters the business as it ot be heard to complain that it might have been made it was conducted in a hazardous manner.

---

s' LIABILITY—RAILROAD COMPANIES.—Section 2701 of Statutes of 1894 of Minnesota, known as the "fellow provides that "every railroad corporation owning or ailroad in this State shall be liable for all damage susagent or servant thereof by reason of the negligence of nt or servant thereof, without contributory negligence hen sustained within this State, and no contract, rule, between such corporation and any agent or servant shall inish such liability."

e court of Minnesota decided on December 9, 1895, in arl Mikkelson v. William H. Truesdale, receiver of the nd St. Louis Railway Company (vol. 65, Northwestern 60), that Mikkelson, who was a wiper in the defendant's d was injured while assisting in coaling an engine by its tly moved, as he claimed, by a coemployee, was injured xposure to the hazards peculiar to the operation of railat a receiver operating a railroad under the appointment of a court of equity is within the provisions of the [illegible] act," and is liable to an employee who is injured by [illegible] a coemployee.

In deciding as to the liability of the receiver under the statute, against which the defendant contended, Chief Justice [illegible], who delivered the opinion of the court, said:

It is true that the word "receiver" is not used in this statute, and that its language is, "every railroad corporation owning or operating a railroad;" but the statute is a police regulation intended to protect life, person, and property, by securing a more careful selection of servants and a more rigid enforcement of their duties by railroad companies, by making them pecuniarily responsible to those of their servants who are injured by the negligence of incompetent or careless fellow-servants. It is remedial in its nature, and must be construed, if not liberally, certainly in accordance with its obvious purpose and spirit. It would be a most unreasonable construction of the statute, if we were to adopt the one claimed for it by the defendant. We are aware that able courts have adopted such a construction of similar statutes, but we are of the opinion that they have taken a narrow view of the statute, and we must decline to follow their conclusions.

If this police regulation does not apply to receivers of railroad corporations, it is difficult to see why such receivers are not absolved from a compliance with each and all of the police regulations made applicable by statute to railroad corporations. Can it be true that a railroad corporation whose road is operated for it by a general manager is, and one whose road is managed for it by a receiver is not, subject to the police regulations of the State? or that the employees of the one have, and those of the other have not, a remedy, when injured by the negligence of a fellow-servant? or that the employees of a corporation whose road is operated by a general manager to-day have such remedy, but if injured to-morrow they will have it not, because a receiver has taken the place of the manager? It would seem that an affirmative answer must be given to these questions, if we held that this statute has no application to receivers of railway corporations. Manifestly, such is not the fair and reasonable construction to be given to the statute. It is only in a technical sense that a receiver manages a railroad for the court appointing him. He operates it subject to the direction of the court, not for its benefit, but for the owners of the road, the corporation and its creditors. In doing so he necessarily exercises the franchises, rights, and powers of the corporation, and discharges its functions as a common carrier, and appropriates the income received from the operation of the road for the benefit of the corporation; and it logically follows that in so operating the road the receiver stands, in respect to duty and liability, just where the corporation would if it was operating the road. A distinction in this respect has been made between the common-law duties and liabilities of the corporation and those imposed on it by statute, but there can be no distinction in principle, for wherein is the duty and liability more imperative or sacred in the one case than in the other? A receiver can not, while exercising the franchises and powers of a corporation, claim immunity from the police regulations and liabilities which have been imposed upon the corporation by the State.

These general considerations lead to the conclusion [illegible] sions of the fellow-servant statute, here in question, [illegible] operating a railroad under the appointment and direction [illegible]

[illegible], page 217, of the Montana Compiled Statutes [illegible] "That in every case the liability of the cor[illegible] employee acting under the orders of his supe[illegible] case of injury sustained by default or wrongful [illegible] or to an employee not appointed or controlled by [illegible] or employee were a passenger." The same pro[illegible] the Montana Codes and Statutes, Sanders Edition, [illegible] section 905.

[illegible] above quoted the supreme court of Montana [illegible] 25, 1895, in the case of Crisswell *v.* Montana [illegible] Company (Pacific Reporter, vol. 42, p. 767), that both [illegible] engineer of a train are the superiors of a brakeman [illegible] and that the railroad company is liable for injuries [illegible]ch brakeman in a collision caused by the negligence of [illegible] running his train into the depot yard at night with[illegible] headlight, or without sending a flagman to see if the [illegible]r, as required by the rules of the company.

[illegible] brought before the supreme court on appeal by the [illegible]y from the judgment of the district court of Cascade [illegible] of the plaintiff, Crisswell, who had been injured while [illegible] the company as a brakeman, under the circumstances

[illegible]ied by the defendant's counsel that the railway corpora[illegible] common law, had performed its whole duty to the [illegible]mployee when it had used ordinary and reasonable care [illegible] safe machinery, (2) furnishing a safe place to work, and [illegible]ellow-servants to prosecute the common employment; [illegible]tute in question does not increase or change the defend[illegible] common law; that it does not change the common law [illegible]llow-servants; that it does not establish the superior[illegible] and enlarge the common-law liability of the defendant [illegible]and was not so intended by the legislature.

[illegible] Pemberton, in delivering the opinion of the supreme [illegible]t considerable length from the opinion of the United [illegible]ourt for the northern district of Iowa, in the case of [illegible]lroad Co. (42 Fed., 383), and from the opinion of the [illegible] circuit court of appeals, eighth circuit, in the case of [illegible]y *v.* Mase (63 Fed., 114), in both of which cases the [illegible] the statute in question was directly involved. In [illegible] opinion in the Ragsdale case the court said, "Under [illegible] corporation is made liable to any one of its employees [illegible]gence on his part, is injured by the default or wrong[illegible] [illegible], even though the latter has no control over the [illegible] Mase case, "It goes without saying that the purpose of this statute was to extend the liability of railroad companies to their servants for the negligence of servants of a higher grade;" also, "The effect of the statute is to give a cause of action against the railroad company to every servant who is himself without fault, for the default or wrongful act of any superior servant, whether or not the latter appointed or exercised any control over the former before or at the time of the infliction of the injury."

After quoting from the cases above referred to Chief Justice Pemberton said:

> We think from the interpretation given to the statute in question by the above authorities that it can not be doubted that the common-law rule contended for by the defendant was modified and changed thereby, and that such was the intention of that legislation. And it is no less plain that this statute establishes the principle that there is a difference in the grade of the employees engaged in a common employment, and gives a right of action to a servant, injured through the negligence of a superior employee or servant, against a master, when such injured servant is without fault or negligence on his part. In view of the extent to which the common-law rule has been carried, the enactment of such legislation is not surprising, nor are we prepared to reprobate the wisdom of the policy of establishing a legislative rule that relaxes the rigor of the common law in such cases.

---

EMPLOYERS' LIABILITY—RAILROAD COMPANIES.—In the case of Missouri, Kansas and Texas Railway Company *v.* Whittaker (vol. 33, Southwestern Reporter, p. 716) the court of civil appeals of Texas, on November 23, 1895, reversed the judgment of the district court of Grayson County, by which Whittaker, a boiler washer in the company's employ, had been awarded $1,000 damages for injuries sustained by him through the negligence of a hostler employed by the company.

The court of civil appeals held, in effect, that under section 2 of chapter 24 of the General Laws of 1891 of Texas, relating to fellow-servants, the provisions of which were reenacted in section 2 of chapter 91 of the General Laws of 1893, a hostler, whose duty it is to bring the engines into the roundhouse and take them out when necessary, and a boiler washer, whose duty it is to clean the boilers of the engine so as to fit them for further service, both being under the orders of the roundhouse foreman and without authority over each other, are, as a matter of law, fellow-servants, and, hence, that one of them was not entitled to recover damages from their common employer, the railroad company, for injuries sustained through the negligence of the other.

Judge Finley, in delivering the opinion of the court of civil appeals, said:

> The evidence, so far as it touches the relation of plaintiff [illegible] hostler who was in charge of the engine and whose negligence [illegible] alleged, caused the injury, to the railway company and to [illegible] and as to the material circumstances under which the injury [illegible] is entirely without conflict.

ndhouse as a boiler washer, and
of engines brought in there to be
on the road. The hostler was
duty was to bring the engines into
gain when they were sent out for
l by, and worked under the super-
ouse. Neither of them had any
intendence over the other in his
in different capacities, about the
engines at the roundhouse when

reads as follows: "That all per-
service of such railway corpora-
e working together at the same
e, of same grade, neither of such
orations with any superintendence
es, are fellow-servants with each
contained shall be so construed
ation, in the service of such cor-
employees of such corporation,
or service of such corporation.
he provisions of this section shall

ve the same master, labor under
rpose, and derive their authority
eneral source, are fellow-servants,
or labor in different and distinct
now fixes the relation of fellow-
ly between those who serve the
are working together at the same
se, in the same department, and
endence or control over his fellow-
rades, or different departments of
power to superintend or control
e statute, the relation of fellow-

tion, unquestionably, the plaintiff
ne master, at the same time and
the same grade, and neither had
w-employees. Were they in the
ere both employed by the foreman
e same special control, their duties
vice at the same time, and their
they were not in actual service
the engines into the roundhouse
y. The plaintiff cleaned out the
rther service. They were clearly
cessary that they should be doing
getting the same compensation
grade or to be employees in the

the statute, and left the jury to
ether the plaintiff and the hostler
dence clearly and without conflict
to be that of fellow-servants, the
atter of law, that they are such.

EMPLOYERS' LIABILITY—RAILROAD COMPANIES—The supreme court of Wisconsin decided, on November [illegible], [illegible], in the case of Smith *v.* Chicago, Milwaukee and St. Paul Railway Company, that, under chapter 220 of the laws of 1893 of Wisconsin, a railway company is not liable for injuries to one of its car repairers caused by a switchman negligently running a train into the stationary car in which the repairer was at work.

The statute referred to provides that "Every railway company operating any railroad or railway, the line of which shall be in whole or in part within this State, shall be liable for all damages sustained within this State by any employee of such company, without contributory negligence on his part, * * * while any such employee is so engaged in operating, running, riding upon or switching, passenger or freight or other trains, engines or cars, and while engaged in the performance of his duties as such employee, and which such injury shall have been caused by the carelessness or negligence of any other employee, officer or agent of such company in the discharge of, or for failure to discharge his duties as such."

The opinion of the supreme court in this case, published in volume 65 of the Northwestern Reporter, page 183, was delivered by Judge Marshall, who in the course of it said:

It has been too long and too well settled that persons working for the same employer, bearing such relations to each other and to the business they are jointly engaged in for such employer as a switchman and a car repairer, are fellow-servants, and that the master is not liable for an injury to one through the negligence of the other, unless made so by statute, to need any discussion of the subject here.

Therefore, if there be any liability of defendant to plaintiff, it is under chapter 220, laws of 1893. The words [in the statute referred to] "while engaged in the performance of his duties as such employee," refer to the words, "while operating, running, riding upon or switching, passenger or freight or other trains, engines or cars." This, we think, is very clear. It is a familiar principle that statutes in derogation of the common law should be strictly construed, and not given any effect beyond the plain legislative intent; but whether the statute under consideration is tested by that rule, or by the more liberal one applicable to purely remedial laws, the result must be the same, for the legislative intent must govern, and that is to be determined "by considering the entire statute, looking at every part, having regard to the legislative idea or purpose of the whole instrument." The legislative idea of that part of the law under consideration plainly is to give a right of action to the class of employees engaged in operating [illegible] trains, engines, and cars, while actually so engaged; and [illegible] used to express such idea are too plain to leave any [illegible] to the rules for judicial construction to determine [illegible]. Plaintiff was not an employee engaged in the branch of [illegible] vice, when injured, entitled to the benefits of the [illegible].

[illegible] COMMON LAW.

[illegible] the United States circuit court, southern dis- [illegible] January 21, 1898, in the case of W. E. Willett v. [illegible] and Indian River Railroad Company, a jury [illegible] the sum of $1,750 as damages against the railroad [illegible] blacklisting him, which amount, together with the costs in [illegible] the company paid.

[illegible] furnished the Department of Labor by the clerk [illegible] in which the case was tried, the facts upon which the [illegible] appear to have been as follows: Willett, while employed [illegible] by the defendant company, sought employment on another [illegible] the Savannah, Florida and Western (formerly the South [illegible]. He was notified that employment would be given him, and [illegible] to report for duty immediately, and passes were sent him to [illegible] to go over the road of the Savannah, Florida and Western [illegible] Company and learn the route before entering regularly upon [illegible] of his new position. He at once telegraphed to the proper [illegible] of the Jacksonville, St. Johns and Indian River Railroad Com- [illegible] to be relieved from duty at a certain station, but was [illegible] by the company to remain in its employ and take out another [illegible] He finished the run he was then making and made the return [illegible] the official that he would leave the employ of the [illegible] upon arrival at its terminus, which he did, and proceeded to [illegible] the line of the Savannah, Florida and Western Railroad Com- [illegible] to learn the route.

[illegible] he had finished the preliminary trip he received a telegram [illegible] the officers of the last-named company directing him to "come [illegible]" He complied with this order, and upon returning was informed [illegible] would not be employed. He subsequently ascertained the rea- [illegible] refusal to employ him to have been that the superintendent [illegible] company whose service he had left, had written a letter to the [illegible] of the company whose service he was about to enter, [illegible] him against Willett, who, the letter stated, had left their [illegible] with certain charges pending against him.

[illegible] principal defense of the railroad company was that the letter [illegible] one, and not written officially, but this defense was of [illegible], as before stated, Mr. Willett successfully prosecuted his [illegible] the company whose superintendent had prevented his [illegible] by another company.

[illegible] Department of Labor is advised, this is the first case of [illegible] in the United States in which an award of damages [illegible] and paid for blacklisting.

EMPLOYERS' LIABILITY.—The supreme judicial court of setts decided on November 29, 1895, in the case of O'Conn that a master, having furnished suitable material, is not resp injuries to a servant resulting from the negligence of anoth in putting a defective plank into a scaffold, though the s erected and the defective plank placed therein before t servant entered the master's service.

The facts in the case are stated in the opinion of the court delivered by Judge Knowlton. Said decision is published 42 of the Northeastern Reporter, page 111, and is as follows

The plaintiff fell and was injured by reason of the breaking in a temporary staging on which he was working in the d building. It is not disputed that the staging was of a kir struction of which is ordinarily left to the servants of the b that the duty of the master concerning it was performed if h a sufficient supply of suitable materials from which to constr this case there was uncontradicted evidence that there wer planks furnished by the defendant from which to build the st the negligence, if there was any, was on the part of the workm the planks in place, in taking one which was not adapted use. Upon these facts, if the plaintiff had been in the d service at the time when the staging was built, it would be that he could not maintain his claim.

But it appears that, although he had previously worked fo erable time upon the building, he was away working for anot four days before the day of the accident, and this staging w a day or two before his last engagement in the defendan began. Under these circumstances, the question is whether t ant is liable to him for the previous negligence of a servan work which may properly be intrusted to servants. We are that an employer, under such circumstances, owes one who i enter his service no duty to inspect all the work which has by his servants previously, and which ordinarily may be in them, without liability to their fellow-servants for their negli he owes no such duty, the risk of accident from previous ne servants in their own field is one of the ordinary risks of th which the employee assumes by virtue of his contract on en service.

---

EMPLOYERS' LIABILITY — RAILROAD COMPANIES.— court of appeals of West Virginia, by decision of Nov reversed the judgment of the circuit court of Braxton case of Skidmore v. West Virginia and Pittsburg Rail by which $6,000 had been awarded Skidmore as dam sustained by him while in the performance of his du in the employ of the railroad company.

The opinion of the supreme court of appeals was English and is published in volume 23 of the Sou page 713. The syllabus of said opinion, prepared

[illegible] to move a tender which lies on its side near [illegible] said track, and said tender inclines a little [illegible] but nothing in its appearance indicates that [illegible] had been broken loose from the body, and [illegible] supervisor could by ordinary diligence dis[illegible], or any separation from the main body, and [illegible] are engaged in moving such tender the bottom [illegible] section hand, he is not entitled to recover from [illegible] damages for the injury so received.

[illegible] and his assistants have equal knowledge of the [illegible] an act about to be done, even if the foreman [illegible], and injury ensues to the assistant, the [illegible] liable. Notwithstanding the request, the [illegible] or not, as he chooses, and if he does comply, he [illegible] of the peril surrounding the situation.

[illegible] the servant is ignorant of the impending danger, [illegible] is not, and the employer fails to warn the servant of [illegible] the master's liability attaches.

[illegible] enters upon a service he assumes to understand it and [illegible]dinary risks that are incident to the employment; and [illegible]loyment presents special features of danger, such as are [illegible], he also assumes the risk of those.

[illegible] danger consists in some latent defect, which is not [illegible] use of ordinary diligence on the part of the master, [illegible] performing his work is thereby injured, when he had [illegible] of observation as the master, no liability attaches to

---

[illegible]NSPIRACY, LABOR COMBINATIONS NOT UNLAWFUL.—[illegible]ourt of Greene County, Ind., Benjamin F. Watson recov[illegible] in an action against Thomas Clemitt and others for [illegible]iven from his employment as workman in a coal mine by [illegible]ngful conspiracy among other workmen in the mine, who [illegible]ch other not to work with him and to quit work unless [illegible]ged, pursuant to which they did quit work upon their [illegible]sing to discharge Watson, by reason whereof the busi[illegible]nded and he was thrown out of work.

[illegible]nts appealed to the appellate court of Indiana, which [illegible] the decision of the circuit court and held that such [illegible] workmen is not actionable in the absence of mal[illegible] violence, or evidence that they were bound to con[illegible] the employer was obliged to retain the plaintiff in

[illegible] appellate court, delivered by Chief Justice Gavin

December 10, 1895, is published in full in volume 42 of the Northeastern Reporter, page 367. In the course of the opinion it is said:

While it is true that, under all civilized forms of government, every man surrenders for the general good a certain amount of that absolute freedom of action which may adhere to the individual in an independent or natural state, yet, under our institutions, it is a cardinal principle that each man retains the greatest freedom of action compatible with the general welfare. The right to control his own labor, and to bestow or withhold it where he will, belongs to every man. Even though he be under contract to render services, the courts will not interfere to compel him to specifically perform them. (Arthur v. Oakes, 11 C. C. A., 209; 63 Fed., 310.)

So far as appears by these instructions [of the circuit court to the jury] none of the appellants were under any continuing contract to labor for their employer. Each one could have quit without incurring any civil liability to him. What each one could rightfully do certainly all could do if they so desired, especially when their concerted action was taken peaceably, without any threats, violence, or attempts at intimidation. There is no law to compel one man or any body of men to work for or with another who is personally obnoxious to them. If they can not be by law compelled to work, I am wholly unable to see how they can incur any personal liability by simply ceasing to do that which they have not agreed to do, and for the performance of which they are under no obligation whatever.

Under our law every workman assumes many risks arising from the incompetency or negligence of his fellow-workmen. It would be an anomalous doctrine to hold that after his fellows have concluded that he was not a safe or even a desirable companion they must continue to work with him under the penalty of paying damages if by their refusal to do so the works are for a time stopped and he thrown out of employment. We can not believe it to be in accordance with the spirit of our institutions or the law of the land to say that a body of workmen must respond in damages because they, without malice or any evil motive, peaceably and quietly quit work which they are not required to continue rather than remain at work with one who is for any reason unsatisfactory to them. To so hold would be subversive of their natural and legal rights, and tend to place them in a condition of involuntary servitude.

---

Strikes, Intimidation, etc.—Right of Court to Interfere by Injunction.—The supreme court of Missouri, on November 26, 1895, affirmed the judgment of the circuit court of the city of St. Louis in the case of Hamilton-Brown Shoe Company *v.* Saxey et al., and adopted as its own the opinion delivered in said case by Hon. L. B. Valliant, one of the judges of the circuit court, by which A. J. Saxey and others were prevented by injunction from attempting by intimidation, threats of personal violence, and other unlawful means to force the employees of the Hamilton-Brown Shoe Company to quit work and join in a strike.

The opinion in this case is published in volume 32 of the Southwestern Reporter, page 1106, and sufficient portions of it are here reproduced to give a clear understanding of the facts in the case and the reasoning upon which the judgment of the court was based:

The amended petition states in substance that the plaintiff conducts a large shoe manufactory in this city, and has in its employ some eight or nine hundred persons, all of whom are earning their living in plaintiff's employment, and are desirous of so continuing; that the defendants, except two of them, were lately in plaintiff's employ, but have gone out of the same, on a strike, and are now, with the other two defendants, engaged in an attempt to force the other employees of plaintiff to quit their work and join in the strike, and that to accomplish this purpose they are intimidating them with threats of personal violence; that among the plaintiff's employees who are thus threatened are about three hundred women and girls and two or three hundred other young persons; that the effect of all this on the plaintiff's business, if the defendants are allowed to proceed, would be to inflict incalculable damage.

The defendants have appeared by their counsel, and, by their demurrer filed, admit that all the statements of the amended petition are true; but they take the position that, even if they are doing the unlawful acts that they are charged with doing, still this court has no right to interfere with them, because they say that what they are doing is a crime, by the State law of this State, and that for the commission of a crime they can only be tried by a jury in a court having criminal jurisdiction. It will be observed that the defendants do not claim to have the right to do what the injunction forbids them doing. Their learned counsel even quotes the statute to show that it is a crime to do so. But he contends that the Constitution of the United States and the constitution of the State of Missouri guarantee them the right to commit crime, with only this limitation, to wit, that they shall answer for the crime, when committed, in a criminal court, before a jury, and that to restrain them from committing crime is to rob them of their constitutional right of trial by jury. If that position be correct, then there can be no valid statute to prevent crime. But that position is contrary to all reason. The right of trial by jury does not arise until the party is accused of having already committed the crime. If you see a man advancing upon another with murderous demeanor and a deadly weapon, and you arrest him—disarm him—you have perhaps prevented an act which would have brought about a trial by jury, but can you be said to have deprived him of his constitutional right of trial by jury? The train of thought put in motion by the argument of the learned counsel for defendants on this point leads only to this end, to wit, that the Constitution guarantees to every man the right to commit crime, so that he may enjoy the inestimable right of trial by jury.

Passing now to the question relating to the particular jurisdiction of a court of equity, we are brought to face the proposition that a court of equity has no criminal jurisdiction, and will not interfere by injunction to prevent the commission of a crime. These two propositions are firmly established; and as to the first, that a court of equity has no criminal jurisdiction, there is no exception. As to the second, that a court of equity will not interfere by injunction to prevent the commission of a crime, that, too, is perhaps without exception, when properly interpreted, but is sometimes misinterpreted. When we say that a court of equity will interfere to prevent that injury, notwithstanding the act may also be a violation of a criminal law. In such case the court does not interfere to prevent the commission of a crime, although that may incidentally result, but it exerts its force to protect the individual's property from destruction, and ignores entirely the criminal portion of the act. There can be no doubt of the jurisdiction of a court of equity in such a case.

Equity will not interfere when there is an adequate remedy at law. But what remedy does the law afford that would be adequate to the plaintiffs' injury? How would their damages be estimated? How compensated? The defendants' learned counsel cites us to the criminal statute, but how will that remedy the plaintiffs' injury? A criminal prosecution does not propose to remedy a private wrong.

What a humiliating thought it would be if these defendants were really attempting to do what the amended petition charges, and what their demurrer confesses—that is, to destroy the business of these plaintiffs, and to force the eight or nine hundred men, women, boys, and girls, who are earning their livings in the plaintiffs' employ, to quit their work against their will—and yet there is no law in the land to protect them. The injunction in this case does not hinder the defendants doing anything that they claim they have a right to do. They are free men, and have a right to quit the employ of the plaintiffs whenever they see fit to do so, and no one can prevent them; and whether their act of quitting is wise or unwise, just or unjust, it is nobody's business but their own. And they have a right to use fair persuasion to induce others to join them in their quitting. But when fair persuasion is exhausted they have no right to resort to force or threats of violence. The law will protect their freedom and their rights, but it will not permit them to destroy the freedom and rights of others. The same law which guarantees the defendants in their right to quit the employment of the plaintiffs at their own will and pleasure also guarantees the other employees the right to remain at their will and pleasure. These defendants are their own masters, but they are not the masters of the other employees, and not only are they not the masters of the other employees, but they are not even their guardians. There is a maxim of our law to the effect that one may exercise his own right as he pleases, provided that he does not thereby prevent another exercising his right as he pleases. This maxim or rule of law comes nearer than any other rule in our law to the golden rule of Divine authority: "That which you would have another do unto you, do you even so unto them." Whilst the strict enforcement of the golden rule is beyond the mandate of a human tribunal, yet courts of equity, by injunction, do restrain men who are so disposed from so exercising their own rights as to destroy the rights of others.

o

# BULLETIN

OF THE

# DEPARTMENT OF LABOR.

---

No. 5—JULY, 1896.

ISSUED EVERY OTHER MONTH.

---

EDITED BY

CARROLL D. WRIGHT,
COMMISSIONER.

OREN W. WEAVER,
CHIEF CLERK.

---

WASHINGTON:
GOVERNMENT PRINTING OFFICE.
1896.

# CONTENTS.

BULLETIN

OF THE

DEPARTMENT OF LABOR.

No. 5. - WASHINGTON. JULY, 1896.

## CONVICT LABOR.

In February, 1887, the Commissioner of Labor published a report on the subject of convict labor in the United States, known as the Second Annual Report of the Commissioner of Labor. The investigation, the results of which were published at that time, was made in accordance with a joint resolution of Congress approved August 2, 1886, authorizing and directing the Commissioner of Labor to make a full investigation as to the kind and amount of work performed in the penal institutions of the various States and Territories of the United States and the District of Columbia, as to the methods under which convicts were employed, and as to all the facts pertaining to convict labor and the influence of the same upon the industries of the country. The inquiry, conducted in accordance with the Congressional authorization, and the results given in a series of sixteen general tables, with an analysis. The results of State investigations previously made were also given, and there was a discussion of the advantages and disadvantages of various systems and plans for the employment of convicts. That report also contained extensive historical notes giving the leading facts of the employment of convicts among the early nations and of the plans and means of utilizing convict labor in recent times in various countries of the world. It also contained all the convict labor laws then in force in the various States and Territories. The present report is confined to convict labor in institutions of the grade of State penitentiaries and an abstract of the various laws relating to convict labor enacted in the different States and Territories since the report of 1885–86.

In the report of ten years ago it was found that the total value of all products of, and work performed in, all the penal institutions of all grades was $28,753,999, while the value of goods produced and work done at that time in the penal institutions now considered was $24,271,078. It will thus be seen that all other institutions, or those of a lower grade than the State penitentiary, were then of little consequence. For this reason the present investigation has been confined to the high-grade institutions.

In 1885 there were four plans or systems followed in the employment of convicts, and they are described in brief as—

1. The contract system, under which a contractor employs convicts at a certain agreed price per day for their labor, the prisoners working under the immediate direction of the contractor or his agents. Under this system the institutions usually furnish to the contractor the power necessary and even the machinery for carrying on the work.

2. The piece-price system, which is simply a modification of the contract system. Under this system the contractor furnishes to the prison the materials in a proper shape for working, and receives from the prison the manufactured articles at an agreed piece price, the supervision of the work being wholly in the hands of the prison officials.

3. The public-account system, under which the institution carries on the business of manufacturing like a private individual or firm, buying raw materials and converting them into manufactured articles, which are sold in the best available market.

4. The lease system, under which the institution leases the convicts to a contractor for a specified sum and for a fixed period, the lessee usually undertaking to clothe, feed, care for, and maintain proper discipline among the prisoners while they perform such labor as may have been determined by the terms of the lease.

These four systems are still followed, with some modifi[illegible] and there, especially in a few States where, by law, con[illegible] employed in the production of things used by the State [illegible] institutions—penal, charitable, etc.

The results of the present investigation are presented [illegible] and a section giving the abstracts of the various [illegible]

[illegible] are so brief and comprehensive in themselves that [illegible] but little analysis.

Table I shows the fiscal years in each State and for each institution for which the facts have been reported.

Table II shows the systems of work in vogue, both in 1885 and in 1895, in each institution comprehended in this report. By examining the last two columns the changes in system are easily seen. It will be noticed that the changes are chiefly from contract to public account or piece price, and from lease to contract or public account. There have not been many changes, however.

Table III states the number of convicts in prisons and penitentiaries in 1895, by systems of work, designating them under those employed in productive labor, those engaged in prison duties, and those who are idle and sick, and the total number, classifying them by sex, under each of these designations. Out of this table is drawn Table IV, which gives a more interesting and important statement.

Table IV consolidates the number in each State under all systems, both for 1885 and 1895. Looking at the total line in Table IV, it is seen that the number of convicts in the prisons of the grade under consideration in 1885 was 41,877, while for 1895 the number rose to 54,244. It is interesting to observe that, in 1885, of these numbers 1,967 were females, while the number of females in 1895 was 1,988, an increase of only 21. In 1885 the number engaged in productive labor was 30,853, being 73.7 per cent of the total number of convicts at that time, while in 1895 the number engaged in productive labor was 38,415, or 70.8 per cent of the total number of convicts. While there was thus a decrease in the proportion of convicts employed in productive labor, it will be seen that there was also a decrease in the proportion of those engaged in prison duties, for in 1885 the total so engaged was 8,391, or 20 per cent of the whole number of convicts, while in 1895 there were 8,804 so engaged, being 16.2 per cent of the whole number. When we come to the idle and sick, an increase in the proportion in the ten years is observed. In 1885 the number was 2,633, or 6.3 per cent of the whole, while in 1895 the number of idle and sick was 7,025, being 13 per cent of the whole. Undoubtedly this proportional increase is in the number of idle, this state of affairs being brought about by changes in the laws relating to the employment of convicts.

Table V is an analytical table giving the kind of goods produced or work done, both in 1885 and 1895, and the value thereof, for each State and for each system. An examination of this table shows the shifting of the industries in the State prisons of the country during the ten years.

Table VI shows the value of goods produced or work done, by systems of work, for the two years under consideration, but the values [illegible] for each State for the various prisons covered by the

Table VII summarizes these totals for the [illegible]
is reproduced here as a part of this text [illegible]

SUMMARY OF VALUE OF GOODS PRODUCED OR WORK DONE [illegible]
WORK, 1885 AND [illegible].

| Systems of work. | [illegible] |
|---|---|
| Public-account system | [illegible] |
| Contract system | 17,071,[illegible] |
| Piece-price system | 1,[illegible] |
| Lease system | [illegible] |
| Total | 24,271,[illegible] |

By this brief table one can note the general changes in v
the public-account system there were produced in the Un
1885 goods to the value of $2,063,892.18, but under this s
there were produced goods to the value of $4,868,563
increase of more than 100 per cent. This system has
popular in recent years, hence the increase. Looking at th
find that under the contract system there has been a decr
50 per cent., the decrease being from $17,071,265.69 to
This system (the contract) has become offensive during
years, and legislatures have sought to change their pl
either to the public-account system or to the piece-price s
the latter the value of goods has increased from $1,484,
to $3,795,483.24 in 1895, an increase of over 150 per cent
lease system the values show the effect of agitation in
States, where that system more generally prevailed in 1
the value of goods produced or work done decreased fr
to $2,167,626.03 in 1895, a decrease of 40.6 per cent.
show a great change, the decrease being from $24,271,67
$19,042,472.33 in 1895, a decrease of 21.5 per cent.

Table VIII summarizes the value of goods produced or
States, without regard to system, and this table shows
communities there has been an increase or decrease in
products in the various State prisons. There has been
16 States and Territories and a decrease in 25 States an
The increase occurs in the following-named States
Alabama, Connecticut, Florida, Maryland, Massach
Missouri, New Hampshire, New Mexico, Rhode Island,
Texas, Vermont, Virginia, Washington, and Wisc
occurs in the following-named States and Territories:
sas, California, Colorado, Georgia, Illinois, Indiana,
Kentucky, Louisiana, Maine, Michigan, Mississi
New Jersey, New York, North Carolina, Ohio,
South Dakota, Tennessee, and West Virgi
no influence arising from geographical [illegible]
decrease. Some effort has been made [illegible]

task performed, and the results are fairly
is ascertained that the prisoners are not
but are employed in prison improvements
been a change of system, which works a
een lowered. In California the convicts
ging the prison and in cultivating a farm
s at the State Prison at Folsom. At the
here has been a discontinuance, by act of
hich in any way entered into competition
there has been a discontinuance, by act
es which entered into competition with
decrease in Georgia has been very large,
on thereof has been given. In the State
contractor threw up his contract in 1893,
. There is a small force in the shoe fac-
s been removed from the prison; and no
made. At the Southern Penitentiary of
ange of system from contract to public
generally results in a loss in the value of
ison (north) in Indiana contractors refused
knitting factory was closed, and the chair
half time. At the State Prison (south) the
ew their contracts. At the State Reform-
the principal industry has been almost
found impossible to compete with those
of the prison by machinery. The small
he State Penitentiary at Anamosa by rea-
zed for goods made, and by the industrial
n Kansas the principal industry was dis-
ion in business. The decrease in Ken-
re of the contractors in 1893, since which
employed in but two industries. In
n account of the low selling value of the
r production and decrease in the selling
ie loss in the State of Maine. In Michi-
causes affecting the State Prison, where
employed in 1895, 200 able convicts being
issippi there was a change from the lease
in 1894, but no positive explanation is
ld have resulted in a loss of $80,000 in
ska the reason given is the depression in
of leasing convicts in several industries,
gricultural implements and clothing, and
ed on before. In Nevada the convicts
ts and repairs of the prison. The manu-
ued, and there was no demand for stone.
is probably owing to a State law restrict-

ing the number of convicts to 100 in any one industry, to smaller production, and to a decrease in the selling value of goods. This was at the State Prison, while at the Essex County Penitentiary the apparent decrease was owing to an error in the report made to the Department in 1885. The decrease in New York of over $2,890,000 in the value of goods produced and work done, was largely owing to changes in legislation. At the Auburn Prison there was no decrease. At the Sing Sing State Prison a change of system, a lowering of the selling value of goods, a less number of convicts employed, and the legislation of 1889, restricting the number of convicts in any one industry, account for the decrease at that point, and the same causes operated at the Clinton Prison, convicts there being employed in less profitable industries. A change in the system of employment accounts for the decrease at the State Reformatory, while at the Albany County Penitentiary a change in the nature of the product—it being from shoes to less profitable industries—satisfies the inquiry there. The decrease at the Erie County Penitentiary and the Kings County Penitentiary resulted from legislation restricting the number of convicts in any one industry, and from a change of system from contract to piece price, while at the Monroe County and Onondaga County Penitentiaries a change of system from contract to public account was responsible. The decrease of about $70,000 in the value of products at the State Penitentiary of North Carolina has not been explained. At the Ohio Penitentiary the decrease came from a lowering of the price of goods and legislation requiring the prison-made goods to be stamped. At the Oregon State Penitentiary the decrease was on account of the general decline in business. In Pennsylvania, at the Eastern Penitentiary, the discontinuance of the hosiery industry because of inventions outside, and a smaller number of convicts employed, account for the loss there; while at the Western Penitentiary the decrease was owing to a change of system from contract to public account, the convicts not being steadily employed. The other prisons of State grade in Pennsylvania which show a loss are those for the counties of Berks, Chester, Delaware, Northumberland, Philadelphia, and Schuylkill. In these, smaller production and a decline in the selling value of goods account for the decrease in the value of products, although in Philadelphia County, in addition to these reasons, there has been a change in the quality of goods made, and on account of dullness of trade the convicts have frequently been idle. The figures for the State of South Dakota for 1895 are shown in connection with those for 1885 for Dakota Territory. In Tennessee the decrease in value of production resulted from smaller output and a decline in selling values. The decrease in West Virginia of about [illegible] explained.

A study of the abstracts of legislation since [illegible] in the different States and Territories, which [illegible] Table VIII, will give a clearer insight [illegible] changes referred to above.

1895, was $19,042,472.33. The same difficulty occurs here that occurs in census work in stating the value of products. The sum just given represents the value of the goods after the materials have been manipulated by the convicts. It does not represent the value which has been added by them, but represents the value of the materials on which work has been bestowed and the work itself. Under the public-account system prison officials purchase the materials and convert them, by the labor of convicts, into goods, which they sell. The value given includes, therefore, the value of the materials purchased plus the labor expended upon them by the prisoners. In purchasing materials the prison officials in many instances purchase the finished products of manufacturers outside of prisons. They purchase everything that is essential for conversion into certain articles. So the contractors working under the contract system take to the prisons the materials, which there, by the labor of the convicts, they convert into finished goods. The value reported to the Department, therefore, covers the value of these materials, whether the finished products of some other manufacturers or purely raw materials plus the labor expended upon them. These explanations, if clearly understood, will prevent any misunderstanding on the part of the reader of this report, and they are equally applicable to the values reported for 1885. The $19,042,472.33 does not represent the labor of the convicts themselves. In 1885 the total wages paid by contractors and lessees to States and counties for the labor of convicts, from which resulted a product of the value of $28,753,999, was only $3,512,970, or $1 of convict-labor wages to $8.19 of finished product of convict labor. There is reason to believe that the ratio at the present time is less than that for 1885. At the present time in all probability the total value of the labor expended by the convicts in the State penitentiaries and prisons of the country, considered in this report, does not exceed $2,500,000.

TABLE I.—DATE OF [illegible]

[The returns made for what has been termed, for brevity's sake, [illegible] really for the dates shown herewith.]

| State. | Locality. | Fiscal year ending— | State. | Locality. | Fiscal year ending— |
|---|---|---|---|---|---|
| Alabama | Wetumpka | Aug. 31, 1894 | New York | Sing Sing | Sept. [illegible] |
| Arizona | Yuma | Dec. 31, 1895 | New York | Dannemora | [illegible] |
| Arkansas | Little Rock | Oct. 31, 1894 | New York | Elmira | [illegible] |
| California | Folsom | June 30, 1895 | New York | Albany | Oct. 31, [illegible] |
| California | San Quentin | June 30, 1895 | New York | Buffalo | Sept. [illegible] |
| Colorado | Cañon City | Nov. 30, 1895 | New York | Brooklyn | July [illegible] |
| Connecticut | Wethersfield | Dec. 31, 1895 | New York | Rochester | Sept. [illegible] |
| Florida | Tallahassee | Dec. 31, 1895 | New York | Syracuse | Oct. [illegible] |
| Georgia | Atlanta | Oct. 1, 1895 | North Carolina | Raleigh | Dec. [illegible] |
| Illinois | Joliet | Sept. 30, 1895 | Ohio | Columbus | Oct. 31, [illegible] |
| Illinois | Chester | Sept. 30, 1895 | Oregon | Salem | Dec. [illegible] |
| Indiana | Michigan City | Oct. 31, 1895 | Pennsylvania | Philadelphia | Dec. 31, [illegible] |
| Indiana | Jeffersonville | Oct. 31, 1895 | Pennsylvania | Allegheny City | Dec. 31, [illegible] |
| Indiana | Indianapolis | Oct. 31, 1895 | Pennsylvania | Philadelphia | Dec. 31, [illegible] |
| Iowa | Fort Madison | June 30, 1895 | Pennsylvania | Reading | Dec. [illegible] |
| Iowa | Anamosa | June 30, 1895 | Pennsylvania | West Chester | Sept. [illegible] |
| Kansas | Lansing | June 30, 1895 | Pennsylvania | Media | Dec. 31, [illegible] |
| Kentucky | Frankfort | Nov. 30, 1895 | Pennsylvania | Lancaster | Nov. [illegible] |
| Louisiana | Baton Rouge | Feb. 28, 1895 | Pennsylvania | Allentown | Dec. 31, [illegible] |
| Maine | Thomaston | Nov. 30, 1895 | Pennsylvania | Norristown | Dec. 31, [illegible] |
| Maryland | Baltimore | Nov. 30, 1895 | Pennsylvania | Easton | Dec. 31, [illegible] |
| Massachusetts | Boston | Sept. 30, 1895 | Pennsylvania | Sunbury | Dec. 31, [illegible] |
| Massachusetts | Concord Junct'n | Sept. 30, 1895 | Pennsylvania | Philadelphia | Dec. 31, [illegible] |
| Massachusetts | Sherborn | Sept. 30, 1895 | Pennsylvania | Pottsville | Dec. [illegible] |
| Michigan | Jackson | June 30, 1895 | Rhode Island | Howard | Dec. 31, [illegible] |
| Michigan | Ionia | June 30, 1895 | South Carolina | Columbia | Oct. 31, [illegible] |
| Minnesota | Stillwater | July 31, 1895 | South Dakota | Sioux Falls | June 30, [illegible] |
| Mississippi | Jackson | Sept. 30, 1895 | Tennessee | Nashville | Dec. 31, [illegible] |
| Missouri | Jefferson City | Dec. 31, 1895 | Texas | Huntsville and Rusk. | Oct. 31, [illegible] |
| Nebraska | Lancaster | Dec. 31, 1895 | | | |
| Nevada | Carson City | Dec. 31, 1895 | Vermont | Windsor | June 30, [illegible] |
| New Hampshire | Concord | Apr. 30, 1896 | Vermont | Rutland | June [illegible] |
| New Jersey | Trenton | Oct. 31, 1895 | Virginia | Richmond | Sept. 30, [illegible] |
| New Jersey | Caldwell | Apr. 30, 1896 | Washington | Walla Walla | Sept. [illegible] |
| New Mexico | Santa Fe | Mar. 4, 1895 | West Virginia | Moundsville | Sept. [illegible] |
| New York | Auburn | Sept. 30, 1895 | Wisconsin | Waupun | Sept. [illegible] |

TABLE II.—SYSTEMS OF WORK IN 1885 AND 1895.

| State. | Institution. | Locality. | 1885. | 1895. |
|---|---|---|---|---|
| Alabama | State Penitentiary | Wetumpka | Lease | Lease and public account. |
| Arizona | Territorial Prison | Yuma | Public account | Public account. |
| Arkansas | State Penitentiary | Little Rock | Lease | Contract and public account. |
| California | State Prison | Folsom | Public account | Public account. |
| California | State Prison | San Quentin | Piece price and public account. | Public account. |
| Colorado | State Penitentiary | Cañon City | Public account | Public account. |
| Connecticut | State Prison | Wethersfield | Contract | Contract. |
| Florida | State Penitentiary | Tallahassee | Lease | Lease. |
| Georgia | State Penitentiary | Atlanta | Lease | Lease. |
| Illinois | State Penitentiary | Joliet | Contract | Contract and public account. |
| Illinois | Southern Penitentiary | Chester | Contract | Public account. |
| Indiana | State Prison (north) | Michigan City | Contract | Contract. |
| Indiana | State Prison (south) | Jeffersonville | Contract | Contract. |
| Indiana | State Reformatory for Women. | Indianapolis | Piece price | Piece price. |
| Iowa | State Penitentiary | Fort Madison | Contract | Contract. |
| Iowa | State Penitentiary | Anamosa | Public account | Public account. |
| Kansas | State Penitentiary | Lansing | Public account and contract. | Public account and contract. |
| Kentucky | State Penitentiary | Frankfort | Lease and public account. | Piece price. |
| Louisiana | State Penitentiary | Baton Rouge | Lease | Lease. |
| Maine | State Prison | Thomaston | Public account | [illegible] |
| Maryland | Penitentiary | Baltimore | Contract | [illegible] |
| Massachusetts | State Prison | Boston | Contract | [illegible] |
| Massachusetts | Reformatory | Concord Junction. | Piece price | [illegible] |
| Massachusetts | Reformatory Prison for Women. | Sherborn | Piece price | [illegible] |
| Michigan | State Prison | Jackson | Contract and piece price. | [illegible] |

TABLE II.—SYSTEMS OF WORK IN 1885 AND 1895—Concluded.

| State. | Institution. | Locality. | 1885. | 1895. |
|---|---|---|---|---|
| Michigan | State House of Correction and Reformatory. | Ionia | Contract | Public account and contract. |
| Minnesota | State Prison | Stillwater | Contract | Contract and public account. |
| Mississippi | State Penitentiary | Jackson | Lease | Public account. |
| Missouri | State Penitentiary | Jefferson City | Contract | Contract. |
| Nebraska | State Penitentiary | Lancaster | Lease | Lease. |
| Nevada | State Prison | Carson City | Public account | Public account. |
| New Hampshire | State Prison | Concord | Contract | Contract. |
| New Jersey | State Prison | Trenton | Piece price | Piece price and public account. |
| New Jersey | Essex County Penitentiary. | Caldwell | Public account | Public account. |
| New Mexico | Territorial Penitentiary. | Santa Fe | Lease | Public account. |
| New York | Auburn Prison | Auburn | Public account and contract. | Piece price and public account. |
| New York | Sing Sing State Prison | Sing Sing | Contract. | Public account and piece price. |
| New York | Clinton Prison | Dannemora | Public account | Public account and piece price. |
| New York | State Reformatory | Elmira | Contract and public account. | Piece price and public account. |
| New York | Albany County Penitentiary. | Albany | Contract | Piece price. |
| New York | Erie County Penitentiary. | Buffalo | Contract | Public account and piece price. |
| New York | Kings County Penitentiary. | Brooklyn | Contract | Piece price. |
| New York | Monroe County Penitentiary. | Rochester | Contract | Piece price and public account. |
| New York | Onondaga County Penitentiary. | Syracuse | Contract | Piece price. |
| North Carolina | State Penitentiary | Raleigh | Public account and lease. | Public account. |
| Ohio | Penitentiary | Columbus | Contract, piece price, and public account. | Contract and piece price. |
| Oregon | State Penitentiary | Salem | Contract and public account. | Contract and public account. |
| Pennsylvania | Eastern Penitentiary | Philadelphia | Public account and piece price. | Public account. |
| Pennsylvania | Western Penitentiary | Allegheny City | Contract | Public account and piece price. |
| Pennsylvania | Philadelphia County House of Correction. | Philadelphia | Public account | Public account. |
| Pennsylvania | Berks County Prison | Reading | Public account | Public account. |
| Pennsylvania | Chester County Prison | West Chester | Public account | Public account. |
| Pennsylvania | Delaware County Prison. | Media | Public account | Public account. |
| Pennsylvania | Lancaster County Prison. | Lancaster | Public account | Public account. |
| Pennsylvania | Lehigh County Prison | Allentown | Public account | Public account. |
| Pennsylvania | Montgomery County Prison. | Norristown | Piece price | Piece price. |
| Pennsylvania | Northampton County Prison. | Easton | Public account | Public account. |
| Pennsylvania | Northumberland County Prison. | Sunbury | Public account | Public account. |

TABLE III.—CONVICTS IN PRISONS AND PENITENTIARIES, BY SYSTEMS OF WORK, [illegible].

PUBLIC-ACCOUNT SYSTEM.

| Marginal number. | Name of institution. | Locality. | Employed in productive labor. | | |
|---|---|---|---|---|---|
| | | | Male. | Female. | Total. |
| | ARIZONA. | | | | |
| 1 | Territorial Prison | Yuma | 158 | 2 | 160 |
| | CALIFORNIA. | | | | |
| 2 | State Prison | Folsom | 742 | .......... | 742 |
| 3 | State Prison | San Quentin | 717 | .......... | 717 |
| | COLORADO. | | | | |
| 4 | State Penitentiary | Cañon City | 168 | .......... | 168 |
| | ILLINOIS. | | | | |
| 5 | Southern Penitentiary | Chester | 517 | .......... | 517 |
| | IOWA. | | | | |
| 6 | State Penitentiary | Anamosa | 363 | .......... | 363 |
| | KANSAS. | | | | |
| 7 | State Penitentiary | Lansing | a 825 | a 16 | a 841 |
| | MAINE. | | | | |
| 8 | State Prison | Thomaston | 89 | .......... | 89 |
| | MICHIGAN. | | | | |
| 9 | State House of Correction and Reformatory. | Ionia | b 378 | .......... | b 378 |
| | MISSISSIPPI. | | | | |
| 10 | State Penitentiary | Jackson | 894 | 10 | 904 |
| | NEVADA. | | | | |
| 11 | State Prison | Carson City | 10 | .......... | 10 |
| | NEW JERSEY. | | | | |
| 12 | Essex County Penitentiary | Caldwell | 212 | .......... | 212 |
| | NEW MEXICO. | | | | |
| 13 | Territorial Penitentiary | Santa Fe | 107 | .......... | 107 |
| | NEW YORK. | | | | |
| 14 | Clinton Prison | Dannemora | c 670 | .......... | c 670 |
| 15 | Erie County Penitentiary | Buffalo | d 230 | .......... | d 230 |
| | NORTH CAROLINA. | | | | |
| 16 | State Penitentiary | Raleigh | 1,000 | 40 | 1,040 |
| | PENNSYLVANIA. | | | | |
| 17 | Eastern Penitentiary | Philadelphia | 194 | 23 | 217 |
| 18 | Western Penitentiary | Allegheny City | a 830 | .......... | a 830 |
| 19 | Philadelphia County House of Correction. | Philadelphia | 508 | 37 | 545 |
| 20 | Berks County Prison | Reading | 103 | .......... | 103 |
| 21 | Chester County Prison | West Chester | 19 | .......... | [illegible] |
| 22 | Delaware County Prison | Media | 38 | 2 | [illegible] |
| 23 | Lancaster County Prison | Lancaster | 51 | .......... | [illegible] |
| 24 | Lehigh County Prison | Allentown | 30 | .......... | [illegible] |
| 25 | Northampton County Prison | Easton | 46 | .......... | [illegible] |
| 26 | Northumberland County Prison | Sunbury | 49 | .......... | [illegible] |
| 27 | Philadelphia County Prison | Philadelphia | 33 | .......... | [illegible] |
| 28 | Schuylkill County Prison | Pottsville | 43 | .......... | [illegible] |
| | SOUTH DAKOTA. | | | | |
| 29 | State Penitentiary | Sioux Falls | [illegible] | .......... | [illegible] |

TABLE III.—CONVICTS IN PRISONS AND PENITENTIARIES, BY SYSTEMS OF WORK, 1895.

PUBLIC-ACCOUNT SYSTEM.

| Engaged in prison duties. | | | Idle and sick. | | | Total. | | | Marginal number. |
|---|---|---|---|---|---|---|---|---|---|
| Male. | Female. | Total. | Male. | Female. | Total. | Male. | Female. | Total. | |
| ........ | ........ | ........ | 5 | ........ | 5 | 163 | 2 | 165 | 1 |
| 160 | ........ | 160 | 24 | ........ | 24 | 927 | ........ | 927 | 2 |
| 447 | 12 | 459 | 108 | 3 | 111 | 1,272 | 15 | 1,287 | 3 |
| 100 | 15 | 115 | 346 | ........ | 346 | 609 | 15 | 624 | 4 |
| 222 | ........ | 222 | 6 | ........ | 6 | 745 | ........ | 745 | 5 |
| 114 | 25 | 139 | 40 | ........ | 40 | 517 | 25 | 542 | 6 |
| ........ | ........ | ........ | 30 | 1 | 31 | *a* 855 | *a* 17 | *a* 872 | 7 |
| 37 | 4 | 41 | 21 | ........ | 21 | 147 | 4 | 151 | 8 |
| 124 | ........ | 124 | 69 | ........ | 69 | *b* 571 | ........ | *b* 571 | 9 |
| 36 | ........ | 36 | 50 | ........ | 50 | 980 | 10 | 990 | 10 |
| 62 | ........ | 62 | 5 | 1 | 6 | 77 | 1 | 78 | 11 |
| 45 | 14 | 59 | ........ | ........ | ........ | 257 | 14 | 271 | 12 |
| 40 | 1 | 41 | 1 | ........ | 1 | 148 | 1 | 149 | 13 |
| 273 | ........ | 273 | 64 | ........ | 64 | *c* 1,007 | ........ | *c* 1,007 | 14 |
| 55 | 40 | 95 | 417 | 11 | 428 | *d* 702 | 51 | *d* 753 | 15 |
| 75 | 35 | 110 | 50 | 8 | 58 | 1,125 | 83 | 1,208 | 16 |
| 97 | 2 | 99 | 1,112 | ........ | 1,112 | 1,403 | 25 | 1,428 | 17 |
| 174 | 23 | 197 | 131 | 1 | 132 | *e* 1,135 | 24 | *e* 1,159 | 18 |
| 319 | 93 | 412 | 201 | 23 | 224 | 1,028 | 203 | 1,231 | 19 |
| 14 | 4 | 18 | ........ | ........ | ........ | 117 | 4 | 121 | 20 |
| 3 | ........ | 3 | 2 | ........ | 2 | 24 | ........ | 24 | 21 |
| 3 | ........ | 3 | 1 | ........ | 1 | 42 | 2 | 44 | 22 |
| 46 | 3 | 49 | 2 | ........ | 2 | 99 | 3 | 102 | 23 |
| 12 | ........ | 12 | 78 | ........ | 78 | 120 | ........ | 120 | 24 |
| 3 | ........ | 3 | 7 | ........ | 7 | 50 | ........ | 50 | 25 |
| 20 | 2 | 22 | 21 | 8 | 29 | 90 | 10 | 100 | 26 |
| 19 | 45 | 64 | 213 | 1 | 214 | 264 | 46 | 310 | 27 |
| 13 | 6 | 19 | 13 | ........ | 13 | 74 | 6 | 80 | 28 |
| 28 | 2 | 30 | 3 | ........ | 3 | 115 | 2 | 117 | 29 |

*d* Including 30 under the piece-price system.
*e* Including 140 under the piece-price system.

IN PRISONS AND PENITENTIARIES, BY SYSTEMS OF WORK, 1895—Continued.

PUBLIC-ACCOUNT SYSTEM—Concluded.

| stitution. | Locality. | Employed in productive labor. | | |
|---|---|---|---|---|
| | | Male. | Female. | Total. |
| XAS. | | | | |
| .................. | Huntsville and Rusk. | a 3,760 | 62 | a 3,822 |
| MONT. | | | | |
| D ................ | Rutland ............ | 42 | .......... | 42 |
| NGTON. | | | | |
| .................. | Walla Walla.......... | 264 | .......... | 264 |
| .................. | .................... | 13,168 | 242 | 13,410 |

CONTRACT SYSTEM.

| stitution. | Locality. | Male. | Female. | Total. |
|---|---|---|---|---|
| NSAS. | | | | |
| .................. | Little Rock .......... | b 749 | b 12 | b 761 |
| TICUT. | | | | |
| .................. | Wethersfield ........ | 303 | .......... | 303 |
| NOIS. | | | | |
| .................. | Joliet................ | b 1,199 | b 43 | b 1,242 |
| ANA. | | | | |
| ) ................ | Michigan City ........ | 655 | .......... | 655 |
| ) ................ | Jeffersonville........ | 475 | .......... | 475 |
| VA. | | | | |
| .................. | Fort Madison......... | 280 | .......... | 280 |

TABLE III.—CONVICTS IN PRISONS AND PENITENTIARIES, BY SYSTEMS OF WORK, 1895—Continued.

PUBLIC ACCOUNT SYSTEM—Concluded.

| Engaged in prison duties. | | | Idle and sick. | | | Total. | | | Marginal number. |
|---|---|---|---|---|---|---|---|---|---|
| Male. | Female. | Total. | Male. | Female. | Total. | Male. | Female. | Total. | |
| 208 | 6 | 214 | 89 | ........... | 89 | a4, 057 | 68 | a4, 125 | 1 |
| 40 | 3 | 43 | ........ | 1 | 1 | 82 | 4 | 86 | 2 |
| 126 | 4 | 130 | 22 | ........... | 22 | 412 | 4 | 416 | 3 |
| 2, 915 | 339 | 3, 254 | 3, 131 | 58 | 3, 189 | 19, 214 | 639 | 19, 853 | |

CONTRACT SYSTEM.

| Engaged in prison duties. | | | Idle and sick. | | | Total. | | | Marginal number. |
|---|---|---|---|---|---|---|---|---|---|
| Male. | Female. | Total. | Male. | Female. | Total. | Male. | Female. | Total. | |
| 46 | ........... | 46 | 80 | 2 | 82 | b875 | b14 | b889 | 4 |
| 108 | 5 | 113 | 14 | ........... | 14 | 425 | 5 | 430 | 5 |
| 311 | 11 | 322 | 51 | 2 | 53 | b1, 561 | b56 | b1, 617 | 6 |
| 144 | ........... | 144 | 53 | ........... | 53 | 852 | ........... | 852 | 7 |
| 200 | ........... | 200 | 140 | ........... | 140 | 815 | ........... | 815 | 8 |
| 105 | ........... | 105 | 115 | ........... | 115 | 500 | ........... | 500 | 9 |
| 64 | 5 | 69 | 3 | ........... | 3 | 677 | 26 | 703 | 10 |
| 163 | 1 | 164 | 198 | 1 | 199 | b808 | 2 | b810 | 11 |
| 125 | 6 | 131 | 12 | 2 | 14 | c494 | 8 | c502 | 12 |
| 448 | 12 | 460 | 553 | 11 | 564 | 2, 106 | 53 | 2, 159 | 13 |
| 13 | 2 | 15 | 1 | ........... | 1 | 181 | 2 | 183 | 14 |
| 387 | 33 | 420 | 627 | ........... | 627 | d2, 052 | 33 | d2, 085 | 15 |
| 221 | ........... | 221 | 17 | 3 | 20 | b396 | 3 | b399 | 16 |
| 12 | 2 | 14 | 5 | ........... | 5 | 153 | 2 | 155 | 17 |
| 55 | 40 | 95 | 21 | 4 | 25 | e970 | 44 | e1, 014 | 18 |

d Including those under the piece-price system.
e Including 247 under the public-account system and 438 under the lease system.

6139—No. 5——2

TABLE III.—CONVICTS IN PRISONS AND PENITENTIARIES, BY SYSTEMS [illegible] [illegible]—Continued.

CONTRACT SYSTEM—Concluded.

| Marginal number. | Name of institution. | Locality. | Employed in p[illegible] labor. Male. | Female. | [illegible] |
|---|---|---|---|---|---|
| | VERMONT. | | | | |
| 1 | State Prison | Windsor | 134 | .......... | [illegible] |
| | VIRGINIA. | | | | |
| 2 | State Penitentiary | Richmond | a 918 | a 88 | a 1,[illegible] |
| | WEST VIRGINIA. | | | | |
| 3 | Penitentiary | Moundsville | 362 | .......... | [illegible] |
| | WISCONSIN. | | | | |
| 4 | State Prison | Waupun | b 422 | c 11 | [illegible] |
| | Total | | 10,409 | 199 | [illegible] |

PIECE-PRICE SYSTEM.

| Marginal number. | Name of institution. | Locality. | Male. | Female. | [illegible] |
|---|---|---|---|---|---|
| | INDIANA. | | | | |
| 5 | State Reformatory for Women | Indianapolis | .......... | 23 | [illegible] |
| | KENTUCKY. | | | | |
| 6 | State Penitentiary | Frankfort | 900 | 30 | [illegible] |
| | MASSACHUSETTS. | | | | |
| 7 | State Prison | Boston | a 513 | .......... | a 513 |
| 8 | Reformatory | Concord Junction | 577 | .......... | 577 |
| 9 | Reformatory Prison for Women | Sherborn | .......... | a 150 | a [illegible] |
| | NEW JERSEY. | | | | |
| 10 | State Prison | Trenton | a 486 | a 10 | a [illegible] |
| | NEW YORK. | | | | |
| 11 | Auburn Prison | Auburn | a 847 | a 50 | a [illegible] |
| 12 | Sing Sing State Prison | Sing Sing | f 1,632 | .......... | f 1,[illegible] |
| 13 | State Reformatory | Elmira | g 596 | .......... | g [illegible] |
| 14 | Albany County Penitentiary | Albany | 790 | .......... | 790 |
| 15 | Kings County Penitentiary | Brooklyn | 825 | 30 | [illegible] |
| 16 | Monroe County Penitentiary | Rochester | h 281 | .......... | h 281 |
| 17 | Onondaga County Penitentiary | Syracuse | 160 | .......... | 160 |
| | PENNSYLVANIA. | | | | |
| 18 | Montgomery County Prison | Norristown | 16 | .......... | 16 |
| | Total | | 7,236 | 301 | 7,537 |

LEASE SYSTEM.

| Marginal number. | Name of institution. | Locality. | Male. | Female. | [illegible] |
|---|---|---|---|---|---|
| | ALABAMA. | | | | |
| 19 | State Penitentiary | Wetumpka | i 1,506 | .......... | i 1,[illegible] |
| | FLORIDA. | | | | |
| 20 | State Penitentiary | Tallahassee | 582 | .......... | [illegible] |
| | GEORGIA. | | | | |
| 21 | State Penitentiary | Atlanta | 2,357 | 67 | 2,[illegible] |
| | LOUISIANA. | | | | |
| 22 | State Penitentiary | Baton Rouge | 1,087 | [illegible] | [illegible] |

a Including [illegible] public-account system.
b Including 77 [illegible] system.
c Under the [illegible] system.
d Including [illegible] system.
e Including 11 under [illegible] system.

TABLE III.—CONVICTS IN PRISONS AND PENITENTIARIES, BY SYSTEMS OF WORK, 1895—Continued.

CONTRACT SYSTEM—Concluded.

| Engaged in prison duties. | | | Idle and sick | | | Total. | | | Marginal number. |
|---|---|---|---|---|---|---|---|---|---|
| Male | Female. | Total. | Male. | Female | Total. | Male. | Female. | Total. | |
| 21 | 6 | 27 | .......... | .......... | .......... | 155 | 6 | 161 | 1 |
| 118 | 10 | 128 | 247 | .......... | 247 | a 1,283 | a 92 | a 1,375 | 2 |
| 54 | 14 | 68 | 105 | .......... | 105 | 511 | 14 | 525 | 3 |
| 73 | 13 | 86 | 95 | .......... | 95 | b 591 | c 24 | d 615 | 4 |
| 2,668 | 160 | 2,828 | 2,337 | 25 | 2,362 | 15,405 | 384 | 15,789 | |

PIECE-PRICE SYSTEM.

| Engaged in prison duties. | | | Idle and sick | | | Total. | | | Marginal number. |
|---|---|---|---|---|---|---|---|---|---|
| Male | Female. | Total. | Male. | Female | Total. | Male. | Female. | Total. | |
| .......... | 4 | 4 | .......... | .......... | .......... | .......... | 36 | 36 | 5 |
| 96 | 9 | 105 | 85 | 12 | 97 | 1,081 | 41 | 1,122 | 6 |
| 164 | .......... | 164 | 23 | .......... | 23 | a 700 | .......... | a 700 | 7 |
| 411 | .......... | 411 | 39 | .......... | 39 | 1,027 | .......... | 1,027 | 8 |
| .......... | 153 | 153 | .......... | 15 | 15 | .......... | a 318 | a 318 | 9 |
| 170 | 12 | 182 | 294 | 5 | 299 | a 950 | a 27 | a 977 | 10 |
| 256 | 32 | 288 | 27 | 8 | 35 | a 1,130 | a 99 | a 1,229 | 11 |
| 295 | .......... | 295 | 46 | 1 | 47 | f 1,374 | 1 | f 1,375 | 12 |
| 316 | .......... | 316 | 43 | .......... | 43 | g 1,257 | .......... | g 1,257 | 13 |
| 201 | 65 | 266 | 12 | 2 | 14 | 913 | 67 | 980 | 14 |
| 95 | 35 | 130 | 132 | 5 | 137 | 1,052 | 70 | 1,122 | 15 |
| 43 | 43 | 86 | 20 | 4 | 24 | h 344 | 47 | h 391 | 16 |
| 50 | 15 | 65 | 100 | 2 | 102 | 310 | 17 | 327 | 17 |
| 15 | 4 | 19 | 49 | .......... | 49 | 80 | 4 | 84 | 18 |
| 2,112 | 372 | 2,484 | 870 | 54 | 924 | 10,218 | 727 | 10,945 | |

LEASE SYSTEM.

| Engaged in prison duties. | | | Idle and sick | | | Total. | | | Marginal number. |
|---|---|---|---|---|---|---|---|---|---|
| Male | Female. | Total. | Male. | Female | Total. | Male. | Female. | Total. | |
| .......... | .......... | .......... | 4 | 65 | 69 | i 1,512 | 65 | i 1,577 | 19 |
| 31 | 19 | 50 | 50 | .......... | 50 | 663 | 19 | 682 | 20 |
| .......... | .......... | .......... | .......... | .......... | .......... | 2,357 | 67 | 2,424 | 21 |
| 46 | 2 | 48 | 7 | .......... | 7 | 1,090 | 37 | 1,127 | 22 |

f Including 862 under the public-account system.
g Including 136 under the public-account system.
h Including 12 under the public-account system.
i Including 536 under the public-account system.

TABLE V.—GOODS PRODUCED OR WORK DONE, 1885 AND 1895.

| 1885. | | 1895. | |
|---|---|---|---|
| Kind. | Value. | Kind. | Value. |
| **ALABAMA—PUBLIC-ACCOUNT SYSTEM.** | | | |
| | | Brick | $3,682.45 |
| | | Erecting and improving buildings. | 38,913.45 |
| | | Farming | 22,316.58 |
| **ALABAMA—LEASE SYSTEM.** | | | |
| Mining, coal | $192,000.00 | Mining, coal | $622,463.68 |
| Farming | 17,400.00 | | |
| Stone, broken | 5,000.00 | | |
| **ARIZONA—PUBLIC-ACCOUNT SYSTEM** | | | |
| Building and repairing prison | $25,000.00 | Building and repairing prison | $6,000.00 |
| **ARKANSAS—PUBLIC-ACCOUNT SYSTEM.** | | | |
| | | Farming | $46,467.82 |
| | | Repairing prison | 3,134.00 |
| | | Wood chopping | 4,417.23 |
| **ARKANSAS—CONTRACT SYSTEM.** | | | |
| | | Farming, etc. | $82,273.34 |
| **ARKANSAS—LEASE SYSTEM.** | | | |
| Brick | $26,000.00 | | |
| Bricklaying, carpentering, etc | 23,250.00 | | |
| Cigars | 50,000.00 | | |
| Farming | 64,000.00 | | |
| Mining, coal | 37,200.00 | | |
| Wood chopping | 30,000.00 | | |
| **CALIFORNIA—PUBLIC-ACCOUNT SYSTEM.** | | | |
| Bags, jute | $101,318.52 | Bags, jute (grain) | $174,114.13 |
| Stone, quarried and dressed | 21,020.00 | Stone, quarried and dressed | 8,652.00 |
| Brick | 4,075.04 | Improvements to prison | 100,000.00 |
| **CALIFORNIA—PIECE-PRICE SYSTEM.** | | | |
| Furniture | $43,277.87 | | |
| Harnesses | 17,500.00 | | |
| Leather tanning | 9,000.00 | | |
| Sashes, doors and blinds | 225,000.00 | | |
| **COLORADO—PUBLIC-ACCOUNT SYSTEM.** | | | |
| Brick | $10,000.00 | Brick | $1,900.00 |
| Lime | 20,000.00 | Lime | 5,420.94 |
| | | Farming | 8,208.50 |
| | | Stone, quarried | 1,691.08 |

TABLE III.—CONVICTS IN PRISONS AND PENITENTIARIES, BY SYSTEMS OF WORK, 1886—Concluded.

LEASE SYSTEM—Concluded.

| Engaged in prison duties. | | | Idle and sick. | | | Total. | | | Marginal number. |
|---|---|---|---|---|---|---|---|---|---|
| Male. | Female. | Total. | Male. | Female. | Total. | Male. | Female. | Total. | |
| 90 | 6 | 96 | 30 | ...... | 30 | 297 | 6 | 303 | 1 |
| 40 | 4 | 44 | 360 | 25 | 385 | 1,300 | 44 | 1,344 | 2 |
| 287 | 31 | 318 | 460 | 90 | 550 | 7,419 | 238 | 7,657 | |

TABLE IV.—CONVICTS IN 1885 AND 1886.

| Engaged in prison duties—Conc'd. | | | Idle and sick. | | | | | | Aggregate. | | | | | | Marginal number. |
|---|---|---|---|---|---|---|---|---|---|---|---|---|---|---|---|
| 1886. | | | 1885. | | | 1886. | | | 1885. | | | 1886. | | | |
| Male. | Female. | Total. | Male. | Female. | Total. | Male. | Female. | Total. | Male. | Female. | Total. | Male. | Female. | Total. | |
| ...... | ...... | ...... | 7 | 3 | 10 | 4 | 65 | 69 | 535 | 29 | 564 | 1,512 | 65 | 1,577 | 3 |
| ...... | ...... | ...... | 15 | ...... | 15 | 5 | ...... | 5 | 157 | ...... | 157 | 163 | 2 | 165 | 4 |
| 46 | ...... | 46 | ...... | ...... | ...... | 80 | 2 | 82 | 553 | 11 | 564 | 875 | 14 | 889 | 5 |
| 607 | 12 | 619 | 56 | ...... | 56 | 122 | 3 | 125 | 1,779 | 29 | 1,808 | 2,199 | 15 | 2,214 | 6 |
| 100 | 15 | 115 | 10 | ...... | 10 | 246 | ...... | 246 | 285 | 5 | 290 | 609 | 15 | 624 | 7 |
| 108 | 5 | 113 | 16 | ...... | 16 | 14 | ...... | 14 | 277 | 4 | 281 | 425 | 5 | 430 | 8 |
| 31 | 19 | 50 | ...... | ...... | ...... | 50 | ...... | 50 | 291 | 5 | 296 | 663 | 19 | 682 | 9 |
| ...... | ...... | ...... | ...... | ...... | ...... | ...... | ...... | ...... | 1,520 | 40 | 1,560 | 2,357 | 67 | 2,424 | 10 |
| 533 | 11 | 544 | 71 | 1 | 72 | 57 | 2 | 59 | 2,243 | 44 | 2,287 | 2,308 | 58 | 2,366 | 11 |
| 314 | 4 | 318 | 33 | 3 | 36 | 193 | ...... | 193 | 1,298 | 177 | 1,475 | 1,667 | 38 | 1,705 | 12 |
| 219 | 25 | 244 | 22 | ...... | 22 | 155 | ...... | 155 | 679 | 11 | 690 | 1,017 | 25 | 1,042 | 13 |
| (b) | (b) | (b) | 24 | ...... | 24 | 30 | 1 | 31 | 856 | 13 | 869 | 855 | 17 | 872 | 14 |
| 96 | 9 | 105 | 12 | ...... | 12 | 85 | 12 | 97 | 1,114 | 28 | 1,142 | 1,081 | 41 | 1,122 | 15 |
| 46 | 2 | 48 | 7 | 2 | 9 | 7 | ...... | 7 | 798 | 43 | 841 | 1,090 | 37 | 1,127 | 16 |
| 37 | 4 | 41 | 8 | ...... | 8 | 21 | ...... | 21 | 168 | 3 | 171 | 147 | 4 | 151 | 17 |
| 64 | 5 | 69 | 21 | ...... | 21 | 2 | ...... | 2 | 513 | 26 | 539 | 677 | 26 | 703 | 18 |
| 575 | 153 | 728 | 128 | 1 | 129 | 62 | 15 | 77 | 1,201 | 311 | 1,512 | 1,727 | 318 | 2,045 | 19 |
| 227 | 1 | 228 | 189 | ...... | 189 | 267 | 1 | 268 | 1,315 | 1 | 1,316 | 1,379 | 2 | 1,381 | 20 |
| 125 | 6 | 131 | 29 | 1 | 30 | 12 | 2 | 14 | 400 | 11 | 411 | 494 | 8 | 502 | 21 |
| 36 | ...... | 36 | ...... | ...... | ...... | 50 | ...... | 50 | 790 | 22 | 812 | 980 | 10 | 990 | 22 |
| 448 | 12 | 460 | 150 | 35 | 185 | 553 | 11 | 564 | 1,620 | 35 | 1,655 | 2,108 | 53 | 2,161 | 23 |
| 90 | 6 | 96 | 12 | ...... | 12 | 30 | ...... | 30 | 204 | 3 | 207 | 297 | 6 | 303 | 24 |
| 62 | ...... | 62 | 21 | 2 | 23 | 5 | 1 | 6 | 128 | 2 | 130 | 77 | 1 | 78 | 25 |
| 13 | 2 | 15 | 4 | ...... | 4 | 1 | ...... | 1 | 128 | 2 | 130 | 181 | 2 | 183 | 26 |
| 315 | 26 | 341 | 100 | ...... | 100 | 294 | 5 | 299 | 1,012 | 38 | 1,050 | 1,207 | 41 | 1,248 | 27 |
| 40 | 1 | 41 | ...... | ...... | ...... | 1 | ...... | 1 | 100 | ...... | 100 | 148 | 1 | 149 | 28 |
| 1,584 | 230 | 1,814 | 799 | 66 | 865 | 861 | 33 | 894 | 6,125 | 375 | 6,500 | 8,089 | 352 | 8,441 | 29 |
| 75 | 35 | 110 | 5 | ...... | 5 | 50 | 8 | 58 | 1,020 | 65 | 1,085 | 1,125 | 83 | 1,208 | 30 |
| 397 | 33 | 430 | 51 | ...... | 51 | 627 | ...... | 627 | 1,950 | 24 | 1,974 | 2,052 | 33 | 2,085 | 31 |
| 221 | ...... | 221 | 9 | ...... | 9 | 17 | 3 | 20 | 272 | ...... | 272 | 396 | 3 | 399 | 32 |
| 738 | 182 | 920 | 822 | 36 | 858 | 1,830 | 33 | 1,863 | 3,362 | 366 | 3,728 | 4,528 | 327 | 4,855 | 33 |
| 12 | 2 | 14 | 3 | ...... | 3 | 5 | ...... | 5 | 238 | ...... | 238 | 153 | 2 | 155 | 34 |
| 55 | 40 | 95 | 20 | 2 | 22 | 21 | 4 | 25 | 802 | 45 | 847 | 970 | 44 | 1,014 | 35 |
| 28 | 2 | 30 | ...... | ...... | ...... | 3 | ...... | 3 | c 90 | c 3 | c 93 | 115 | 2 | 117 | 36 |
| 40 | 4 | 44 | ...... | ...... | ...... | 360 | 25 | 385 | 1,278 | 45 | 1,323 | 1,500 | 44 | 1,544 | 37 |
| 208 | 6 | 214 | ...... | ...... | ...... | 89 | ...... | 89 | 2,877 | 45 | 2,922 | 4,057 | 68 | 4,125 | 38 |
| 61 | 9 | 70 | 5 | ...... | 5 | ...... | 1 | 1 | 156 | 9 | 165 | 237 | 10 | 247 | 39 |
| 112 | 16 | 128 | ...... | ...... | ...... | 247 | ...... | 247 | 956 | 68 | 1,024 | 1,283 | 92 | 1,375 | 40 |
| 126 | 4 | 130 | 7 | ...... | 7 | 22 | ...... | 22 | 82 | ...... | 82 | 412 | 4 | 416 | 41 |
| 54 | 14 | 68 | 15 | ...... | 15 | 105 | ...... | 105 | 255 | 6 | 261 | 511 | 14 | 525 | 42 |
| 73 | 13 | 86 | 16 | ...... | 16 | 95 | ...... | 95 | 443 | 13 | 456 | 591 | 24 | 615 | 43 |
| 7,902 | 902 | 8,804 | 2,481 | 152 | 2,633 | 6,798 | 227 | 7,025 | 39,910 | 1,967 | 41,877 | 52,256 | 1,988 | 54,244 | |

c Dakota Territory.

TABLE V.—GOODS PRODUCED OR WORK DONE, 1885 AND 1895—Continued.

IOWA—PUBLIC-ACCOUNT SYSTEM.

| 1885. | | 1895. | |
|---|---|---|---|
| Kind. | Value. | Kind. | Value. |
| Farming | $2,000.00 | Farming | $1,860.00 |
| Stone, dressed | 15,000.00 | Stone, dressed | 80,768.40 |
| | | Masonry work | 11,175.00 |
| | | Stone, broken | 1,560.00 |

IOWA—CONTRACT SYSTEM.

| 1885. | | 1895. | |
|---|---|---|---|
| Kind. | Value. | Kind. | Value. |
| Agricultural implements | $120,590.00 | Agricultural implements | $150,870.00 |
| Chairs | 100,000.00 | Chairs | 90,216.00 |
| Boots and shoes | 161,000.00 | Erecting buildings | 1,550.00 |

KANSAS—PUBLIC-ACCOUNT SYSTEM.

| 1885. | | 1895. | |
|---|---|---|---|
| Kind. | Value. | Kind. | Value. |
| Building and repairing prison | $158,000.00 | Building and repairing prison | $59,000.00 |
| Clothing (for convicts) | 24,964.72 | Clothing (for convicts) | 14,708.00 |
| Mining, coal | 85,630.05 | Mining, coal | 92,624.25 |

KANSAS—CONTRACT SYSTEM.

| 1885. | | 1895. | |
|---|---|---|---|
| Kind. | Value. | Kind. | Value. |
| Boots and shoes | $70,125.00 | Shoes | $26,361.00 |
| Wagons | 720,000.00 | Furniture | 24,380.00 |
| | | Horse collars | 20,040.00 |

KENTUCKY—PUBLIC-ACCOUNT SYSTEM.

| 1885. | | 1895. | |
|---|---|---|---|
| Kind. | Value. | Kind. | Value. |
| Building prison | $37,200.00 | | |

KENTUCKY—PIECE-PRICE SYSTEM.

| 1885. | | 1895. | |
|---|---|---|---|
| Kind. | Value. | Kind. | Value. |
| | | Bedsteads | $17,801.00 |
| | | Chairs | 152,725.00 |

KENTUCKY—LEASE SYSTEM.

| 1885. | | 1895. | |
|---|---|---|---|
| Kind. | Value. | Kind. | Value. |
| Brooms | $30,000.00 | | |
| Building railroad | 152,000.00 | | |
| Chairs, tables, etc | 18,000.00 | | |
| Laundering | 1,380.00 | | |
| Mining, coal | 175,000.00 | | |
| Shoes | 24,900.00 | | |
| Wagon driving | 10,000.00 | | |

LOUISIANA—LEASE SYSTEM.

| 1885. | | 1895. | |
|---|---|---|---|
| Kind. | Value. | Kind. | Value. |
| Farming | $56,000.00 | Farming, repairing levee, repairing railroad, and pantaloons. | $165,647.85 |
| Repairing levee | 42,000.00 | | |
| Repairing railroad | 254,000.00 | | |

MAINE—PUBLIC-ACCOUNT SYSTEM.

| 1885. | | 1895. | |
|---|---|---|---|
| Kind. | Value. | Kind. | Value. |
| Carriages and sleighs | $45,000.00 | Carriages, wagons, and sleighs | $14,590.00 |
| Harnesses | 27,000.00 | Harnesses | 9,572.00 |
| | | Brooms and brushes | 18,764.13 |
| | | Furniture | 900.00 |

TABLE V.—GOODS PRODUCED OR WORK DONE, 1885 AND 1886—Continued.

CONNECTICUT—CONTRACT SYSTEM.

| 1885. | | 1886. | |
|---|---|---|---|
| Kind. | Value. | Kind. | Value. |
| Boots and shoes | $[illegible] | Boots and shoes | $[illegible] |

FLORIDA—LEASE SYSTEM.

| Kind. | Value. | Kind. | Value. |
|---|---|---|---|
| Naval stores | $1[illegible] | Naval stores | $[illegible] |
| | | Building railroad | 24,[illegible] |
| | | Improvements, etc | 2,[illegible] |
| | | Mining, phosphate | 1[illegible] |

GEORGIA—LEASE SYSTEM.

| Kind. | Value. | Kind. | Value. |
|---|---|---|---|
| Brick | $172,000.00 | Brick | $[illegible] |
| Farming | 16,000.00 | Farming | 21,426.00 |
| Lumber | 34,000.00 | Lumber | 13[illegible] |
| Building railroad | 63,000.00 | Ditching and clearing land, wood chopping, etc. | 3,000.00 |
| Lime | 7,000.00 | | |
| Mining, coal and iron ore (and making pig iron). | 142,000.00 | | |
| Mining, iron ore | 26,000.00 | | |

ILLINOIS—PUBLIC-ACCOUNT SYSTEM.

| Kind. | Value. | Kind. | Value. |
|---|---|---|---|
| | | Brick | $16,140.00 |
| | | Brooms | 24,500.00 |
| | | Chairs | 153,155.00 |
| | | Cigars | 22,507.00 |
| | | Cooperage | 39,105.00 |
| | | Harnesses | 158,642.00 |
| | | Hollow ware | 128,267.40 |
| | | Hosiery | [illegible] |
| | | Pearl button blanks | 7,501.10 |
| | | Stone | 56,272.16 |

ILLINOIS—CONTRACT SYSTEM.

| Kind. | Value. | Kind. | Value. |
|---|---|---|---|
| Boots and shoes | $1,530,000.00 | Boots and shoes | $156,000.00 |
| Barrels, etc | 375,000.00 | Willow ware | 97,000.00 |
| Brick | 25,000.00 | | |
| Fence wire, barbed | 318,000.00 | | |
| Harnesses and saddlery | 148,000.00 | | |
| Hollow ware | 10,000.00 | | |
| Hosiery and overalls | 98,000.00 | | |
| Stone and marble (dressed) and monuments. | 500,000.00 | | |

INDIANA—CONTRACT SYSTEM.

| Kind. | Value. | Kind. | Value. |
|---|---|---|---|
| Boots and shoes | $275,000.00 | Boots and shoes | $611,205.00 |
| Saddletrees | 10,000.00 | Saddletrees | 15,187.00 |
| Chairs and baby cradles | 168,000.00 | Chairs | 91,566.50 |
| Boots and shoes, men's and women's | 297,716.40 | Bicycles | 58,365.00 |
| Brooms | 31,300.00 | Brushes and wire goods | 72,440.00 |
| Hardware, fancy | 390,600.00 | Cooperage | 73,572.00 |
| Hosiery and cloth goods | 305,298.92 | Hollow ware | 153,364.50 |
| Tierces, pork and lard | 174,497.50 | | |

INDIANA—PIECE-PRICE SYSTEM.

| Kind. | Value. | Kind. | Value. |
|---|---|---|---|
| Cane-seating chairs | $5,460.00 | Cane-seating chairs | $1,122.00 |
| Family sewing | 2,000.00 | Family sewing | 2,376.30 |
| Laundering | 5,300.00 | Laundering | 1,134.78 |
| Overalls and shirts | 5,605.50 | Overalls and jumpers | 3,123.80 |
| Toeing stockings | 147.75 | | |

Table V.—[illegible]

IOWA—[illegible]

| [illegible]. | | [illegible] | |
|---|---|---|---|
| Kind. | Value. | [illegible] | [illegible] |
| Farming | [illegible] | [illegible] | [illegible] |
| Stone, dressed | 15, [illegible]. 00 | [illegible] | [illegible] |
| | | Masonry work | [illegible] |
| | | Stone, [illegible] | [illegible] |

IOWA—Contract System.

| Kind. | Value. | Kind. | Value. |
|---|---|---|---|
| Agricultural implements | $126, 590. 00 | Agricultural implements | [illegible] |
| Chairs | 168, 000. 00 | Chairs | [illegible] |
| Boots and shoes | 161, 000. 00 | Erecting buildings | [illegible] |

KANSAS—Public-Account System.

| Kind. | Value. | Kind. | Value. |
|---|---|---|---|
| Building and repairing prison | $156, 000. 00 | Building and repairing prison | [illegible] |
| Clothing (for convicts) | 24, 964. 72 | Clothing (for convicts) | [illegible] |
| Mining, coal | 85, 680. 06 | Mining, coal | [illegible] |

KANSAS—Contract System.

| Kind. | Value. | Kind. | Value. |
|---|---|---|---|
| Boots and shoes | $70, 125. 00 | Shoes | [illegible] |
| Wagons | 720, 000. 00 | Furniture | [illegible] |
| | | Horse collars | [illegible] |

KENTUCKY—Public-Account System.

| Kind. | Value. | Kind. | Value. |
|---|---|---|---|
| Building prison | $37, 200. 00 | | [illegible] |

KENTUCKY—Piece-Price System.

| Kind. | Value. | Kind. | Value. |
|---|---|---|---|
| | | Bedsteads | [illegible] |
| | | Chairs | [illegible] |

KENTUCKY—Lease System.

| Kind. | Value. | Kind. | Value. |
|---|---|---|---|
| Brooms | $30, 000. 00 | | |
| Building railroad | 152, 000. 00 | | |
| Chairs, tables, etc | 18, 000. 00 | | |
| Laundering | 1, 380. 00 | | |
| Mining, coal | 175, 000. 00 | | |
| Shoes | 24, 000. 00 | | |
| Wagon driving | 10, 000. 00 | | |

LOUISIANA—Lease System.

| Kind. | Value. | Kind. | Value. |
|---|---|---|---|
| Farming | [illegible] | Farming, repairing levee, repairing railroad, and [illegible] | [illegible] |
| Repairing levee | 42, [illegible]. 00 | | |
| Repairing railroad | 254, 000. 00 | | |

MAINE—Public-Account System.

| Kind. | Value. | Kind. | Value. |
|---|---|---|---|
| Carriages and sleighs | [illegible] | Carriages, wagons, and sleighs | [illegible] |
| Harnesses | [illegible] | [illegible] | [illegible] |

TABLE V.—GOODS PRODUCED OR WORK DONE, 1885 AND 1895—Continued.

MARYLAND—CONTRACT SYSTEM.

| 1885. | | 1895. | |
|---|---|---|---|
| Kind. | Value. | Kind. | Value. |
| Marble, dressed | $150,000.00 | Marble, dressed | $140,000.00 |
| Shoes, women's and girls' | 125,000.00 | Shoes, men's, boys', and youths' | 428,008.20 |
| Stoves and hollow ware | 120,000.00 | Iron hollow ware | 130,000.00 |

MASSACHUSETTS—PUBLIC-ACCOUNT SYSTEM.

| 1885 Kind | Value | 1895 Kind | Value |
|---|---|---|---|
| | | Brushes | $14,118.00 |
| | | Butter | 874.34 |
| | | Harnesses | 18,946.00 |
| | | Laundering | 3,366.00 |
| | | Poultry, eggs, etc | 115.00 |
| | | Shoes, men's | 128,120.20 |
| | | Trunks | 3,344.00 |

MASSACHUSETTS—CONTRACT SYSTEM.

| 1885 Kind | Value | 1895 Kind | Value |
|---|---|---|---|
| Beds, spring and mantel | $71,415.70 | | |
| Moldings, wooden | 7,584.30 | | |

MASSACHUSETTS—PIECE-PRICE SYSTEM.

| 1885 Kind | Value | 1895 Kind | Value |
|---|---|---|---|
| Boots and shoes, men's and boys' | $199,720.00 | Buttons | $7,930.77 |
| Clothing, knit goods, and laundering. | 22,250.00 | Chairs | 384,120.00 |
| | | Shirts | 106,607.75 |
| Harnesses and saddlery | 66,250.00 | Shoes, men's | 384,570.00 |
| Pantaloons | 52,512.00 | Shoes, women's | 72,666.75 |

MICHIGAN—PUBLIC-ACCOUNT SYSTEM.

| 1885 Kind | Value | 1895 Kind | Value |
|---|---|---|---|
| | | Boxes, etc | $7,040.61 |
| | | Brooms, etc | 31,890.00 |
| | | Chairs, etc., caned | 4,295.00 |
| | | Clothing | 2,148.35 |
| | | Furniture | 76,692.00 |

MICHIGAN—CONTRACT SYSTEM.

| 1885 Kind | Value | 1895 Kind | Value |
|---|---|---|---|
| Agricultural implements | $280,000.00 | Agricultural implements | $125,000.00 |
| Boots and shoes | 45,000.00 | Hosiery, gloves, and mittens | 36,500.00 |
| Chairs | 28,778.00 | Monuments | 46,500.00 |
| Cigars | 165,000.00 | Wagons | 72,000.00 |
| Wagons | 300,000.00 | | |

MICHIGAN—PIECE-PRICE SYSTEM.

| 1885 Kind | Value | 1895 Kind | Value |
|---|---|---|---|
| Brooms | $35,000.00 | | |

MINNESOTA—PUBLIC-ACCOUNT SYSTEM.

| 1885 Kind | Value | 1895 Kind | Value |
|---|---|---|---|
| | | Binding cord | $150,000.00 |

MINNESOTA—CONTRACT SYSTEM.

| 1885 Kind | Value | 1895 Kind | Value |
|---|---|---|---|
| Sashes, doors, and blinds | $50,000.00 | Boots and shoes | $175,000.00 |
| Threshing machines | 195,500.00 | | |

MISSISSIPPI—PUBLIC-ACCOUNT SYSTEM.

| 1885 Kind | Value | 1895 Kind | Value |
|---|---|---|---|
| | | Clothing (for convicts) | $9,506.00 |
| | | Farming | 282,468.00 |
| | | Shoes (for convicts) | 2,[illegible] |

TABLE V.—GOODS PRODUCED OR WORK DONE [illegible]

MISSISSIPPI—LEASE SYSTEM.

| 1885. | | [illegible] | |
|---|---|---|---|
| Kind. | Value. | Kind. | [illegible] |
| Building railroad | [illegible] | [illegible] | [illegible] |
| Farming and clearing land | [illegible] | [illegible] | [illegible] |
| Gravel digging | [illegible] | [illegible] | [illegible] |
| Lumber | 16, 000. 00 | [illegible] | [illegible] |
| Wagons, furniture, brick, etc | 66, 000. 00 | [illegible] | [illegible] |

MISSOURI—CONTRACT SYSTEM.

| Kind. | Value. | Kind. | Value. |
|---|---|---|---|
| Harnesses and saddlery | [illegible] | Harnesses, saddlery, and whips | [illegible] |
| Overalls | 45, 000. 00 | Overalls | [illegible] |
| Saddletrees | 175, 000. 00 | Saddletrees | [illegible] |
| Boots and shoes | 765, 000. 00 | Brick | [illegible] |
| | | Coats | [illegible] |
| | | Erecting buildings | [illegible] |
| | | Jackets | [illegible] |
| | | Shoes | [illegible] |
| | | Stone, dressed | [illegible] |
| | | Stone, quarried | [illegible] |

NEBRASKA—LEASE SYSTEM.

| Kind. | Value. | Kind. | Value. |
|---|---|---|---|
| Harnesses and collars | $27, 000. 00 | Harnesses, collars, etc | [illegible] |
| Agricultural implements | 68, 000. 00 | Cooperage | [illegible] |
| Brooms and trunks | 4, 000. 00 | Stoves | [illegible] |
| Clothing | 15, 000. 00 | | [illegible] |
| Laundering | 6, 000. 00 | | [illegible] |
| Stone, dressed | 28, 000. 00 | | [illegible] |

NEVADA—PUBLIC-ACCOUNT SYSTEM.

| Kind. | Value. | Kind. | Value. |
|---|---|---|---|
| Stone, quarried and dressed | $7, 768. 37 | Stone, quarried and dressed | [illegible] |
| Boots and shoes | 12, 605. 65 | Improvements to prison | [illegible] |

NEW HAMPSHIRE—CONTRACT SYSTEM.

| Kind. | Value. | Kind. | Value. |
|---|---|---|---|
| Bedsteads | $100, 000. 00 | Chairs | [illegible] |

NEW JERSEY—PUBLIC-ACCOUNT SYSTEM.

| Kind. | Value. | Kind. | Value. |
|---|---|---|---|
| Clothing (for convicts) | $970. 20 | Clothing, shoes, etc. (for convicts) | [illegible] |
| Stone, quarried and crushed | 2, 317. 90 | Stone, broken | [illegible] |
| | | Erecting buildings | [illegible] |

NEW JERSEY—PIECE-PRICE SYSTEM.

| Kind. | Value. | Kind. | Value. |
|---|---|---|---|
| Brushes, scrub, shoe, and stove | $34, 560. 00 | Brushes | [illegible] |
| Hosiery | 10, 800. 00 | Hosiery, cotton | [illegible] |
| Collars, cuffs, shirts, and laundering | 232, 084. 40 | Mats and matting | [illegible] |
| Pantaloons (coarse) and working shirts. | 129, 000. 00 | Pantaloons, common | [illegible] |
| | | Shirts, men's coarse | [illegible] |
| Shoes, men's, girls', and children's | 180, 000. 00 | Shoes, infants' | [illegible] |

NEW MEXICO—PUBLIC-ACCOUNT SYSTEM.

| Kind. | Value. | Kind. | Value. |
|---|---|---|---|
| | | Brick | [illegible] |
| | | Clothing | [illegible] |
| | | Erecting and improving buildings | [illegible] |
| | | Farming | [illegible] |
| | | Lime | [illegible] |
| | | Road making | [illegible] |
| | | Sewer pipe | [illegible] |
| | | Shoes | [illegible] |
| | | Well digging | [illegible] |

NEW MEXICO—Lease System.

| [illegible]. | | [illegible]. | |
|---|---|---|---|
| Kind. | Value. | Kind. | Value. |
| Stone, quarried, ditch digging, etc. | $16,000.00 | | |

NEW YORK—Public-Account System.

| Kind. | Value. | Kind. | Value. |
|---|---|---|---|
| Brooms | $19,238.54 | Brooms | [illegible] |
| Clothing, men's and boys' | 228,714.19 | Clothing, men's and boys' | [illegible] |
| Brushes, scrub and shoe | 46,022.90 | Baskets, willow and rattan | [illegible] |
| Shoes, men's | 226,785.76 | Boots and shoes, men's | [illegible] |
| | | Brush fiber and curled hair products. | 74,[illegible] |
| | | Candy and tobacco pails | [illegible] |
| | | Clothing for State institutions | [illegible] |
| | | Clothing, women's | [illegible] |
| | | Erecting and improving buildings | [illegible] |
| | | Farming | [illegible] |
| | | Hardware | [illegible] |
| | | Pins and combs, horn | [illegible] |
| | | Shirts, men's white | [illegible] |
| | | Shoes for State institutions | [illegible] |
| | | Toys and novelties | [illegible] |
| | | Wooden shovels | [illegible] |

NEW YORK—Contract System.

| Kind. | Value. | Kind. | Value. |
|---|---|---|---|
| Bolts, iron | $45,000.00 | | |
| Boots and shoes, men's | 862,400.00 | | |
| Boots and shoes, men's and women's | 226,[illegible].00 | | |
| Brushes, scrub, shoe, etc | 54,000.00 | | |
| Hames, wooden | 25,000.00 | | |
| Hardware, saddlery | 445,000.00 | | |
| Hollow ware | 130,000.00 | | |
| Horse collars | 48,215.00 | | |
| Laundering | a 300,000.00 | | |
| Shoes, men's and women's | 1,587,500.00 | | |
| Shoes, women's and boys' | 377,500.00 | | |
| Stoves | 608,980.00 | | |

NEW YORK—Piece-Price System.

| Kind. | Value. | Kind. | Value. |
|---|---|---|---|
| | | Bags | $[illegible] |
| | | Bolts, iron | 22,000.00 |
| | | Brooms | 60,000.00 |
| | | Brushes, scrub, shoe, and stove | 61,067.50 |
| | | Buttons | 34,371.60 |
| | | Cane-seating chairs | 103,706.80 |
| | | Chairs | 81,282.00 |
| | | Clothing, men's, youths', and boys' | 50,000.00 |
| | | Clothing, women's | 110,[illegible] |
| | | Curled hair and brush fiber | 23,602.14 |
| | | Furniture | 90,737.00 |
| | | Hollow ware | 60,000.00 |
| | | Hosiery | 119,000.00 |
| | | Iron castings | 48,000.00 |
| | | Machinery and tools | 10,[illegible].31 |
| | | Marble, dressed and polished | 29,462.00 |
| | | Pantaloons | 62,080.00 |
| | | Pins, horn | 10,325.00 |
| | | Plumbers' supplies, etc | 130,814.00 |
| | | Rags, seamed, skirted, and assorted. | 45,000.00 |
| | | Shirts, men's | 232,702.00 |
| | | Shirts, men's laundered | 134,205.75 |
| | | Shoes, men's and boys' | 176,807.70 |
| | | Stone, crushed | 2,500.00 |
| | | Umbrellas | 90,000.00 |
| | | Wooden goods | 102,191.22 |

a Value of shirts made and laundered.

**OODS PRODUCED OR WORK DONE, 1885 AND 1895—Conti**

NORTH CAROLINA—PUBLIC-ACCOUNT SYSTEM.

| 1885. | | 1895. |
|---|---|---|
| | Value. | Kind. |
| .............. | $13,725.40 | Brick .............................. |
| .............. | 18,714.41 | Farming .............................. |
| ansion....... | 11,000.00 | Wagons, etc .............................. |
| .............. | 6,500.00 | .............................. |
| s .............. | 6,347.13 | .............................. |
| .............. | 4,785.19 | .............................. |

NORTH CAROLINA—LEASE SYSTEM.

| | | |
|---|---|---|
| .............. | $200,000.00 | .............................. |

OHIO—PUBLIC-ACCOUNT SYSTEM.

| | | |
|---|---|---|
| .............. | $13,403.39 | .............................. |
| .............. | 50,836.13 | .............................. |
| .............. | 30,510.32 | .............................. |
| .............. | 4,332.67 | .............................. |

OHIO—CONTRACT SYSTEM

| | | |
|---|---|---|
| .............. | $175,939.00 | Hardware, saddlery .............. |
| .............. | 60,060 00 | Stoves .............................. |
| .............. | 36,000 00 | Brooms and brushes .............. |
| .............. | 99,910.00 | Farm and garden tools.............. |
| s.............. | 62,000.00 | Hardware, wagon, and foundry work. |
| .............. | 18,000 00 | |
| s, etc......... | 75,700.00 | Shafts and spokes .............. |
| .............. | 31,500.00 | .............................. |
| .............. | 2,681.00 | .............................. |
| .............. | 46,500 00 | .............................. |
| ngs.......... | 209,884.00 | .............................. |
| joiners'...... | 44,558 00 | .............................. |

TABLE V.—GOODS PRODUCED OR WORK DONE, 1885 AND 1886—Continued.

PENNSYLVANIA—PUBLIC-ACCOUNT SYSTEM.

| 1885. | | 1886. | |
|---|---|---|---|
| Kind. | Value. | Kind. | Value. |
| Brooms | $694.50 | Brooms, etc. | $[illegible] |
| Cane-seating chairs | 7,416.58 | Cane-seating chairs | 6,[illegible] |
| Carpeting | 17,128.90 | Carpeting | 4,158.48 |
| Carpeting, rag | 37,094.60 | Carpeting, rag | 44,718.70 |
| Carpeting, rag and ingrain | 28,479.40 | Carpeting, rag and ingrain | 7,[illegible] |
| Cigars | 16,342.00 | Cigars | 2,[illegible] |
| Hosiery | 320.75 | Hosiery | 53,951.73 |
| Baskets | 121.34 | Blacksmithing | 4,[illegible] |
| Boots and shoes | 1,495.94 | Blankets | [illegible] |
| Boots and shoes, men's and women's | 60,442.00 | Brushes, etc. | 1,[illegible] |
| Carpeting, rag and jute | 7,816.14 | Brushes, scrubbing | 2,[illegible] |
| Checks, cotton | 6,225.94 | Carpentering | 6,[illegible] |
| Hosiery, cotton | 1,560.00 | Carpeting, ingrain | [illegible] |
| Hosiery, woolen and cotton | 3,965.00 | Checks | 62.34 |
| Nets, fishing | 381.81 | Clothing | 10,[illegible].75 |
| Shoes, men's | 15,757.50 | Drillings | 11.04 |
| Stone | 23,532.40 | Farming | 10,114.25 |
| | | Flannels | 5,281.68 |
| | | Ginghams | 197.[illegible] |
| | | Illuminating gas | 25,[illegible] |
| | | Jeans | 1,420.[illegible] |
| | | Machine-shop products | 1,354.20 |
| | | Mats and matting | 165,[illegible].51 |
| | | Muslins | [illegible] |
| | | Sheetings | [illegible] |
| | | Shirts | [illegible] |
| | | Shoes | 12,[illegible] |
| | | Shoes, men's and boys' | 5,705.[illegible] |
| | | Shoes, men's brogans | 2,[illegible] |
| | | Shoes, women's | 3,745.00 |
| | | Soap | 2,951.04 |
| | | Stone masonry | 2,867.00 |
| | | Stone, quarried | 9,128.00 |
| | | Tickings | 286.08 |
| | | Tinware | 1,397.[illegible] |

PENNSYLVANIA—CONTRACT SYSTEM.

| Kind. | Value. | Kind. | Value. |
|---|---|---|---|
| Brooms | $66,000.00 | | |
| Cigars | 64,000.00 | | |
| Iron, architectural | 20,000.00 | | |
| Shoes, men's, women's, and girls' | 354,452.51 | | |

PENNSYLVANIA—PIECE-PRICE SYSTEM.

| Kind. | Value. | Kind. | Value. |
|---|---|---|---|
| Hosiery, cotton | $158,085.00 | Hosiery, cotton | $10,040.40 |
| Hosiery, woolen and cotton | 6,050.00 | Brooms, etc | 64,091.91 |
| | | Mops | 1,086.37 |

RHODE ISLAND—CONTRACT SYSTEM.

| Kind. | Value. | Kind. | Value. |
|---|---|---|---|
| Boots and shoes | $75,000.00 | Shoes, men's and boys' | $151,482.40 |
| Wire goods (screens and railings) | 6,000.00 | Wire goods (screens and railings) | 8,000.00 |

SOUTH CAROLINA—PUBLIC-ACCOUNT SYSTEM.

| Kind. | Value. | Kind. | Value. |
|---|---|---|---|
| Building State canal | $80,000.00 | Brick | $15,000.00 |
| Clothing (for convicts) | 4,457.09 | Clothing (for convicts) | 2,008.00 |
| Farming | 10,000.00 | Farming | 68,010.00 |
| Repairing prison | 5,000.00 | General labor | 17,114.36 |

SOUTH CAROLINA—CONTRACT SYSTEM.

| Kind. | Value. | Kind. | Value. |
|---|---|---|---|
| Hosiery | $60,000.00 | Hosiery | $270,000.00 |
| Boots and shoes | 150,000.00 | | |

TABLE V.—GOODS PRODUCED OR WORK [illegible]

SOUTH CAROLINA—[illegible]

| [illegible] | | [illegible] | |
|---|---|---|---|
| Kind. | Value. | Kind. | [illegible] |
| Mining, phosphate | [illegible] | Farming, etc. | [illegible] |
| **SOUTH DAKOTA—Public-Account System.** | | | |
| | | Farming | [illegible] |
| | | Stone, dressed | [illegible] |
| | | Stone, quarried | [illegible] |
| **SOUTH DAKOTA—Contract System.** | | | |
| Stone, dressed | a $11,377.36 | | |
| **TENNESSEE—Lease System.** | | | |
| Farming | $6,500.00 | Farming | [illegible] |
| Mining, coal | 451,500.00 | Mining, coal | [illegible] |
| Mining, iron ore | 134,000.00 | Coke | [illegible] |
| Wagons | 550,000.00 | Grading railroad, etc. | [illegible] |
| | | Labor and camp duty at mines | [illegible] |
| | | Saddlery | [illegible] |
| | | Stoves and hollow ware | [illegible] |
| | | Work for State | [illegible] |
| **TEXAS—Public-Account System.** | | | |
| Engines, boilers, pumps, etc. | $16,187.00 | Engines, boilers, pumps, etc. | [illegible] |
| Furniture and lumber | 11,219.00 | Furniture and lumber | [illegible] |
| Cloth (for prison) | 16,480.00 | Cotton and woolen cloth and hosiery (for convicts). | [illegible] |
| Pig iron and castings | 129,000.00 | Pig iron | [illegible] |
| Shoes (for convicts) | 10,800.00 | Boots and shoes (for convicts) | [illegible] |
| Wagons and cotton presses | 48,965.00 | Wagons, buggies, etc | [illegible] |
| Mining, iron ore (and burning charcoal), etc. | 85,000.00 | Mining, iron ore | [illegible] |
| Stone, quarried | 40,000.00 | Burning charcoal | [illegible] |
| | | Blacksmith-shop products | [illegible] |
| | | Boilers | [illegible] |
| | | Building railroad | [illegible] |
| | | Building reservoir | [illegible] |
| | | Carpentering and planing-mill products. | [illegible] |
| | | Cast-iron water pipes | [illegible] |
| | | Chewing tobacco (for convicts) | [illegible] |
| | | Clothing, mattresses, and shoes | [illegible] |
| | | Farming | [illegible] |
| | | Foundry and machine-shop products. | [illegible] |
| | | Hollow ware, castings, etc | [illegible] |
| | | Paint shop | [illegible] |
| | | Pattern shop | [illegible] |
| | | Tin shop | [illegible] |
| | | Miscellaneous, labor | [illegible] |
| **TEXAS—Contract System.** | | | |
| Farming | b $220,000.00 | Farming | [illegible] |
| Building railroad | b 45,000.00 | Railroad labor | [illegible] |
| Saddletrees and stirrups | 30,000.00 | | |
| **VERMONT—Public-Account System.** | | | |
| | | Marble, dressed | [illegible] |

a Dakota Territory.

b [illegible]

TABLE V.—GOODS PRODUCED OR WORK DONE, 1885 AND 1895—Concluded.

VERMONT—CONTRACT SYSTEM.

| 1885. Kind. | 1885. Value. | 1895. Kind. | 1895. Value. |
|---|---|---|---|
| Marble (dressed) and monuments | $30,000.00 | Shoes, boys' and youths' | $16,885.00 |
| Shoes, women's | 90,837.75 | Shoes, children's | 51,269.00 |
| | | Shoes, women's and misses' | 245,446.50 |

VIRGINIA—PUBLIC-ACCOUNT SYSTEM.

| 1885. Kind. | 1885. Value. | 1895. Kind. | 1895. Value. |
|---|---|---|---|
| | | Farming | $12,000.00 |
| | | Roadmaking | 13,200.00 |

VIRGINIA—CONTRACT SYSTEM.

| 1885. Kind. | 1885. Value. | 1895. Kind. | 1895. Value. |
|---|---|---|---|
| Shoes, women's | $631,289.95 | Shoes, women's and misses' | $1,054,421.91 |
| Tobacco, plug and twist | 60,000.00 | Tobacco, plug and twist | 45,000.00 |
| Barrels, etc | 30,000.00 | | |
| Building railroad | 65,000.00 | | |

WASHINGTON—PUBLIC-ACCOUNT SYSTEM.

| 1885. Kind. | 1885. Value. | 1895. Kind. | 1895. Value. |
|---|---|---|---|
| | | Building and improving prison | $23,657.45 |
| | | Burlap | 439.95 |
| | | Grain bags, jute | 93,130.83 |
| | | Hop cloth | 8,021.34 |
| | | Kiln cloth | 27.11 |
| | | Matting | 225.60 |
| | | Ore bags, jute | 76.72 |
| | | Sewing twine | 63.12 |
| | | Wool bags | 1,578.21 |

WASHINGTON—LEASE SYSTEM.

| 1885. Kind. | 1885. Value. | 1895. Kind. | 1895. Value. |
|---|---|---|---|
| Sashes, doors, and blinds | $30,000.00 | | |

WEST VIRGINIA—CONTRACT SYSTEM.

| 1885. Kind. | 1885. Value. | 1895. Kind. | 1895. Value. |
|---|---|---|---|
| Brooms and leather whips | $125,000.00 | Brooms | $65,000.00 |
| Wagons | 150,000.00 | Fly nets, etc | 50,000.00 |
| | | Leather whips | 83,300.00 |
| | | Pantaloons | 43,750.00 |

WISCONSIN—PUBLIC-ACCOUNT SYSTEM.

| 1885. Kind. | 1885. Value. | 1895. Kind. | 1895. Value. |
|---|---|---|---|
| | | Hosiery | $29,741.50 |
| | | Overalls | 47,385.00 |

WISCONSIN—CONTRACT SYSTEM.

| 1885. Kind. | 1885. Value. | 1895. Kind. | 1895. Value. |
|---|---|---|---|
| Boots and shoes | $360,000.00 | Boots and shoes | $600,000.00 |

TABLE VI.—VALUE OF GOODS PRODUCED [illegible] 1885 AND 1895.

PUBLIC-ACCOUNT SYSTEM.

| State. | Value. 1885. | Value. 1895. | State. | Value. 1885. | Value. 1895. |
|---|---|---|---|---|---|
| Alabama | .......... | $64,912.81 | New Mexico | .......... | [illegible] |
| Arizona | $25,000.00 | 6,000.00 | New York | [illegible] | [illegible] |
| Arkansas | .......... | 54,018.65 | North Carolina | [illegible] | [illegible] |
| California | 136,412.56 | 283,768.13 | Ohio | [illegible] | [illegible] |
| Colorado | 30,000.00 | 12,230.52 | Oregon | [illegible] | [illegible] |
| Illinois | .......... | 773,546.62 | Pennsylvania | [illegible] | [illegible] |
| Iowa | 17,000.00 | 25,242.40 | South Carolina | [illegible] | [illegible] |
| Kansas | 268,594.77 | 166,324.25 | South Dakota | .......... | [illegible] |
| Kentucky | 37,200.00 | .......... | Texas | [illegible] | [illegible] |
| Maine | 72,000.00 | 43,826.13 | Vermont | .......... | [illegible] |
| Massachusetts | .......... | 175,347.69 | Virginia | .......... | [illegible] |
| Michigan | .......... | 121,646.55 | Washington | .......... | [illegible] |
| Minnesota | .......... | 150,000.00 | Wisconsin | .......... | [illegible] |
| Mississippi | .......... | 344,066.00 | | | |
| Nevada | 21,272.02 | 16,607.96 | Total | 2,6[illegible].25 | [illegible] |
| New Jersey | 8,288.10 | 10,582.60 | | | |

CONTRACT SYSTEM.

| State. | Value. 1885. | Value. 1895. | State. | Value. 1885. | Value. 1895. |
|---|---|---|---|---|---|
| Arkansas | .......... | $82,273.24 | Ohio | $[illegible] | [illegible] |
| Connecticut | $109,000.00 | 242,375.00 | Oregon | [illegible] | [illegible] |
| Illinois | 2,006,000.00 | 255,000.00 | Pennsylvania | [illegible] | .......... |
| Indiana | 1,551,807.82 | 799,700.00 | Rhode Island | [illegible] | [illegible] |
| Iowa | 381,590.00 | 251,636.00 | South Carolina | [illegible] | [illegible] |
| Kansas | 790,125.00 | 74,761.00 | South Dakota | a[illegible] | .......... |
| Maryland | 395,000.00 | 680,000.30 | Texas | [illegible] | [illegible] |
| Massachusetts | 79,000.00 | .......... | Vermont | [illegible].75 | [illegible] |
| Michigan | 818,778.00 | 286,330.00 | Virginia | [illegible] | 1,[illegible] |
| Minnesota | 245,500.00 | 175,000.00 | West Virginia | 275,[illegible] | [illegible] |
| Missouri | 1,135,000.00 | 1,184,062.47 | Wisconsin | [illegible] | [illegible] |
| New Hampshire | 100,000.00 | 138,862.50 | | | |
| New York | 4,903,575.00 | .......... | Total | 17,071,[illegible] | [illegible] |

PIECE-PRICE SYSTEM.

| State. | Value. 1885. | Value. 1895. | State. | Value. 1885. | Value. 1895. |
|---|---|---|---|---|---|
| California | $294,777.87 | .......... | New York | .......... | $[illegible] |
| Indiana | 16,493.25 | $7,720.88 | Ohio | 945,642.00 | [illegible] |
| Kentucky | .......... | 170,528.00 | Pennsylvania | 164,136.00 | [illegible] |
| Massachusetts | 341,732.00 | 859,596.17 | | | |
| Michigan | 35,000.00 | .......... | Total | 1,484,236.88 | 3,70[illegible] |
| New Jersey | 586,444.40 | 394,685.51 | | | |

LEASE SYSTEM.

| State. | Value. 1885. | Value. 1895. | State. | Value. 1885. | Value. 1895. |
|---|---|---|---|---|---|
| Alabama | $214,400.00 | $622,463.00 | New Mexico | $16,000.00 | .......... |
| Arkansas | 230,450.00 | .......... | North Carolina | 308,000.00 | .......... |
| Florida | 100,000.00 | 283,172.00 | South Carolina | 23,560.00 | $[illegible] |
| Georgia | 460,000.00 | 177,416.00 | Tennessee | 1,142,000.00 | [illegible] |
| Kentucky | 411,280.00 | .......... | Washington | 30,000.00 | .......... |
| Louisiana | 352,000.00 | 165,647.85 | | | |
| Mississippi | 324,000.00 | .......... | Total | 3,061,690.00 | 3,1[illegible] |
| Nebraska | 148,000.00 | 72,436.25 | | | |

TABLE VII.—SUMMARY OF VALUE OF GOODS PRODUCED OR WORK DONE, BY [illegible]TEMS OF WORK, 1885 AND 1895.

| Systems of work. | Value. 1885. | Value. 1895. |
|---|---|---|
| Public-account system | $2,062,[illegible].18 | $4,[illegible] |
| Contract system | 17,071,[illegible] | [illegible] |
| Piece-price system | 1,484,[illegible] | 3,7[illegible] |
| Lease system | 3,061,690.00 | 2,[illegible] |
| Total | 24,[illegible] | 2[illegible] |

a Dakota Territory.

TABLE VIII.—SUMMARY OF VALUE OF GOODS PRODUCED OR WORK DONE, BY STATES.

| State. | Value. | | State. | Value. | |
|---|---|---|---|---|---|
| | 1885. | 1895. | | 1885. | 1895. |
| Alabama | $314,400.00 | $657,976.41 | Nevada | $631,372.02 | $410,607.96 |
| Arizona | 35,606.00 | 6,000.00 | New Hampshire | 100,000.00 | 128,282.90 |
| Arkansas | 230,450.00 | 138,[illegible].29 | New Jersey | 509,782.00 | 406,306.11 |
| California | 421,191.42 | 282,766.13 | New Mexico | 16,000.00 | 22,281.00 |
| Colorado | 30,000.00 | 12,230.52 | New York | 5,828,482.48 | 2,708,972.06 |
| Connecticut | 100,000.00 | 342,375.00 | North Carolina | 261,073.13 | 130,000.00 |
| Florida | 100,000.00 | 283,173.00 | Ohio | 987,499.51 | 728,172.28 |
| Georgia | 400,000.00 | 177,416.00 | Oregon | 120,000.00 | 44,[illegible].50 |
| Illinois | 3,065,000.00 | 1,026,540.62 | Pennsylvania | 884,400.22 | 465,740.30 |
| Indiana | 1,562,301.07 | 607,420.88 | Rhode Island | 81,000.00 | 192,402.40 |
| Iowa | 300,000.00 | 246,679.40 | South Carolina | 323,017.00 | 347,312.00 |
| Kansas | 1,058,719.77 | 241,086.25 | South Dakota | a 11,577.36 | 7,[illegible].60 |
| Kentucky | 448,400.00 | 170,322.60 | Tennessee | 1,168,000.00 | 690,300.00 |
| Louisiana | 352,000.00 | 165,647.88 | Texas | 682,742.00 | 1,664,615.20 |
| Maine | 72,000.00 | 43,828.13 | Vermont | 129,897.75 | 228,000.00 |
| Maryland | 385,000.00 | 690,060.30 | Virginia | 796,209.96 | 1,124,021.91 |
| Massachusetts | 420,782.00 | 1,034,257.26 | Washington | 30,000.00 | 127,880.33 |
| Michigan | 853,775.00 | 407,976.55 | West Virginia | 275,000.00 | 242,050.00 |
| Minnesota | 245,500.00 | 325,000.00 | Wisconsin | 300,000.00 | 677,126.50 |
| Mississippi | 324,000.00 | 244,066.00 | | | |
| Missouri | 1,135,000.00 | 1,184,062.47 | Total | 24,371,078.39 | 19,042,472.33 |
| Nebraska | 148,000.00 | 72,436.25 | | | |

a Dakota Territory.

## ABSTRACT OF LAWS RELATING TO CONVICT LABOR PASSED SINCE 1885.

ALABAMA.—Act of February 18, 1895, provides in substance as follows: State convicts shall be hired or employed within the State as may be determined by the board of inspectors with the approval of the governor. All hiring made must be per capita. Not less than 50 State convicts shall be hired to one person or kept at one prison, except that where convicts are worked in the county where convicted, less than 50 may be worked at one place. The board of inspectors may make contracts for the hire of State convicts by the day, month, or year, or a term of years, the State in such cases controlling and supporting the convicts.

County convicts may be worked on public roads, bridges, and other public works in the county, and may be hired to labor anywhere within the State.

ARIZONA.—In section 2424 of the Revised Statutes of 1887, originally part of an act approved March 10, 1887, it is provided that the board of commissioners shall lease the labor of the convicts in the Territorial Prison to be employed within the prison walls in such manufacturing enterprises as the board may deem proper.

Section 9 of act No. 19 of the acts of 1895, approved March 8, 1895, provides that the board of control of the Territorial institutions shall have power to lease on shares or for cash the property, buildings, and lands belonging to the Territory for the purpose of furnishing employment for the inmates of the Territorial Prison and reform school, and to make contracts to furnish the labor of the inmates of said institutions either within or outside the walls of the prison or the confines of the reform school. The labor of said inmates shall not be leased when it is required upon the buildings or properties of said institutions.

ARKANSAS.—Sections 5499 to 5506, inclusive, of the Digest of 1894, originally forming part of an act approved March 21, 1893, direct that the board of commissioners of the penitentiary shall employ the convicts either within or outside the walls of the penitentiary, and shall purchase or lease and equip a farm or farms, upon which convicts who are not suitable for contract labor and who can not be made self-supporting within the walls shall be worked on State account; that the system of labor shall be the State account system or the contract system, or partly one and partly the other, but no contract shall be let if equally remunerative employment can be furnished by the State and worked on State account, and that said board shall establish within the walls of the penitentiary such industries as are for the best interests of the State, and as will furnish the charitable institutions of the State with such articles as are necessary to be used therein.

The board is authorized to have State coal lands opened and operated by convict labor on the State account system. It is also authorized to employ convicts on State timber lands in clearing and fencing the same and cutting the timber, and to purchase a tract of land on which there is an abundant supply of good building stone, and to employ convicts thereon in cutting, quarrying, and otherwise preparing the stone for use.

CALIFORNIA.—In chapter 208 of the acts of 1895, approved March 28, 1895, provision is made for establishing at one or both of the State prisons a rock or stone crushing plant, to be operated by convict labor, with such free labor as is necessary for superintendence and direction.

COLORADO.—Section 3447 of the Annotated Statutes of 1891, originally part of an act approved April 2, 1887, forbids the employment of the penitentiary or prison convicts outside the prison walls or grounds in the vicinity of such penitentiary or prison, and provides that the board of penitentiary commissioners shall not hire out any convict for the purpose of carrying on an industry that comes in competition with free labor in the State.

Section 4163, originally part of an act approved April 19, 1889, provides that so far as practicable the industries upon which the convicts of the State reformatory shall be employed shall be the manufacture of articles not elsewhere manufactured in the State.

CONNECTICUT.—Chapter 153 of the acts of 1895, approved May 23, 1895, provides that no convict shall be employed in or about the manufacture or preparation of any drugs, medicines, food or food material, cigars or tobacco, or any preparation thereof, pipes, chewing gum, or any other article or thing used for eating, drinking, chewing, or smoking, or for any other use within or through the mouth of any human being.

IDAHO.—Section 3 of article 13 of the constitution of the State, adopted in 1889, provides that all labor of convicts shall be done within the prison walls, except where the work is done on public works under the direct control of the State.

Section 2 of an act approved February 3, 1891, as amended by an act approved March 6, 1893, empowers the board of State prison commissioners, either by direct expenditure or by contract, to provide for the employment of all convicts confined in the penitentiary, and provides that no contract shall be let to perform any labor which will conflict with any existing manufacturing industries in the State.

KANSAS.—In sections 35 to 43, inclusive, of chapter 152 of the acts of 1891, approved March 11, 1891, the following provisions are made: The party hiring the labor of the convicts in the penitentiary shall be required, so far as practicable, to teach the prisoner as much of the trade at which he is employed as will enable him to work at the same when discharged. No contract shall be made for the employment of the prisoners outside of the prison grounds. The warden of the penitentiary is authorized to employ the labor of such convicts as are not required in other departments in mining coal upon the lands belonging to the State upon which the penitentiary is located and adjacent thereto, and to use such portion of the convict labor as may be necessary to keep the wagon road from the penitentiary to the city of Leavenworth in repair.

Section 5 of chapter 46 of the acts of 1895, approved March 8, 1895, makes it unlawful to allow convicts in the penitentiary to perform labor for private citizens outside the prison grounds, for hire or otherwise, and makes it the duty of the warden to employ the surplus convict labor in extending and repairing the State road and upon other work exclusively for the benefit of the State.

KENTUCKY.—Sections 253 and 254 of the constitution, adopted in 1891, provide that the labor only of penitentiary convicts may be leased and must be performed within the prison walls. Also, that the employment of convicts outside the walls can not be authorized except upon public works, or when, during pestilence or in case of the destruction of the prison building, they can not be confined in the penitentiary.

LOUISIANA.—Act No. 114 of the acts of 1890, approved July 10, 1890, provides for the extending of the lease for the labor of the penitentiary convicts for a period of ten years from March 3, 1891, but limits the employment of such labor to the levees, railroads, and other works of public improvement.

It absolutely prohibits the subletting, rental-out, or use by the lessee himself of the convict labor in the cultivation, planting, or gathering of any agricultural crop, such as rice, sugar, cotton, or corn.

Chapter 134 of the acts of 1894, approved July 11, 1894, permits the employment of the penitentiary convicts on plantation or farm work, or any other work not provided for in act No. 114 of the acts of 1890, when it shall have been demonstrated to the governor that employment for the convicts can not be obtained on the public levees of the State.

MASSACHUSETTS.—In chapter 447 of the acts of 1887, approved June 16, 1887, the following provisions were made:

No contract shall hereafter be made for the labor of prisoners confined in the State Prison, reformatories, or any of the houses of correction. Such prisoners shall be employed by the head officers of said institutions in such industries as shall from time to time be fixed upon. No new machinery to be propelled by other than hand or foot power shall be used in any such institution. The number of prisoners employed in a single industry at the same time in any institution shall not exceed one-twentieth of the number of persons employed in such industry in the State, except that county institutions doing business at the date of passage of this act on public account may continue such industries as they maintain, but can not employ more than 250 persons in any one industry at the same time. The goods to be manufactured in said institutions shall be, as far as may be, such articles as are in use in the several State and county institutions and as are necessary to their maintenance.

Chapter 22 of the acts of 1888, approved February 9, 1888, provides that the words "contract for the labor of prisoners," used in chapter 447 of the acts of 1887, shall not be construed as applying to a contract for the manufacture of articles by the piece, under what is known as the "piece-price system" with persons who furnish the materials used in such manufacture.

Section 2 of chapter 403 of the acts of 1888, as amended by chapter 371, acts of 1891, provides that the number of prisoners employed in any industry in the State Prison, Reformatory, Reformatory Prison for Women, or in any house of correction, shall not exceed one-twentieth of the number of persons employed in such industry in the State, unless a larger number is needed to produce articles to be supplied to State and county institutions. Fifty prisoners may be employed in the manufacture of brushes at the House of Correction at Cambridge, upon the public-account system.

Chapter 209 of the acts of 1891, approved April 17, 1891, provides that prisoners in any State institution shall not be employed outside the precincts of the same in any mechanical or skilled labor for private parties.

Chapter 460 of the acts of 1894, approved June 9, 1894, provides that no new contract for the employment of prisoners in the manufacture of reed or rattan goods shall be made until the number so employed is reduced to 75, and that thereafter no contract for the employment of more than 75 in the manufacture of such goods shall be approved.

MINNESOTA.—Section 3598 of the statutes of 1894, originally part of chapter 208 of the acts of 1887, as amended by chapter 112 of the acts of 1891, prohibits the contract system of convict labor in the State Reformatory, and provides that the board of managers shall retain control of the labor of the convicts. It also provides that during any year the board shall not employ or engage, on the average, to exceed 33 per cent of the prisoners in quarrying, manufacturing, and cutting of granite for sale.

In chapter 154 of the acts of 1895, approved [illegible] were made: The prisoners in the State Prison and [illegible] some trade or handicraft. No contracts shall be [illegible] the prisoners in said institutions at a certain rate [illegible] control of the labor, but they shall be employed by the [illegible] other chief officer having charge thereof, in such industrial [illegible] be fixed upon, or in the manufacture of articles by the piece [illegible] system" by contracts with persons who furnish the materials [illegible] ture. The number of prisoners employed in a single industry [illegible] any institution shall not exceed 10 per cent of the total number [illegible] such industry in the State unless a greater number is necessary [illegible] or articles to be supplied to State and other municipal institutions [illegible] table. This does not apply to the number of prisoners employed in [illegible] of binding twine at the State Prison at Stillwater.

The board of managers of these institutions shall, as far as may be, [illegible] tured in said prison and reformatory such articles as are in common [illegible] State institutions, whether penal or otherwise, and the officers of [illegible] shall purchase such of said articles as are necessary for the [illegible] institution they may represent.

MISSISSIPPI.—Sections 223 to 226, inclusive, of the State constitution, [illegible] 1890, provide that after the year 1894 no penitentiary convict shall ever [illegible] hired to any person or persons or corporation, private or public or [illegible] board; that said convicts may be employed under State supervision on [illegible] or other public works, or by any levee board on any public levees; that the [illegible] lature may place the penitentiary convicts on State farms, and have [illegible] thereon, under State supervision exclusively, in tilling the soil or [illegible] both; and that convicts in the county jails shall not be hired or leased to [illegible] or corporation outside the county of their conviction after 1892.

Section 3201 of the code of 1892 provides that after the year 1894 penitentiary [illegible] victs shall not be hired or leased out, but shall be employed in the [illegible] on a farm or farms leased for that purpose, under the sole control, [illegible] management of the officers and employees of the penitentiary; and that the [illegible] of control may work said convicts on public roads, works, and levees.

Chapter 75 of the acts of 1894 provides for the establishment of penitentiary [illegible], and that the board of control of the penitentiary may carry on in connection [illegible] said farms such industrial enterprises as may be deemed advisable, including [illegible] manufacture of drainage tile, brick, wagons, agricultural implements, [illegible] shoes, clothing, and other articles for the convicts.

Chapter 76 of the acts of 1894, approved February 10, 1894, provides that jail [illegible] victs shall be worked either by hiring them out to the best bidder or by [illegible] them, under the directions of the board of supervisors, on public roads or [illegible] on a farm or farms which the board may buy or lease or work on [illegible] delivering them to the county contractor; that the sheriff may hire out the [illegible] with their consent, or, if they do not consent, at public outcry, at the [illegible] courthouse, to persons who will pay in cash the amount of their several [illegible] and jail fees, or give bond for the payment of the same; and that if the [illegible] not hired out and the board of supervisors do not work them on the public [illegible] works, or farms, said board may agree with a person, as convict contractor, [illegible] them at a price to be agreed upon by the board.

MONTANA.—Section 2 of article 18 of the constitution of the State, [illegible] 1889, provides that the labor of the convicts in the reformatory [illegible] penitentiaries of the State shall not be let by contract.

Section 2960 of the Penal Code (Codes and Statutes, 1895, Sanders' [illegible] nally part of an act passed in 1895, provides that the [illegible] may be employed within the prison in any mechanical [illegible] without the walls or inclosure of the [illegible] otherwise, but that such employment must be under the exclusive control of the board of prison commissioners; and that neither the board nor the warden must let by contract to any person the labor of any convict in the prison.

NEBRASKA.—Section 5202 of the Compiled Statutes of 1895, originally part of chapter 66 of the acts of 1895, authorizes and empowers the board of public lands and buildings to lease the labor of convicts in the penitentiary to responsible persons. when in its judgment the best interests of the State will be subserved thereby.

NEVADA.—Chapter 91 of the acts of 1887, approved February 28, 1887, provides that all the boots and shoes required to be used by the prisoners in the State Prison shall be made in the prison shop, and that the managers of other State institutions shall be supplied with all they need for the use of those under their charge; also that the surplus product of the shop may be sold by the warden in open market, but only by wholesale in full cases and unbroken packages of not less than one dozen pairs of boots or shoes each.

NEW JERSEY.—Chapter 357 of the acts of 1895, approved March 28, 1895, provides that the convicts in the reformatory may be employed in agricultural or mechanical labor, and that the system employed shall be either the "piece-price plan" or the "public-account system," or partly one and partly the other, in the discretion of the board of managers.

NEW MEXICO.—Sections 39 and 59 of chapter 76 of the acts of 1889, passed over veto February 22, 1889, provides that the labor of the penitentiary convicts may be hired out to the best advantage, and that those who are not hired out may be employed in and about any work, labor, or improvement on the capitol building or grounds, in grading, repairing, opening, cleaning, or leveling the streets, alleys, roads, and bridges in and near the city of Santa Fe; in quarrying and hauling stone, and in securing, bettering, and protecting the banks of the Santa Fe River from overflowing or destruction, so as to prevent damage to the city of Santa Fe.

NEW YORK.—Section 29 of article III of the constitution of the State, adopted in 1894, provides that after January 1, 1897, no person sentenced to the several State prisons, penitentiaries, jails, and reformatories shall be required or allowed to work at any trade, industry, or occupation wherein or whereby his work, or the product or profit of his work, shall be farmed out, contracted, given, or sold to any person, firm, association, or corporation, and that convicts may work for, and the products of their labor may be disposed of to, the State or any political division thereof, or for or to any public institution owned or managed and controlled by the State or any political division thereof.

Chapter 586 of the acts of 1888, approved August 1, 1888, provides that no motive-power machinery for manufacturing purposes shall be placed or used in any of the penal institutions of the State, and that no convict in such institutions shall be required or allowed to work at any trade or industry where his labor, or the production or profit of his labor, is farmed out, contracted, given, or sold to any person or persons whomsoever; also, that only such articles as are commonly needed and used in public institutions of the State for clothing and other necessary supplies shall be manufactured in the penal institutions.

Chapter 382 of the acts of 1889, amending article 3 of title 2 of chapter 3 of part 4 of the Revised Statutes of 1881, and approved June 6, 1889, provides that no contract shall be made by which the labor or time of any prisoner in the State prisons shall be contracted, let, or hired to contractors at a price per day or for other period of time; that the labor of the prisoners shall be either for the purpose of production and profit or for the purpose of industrial training and instruction, or partly for one and partly for the other; that the system of productive training in the State prisons shall be either the "public-account system" or the "piece-price system," or partly one and partly the other; that none of the product of the labor of the convicts shall be sold for less than 10 per centum in excess of the cost of the materials used in the manufacture of such products; that in determining the line of productive labor to be pursued the superintendent shall select diversified lines of industry with reference to interfering as little as possible with the same lines of industry carried on by the citizens of the State; that the number of convicts employed at one time in manufacturing one kind of goods shall not exceed 5 per centum of the number of all persons within the State employed in manufacturing the same kind of goods, except in industries in which not to exceed 50 free laborers are employed, and provided that not more than 100 convicts shall be employed in all the prisons of the State in the manufacture of stoves, iron hollow ware, and boots and shoes, and that no convict shall be employed upon any of said specified industries in any of the penitentiaries, reformatories, or houses of correction in the State; that there shall be manufactured in the State prisons such articles as are commonly needed and used in the public institutions of the State, for clothing and other necessary supplies, and that such of said articles as are not needed in the State prisons shall be furnished to the several public institutions supported in whole or in part by the State; also that the labor of the prisoners in the State Reformatory at Elmira, and in the penitentiaries and other penal institutions of the State other than the State prisons, shall be conducted on the "public-account system" or the "piece-price system."

Chapter 237 of the acts of 1894, approved April 2, 1894, prohibits the manufacture of brushes, in whole or in part, by the convicts confined in the Albany Penitentiary.

Chapter 737 of the acts of 1894, approved May 21, 1894, restricts the total number of convicts employed in the State in manufacturing brooms and brushes from broom corn to 5 per centum of the total number of persons employed within the State in manufacturing such goods.

NORTH CAROLINA.—Section 4 of chapter 283 of the acts of 1893, ratified March 6, 1893, provides for the employment of the penitentiary convicts either within the walls of said institution or on farms leased or owned by the penitentiary, and makes it the duty of the superintendent to make contracts with persons or corporations, in order to employ and support as many of the able-bodied convicts on public works as the interests of the State and the constitution will permit

OHIO.—Section 7436-5 of the Revised Statutes, seventh edition, and originally part of an act passed in 1889, provides that the managers of the penitentiary shall employ all convicts directly for the State whenever the legislature shall provide means for machinery, materials, etc.; that said board shall have power to agree with manufacturers and others to furnish machinery, materials, etc., for the employment of the convicts, under the control of the managers and their officers; that such employment shall be on the "piece or process plan," and that bids for the product of such labor on said plan shall be advertised for; that the board may arrange with the employers of prisoners, so obtained, to pay for the labor of such number of prisoners necessary to the conduct of the general business (when they are employed in connection with larger numbers of other prisoners working by the piece or process plan), by the day or week, or otherwise, and that with this exception none of the labor of the prisoners in the penitentiary shall be let on contract by the day; that as far as practicable the board must employ all prisoners necessary in making all articles for the various State institutions not manufactured by such institutions, and said institutions shall purchase and pay to the penitentiary the market price for all such articles.

An act passed April 16, 1892, directs that the total number of prisoners and inmates employed at one time in the penitentiaries, workhouses, and reformatories of the State in the manufacture of any one kind of goods shall not exceed 5 per cent of the number of all persons in the State employed outside of said penitentiaries, etc., in manufacturing the same kind of goods, except in industries in which not more than 50 free laborers are employed.

An act passed April 21, 1893, prohibits any board or authority having the control or management of any penal, reformatory, or charitable institution, or asylum from contracting with any person, firm, or corporation for the manufacture of knit or woolen goods, or from establishing any mill or manufactory for the manufacture of such goods.

OKLAHOMA.—Section 5436 of the statutes of 1893 provides that the sheriff may employ jail convicts sentenced to hard labor within the jail, and the county shall be entitled to their earnings; also that he may employ said convicts outside of the jail or yard either for the county or for any municipality in the county in work on public streets, or highways, or otherwise.

OREGON.—Section 1 of an act approved February 23, 1895, authorizes the governor to contract and lease to any person, firm, or corporation the whole or any part of the labor of the convicts at any time confined in the penitentiary for any period or periods of time not exceeding ten years. It provides that the officers of the penitentiary shall have general charge and custody of the convicts while they are engaged in such labor, and that the labor must be performed within the penitentiary building or within the yard or inclosure thereof.

SOUTH CAROLINA.—Sections 662 and 663 of the Civil Statute Laws, originally part of chapter 21 of the acts of 1893, section 554 of the Criminal Statute Laws, originally part of chapter 481 of the acts of 1893, section 33 of article 5, and section 6 of article 12 of the constitution of 1895 provide for the labor of jail convicts upon the public works, roads, highways, and bridges, under control of the county and municipal authorities.

Section 578 of the Criminal Statute Laws, originally part of chapter 20 of the acts of 1889, provides that no hiring or leasing of convicts in phosphate mining shall be made by the board of directors of the penitentiary.

Section 9 of article 12 of the constitution of 1895 directs that in case penitentiary convicts are hired or farmed out, they shall be under the supervision and control of officers of the State.

SOUTH DAKOTA.—Chapter 11 of the acts of 1890, approved March 8, 1890, provides for the working of the undeveloped stone quarries belonging to the State, and situated upon the penitentiary grounds at Sioux Falls, by the labor of the penitentiary convicts.

Chapter 131 of the acts of 1893, approved March 6, 1893, provides for the establishment of a plant in the State Penitentiary for the manufacture of binding twine from hemp or flax fiber by the use of convict labor, and directs that in the manufacture of the twine preference shall be given to the fiber grown in the State.

TENNESSEE.—Chapter 204 of the acts of 1889, approved April 4, 1889, provides for the leasing of the penitentiary buildings, quarry, grounds, fixtures, etc., and the labor of the convicts for the term of six years from January 1, 1890.

Chapter 78 of the acts of 1893, approved April 4, 1893, provides that, for the purpose of removing, as far as possible, convict labor from competition with free labor and to utilize, under the exclusive control and management of the State and for and on the State's account, the labor of the convicts, a new penitentiary shall be erected, with such buildings and workshops within its outer walls as may be deemed necessary to utilize the labor of the convicts in diversified industries; that in connection with said penitentiary there shall be a farm, not exceeding 1,500 acres of land, and that coal lands shall be purchased of not more than 10,000 acres for the purpose of opening and operating a coal mine or mines to be worked by convict labor. Said chapter also provides that "from and after the expiration of the lease now in force between the State and the lessees of the hire and labor of the convicts of the State," the contract lease system shall be forever abolished.

TEXAS.—Chapter 82 of the acts of 1891, approved April 15, 1891, provides that the system of labor in the State penitentiaries shall be either the State account system or the contract system, or partly the one and partly the other, but no contract shall be let for any of such convict labor if equally remunerative employment can be furnished by the State and worked on State account, and that no contract shall be made by which the control of the convicts shall pass from the State or its officers. It also provides that all convicts shall be placed within the walls of the penitentiaries or on State farms and worked on State account as soon and speedily as possible.

UTAH.—Section 3 of article [illegible] that the legislature shall prohibit [illegible] convicts outside prison grounds, except [illegible] the State.

WASHINGTON.—Section 29 of article 2 of the [illegible] 1889, directs that after January 1, 1890, the [illegible] contract to any person, copartnership, company, or [illegible] ture shall provide for the working of convicts for the [illegible]

Section 1158 of the statutes and codes of 1891, [illegible] March 9, 1891, provides that all convicts may be employed [illegible] board of directors of the penitentiary in the performance [illegible] the manufacture of any article or articles for the State, or [illegible] is sanctioned by law.

It also provides that at the State Penitentiary no articles [illegible] for sale, except jute fabrics and brick.

WEST VIRGINIA.—Chapter 46 of the acts of 1893, approved [illegible] provides that the directors of the penitentiary shall be required [illegible] labor of the convicts, upon such branches of business and for [illegible] such articles, as in their judgment will best accomplish the [illegible] interests of the State; that all convicts not employed on contracts shall be in the performance of work for the State or temporarily hired; that a port[illegible] convicts may be employed in the manufacture and repair of articles us[illegible] State in carrying on the penitentiary or articles used by any of the ot[illegible] institutions; that any convicts not employed under contract may be em[illegible] let to contract on the piece-price system or employed in manufacturing for such articles as may be selected by the board of directors of the peniten[illegible] that convicts may be furnished to counties to work on public roads.

WISCONSIN.—Section 567d of the Annotated Statutes of 1889, originall[illegible] 437 of the acts of 1887, provides that the State board of supervision of c[illegible] reformatory, and penal institutions shall, whenever in the opinion of such[illegible] is best for the interest of the State, establish in the State prison the b[illegible] manufacturing.

WYOMING.—Chapter 37 of the acts of 1890–91, approved January 8, 1891 that it shall be the duty of the board of charities and reform, either expenditure or contract, to provide for the care, maintenance, and emplo[illegible] all inmates of the penitentiary, reform school, or any penal or reformator[illegible] tions in the State, but that no convict shall be used or contracted to be us[illegible] coal mine or occupation when the products of his labor may be in competi[illegible] that of any citizen of the State. It also provides that when the cost of ma[illegible] said convicts can be reduced to the State by their employment in some oc[illegible] not unreasonably laborious or unhealthy, or when they can be employed to or repair the place or surroundings of the place in which they are confi[illegible] shall be so employed.

UNITED STATES.—Chapter 213 of the acts of 1886–87 (second session, Fo[illegible] Congress), approved February 23, 1887, prohibits any officer, servant, or ag[illegible] United States Government from hiring or contracting out the labor, or [illegible] mitting any warden, agent, or official of any State prison, penitentiar[illegible] house of correction from hiring or contracting out the labor of any Unit[illegible] criminals confined in said institutions.

Chapter 529 of the acts of 1890–91 (second session, Fifty-first Congress), March 3, 1891, provides for the erection of three United States prisons in sections of the country, and also provides that the convicts therein be exclusively in the manufacture of such supplies for the Government as can be factured without the use of machinery, and that they be not worked [illegible] prison inclosure.

# INDUSTRIAL COMMUNITIES. (a)

BY W. F. WILLOUGHBY.

## CHAPTER IV.

## IRON AND STEEL WORKS OF FRIEDRICH KRUPP, ESSEN, GERMANY.

There is no industrial center whose development, in relation to the condition of labor, can be studied with greater profit than that of Essen, Germany, the seat of the great iron and steel works of Friedrich Krupp.

The business of Krupp was founded in 1810, before the age of steel had fairly begun, and in its career is presented a type of the development of modern industry during the nineteenth century. From a beginning of a small shop, employing scarcely a dozen workingmen, it has increased, at first slowly and then by leaps and bounds, to its present huge proportions—the largest single manufacturing establishment in the world.

The study of such an example of industrial evolution would of itself possess great interest, but accompanying this growth has been the creation of social institutions for the benefit of the employees of the works on a scale and in a variety existing to the same extent nowhere else—institutions that have had a profound influence in shaping public thought, both in Germany and elsewhere, concerning the best means of improving the condition of labor in strictly manufacturing districts. It is the existence of these institutions in connection with the industrial prosperity of Krupp's works that makes Essen a center that can be studied historically and analytically with especial profit. The village of Essen was in existence previous to the founding of the Krupp works. After a slow development during at least a thousand years, as far as records are in existence, Essen at the beginning of the present century was a small agricultural village comprising not over 3,000 inhabitants. The cultivation of the soil was almost the only industry. Though the mineral resources of the surrounding country were known, coal and iron were mined only in sufficient quantities to satisfy local household needs. This rural character changed suddenly with the rapid growth in the use of steam power. Mines were opened in rapid succession, and an influx of labor set in that has been likened to the California gold fever of 1849.

The history of the great iron and steel works of Krupp is but a repetition of that of the district in which they are situated. Founded in 1810, for nearly forty years the establishment progressed in the most

a See footnote to the beginning of this series of articles in Bulletin No. 3.

leisurely manner. In 1848, after an exi[illegible] employed but 72 workingmen all told. From [illegible] the present time its development has been with[illegible] trial annals. The following table, giving the popula[illegible] year to year and the total number of employees of the [illegible] Krupp in Essen, shows the relation between the city and [illegible]

POPULATION OF ESSEN AND EMPLOYEES OF THE KRUPP IRON AND [illegible] LIVING IN ESSEN, 1803 TO 1894.

| Year. | Population of Essen. | Employees of Krupp firm in Essen. | Year. | Population of Essen. | Employees of Krupp firm in Essen. | Year. | Population of Essen. | [illegible] |
|---|---|---|---|---|---|---|---|---|
| 1803 | 3,480 | | 1856 | | 970 | 1876 | [illegible] | [illegible] |
| 1813 | 4,000 | | 1857 | | 992 | 1877 | [illegible] | [illegible] |
| 1820 | 4,636 | | 1858 | 17,165 | 1,047 | 1878 | [illegible] | [illegible] |
| 1830 | 5,457 | | 1859 | | 1,391 | 1879 | [illegible] | [illegible] |
| 1832 | | 10 | 1860 | | 1,764 | 1880 | [illegible] | [illegible] |
| 1833 | | 9 | 1861 | 20,766 | 2,082 | 1881 | [illegible] | [illegible] |
| 1840 | 6,325 | | 1862 | | 2,512 | 1882 | [illegible] | [illegible] |
| 1843 | 7,119 | 99 | 1863 | | 4,185 | 1883 | [illegible] | [illegible] |
| 1844 | | 107 | 1864 | 31,327 | 6,693 | 1884 | [illegible] | [illegible] |
| 1845 | | 122 | 1865 | | 8,187 | 1885 | [illegible] | [illegible] |
| 1846 | 7,841 | 120 | 1866 | | 6,350 | 1886 | [illegible] | [illegible] |
| 1847 | | 93 | 1867 | 40,695 | 6,869 | 1887 | [illegible] | [illegible] |
| 1848 | | 72 | 1868 | | 6,317 | 1888 | [illegible] | [illegible] |
| 1849 | 8,734 | 107 | 1869 | | 6,318 | 1889 | [illegible] | [illegible] |
| 1850 | | 237 | 1870 | | 7,084 | 1890 | [illegible] | [illegible] |
| 1851 | | 192 | 1871 | 51,840 | 8,810 | 1891 | | [illegible] |
| 1852 | 10,475 | 340 | 1872 | | 10,394 | 1892 | | [illegible] |
| 1853 | | 352 | 1873 | | 11,671 | 1893 | | [illegible] |
| 1854 | | 360 | 1874 | | 11,543 | 1894 | | [illegible] |
| 1855 | 12,891 | 693 | 1875 | 55,045 | 9,743 | | | |

It should be understood that the figures just given for the number of employees of Krupp by no means include the total number. A constantly increasing number of employees have found homes just outside of the city limits. This is especially true of the period since 1873–74, when the largest group of workingmen's houses was erected. In the prosecution of its business the firm has found it advisable to acquire and manage extensive works outside of Essen. It now owns, in addition to its works at Essen, several iron works elsewhere, three coal mines—which, however, do not supply all the coal required—various iron mines in Germany and Spain, etc. The total number of employees in 1888 was 20,960, of whom 13,198 were in Essen. In 1891 the total number had increased to over 24,000, of whom 16,161 belonged to Essen, and in 1892 to over 25,000, of whom 16,865 were in Essen.

These figures stand for the actual number of employees only. The firm employs no women. The total number of employees and their families reached in 1888 the large number of 73,769, and in 1892, 85,591, all dependent upon this single manufacturing concern. It is unnecessary to more than mention these figures. The establishment of Krupp constitutes a state in itself.

Naturally such a rapid increase in population in a district unpre[illegible] pared to receive it introduced a number of evils. The greatest of the[illegible] was that of insufficient housing. The price of land and the rent [illegible] houses went up enormously. The following figures, giving the [illegible] number of houses and the average number of inmates per house in Essen at different periods, show how largely the increase in the number of houses failed to keep pace with the increase in the population:

NUMBER OF HOUSES AND PERSONS PER HOUSE AT ESSEN, 1820 TO 1890.

| Year. | Houses. | Persons per house. | Year. | Houses. | Persons per house. | Year. | Houses. | Persons per house. |
|---|---|---|---|---|---|---|---|---|
| 1820...... | ........... | 6.30 | 1852..... | 1,024 | 10.22 | 1871..... | 3,222 | 16.61 |
| 1830...... | ........... | 6.80 | 1855..... | 1,106 | 11.67 | 1875..... | ........... | 12.75 |
| 1840...... | 846 | 7.53 | 1858..... | 1,319 | 12.01 | 1880..... | 4,214 | 12.82 |
| 1843...... | 872 | 8.16 | 1861..... | 1,636 | 12.60 | 1885..... | 4,296 | 15.14 |
| 1846...... | 923 | 8.50 | 1864..... | 2,045 | 15.31 | 1890..... | 4,853 | 14.22 |
| 1849...... | 941 | 9.28 | 1867..... | 2,970 | 13.70 | | | |

In the distinctively workingmen's quarter the density of population was even greater. In 1864 in two streets of this quarter there were 1,443 persons living in 77 houses, or an average of 18.74 persons per house, and in another street 2,962 persons living in 124 houses, or an average of 23.89 per house.

A natural result of these conditions was a great rise in rents. According to official investigations the annual rent of a two-room house in 1855 was from 24 to 30 thalers ($17.14 to $21.42), and gradually rose until it was from 36 to 50 thalers ($25.70 to $35.70). The crowding also gave rise to insanitary conditions, so that, in the first of the two streets mentioned as examples of overcrowding, the death rate was 4.24 per cent of the population and in the second street 4.59 per cent, while the average for all Essen was but 3.41. The cholera epidemic of 1866 was particularly fatal in these quarters, many squares being almost depopulated.

The second result of this increase in population was the great rise in the prices of all commodities consumed by workingmen. At the same time the workingmen became thoroughly demoralized through various trade practices on the part of the merchants. Trade was solicited through the granting of prizes and portions of whisky with purchases, and the workingmen were encouraged by the merchants to run in debt in order that they might be more securely held in their power. In a word, Essen during this period presented all the evils found in one of our new mining or other rapidly growing industrial communities.

These evils the firm of Krupp sought to remove through its extensive building operations and the creation of the vast system of Consum-Anstalten or cooperative distributive stores. To these two institutions, corresponding to the two main classes of material wants of the laboring population, was joined a whole system of institutions, the object of which was to provide assistance to the workingmen in times of special need. These institutions took the form of separate funds for the insurance of the men against sickness and accident, of pension funds for old employees, of the encouragement and provision of facilities for life insurance, of savings institutions, etc. Such a community, if it was to advance in prosperity and welfare, needed scholastic institutions. Primary and secondary schools for general education were provided.

but special effort was made to provide instruction in practical in the way of industrial schools for boys and housekeeping and schools for girls.

Without attempting to state all the forms of activity adopted firm, the workingmen's institutions organized by it can be group the following main classes:

1. The housing of employees.
2. Relief and pension funds.
3. Workingmen's life insurance association.
4. Cooperative distributive stores.
5. Funds for the benefit of workingmen other than the and pension funds.
6. Schools.
7. Health service.
8. Other institutions.

## THE HOUSING OF EMPLOYEES.

The building operations of the Krupp firm for the housing employees have been conducted on a vast scale. As has been the conditions at Essen were such as to render such action imp These building operations have taken several forms to meet th ing needs of the different classes of the firm's employees. Th be grouped under the following heads:

1. Labor colonies or distinct communities of workingmen's dw
2. Maintenance of a building fund to aid employees to buil own houses.
3. Ménage or boarding house for unmarried employees.
4. Workingmen's Cooperative Boarding House for small gr single workingmen.

### GROUPS OF WORKINGMEN'S HOUSES.

The report of this Department on the Housing of the W People gives a brief account of the technical details of several better types of houses erected by the Krupp firm at Essen. It building up of the whole community and the policy pursued reg the provision of workingmen's houses that is the feature of inte the present report.

The first efforts of the firm date from the year 1861, when the of ten houses for boss workmen and foremen was commenced. then the efforts of the firm have never ceased, and every few new group of houses has been erected, though even then it h difficult to keep pace with the rapid increase in the number of em of the firm.

The selection of a building system was limited from the begin existing conditions. The firm preferred the cottage system, wh house is isolated and surrounded by a garden. This system, was utterly impracticable at Essen. In fact, there was land to be had in the vicinity of the works to accommodate the 2,358 family tenements built during the years 1871 to 1873, if erected on that plan. But even if the land could have been acquired, the prices were such as to have made the dwellings too expensive for workingmen. The distance that many of the men would have been compelled to live from their work would have also been a great inconvenience, and would have prevented them from taking their midday meal at home.

Another objection would have been the lack of water facilities in the vicinity of Essen, as, owing to the mining operations, all well water had been drawn off. A water service by means of water mains to the widely extended cottage district would necessarily have increased the rents considerably. For these reasons it was necessary to adopt a plan whereby the dwellings could be somewhat more closely concentrated. In every case, however, the effort has been made to provide for the dwellings a healthy location, free access to light and air, and an abundant supply of good water.

Though the majority of houses contain a number of tenements, they are completely detached from one another; there are numerous streets and open spaces; the Krupp waterworks supply ample drinking water, and the streets are well lighted by the Krupp gas works. Though the tenements in a house have a common outside entrance, inside they are completely isolated from each other.

The houses erected at different periods differ greatly. The later built houses are much superior, and a remarkable opportunity is given for the study of the evolution of the workingman's home.

In 1892 Mr. Gussman, an official of the Krupp firm, was invited to deliver an address before the Centralstelle für Wohlfahrts-Einrichtungen, in Berlin, concerning the housing operations of the Krupp firm. In this address is given, in the most direct and chronological order, an account of the erection of workingmen's houses at Essen. The following account is largely a reproduction of this address, supplemented by other information furnished by the firm (*a*):

The first tenement houses for working people were constructed in 1861–62. Two rows of houses, one with 6 tenements and the other with 4, were built for the foremen of the factory. Each tenement contains 3 rooms on the first story, 3 in the attic, and cellar space. The outer walls are of heavy stone and plaster work, and the inner partition walls of a heavy wooden framework filled in with broken stone and plastered over.

The first workingmen's colony, known as Alt-Westend, was built during the three summer months of 1863. It consists of one row of houses two stories high, containing 16 four-room tenements, and eight rows containing tenements of three or four rooms each, and so arranged that the four-room tenements can be divided into two-room tenements.

---

*a* Free use has been made of the translation of a part of this address, as given by Dr. Lindsay in the Annals of the American Academy of Political and Social [illegible] November, 1892.

The outer walls of the buildings are of plastered stonework in the first story, surmounted by plastered wood framework. Each tenement has a cellar. The houses are very plain, and, as Mr. Krupp expresses it, "were intended for poor families who must save, but who desire a healthful dwelling, and not for those to whom a few more thalers a year make no difference, when it means that they could live more comfortably."

During the winter of 1871-72 a second colony, called Neu-Westend, was completed. This colony consists of 10 double houses three stories high, each containing 2 two-room tenements on each floor—that is, 6 tenements to each house and 60 in all; also 8 double houses three stories high, with 2 three-room tenements per floor, or 6 tenements per house. This makes a total of 48 tenements of this kind and a grand total of 108 two and three room tenements for the colony. The houses in this colony are built of brick.

Another colony, known as Nordhof, was completed in the fall of 1871. This consists of 36 tenements of three and four rooms each, and 126 tenements of two rooms each, the latter built on the so-called "Baracken" system—that is, in solid rows. The latter houses are plain wood structures, two stories high. Each tenement has a separate entrance from the street. All sanitary arrangements, closets, etc., are outside of the house. The other 36 tenements have three and four rooms each and are built of brick. The houses are three stories high.

The next colony, called Baumhof (Dreilinden), was built in 1871 and enlarged in 1890. These houses are built in a more rural style, each having a garden, and some having stables. Eight houses have 4 four-room tenements each, 23 have 4 three-room tenements each, 6 have 3 four-room tenements each, and 4 have 3 five-room tenements each, making a total of 41 houses containing 154 tenements. The population in 1892 was 910 persons.

During the years 1872 and 1873 the Schederhof colony was built. This colony consists of 2 houses, each containing 6 four-room tenements; 44 houses, each containing 6 three-room tenements, and 36 houses, each containing 6 two-room tenements, or a total of 82 houses containing 492 tenements. Each house is three stories high and has a cellar and attic.

In addition to the above there are 70 houses containing 4 two-room tenements each, or a total of 280 tenements. These houses are built in solid rows. The population of this colony was 4,002 persons in 1892.

The largest of all the colonies is known as Cronenberg. It was built in the years 1872-1874, but has since then been enlarged from time to time. The colony is situated within 100 steps from the west boundary of the works. It covers over 50 acres of land, and, according to the census of 1892, had a population of 8,001 persons.

The colony consists of 226 three-story houses built with walls partly of rough brick and partly of rough stone. Some are in rows and some in pairs. They are built on eight streets running the length of the colony and ten cross streets, ranging in width between curbs from 8½ to 12½ meters (27.9 to 41 feet) and with sidewalks 2½ meters (8 feet) wide. In 1892 there were in all 1,437 tenements. Of these 720 have two rooms each, 600 have three rooms each, 104 have four rooms each, and 13 have five and six living rooms. Each house has a garden, a cellar, and a drying space.

In this colony are located dwellings for several officials and school teachers connected with the works. A parsonage, two school buildings, a Protestant church, several branches of the cooperative store, an apothecary shop, a post-office, a market place over three-fourths of an acre in size, a restaurant with games, bowling alley, etc., and a library with a large hall for workingmen's meetings have been established by the firm within the limits of Cronenberg.

In 1892 plans were prepared for another colony, larger than the preceding, to be known as Holsterhausen. According to the plans this colony consists of 280 tenement houses, 180 of which are built in pairs and 100 completely detached. Each tenement house has a large garden space at the side and rear. Near the center are a large market place and a public park. The 90 double houses accommodate four families each or two families per house, and the 100 detached houses contain one tenement each. The tenements in the former contain three rooms and an attic each. The detached houses, of which there are two types, contain five rooms and an attic each. The colony is planned to accommodate 460 families.

The annual rents of workingmen's tenements range (1892) as follows:

| | |
|---|---|
| Two-room tenements in rows (*Baracken*) | $14.28 to $21.42 |
| Other two-room tenements, with cellar | 21.42 to 25.70 |
| Three-room tenements, with cellar | 28.56 to 38.56 |
| Four-room tenements, with cellar | 42.84 to 47.60 |
| Five-room and larger tenements, with cellar | 49.98 to 78.54 |

The following table shows the financial status of the firm's housing operations, July 1, 1891:

CAPITAL INVESTED AND ANNUAL RECEIPTS AND EXPENDITURES OF THE KRUPP IRON AND STEEL WORKS FOR HOUSING OF WORKING PEOPLE, JULY 1, 1891.

| Items. | Essen. | Outside of Essen. | Total. |
|---|---|---|---|
| Tenements | a 3,659 | b 523 | c 4,182 |
| Capital invested: | | | |
| Buildings | $2,628,103.34 | $366,924.60 | $2,995,027.94 |
| Ground | 288,842.51 | 47,751.13 | 336,593.64 |
| Total | 2,916,945.85 | 414,675.73 | 3,331,621.58 |
| Total receipts from rents | 115,352.65 | 15,956.71 | 131,309.36 |
| Expenses: | | | |
| Repairs | 26,817.36 | 2,663.46 | 29,480.82 |
| Lighting, water, roads, etc | 16,100.70 | 719.95 | 16,820.65 |
| Taxes and insurance | 11,424.00 | 1,881.15 | 13,305.15 |
| Total | 54,342.06 | 5,264.56 | 59,606.62 |
| Net receipts | 61,010.59 | 10,692.15 | 71,702.74 |
| Per cent of net receipts of capital | 2.09 | 2.58 | 2.15 |

a Not including 43 rent free. b Not including 131 rent free. c Not including 174 rent free.

This shows a net income of 2.69 per [illegible] housing operations at Essen, and 2.58 per cent [illegible] where, or a net income of 2.15 per cent on the [illegible] such property.

In conclusion, it may be proper to present the [illegible] show the relation between the total number of [illegible] support upon the Krupp Company and these [illegible] In March, 1892, the number of employees was 25,200 [illegible] of families depending on the same 62,700, a total of [illegible]

Of the total number of persons depending for [illegible] Krupp Company, as just shown, 15,300 lived in [illegible] 25,800 rented from the firm, and 46,800 rented from [illegible]

## BUILDING FUND.

None of the houses built by the firm in these labor [illegible] been sold. The firm, however, is very desirous that [illegible] should become, as far as possible, owners of their [illegible] encourage this Mr. Krupp in 1889 set aside a sum of [illegible] ($119,000) to be employed in making loans to workingmen [illegible] to build. This money is loaned under the following conditions [illegible]

Applicants must be married and between 25 and 50 [illegible] have been at least three years in the service of the firm, and [illegible] than 3,000 marks ($714) per year. A payment of at least 300 [illegible] ($71.40) must be made from the applicant's own resources. If the [illegible] is demanded for the purchase of a house already built, the [illegible] appraised by experts and the loan made according to their [illegible] ment. If the loan is for the purpose of building a house, the [illegible] must be furnished and the name of the builder, and then, if the [illegible] approves, the loan will be paid in regular installments during the [illegible] ress of the work. The services of the firm's experts and architect will be furnished gratis to the borrower. The loan is secured by a [illegible] gage on the house and ground. Three per cent interest is [illegible] The capital is paid off in monthly or fortnightly installments, but [illegible] tional payments can be made to hasten the liquidation of the debt. In cases of illness payments may be temporarily suspended. The [illegible] ments and interest payments together rarely exceed ordinary [illegible] payments.

In 1891 seventy-five houses, varying in value from 1,000 to [illegible] marks ($238 to $3,094), had been erected through the medium of [illegible] loans. In that year between one-fifth and one-sixth of all the [illegible] of the firm lived in houses owned by themselves.

## BOARDING HOUSE FOR UNMARRIED EMPLOYEES.

The firm found it very desirable to make some provision for the housing of its unmarried employees. In 1856 it erected its first [illegible] or single men's boarding and lodging house, a building with [illegible] dations for 200 men. [illegible] largely increased. In 1891 the ménage was occupied by about 800 men. Since 1884 all married workingmen who are not skilled laborers, and therefore have small wages, and who are separated from their families, are obliged to become inmates of the ménage.

In the beginning the cost per day for lodging, dinner (meat four times per week), supper, washing, etc., was 58 pfennigs (13.8 cents). On January 1, 1862, the rate was increased to 66 pfennigs (15.7 cents), and in 1869 further increased to 70 pfennigs (16⅔ cents). Since the latter date, however, meat has been served at dinner daily. In 1874, owing to the increase in the prices of food products, another increase was made to 80 pfennigs (19 cents), which is the present rate charged. At present, however, meat or fish is also served in the evenings at supper three times a week.

The management of the ménage is exactly that of a military barrack. There are a number of beds in each room, meals are served at regular hours, and regulations prescribe the use to be made of the building. The club feature is introduced to a considerable extent. There is a special room for a small library, where periodicals and papers can be consulted, and there are rooms for a billiard table, a bowling alley, a restaurant, etc.

### COOPERATIVE BOARDING HOUSE FOR UNMARRIED EMPLOYEES.

It is unnecessary to say that there are a good many of the higher paid and more intelligent workingmen to whom the life in a ménage such as has been described is extremely distasteful. Mr. Krupp has appreciated the desires of this class, and in 1893 he inaugurated an extremely interesting plan of erecting small compact houses specially constructed for the accommodation of 30 unmarried men. As yet only one such house has been erected as an experiment, but if successful the system will be further extended.

The idea, as exemplified in the existing house, is to provide a house that can be taken by a club of 30 men who will run it as a sort of living club house and share the payment of the expenses. The house that has been built contains three stories. On the ground floor are the dining room and kitchen. There are three reading and studying rooms, dressing rooms, a bath, lavatories, etc. On the upper floors are the living rooms for 30 men. Part of the rooms are single and part double-bedded rooms. The rooms are cheerful and contain a bed, a bureau, a wardrobe, a table, and chairs. In the writing rooms each member has a special drawer in the table provided with lock and key.

Thirty men club together and agree to take this house. The rent paid is 10 marks ($2.38) per month by those occupying single, and 8 marks ($1.90) by those occupying double-bedded rooms. A general manager is then elected by the members from among their number, who has entire charge of the building. He appoints a housekeeper, who takes care of the building, does the cooking, etc. The determination of the cost of the meals rests entirely with the members themselves.

[illegible] in the morning a first [illegible] of two meat and [illegible] of meat, vegetables, [illegible] and in the evening a [illegible] than the dinner. Beer [illegible] a bottle.

[illegible] the cost to each member for [illegible] for maintenance of the [illegible] marks ($2.38) per month, [illegible] the cost of board is 1.25 marks (29¾ [illegible] board and lodging being, therefore, [illegible] per day. The members are mostly [illegible] earning from 4 to 5 marks (95 cents [illegible]

[illegible] of the firm, if the house is full, a yearly rent of [illegible] will be obtained. The cost of the ground and [illegible] marks ($15,470). The maximum interest [illegible] investment is therefore 4.17 per cent. It should be [illegible] that gas is supplied free by the firm, and the house [illegible] daughter who assists her are able to live on a small [illegible] husband is pensioned by the firm. A visitor to this [illegible] house is struck with the admirable arrangement [illegible] full provision for the comfort and pleasure of the [illegible] at the same time the absolute freedom left to each, [illegible] upon his own room as his home. A vacancy is [illegible] for by a great many workingmen.

## [illegible] RELIEF AND PENSION FUNDS.

### [illegible] GENERAL RELIEF AND PENSION FUND.

[illegible] relief and pension funds at Essen in its present form [illegible] This is due to the radical changes introduced in [illegible] all such institutions by the series of laws enacted [illegible] during the years 1883 to 1891 concerning the [illegible] workingmen.

[illegible] such funds in connection with the Krupp Works, [illegible] the year 1853, when a sick and death benefit fund [illegible] kasse) for the benefit of workingmen of the firm [illegible] entering into details concerning the fund, for [illegible] another system, it is sufficient to say that it com [illegible] the various services of a sick, accident, and [illegible] The following table, showing the receipts and [illegible] extended comment unnecessary. Thus for [illegible] of the compulsory features in

... insurance by the State, a system had been organized ... not quite as liberal benefits to workingmen as those ... the new series of laws.

... EXPENDITURES OF THE OLD GENERAL RELIEF AND PENSION ... KRUPP IRON AND STEEL WORKS, BY FIVE YEAR PERIODS, 1856 TO

| ...ns. | 1856-1860 | 1861-1865 | 1866-1870 | 1871-1875 | 1876-1880 | 1881-1884 | Total. |
|---|---|---|---|---|---|---|---|
| ...CEIPTS. | | | | | | | |
| ... | ......... | $11,314.85 | $35,818.39 | $66,564.91 | $177,662.80 | $253,207.53 | ...... |
| ... fees | $1,182 22 | 4,367.14 | 3,324.37 | 8,222 82 | 2,316 76 | 2,485.75 | $21,890.06 |
| ...utions of members | 16,332 66 | 79,211 03 | 128,065 18 | 254,489.26 | 232,015.16 | 237,008 69 | 948,021.98 |
| ...utions of firm | 8,166 46 | 39,605.52 | 64,482.59 | 127,255 39 | 116,007.57 | 118,504 34 | 474,021.87 |
| ... | 867.75 | 5,563.76 | 6,290 56 | 9,437 24 | 6,103 68 | 6,306.85 | 34,589 84 |
| ...est | 914.34 | 3 894.46 | 9,151.22 | 22,656 22 | 42,885.29 | 57,066.45 | 136,567.98 |
| ...cellaneous receipts | 645 85 | 152.12 | 844.33 | 4,838.76 | 2,060 90 | 731.05 | 9,27[illegible].01 |
| Total | 28,109.28 | 144 108 88 | 248,876.64 | 493 484.60 | 579,052.16 | 675,310 66 | 1,624,373.74 |
| EXPENDITURES. | | | | | | | |
| Hospital service | a8,490.40 | a45,824 75 | 17,189.52 | 78,184.23 | 22,051 50 | 24,467 74 | 156,208.14 |
| Sick benefits, cash | (b) | (b) | 52,542.24 | 120,363 18 | 136,359.64 | 119,299.40 | 428,564.56 |
| Physicians, medicines, nurses, etc. | 6,144 29 | 35,306.77 | 63,090.19 | 87,942 04 | 87,326.84 | 85,109.44 | 365,119.57 |
| Voluntary and other irregular relief | 813.49 | 21,776.09 | 35,465 24 | 47,040 43 | 44,392.88 | 24,568.77 | 174,656.90 |
| Pensions, cash | 687.49 | 1,895.02 | 4,719 85 | 9,057.62 | 23,355.34 | 44,508.74 | 84,224.06 |
| Death benefits, cash | 564.39 | 3 028.79 | 7,505.33 | 10,225.18 | 9,544.32 | 10,108.32 | 40,976.28 |
| Miscellaneous expenditures | 94 37 | 259 07 | 1,799.36 | 2,409.17 | 2,814.11 | 1,499.07 | 8,875.15 |
| Balance | 11,314 85 | 35,818.39 | 66,564.91 | 177,662.80 | 253,207.53 | 365,749.09 | 365,749.09 |
| Total | 28,109 28 | 144,108 88 | 248,876 64 | 493,484.60 | 579,052.16 | 675,310 66 | 1,624,373.74 |

a Including sick benefits   b Included in hospital service.

The following table shows in detail the average number of members, the number of cases of sickness, and the expenditure per member during the entire existence of the old general relief and pension fund. It should be noted that the item of expense of administration is not included.

AVERAGE MEMBERS, CASES OF SICKNESS, AND EXPENDITURE PER MEMBER OF THE OLD GENERAL RELIEF AND PENSION FUND OF THE KRUPP IRON AND STEEL WORKS, 1856 TO 1884

[It will be noticed that the average members from 1856 to 1882 as shown in this table agree with the total employees shown for the same years on page 480, while for 1883 and 1884 the figures are different. The explanation is not known. The figures are given as published by the company.]

| Year. | Average members | Cases of sickness | Cases of sickness per member. | Expenditure per member. | Year. | Average members. | Cases of sickness. | Cases of sickness per member. | Expenditure per member. |
|---|---|---|---|---|---|---|---|---|---|
| 1856 | 970 | 1,750 | 1.80 | $1.83 | 1871 | 8,810 | 20,651 | 2.34 | $5.82 |
| 1857 | 992 | 1,871 | 1.89 | 2 62 | 1872 | 10,394 | 13,280 | 1.28 | 5.68 |
| 1858 | 1,047 | 1,837 | 1.75 | 3.27 | 1873 | 11,671 | 22,874 | 1.96 | 5.19 |
| 1859 | 1,391 | 1,882 | 1.35 | 2.94 | 1874 | 11,543 | 22,474 | 1.95 | 5.98 |
| 1860 | 1,764 | 2,343 | 1.33 | 2.78 | 1875 | 9,743 | 19,828 | 2.04 | 7.80 |
| 1861 | 2,082 | 3,496 | 1.68 | 3.81 | 1876 | 8,998 | 16,664 | 1.86 | 7.17 |
| 1862 | 2,512 | 4,345 | 1.73 | 4.18 | 1877 | 8,586 | 15,150 | 1.76 | 7 25 |
| 1863 | 4,185 | 8,373 | 2.00 | 4.18 | 1878 | 9,414 | 17,886 | 1.90 | 7.02 |
| 1864 | 6,693 | 16,894 | 2.52 | 4.80 | 1879 | 7,964 | 14,334 | 1.80 | 8.24 |
| 1865 | 8 187 | 21 215 | 2.59 | 4.91 | 1880 | 8,806 | 16,851 | 1.91 | 7.65 |
| 1866 | 6,350 | 15,840 | 2.49 | 7.50 | 1881 | 10,598 | 20,774 | 1.96 | 7.28 |
| 1867 | 6,869 | 15,490 | 2.26 | 5.04 | 1882 | 11,011 | 21,139 | 1.92 | 7.03 |
| 1868 | 6,217 | 14,275 | 2.30 | 4.87 | 1883 | 10,402 | 19,629 | 1.89 | 7.32 |
| 1869 | 6,318 | 15,003 | 2.37 | 4.96 | 1884 | 10,214 | 18,715 | 1.83 | [illegible] |
| 1870 | 7,084 | 15,100 | 2.14 | 5.43 | | | | | |

They determine how much they will pay, [illegible] then makes the best use of this money that he [illegible] pays 1.25 marks (29¾ cents) per member per day. [illegible] including the housekeeper and her assistants, [illegible]

Meals of the following character are furnished: [illegible] breakfast of coffee and bread, and a second breakfast [illegible] cheese sandwiches; at noon a warm meal consisting of [illegible] bread, etc.; in the afternoon sandwiches again, and in [illegible] warm supper of meat, somewhat less substantial than [illegible] can be had at the house for 15 pfennigs (3.6 cents) a [illegible]

From the foregoing it can be seen that the cost to each [illegible] rent, including 1 mark (23.8 cents) per month for [illegible] furnishings of the house, averages about 10 marks [illegible] or 33⅓ pfennigs (7.9 cents) per day; the cost of board is 1.25 [illegible] cents) per day; the total cost for board and lodging being [illegible] about 1.58⅓ marks (37.7 cents) per day. The members [illegible] skilled iron workers and turners, earning from 4 to 5 marks [illegible] to $1.19) per day.

From the standpoint of the firm, if the house is full, a yearly [illegible] 2,708 marks ($644.50) will be obtained. The cost of the [illegible] building was about 65,000 marks ($15,470). The maximum [illegible] returns on the investment is therefore 4.17 per cent. It should [illegible] noted, however, that gas is supplied free by the firm, and the [illegible] keeper and her daughter who assists her are able to live on [illegible] salary, as her husband is pensioned by the firm. A visitor [illegible] cooperative lodging house is struck with the admirable arrange[illegible] here devised, the full provision for the comfort and pleasure [illegible] inhabitants, and at the same time the absolute freedom left to [illegible] so that he can look upon his own room as his home. A vacan[illegible] immediately competed for by a great many workingmen.

## RELIEF AND PENSION FUNDS.

### THE OLD GENERAL RELIEF AND PENSION FUND.

The system of relief and pension funds at Essen in its present [illegible] is of recent origin. This is due to the radical changes introduce[illegible] the organization of all such institutions by the series of laws ena[illegible] by the German Empire during the years 1883 to 1891 concerning [illegible] insurance of workingmen.

The existence of such funds in connection with the Krupp [illegible] however, dates from the year 1853, when a sick and death bene[illegible] (Kranken- und Sterbekasse) for the benefit of workingmen of [illegible] was created. Without entering into details concerning the [illegible] it is now replaced by another system, it is sufficient to say that [illegible] bined in one institution the various services of a sick, accid[illegible] old-age pension fund. The following table, showing the receipt[illegible] disbursements, renders extended comment unnecessary. [illegible] thirty years previous to the introduction of the compulsory [illegible] provided for by the new series of laws.

RECEIPTS AND EXPENDITURES OF THE OLD GENERAL RELIEF AND PENSION FUND OF THE KRUPP IRON AND STEEL WORKS, BY FIVE-YEAR PERIODS, 1856 TO 1884.

| Items. | 1856-1860 | 1861-1865 | 1866-1870 | 1871-1875 | 1876-1880 | 1881-1884 | Total. |
|---|---|---|---|---|---|---|---|
| RECEIPTS. | | | | | | | |
| Balance | | [illegible]1,314.[illegible] | [illegible],818.[illegible] | [illegible],564.91 | $177,[illegible] | $[illegible],307.[illegible] | |
| Entrance fees | $1,[illegible] | 4,[illegible].14 | 3,394.37 | 8,[illegible] | 3,216.7[illegible] | 2,4[illegible].73 | [illegible] |
| Contributions of members | 16,[illegible] | 79,211.6[illegible] | 1[illegible] | [illegible]4,490.36 | 2[illegible],076.16 | 257,002.[illegible] | 946,[illegible] |
| Contributions of firm | 8,[illegible] | 20,605.[illegible] | 64,[illegible] | 127,236.30 | 116,[illegible].57 | 112,504.34 | 474,021.[illegible] |
| Fines | 257.7[illegible] | 5,[illegible] | 6,290.5[illegible] | 9,457.3[illegible] | 6,10[illegible] | 6,[illegible] | 34,[illegible] |
| Interest | 304.34 | 3,894.46 | 9,151.32 | 22,656.22 | 42,[illegible] | 57,[illegible] | 134,567.[illegible] |
| Miscellaneous receipts | 645.[illegible] | 152.19 | 844.[illegible] | 4,[illegible].76 | 2,[illegible].90 | 731.05 | 9,3[illegible].81 |
| Total | 28,100.28 | 144,108.88 | 248,876.64 | 493,494.60 | 579,052.16 | 675,310.66 | 1,024,373.74 |
| EXPENDITURES. | | | | | | | |
| Hospital service | a8,490.40 | a45,824.78 | 17,183.82 | 38,184.33 | 33,051.[illegible] | 34,467.74 | 156,[illegible].14 |
| Sick benefits, cash | (b) | (b) | 52,542.24 | 120,[illegible].18 | 136,[illegible].64 | 119,[illegible].40 | 428,[illegible] |
| Physicians, medicines, nurses, etc. | 6,144.39 | 35,606.77 | 63,090.19 | 87,942.04 | 67,[illegible].84 | 66,109.46 | 323,[illegible].67 |
| Voluntary and other irregular relief | 812.49 | 21,776.09 | 36,465.24 | 47,648.43 | 44,382.68 | 24,[illegible].77 | 174,[illegible].90 |
| Pensions, cash | 687.48 | 1,806.02 | 4,719.83 | 9,057.62 | 22,255.34 | 44,508.74 | 84,224.06 |
| Death benefits, cash | 964.39 | 3,038.79 | 7,505.33 | 10,225.13 | 9,844.33 | 10,106.32 | 40,876.36 |
| Miscellaneous expenditures | 94.87 | 289.67 | 1,798.36 | 2,409.17 | 2,814.11 | 1,490.07 | 8,875.18 |
| Balance | 11,314.85 | 35,618.3[illegible] | 96,564.91 | 177,662.80 | 253,307.58 | 365,749.09 | 365,749.09 |
| Total | 28,100.28 | 144,108.88 | 348,876.64 | 493,484.60 | 579,052.16 | 675,310.66 | 1,624,373.74 |

a Including sick benefits. b Included in hospital service.

The following table shows in detail the average number of members, the number of cases of sickness, and the expenditure per member during the entire existence of the old general relief and pension fund. It should be noted that the item of expense of administration is not included.

AVERAGE MEMBERS, CASES OF SICKNESS, AND EXPENDITURE PER MEMBER OF THE OLD GENERAL RELIEF AND PENSION FUND OF THE KRUPP IRON AND STEEL WORKS, 1856 TO 1884.

[It will be noticed that the average members from 1856 to 1882 as shown in this table agree with the total employees shown for the same years on page 480, while for 1883 and 1884 the figures are different. The explanation is not known. The figures are given as published by the company.]

| Year. | Average members. | Cases of sickness. | Cases of sickness per member. | Expenditure per member. | Year. | Average members. | Cases of sickness. | Cases of sickness per member. | Expenditure per member. |
|---|---|---|---|---|---|---|---|---|---|
| 1856 | 970 | 1,760 | 1.80 | $1.83 | 1871 | 8,810 | 20,651 | 2.34 | $3.82 |
| 1857 | 982 | 1,871 | 1.90 | 2.63 | 1872 | 10,394 | 13,280 | 1.28 | 5.64 |
| 1858 | 1,047 | 1,837 | 1.75 | 3.27 | 1873 | 11,671 | 22,874 | 1.96 | 5.19 |
| 1859 | 1,391 | 1,882 | 1.35 | 2.94 | 1874 | 11,543 | 22,474 | 1.95 | 5.98 |
| 1860 | 1,764 | 2,343 | 1.33 | 2.78 | 1875 | 9,743 | 19,828 | 2.04 | 7.90 |
| 1861 | 2,062 | 3,490 | 1.68 | 3.81 | 1876 | 8,908 | 16,064 | 1.80 | 7.17 |
| 1862 | 2,512 | 4,348 | 1.73 | 4.18 | 1877 | 8,566 | 15,150 | 1.76 | 7.25 |
| 1863 | 4,185 | 8,373 | 2.00 | 4.18 | 1878 | 9,414 | 17,880 | 1.90 | 7.02 |
| 1864 | [illegible] | 16,[illegible] | 2.52 | 4.80 | 1879 | 7,964 | 14,334 | 1.80 | 8.[illegible] |
| 1865 | 8,157 | 21,215 | 2.59 | 4.91 | 1880 | 8,806 | 16,851 | 1.91 | 7.65 |
| 1866 | 6,380 | 15,840 | 2.49 | 7.30 | 1881 | 10,568 | 20,774 | 1.96 | 7.28 |
| 1867 | 6,[illegible] | 15,490 | 2.36 | 5.04 | 1882 | 11,011 | 21,130 | 1.92 | 7.03 |
| 1868 | 6,217 | 14,275 | 2.30 | 4.87 | 1883 | 10,402 | 19,629 | 1.89 | 7.32 |
| 1869 | 6,[illegible] | 15,[illegible] | 2.37 | 4.98 | 1884 | 10,214 | 18,715 | 1.83 | 7.76 |
| 1870 | 7,[illegible] | 15,190 | 2.14 | 5.[illegible] | | | | | |

## NEW SICK FUND.

The old fund was forced to wind up its operations in accordance with the national law of June 15, 1883, which, while making the maintenance of sick and pension funds obligatory, prohibited their union in a single fund. To replace this old fund, therefore, there were organized by the Krupp firm two independent funds, a sick fund (Krankenkasse) and a pension fund (Arbeiter-Pensions-, Wittwen- und Waisenkasse).

On winding up the affairs of the old fund, a net surplus of 1,536,760.87 marks ($365,749.09) remained, as will be seen by the table. Of this sum it was determined to apportion 10,000 marks ($2,380) to the new sick fund, with the proviso that in case an extraordinary demand should be made on the sick fund, as in case of an epidemic, before the legal reserve fund could be accumulated, a further sum of 100,000 marks ($23,800) could be claimed. This, however, it never became necessary to do. The remainder of the old fund's assets was transferred to the new pension fund, for it was for the payment of these pensions that the large reserve fund had been accumulated.

Membership in the new sick fund is obligatory for all persons employed by the firm with the exception of officials receiving more than 6⅔ marks ($1.59) per day, or 166⅔ marks ($39.67) per month, or 2,000 marks ($476) per year. Provision is also made for voluntary membership.

The resources of the fund, in addition to the 10,000 marks ($2,380) received from the old fund, consist practically of contributions by members of an amount equal to 2 per cent of their average wages as far as the latter do not exceed 4 marks (95 cents) per day, and contributions by the firm of a sum equal to one-half of that amount. Other receipts are unimportant, and consist of interest on funds invested, entrance fees, fines, and other miscellaneous items. If the state of the funds permit, the contribution of the members can be reduced, provided that their contributions to the pension fund are augmented to an equal amount. In consequence of this provision, the contributions of members were reduced in 1889 to 1.7 per cent of their average wages. Considerable latitude is also given regarding the power to raise and lower the rate of benefits granted in order to conform to the needs of the fund's budget. A reserve fund is provided for, the minimum amount of which can not be less than the average, of the last three years' expenditures.

From this fund provision is made for the granting of the following benefits:

1. Free medical attendance by a physician employed by the fund, free medicine, free medical supplies, etc. This includes hospital treatment and nursing for members who are married or members of families, if the disease is of such a nature as to require treatment outside their homes, and to all other persons unconditionally. In this case, however, the cash benefits paid to sick members are somewhat reduced.

2. Cash benefits to workingmen when sick, after the third day of their illness, equal to one-half their average wages. If they are married or widowed and have children under 15 years of age who are not working, an additional allowance of 5 per cent for each child is accorded. Such total additional allowances can, however, in no case exceed 16⅔ per cent of the wages, so that the regular and additional benefits can never exceed two-thirds of the recipient's wages. If the member has been in the employ of the firm less than five years, these benefits are paid during thirteen weeks only; but if they have been so employed five or more years, during a period of twenty-six weeks. Liberal power, however, is given to the management of the fund to continue the payments in worthy cases for a year if necessary.

3. In case of death a cash benefit equal to 20 times the amount of the decedent's average daily wages, but not to exceed 80 marks ($19.04) nor be less than 20 times the average daily wages of ordinary laborers. A death benefit equal to two-thirds of this benefit is also paid in case of the death of the wife of a member.

The sick-benefit fund employs 16 physicians, 10 of whom live in Essen and the others in neighboring towns where the firm employs labor. Two oculists are among the physicians who reside in Essen. Arrangements also exist whereby the services of specialists for throat and other diseases may be secured whenever recommended by the attending physicians. The choice of physicians is optional with the patient. The compensation of the physicians depends upon the number of members of the fund and the number treated by each, respectively. For instance, the total for all physicians is determined in a lump sum, according to the membership. This amount is then apportioned to the physicians according to the number of persons treated by each. In order to facilitate the rendering of aid in case of injury, three special attendants (Heildiener) are stationed at two bandage dispensaries (Verbandstationen) within the works. Medicines, bandages, etc., are also placed in six portable medicine and bandage chests distributed at convenient points. Three special attendants are employed for performing minor surgical operations, such as cupping, leeching, tooth extracting, etc., whenever directed by the attending physicians. They receive compensation according to a fixed schedule of rates.

All apothecaries in Essen, Altendorf, Altenessen, Borbeck, Berge-Borbeck, Rellinghausen, Werden, and Muhlheim-on-the-Ruhr participate in furnishing medicines. The sick-benefit fund enjoys a rebate of from 10 to 25 per cent on all medicines. The charges allowed for medicines are determined by a paid pharmacist in accordance with the usual rates charged for medicines, and are verified by a factory physician. The bandage material is bought at wholesale directly from the manufacturer. All bandages, as well as orthopedic instruments, artificial limbs, etc., are furnished gratis to members of the fund.

The claims for assistance are usually made by a representative of the working people who is on the board of directors. He examines the applications, usually certified to by [illegible] result to the board. Two inspectors [illegible] the patients receiving aid and to keep a [illegible] condition of the recipients. They report [illegible] of the sick fund any facts that may be of [illegible] assistance to be granted.

Another feature of the sick fund is one in [illegible] members. By its provisions every member may [illegible] (29¾ cents) quarterly to the so-called "family fund," [illegible] to the services of a physician for members of his [illegible] not already connected with some sick fund or are not [illegible] to aid in case of sickness. Since April 1, 1890, this [illegible] extended to the families of pensioned invalids. A [illegible] tion, which at present amounts to 1,200 marks ($285.[illegible] of the workingmen's fund (Arbeiter-Stiftung) in aid of [illegible]

The administration of the fund is in the hands of a [illegible] and board of directors elected by the firm and by the [illegible] fund. For the duties and powers of these bodies reference mus[illegible] to the constitution.

The operations of the sick-benefit fund extend far beyond th[illegible] scribed by law. They include, among other things, maintenan[illegible] long a period as fifty-two weeks for persons over five years in the [illegible] of the firm, and extra assistance in addition to regular sick [illegible] equal to 5 per cent of the wages for each child under 15 years [illegible] and not working, the total not exceeding two-thirds of a [illegible] wage of 4 marks (95 cents) per day.

Notwithstanding the liberality of this fund the firm of Kr[illegible] done a great deal in other ways for the relief of suffering [illegible] from sickness among its employees.

In 1879 Mr. Alfred Krupp donated a sum of 6,000 marks ($1,[illegible] relief in cases of sickness and distress in families, in comme[illegible] of the golden wedding of the late Emperor William I. Later [illegible] of Friedrich Krupp placed at the disposal of this fund an annu[illegible] 3,000 marks ($714). A careful statement is kept of the dispo[illegible] these funds.

The directors of the fund have, furthermore, at their dispos[illegible] per cent interest on a sum of 40,000 marks ($9,520) donated [illegible] Friedrich Alfred Krupp. This interest, namely, 2,000 marks ($[illegible] year, is intended for the payment of a part of the expenses [illegible] needy employees who have members of their families under t[illegible] in the women's and children's division of the Krupp Hospital.

Lastly the trustees of the workingmen's fund (Arbeiter-Stiftu[illegible] turned over to the directors of the sick fund a sum of 1,50[illegible] ($357) for the last half of the year 1889 and 3,000 marks ($[illegible] the year 1890, to be used for assisting members and their fa[illegible]

The directors of the sick fund [illegible] amounts of money for assisting [illegible] own illness but also in case of indigence in their families. The operations of the fund proper are shown in the table that follows. It is unnecessary to make any comment. The salient features may be seen at a glance. It is of interest, however, to know what is the average expenditure per member, and whether this average tends to increase or diminish during the year. The necessary calculations have been made in the following statement. The same information was given for the old fund, embracing the years 1856 to 1884. The two tables together, therefore, show the average annual cost per member of a general sick fund.

AVERAGE MEMBERS AND EXPENDITURE PER MEMBER OF THE NEW SICK FUND OF THE KRUPP IRON AND STEEL WORKS, 1885 TO 1894.

| Year. | Average members. | Total expenditure. | Average expenditure per member. |
|---|---|---|---|
| 1885 | 10,686 | $65,944.89 | $6.[illegible] |
| 1886 | 11,784 | 74,172.21 | 6.29 |
| 1887 | 12,694 | 86,266.39 | 6.79 |
| 1888 | 13,298 | 81,680.59 | 6.16 |
| 1889 | 14,061 | 87,232.12 | 6.20 |
| 1890 | 15,572 | 92,350.00 | 5.93 |
| 1891 | 16,065 | 101,858.64 | 6.33 |
| 1892 | 16,779 | 109,667.82 | 6.54 |
| 1893 | 17,074 | 108,237.56 | 6.34 |
| 1894 | 16,705 | 97,370.17 | 5.83 |

RECEIPTS AND EXPENDITURES OF THE NEW SICK FUND OF THE KRUPP IRON AND STEEL WORKS, 1885 TO 1894.

| Items. | 1885. | 1886. | 1887. | 1888. | 1889. | 1890. |
|---|---|---|---|---|---|---|
| RECEIPTS. | | | | | | |
| Balance on hand at beginning | $2,380.00 | $34.74 | $186.98 | $281.89 | $4,514.26 | $9,833.38 |
| Contributions of members | 58,686.02 | 57,384.29 | 61,759.61 | 66,469.86 | 64,366.21 | 65,568.15 |
| Contributions of firm | 26,348.01 | 28,642.14 | 30,879.81 | 33,234.93 | 31,480.03 | 32,788.53 |
| Interest | 511.31 | 1,347.06 | 1,904.79 | 3,436.72 | 3,351.04 | 3,351.04 |
| Restitution by others for aid granted sick | 628.45 | 1,826.41 | 2,436.32 | 1,785.82 | 2,864.90 | 3,163.06 |
| Miscellaneous receipts | 2,640.84 | 6.48 | 32.81 | 5.63 | 6.96 | 20.99 |
| Total | 90,205.63 | 88,640.14 | 97,200.27 | 105,214.85 | 106,585.50 | 114,725.15 |
| EXPENDITURES. | | | | | | |
| Expenses of administration | 680.73 | 764.84 | 692.70 | 750.20 | 725.35 | 747.89 |
| Medical attendance | 8,257.10 | 9,074.72 | 10,096.27 | 10,259.86 | 10,789.07 | 12,084.97 |
| Medicines | 10,977.32 | 13,443.59 | 14,073.13 | 13,020.88 | 14,111.07 | 15,150.60 |
| To hospitals for attendance and nursing | 4,144.78 | 4,481.15 | 5,633.11 | 5,949.32 | 5,799.93 | 7,614.66 |
| Sick benefits paid to members | 26,343.95 | 38,353.83 | 46,547.50 | 37,145.65 | 40,856.06 | 41,055.79 |
| Sick benefits paid to dependents of members | 557.13 | 587.14 | 977.59 | 7,212.84 | 7,601.64 | 7,818.69 |
| Death benefits | 2,327.20 | 3,006.62 | 3,456.55 | 3,060.04 | 3,433.78 | 3,571.43 |
| Miscellaneous relief | 1,317.69 | 1,517.49 | 1,552.96 | 1,511.06 | 1,885.80 | 1,768.77 |
| Miscellaneous expenditures | 1,350.09 | 2,941.83 | 3,178.58 | 2,750.74 | 1,929.43 | 2,537.21 |
| Contributed to reserve fund | 30,226.00 | 14,280.00 | 10,710.00 | 19,040.00 | 9,520.00 | 14,280.00 |
| Balance on hand at end | 34.74 | 186.98 | 281.89 | 4,514.26 | 9,833.38 | 8,095.06 |
| Total | 90,205.63 | 88,640.14 | 97,200.27 | 105,214.85 | 106,583.50 | 114,725.15 |
| Reserve fund at beginning | .......... | 30,226.00 | 44,506.00 | 55,216.00 | 74,256.00 | 83,776.00 |
| Contributed to reserve fund | 30,226.00 | 14,280.00 | 10,710.00 | 19,040.00 | 9,520.00 | 14,280.00 |
| Reserve fund at end | 30,226.00 | 44,506.00 | 55,216.00 | 74,256.00 | 83,776.00 | 98,056.00 |

RECEIPTS AND EXPENDITURES OF THE [illegible] AND STEEL WORKS, [illegible]

| Items. | 1881. | [illegible] | [illegible] | [illegible] |
|---|---|---|---|---|
| RECEIPTS. | | | | |
| Balance on hand at beginning | [illegible] | [illegible] | [illegible] | [illegible] |
| Contributions of members | 69,447.67 | 73,[illegible] | [illegible] | [illegible] |
| Contributions of firm | 34,704.80 | [illegible] | [illegible] | [illegible] |
| Interest | 2,769.80 | 5,[illegible] | [illegible] | [illegible] |
| Restitution by others for aid granted sick | 3,215.11 | 1,[illegible] | [illegible] | [illegible] |
| Miscellaneous receipts | 7.06 | 170.72 | 1,[illegible] | [illegible] |
| Total | 118,230.44 | 125,567.03 | [illegible] | [illegible] |
| EXPENDITURES. | | | | |
| Expenses of administration | 751.24 | 824.72 | 911.[illegible] | [illegible] |
| Medical attendance | 12,898.40 | 14,[illegible] | [illegible] | [illegible] |
| Medicines | 16,466.56 | 20,324.81 | [illegible] | [illegible] |
| To hospitals for attendance and nursing | 9,055.99 | 3,128.14 | 6,431.78 | [illegible] |
| Sick benefits paid to members | 45,246.62 | 48,750.57 | 51,216.29 | [illegible] |
| Sick benefits paid to dependents of members | 8,366.65 | 8,665.86 | 4,726.67 | [illegible] |
| Death benefits | 4,536.09 | 4,[illegible] | 4,[illegible] | [illegible] |
| Miscellaneous relief | 2,273.90 | 2,404.87 | 71.78 | [illegible] |
| Miscellaneous expenditures | 2,493.17 | 1,190.54 | 2,[illegible] | [illegible] |
| Contributed to reserve fund | 8,330.00 | .......... | 12,961.60 | .......... |
| Balance on hand at end | 8,041.80 | 16,519.41 | 2,162.66 | [illegible] |
| Total | 118,230.44 | 125,567.03 | [illegible] | [illegible] |
| Reserve fund at beginning | 98,066.00 | 106,396.00 | 106,396.00 | [illegible] |
| Contributed to reserve fund | 8,330.00 | .......... | 12,961.60 | .......... |
| Reserve fund at end | 106,396.00 | 106,396.00 | a118,296.60 | [illegible] |

a It will be noticed that $12,961.60 is carried to the reserve fund, while the reserve [illegible] only $11,900. The explanation of this is not known; the figures are as furnished by [illegible]

## NEW PENSION FUND.

The imperial law of June 15, 1883, regarding workingmen's [illegible] necessitated, as has been shown, the abolition of the old sick and [illegible] sion fund and the creation in its place of two independent [illegible] constitution of the new pension fund (Pensionskasse für die [illegible] fabrik der Firma Fried. Krupp) was adopted October 22, 188[illegible] bership in this fund is obligatory, and applies to the same [illegible] are included in the sick fund.

The resources of the fund were made to consist of (1) the entire [illegible] of the old sick and pension fund, excepting the 10,000 marks ([illegible] previously mentioned, that were turned over to the sick-benefit [illegible] (2) an entrance fee from each member of an amount equal to 1[illegible] his daily wages, where such wages do not exceed 4 marks (95 [illegible] (3) a contribution by each member of an amount equal originally [illegible] per cent, but increased July 1, 1889, to 1.3 per cent of his annual [illegible] at the same time that his contribution to the sick fund was [illegible] from 2 per cent to 1.7 per cent; (4) a contribution by the firm of [illegible] amount equal originally to one-half, but in 1891 increased to the [illegible] amount as that paid by the workingmen; (5) miscellaneous [illegible] such as interest on funds invested, fines, [illegible] of the contributions can, within certain limits, be increased or diminished if the condition of the fund is such as to demand or warrant such a change.

From the fund thus constituted provision was made for the payment of pensions as follows:

1. To workingmen who have been in the employ of the company for twenty years continuously, or who have done specially difficult work for the firm during at least fifteen years, an annual pension equal to 40 per cent of their average wages during the preceding three years. For every additional year of service the pension is increased by an amount equal to 1½ per cent of the average wages. Wages over 4 marks (95 cents) per day are not considered in making this calculation.

2. To widows of pensioners, or those who would have been entitled to receive a pension in case of disability until their death or remarriage, 33⅓ per cent of the pension of their husbands.

3. To children of these fathers until the age of 15 years or death, 10 per cent, or in case they have no mother, 15 per cent of the pensions of their fathers. In no case, however, can the total of pensions to survivors exceed 90 per cent of the pension to which the husband or father was entitled.

4. Partial pensions are also given to those entitled to pensions, but who are capable of doing light work. The partial pension can not exceed in amount one-half of that granted to totally disabled invalids, nor can the total amount paid for all partial pensions exceed 10 per cent of that for all pensioners.

The imperial law of June 22, 1889, regarding invalidity and old-age pensions necessitated several changes in the constitution of this fund. These modifications, which are very comprehensive in character, were incorporated in an amendment to the constitution adopted July 5 and approved August 18, 1890.

The imperial law, establishing a system of old-age and invalidity pensions under the auspices of the State, provided that an employer already maintaining a fund for the granting of such pensions to his employees could reduce the amount of such private pension by the amount granted under the national system, or, indeed, do away with the private fund altogether, provided, of course, no obligations already incurred were violated. The Krupp firm had been maintaining a private fund, which granted pensions in excess of those required by the imperial law. In order that its employees might receive some additional advantages from the law, the firm amended the constitution of its fund so that the pension received from this source would be reduced by an amount equal to only one-half the pension received from the general system. The effect of this was to increase the total pension received from the national and private funds by an amount equal to one-half the pension provided for in the law, the total amount received thus being far greater than that required by law.

The maximum wage for eligibility to pensions was raised from [illegible] to 2,000 marks ($285.60 to [illegible] for all those employed who earn over [illegible] ment also provided that the widow's [illegible] per cent of that received by the husband [illegible] this additional expense the firm raised its [illegible] an amount equal to that contributed by the [illegible]

The significance of these changes can be [illegible] showing the total amount of the pensions [illegible] earning 1,200 marks ($285.60) per annum, or by [illegible]

SCHEDULE OF PENSIONS PAID FOR INVALIDITY TO [illegible] SION FUND OF THE KRUPP IRON AND STEEL WORKS [illegible] ANNUM, AND TO THEIR WIDOWS AND CHILDREN.

[The figures in the table indicate that the child in each case receives [illegible] provided by the constitution. The explanation is not known. The [illegible] by the company.]

| Length of service of members. | Pension paid [illegible] | | | |
|---|---|---|---|---|
| | Members before passage of law. | Members since passage of law. | Widows. | [illegible] |
| 20 years | $114.24 | $142.42 | [illegible] | [illegible] |
| 25 years | 135.66 | 168.86 | [illegible] | [illegible] |
| 30 years | 157.08 | 194.30 | [illegible] | [illegible] |
| 35 years | 178.50 | 219.75 | [illegible] | [illegible] |
| 40 years | 199.92 | 245.30 | [illegible] | [illegible] |

a Calculated on the same basis as the other figures in the column this should be $13[illegible] are given, however, as published by the company.

The following tables give all the desirable information the operations of the fund since its creation:

RECEIPTS AND EXPENDITURES OF THE NEW PENSION FUND OF THE [illegible] AND STEEL WORKS, 1885 TO 1894.

| Items. | 1885. | 1886. | 1887. | 188[illegible]. | [illegible] |
|---|---|---|---|---|---|
| RECEIPTS. | | | | | |
| Balance on hand at beginning | $363,536.57 | $20,844.38 | $21,830.01 | $8,490.73 | $12,[illegible] |
| Contributions of members | 26,317.02 | 28,585.42 | 30,[illegible].71 | 32,260.29 | [illegible] |
| Contributions of firm | 13,158.52 | 14,292.73 | 15,416.87 | 16,130.16 | [illegible] |
| Entrance fees | 2,175.99 | 2,830.58 | 3,273.39 | 2,[illegible]1.70 | 5,1[illegible] |
| Fines | 1,465.97 | 1,762.59 | 1,763.77 | 2,030.48 | 2,231 |
| Interest | 14,619.88 | 16,127.37 | 17,566.73 | 18,851.70 | 20,023 |
| Miscellaneous | | | [illegible] | | [illegible] |
| Total | 421,273.95 | 84,[illegible].01 | 90,972.37 | 80,[illegible] | 10[illegible] |
| EXPENDITURES. | | | | | |
| Pensions | 19,012.26 | 24,7[illegible].51 | 31,7[illegible] | 41,244.[illegible] | [illegible] |
| Premiums, expenses incident to purchase of securities, etc. | 12,516.31 | 2,1[illegible] | 3,120.[illegible] | 1,[illegible] | 2,[illegible] |
| Miscellaneous | | | | 13.47 | 5[illegible] |
| Invested in securities | 368,900.00 | [illegible] | 47,[illegible] | [illegible] | [illegible] |
| Balance | 20,844.38 | 21,[illegible] | [illegible] | 13,[illegible] | 14,[illegible] |
| Total | 421,273.95 | [illegible] | [illegible] | [illegible] | [illegible] |
| Reserve invested in 4 per cent securities— | | | | | |
| On hand at beginning | | [illegible] | [illegible] | [illegible] | [illegible] |
| Invested | [illegible] | [illegible] | [illegible] | [illegible] | [illegible] |
| On hand at end | [illegible] | [illegible] | [illegible] | [illegible] | [illegible] |

RECEIPTS AND EXPENDITURES OF THE NEW PENSION FUND OF THE KRUPP IRON AND STEEL WORKS, 1885 TO 1894—Concluded.

| Items. | 1891. | 1892. | 1893. | 1894. | Total. |
|---|---|---|---|---|---|
| RECEIPTS. | | | | | |
| Balance on hand at beginning | $23,781.94 | $45,574.53 | [illegible] | [illegible] | [illegible] |
| Contributions of members | 59,[illegible] | 65,149.30 | 64,097.61 | [illegible] | 457,[illegible] |
| Contributions of firm | 59,330.59 | 65,149.30 | 64,097.61 | [illegible] | [illegible] |
| Entrance fees | 4,045.39 | 3,949.00 | 3,524.97 | 2,051.21 | [illegible] |
| Fines | 2,390.36 | 2,363.56 | 2,425.08 | 2,004.27 | 31,377.[illegible] |
| Interest | 28,[illegible].73 | 27,388.02 | 30,451.41 | 32,[illegible] | [illegible] |
| Miscellaneous | 307.15 | 236.05 | 457.90 | 245.05 | 2,[illegible] |
| Total | 173,612.06 | 205,909.54 | 194,178.96 | 219,033.31 | 1,457,207.16 |
| EXPENDITURES. | | | | | |
| Pensions | 65,280.04 | 74,576.04 | 88,166.22 | 116,237.31 | 862,181.00 |
| Premiums, expenses incident to purchase of securities, etc. | 3,440.54 | 6,784.12 | 3,067.06 | 1,856.16 | 30,[illegible] |
| Miscellaneous | 37.96 | 43.39 | 45.03 | 69.59 | 391.40 |
| Invested in securities | 59,580.00 | 95,280.00 | 47,650.00 | 23,880.00 | 773,[illegible] |
| Balance | 45,574.53 | 29,305.29 | 55,299.65 | 77,008.25 | 77,008.25 |
| Total | 173,612.06 | 205,909.54 | 194,178.96 | 219,033.31 | 1,457,207.16 |
| Reserve invested in 4 per cent securities— | | | | | |
| On hand at beginning | 547,400.00 | 606,980.00 | 702,100.00 | 749,700.00 | .............. |
| Invested | 59,580.00 | 95,280.00 | 47,600.00 | 23,880.00 | 773,580.00 |
| On hand at end | 606,980.00 | 702,100.00 | 749,700.00 | 773,580.00 | 773,580.00 |

PENSIONERS AND AVERAGE PENSIONS UNDER THE NEW PENSION FUND OF THE KRUPP IRON AND STEEL WORKS, 1885 TO 1894.

| Year. | Men. | | Widows. | | Orphans. | | Partial pensions. | | Total pensions. | |
|---|---|---|---|---|---|---|---|---|---|---|
| | Number. | Average. | Number. | Average. | Number. | Average. | Number. | Average. | Number. | Average. |
| 1885 | 110 | $89.64 | 126 | $63.57 | 4 | $11.30 | ........ | ........ | 240 | $79.22 |
| 1886 | 155 | 95.87 | 156 | 62.06 | 11 | 7.81 | ........ | ........ | 322 | 78.[illegible] |
| 1887 | 204 | 92.51 | 211 | 60.17 | 12 | 11.55 | 5 | $7.34 | 432 | 73.48 |
| 1888 | 247 | 103.54 | 247 | 63.14 | 18 | 11.37 | 6 | 19.66 | 518 | 79.62 |
| 1889 | 281 | 106.45 | 296 | 62.15 | 26 | 11.51 | 10 | 24.41 | 613 | 79.69 |
| 1890 | 312 | 109.89 | 357 | 62.09 | 35 | 11.47 | 15 | 30.72 | 719 | 78.71 |
| 1891 | 344 | 109.75 | 411 | 64.36 | 43 | 11.24 | 17 | 33.52 | 815 | 80.07 |
| 1892 | 395 | 104.64 | 488 | 65.70 | 65 | 13.18 | 17 | 32.94 | 965 | 77.28 |
| 1893 | 478 | 104.19 | 538 | 68.04 | 88 | 11.52 | 26 | 27.46 | 1,130 | 78.03 |
| 1894 | 650 | 110.97 | 587 | 69.31 | 96 | 13.17 | 52 | 23.96 | 1,385 | 83.32 |

AVERAGE MEMBERS AND EXPENDITURES PER MEMBER OF THE NEW PENSION FUND OF THE KRUPP IRON AND STEEL WORKS, 1885 to 1894.

[It will be noticed that the figures in this table showing average members do not agree with the figures on page 480. The explanation is not known. The figures are given as furnished by the company.]

| Year. | Average members. | Expenditures. | | | | | |
|---|---|---|---|---|---|---|---|
| | | Pension. | Pension per member. | Other expenses. | Other expenses per member. | Total. | Total per member. |
| 1885 | 10,673 | $19,013.26 | $1.78 | $12,516.31 | $1.17 | $31,529.57 | $2.95 |
| 1886 | 11,707 | 24,782.51 | 2.12 | 2,150.49 | .18 | 26,933.00 | 2.30 |
| 1887 | 12,642 | 31,743.50 | 2.51 | 3,130.05 | .25 | 34,873.64 | 2.76 |
| 1888 | 13,168 | 41,244.64 | 3.13 | 1,899.93 | .15 | 43,144.57 | 3.28 |
| 1889 | 14,187 | 48,852.23 | 3.44 | 2,416.91 | .17 | 51,269.14 | 3.61 |
| 1890 | 15,482 | 57,312.96 | 3.70 | 1,987.25 | .13 | 59,300.21 | 3.83 |
| 1891 | 15,980 | 65,280.04 | 4.08 | 3,478.50 | .22 | 68,758.54 | 4.30 |
| 1892 | 16,741 | 74,576.04 | 4.45 | 6,827.41 | .41 | 81,404.25 | 4.86 |
| 1893 | 16,906 | 88,166.22 | 5.21 | 3,112.70 | .18 | 91,280.92 | 5.39 |
| 1894 | 16,509 | 116,237.31 | 7.04 | 1,927.76 | .12 | 118,165.07 | 7.[illegible] |

From the experience thus far it is impossible to judge how long, on an average, an invalid of the steel works remains on the pension rolls. Cases have occurred in which quite considerable demands have been made upon the fund. Thus, for instance, three invalids died recently who lived $18\frac{3}{12}$, $19\frac{11}{12}$, and $13\frac{2}{12}$ years after having been placed on the pension list, and these invalids drew altogether 11,497.50, 14,937.50, and 6,912.50 marks ($2,736.41, $3,555.13, and $1,645.18), respectively, in pension allowances.

## FUNDS FOR THE INSURANCE OF OFFICIALS AND HIGHER PAID EMPLOYEES.

The pension funds that have been considered are for the laboring classes, strictly speaking. Their provisions apply only to those employees who are earning less than 2,000 marks ($476) per annum. For those of its employees who are earning over that amount the firm has organized a parallel series of funds. The first of these, that for the granting of pensions to this class of employees, their widows, and orphans, was created July 1, 1890. As a nucleus for the fund Mr. Friedrich Alfred Krupp donated the sum of 500,000 marks ($119,000). The regular sources of income consist of entrance fees equal to one-twelfth of the members' yearly earnings, dues equal to 3 per cent of the earnings, and a contribution by the firm equal to the amount paid by the members as dues. Membership is obligatory.

Members of the fund who have been at least five years in the employ of the firm receive in case of disability a pension of fifteen-sixtieths of their salary after five years' service, and one-sixtieth for each additional year of service. The pension can not in any case, however, exceed forty-five sixtieths of the salary received at the time of retirement. In case of death the widow of the member receives one-half of the husband's pension and each child under 18 years of age gets one-twentieth of the pension. The total paid to the heirs can not, however, exceed three-fourths of the entire pension. Any member who has been in the employ of the firm thirty-five years, or who has completed his sixty-fifth year of age, may apply for a pension. In all other cases, except when specially directed by the firm of Friedrich Krupp, a physician's certificate of disability must accompany the request for a pension.

The fund is administered by a board of directors. This consists of the president, treasurer, and secretary, who are appointed by the firm, and three directors elected by the members of the fund. The auditing committee consists of three members elected every three years by the members of the fund.

As the particular funds of the Krupp Works, as well as the provisions of the general law regarding the insurance of workmen against accident, do not apply to the insurance against accidents of employees earning over 2,000 marks ($476) per annum, the firm has created a fund for the latter's insurance. The expense of the fund is borne entirely by the firm. The schedule of pensions provides for the following indemnities:

In the case of death by accident the survivors receive the full wages which the member would have drawn at the time of the accident for the month when he died and the two months following. As soon as the payment of the full wages ceases, an annuity of 20 per cent of the wages is paid to the widow until her death or remarriage, and 15 per cent to each child until the attainment of 15 years of age. The total paid to survivors, however, can not exceed 60 per cent of the member's wages.

In case of accidents, not fatal, the injured person receives his full salary while under surgical treatment. If after surgical treatment he remains unable to perform his full duties, he receives an annuity of 66⅔ per cent of his average wages in case of total disability; in case of partial disability he receives, during such disability, a fraction of the annuity mentioned above, to be measured by the extent of the partial disability.

The general workings of these two funds are the same as in the case of similar funds for workingmen proper, and it is scarcely necessary to follow the details of their operations.

## WORKINGMEN'S LIFE INSURANCE ASSOCIATION.

The desirability of life insurance is unquestionable. But among the working classes large numbers are deterred by their inability to understand the mechanism of insurance, the difference between different kinds of policies, and the real cost of carrying insurance; and in these matters the agents of the insurance companies are not always unbiased advisers.

To obviate the obstacles in the way of life insurance for its employees, and at the same time to cheapen the cost, the firm of Krupp organized in 1877 the present life insurance association. The objects of this association are not to itself offer insurance, but, as stated in its constitution—

1. To encourage insurance among the employees of the firm.

2. By making contracts with companies and undertaking the payment of premiums to secure to employees and their families, under advantageous conditions, the acquisition of life insurance of any kind, to be paid for in full or by annual premiums.

3. To act in general as an intermediary between the insurance companies and the insured.

4. To create a fund for the purpose of insuring the payment of premiums by members and to extend other extraordinary aid as far as its resources permit.

To accomplish these objects the association entered into special contracts with the principal life insurance companies, according to which the association agreed that thereafter it would act as the agent of the companies, that it would encourage insurance in every way, that it would collect the premiums and

pay to those entitled to it the insurance

icy holders; in a word, that it would act

sary business transactions.

In return, the companies agreed to pay to the

sions that they usually paid to their local agents,

that they would charge no fee for the making

medical examination.

The contract was made with a number of

greater freedom of choice might be left to those wishing

insured as to the kind of policies they desired to take out.

To those wishing to become insured, the advantages

association were numerous. First of all, all the

necessary for securing a policy are managed by the

fees for examination or otherwise are abolished.

Financially, the members are benefited through their partic in the funds accumulated by the association through the paym the insurance companies of sums in lieu of commissions and fro and donations received by it. When the association was first f Mr. Alfred Krupp donated to it 50,000 marks ($11,900), to wl subsequently added 4,000 marks ($952). The firm also makes a quarterly contribution equal to one-half the amount received fr companies in lieu of commissions. The expenses of administrat borne entirely by the firm, so that the association has practic running expenses. The funds of the association are used in the ing ways:

One-half of the amount received by the association from the nies is credited to the policy holders as deductions on the am their premiums. Premium payments are also reduced by the application of the funds of the association. During recent yea reduction has amounted to nearly 8 per cent of the annual pre The remaining sum is used to aid members to pay their prem times of temporary distress and to grant loans to members. In sickness or accident the necessary premium payments of a mem be made out of this fund. In these cases, however, the assc should be reimbursed within a year. In particularly worthy and cases members can be permanently released from the nece making these repayments.

The system of granting loans enables the members to use the cies as collateral security for the raising of loans in times of eme No interest whatever is charged. The amount of the loan is lin that which the private companies would grant in similar cases, loan must be repaid within one year in equal monthly or fort installments. If repayment is not promptly made, interest at t of 6 per cent is charged. The policy is always transferred to t ciation as collateral security.

The affairs of the association are managed by a board of directors, consisting of nine members, of whom six are elected by the general assembly of all members, and three, including the presiding officer, are selected by the firm. The board of directors has at its service a number of confidential agents (Vertrauensmänner), of whom one or more are selected by the board from the membership of each of the forty-two districts into which the workingmen are divided. These officials hold office as an honorary post, and have the functions of insurance agents as far as they relate to efforts to encourage insurance and to preparing applications. The latter, when made, are acted upon by the board of directors. The agents are furnished with special instructions as to their duties and functions.

The workingmen, on taking out insurance, may select any of the policies offered in the rules and prospectuses of the different companies. The amount of the premium is therefore determined in each individual case according to the kind of policy and age, health, and occupation of the applicant. The following table, however, gives an average rate for 1,000 marks ($238), payable at death:

AVERAGE FORTNIGHTLY PREMIUMS ON AN ORDINARY LIFE INSURANCE POLICY FOR $238 IN THE WORKINGMEN'S LIFE INSURANCE ASSOCIATION OF THE KRUPP IRON AND STEEL WORKS.

| Age of applicant. | Premium payable fortnightly, fully paid up in— | | | |
|---|---|---|---|---|
| | 20 years. | 25 years. | 30 years. | 35 years. |
| 20 years | $0.29 | $0.26 | $0.24½ | $0.23 |
| 25 years | .31 | .28 | .26 | .24½ |
| 30 years | .34 | .30½ | .28½ | .27 |
| 35 years | .37½ | .34 | .31½ | .30½ |
| 40 years | .42 | .38 | .36 | .34½ |

The system of the joint insurance of man and wife is one often preferred by workingmen, by which the survivor receives the insurance money. The following table shows the average premium rate per fortnight, fully paid up in 25 years:

AVERAGE FORTNIGHTLY PREMIUMS ON A JOINT LIFE INSURANCE POLICY FOR $238 IN THE WORKINGMEN'S LIFE INSURANCE ASSOCIATION OF THE KRUPP IRON AND STEEL WORKS.

| Age of older person. | Age of younger person. | | | |
|---|---|---|---|---|
| | 20 years. | 25 years. | 30 years. | 35 years. |
| 25 years | $0.39 | $0.40½ | .......... | .......... |
| 30 years | .41 | .42 | $0.44 | .......... |
| 35 years | .44 | .45 | .46½ | $0.49 |
| 40 years | .48 | .49 | .50½ | .52½ |

The operations of the association are fully set forth in the following tables (pages 502–504).

[illegible] OF THE WORKINGMEN'S LIFE INSURANCE [illegible] IRON AND STEEL WORKS, [illegible] TO [illegible].

| Year. | [illegible] | [illegible] |
|---|---|---|
| 1877 | [illegible] | [illegible] |
| 1878 | [illegible] | [illegible] |
| 1879-1884 | [illegible] | [illegible] |
| 1885-1889 | [illegible] | [illegible] |
| 1890 | [illegible] | [illegible] |
| 1891-1894 | [illegible] | [illegible] |

a Should be 2,1[illegible] in order to balance. The figures are given, however, as publish[illegible]

POLICIES HELD BY MEMBERS OF THE WORKINGMEN'S LIFE [illegible] CIATION OF THE KRUPP IRON AND STEEL WORKS, BY FACE VAL[illegible] 1878 TO 1894.

| Face value of policy. | 1878. | | | | [illegible] | | |
|---|---|---|---|---|---|---|---|
| | Policies. | Per cent. | Amount. | Per cent. | Policies. | Per cent. | Amount. |
| $35.70 | ..... | ..... | ..... | ..... | ..... | ..... | [illegible] |
| $47.60 | ..... | ..... | ..... | ..... | ..... | ..... | [illegible] |
| $71.40 | 118 | 8.85 | [illegible],425.20 | 1.[illegible] | [illegible] | [illegible] | [illegible] |
| $71.64 to $142.80 | [illegible] | 36.91 | [illegible],472.60 | 15.3[illegible] | [illegible] | [illegible] | [illegible] |
| $143.04 to $214.20 | 295 | 22.13 | [illegible] | 13.[illegible] | [illegible] | [illegible] | [illegible] |
| $214.44 to $285.60 | 129 | 9.68 | [illegible],231.90 | 7.46 | 126 | [illegible] | [illegible] |
| $285.84 to $476 | 153 | 11.48 | [illegible],578.60 | 12.14 | 155 | [illegible] | [illegible] |
| $476.24 to $714 | 72 | 5.40 | [illegible] | 11.16 | 81 | [illegible] | [illegible] |
| $714.24 to $1,428 | 41 | 3.08 | 58,912.[illegible] | 12.[illegible] | 47 | [illegible] | [illegible] |
| Over $1,428 | 33 | 2.47 | 110,985.27 | 24.[illegible] | 37 | [illegible] | [illegible] |
| Total | 1,333 | 100.00 | 445,832.97 | 100.00 | [illegible] | 100.00 | [illegible] |

POLICIES HELD BY MEMBERS OF THE WORKINGMEN'S LIFE INSURANC[illegible] TION OF THE KRUPP IRON AND STEEL WORKS, BY KIND OF POLICY, 1[illegible]

| Kind of policy. | 1878. | | | | [illegible] | | |
|---|---|---|---|---|---|---|---|
| | Policies. | Per cent. | Amount. | Per cent. | Policies. | Per cent. | Amount. |
| Payable at death or specified age | 751 | 56.34 | $328,165.77 | 73.61 | 737 | [illegible] | [illegible] |
| For two lives conjointly | 577 | 43.29 | 115,525.20 | 25.91 | [illegible] | 43.27 | [illegible] |
| Endowment | 5 | .37 | 2,142.00 | .48 | 16 | [illegible] | [illegible] |
| Total | 1,333 | 100.00 | 445,832.97 | 100.00 | 1,352 | 100.00 | 478,[illegible] |
| Insurance per member | ...... | ...... | [illegible] | ...... | ...... | ...... | [illegible] |

[illegible] OF THE WORKINGMEN'S LIFE INSURANCE ASSOCIATION OF THE KRUPP IRON AND STEEL WORKS, 1877 TO 1894.

| Policies canceled. | | | Condition of business at end of year. | | | | | |
|---|---|---|---|---|---|---|---|---|
| | | | | Policies. | | | Annual premiums on policies. | |
| Number. | Amount. | Average. | Persons insured. | Number. | Amount. | Average. | Amount. | Per cent of policies. |
| ........ | ........ | ........ | 35 | 37 | [illegible] | $1,744.[illegible] | [illegible].12 | ........ |
| 44 | $12,114.60 | $[illegible].60 | 1,276 | 1,[illegible] | 445,[illegible].97 | [illegible].40 | 8,[illegible].55 | ........ |
| [illegible] | 1[illegible],[illegible].43 | [illegible].67 | 1,[illegible] | 1,904 | 666,[illegible].30 | [illegible].20 | 35,[illegible].44 | 2.[illegible] |
| [illegible] | 177,182.45 | 345.10 | 1,[illegible] | 2,006 | 804,[illegible].50 | 401.22 | 27,[illegible] | 2.[illegible] |
| 107 | [illegible],201.90 | 244.[illegible] | 2,045 | a 2,190 | [illegible],672.10 | 394.37 | 29,672.96 | 2.43 |
| 404 | 165,[illegible].40 | [illegible].79 | 2,[illegible] | 2,466 | 1,081,460.10 | 423.55 | 28,[illegible].65 | 2.67 |

POLICIES HELD BY MEMBERS OF THE WORKINGMEN'S LIFE INSURANCE ASSOCIATION OF THE KRUPP IRON AND STEEL WORKS, BY FACE VALUE OF POLICY, 1878 TO 1894.

| 1884. | | | | 1889. | | | | 1894. | | | |
|---|---|---|---|---|---|---|---|---|---|---|---|
| Policies. | Per cent. | Amount. | Per cent. | Policies. | Per cent. | Amount. | Per cent. | Policies. | Per cent. | Amount. | Per cent. |
| 3 | .16 | $107.10 | .02 | 3 | .15 | $107.10 | .01 | 3 | .12 | [illegible].30 | .01 |
| 12 | .63 | 583.10 | .09 | 11 | .55 | 533.60 | .07 | 12 | .49 | 571.30 | .05 |
| 139 | 6.[illegible] | 9,472.40 | 1.42 | 117 | 5.83 | 8,[illegible].10 | 1.08 | 93 | 3.77 | 6,[illegible].50 | .61 |
| 614 | 32.25 | 82,871.60 | 12.43 | 539 | 26.87 | 72,173.50 | 8.97 | 491 | 19.91 | 65,200.10 | 6.03 |
| 270 | 14.18 | 56,417.90 | 8.46 | 228 | 11.37 | 47,[illegible].00 | 5.89 | 183 | 7.38 | 87,748.80 | 8.40 |
| 396 | 20.80 | 97,127.80 | 14.56 | 508 | 25.32 | 128,273.10 | 15.92 | 894 | 36.25 | 214,783.10 | 19.86 |
| 242 | 12.71 | 94,802.20 | 14.22 | 283 | 14.11 | 111,610.20 | 13.89 | 378 | 15.33 | 151,803.20 | 14.04 |
| 117 | 6.14 | 81,229.40 | 12.18 | 143 | 7.13 | 100,186.10 | 12.45 | 179 | 7.26 | 127,294.30 | 11.77 |
| 67 | 3.52 | 84,020.80 | 12.60 | 104 | 5.18 | 129,350.80 | 16.07 | 130 | 5.27 | 160,792.80 | 14.87 |
| 50 | 2.63 | 160,191.00 | 24.03 | 70 | 3.49 | 211,718.00 | 26.30 | 104 | 4.22 | 316,560.80 | 29.27 |
| 1,904 | 100.00 | 666,823.30 | 100.00 | 2,006 | 100.00 | 804,839.50 | 100.00 | 2,466 | 100.00 | 1,081,460.10 | 100.00 |

POLICIES HELD BY MEMBERS OF THE WORKINGMEN'S LIFE INSURANCE ASSOCIATION OF THE KRUPP IRON AND STEEL WORKS, BY KIND OF POLICY, 1878 TO 1894.

| 1884. | | | | 1889. | | | | 1894. | | | |
|---|---|---|---|---|---|---|---|---|---|---|---|
| Policies. | Per cent. | Amount. | Per cent. | Policies. | Per cent. | Amount. | Per cent. | Policies. | Per cent. | Amount. | Per cent. |
| 1,211 | 63.60 | $512,099.50 | 76.80 | 1,295 | 64.56 | $633,658.00 | 78.73 | 1,611 | 65.33 | $858,930.10 | 79.42 |
| 612 | 32.14 | 129,483.90 | 19.42 | 599 | 29.86 | 129,210.20 | 16.05 | 590 | 23.92 | 131,756.80 | 12.18 |
| 81 | 4.26 | 25,239.90 | 3.78 | 112 | 5.58 | 41,971.30 | 5.22 | 265 | 10.75 | 90,773.20 | 8.40 |
| 1,904 | 100.00 | 666,827.30 | 100.00 | 2,006 | 100.00 | 804,839.30 | 100.00 | 2,466 | 100.00 | 1,081,460.10 | 100.00 |
| ...... | ...... | 365.38 | ...... | ...... | ...... | 385.09 | ...... | ...... | ...... | 468.57 | ...... |

BENEFITS OBTAINED FOR MEMBERS OF THE WO[illegible] ASSOCIATION OF THE KRUPP IRON AND [illegible]

| Year. | Reduction of premiums by use of— | | Premiums paid in cases of— | | Interest saved. | Other benefits. | [illegible] |
|---|---|---|---|---|---|---|---|
| | One-half of commissions paid by companies. | Funds of society (limited to 8 per cent of premiums). | Sickness. | Poverty. | | | |
| 1877 | ........ | ........ | ........ | ........ | ........ | ........ | [illegible] |
| 1878 | $354.82 | ........ | ........ | ........ | ........ | [illegible] | [illegible] |
| 1879–1884 | 3,571.50 | ........ | $1,565.68 | $941.51 | $1,074.[illegible] | [illegible] | [illegible] |
| 1885–1889 | 2,554.26 | $7,322.99 | 1,958.93 | 627.58 | 1,762.71 | [illegible] | [illegible] |
| 1890 | 886.99 | 2,177.28 | 480.50 | 84.51 | [illegible] | [illegible] | [illegible] |
| 1891–1894 | 3,470.11 | 10,537.59 | 2,115.67 | [illegible] | 1,[illegible] | [illegible] | [illegible] |
| Total | 11,837.68 | 20,037.86 | 6,120.78 | 1,977.65 | 4,792.65 | 2,771.66 | [illegible] |

## COOPERATIVE DISTRIBUTIVE STORES.

During the first period of the rapid development of Essen [illegible] under which the workingmen made their purchases was, as [illegible] shown, demoralizing in the extreme. Not only were the articles purc[illegible] poor in quality and the prices exorbitant, but the workingmen demoralized through the system of buying on credit that was encou[illegible] by the storekeepers in order better to keep their customers in power, through various lottery and prize schemes in connection wit[illegible] chases, and through the prevalence of liquor selling in ordinary s[illegible]

To remedy this the Krupp firm, in 1868, at the request of the bers of a cooperative distributive society, took over its busines[illegible] with this as a beginning inaugurated a vast system of distrib[illegible] stores (Consum-Anstalten) in which almost every article desired [illegible] workingmen could be purchased.

As regards profits, the original policy of the firm was to so reg[illegible] prices that the stores should pay for themselves and neither prof[illegible] loss be sustained. Though at this time there was no true divis[illegible] profits the stores were never run on the truck-store principle. [illegible] cases where a profit was realized it was applied in some way fo[illegible] benefit of the workingmen. On January 1, 1890, however, in ord[illegible] remove any suspicion that selfish interests were allowed to hav[illegible] play, the firm introduced the true principle of cooperation wherel[illegible] profits were to be distributed among purchasers in proportion t[illegible] value of their purchases. Sales, as has been said, are made on a[illegible] solutely cash basis, but each purchaser is provided with a pass bo[illegible] which is entered the amount of his purchases in order that his [illegible] in the profits can be determined. Anyone can trade at the store[illegible] only employees of the firm are entitled to share in profits.

Under this régime prices are kept as near as possible to current p[illegible] elsewhere, though care is exercised that the latter be not exce[illegible] Printed posters are displayed showing the prices of the principal ar[illegible] of consumption and indicating any changes in prices. The follo[illegible] table, compiled by the firm from their quotations, forms an inter[illegible] exhibit of the variation in prices of the principal commodities since[illegible] The table can be used as a fair statement of the range of prices in [illegible] many during these years, and has, [illegible]

## PRICES OF COMMODITIES IN THE COOPERATIVE DISTRIBUTION STORES OF THE KRUPP IRON AND STEEL WORKS, 1872 TO 1893.

| Articles. | Unit. | 1872. | 1873. | 1874. | 1875. | 1876. | 1877. | 1878. | 1879. | 1880. | 1881. | 1882. |
|---|---|---|---|---|---|---|---|---|---|---|---|---|
| Potatoes | Bushel | [illegible] | [illegible] | [illegible] | [illegible] | [illegible] | [illegible] | $0.515 | $0.514 | $0.515 | $0.418 | $0.417 |
| Rye bread | Pound | .017 | .017 | .019 | .017 | .018 | .017 | .018 | .018 | .020 | .022 | .018 |
| Beef: | | | | | | | | | | | | |
| First quality | Pound | | | | .140 | .140 | .140 | .140 | .139 | .139 | .136 | .135 |
| Second quality | Pound | | | | .119 | .119 | .128 | .128 | .128 | .128 | .128 | .134 |
| Veal: | | | | | | | | | | | | |
| First quality | Pound | | | | .128 | .128 | .129 | .131 | .140 | .140 | .128 | .140 |
| Second quality | Pound | | | | .124 | .124 | .116 | .121 | .130 | .130 | .128 | .130 |
| Mutton: | | | | | | | | | | | | |
| First quality | Pound | | | | .136 | .135 | .132 | .144 | .148 | .140 | .142 | .140 |
| Second quality | Pound | | | | .124 | .124 | .127 | .124 | .128 | .128 | .130 | .128 |
| Pork | Pound | | | | .147 | .160 | .162 | .151 | .144 | .151 | .161 | .159 |
| Pork sausage | Pound | .165 | .157 | .159 | .162 | .171 | .178 | .178 | .164 | .162 | .162 | .162 |
| Smoked bacon | Pound | .151 | .156 | .154 | .162 | .180 | .173 | .158 | .162 | .166 | .172 | .190 |
| American lard | Pound | .134 | .117 | .138 | .152 | .142 | .129 | .100 | .090 | .128 | .128 | .167 |
| Butter, first quality | Pound | .246 | .258 | .284 | .288 | .270 | .245 | .226 | .234 | .255 | .267 | .262 |
| Wheat flour | Pound | .041 | .044 | .042 | .039 | .034 | .030 | .035 | .034 | .037 | .036 | .036 |
| Groats | Pound | .054 | .054 | .054 | .054 | .047 | .045 | .045 | .041 | .044 | .042 | .042 |
| Buckwheat | Pound | .031 | .036 | .036 | .034 | .033 | .030 | .026 | .027 | .032 | .035 | .032 |
| Beans, white | Pound | .039 | .030 | .036 | .032 | .036 | .030 | .028 | .025 | .030 | .033 | .030 |
| Pease | Pound | .027 | .029 | .026 | .024 | .032 | .031 | .026 | .028 | .030 | .032 | .031 |
| Lentils | Pound | .035 | .032 | .038 | .045 | .043 | .039 | .035 | .034 | .045 | .050 | .051 |
| Barley | Pound | .037 | .038 | .040 | .039 | .033 | .037 | .034 | .038 | .035 | .032 | .030 |
| Rice | Pound | .039 | .037 | .037 | .035 | .041 | .038 | .038 | .037 | .037 | .037 | .035 |
| Vermicelli | Pound | .073 | .078 | .066 | .069 | .068 | .065 | .066 | .065 | .065 | .065 | .065 |
| Cheese, Holland | Pound | .150 | .149 | .137 | .137 | .145 | .142 | .140 | .145 | .158 | .162 | .155 |
| Turnip tops | Pound | .041 | .036 | .034 | .036 | .030 | .040 | .033 | .025 | .030 | .034 | .031 |
| Coffee, Java | Pound | .212 | .262 | .272 | .272 | .266 | .254 | .241 | .236 | .217 | .202 | .184 |
| Salt | Pound | .019 | .019 | .019 | .019 | .019 | .019 | .019 | .019 | .019 | .019 | .019 |
| Prunes, Turkish | Pound | .080 | .073 | .086 | .077 | .080 | .065 | .076 | .080 | .072 | .062 | .072 |
| Sugar: | | | | | | | | | | | | |
| Refined | Pound | .127 | .121 | .119 | .114 | .109 | .116 | .100 | .097 | .097 | .096 | .101 |
| Loaf | Pound | .162 | .165 | .166 | .158 | .144 | .146 | .128 | .130 | .130 | .130 | .130 |
| Soap: | | | | | | | | | | | | |
| Bar | Pound | .065 | .065 | .065 | .062 | .060 | .060 | .060 | .060 | .057 | .052 | .052 |
| Soft | Pound | .048 | .048 | .047 | .042 | .042 | .042 | .042 | .040 | .039 | .039 | .039 |
| Rape-seed oil | Quart | .184 | .171 | .145 | .145 | .160 | .166 | .152 | .130 | .125 | .123 | .131 |
| Petroleum | Quart | .064 | .060 | .048 | .048 | .053 | .050 | .042 | .040 | .048 | .045 | .041 |

| Articles. | Unit. | 1883. | 1884. | 1885. | 1886. | 1887. | 1888. | 1889. | 1890. | 1891. | 1892. | 1893. |
|---|---|---|---|---|---|---|---|---|---|---|---|---|
| Potatoes | Bushel | $0.443 | $0.368 | $0.386 | $0.373 | $0.350 | $0.442 | $0.443 | $0.388 | $0.549 | $0.484 | $0.315 |
| Rye bread | Pound | .017 | .016 | .016 | .015 | .015 | .014 | .015 | .016 | .019 | .019 | .015 |
| Beef: | | | | | | | | | | | | |
| First quality | Pound | .145 | .140 | .140 | .135 | .130 | .127 | .140 | .148 | .151 | .151 | .142 |
| Second quality | Pound | .124 | .130 | .130 | .124 | .109 | .112 | .129 | .137 | .140 | .140 | .131 |
| Veal: | | | | | | | | | | | | |
| First quality | Pound | .140 | .140 | .140 | .140 | .140 | .128 | .140 | .145 | .147 | .151 | .144 |
| Second quality | Pound | .130 | .130 | .130 | .130 | .130 | .124 | .130 | .132 | .135 | .130 | .132 |
| Mutton: | | | | | | | | | | | | |
| First quality | Pound | .140 | .140 | .140 | .140 | .140 | .140 | .140 | .148 | .151 | .151 | .141 |
| Second quality | Pound | .130 | .130 | .130 | .123 | .119 | .119 | .123 | .130 | .130 | .127 | .108 |
| Pork | Pound | .156 | .125 | .136 | .130 | .130 | .125 | .154 | .161 | .145 | .149 | .154 |
| Pork sausage | Pound | .169 | .162 | .162 | .162 | .162 | .153 | .169 | .175 | .173 | .171 | .168 |
| Smoked bacon | Pound | .176 | .154 | .156 | .151 | .151 | .141 | .174 | .185 | .162 | .154 | .164 |
| American lard | Pound | .137 | .115 | .103 | .093 | .095 | .103 | .108 | .108 | .105 | .112 | .141 |
| Butter, first quality | Pound | .254 | .254 | .235 | .236 | .232 | .245 | .257 | .257 | .261 | .262 | .258 |
| Wheat flour | Pound | .032 | .028 | .028 | .027 | .028 | .029 | .031 | .032 | .036 | .033 | .027 |
| Groats | Pound | .043 | .038 | .035 | .035 | .035 | .035 | .035 | .039 | .041 | .042 | .039 |
| Buckwheat | Pound | .030 | .027 | .026 | .028 | .025 | .026 | .028 | .028 | .030 | .031 | .028 |
| Beans, white | Pound | .030 | .030 | .029 | .026 | .025 | .029 | .028 | .027 | .030 | .024 | .024 |
| Pease | Pound | .032 | .031 | .028 | .024 | .024 | .024 | .027 | .029 | .033 | .033 | .030 |
| Lentils | Pound | .049 | .045 | .040 | .051 | .049 | .057 | .053 | .051 | .050 | .051 | .065 |
| Barley | Pound | .028 | .028 | .028 | .028 | .028 | .028 | .028 | .030 | .033 | .032 | .027 |
| Rice | Pound | .035 | .036 | .035 | .034 | .032 | .032 | .032 | .033 | .035 | .035 | .036 |
| Vermicelli | Pound | .065 | .064 | .060 | .060 | .060 | .060 | .060 | .060 | .063 | .064 | .060 |
| Cheese, Holland | Pound | .147 | .150 | .136 | .160 | .159 | .164 | .171 | .179 | .172 | .173 | .173 |
| Turnip tops | Pound | .027 | .024 | .024 | .028 | .026 | .029 | .030 | .029 | .027 | .029 | .030 |
| Coffee, Java | Pound | .182 | .184 | .172 | .174 | .232 | .222 | .245 | .268 | .286 | .285 | .285 |
| Salt | Pound | .019 | .019 | .019 | .019 | .019 | .019 | .019 | .021 | .022 | .022 | .022 |
| Prunes, Turkish | Pound | .070 | .069 | .049 | .048 | .049 | .047 | .046 | .056 | .074 | .049 | .063 |
| Sugar: | | | | | | | | | | | | |
| Refined | Pound | .091 | .081 | .073 | .071 | .063 | .069 | .078 | .070 | .069 | .070 | .071 |
| Loaf | Pound | .126 | .111 | .106 | .108 | .108 | .108 | .110 | .108 | .108 | .108 | .108 |
| Soap: | | | | | | | | | | | | |
| Bar | Pound | .066 | .052 | .049 | .045 | .042 | .043 | .042 | .043 | .048 | .046 | .048 |
| Soft | Pound | .039 | .037 | .035 | .035 | .036 | .035 | .032 | .041 | .036 | .037 | .039 |
| Rape-seed oil | Quart | .154 | .123 | .112 | .101 | .108 | .116 | .137 | .150 | .137 | .121 | .118 |
| Petroleum | Quart | .042 | .043 | .043 | .041 | .041 | .043 | .043 | .042 | .041 | .039 | .035 |

The system developed with extraordinary [illegible] operations are conducted on a vast scale. In 18[illegible] [illegible]prised 68 shops. Of these, 15 were general stores, [illegible] clothing stores, 3 shoe stores, 1 hardware and house-f[illegible] 6 bake shops, 7 meat shops, 2 tailor shops, 7 restaurants, 9 [illegible] 2 coffeehouses, etc. In connection with these are numerous [illegible]ments for the production or manufacture of various articles. [illegible] the chief are 1 slaughterhouse and sausage factory, 2 bak[illegible] shop, 1 flour mill, 1 brush factory, 1 paper-bag factory, and [illegible]tory. As part of the system, there is also a hotel, a club room, [illegible] markets, and a laundry.

The stores are all well housed. A central magazine was [illegible] July 1, 1874. It is a three-story structure, comprising a central [illegible]ing 22.4 meters (73½ feet) by 31 meters (101.7 feet), and 19 meters [illegible] feet) high, and two wings, each 18 meters (59 feet) by 15.2 [illegible] (49.9 feet), and the same height as the central building. It cont[illegible] general store, a dry goods and clothing store, a hardware store, [illegible] store, offices, a storage room, 2 tailor shops, 1 dining room for [illegible]sonnel employed in the building, and dwellings for the janitress, [illegible] women, servants, and porter. It also contains cellars for wine, [illegible] and leather. The building is provided with hot-air heating [illegible] and a hydraulic elevator. There is also a three-story warehouse [illegible] 15.7 meters (203.4 by 51½ feet), with a cellar and attic for the storage [illegible] supplies destined for the general stores. It contains also a [illegible] roasting room and a spice mill. A gas motor supplies power to the elevator, spice mill, etc. A petroleum storage reservoir, with a capacity of 50,000 liters (13,208½ gallons), tank wagons, and iron receptacles at the stores are also provided.

The articles sold at the general stores are groceries, bakery and meat products, bottled beer, wines, liquors, mineral waters, tobacco, cigars, brushes, glass, porcelain and earthenware, stationery, school books, etc. The following list shows the quantity of the principal grocery products sold at these stores during the year 1890:

| Article | Unit | Quantity | Article | Unit | Quantity |
|---|---|---|---|---|---|
| Wheat flour | pounds | 2,059,322 | Rape-seed oil | pounds | [illegible] |
| Groats | do | 55,675 | Turnip tops | do | [illegible] |
| Buckwheat flour | do | 120,813 | Coffee | do | [illegible] |
| Beans | do | 201,178 | Substitutes for coffee | do | [illegible] |
| Pease | do | 243,461 | Salt | do | [illegible] |
| Lentils | do | 39,403 | Prunes | do | [illegible] |
| Barley | do | 79,221 | Coarse sugar | do | [illegible] |
| Rice | do | 246,611 | Lump sugar | do | [illegible] |
| Vermicelli | do | 68,685 | Bar soap | do | [illegible] |
| Cheese | do | 19,013 | Soft soap | do | [illegible] |
| American lard | do | 249,111 | Petroleum | do | [illegible] |
| Butter | do | 246,413 | Potatoes | bushels | [illegible] |
| Coal | tons of 2,240 lbs | 9,890 | | | |

The personnel of the general stores consists of 15 man[illegible] assistants, 2 apprentices, 71 saleswomen, and 35 laborers.

In the dry goods and clothing stores are sold dry goods, clothing, underwear, hats, umbrellas, sewing machines, etc. Forty-two persons are employed in these stores.

In the tailor shops clothing is made to order for men and boys, and such articles as bedding, underwear, overalls, etc. Repair work of all sorts is done. The personnel comprises a master tailor, a cutter, 21 journeymen tailors, and 2 seamstresses.

In the shoe stores foot wear of all kinds is kept in stock. Boots and shoes are made to order and repaired at the shoe shops. The personnel of the shops and stores consists of 1 master shoemaker, 11 journeymen shoemakers, 1 apprentice, 1 clerk, and 3 saleswomen.

The bakeries are furnished with 14 ovens, 4 kneading machines, and 1 cutting machine, operated by steam power. There is a mill for rough-grinding rye. The personnel of the mill and bakeries comprises 29 employees. The production during 1890 was 1,184,886 kilograms (2,612,223 pounds) of rye bread, 548,108 kilograms (1,208,370 pounds) of mixed (rye and wheat) bread, 233,523 kilograms (514,829 pounds) of wheat bread, besides quantities of rolls, toast, etc.

The abattoir comprises stables and pens for beeves, hogs, and sheep, a slaughterhouse, meat-storage room, sausage factory, sales room, smokehouse, 2 cold-storage rooms ventilated by water power, 3 cellars for pickling meat, drying rooms, warerooms, etc. Since July 1, 1885, the slaughtering and inspecting of meat is done at public abattoirs as required by law. In 1890 1,335 beeves, 1,343 calves, 4,907 hogs, and 817 sheep were slaughtered for the cooperative stores. The personnel of the abattoirs, including clerks, salesmen, cashiers, and butchers, comprised 52 persons.

The restaurants all have gardens and some have tenpin alleys. One, at Bredeney, has lodgings for transient guests. The restaurant in the central building is for persons employed there. One of the restaurants at Cronenberg has a hall with a seating capacity of 1,500 persons, a stage, galleries, and bookcases. It is intended for society meetings and festivities. In the wintertime the personnel of the Essen theater give fortnightly theatrical performances at this hall. The total amount of beer sold at the various bars and restaurants during 1890 was 1,424,539 liters (376,328 gallons). The personnel of the restaurants comprises 16 employees, such as managers, assistants, and laborers.

The hotel at Essen, the "Essener Hof," is intended for persons doing business with the Krupp establishment. Strangers are admitted at all times without any formalities. The hotel has a clubroom for the use of officials of the firm. The hotel contains twenty-five bedrooms, breakfast and dining halls, a billiard parlor, a tenpin alley, and a large garden. All beverages used at the hotel and restaurants must be obtained from the cooperative stores, as well as all food, which is sold at prices fixed by the management of the cooperative stores.

The wines and liquors for the restaurants and general stores are kept in five cellars, where 1 master cooper and 8 [illegible] employed. In 1890 there were sold 167,000 [illegible] taining three-fourths of a liter (1½ pints).

All the ice which is not required for the steel [illegible] stores, and restaurants, is sold to customers on [illegible] (May 1 to September 30), the total cost of ice for the [illegible] subscribers was 27 marks ($6.43) for 15 kilograms (33 [illegible] 20 marks ($4.76) for 10 kilograms (22 pounds), and 11 [illegible] 5 kilograms (11 pounds) daily, delivered at the house. [illegible] sale at the central magazine. It is delivered gratis to the [illegible] the sick fund, when required on account of sickness.

Coffee stands are located at the main entrances to the [illegible] which coffee and rolls are sold in the morning before [illegible]. The average daily consumption in 1890 was 400 cups of coffee and 800 rolls.

The brush factory supplies all the brushes needed in the steel works and at the various stores. It employs 1 master brush maker, 2 skilled laborers, 12 partial invalids, and 2 assistants.

The paper-bag factory employs seven young women, daughters of widows.

Clothing is ironed at a fixed rate both for employees and others. Eighteen widows and daughters of former employees are employed in this work.

Weekly markets were established at the Cronenberg and Schederhof colonies for the sale of vegetables, bakery products, fish, and other food stuffs on squares belonging to the firm. Market masters employed by the firm keep the places in order, rent the chairs and stands, and attend to the weighing of products. Dealers pay a fee for the use of their stands.

Widows and dependents of deceased workingmen often secure employment through these various institutions by sewing overalls, cartridge bags, etc., for the steel works; shirts, quilts, etc., for the drygoods department; and by cleaning up stores, offices, etc. Sewing machines are sold to them upon payment of small installments. During the year 1890, 449 widows and daughters of deceased workingmen [illegible] employed in this way, earning altogether 43,031.74 marks ($10,24[illegible]).

The development of these stores has introduced entirely new [illegible] regarding savings and the avoidance of debt. The small [illegible] stores have not been entirely crushed out by its competition, but [illegible] the system of apportioning profits, adopted in 1890, a great [illegible] yearly made. During the fiscal year 1890–91 there was a tot[illegible] 11,154 pass books in use, showing about that number of families [illegible] ronizing the stores. It is impossible to tell how many persons [illegible] than employees of the firm were also customers, but the number [illegible] said to be very large. The average profit distributed on sales [illegible] 5 per cent.

## FUNDS FOR THE BENEFIT OF WORKINGMEN OTHER THAN THE REGULAR RELIEF AND PENSION FUNDS.

### WORKINGMEN'S AID FUND.

Complete as are the provisions of the different benefit and pension funds, they can not provide for all cases of distress, even though the cases are such as are quite worthy of outside aid. Such, for instance, are those of workingmen who are sick, but can not fulfill the conditions entitling them to relief from the different relief funds, or when an unusual amount of sickness is experienced in a workingman's family. To meet these and other cases where general aid is desirable, Mr. Friedrich Alfred Krupp, in accordance with the wishes of his late father, set aside 1,000,000 marks ($238,000) to constitute a workingmen's fund (Arbeiter-Stiftung). The administration of the fund is determined by a constitution adopted November 19, 1888.

Briefly recapitulated, the provisions of the constitution are the creation of a fund of 1,000,000 marks ($238,000), to be known as the Krupp's workingmen's fund. Only the interest of the fund can be used, the capital must remain undiminished. The earnings must be applied exclusively to the benefit of the workingmen of the steel works at Essen and of the branches elsewhere, and for the dependents of the workingmen. By workingmen in this sense is meant not only those in actual service, but also those who have left the employ of the works or its branches on account of invalidity. Dependents include surviving members of the family after the death of a workingman.

The income of the fund can not be used for any expenditures resulting from present or future regulations. This applies as well to expenditures imposed by law upon the proprietor as to such expenditures as are obligatory upon such institutions as sick and pension funds, poor associations, etc., on account of statutory requirements.

The income is applied, first of all, to the rendering of aid in money or money value in cases of merited distress. It may also be used for the erection of institutions having for their object the physical and spiritual welfare of Krupp's employees or for assisting such institutions, already existing, in the form of subsidies. The assistance in the form of money or money value is especially granted:

1. To workingmen who were totally disabled but who have no claim to pensions—that is, such as have become incapacitated for work before having attained the time of service required by the pension regulations.
2. To widows and orphans of deceased workingmen who have no claim to pensions, namely, widows and orphans of workingmen who died or became permanently disabled before having attained the time of service required by the regulations.
3. To workingmen who remained ill and unable to work after the expiration of the time allowed, by the sick fund, for granting sick benefits, and who have therefore lost their claims upon that fund. Assistance may be given during the continuance of such sickness and disability.

Workingmen's schools (Fortbildungsschulen) have existed in Essen since 1860. The course of instruction covers drawing, French, German, natural philosophy, mathematics, history, mensuration, mechanics, and the science of machinery and construction. Attendance is compulsory for all apprentices. An average tuition fee of 18 marks ($4.28) per year is charged. The schools at Essen have 21 classes, 45 teachers, and about 900 pupils; a similar one at Altendorf has 7 evening and 7 drawing classes, 22 teachers, and about 300 pupils. The firm gives material aid to these schools, as well as to mining schools in the mining districts.

Industrial schools are of two kinds, those for women and those girls of school age. Instruction in the former consists of sewing

per 3 months, and 6 hours daily, 15 marks ($3.57) per 3 months; embroidery, 3 hours daily, 3 marks (71 cents) per month, and 6 hours daily, 5 marks ($1.19) per month; dressmaking, 3 hours daily, 10 marks ($2.38) per 3 months; ironing, twice per week, 4 marks (95 cents) per month.

These fees are doubled for persons not members of employees' families. In 1890 the average attendance was 77 pupils in sewing by hand, 46 in sewing by machine, 29 in dressmaking, 28 in embroidery, and 6 in ironing, or a total of 186 pupils.

There are three industrial schools for girls of school age, situated in the three principal groups of workingmen's houses. The object of these schools is to give instruction in knitting, sewing, and crocheting. Only children of employees are admitted. The attendance in 1890 was 1,897 pupils, of whom 62 per cent were instructed in knitting, 30 per cent in crocheting, and 8 per cent in sewing. The pupils furnish their own materials, and when the articles are finished they can take them home for the use of the family. The tuition fee is 20 pfennigs (4.8 cents) per month in advance, but if the pupil attends regularly during the 15 months of the course and has a good record all the tuition money is refunded. The principal of the school for adults has supervision over these classes. The other teachers, 38 in number, are mostly widows and dependents of deceased employees.

The object of the housekeeping school is the education of girls over 14 years of age in the duties of an ordinary household. The instruction comprises the preparation of meals, preserving of fruits and vegetables, buying victuals, cultivating garden vegetables, washing, wringing, ironing, mending, darning, and all sorts of housework. A trained matron and two assistants have charge of the instruction. During attendance girls must board at the school. No tuition is charged, and the board and lodging cost 6 marks ($1.43) per month.

In the case of poor persons this is frequently remitted in part. The course covers 4 months. Twelve new pupils are admitted every 2 months, making thus an attendance of 24 at any one time. Certificates are given to all who complete the course, and in meritorious cases money prizes are awarded.

A midday meal is served daily at the housekeeping school, for 35 pfennigs (8.3 cents), for widowers and their children, unmarried invalids, married men whose wives are ill, and widows who are prevented by work or sickness from doing their own cooking. In cases where, on account of the illness of the housewife, there is real destitution in the family, the physician may order that meals be furnished gratuitously.

The industrial and the housekeeping schools are under the management of the cooperative stores. The firm of Krupp furnishes the school accommodations and bears all expenses of maintenance, heating supplies and apparatus, salaries, etc. The housekeeping school cost in this manner a sum of 14,800 marks ($3,522.40) during the first year of its operation, 1890–91.

It is unnecessary to comment upon the practical and useful character of the educational work of the firm. It seems to be perfectly adapted to the needs of the people.

## HEALTH SERVICE.

To cope with the problem of public health consequent upon the aggregation of so many employees in one place, the firm has been under the compulsion of organizing a health service as elaborate and complete as that maintained by a government of a principality. Its service includes an elaborate board of health code, the maintenance of a corps of physicians, and the organization of a complete hospital service, and, finally, in order that exact information may always be at hand concerning the condition of the health of workmen, complete statistics concerning the health not only of employees but of their families as well.

The hospital service is naturally the most important and the central feature of the work. The service was first inaugurated in 1872, when buildings originally intended for the care of the wounded in the Franco-Prussian war were converted into hospital quarters. In 1888 the three original pavilions were supplemented by two new ones for women and children. In 1890 the three old ones were thoroughly renovated and partly reconstructed. The hospital is situated near the center of the town of Essen. The buildings are well isolated, thus affording plenty of light and air. The grounds cover an area of 1.7302 hectares (about 4¼ acres), of which 2,924 square meters (31,474 square feet) are built upon. The vacant ground is used for garden purposes for patients only.

Every attention is paid to the care of the buildings. All the walls are painted in light oil colors and the floors are of oiled hard wood. The bathrooms and water-closets have tiled wainscoting and tiled floors. The bedsteads are of iron and have wire mattresses. Every precaution is thus taken to insure cleanliness. Lighting is done by gas and heating by hot-air furnaces. The operating room, dispensary, and the room for operating upon diphtheria patients are heated by gas stoves. The hospital is entirely owned by the firm, and its management is in the hands of the chief physician. The sick-fund management has no authority over it. The fund simply turns over its patients, for which it pays the firm 1.50 marks (35.7 cents) per day for men, 1.20 marks (28.6 cents) for women, 80 pfennigs (19 cents) for children, and 40 pfennigs (9½ cents) for infants. Any deficit is met by the firm.

The firm has also two hospitals for the treatment of epidemic cases. The larger of these is situated to the north and the other to the south of the steel works. They are both well isolated. The former consists of six barracks arranged in V shape and an administration building at the open end of the V. Each barrack contains four sick wards and a sitting room, lavatory, and bathroom. Each ward has four beds. The cubic air space is 40 cubic meters (1,413 cubic feet) per bed. Each ward is completely separated by wall partitions from every other. Each has its own stove. Oil is used for lighting. The administration building is one story high and contains nine rooms for the office and dwelling of the physician and his help.

The smaller hospital consists of two barracks, an administration building, and a deadhouse. Each barrack contains two large sick wards, a hall, sitting room, and water-closet. Each ward has fifteen beds. The air space per bed is 40 cubic meters (1,413 cubic feet). Special attention is given in both hospitals to light and ventilation.

The following tables give some idea of the scope of the work done by the hospital service in Essen:

STATISTICS OF DISEASES TREATED AT THE HOSPITALS OF THE KRUPP IRON AND STEEL WORKS, 1872-73 TO 1893-94.

| Year. | Internal diseases. | | Surgical cases. | | Skin diseases. | | Venereal diseases. | | Total diseases. | |
|---|---|---|---|---|---|---|---|---|---|---|
| | Cases. | Per cent of deaths. | Cases. | Per cent of deaths. | Cases. | Per cent of deaths. | Cases. | Per cent of deaths. | Cases. | Per cent of deaths. |
| 1872-73.. | 533 | 6.1 | 466 | 3.2 | 162 | ........ | 47 | ........ | a1,228 | 3.6 |
| 1873-74.. | 355 | 7.0 | 344 | 3.1 | 117 | ........ | 28 | ........ | 844 | 4.3 |
| 1874-75.. | 223 | 8.1 | 225 | 1.3 | 59 | ........ | 22 | ........ | 529 | 4.9 |
| 1875-76.. | 217 | 6.5 | 194 | 2.0 | 50 | ........ | 18 | ........ | 479 | 3.8 |
| 1876-77.. | 129 | 9.3 | 134 | 3.7 | 23 | ........ | 14 | ........ | 300 | 5.6 |
| 1877-78.. | 129 | 13.1 | 183 | 0.5 | 59 | ........ | 27 | ........ | a401 | 4.4 |
| 1878-79.. | 115 | 13.0 | 175 | ........ | 35 | ........ | 32 | ........ | 357 | 4.2 |
| 1879-80.. | 107 | 9.3 | 190 | 2.6 | 36 | ........ | 35 | ........ | 368 | 4.0 |
| 1880-81.. | 289 | 13.7 | 289 | 2.8 | 86 | ........ | 62 | ........ | 726 | 5.2 |
| 1881-82.. | 226 | 9.7 | 357 | 1.7 | 88 | ........ | 90 | ........ | 761 | 3.0 |
| 1882-83.. | 222 | 2.6 | 228 | 0.8 | 65 | ........ | 24 | ........ | 539 | 5.5 |
| 1883-84.. | 150 | 14.0 | 169 | 2.3 | 36 | ........ | 16 | ........ | 371 | 6.5 |
| 1884-85.. | 152 | 13.8 | 227 | 0.9 | 22 | ........ | 17 | ........ | 418 | 5.8 |
| 1885-86.. | 256 | 8.6 | 285 | 0.7 | 23 | ........ | 23 | ........ | 587 | 4.1 |
| 1886-87.. | 368 | 10.1 | 423 | 0.9 | 18 | ........ | 20 | ........ | 828 | 4.8 |
| 1887-88.. | 464 | 9.1 | 532 | 1.1 | 31 | ........ | 18 | ........ | 1,045 | 4.9 |
| 1888-89.. | 484 | 8.1 | 623 | 0.6 | 66 | ........ | 21 | ........ | 1,194 | 3.6 |
| 1889-90.. | 655 | 10.0 | 860 | 1.0 | 88 | ........ | 20 | ........ | 1,623 | 6.2 |
| 1890-91.. | 528 | 9.8 | 979 | 1.1 | 159 | ........ | 28 | ........ | a1,695 | 6.1 |
| 1891-92.. | 617 | 12.6 | 1,137 | 1.3 | 197 | ........ | 34 | ........ | 1,985 | 4.7 |
| 1892-93.. | 616 | 18.3 | 1,057 | 1.9 | 225 | ........ | 63 | ........ | 1,961 | 6.7 |
| 1893-94.. | 506 | 15.4 | 1,041 | 2.0 | 208 | ........ | 53 | ........ | 1,808 | 5.4 |

a This total does not agree with the sum of the items. The figures are given, however, as published by the company.

## HOSPITAL PATIENTS AND COST OF HOSPITAL SERVICE OF THE KRUPP IRON AND STEEL WORKS, 1875-76 TO 1893-94.

| Year. | Average employees of Krupp firm. | Hospital patients. | | | Days of patients in hospital. | | Hospital expenditure (including depreciation). | | Per cent of employees in hospitals. |
|---|---|---|---|---|---|---|---|---|---|
| | | Men. | Women and children. | Daily average. | Total. | Per patient. | Total. | Per employee. | |
| 1875-76 | 9,720 | 479 | .......... | 28.9 | 10,552 | 22.03 | $5,903.71 | $0.51 | 4.93 |
| 1876-77 | 8,510 | 300 | .......... | 22.8 | 8,306 | 27.69 | 3,946.28 | .46 | 3.53 |
| 1877-78 | 9,255 | 401 | .......... | 26.4 | 9,644 | 24.05 | 4,276.38 | .46 | 4.33 |
| 1878-79 | 8,655 | 357 | .......... | 27.3 | 9,948 | 27.87 | 5,643.93 | .65 | 4.12 |
| 1879-80 | 8,100 | 368 | .......... | 29.7 | 10,838 | 29.45 | 4,970.15 | .61 | 4.49 |
| 1880-81 | 9,767 | 726 | .......... | 51.9 | 18,950 | 26.10 | 6,629.01 | .68 | 7.43 |
| 1881-82 | 11,021 | 813 | .......... | 56.2 | 20,508 | 25.23 | 6,842.98 | .62 | 7.38 |
| 1882-83 | 10,753 | 539 | .......... | 40.7 | 14,871 | 27.59 | 5,573.96 | .52 | 5.01 |
| 1883-84 | 10,207 | 371 | .......... | 26.1 | 9,519 | 25.66 | 4,072.89 | .40 | 3.63 |
| 1884-85 | 10,402 | 418 | .......... | 30.5 | 11,138 | 26.65 | 4,465.80 | .43 | 4.02 |
| 1885-86 | 11,138 | 587 | .......... | 28.0 | 10,216 | 17.40 | 4,930.17 | .44 | 5.27 |
| 1886-87 | 12,257 | 697 | 131 | 41.0 | 14,959 | 18.07 | 6,974.40 | .57 | 5.69 |
| 1887-88 | 13,057 | 734 | 311 | 52.8 | 19,263 | 18.43 | 10,216.15 | .78 | 5.62 |
| 1888-89 | 13,403 | 730 | 464 | 60.0 | 21,900 | 18.34 | 14,600.59 | 1.09 | 5.45 |
| 1889-90 | 14,967 | 1,135 | 567 | 85.0 | 31,022 | 18.23 | 19.608.34 | 1.31 | 7.58 |
| 1890-91 | 15,918 | 1,192 | 575 | 95.5 | 34,868 | 19.73 | 28,827.75 | 1.81 | 7.40 |
| 1891-92 | 16,511 | 1,384 | 601 | 92.3 | 33,683 | 16.97 | 23,118.37 | 1.40 | 8.38 |
| 1892-93 | 16,808 | 1,236 | 729 | 76.8 | 28,049 | 14.27 | 22,562.40 | 1.34 | 7.35 |
| 1893-94 | 17,168 | 1,148 | 660 | 74.4 | 27,164 | 15.02 | 20,798.34 | 1.21 | 6.69 |

## MORTALITY OF EMPLOYEES OF THE KRUPP IRON AND STEEL WORKS LIVING IN ESSEN, 1870 TO 1894.

| Year. | Employees. | Deaths. | Deaths per 1,000. | Year. | Employees. | Deaths. | Deaths per 1,000. | Year. | Employees. | Deaths. | Deaths per 1,000. |
|---|---|---|---|---|---|---|---|---|---|---|---|
| 1870 | 7,084 | 82 | 12 | 1879 | 7,964 | 104 | 13 | 1887 | 12,674 | 152 | 12 |
| 1871 | 8,810 | 152 | 17 | 1880 | 8,806 | 145 | 16 | 1888 | 13,198 | 110 | 8 |
| 1872 | 10,394 | 142 | 14 | 1881 | 10,598 | 148 | 14 | 1889 | 14,223 | 147 | 10 |
| 1873 | 11,671 | 150 | 13 | 1882 | 11,011 | 113 | 10 | 1890 | 15,519 | 158 | 10 |
| 1874 | 11,543 | 137 | 12 | 1883 | 10,491 | 159 | 15 | 1891 | 16,161 | 173 | 11 |
| 1875 | 9,743 | 121 | 12 | 1884 | 10,213 | 139 | 14 | 1892 | 16,865 | 173 | 10 |
| 1876 | 8,998 | 100 | 11 | 1885 | 10,656 | 120 | 11 | 1893 | 17,100 | 145 | 8 |
| 1877 | 8,586 | 86 | 10 | 1886 | 11,723 | 125 | 11 | 1894 | 16,706 | 107 | 6 |
| 1878 | 9,414 | 119 | 13 | | | | | | | | |

## OTHER INSTITUTIONS.

It would be almost impossible to describe the thousand and one ways in which the firm manifests its solicitude for the permanent welfare of its employees in addition to the ways already described. Contributions are made to almost every institution having for its object the elevation of the working classes. At the works provision is made for the furnishing of coffee and rolls at a minimum cost. A cup of coffee with sugar and a roll is provided for 7 pfennigs (1⅔ cents). Several eating houses with gardens are also provided for workingmen who find it too far to return home for meals.

The unusual extent to which bathing facilities have been provided for employees should be especially commented upon. Bathrooms are provided at the exits of most of the shops. In addition there is a central bath house containing seven bathrooms with tubs, hot and cold water, and shower appliances, and a steam bath in which six persons can be accommodated simultaneously. The object of this central bath house is, first of all, the accommodation of patients who are not inmates of the hospital. Where baths are ordered by the physician the fees are paid by the sick fund. Other employees may also use the baths when not required by patients. The fees for employees are 15 pfennigs (3.6 cents) for tub baths and 1 mark (23.8 cents) for steam baths. The same fees are charged to the sick fund for patients. Free baths are allowed to workingmen whose work is of such a nature as to make baths very desirable.

At the Hanover mines a bath house containing 28 cells with shower-bath appliances was erected at a cost of 20,000 marks ($4,760). The daily attendance at these baths is about 1,100 persons.

Another bath house containing 16 cells for shower baths and one tub and shower bath was erected at the smelting works near Duisburg, at a cost of 10,000 marks ($2,380). Out of 491 men employed there in 1890, the average daily attendance at the baths was 107 persons. No fees are charged for the use of these baths.

## CONCLUSION.

The magnitude of the enterprise of Krupp at Essen, and the variety of social institutions that are found there, almost preclude any attempt at a general résumé of results. There are, however, certain general principles underlying the management of all these institutions that, though it is impossible to incorporate them in constitutions, yet determine the real spirit in which the institutions are carried on. They may be said to constitute the soul of the institutions. A study of these institutions in their practical workings shows, first of all, that they have been conceived in the most liberal spirit as regards the participation of the workingmen themselves in their management. In spite of the great prominence of the firm, the independence of the individual has been sacrificed as little as possible.

There can be no doubt that the firm has succeeded in gaining the respect and good wishes of its employees. The feeling that the firm has the true welfare of the latter at heart seems to be universal. At the same time Essen is not the result of any sentimental effort for reform. To the visitor the first serious impression is that here there has been no carrying out of a caprice, or a personal desire to do this or that for the workingmen. Everything has the appearance of having been the result of stern necessity. Each institution has developed in response to a distinct demand. Economy is everywhere. The laborers are not given china where tin or iron will suffice. The schools are especially plain, but they have the appearance of being of a character suitable to a laboring population.

That the laborers constitute a contented class is shown by the almost absolute absence of labor difficulties, and the high degree of stability of employment. Twenty-one per cent of all employees have been continuously employed over fifteen years, and 23 per cent have been employed more than five but less than fifteen years, or a total of 44 per cent that have been in the employ of the firm more than five years. It should be remembered, moreover, that the rapid increase in the number of employees within recent years has necessitated the constant entrance of men to swell the number of those employed but a short time.

As regards the effect of the expense entailed upon the firm by its various social enterprises, the firm is emphatic in the statement that it has been more than repaid by the better class of workingmen that they have been able to obtain and retain, and the absence of friction between the management and its personnel. All improvements in the condition of its employees have been followed by improvements in the character of the work performed by them, and by increased faithfulness to the interests of the establishment.

## RECENT REPORTS OF STATE BUREAUS OF [illegible]

### MARYLAND.

*Fourth Annual Report of the Bureau of Industrial* [illegible] 1896. A. B. Howard, jr., Chief of Bureau. [illegible]

This report treats of the following subjects: Personal [illegible] 95 pages; building and loan associations, 56 pages; [illegible] 10 pages.

The presentation concerning building and loan [illegible] principally of quotations from the Ninth Annual Report [illegible] ment of Labor. Under the title of "Strikes and Lockouts" [illegible] torical statement and the estimated loss in wages are given [illegible] disturbance that occurred in the State during the year. [illegible] loss in wages is fixed at $25,000.

Personal Property Values.—This presentation is a continuat[illegible] and completion of an investigation commenced in 1894 and publis[illegible] in the report for that year. The inquiry of 1894 was confined to [illegible] city of Baltimore. The statistics for 1895 show, for each county in [illegible] State, the number of estates probated and the value of personal pr[illegible] erty belonging to them, for each year of two periods of five years eac[illegible] 1875 to 1879 and 1890 to 1894, inclusive. The estates for each cou[illegible] are arranged in nine classes according to value. Estates which w[illegible] not of sufficient value to pay off all the debts charged against them [illegible] not included.

The following statement gives the State totals, not including Ba[illegible] more City, for the two periods:

ESTATES PROBATED AND VALUE OF PERSONAL PROPERTY.

| Estates having personal property valued— | 1875 to 1879, inclusive. | | 1890 to 1894, [illegible] | |
|---|---|---|---|---|
| | Number. | Value. | Number. | [illegible] |
| Under $500 | 1,724 | $370,810 | [illegible] | [illegible] |
| $500 to $1,000 | 895 | 503,896 | [illegible] | [illegible] |
| $1,000 to $2,500 | 1,063 | 1,490,010 | 1,135 | [illegible] |
| $2,500 to $5,000 | 535 | 1,703,783 | [illegible] | [illegible] |
| $5,000 to $10,000 | [illegible] | [illegible] | [illegible] | [illegible] |
| $10,000 to $25,000 | 256 | [illegible] | [illegible] | [illegible] |
| $25,000 to $50,000 | 65 | [illegible] | [illegible] | [illegible] |
| $50,000 to $100,000 | 37 | [illegible] | [illegible] | [illegible] |
| Over $100,000 | 10 | [illegible] | 16 | [illegible] |
| Total | a4,[illegible] | b$16,1[illegible] | [illegible] | [illegible] |

a Figures here apparently should be 4,9[illegible]; those given are, however, according to the [illegible]
b Figures here apparently should be $16,[illegible]; those given are, however, according [illegible]

## MICHIGAN.

*Thirteenth Annual Report of the Bureau of Labor and Industrial Statistics of Michigan.* Year ending February 1, 1896. Charles H. Morse, Commissioner; H. R. Dewey, Deputy Commissioner. xxvi, 402 pp.

This report consists of an introduction of 26 pages, which includes statistics of street railway companies, and 6 parts which treat, respectively, of the following subjects: Laborers engaged in transportation, 227 pages; organized labor, 44 pages; miscellaneous statistics, 63 pages; penal and reformatory institutions, 9 pages; strikes, 30 pages; Michigan laws of 1895 affecting labor, etc., 26 pages.

The statistics presented under the first two titles are the results of original investigations by the bureau. The other presentations consist principally of compilations from official reports.

LABORERS ENGAGED IN TRANSPORTATION.—The individual reports from employees on street railways, hack and bus lines, etc., and from owners who drive their own hacks, buses, drays, or teams are published in detail. The following statement indicates the character of the questions asked and gives some of the principal facts brought out by the analysis of the statistics:

PERSONS ENGAGED IN TRANSPORTATION IN 1896.

| Items. | Employees on— | | Owners who drive their own hack, bus, dray, or team. |
|---|---|---|---|
| | Street railways. | Hack and bus lines, etc. | |
| Total number considered | 1,865 | 3,127 | 1,943 |
| Native born | 1,070 | 1,930 | .......... |
| Foreign born | 795 | 1,196 | .......... |
| Married | 1,285 | 1,721 | 1,614 |
| Single | 557 | 1,376 | 306 |
| Widowed | 23 | 30 | 21 |
| Average hours in day's work | 10½ | 10½ | 10½ |
| Number reporting hours of work increased during past year | 577 | 119 | .......... |
| Number reporting hours of work not increased during past year | 958 | 2,846 | .......... |
| Number reporting hours of work decreased during past year | 95 | 31 | .......... |
| Number who work overtime | 520 | 1,157 | .......... |
| Number who do not work overtime | 1,300 | 1,829 | .......... |
| Number who receive extra pay for overtime | 398 | 345 | .......... |
| Number who do not receive extra pay for overtime | 121 | 783 | .......... |
| Average daily wages | $1.69 | $1.35 | (a) $2.16 |
| Number reporting wages increased during past year | 748 | 224 | 322 |
| Number reporting no change during past year | 824 | 2,588 | 710 |
| Number reporting wages decreased during past year | 45 | 235 | 850 |
| Number who lost time during past year | 1,614 | 1,552 | 1,202 |
| Number who lost no time during past year | 247 | 1,557 | 731 |
| Average number of days lost for those who lost time | .......... | 56½ | 78+ |
| Number who saved money during past year | 796 | 774 | 592 |
| Number who did not save money during past year | 994 | 2,222 | 1,301 |
| Average savings of those who saved | $123.28 | $100.84 | $209.73 |
| Number who say times are better than one year ago | 1,087 | 827 | 498 |
| Number who say times are worse than one year ago | 344 | 820 | 1,007 |
| Number who own a home | 460 | 619 | 1,112 |
| Number who own a home clear of incumbrance | 254 | 322 | 631 |
| Number who rent homes | 821 | 1,149 | 607 |
| Average rent paid per month | $7.27 | $6.46 | $7.53 |
| Number who say cost of living increased during past year | 843 | 676 | 765 |
| Number who say cost of living decreased during past year | 79 | 142 | 154 |
| Number who buy beer or spirituous liquors | 496 | 1,305 | 710 |
| Average expenditure per month for beer or spirituous liquors by those who admit their use | $1.10 | $1.05 | $1.96 |
| Number who belong to labor organizations | 1,074 | 277 | 166 |

a Average per vehicle per day.

In some of the returns answers were not given [illegible] It therefore does not follow that the difference [illegible] given for any particular item in the above summary [illegible] ber considered represents the number reporting the [illegible] shown.

ORGANIZED LABOR.—Reports were received from 237 [illegible] tions in the State, which are published in detail. The [illegible] reported 19,192 male and 302 female members. There were [illegible] bers initiated and 1,256 suspended during the year. Twenty-one [illegible] zations gave out-of-work—73 sick, 107 strike, and 93 burial—[illegible] Fifty-eight furnished life insurance, and 194 reported the [illegible] daily wages received by members, the average wage being [illegible] day. The average hours in a day's work, as reported by 196 [illegible] tions, was $9\frac{3}{8}$. There were 23 organizations that reported 31 [illegible] involving 5,956 men, as having occurred during the year. The [illegible] ployees were successful in 16 strikes, failed in 2, and compromised 7, no information as to settlement being furnished for 6.

## NORTH CAROLINA.

*Ninth Annual Report of the Bureau of Labor Statistics of the [illegible] North Carolina for the year 1895.* B. R. Lacy, Commissioner; [illegible] Terrell, Chief Clerk. v, 408 pp.

This report treats of the following subjects: Cotton and woolen factories, 76 pages; agricultural statistics, 146 pages; reports of laboring men, 80 pages; tobacco and miscellaneous factories, 33 pages; railroads, 10 pages; organized labor, 25 pages; fishing industry, 15 pages; newspapers, 13 pages; bureaus of labor, 10 pages.

The presentation concerning the first four subjects treated consists of statistics as to capital, machinery, materials and products, employees and wages, hours of work, and the general social and financial condition of employees in the cotton and woolen, tobacco, and miscellaneous factories; also the financial, social, and moral condition of farmers, and the wages, hours of labor, educational, moral, and financial condition of mechanics and laboring men in various industries. These statistics are shown in detail, the individual reports being given for numerous factories and workingmen in different sections of the State. The reports are arranged by counties, and are followed by letters from employers and employees giving their personal views on the various subjects treated.

Totals and general averages for the State are shown for cotton and woolen factories and for farmers.

The following statement gives statistics of cotton and woolen factories for 1895:

| | |
|---|---|
| Number of mills | [illegible] |
| Capital | [illegible] |
| Cotton and wool consumed, pounds | [illegible] |

| | |
|---|---|
| Products: | |
| Yarn, pounds | 79,473,949 |
| Domestics, yards | 87,742,655 |
| Plaids, yards | 51,737,547 |
| Woolen goods, yards | 18,424,200 |
| Number of employees: | |
| Men | 4,888 |
| Women | 6,175 |
| Children | 3,311 |
| Children under 14 years of age— | |
| Boys | 778 |
| Girls | 780 |
| Total employees | 15,932 |
| Average wages per day of— | |
| Machinists | $1.93½ |
| Engineers | 1.61½ |
| Firemen | .89¼ |
| Skilled men | 1.10 |
| Unskilled men | .70 |
| Skilled women | .65 |
| Unskilled women | .50 |
| Children | .30 |
| Average number of days in operation during year | 286½ |
| Average number of hours constituting day's work | 11¼ |
| Number of spindles | 913,458 |
| Number of looms | 24,853 |

The reports show that almost invariably wages were paid weekly and in cash, and that wages had neither increased nor decreased as compared with 1894; also that the sanitary condition of the factories and of the houses of employees was good, that the employees had religious and educational facilities of which they availed themselves and were improving mentally and morally, and that a large percentage of them could read and write.

Following are statistics relating to farm laborers:

AVERAGE WAGES AND VALUE OF RATIONS PER MONTH OF FARM LABORERS.

| | 1893 | 1894. | 1895. |
|---|---|---|---|
| Men | $9.50 | $9.00 | $8.75 |
| Women | 5.50 | 5.00 | 4.65 |
| Children | 3.20 | 3.00 | 2.90 |
| Rations | 4.25 | 4.00 | 3.84 |
| Rent and pasturage | ....... | 3.00 | 2.57 |

In explanation of the comparatively low wages of farm laborers it is stated that they have no house rent to pay, "their fire wood is obtained by simply going out and gathering it up—it is free; and in most every instance the landlord gives them a team to haul it up with, and charges nothing for it. Gardens, truck patches, and places to raise pigs and poultry they have free of charge. We find, too, that the majority of tenant farmers are furnished with horse and plow to work their patches with free, and often work them in the landlord's time."

are published under this title.

## REPORT OF THE MASSACHUSETTS BOARD TO INVESTIGATE THE SUBJECT OF THE UNEMPLOYED.

*Report of the Massachusetts Board to Investigate the Subject of the Unemployed.* January 1, 1895, and March 13, 1895. Davis R. Dewey, David F. Moreland, and Haven C. Perham, Commissioners. ccxx, 582 pp.

This report was prepared under authority of an act of the legislature approved April 12, 1894. The subject is treated under the following titles: Part I, Relief measures, 264 pages; Part II, Wayfarers and tramps, 123 pages; Part III, Public works, 135 pages; Part IV, The amount of nonemployment and causes thereof, 87 pages; Part V, Final report, 193 pages.

RELIEF MEASURES.—The various agencies, both public and private, that were at work in Massachusetts during the winter of 1893–94 to relieve or prevent distress among the unemployed are grouped in five classes.

1. Special citizens' relief committees, organized primarily to aid the unemployed.

2. Municipal departments having charge of public works upon which it was possible to give employment.

3. Labor organizations giving aid either by usual out-of-work benefits or by extraordinary methods.

4. Private charities, including all permanent relief-giving organizations not connected with the State or municipal government on the one hand, or with labor organizations on the other.

5. The permanently established public relief agencies administered for the State and for municipalities, such as poor departments.

The information for the entire State is summarized under these groups and then presented in detail for the different municipalities. Some of the features discussed are as follows: Methods of obtaining funds, character of the recipients of relief, distribution of relief, value of the relief work, characteristics of relief by public work, wages offered, and economic results.

The thirteen citizens' relief committees of the State raised about $147,000, of which two-thirds was raised in Boston and one-half of the remainder in Lynn. The amount of extra appropriation to give work to the unemployed upon public works was $352,000. The out-of-door aid granted by public poor departments in all cities and towns of the State was $700,000 for the year 1893–94.

The amount of relief afforded was much [illegible] normal years. The whole relief afforded by [illegible] and employment upon public works may be [illegible] the usual amounts. Most of that afforded by trade [illegible] usual. The increase of relief by private charities [illegible] 50 per cent. The increase by public poor department [illegible] about one-eighth.

It is impossible to ascertain definitely the number of [illegible] by means of all the methods referred to. Five of the [illegible] societies aided 6,462 families in 1893-94, as compared [illegible] 1892-93. The number aided by the citizens' relief committee [illegible] upon the public works and by the public poor department, rep about 85,000 families, being an increased or unusual aid for m 40,000 persons.

WAYFARERS AND TRAMPS.—Methods of reducing the nu wayfarers and tramps and of determining those worthy of a are discussed. The commissioners, in conclusion, recommend tion designed to give effect to the following principles:

It should be easier to convict vagrants and tramps.

Overseers of the poor in every town shall provide decen modations of food and lodging for wayfarers, and in return shall demand work. Refusal on the part of wayfarers to com this demand shall constitute prima facie evidence of tramping. farer shall be lodged in police stations or in tramp rooms c with such station. These stations shall be reserved solely f under criminal charge or sentence. Failure on part of the ove the poor to demand work shall be subject to penalties.

All persons found riding on freight trains without author mission should be punished with the penalties against tramps.

There should be uniform methods of treating wayfarers th the State.

It would be desirable for this Commonwealth to establish a institution for the care and training of tramps and vagran 30 years of age.

PUBLIC WORKS.—The following are extracts from the co derived from a careful consideration of the large body of co evidence presented under this title:

That, as a rule, the city does not do construction work di cheaply as can a contractor to whom the work is intrusted.

That, in exceptional instances, where civil-service rules are and uniformly followed, and where the city is not too strictly li ordinances as to the minimum rates of wages and other cond labor, the city can do its work as cheaply as any private em labor.

That the quality of the work done by direct municipal em is generally better than that done by contractors.

Nonemployment is frequently aggravated by the influx of a large number of nonresident and ofttimes alien laborers, brought in by contractors.

Greater care should be taken in the letting of contracts to prevent the introduction of large gangs of nonresident and particularly alien labor, unless there is clear proof that there is a scarcity in the vicinity of labor to be hired at a fair market price.

The plans for the establishment of factories or farms on State initiative appear impracticable.

THE AMOUNT OF NONEMPLOYMENT AND CAUSES THEREOF.—There is but little statistical material available which will show the amount of nonemployment in the different trades and occupations throughout the State for any series of years. The commission made special inquiries into conditions of employment in a few selected industries, with the purpose of determining whether the amount of nonemployment has been increasing or decreasing during the past ten years. The information is presented in detail for each of the trades investigated, it being impracticable to make a summarization.

FINAL REPORT.—The entire subject is discussed under the two general heads of temporary or emergency relief and permanent measures. The measures for temporary relief are treated under five subdivisions, as follows:

1. The permanently established relief agencies of town and State, such as the poor departments.
2. Municipal departments of public works, temporarily used for furnishing work relief.
3. Private charities.
4. Special relief committees.
5. Labor organizations.

The measures for permanent relief are treated as follows:

1. Removal of residents of the cities to the country and farms.
2. Removing the competition, and hence displacement of free labor occasioned by the labor of inmates of reformatory and penal institutions.
3. Reducing the hours of a day's labor.
4. Restriction of immigration.
5. An extension of industrial education.
6. Improving the intelligence and employment offices, or establishing free employment offices.

*Systematisches Verzeichnis der Gewerbe für statistische Zwecke* [illegible] *dels- und Gewerbekammern in den im Reichsrathe* [illegible] *reichen und Ländern.* 1896. 87 pp.

This report comprises a list of all the skilled trades in [illegible] represented in the Austrian Parliament (Reichsrath) and [illegible] within the scope of the trade regulations established by [illegible] beordnung). It is intended for the use of the chambers of [illegible] and industry in their statistical work. The list comprises 4,[illegible] trial and 2,101 mercantile trades. The trades are divided, [illegible] to their nature, into 25 classes, comprising 363 groups. The [illegible] trades in a second list are arranged alphabetically.

*Die gewerblichen Genossenschaften in Oesterreich. Verfasst und* [illegible] *gegeben vom statistischen Departement im k. k. Handels*[illegible] 1895. 1,480 pp.

This is the first complete statistical report of trades guilds in Aus[illegible] tria published by the Government. The guilds herein reported [illegible] all the trades associations that are organized under the provisions of the act of March 15, 1883, and also those whose constitutions have not yet been approved as being in conformity with that act. Out of a total of 5,317 trades guilds reported, only 180 come under the latter class.

By the provisions of the act of 1883, all persons carrying on similar skilled trades on an independent basis in the same or neighboring [illegible], together with their helpers and apprentices, are required to [illegible] themselves into trades guilds. Under certain conditions the guilds may also be composed of persons of different trades. Of the [illegible] number of guilds, 992 are composed of persons of the same and allied trades, and 4,325 of persons of different trades. Owners and employees of factories are not included within the provisions of this act. Owners of shops are known as members proper (Mitglieder), while their employees are regarded as associate members (Angehörige) of the guilds.

The object of the Austrian trades guilds is, among other things, the regulation of the relations between employers and employees, the systematic training of apprentices, the care of workingmen in cases of illness by means of sick insurance funds, and the establishment [illegible] maintenance of journeymen's homes (Gesellen-Herbergen), employment agencies, arbitration commissions, trade schools, etc.

The statistics contained in this report cover upper Austria, [illegible] Austria, Salzburg, Styria, Carinthia, [illegible], Trieste and [illegible]

Goritz and Gradiska, Istria, Tyrol, Vorarlberg, Bohemia, Moravia, Silesia, Galicia, Bukowina, and Dalmatia. The date of the enumeration of the individual guilds was December 31, 1894, except in the case of 20, which were founded during the first half of the year 1895, and are also included.

The 5,317 trades guilds within the territory enumerated contained 554,335 members proper (owners of shops) and 692,753 associate members (employees), making a total of 1,247,088 members and associate members. This is equal to 5.3 per cent of the entire resident population.

The following titles of the various tables presented in the report will give a fair idea of the scope of the enumeration:

1. Number of trades guilds and number of members proper and of associate members in the various classes of guilds.

2. Trades guilds classified according to territorial extent and membership.

3. Trades guilds classified according to institutions (dues, membership qualifications, sick funds, etc.).

4. Trades guilds classified according to apprenticeship conditions and regulations.

5. Relief funds of the trades guilds.

6. Dates of approval of the constitutions of the guilds, of employees' assemblies, and of the arbitration commissions.

7. Relation between the number of apprentices and the number of journeymen.

8. Trades guilds classified according to the dates of their creation.

9. Trades guilds classified according to the population of the localities in which they are situated.

10. Trades guilds for single trades, by industries.

11. Provinces and minor divisions and the number of trades guilds in each.

12. Statistics showing population, number of proprietors of establishments, number of trades guilds, and number of members proper.

The tables present the statistics by provinces and minor divisions. Information is also given in detail for each individual guild.

The following statement shows the guilds, classified according to the number of members in each:

MEMBERSHIP OF TRADES GUILDS.

| Trades guilds having a membership of— | Number. | Per cent. |
|---|---|---|
| 1 to 30 members and associates | 195 | 3.7 |
| 31 to 150 members and associates | 2,879 | 54.2 |
| 151 to 300 members and associates | 1,352 | 25.4 |
| 301 to 600 members and associates | 586 | 11.0 |
| 601 to 1,500 members and associates | 187 | 3.5 |
| Over 1,500 members and associates | 74 | 1.4 |
| Not reported | 44 | .8 |
| Total | 5,317 | 100.0 |

It appears from the report that [illegible] have no journeymen and 11.8 per cent, [illegible] membership. There are, on an average, in [illegible] bers proper, 97.5 journeymen, and 32.8 appr[illegible] ship of 234.6 per guild.

In connection with these guilds there are [illegible] tions created for the purpose of carrying [illegible] organization, as contemplated by the act of 18[illegible] number of guilds enumerated, 3,196, or 60.1 per [illegible] men's assemblies; 3,049, or 57.3 per cent, have arb[illegible] 122, or 2.3 per cent, have trade and continuatio[illegible] 7.5 per cent, have journeymen's homes (Gesellen-Her[illegible] or 28.1 per cent, of the guilds journeymen may be [illegible] executive councils.

A very important feature of the guilds is the ex[illegible] funds. Prior to the act of 1859 there was no regular [illegible] relief organization in trades guilds. Although there [illegible] journeymen's funds, neither these nor the regular guild [illegible] regarded as actual sick funds. The act of 1859 required [illegible] either establish funds for the relief of journeymen in [illegible] or to participate in existing sick funds. This act was so [illegible] that of March 30, 1888, that all existing sick funds came [illegible] provisions of the latter act, and as a result they now [illegible] resemblance to the sick-insurance funds of Germany.

The report shows that there were 1,030 special sick funds p[illegible] in by 1,475 guilds. This leaves 3,842, or 72.3 per cent, of [illegible] without special relief funds. Seven hundred and thirty-four, [illegible] per cent, of the sick funds pay benefits equal to the legal [illegible] required in the case of district sick funds, while 32, or 3.1 per [illegible] larger amounts. In 264, or 25.6 per cent, of the journeym[illegible] funds the benefits are of such a nature as not to permit of com[illegible]

By an act of April 4, 1889, apprentices who are members [illegible] guilds are permitted to participate in special sick-relief fun[illegible] the time of the enumeration there were 313 such funds, partici[illegible] by 388 trades guilds. There were, in addition, 27 funds for bo[illegible] neymen and apprentices, and 42 proprietors' sick funds. The [illegible] shows the existence of 23 special funds for other forms of relief.

The dues required of participants in the benefits of the journey[illegible] sick-relief funds vary in the different guilds. In 762, or 74 per c[illegible] the funds, the dues are equal to 2 per cent of the wages, and in [illegible] 1.5 per cent, they are over 2 per cent of the wages. In 15[illegible] per cent of the funds, the dues are fixed at 10 kreutzer [illegible] or under per week, while in 108, or 10 per cent of the fu[illegible] over 10 kreutzer (nearly 5 cents) per week.

Each trades guild makes its ow[illegible] ing the employment and educa[illegible] that 4,797, or 90.2 per cent, of the guilds have regulations limiting the number of apprentices, and in 85.8 per cent of these cases the limitations relate to the employment of apprentices by masters who have no journeymen employees.

Following are the terms of apprenticeship required by the constitutions of the various guilds: 2 years or under, in 3.1 per cent of the guilds; 3 years in 23 per cent; 4 years in 6.5 per cent; 2 to 3 years in 4.7 per cent; 3 to 4 years in 17.3 per cent; from 2 to 4 years in 45.4 per cent.

Special regulations governing the examination of apprentices are provided for in 4,282, or 80.5 per cent, of the guilds.

The following statement shows the relation between the number of journeymen and the number of apprentices in the trades guilds:

JOURNEYMEN AND APPRENTICES IN TRADES GUILDS.

| Trades guilds having— | Number. | Per cent. |
|---|---|---|
| Neither journeymen nor apprentices | 280 | 5.3 |
| Journeymen, but no apprentices | 269 | 5.1 |
| Apprentices, but no journeymen | 161 | 3.0 |
| As many apprentices as journeymen | 115 | 2.2 |
| More journeymen than apprentices | 3,396 | 63.8 |
| More apprentices than journeymen | 1,012 | 19.0 |
| Not reported | 84 | 1.6 |
| Total | 5,317 | 100.0 |

The following statement shows the number of the present trades guilds founded during each of the specified periods:

DATE OF ESTABLISHMENT OF TRADES GUILDS.

| Period. | Number. | Per cent. |
|---|---|---|
| Eighth century | 1 | 0.02 |
| Eleventh century | 1 | .02 |
| Thirteenth century | 2 | .04 |
| Fourteenth century | 2 | .04 |
| Fifteenth century | 14 | .26 |
| Sixteenth century | 37 | .70 |
| Seventeenth century | 97 | 1.82 |
| First half of eighteenth century | 53 | 1.00 |
| Second half of eighteenth century | 51 | .96 |
| 1801 to 1859 | 60 | 1.13 |
| 1860 to 1882 | 372 | 7.00 |
| 1883 to 1895 | 4,593 | 86.38 |
| Unknown | 34 | .63 |
| Total | 5,317 | 100.00 |

The periods during the present century were divided, as indicated, in order to show the development as affected by legislation. The first period, namely, 1801 to 1859, was prior to the act of 1859. During the second period, 1860 to 1882, the guilds operated under that act. The

*Statistique des Grèves et des*
*Survenus Pendant l'Année 1895.*
merce, de l'Industrie, des Postes

In Bulletin No. 1 was given a brief
during the year 1894, with a short
ures concerning strikes during the
annual volumes concerning strikes
bureau. The present volume relates
principal results are summarized in the
aration of this statement the same
sentation have, with a single exception,
easy matter to compare the information for
prior years.

In 1895 there were reported a total of 405
establishments, in which 45,801 workingmen
and resulting in a loss of 617,469 days labor,
ever, includes 61,597 days lost by 5,899 persons
but were thrown out of employment as the result

In 1894 there were but 391 strikes, but they
lishments, 54,576 strikers, and caused a loss of
labor. The number of strikers in 1895, moreover,
12.83 out of every 1,000 persons productively employed
19.83 in 1894. Strikes, therefore, on the whole, were
severe in 1895 than in the preceding year.

But little difference is discernible as regards the
achieved by the strikers in the two years. In 1895 out of
403 strikes (2 strikes not having been terminated when
closed) 100, or 24.81 per cent, were successful, 117, or 29.03
were partly successful, and 186, or 46.16 per cent, resulted
The percentages for 1894 were 21.48, 32.90, and 45.53,
really better criterion of results would be to take the
ers as a basis. Doing this, and eliminating the number
involved in strikes not yet terminated, it will be found
18.72 per cent, of strikers succeeded, 20,672, or 45.18 per
succeeded, and 16,521, or 36.10 per cent, failed to
demands. In 1894 the percentages were 23.63, 45.41,
respectively.

The great majority of strikes in 1895, 320 out of 405,
establishment, 30 involved from 2 to 5 establishments, 20
27 from 11 to 25, and 8 from 26 to 50 establishments.

The two tables that follow show the number of
establishments involved according to the
as the number of days work lost
of strikers represent of the total
to 17 main groups of industries.

a The report for

| Industry. | Succeeded. | | Succeeded partly. | | Failed. | | Total. | |
|---|---|---|---|---|---|---|---|---|
| | Strikes. | Establishments. | Strikes. | Establishments. | Strikes. | Establishments. | Strikes. | Establishments. |
| Agriculture, forestry, and fisheries | ........ | ........ | 2 | 2 | 2 | 2 | 4 | 4 |
| Mining | 2 | 2 | 4 | 4 | 3 | 3 | 9 | 9 |
| Quarrying | 4 | 17 | 8 | 22 | 6 | 8 | 19 | [illegible] |
| Food products | 4 | 22 | ........ | ........ | 3 | 52 | 7 | [illegible] |
| Chemical industries | 1 | 1 | 2 | 5 | 5 | 10 | 8 | [illegible] |
| Printing | 1 | 1 | 1 | 1 | 13 | 17 | 15 | [illegible] |
| Hides and leather | 11 | 31 | 9 | 9 | 18 | 21 | a [illegible] | [illegible] |
| Textiles proper | 32 | 66 | 45 | 82 | 63 | 58 | a 141 | a 212 |
| Clothing and cleaning | 1 | 1 | 2 | 2 | 4 | 5 | 7 | 8 |
| Woodworking | 3 | 7 | 4 | 6 | 7 | 22 | 14 | [illegible] |
| Building trades (woodwork) | 3 | 9 | 8 | 110 | 6 | 68 | 18 | 187 |
| Metal refining | 1 | 1 | 1 | 1 | 3 | 3 | 5 | 5 |
| Metallic goods | 16 | 39 | 7 | 66 | 21 | 23 | 44 | 128 |
| Precious-metal work | ........ | ........ | ........ | ........ | ........ | ........ | ........ | ........ |
| Stone cutting and polishing, glass and pottery work | 4 | 4 | 2 | 14 | 8 | 8 | 14 | 26 |
| Building trades (stone, earthenware, glass, etc.) | 14 | 56 | 24 | 226 | 15 | 20 | 53 | [illegible] |
| Transportation and handling | 4 | 21 | 3 | 16 | 9 | 43 | 16 | [illegible] |
| Total | 109 | 277 | 117 | 598 | 186 | 483 | b 405 | b 1,2[illegible] |

a Including 1 strike not yet terminated.
b Including 2 strikes not yet terminated.

STRIKERS AND DAYS OF WORK LOST IN 1896, BY INDUSTRIES.

| Industry. | In successful strikes. | In partly successful strikes. | In strikes which failed. | Total strikers. | Strikers per 1,000 work people. (a) | Days of work lost. |
|---|---|---|---|---|---|---|
| Agriculture, forestry, and fisheries | ........ | 53 | 8 | 61 | 0.02 | 53 |
| Mining | 265 | 1,506 | 738 | 2,509 | b 21.13 | 51,019 |
| Quarrying | 705 | 740 | 421 | 1,866 | (c) | 8,997 |
| Food products | 528 | ........ | 365 | 893 | 6.97 | 1,165 |
| Chemical industries | 305 | 1,564 | 2,042 | 3,911 | 65.18 | 61,956 |
| Printing | 150 | 12 | 210 | 372 | 3.67 | 1,720 |
| Hides and leather | 796 | 459 | 873 | d 2,129 | d 16.66 | d 16,412 |
| Textiles proper | 3,101 | 5,899 | 5,610 | d 14,641 | d 20.27 | d 190,655 |
| Clothing and cleaning | 16 | 32 | 97 | 145 | .20 | 3,000 |
| Woodworking | 201 | 337 | 259 | 897 | 6.20 | 11,986 |
| Building trades (woodwork) | 440 | 407 | 417 | 1,264 | (e) | 20,503 |
| Metal refining | 420 | 300 | 597 | 1,317 | 12.86 | 12,296 |
| Metallic goods | 650 | 689 | 967 | 2,306 | 7.61 | 28,320 |
| Precious-metal work | ........ | ........ | ........ | ........ | ........ | ........ |
| Stone cutting and polishing, glass and pottery work | 148 | 645 | 1,762 | 2,555 | 34.50 | 126,483 |
| Building trades (stone, earthenware, glass, etc.) | 585 | 5,613 | 826 | 7,024 | f 19.27 | 48,550 |
| Transportation and handling | 175 | 2,416 | 1,330 | 3,921 | 16.61 | 23,162 |
| Total | 8,565 | 20,673 | 16,521 | g 45,801 | g 12.83 | g 617,469 |

a Census of 1891.
b Including quarrying.
c Included in mining.
d Including 1 strike not yet terminated.
e Included in building trades (stone, earthenware, glass, etc.).
f Including building trades (woodwork).
g Including 2 strikes not yet terminated.

According to the number of strikes the industry most affected was that of textiles proper. A total of 141 strikes, involving 212 establishments and 14,641 strikers, and resulting in a loss of 190,655 days labor, occurred in this industry. Next in importance came that of the building trades, the two groupings together having had 69 strikes, involving 567 establishments and 8,288 workingmen, and causing a loss of 69,053 days labor. Metallic goods came third with 44 strikes, 128 establishments, 2,396 strikers, and [illegible] occupied the first three [illegible] in 1894.

As regards the relative number [illegible] part in strikes, however, the [illegible] especially affected by strikes, over [illegible] that industry participating in 1895; [illegible] and pottery work coming second with [illegible] engaged in strikes.

The information given in the two [illegible] tables which follow, according to the causes [illegible] were undertaken instead of according to [illegible]

STRIKES IN 1895, BY CAUSES.

[A considerable number of strikes were due to two or three causes, [illegible] been tabulated under each cause. Hence the totals for this table [illegible] those for the preceding tables.]

| Cause or object. | Succeeded. | | Succeeded partly. | | [illegible] | |
|---|---|---|---|---|---|---|
| | Strikes. | Establishments. | Strikes. | Establishments. | Strikes. | [illegible] |
| For increase of wages | 48 | 180 | 78 | 484 | [illegible] | [illegible] |
| Against reduction of wages | 13 | 14 | 13 | 13 | [illegible] | [illegible] |
| For reduction of hours of labor with present or increased wages | 23 | 170 | 7 | 63 | [illegible] | [illegible] |
| Relating to time and method of payment of wages, etc | 13 | 60 | 1 | 13 | 7 | [illegible] |
| For or against modification of conditions of work | 10 | 17 | 4 | 4 | 14 | [illegible] |
| Against piecework | 2 | 2 | 2 | 2 | [illegible] | [illegible] |
| For or against modification of shop rules | 3 | 3 | 1 | 1 | 15 | [illegible] |
| For abolition or reduction of fines | 3 | 3 | 2 | 2 | 7 | [illegible] |
| Against discharge of workmen, foremen, or directors, or for their reinstatement | 7 | 18 | 3 | 3 | 18 | [illegible] |
| For discharge of workmen, foremen, or directors | 16 | 18 | 5 | 5 | 35 | [illegible] |
| Against employment of women | .......... | .......... | .......... | .......... | 3 | [illegible] |
| For discharge of apprentices or limitation in number | 1 | 5 | .......... | .......... | 1 | 1 |
| Relating to deduction from wages for the support of insurance and aid funds | 3 | 29 | .......... | .......... | .......... | .......... |
| Other | 2 | 2 | 1 | 1 | 8 | [illegible] |

*a* Including 1 strike not yet terminated.

STRIKERS AND DAYS OF WORK LOST IN 1895, BY CAUSES.

[A considerable number of strikes were due to two or three causes, and the facts in such cases have been tabulated under each cause. Hence the totals for this table necessarily would not agree with those for the preceding tables.]

| Cause or object. | In successful strikes. | In partly successful strikes. | In strikes which failed. | Total strikers. | Days of work lost. |
|---|---|---|---|---|---|
| For increase of wages | 4,224 | 15,781 | 8,883 | 28,888 | 262,236 |
| Against reduction of wages | 682 | 842 | 1,649 | a 3,204 | a [illegible] |
| For reduction of hours of labor with present or increased wages | 3,602 | 777 | 1,727 | 6,106 | 67,461 |
| Relating to time and method of payment of wages, etc | 1,734 | 810 | 134 | 2,678 | 13,281 |
| For or against modification of conditions of work | 1,109 | 2,707 | 1,046 | 4,862 | 55,143 |
| Against piecework | 38 | 506 | 519 | 1,063 | 12,308 |
| For or against modification of shop rules | 208 | 35 | 1,257 | 1,500 | 14,150 |
| For abolition or reduction of fines | 563 | 56 | 1,200 | 1,819 | 9,732 |
| Against discharge of workmen, foremen, or directors, or for their reinstatement | 938 | 747 | 3,530 | a 5,217 | a 177,761 |
| For discharge of workmen, foremen, or directors | 1,318 | 837 | 2,798 | 4,953 | 32,583 |
| Against employment of women | ........ | ........ | 83 | 83 | 1,236 |
| For discharge of apprentices or limitation in number | 164 | ........ | 21 | 185 | 6,694 |
| Relating to deduction from wages for the support of insurance and aid funds | 378 | ........ | ........ | 378 | 5,279 |
| Other | 85 | 142 | 244 | 471 | 3,862 |

a Including 1 strike not yet terminated.

The demand for higher wages or the refusal to accept a reduction of wages alone or in connection with other causes still continues, as in former years, the chief cause of strikes. This cause accounts for 62.47 per cent of all strikes, 70 per cent of all strikers, and 68.33 per cent of days of labor lost. The question of the employment or nonemployment of workingmen, foremen, or directors figures as the second important cause, having produced 85 strikes in 1895 as against 78 in 1894. Demand for shortening the hours of labor caused 30 strikes in 1894 and 49 in 1895. These strikes resulted in the substitution of 11 hours of labor in place of 12 in 14 cases, of 10 hours in place of 11 in 8 cases, and of 10 hours in place of 12 in 4 cases.

The results of strikes according to their importance and severity—that is, according to the number of persons involved and the duration of the strikes—is shown in the two tables that follow:

STRIKES AND STRIKERS, BY DURATION OF STRIKES, IN 1895.

| Days of duration. | Strikes. | | | | Strikers. | | | |
|---|---|---|---|---|---|---|---|---|
| | Succeeded. | Succeeded partly. | Failed. | Total. | Succeeded. | Succeeded partly. | Failed. | Total. |
| 7 or under | 72 | 73 | 131 | 276 | 6,441 | 11,405 | 7,182 | 25,028 |
| 8 to 15 | 16 | 20 | 25 | 61 | 1,258 | 5,293 | 2,173 | 8,724 |
| 16 to 30 | 4 | 13 | 16 | 33 | 350 | 1,502 | 3,190 | 5,042 |
| 31 to 100 | 8 | 11 | 11 | a 31 | 521 | 2,467 | 2,719 | a 5,738 |
| 101 or over | ........ | ........ | 3 | a 4 | ........ | ........ | 1,257 | a 1,269 |
| Total | 100 | 117 | 186 | b 405 | 8,565 | 20,672 | 16,521 | b 45,801 |

a Including 1 strike not yet terminated. b Including 2 strikes not yet terminated.

| Strikers involved. | Strikes. | | | | | |
|---|---|---|---|---|---|---|
| | Succeeded. | Succeeded partly. | Failed. | Total. | [illegible] | [illegible] |
| 25 or under | 25 | 27 | 84 | a137 | [illegible] | [illegible] |
| 26 to 50 | 29 | 20 | 31 | a81 | [illegible] | [illegible] |
| 51 to 100 | 18 | 24 | 26 | 68 | [illegible] | [illegible] |
| 101 to 200 | 16 | 21 | 18 | 55 | [illegible] | [illegible] |
| 201 to 500 | 12 | 18 | 10 | 40 | [illegible] | [illegible] |
| 501 to 1,000 | ........ | 3 | 3 | 6 | [illegible] | [illegible] |
| 1,001 or over | ........ | 4 | 2 | 6 | [illegible] | [illegible] |
| Total | 100 | 117 | 186 | b405 | [illegible] | [illegible] |

a Including 1 strike not yet terminated. b Including 3 strikes [illegible]

The great majority of strikes were comparatively [illegible] Two hundred and seventy-six, or over two-thirds, lasted [illegible] and but 35 lasted a month or over, while 137, or over [illegible] less than 26 persons and but 12 involved over 500 persons [illegible] the 12 large strikes were completely successful, 7 having [illegible] compromise and 5 having been failures.

There are two classes of information contained in these [illegible] the French bureau that were not touched upon in the first [illegible] concerning which it will perhaps be well to make some [illegible] considerable portion of each report, in the present case 16[illegible] devoted to giving an account of the more important strikes of [illegible] and secondly, beginning with the volume for 1893, information [illegible] concerning the extent to which use has been made of the [illegible] December 27, 1892, relating to the arbitration of labor disputes.

Briefly stated, this law provides that on the arising of any [illegible] between an employer and his employees the question in dispute [illegible] both parties agree, be submitted to a council of conciliation, [illegible] actual strike has been begun, to a council of arbitration, which [illegible] the presidency of the local justice of the peace, attempts to [illegible] a solution of the difficulty, or the justice of the peace may him[illegible] the initiative and request the parties to submit their differences to [illegible] a council. This council consists of delegates, not exceeding [illegible] number, chosen by each party. If they fail to reach an agreement [illegible] can appoint one or more arbitrators. The submission of a dispute [illegible] arbitration, however, is entirely voluntary, and the decision, no [illegible] how arrived at, can not be legally enforced, its acceptance being [illegible] matter to be determined by the parties as they deem best.

Demands for the application of this law were made in the case [illegible] of the 405 strikes occurring during 1895, or in 20.74 per cent of all [illegible] disputes. The initiative in making these demands was taken [illegible] times by the workingmen, twice by the employers, 3 times by [illegible] the employers and employees, and 34 (a) times by the justice [illegible]

a In the case of 1 strike extending to [illegible] the initiative in 1 department was taken [illegible] justice of the peace.

peace. In 4 cases work was resumed without waiting for the constitution of councils of arbitration, in 3 of which the strikers abandoned their claims and in the fourth a compromise was effected. In the remaining 80 cases submission to arbitration was refused in 34, of which 32 were by the employers and 2 by the workingmen. In 2 of these cases the intervention of the justice of the peace led to a settlement, and in a third the employer was compelled to accept arbitration by those of his employees who had not struck threatening to join the others. Arbitration was thus definitely refused in 31 cases. In these 31 cases the refusal to arbitrate was followed in 3 cases by the immediate resumption of work. The remaining 28 cases were fought out, resulting in success for the strikers in 4 cases, in partial success in 9, and in failure in 15.

Councils of arbitration were constituted for the adjustment of the remaining 49 strikes. Twenty-four of these were immediately adjusted by the council, and 5 others later on as the result of further negotiations. Of these 29 thus settled, 4 were in favor of the strikers, 24 were compromised, and 1 was in favor of the employer. It is a matter of interest to notice the large number of strikes that were thus settled by mutual concessions on the part of both parties. The 20 strikes remaining, in which the constitution of councils proved of no avail, resulted in success for the workingmen in 3 cases, in compromise in 8 cases, and failure in 9 cases.

In addition to these 84 strikes in which the constitution of councils of arbitration was asked, 5 demands were made by the employees for the submission of differences to councils of conciliation before the actual outbreak of the strikes. In one of these cases the employers refused to discuss the matter and the workingmen continued their work; in 4 cases a council was constituted and a settlement not being achieved, strikes resulted, the outcome being 2 compromises and 2 failures for the strikers.

The year 1895 was the third year that the law concerning arbitration had been in force. In the first year, 1893, arbitration was requested in 109 of the 634 strikes, or in 17.19 per cent of all strikes; in 1894, in 101 of the 391 strikes, or 25.83 per cent of all strikes. In 1893, 51 of these cases were finally adjusted by the councils, resulting 12 times in success to the strikers, 26 times in a compromise, and 13 times in failure. In 1894, 53 cases were thus adjusted, the results being 13 successes, 24 compromises, and 16 failures. In 1895, if to the 29 cases settled by the councils as the result of formal meetings there be added the 4 strikes adjusted before the councils could be formed and the 3 strikes terminated as soon as the decision of the employers concerning arbitration was known, there were adjusted a total of 36 strikes, 4 of which resulted in success, 25 in compromises, and 7 in failures.

Since the law went into effect, therefore, arbitration was requested in 295 out of 1,430 strikes, or in 20.63 per cent of all strikes. Owing to the refusal to arbitrate, or for other reasons, but 140 of these strikes were actually adjusted by councils of arbitration, their decisions result[illegible] 29 successes to the strikers, in 75 compromises, and in 36 failures.

*Report of the Strikes and Lockouts of 1894 in* [illegible] 1895. 345 pp. (Published by the Labor Department [illegible] Board of Trade.)

This report treats of the state of the labor market [illegible] character, magnitude, and method of settlement of [illegible] lockouts that occurred in the United Kingdom during [illegible] statistics are presented in detail for each dispute, and [illegible] statements include comparative data for previous years.

The state of the labor market in 1894 and in the seven [illegible] years is indicated by the percentage of the members of trade [illegible] that were reported as unemployed at the close of each month [illegible] the period. These percentages are shown in the following [illegible]

PERCENTAGE OF MEMBERS OF TRADE UNIONS REPORTED AS UNEMPLOYED AT THE END OF EACH MONTH, 1887 TO 1894.

| Month. | 1887. | 1888. | 1889. | 1890. | 1891. | 1892. | 1893. | 1894. |
|---|---|---|---|---|---|---|---|---|
| January | 10.3 | 7.8 | 3.1 | 1.4 | 2.4 | [illegible] | 10.[illegible] | [illegible] |
| February | 8.5 | 7.0 | 2.8 | 1.6 | 2.6 | [illegible] | [illegible] | [illegible] |
| March | 7.7 | 5.7 | 2.2 | 1. | 2.8 | [illegible] | [illegible] | [illegible] |
| April | 6.8 | 5.2 | 2.0 | 2. | 2.7 | [illegible] | [illegible] | [illegible] |
| May | 6.5 | 4.8 | 2.0 | 2. | 2.9 | [illegible] | [illegible] | [illegible] |
| June | 8.0 | 4.6 | 1.8 | 1. | 2.9 | [illegible] | [illegible] | [illegible] |
| July | 8.5 | 3.9 | 1.7 | 2. | 3.3 | [illegible] | [illegible] | [illegible] |
| August | 8.3 | 4.8 | 2.5 | 2. | 4.2 | [illegible] | [illegible] | [illegible] |
| September | 7.5 | 4.4 | 2.1 | 2. | 4.5 | [illegible] | [illegible] | [illegible] |
| October | 8.6 | 4.4 | 1.8 | 2. | 4.4 | 7.3 | 7.3 | [illegible] |
| November | 8.5 | 3.1 | 1.5 | 2. | 3.8 | [illegible] | [illegible] | [illegible] |
| December | 6.9 | 3.3 | 1.7 | 3. | 4.4 | 10.2 | [illegible] | [illegible] |

At the end of 1888 the number of unions reporting the number of their unemployed seldom exceeded 20, with an aggregate membership of 200,000, while at the end of 1894 there were 62 unions reporting with a membership of 362,000. While the increase in the number reported enhances the value of the figures for the later years, the percentages given in the above statement are not absolutely comparable for the entire period of eight years, the general effect being that the percentages for the earlier years are slightly too high.

The statistics concerning the number of strikes and lockouts, and the persons affected by them, in the United Kingdom during 1894 are summarized in the following statement:

STRIKES AND LOCKOUTS AND PERSONS AFFECTED IN 1894.

[Persons affected means persons thrown out of work.]

| Division. | Total strikes and lockouts. | Strikes and lockouts for which persons affected were reported. | |
|---|---|---|---|
| | | Number. | Persons affected. |
| England | [illegible] | [illegible] | [illegible] |
| Wales | 78 | [illegible] | [illegible] |
| Scotland | 158 | [illegible] | [illegible] |
| Ireland | [illegible] | [illegible] | [illegible] |
| Total | [illegible] | [illegible] | [illegible] |

the following statements:

RESULTS OF STRIKES AND LOCKOUTS, BY CAUSES, IN 1894.

| Cause or object. | Succeeded. | Succeeded partly. | Failed. | Not reported. | Total. |
|---|---|---|---|---|---|
| Wages | 182 | 161 | 190 | 31 | [illegible] |
| Hours of labor | 6 | 3 | 12 | 3 | 33 |
| Working arrangements | 79 | 50 | 79 | 11 | 219 |
| Class disputes | 29 | 8 | 25 | 3 | 65 |
| Unionism | 37 | 6 | 28 | 3 | 74 |
| Other causes or objects | 30 | 16 | 66 | 4 | 116 |
| Cause not known | ...... | ...... | ...... | 2 | 2 |
| Total | 372 | 244 | 389 | 56 | 1,061 |

PERSONS AFFECTED BY STRIKES AND LOCKOUTS, BY CAUSES AND RESULTS, IN 1894.

[Persons affected means persons thrown out of work.]

| Cause or object. | Succeeded. | | Succeeded partly. | | Failed. | | Not reported. | | Total. | |
|---|---|---|---|---|---|---|---|---|---|---|
| | Strikes and lock-outs. | Persons affected. | Strikes and lock-outs. | Persons affected. | Strikes and lock-outs. | Persons affected. | Strikes and lock-outs. | Persons affected. | Strikes and lock-outs. | Persons affected. |
| Wages | 175 | 31,150 | 156 | 93,531 | 178 | 107,112 | 17 | 3,110 | 526 | 224,908 |
| Hours of labor | 6 | 2,650 | 2 | 1,560 | 12 | 1,895 | ...... | ...... | 20 | 6,105 |
| Working arrangements | 78 | 15,042 | 47 | 9,137 | 77 | 12,128 | 5 | 1,456 | 207 | 37,763 |
| Class disputes | 27 | 1,612 | 8 | 813 | 24 | 1,248 | 2 | 26 | 61 | 3,699 |
| Unionism | 36 | 12,570 | 5 | 705 | 28 | 2,202 | 2 | 42 | 71 | 15,519 |
| Other causes or objects | 38 | 8,627 | 16 | 5,332 | 54 | 11,788 | 4 | 499 | 112 | 26,256 |
| Total | 360 | 71,661 | 234 | 111,078 | 373 | 136,373 | 30 | 5,133 | 997 | 324,245 |

The number of persons thrown out of employment during 1894 is shown for 997 strikes and lockouts; the remaining 64 are known to have been insignificant. Of the 324,245 persons thrown out of employment, 257,937 were directly engaged in the disputes and 66,308 were indirectly engaged. The number of persons affected was less by 312,141 than the number reported affected for 1893, and the average number per dispute was 306, as compared with 814 in 1893. The average duration in working days was 24.6, as compared with 28 days in 1893.

The information concerning the strikes and lockouts for which the number of persons affected and the working days lost were reported is summarized in the following statement. The total number of persons affected is also shown. The strikes and lockouts are grouped according to the number of persons affected.

WORKING DAYS LOST AND PERSONS AFFECTED BY STRIKES AND LOCKOUTS IN 1894.

[Persons affected means persons thrown out of work.]

| Groups. | Strikes and lockouts for which both persons affected and working days lost were reported. | | | | Strikes and lockouts for which persons affected were reported. | |
|---|---|---|---|---|---|---|
| | Number. | Persons affected. | Working days lost. | | Number. | Persons affected. |
| | | | Number. | Average per person affected. | | |
| 5,000 persons and upward | 4 | 118,000 | 5,995,000 | 50.8 | 4 | 118,000 |
| 1,000 to 5,000 persons | 48 | 80,133 | 910,329 | 11.4 | 49 | 81,133 |
| 500 to 1,000 persons | 59 | 40,598 | 378,231 | 9.3 | 59 | 40,598 |
| 100 to 500 persons | 276 | 61,152 | 1,632,486 | 26.7 | 297 | 65,811 |
| Under 100 persons | 490 | 16,160 | 406,050 | 25.1 | 588 | 18,703 |
| Total | 877 | 316,043 | 9,322,096 | 29.5 | 997 | 324,245 |

In the following statement the strikes and lockouts are grouped according to the method of settlement:

METHOD OF SETTLEMENT OF STRIKES AND LOCKOUTS IN 1894.

[Disputes settled by a combination of two or more of the methods enumerated have been classed under the most important one. Disputes settled partly by arbitration and partly by other methods are classed under arbitration. Persons affected means persons thrown out of work.]

| Method of settlement. | Total strikes and lockouts. | Strikes and lockouts for which persons affected were reported. | |
|---|---|---|---|
| | | Number. | Persons affected. |
| Negotiation or conciliation between the parties | 607 | 563 | 144,125 |
| Mediation or conciliation by third parties | 18 | 18 | 8,399 |
| Arbitration | 32 | 12 | 10,785 |
| Submission of work people | 170 | 162 | 147,044 |
| Replacement of hands | 150 | 155 | 6,451 |
| Closing of works or establishments | 13 | 11 | 985 |
| Withdrawal or disappearance of cause without mutual agreement | 6 | 6 | 1,323 |
| Indefinite, or no information | 56 | 30 | 5,133 |
| Total | 1,061 | 997 | 324,245 |

Although the largest number of disputes during the year were settled by negotiation between the parties or by some other conciliatory method, the largest proportion of work people affected had their disputes terminated by submission.

The settlement of disputes by conciliation and arbitration is treated separately. The modes of settlement considered under this head are only those in which an independent individual or permanent body intervened or took part. Settlements due to the mediation of a trade union or trades council on the one side or association of employers or chamber of commerce on the other are not considered in this connection.

Including disputes which began in 1893 and were referred to arbitration or settlement by conciliation in 1894, but excluding those commenced in 1894 and referred to settlement in 1895, there were 42, affecting 18,325 persons, as compared with 25 in 1893, affecting 312,000 persons. The decrease in the number affected is explained by the fact that in 1893 the greatest dispute of the year, viz, that in the coal trade, involved no less than 300,000 persons. This dispute was settled by the mediation of Lord Rosebery.

In the following statement the strikes and lockouts settled by conciliation and arbitration are classified according to the agency employed in their settlement:

STRIKES AND LOCKOUTS SETTLED BY CONCILIATION AND ARBITRATION IN 1894.

[Persons affected means persons thrown out of work.]

| Agency employed. | Conciliation. | | Arbitration. | | Total. | |
|---|---|---|---|---|---|---|
| | Strikes and lock-outs. | Persons affected. | Strikes and lock-outs. | Persons affected. | Strikes and lock-outs. | Persons affected. |
| Trade boards | 7 | 2,572 | 3 | 2,957 | 10 | 5,529 |
| Individuals | 11 | 2,799 | 16 | 9,187 | 27 | 11,986 |
| Trades councils and federations of trade unions (disputes between groups of work people) | 1 | 312 | 4 | 498 | 5 | 810 |
| Total | 19 | 5,683 | 23 | 12,642 | 42 | 18,325 |

It is not to be supposed that permanent boards of conciliation and arbitration are ineffective because so few disputes were settled by them. The greater part of their work consisted in dealing with questions, any or all of which might, if unsettled, have terminated in strikes. Their work, in fact, is rather preventive than remedial.

In all 64 trade boards are believed to have existed in 1894. Ten of these boards dealt with no questions during the year. The number of cases reported as dealt with by the remaining 54 (a) boards was 1,733, of which 368 were withdrawn by one or both of the parties, or referred back or ruled out of order by the boards. Thus, 1,365 cases were settled in 1894, as compared with 1,228 in 1893. Of these 1,365 the boards settled 1,142, the remaining 223 being referred to arbitrators by the boards or settled by independent chairmen of the boards.

Out of 22 district boards believed to have been in existence during the year, only 7 are known to have offered their services in any dispute, and only 3 actually dealt with any questions.

The loss to employers and work people caused by strikes and lockouts is shown by statistics taken from the returns received from employers and trade unions.

a In the case of 6 of these boards no information was reported; in the case of 2 only the principal questions dealt with were reported.

The information contained in the following ... the returns received from employers:

COST OF STRIKES AND LOCKOUTS IN 1894, AS ...

| Items. | [illegible] |
|---|---|
| Estimated value of fixed capital laid idle | [illegible] |
| Estimated value of capital where number of persons is not known | [illegible] |
| Estimated annual ratable value of property laid idle | [illegible] |
| Estimated actual outlay by employers in stopping and reopening works, and in payment of fixed charges, salaries, etc | [illegible] |
| Estimated actual outlay in cases where number of persons is not known | [illegible] |
| Amount paid in defense against strikes or in support of lockouts by organizations of employers | [illegible] |

Compiled from partial returns, the trade unions repor expended during 1894 from trade-union funds in suppo defense against lockouts in 329 disputes involving 35,94 5 of these disputes, however, the persons involved wer The amount expended in this manner from funds othe trade unions is reported as $33,944 for 142 disputes in persons.

## DECISIONS OF COURTS AFFECTING LABOR.

[This subject, begun in Bulletin No. 2, will be continued in successive issues, dealing with the decisions as they occur. All material parts of the decisions are reproduced in the words of the courts, indicated, when short, by quotation marks, and when long by being printed solid. In order to save space immaterial matter, needed simply by way of explanation, is given in the words of the editorial reviser.]

### DECISIONS UNDER STATUTORY LAW.

**Employers' Liability—Railroad Companies**—*Culver v. Alabama R. R. Co. 18 Southern Reporter, page 827.*—Section 2590 of the Code of Alabama provides that "when a personal injury is received by a servant or employee in the service or business of the master or employer, the master or employer is liable to answer in damages to such servant or employee as if he were a stranger, and not engaged in such service or employment, in the cases following:

* * * * * * *

"5. When such injury is caused by reason of the negligence of any person in the service or employment of the master or employer, who has the charge or control of any signal, points, locomotive, engine, switch, car, or train upon a railway, or of any part of the track of a railway."

Under the above provision suit was brought against the Alabama Midland Railway Company by Levin L. Culver as administrator of Virgil Mowdy, deceased, to recover damages for injuries sustained by Mowdy, which resulted in his death, caused by the alleged negligence of an engineer in charge of a locomotive. Judgment was given by the circuit court of Dale County, Ala., for the railroad company, whereupon Culver appealed to the supreme court, which reversed the judgment of the circuit court and remanded the case by decision rendered December 19, 1895.

In delivering the opinion of the supreme court Judge Coleman said:

The employer is liable for an injury inflicted upon an employee by the negligence of a coemployee when such negligence comes within the provisions of the employer's act [section 2590, Code of Alabama], and that without reference to the care and diligence used by the employer in the selection of his servants or employees. The employer's act in no wise relieves the employer from the duty of selecting with reasonable care his servant. The act imposes a further liability, and makes him responsible for injuries sustained by an employee in consequence of any neglect by the employer or his servants, specified in the act itself.

EMPLOYERS' LIABILITY—RAILROAD COMPANIES [illegible] *sota Belt Line and Transfer Co. 65 Northwestern Reporter* [illegible] In this case the allegations of the complaint of the [illegible] effect that he had been employed in the defendant's [illegible] that, when a stock train arrived, his duty was to step [illegible] platform up on top of the cars as they drew up opposite [illegible] and pull bundles of hay from the platform up on the top of [illegible] that the conductor of the train negligently ordered him [illegible] the platform up on the top of a passing car while it [illegible] great a rate of speed to enable him to do so with safety [illegible] was unknown to him, and that owing to the dangerous [illegible] the car, he, while stepping upon it, was thrown to the ground [illegible] arm run over by the wheels of a car.

From an order by the district court of Hennepin County, Minn., over-ruling the defendant's demurrer to Leier's complaint, appeal [illegible] to the supreme court of the State, which tribunal, on December [illegible], sustained the action of the district court, and decided that, [illegible] to the complaint, the plaintiff was injured by reason of [illegible] hazards peculiar to the operation of railroads, and that the [illegible] within the purview of section 2701 of the General Statutes of [illegible] Minnesota, making railroad companies liable to their servants for inju-ries caused by the negligence of their fellow-servants.

The opinion of the supreme court was delivered by Judge Mitchell, who, after summarizing the allegations of the complaint, said:

We think the fair construction of these allegations is: First, that [illegible] was usual and customary for defendant's servants to do this work under the directions of the conductor, and, hence, that in giving such instruc-tions the conductor was acting within the scope of his duty; second, that the conductor knew, or, in the exercise of ordinary care, ought to have known, that the car was moving too fast for the plaintiff to step upon it without exposing himself to great danger of personal injury. If this was so, then the conductor was guilty of negligence in giving the order. It does not appear—certainly not conclusively—from the allegations of the complaint that defendant [plaintiff] was guilty of negligence in obeying the order. It must be remembered that con-tributory negligence is a matter of defense and that a plaintiff is not required to negative it in his complaint. In doing the work which he was doing, in getting upon a moving car, plaintiff was exposed to an element of hazard or condition of danger which is peculiar to railroad business, and, as this element of danger caused or contributed to his injury, the statute (Gen. St. 1894, Sec. 2701) applies, and the railway company would be liable if the injury was caused by the negligence of a fellow-servant.

---

EMPLOYERS' LIABILITY—RAILROAD COMPANIES—*Pennsylvania* [illegible] *v. McCann. 42 Northeastern Reporter, page 768.*—The following [illegible] facts in this case: McCann, who was a brakeman in the service of [illegible] Pennsylvania Railroad Company, in attempting, in the State of [illegible] sylvania, to board one of its moving cars, put his foot in a stirrup that was suspended from the sill of the car and used as a step in mounting the car. The stirrup yielded to the pressure of his foot, causing him to be thrown under the car, whereby a wheel of the locomotive, which was backing, ran over one of his legs, inflicting the injury of which he complained. The railroad company was operating a line running from Youngstown, in Ohio, to a point in the State of Pennsylvania. Suit was brought against the railroad company in the court of common pleas in the State of Ohio. After the evidence had been presented for McCann the attorneys for the railroad company moved the court to take the case from the jury and to render a judgment in their favor, which was done. McCann then carried the cause to the circuit court in Mahoning County, Ohio. The circuit court reversed the judgment rendered in the court below on the sole ground that the act of April 2, 1890 (87 Ohio Laws, p. 149), was applicable, by force of which the fact that the stirrup was defective made a prima facie case of negligence against the railroad company.

The railroad company then brought the case on error to the supreme court of Ohio, which court on January 21, 1896, gave its decision affirming the judgment of the circuit court. From the opinion of the court, read by Judge Bradbury, the following is quoted:

The only question arising upon the record of sufficient importance to be worthy of extended consideration is whether the act of general assembly of this State, passed April 2, 1890 (87 Ohio Laws, p. 149), is applicable to the case or not, the injury complained of having been sustained beyond the limits of this State. The second section of the act in question prescribes the effect that shall be given to evidence which establishes a defect in the locomotives, cars, machinery, or attachments of certain railroads, in actions for injuries to its [their] employees, caused by such defects, and declares that, when such defects are made to appear, the same shall be prima facie evidence of negligence. There can be no doubt respecting the general power of a State to prescribe the rules of evidence which shall be observed by its judicial tribunals. It is a matter concerning its internal policy, over which its legislative department necessarily has authority, limited only by the constitutional guaranties respecting due process of law, vested rights, and the inviolability of contracts. The rules of evidence pertain to the remedy, and usually are the same, whether the cause of action in which they are applied arises within or without the State whose tribunal is investigating the facts in contention between the parties before it. Nor is it material, in this respect, whether the parties are residents or nonresidents of the State. The law of evidence, in its ordinary operation, is no more affected by one of these considerations than the other. No extraterritorial effect is given to a statute creating a rule of evidence by the fact that the rule is applied to the trial of a cause of action arising in another State, or to the trial of an action between parties who are nonresidents. If the tribunal of a State obtains jurisdiction of the parties and the cause, it will conduct the investigation of the facts in controversy between them according to its own rules of evidence, which is simply to follow its own laws within its own borders. The second section [of the act in question], in forbidding the use of defective cars and locomotives by railroad companies, refers to them as "such corporations," manifestly including every corporation owning or operating a railroad any part of which extends into this State.

Here, again, the prohibitive language employed is broad enough to include acts or conduct occurring in other States. In the subsequent clause of the second section of the act, wherein the general assembly sought to prescribe the rule of evidence, before referred to, applicable to the trial of actions in the courts of this State brought by employees of railroad companies on account of injuries sustained by reason of defective cars, locomotives, machinery, or attachments, it approached the question of procedure in our judicial tribunals, over which, as we have seen, the authority of the general assembly is practically supreme. This clause of the statute is purely remedial, and should receive a liberal construction. The language employed by the act in this connection is consistent with a legislative purpose to extend the remedy to all actions of the character named in the act against all railroad companies, and no sufficient reason has been assigned for limiting its operation to causes of action that arose within the State. Indeed, it would be somewhat anomalous to prescribe to the courts of the State rules of evidence depending upon the question whether the cause of action arose within or without the State; and an intent to create this distinction should not be imputed to the legislative power unless it is fairly inferable from the language it has used. That language is as follows: "And when the fact of such defect shall be made to appear in the trial of any action in the courts of this State brought by such employee or his legal representatives against any railroad corporation for damages on account of such injuries so received the same shall be prima facie evidence of negligence on the part of such corporations." This language contains nothing indicating a purpose to confine the rule of evidence it creates to causes of action that should arise in this State. On the contrary, it expressly extends the rule to "any action in the courts of this State brought by such employee * * * against any railroad corporation." In fact the language is comprehensive enough to apply the rule to a railroad company, in this class of actions, whether any part of its line extended into Ohio or not; and if the courts of our State should acquire jurisdiction over the person of a railroad company whose line lay wholly without the State, no reason is perceived why the rule should not be applied. Judgment affirmed.

---

THE FELLOW-SERVANT ACT OF TEXAS—*San Antonio and Aransas Pass R. R. Co. v. Harding et al.* 33 *Southwestern Reporter, page 373.*—In the district court of Harris County, Tex., judgment was rendered awarding $16,000 damages against the San Antonio and Aransas Pass Railway Company in favor of Laura Harding and others, the widow and minor children of Edward Harding, who was killed in a collision between the engine in which Harding was engineer and another engine used in switching in the company's yard at Waco, Tex. The case was carried, on appeal by the company, to the court of civil appeals, which tribunal affirmed the judgment of the district court by decision rendered November 28, 1895.

The circumstances under which Harding was killed were as follows: Deceased was an engineer, in the service of the company, in charge of a train going from Yoakum to Waco, and was under the control of the train master at Yoakum. In the company's yard at Waco was a regular yard crew, consisting of a night yard-master or foreman, a yard engineer or "hostler," a fireman, and other employees, and these were engaged in switching cars in the yard with engine No. 53. This yard crew was under the immediate supervision of one Hall, the foreman, who had no control over Harding. When engine No. 53 was taken to the yard to be used in switching cars its lamp was in a defective and leaking condition and was found empty. It was refilled and relighted by the yard engineer and fireman, who, it seems, had not been notified of its defective condition. The evidence was sufficient to show that when Harding arrived in sight the defective lamp had gone out, and nothing was done to give Harding notice of the switching engine's presence on the track upon which he was approaching, or to prevent a collision, except that when he had approached so close that he had not time to stop and avoid the danger, the yard engineer gave him a signal with his lantern to stop, and then endeavored to back the switch engine out of the way, but was prevented from doing so by the number of cars already occupying the side tracks. Deceased failed to discover the switching engine because of the absence of the headlight and received no other sufficient warning. A collision ensued, which resulted in his death.

In delivering the opinion of the court of civil appeals, Judge Williams said:

As negligence of the defendant in failing to exercise proper care to see that the headlight was in good condition was one of the causes contributing to the death of Harding, defendant is liable, even if it were true that the negligence of employees who were fellow-servants of the deceased also contributed. There can be little doubt that, if the headlight had been kept in proper condition, it would have continued to burn, and would have notified Harding of the presence of the switch engine in time to have enabled him to avoid danger. No other cause for the extinguishment of the light is suggested by the evidence but that the oil had leaked out and that none remained to feed the light. The company is responsible for the omissions of servants, to whom it left the performance of the duty of seeing after the condition of the lamp.

Under our fellow-servants' act the employees working with the switch engine were not the fellow-servants of Harding. (Laws of 1893, p. 120.) The employees in the yard, under the supervision and control of the yard master, were in a different department from engineers running trains on the road, under the supervision and control of the train master at another place. It is contended that the two engineers were in the common service of the company, were in the same department, were of the same grade, and were working together at the same time and place, and to a common purpose, and, therefore, come within the definition of "fellow-servants" as given in the statute. If this were conceded, we do not think it could relieve appellant, even if no negligence but that of its servants were shown, because [illegible] have resulted from the negligence of the [illegible] was guilty of negligence, the foreman was also [illegible] the negligence of a fellow servant merely contributed [illegible] not relieve the company, if its own negligence, or [illegible] who are not fellow-servants with the injured [illegible] But we are not prepared to concede that the "[illegible] servant under the statute. In a sense, as stated by [illegible] the two engineers were in the same department, the [illegible] department," but this has reference to the divisions of [illegible] branches made by the company. Under its regulations [illegible] be in the same department as named by it, and yet in [illegible] ments as intended by the statute. Such questions [illegible] by the relations which the employees actually bear [illegible] not by the mere names that are given by the company [illegible] branches of the service.

Nor do we think that the engineers were in the [illegible] "in the common service," or that they were "working [illegible] mon purpose." Their superiors, to whose authority they [illegible] were vice principals of the corporation, and stood to the [illegible] their control in the relation of master. This the [illegible] declares, and this provision, we think, enables us to [illegible] meant by the words "departments," "common service," [illegible] purpose." As pointed out in the Ross case (112 U. S., 3[illegible] 184), there is a line of decisions holding that employees are [illegible] department, and in a common employment, only when they [illegible] to the same immediate supervision and control. This view [illegible] generally prevailed, and was not adopted by the courts [illegible] and it seems to us, from the whole of the statute, that it [illegible] to substantially adopt it. The servant having control of [illegible] declared to stand in the relation of master to those under him, [illegible] in defining the relation of other employees to each other, it is [illegible] that, in order to be fellow-servants, they must be in the [illegible] in the same department, of the same grade, working [illegible] same time and place, and to a common purpose. The [illegible] to the control of different supervisors are thus treated as [illegible] rate departments and different service. When we consider [illegible] authorities, including some of the later opinions of our [illegible] had expressed the view that sound reason for the existence of [illegible] as to fellow-servants could only be found in cases where the [illegible] were so situated with reference to each other as to be enabled [illegible] cise over the conduct of each other that watchfulness regarded [illegible] tial to the efficiency of the service and the safety of the public, [illegible] that the legislature has adopted that view, and intended [illegible] in the provisions referred to. Under our construction of the [illegible] none of the employees in the Waco yard were fellow-[illegible] Harding, unless indeed, in the performance of his duties, he [illegible] temporarily subject, while operating in the yard, to the [illegible] the yard foreman. Then he might be considered for the time [illegible] servant with the others, subject to the same authority, but not [illegible] foreman himself.

The court, in its charge, gave to the jury all of the [illegible] statute, leaving them only to apply the evidence. Contention [illegible] that the rule applicable when one servant is intrusted with [illegible] others should not have been given, because there was no [illegible] support it. As before noted, there was evidence tending [illegible] engineers, while in the yards, were subject to the control of the yard master, and, if for no other reason than to prevent confusion in the minds of the jury, it was not improper for the court to tell them that, even in that view, Harding could not be a fellow-servant with the yard master. We think it evident that Harding on the occasion in question, never became subject to the authority of the yard master, but it could have done no harm for the court to inform the jury that, if he did, they were not fellow-servants. If we are correct in our view that none of the employees in the yard were fellow-servants of the deceased, then, even if the court committed error in defining those who might be fellow-servants, it is immaterial.

While this verdict is large, and may be for a greater amount than this court would allow if trying the case, it is not so clearly excessive as to authorize us to disregard the opinion of the jury and of the court below. In refusing to reverse such verdicts, we are not to be understood as approving them, but simply as adhering to the rules governing appellate courts in such matters.

---

CHINESE EXCLUSION ACTS—*United States v. Wong Hong. 71 Federal Reporter, page 283.*—The facts in this case are as follows: Prior and up to November 9, 1893, the defendant had resided continuously in the State of California for a period of sixteen years. On said date he departed for China and did not return until May 27, 1895. For a period of seven years preceding and up to August 1, 1893, the defendant was a merchant as defined by act of Congress passed November 3, 1893, being chapter 14 of the first (extra) session of the Fifty-third Congress. On August 1, 1893, his store was destroyed by fire. About six weeks or two months after the fire another store was built on the same lot where the original store stood. The firm of Duey Lee & Co. opened business in this new store two weeks or more before the defendant left for China. Defendant was a member of this firm and put $800 into the business. After the fire and up to his departure for China the defendant devoted himself to the business. After his return from China and up to the time of his arrest he stayed in the store and aided in the transaction of its business and retained his interest in the firm continuously up to the rendering of the decision in this case. The defendant was charged with being a Chinese laborer unlawfully within the United States. The case was heard in the district court for the southern district of California, and on December 2, 1895, the decision was given by Judge Wellborn. In the course of his opinion the following language was used:

The defendant's right to be in the United States must depend upon his having been a merchant at the time of his departure therefrom, November 9, 1893. If at that time he was a laborer, his return to the United States was in contravention of the act of October 1, 1888 (chapter 1015, acts of 1887–88), and unlawful. The defendant, having departed from the country in 1893, can not now be lawfully here, unless the facts sustain his contention that he was a merchant at the time of such departure. The act of November 3, 1893 (chapter 14, acts of 1893, extra session), provides as follows: "The term 'merchan herein and in the acts of which this is amendatory, sha lowing meaning and none other: A merchant is a per buying and selling merchandise, at a fixed place of b business is conducted in his name, and who during the to be engaged as a merchant, does not engage in the any manual labor except such as is necessary in the business as such merchant." An analysis of this provis in order to constitute a person a merchant, four things First, such person must be engaged in buying and sellin second, he must be engaged at a fixed place of busin business must be conducted in his name; fourth, he m the time he claims to be engaged as a merchant engage ance of any manual labor, except such as is necessary of his business as such merchant. With reference to ents, it is only necessary to say, that the defendant has to establish the third constituent, but the evidence sho ence. The evidence is uncontradicted and positive to the firm name was Duey Lee & Co., and there is no evidence that the defendant's name appeared in any duct of said business. It is impossible, therefore, to ho that the defendant is a merchant, without an utter d act of Congress above mentioned. The circuit court of circuit has decided that, in order to constitute a pers within the meaning of said act, it is not necessary that h in the firm designation, but it is sufficient if his inter appear in the business and partnership articles in his the present case there is no proof that the defendant's in the partnership articles or elsewhere in the business, is positive that the business subsequent to August l conducted in the defendant's name. My conclusion is t ant, Wong Hong, is a Chinese laborer, and unlawfully w diction of the United States, and the judgment of th that the said defendant, Wong Hong, be removed fr States to China.

---

CONSTITUTIONALITY OF LAW REQUIRING BLOWER WHEELS—*People v. Smith. 66 Northwestern Reporter*, No. 136 of the session laws of Michigan of 1887, now 1690z2, and 1690z3 of the third volume of Howell's Anno as amended by act No. 111 of the session laws of 1893, r

SECTION 1. All persons, companies, or corporations, factory or workshop where emery wheels or emery belts tion are used, either solid emery, leather, leather cover linen, paper, cotton, or wheels or belts rolled or coated corundum, or cotton wheels used as buffs, shall provid blowers, or similar apparatus, which shall be placed under such wheels or belts in such a manner as to pro or persons using the same from the particles of dust caused thereby, and to carry away the dust arising fro by such wheels or belts while in operation, directly to the building or to some receptacle placed so as to

used at the point of the grinding contact shall be exempt from the conditions of this act.

SEC. 2. Any such person or persons and the managers or directors of any such corporation who shall have [the] charge or management of such factory or workshop, who shall fail to comply with the provisions of this act, shall be deemed guilty of a misdemeanor, and upon a conviction thereof before any court of competent jurisdiction shall be punished by a fine not less than twenty-five dollars and not exceeding one hundred dollars, or imprisonment in the county jail not less than thirty days or exceeding ninety days, or both such fine and imprisonment in the discretion of the court.

SEC. 3. Nothing in this act shall apply to factories, sawmills, shingle mills, and workshops in which such wheels or belts are occasionally used and only by men not especially employed for that purpose.

Joseph N. Smith was convicted of a violation of this act and appealed to the supreme court of Michigan from judgment on a writ of certiorari from the justice's court in Detroit rendered in the circuit court of Wayne County. The supreme court rendered its decision March 3, 1896, sustaining the constitutionality of the act and affirming the judgment of the court below. The following is quoted from the opinion of the supreme court, delivered by Judge Hooker:

Counsel for the defendant assert that they care to raise but one question, viz, the constitutionality of this law. It is not disputed that the State may regulate the use of private property, when the health, morals, or welfare of the public demands it. Such laws have their origin in necessity.

Counsel say that this law is invalid because it does not apply to all, not even to all who have emery wheels, because some may use with water, and others may not work continuously. For the purposes of this case, it may be said that all persons who are given continuous employment over dry emery wheels are within the provisions of this act. This singles out no class, as it applies to all persons who use emery wheels in that manner. Necessarily the practical application is limited to those who engage in such business, but such is the case with many laws. All criminal laws apply only to those who choose to break them. This law applies to all who choose to use the emery wheel.

The legislature has seen fit to permit certain uses of the dry wheel without a blower, while in other cases it is required. This is competent, and is not class legislation as between operatives. It fixes the limits of use without a blower, and requires it after such limits are passed; but the rules apply to all.

The vital question in this case is the right of the State to require the employer to provide, and the employee to use, appliances intended for the protection of the latter. Laws of this class embrace provisions for the safety and welfare of those whom necessity may compel to submit to existing conditions involving hazards which they would otherwise be unwilling to assume. Among them are provisions for fire escapes, the covering or otherwise rendering machinery safe, the condition of buildings, ventilation, etc. In the main, where the necessity is obvious, they commend themselves to those who have at heart the welfare of their fellows, and should be upheld if they do not contravene private rights. The constitution secures to the citizen the rights of life, liberty, and private property, and, as the only value in the latter consists in its use, it follows that the right to use private prop-

...ditions which threaten conta-...quarantine is required,—the ...h laws are obvious. But at ...where there is no direct danger ...s sought to be regulated, and

...r the public welfare nor health ...sought to be afforded is limited ...contract of employment, signi-...s its dangerous condition, and

...y provide inferior and even dan-...t and enterprises more or less ...en may contract to use such ...service, and have no remedy if ...e police power is limited by such

..., the contract may cut off legal ...e not satisfied that the authority ...n of those who do not sustain ...n the absence of a law requiring ...building, and is injured may be ...to his employment, but we know ...m making a regulation requir-...buildings, though the only object ...s who are under contract to work ...ge. Fire escapes in hotels and ...ls are required by law for the ...irtue of contract relations with ...general, and the danger to the ...who are subjected to the dan-...legislative intervention.

...ject under discussion are here ...n to say:

...ver the inability of the courts to ...such laws may be tested. Each ...e (2) Is there a threatened dan-...stitutional right? (3) Is the ...e controversy is raised over ...upon to discuss it. As is ...al right to use property ...welfare requires its regu-...the right must yield to

...blic exigency. And it is upon this question of necessity that the ...question depends.

..., then, seems to be embraced in the question of necessity. Unless ...mery wheel is dangerous to health, there is no necessity, and con...uently no power, to regulate it. Unless the blower is a reasonable ... proper regulation, it is not a necessary one. Who shall decide the ...estion and by what rule? Shall it be the legislature or the courts? ...nd, if the latter, is it to be determined by the evidence in the case that happens to be the first brought, or by some other rule? Does it become a question of fact to be submitted to the jury or decided by the court?

There is a manifest absurdity in allowing any tribunal, either court or jury, to determine from the testimony in the case the question of the constitutionality of the law. Whether this law invades the rights of all the persons using emery wheels in the State is a serious question. If it is a necessary regulation, the law should be sustained, but, if an unjust law, it should be annulled. The first case presented might show, by the opinions of many witnesses, that the use of the dry emery wheel is almost necessarily fatal to the operative, while the next might show exactly the opposite state of facts. Manifestly, then, the decision could not settle the question for other parties, or the fate of the law would depend upon the character of the case first presented to the court of last resort, which would have no means of ascertaining whether it was a collusive case or not, or whether the weight of evidence was in accord with the truth.

It would seem, then, that the questions of danger and reasonableness must be determined in another way. The legislature, in determining upon the passage of the law, may make investigations which the courts can not. As a rule, the members (collectively) may be expected to acquire more technical and experimental knowledge of such matters than any court can be supposed to possess, both as to the dangers to be guarded against and the means of prevention of injury to be applied; and hence, while under our institutions the validity of laws must be finally passed upon by the courts, all presumptions should be in favor of the validity of legislative action. If the courts find the plain provisions of the constitution violated, or if it can be said that the act is not within the rule of necessity in view of facts of which judicial notice may be taken, then the act must fall; otherwise it should stand. Applying this test, we think the law constitutional, and the judgment is therefore affirmed.

---

EMPLOYERS' LIABILITY—RAILROAD COMPANIES—*Pennsylvania Co. v. Finney. 42 Northeastern Reporter, page 816.*—This action was brought by Michael Finney, administrator, against the Pennsylvania Company to recover damages for the killing of the plaintiff's intestate, Patrick J. Finney, a brakeman in its employ. A judgment was rendered in favor of the plaintiff, and the defendant appealed from the superior court of Allen County, Ind., where the trial was had, to the supreme court of the State. From the evidence it appeared that the decedent was 22 years old and had been in the employ of the defendant for six months as a brakeman on a freight train; that near Columbia City the defendant maintained a water plug so near its track that a person descending a passing car on that side could not avoid it; th... decedent was familiar with its location and had passed it almost d...

during his employment; [illegible]
of a train while passing through [illegible]
through Columbia City, he walked [illegible]
the plug, while within 200 feet of [illegible]
ascertain the attendant danger, [illegible]
was carried against the plug; that he [illegible]
particular time, but attempted to do [illegible]
opinion of the supreme court, delivered by [illegible]
1896, the following statements are made:

Considered in the light of the law which [illegible] we are of the opinion, under the facts, that [illegible] ized in finding a verdict in favor of the [illegible] deciding, that the appellant was chargeable with [illegible] in maintaining the water crane in the manner [illegible] shown, still there is an absence of evidence showing [illegible] tributory negligence upon the part of the deceased [illegible] which appellee complains. The rule is settled that [illegible] a case as this, must affirmatively show, by the evidence, [illegible] gence upon the part of the master, but freedom [illegible] part of the servant. The freedom from fault or negligence [illegible] part of the latter being under the law an essential [illegible] which must be found to exist in order to warrant a [illegible] to establish the same results in defeating the action; [illegible] dence in the record fails to prove this material fact, the [illegible] appeal to this court, must necessarily be reversed.

After reviewing the facts in the case the court uses the [illegible] language:

We may affirm that appellee's decedent did not, [illegible] observe his surroundings, or exert the care required of him [illegible] law; and hence, in the eye of the latter, he was chargeable [illegible] tributory negligence, and the allegations in the compl[illegible] extent, at least, are not sustained by the evidence. As [illegible] tofore stated, the accident occurred as the deceased was [illegible] descend to the caboose upon his own volition, and not under [illegible] direction of the appellant. We are unable to discover in this [illegible] evidence in the record from which a reasonable inference [illegible] arise that appellee's decedent was in the exercise of due [illegible] care at the time of the fatal accident. The jury was not [illegible] arbitrarily, without evidence, to infer the absence of contri[illegible] ligence upon the part of the deceased servant. The judg[illegible] reversed and the cause is remanded, with instructions to [illegible] court to sustain the motion for a new trial.

## DECISIONS UNDER COMMON LAW.

EMPLOYERS' LIABILITY—RAILROAD COMPANIES—*Central [illegible] of New Jersey v. Keegan. 16 Supreme Court Reporter, page [illegible]* action brought by one Keegan against the Central Railroad [illegible] of New Jersey judgment was rendered in favor of Keegan [illegible] verdict of a jury awarding him damages for injuries [illegible] while acting as brakeman in the [illegible] injuries having been caused by [illegible] foreman of a drill crew, of which Keegan was a member, which was employed in the company's yard at Jersey City, N. J., in taking cars from the tracks on which they had been left by incoming trains and placing them upon floats by which they were transported across the North river to the city of New York. The negligence of the foreman, resulting in the injury to Keegan, consisted in his failure to place himself or some one else at the brake of certain backwardly-moving cars, so that there was no one to check their motion by applying the brakes, in consequence of which the rear wheel passed over Keegan's leg, who, while in the performance of his dut , had caught his right foot in the guard rail of a switch, and was thereby prevented from moving out of the way of the cars.

The case was carried by the railroad company to the United States circuit court of appeals for the second circuit, where two judges, sitting as the court, differed in opinion upon questions of law, and certified the two following questions to the Supreme Court of the United States for instructions, to enable them to render a proper decision: "(1) Whether the defendant in error [Keegan] and O'Brien were or were not fellow-servants; and (2) whether, from negligence of O'Brien in failing to place himself or some one else at the brake of the backwardly-moving cars, the plaintiff in error, the railroad company, is responsible."

The United States Supreme Court, through Mr. Justice White, decided, December 23, 1895, that Keegan and O'Brien were fellow-servants and that the railroad company was not responsible for the injuries sustained by the former through the negligence of the latter; but Mr. Chief Justice Fuller, Mr. Justice Field, and Mr. Justice Harlan dissented.

The following extract is made from the opinion in the case:

We held in Railroad Co. *v.* Baugh (149 U. S., 368; 13 Sup. Ct., 914) that an engineer and fireman of a locomotive engine running alone on a railroad, without any train attached, when engaged on such duty, were fellow-servants of the railroad company; hence, that the fireman was precluded from securing damages from the company for injuries caused, during the running, by the negligence of the engineer. In that case it was declared that: "Prima facie, all who enter the employment of a single master are engaged in a common service, and are fellow-servants. * * * All enter in the service of the same master to further his interests in the one enterprise." And while we in that case recognized that the heads of separate and distinct departments of a diversified business may, under certain circumstances, be considered, with respect to employees under them, vice principals or representatives of the master, as fully and completely as if the entire business of the master was by him placed under the charge of one superintendent, we declined to affirm that each separate piece of work was a distinct department, and made the one having control of that piece of work a vice principal or representative of the master. It was further declared that "the danger from the negligence of one specially in charge of the particular work was as obvious and as great as from that of those who were simply coworkers with him upon it. Each is equally with the other an ordinary risk of the employment," which the employee assumes whether defendant is charged with any liability ... for the evidence makes it clear that the fall ... defects in the material, but to defective construction ... appears that defendant personally took no part in ... that it was constructed by the masons, in accordance ... the trade. As the error of construction which ... injury was committed by workmen with whom he ... common employment, subject to a common danger ... knew must result from a negligent construction of ... which denies the liability of the master for injury ... negligence of a fellow-servant was plainly applicable ... no evidence that defendant was negligent in the selection ... ment of the masons engaged in the work, there was ... defendant's liability sufficient to be submitted to a jury. ... reasons I think the rule should be made absolute.

---

EMPLOYERS' LIABILITY—VICE PRINCIPAL—*Carlson v. ... ern Telegraph Exchange Co. 65 Northwestern Reporter*, page ... case was brought before the supreme court of Minnesota on ... from the district court of Hennepin County. Suit had been b... the district court to recover damages for personal injuries ... the plaintiff by reason of the defendant's negligence. Verdict ... rendered for the plaintiff, and the defendant appealed from ... denying its motion for a new trial. January 20, 1896, the ... court gave its decision, affirming the order of the court below. ... evidence showed that the plaintiff, with some eighty other la... employed by the defendant, a corporation, was engaged on Aug... 1893, in the work of excavating a ditch in which telephone wires ... to be laid. The work of making this ditch was in charge of a ... of the name of Purvey, who had control of the work and of ... men engaged thereon, with power to employ and discharge them, ... to direct them what to do and where to work. He was the ... authority there present, and all of the men were subject to his ... in every particular and no one present had any authority over him. ... curbing had been put in to hold the side of the ditch in place, ... was insufficient for the purpose, and the complaint alleged the ... defendant knew that the walls of the ditch were unsafe and were ... to cave in and injure persons there working. The plaintiff ... knowledge of the condition of the ditch at the point where the ... was placed. When it was completed, he was ordered by the ... to go into the ditch and clean out the loose sand at the bottom. ... obeyed the order and commenced the work as directed, when a ... earth, constituting the side of the ditch, settled down ... ing into the bottom of the ditch, catching the plaintiff and ... his foot and ankle. The foreman did not caution or advise ... as to the unsafe condition of the ditch at the place ...

into. The following is the essential part of the opinion of the court, which was delivered by Chief Justice Start:

This brings us to the important, and practically the only real, question presented by the record for our decision, viz: Was the foreman discharging a duty which rested upon the defendant as master, when, under the particular circumstances and conditions of this case, he ordered the plaintiff into the ditch at the point where the accident occurred?

We answer the question in the affirmative. In the case of Leindvall v. Wood (41 Minn., 212) this court had the question under consideration, and as a result of a review of its previous decisions, and upon principle, reached the conclusion: "That it is not the rank of the employee, or his authority over other employees, but the nature of the duty or service which he performs, that is decisive; that whenever a master delegates to another the performance of a duty to his servant which rests upon himself as an absolute duty, he is liable for the manner in which that duty is performed by the middleman whom he has selected as his agent; and to the extent of a discharge of those duties by the middleman, however high or low his rank, or however great or small his authority over other employees, he stands in the place of the master, but as to all other matters he is a mere coservant. It follows that the same person may occupy a dual capacity of vice principal as to some matters and of fellow-servant as to others." We adhere to this conclusion. It is correct in principle, and furnishes a just and rational test for determining whether the act or default of an employee in a given case is that of a fellow-servant or of a vice principal. The decisive test is not the conventional title, grade, or rank given to the employee, but the character of what he is authorized to and does do. Applying the rule to the facts of this case, it is clear upon principle that the foreman, Purvey, in ordering the plaintiff into the ditch at the point of the accident, must be regarded as a vice principal. While the employee assumes for himself the ordinary and obvious dangers of the work or business in which he engages, yet the master is bound to use ordinary care to warn and protect the employee from unusual and unnecessary dangers and risks. If the nature and magnitude of the master's work, whether it be that of construction or otherwise, and the number of men engaged in its execution, are such that the exercise of ordinary care for the safety and protection of the workmen from unusual and unnecessary dangers requires that they be given reasonable orders, and that they be not ordered from one part of the work to another, without warning, into places of unusual danger and risks, which are not obvious to the senses and known to them, but which might be ascertained by the master by a proper inspection, the absolute duty rests upon the master to give such reasonable orders. Considerations of justice and a sound public policy impose this duty upon the master as such, which he can not delegate so as to relieve himself from the consequences of a negligent discharge of it. A workman, when ordered from one part of the work to another, can not be allowed to stop, examine, and experiment for himself, in order to ascertain if the place assigned to him is a safe one; and therefore, in obeying the order, while he assumes obvious and ordinary risks, he has a right to rely upon a faithful discharge of the master's duty to use ordinary care to warn and protect him against unusual dangers. In the present case the foreman, Purvey, was the supreme authority in charge of the work, with power to give all orders directing the places where the employees should work, and all reasonable and necessary [illegible] safety, which orders the plaintiff was bound to [illegible] was injured by reason of the negligence of the [illegible] ordering him into a place of unusual danger without [illegible] the risks incurred in obeying the order. In giving [illegible] the special facts of this case, the foreman represented [illegible] defendant.

---

RAILROAD AID ASSOCIATION—ACCEPTANCE OF [illegible]LEASE OF CLAIM—*Otis v. Pennsylvania Co. 71 Federal* [illegible] *136.*—In this case the United States circuit court, district of [illegible] decided, on January 3, 1896, that where a railroad relief [illegible] composed of associated companies and their employees, is in [illegible] the companies, who guaranty the obligations, supply the facilities for the business, pay the operating expenses, take charge of and are [illegible]sible for the funds, make up deficits in the benefit fund, and [illegible] surgical attendance for injuries received in their service, an [illegible] agreement, in his voluntary application for membership, that [illegible]ance of benefits from the association for an injury shall release the railroad company from any claim for damages therefor, is not invalid as being against public policy or for want of consideration or mutuality.

The opinion of the court, delivered by District Judge Baker, is as follows:

This is an action by the plaintiff, Eugene V. Otis, for the recovery of damages from the defendant, the Pennsylvania Company, for injuries received by him through the negligence of the defendant in employing and retaining in its service a careless and drunken engineer, with full knowledge of his habits, by whose carelessness the plaintiff sustained serious and permanent injuries, without fault on his part. The defendant has answered in two paragraphs. The first is a general denial. The second sets up matter in confession and avoidance. To this paragraph of answer the plaintiff has interposed a demurrer, and the question for decision is, Does this paragraph of answer set up facts sufficient to constitute a defense? The gist of this paragraph of answer is the payment to and acceptance by the plaintiff of benefits to the amount of $660 from the relief fund of the defendant's "voluntary relief department" on account of the injuries for which the action is brought, in full payment and satisfaction thereof.

It is alleged in the paragraph under consideration that the plaintiff was a member of the relief department mentioned, which is composed of the different corporations forming the lines of the Pennsylvania Company west of Pittsburg, to which such of their employees as voluntarily become members contribute monthly certain agreed amounts. This department has for its object the relief of such employees as become members thereof in cases of sickness or disability from accident, and the relief of their families in case of death, by the payment to them of definite amounts out of a fund "formed by voluntary contributions from employees, contributions, when necessary to make up any deficit, by the several companies respectively, and income or profit derived from investments of the moneys of the fund, and such gifts as may be made for the use of the fund." The associated companies have [illegible] charge of the department, guaranty the full amount of the obligations assumed by them, and for this purpose annually pay into the funds of the department the sum of $30,000 in conformity with established regulations, furnish the necessary facilities for conducting the business of the department, and pay all the operating expenses thereof, amounting annually to the sum of $25,000. The associated companies have charge of the funds, and are responsible for their management and safe-keeping. Employees of the Pennsylvania Company are not required to become members of the relief department, but are at liberty to do so if admitted on their voluntary written application; and may continue their membership by the payment of certain monthly dues, the amount of which depends upon the respective classes to which they may be admitted; and the benefits to which they may become entitled are determined by the class to which they belong. A disabled member is also entitled to surgical attendance at the company's expense, if injured while in its employ. The plaintiff agreed in his application for membership:

"That the acceptance of benefits from the said relief fund for injury or death shall operate as a release of all claim for damages against said company arising from such injury or death which may be made by or through me, and that I or my legal representatives will execute such further instrument as may be necessary formally to evidence such acquittance."

Each company to the contract also agreed in behalf of itself and employees to appropriate its ratable proportion of the joint expense of administration and management, and the entire outlay necessary to make up deficits for benefits to its employees. It is further alleged that the member was a member of the relief department when injured, and that there was paid to him by the defendant through such department, on account of the injuries so received, and in accordance with his application therefor, and in accordance with the certificate of membership so issued to him, and the rules and regulations of the relief department, the sum of $660, being at the rate of $60 per month for eleven months, which he accepted and received as the benefits due to him from the said relief department under his said application and certificate and the rules and regulations of said relief department.

It is strenuously insisted by the learned counsel for the plaintiff that the contract is void, because it is repugnant to sound public policy, and is an attempt by the defendant to exempt itself, by contract, from the consequences of its own negligence; and because the agreement that the payment and acceptance of the benefits should operate to release the company from responsibility for its wrongful act is without consideration, for the reason that the plaintiff, by the payment of his monthly dues, became entitled as a matter of legal right to receive the stipulated benefits as fully as he was entitled to the payment of his monthly wages.

As a general proposition, it is unquestionably true that a railroad company can not relieve itself from responsibility to an employee for an injury resulting from its own negligence by any contract entered into for that purpose before the happening of the injury, and, if the contract under consideration is of that character, it must be held to be invalid. But upon a careful examination it will be seen that it contains no stipulation that the plaintiff should not be at liberty to bring an action for damages in case he sustained an injury through the negligence of the defendant. He still had as perfect a right to sue for his injury as though the contract had never been entered into. Before the contract was entered into, his right of action for an injury resulting from the defendant's negligence was limited to a suit against it for the recovery of damages therefor. By the contract he was given an election either to receive the benefits stipulated for, or to waive his right to the benefits, and to pursue his remedy at law. He voluntarily agreed that, when an injury happened to him, he would then determine whether he would accept the benefits secured by the contract, or waive them and retain his right of action for damages. He knew, if he accepted the benefits secured to him by the contract, that it would operate to release his right to the other remedy. After the injury happened, two alternative modes were presented to him for obtaining compensation for such injury. With full opportunity to determine which alternative was preferable, he deliberately chose to accept the stipulated benefits. There was nothing illegal or immoral in requiring him so to do. And it is not perceived why the court should relieve him from his election in order to enable him now to pursue his remedy by an action at law, and thus practically to obtain double compensation for his injury. Nor does the fact that the fund was in part formed by his contributions to it alter the case. The defendant also contributed largely to the fund under its agreement to make up deficits, to furnish surgical aid and attendance, to pay expenses of administration and management, and to be responsible for the safe-keeping of the funds of the relief department. It had a large pecuniary interest in the very money which the plaintiff received. We are not concerned with the question whether the plaintiff might not have secured a larger sum of money if he had prosecuted his legal remedy for the recovery of damages for his injury. After the injury, the plaintiff was at liberty to compromise his right of action with the defendant for any valuable consideration, however small; and, if he chose to accept a less amount than that which he might have recovered by action, such settlement, if fairly entered into, constitutes a full accord and satisfaction, from which the court can not, and ought not to, relieve him.

The question of the validity of such a contract as that relied upon in the paragraph of answer under consideration is a new one in this court, but it has been considered by a number of reputable courts in other jurisdictions, and, with a single exception, so far as I am advised, it has been uniformly held that such a contract is not invalid for repugnancy to sound public policy, or for want of consideration, or for want of mutuality. In the views expressed in these cases I entirely concur.

---

INJUNCTIONS AGAINST LABOR ORGANIZATIONS—*Silver State Council, No. 1, of American Order of Steam Engineers v. Rhodes et al. 43 Pacific Reporter, page 451.*—This case was heard in the [illegible] of Arapahoe County, Colo., the injunction asked for being [illegible] judgment rendered for the defendants. The plaintiff [illegible] on error to the court of appeals of Colorado, and on [illegible] the decision of said court was given, in which the [illegible] in the court below was affirmed. The facts in the case [illegible] set out in the opinion of the court delivered by Judge [illegible] is as follows:

The purpose for which this proceeding was [illegible] the prayer of the complaint, which we quote: "[illegible] out of this honorable court, enjoining, restraining, and prohibiting the above-named defendants, each and all of them, and their said organizations, their servants, agents, and employees, both as individuals and organizations, in any manner interfering with or trying by threats, boycotts, strikes, or intimidations to break up and destroy, or cause the resignation of any member by threats, boycotts, strikes, or intimidations, of Silver State Council, No. 1, American Order of Steam Engineers, plaintiff herein, or by strikes, boycotts, or any other threats to compel it or its members to throw up its certificate, articles, or charter of incorporation or organization, or to in any manner interfere with the rights and privileges of Silver State Council, No. 1, of the American Order of Steam Engineers, plaintiff herein, or its right to exist and enjoy its rights, privileges, and freedom under the laws under which it was created; for costs herein expended, and will ever pray."

The complaint avers the capacity in which the plaintiff sues, and the objects of its corporate existence as follows: "That plaintiff is a corporation organized and existing by and under the laws of the State of Colorado, for the purpose of promoting a thorough knowledge in its members of theoretical and practical steam engineering, to help each other to obtain employment, bury the dead, extend the license law throughout the United States as well as the State of Colorado, and for the further purpose of helping its members according to the terms set forth in its certificate of incorporation, reference to which is hereto made." The complaint further states that the plaintiff is a "nonstriking labor organization;" that certain of the defendants are trustees of an organization called the "Trades and Labor Assembly," which is composed of various labor unions of Denver and vicinity, and was organized for the purpose (among other things) "of enforcing the rights of their several component parts by ordering a strike against all other organizations, employers, or individuals against whom it or they may have a grievance, and can not enforce their rights upon which they base their demands;" that certain other defendants are officers and members of an organization known as "Steam Engineers' Protective Union, No. 5703, of the American Federation of Labor," whose objects are to compel all stationary steam engineers to join their order, "and to resort to force by boycotting anyone who employs stationary steam engineers not members of said organization," or not subject to its orders or those of the Federation of Labor; and that the members of this organization are also members of the Trades and Labor Assembly.

It is also alleged that in March, 1892, the plaintiff was admitted into, and became a member of, the Trades and Labor Assembly, and in

and to entitle it to relief, it must appear that its corporate rights are threatened with some injury of a kind which may be made the subject of an action, and for which courts have the power to afford redress. The complaint is that the defendants have banded together and conspired to "exterminate" the plaintiff; and that they propose to accomplish their purpose by compelling its members to leave it. Of course, when its members have all withdrawn, it will be extinct. We need not discuss the character of the means to be employed for its disintegration. Whether they are legal or illegal, they can not be made the subject of an action in favor of the plaintiff. It has no property in its members, and, in losing them, it sustains no damage which the law recognizes as damage. It can not compel its members to remain with it; and, if they are violently driven out of it,—if they are forced to relinquish their membership against their will,—the grievance is theirs, and not the plaintiff's. Or if, for the purpose of forcing their withdrawal, others, by means of "boycotts" or "strikes," are made to suffer, the latter must fight their own battles. The law does not make the plaintiff their champion. The disorganization and resulting extinction of the plaintiff would, doubtless, be a calamity; but it is one which the law is powerless to avert. We have cited no authorities because we can find none which are of any use. If a case bearing the remotest analogy to this was ever the subject of adjudication, our most diligent effort has failed to unearth any record of it. The judgment will be affirmed.

EMPLOYERS' LIABILITY—*Durst v. Carnegie Steel Co., Limited.* 33 *Atlantic Reporter, page 1102.*—This was an action to recover damages for the death of Andrew Durst, who was an employee of the defendant, and lost his life by the fall of an embankment of earth into a ditch, which he was engaged with others in digging on October 18, 1893. The plaintiff was nonsuited in the court of common pleas of Allegheny County, Pa., and carried the case on appeal to the supreme court. In the opinion of the court of common pleas the following points were decided: (1) When a master intrusts to the superintendent in charge of an excavation the matter of notifying the employees of any latent danger, the foremen in charge of the gangs engaged in the work of excavation are not vice principals in the absence of the superintendent, so as to render the employer liable for their failure to notify the employee of such danger. (2) When the only possible danger to an employee engaged in making an excavation is such as may arise during the progress of the work, the employer is not bound to stand by during the work to see if a danger arises, it being sufficient if he provides against such dangers as may possibly arise and gives the workmen the means of protecting themselves. In arriving at these conclusions the following language was used by said court:

The principles of the law governing such cases are well established. The difficulty arises in their application to particular cases. An employee assumes all the ordinary risks of the business in which he may be employed. This includes accidents caused by the negligence of coemployees. But the master owes certain duties to the employee,

excavations. The company was therefore responsible for any neglect by him to perform the duty imposed by law upon them. It is contended that Patterson and McMillan, who were foremen of the gangs engaged at this work, were charged with the duty of protecting it in absence of Molamphy, and were therefore vice principals for the time being. We do not so understand it. They were mere foremen in charge of the men, supervising and directing their labor. Under all the cases they were coemployees with the men under their charge. We do not understand that Molamphy turned over any of his duties to them, but merely instructed them, in cases of necessity, to call him or the carpenter to provide against danger. There was no evidence to show contributory negligence on part of deceased. The only question, therefore, is whether there was evidence of neglect on the part of Molamphy to perform any of the duties imposed by law upon the defendants. The duty which it is alleged was neglected is that of furnishing a reasonably safe place to work. It will be observed that the place as it stood when the work commenced was perfectly safe. The danger could only arise as the work progressed, and be caused by the work done. In such a case we do not think it is the duty of the employer to stand by during the progress of the work to see when a danger arises. It is sufficient if he provides against such dangers as may possibly or probably arise, and gives the workmen the means of protecting themselves. They should look out for such dangers and use the means provided.

The supreme court, on January 6, 1896, approved the decision of the court of common pleas and affirmed its judgment in the following terms:

The opinion of the learned court below on the motion to take off the nonsuit is so full, clear, and convincing that for the reasons there stated, and upon the authorities cited, we affirm the judgment in this case.

## PROTECTION OF GARMENT WORKERS IN SWEAT [illegible] MARYLAND.

[January session, 1896, chapter [illegible], public local laws.]

SECTION 1. *Be it enacted by the General Assembly of Marylan* two new sections be and the same are hereby added to article Code of Public Local Laws, title "City of Baltimore," under t title "Buildings," to follow section 131, to be designated as 131A and 131B, and to read as follows:

131A. It shall not be lawful for any person, agent, owner or pr of any sweat shop or factory where four or more persons are en to use any coal oil, gasoline, or any other explosive or infla compound for the purpose of lighting or heating in any form; son, agent, owner or proprietor violating the provisions of this shall be guilty of a misdemeanor, and, on conviction thereof, by the court before whom such conviction is had for every viola sum of one hundred dollars and costs, and stand committed un fine and costs be paid.

131B. The owner or owners of any such house or building u sweat shop or factory where four or more persons are employed ment workers, on other than the first floor of such house or b shall provide fire escapes for the same, and if any owner or ow any such house or building so used, fail to make or provide a fir within six months after the passage of this act, upon conviction shall pay a fine of two hundred dollars to be recovered as oth in this State, or imprisonment in the city jail for sixty days, or b and imprisonment, in the discretion of the court.

SEC. 2. *And be it enacted*, That this act shall take effect from of its passage.

Approved April 4, 1896.

## RECENT GOVERNMENT CONTRACTS.

[It is proposed to publish in the Bulletin from time to time statements of the contracts for constructions entered into by the Treasury, War, and Navy Departments.]

The following contracts have been made by the office of the Supervising Architect of the Treasury:

FISHERMANS ISLAND, VA.—June 11, 1896. Contract with William H. Virden, of Lewes, Del., for work on new toilet rooms, plumbing, piping, sewer, miscellaneous work, etc., in connection with extensions to barracks at quarantine station, for $2,032.60. Work to be completed within fifty-five days.

NEWBERN, N. C.—June 24, 1896. Contract with Gude & Walker, Atlanta, Ga., for interior finish, plumbing, approaches, etc., for post office, courthouse, and customhouse, $20,080. Work to be completed within seven months.

WASHINGTON, D. C.—June 25, 1896.—Contract with Crook, Horner & Co., Baltimore, Md., for low-pressure steam-heating apparatus for extension to United States Bureau of Engraving and Printing, $1,638.20. Work to be completed within thirty days.

NEWARK, N. J.—June 26, 1896. Contract with the Standard Paving Company, Newark, N. J., for approaches to customhouse and post office, $5,000. Work to be completed within thirty days.

PHILADELPHIA, PA.—June 30, 1896. Contract with Morse, Williams & Co., Philadelphia, Pa., for two hydraulic passenger elevators in the courthouse and post office, $22,097. Work to be completed within four months.

CHICAGO, ILL.—July 1, 1896. Contract with D. H. Hayes, Chicago, Ill., for the erection and completion of an operating wing to the United States Marine Hospital, $6,948. Work to be completed within three months.

PAWTUCKET, R. I.—July 6, 1896. Contract with L. L. Leach & Son, Chicago, Ill., for erection and completion (except heating apparatus) of post office, $40,774. Work to be completed within ten months.

MILWAUKEE, WIS.—Contract with Empire Fire Proofing Company, Chicago, Ill., for terra-cotta fireproofing, miscellaneous ironwork, etc., on post office, courthouse, and customhouse, $39,800. Work within building proper to be completed within ninety days, in tower within fifteen days from the time roofs are on.

The following contracts have been made by the Navy Department:

PITTSBURG AND SOUTH BETHLEHEM, PA.—June 1, 1896. Contract with the Carnegie Steel Company, of Pittsburg, and the Bethlehem Iron Company, of South Bethlehem, for nickel-steel armor plates and appurtenances for battleships Nos. 5 and 6, the "Kearsarge" and the "Kentucky," the former amounting to $1,660,518.20 and the latter to $1,462,191.80.

# BULLETIN

OF THE

# DEPARTMENT OF LABOR.

No. 6—SEPTEMBER, 1896.

ISSUED EVERY OTHER MONTH.

EDITED BY

CARROLL D. WRIGHT,

COMMISSIONER.

OREN W. WEAVER,

CHIEF CLERK.

WASHINGTON:

GOVERNMENT PRINTING OFFICE.

1896.

# CONTENTS.

BULLETIN

OF THE

DEPARTMENT OF LABOR.

No. 6. WASHINGTON. SEPTEMBER, 1896.

# INDUSTRIAL COMMUNITIES. (a)

BY W. F. WILLOUGHBY.

## CHAPTER V.

## FAMILISTÈRE SOCIETY OF GUISE, FRANCE.

The study of the history and practical operations of the Familistère Society of Guise is of interest for reasons quite different from those that render interesting the study of the other industrial centers that have been described. In the latter cases the industrial undertakings have been organized and conducted on the same basis as that existing for the conduct of business generally. The industrial organization of the Familistère of Guise, on the other hand, is unlike that existing anywhere else in the world. This organization has been the result, not of a historical growth, but of a deliberate putting into execution of a previously elaborated scheme by which it was intended to build up a special industrial community on a distinctly communistic basis. The object sought was to secure a perfect mutuality of interests of all concerned in production. This mutuality was to be obtained not only by the introduction of an advanced type of cooperative production and distribution, but by the erection of large tenement houses, called "familistères," in which the members of the society should reside almost as one big family, by the common education of the children of members, by the erection of a theater and bath and wash houses for general use, and by the development of mutuality in every way through the organization of mutual insurance and aid funds and kindred institutions.

The present society dates from 1880, when it took over the business until then carried on by M. Godin as an individual enterprise. To

a See footnote to the beginning of this series of articles in Bulletin No. 3.

understand the present [illegible] necessary to trace its rise in the [illegible] for several decades carried on by M. [illegible] Familistère of Guise embraces, [illegible]

1. History of the business of M. [illegible] until the organization of the Société du [illegible]
2. Organization and description of the [illegible]
3. Results of the practical operations of [illegible] the present time.

## HISTORY OF THE BUSINESS [illegible]

M. Jean-Baptiste-André Godin, the founder [illegible] carried on by the Société du Familistère de Guise, [illegible] society, and until his death, in 1888, its general [illegible] January 26, 1817, at Esquehéries, a small village [illegible] Aisne, France. His father was a simple artisan [illegible] schooling was of the most rudimentary description [illegible] respect from that received by the children of other [illegible] peasants. At 11 years of age he commenced his appren[illegible] locksmith in his father's workshop. Here he remained [illegible] teenth year, when, following the custom prevalent at [illegible] undertook a tour of France in order to perfect himself [illegible] In 1840, being then 23 years of age, he returned to his native [illegible] opened a small workshop for the manufacture of stoves and [illegible] utensils. In this he made a radical departure from ex[illegible] of manufacture. As the result of inventions, for which [illegible] letters patent, he commenced the manufacture of stoves [illegible] instead of sheet iron, which had previously been employed.

He continued to invent new devices, for which he secured [illegible] fifty letters patent. As a result of these improvements [illegible] increased constantly in importance, and in 1846 he found [illegible] to move his works to Guise, a neighboring village more [illegible] uated for the conduct of a large manufacturing enterprise [illegible] date he employed about thirty workingmen. After its remo[illegible] the business continued to grow, until to-day it is one of the [illegible] tant manufacturing establishments in France. Until 18[illegible] of the industry differs little from that of any growing [illegible] In that year was commenced the erection of the left wing [illegible] listère, a large building for housing the employees of the [illegible] inauguration of a system for the organization of labor that [illegible] was to give to the life of the employees of the establishment [illegible] iar character.

Before entering upon a description of this system [illegible] go back and trace what were the [illegible] realize through it. From his [illegible] of a philosopher who [illegible] him. When but 11 years of age, while at school, he was harboring plans for improving methods of instruction. But from the moment of his entrance into active industrial life he devoted all his energies toward devising methods for the improvement of the conditions of the industrial classes. The development of the communistic ideas of St. Simon, Cabet, and Fourier gave a final shaping to his thoughts, and in him Fourier found his most ardent disciple. It is unnecessary to enter deeply into this period of Godin's life. In his book entitled Solutions Sociales, published in 1871, an English translation of which was made by Marie Howland in 1886, Godin has given us his autobiography and a full statement of the development of his theories and his connection with Fourierism. In the abortive attempt of M. Victor Considérant, in 1853, to establish a community in the State of Texas to be conducted on Fourier's communistic plans Godin lost 100,000 francs ($19,300), or one-third of his fortune at that time. Nothing daunted by this failure, Godin determined to introduce into his own industry his ideas regarding mutuality, and by the gradual reconstruction of the organization of the industry transform it into one conducted upon a strictly communistic basis. The ideas that he wished to realize were, briefly, these:

First, and above all, the principle of mutuality was to be developed in every possible way. To do this there were to be created institutions answering to almost every need of his employees, by means of which the lives of the employees were to be lived largely in common. The keystone to the whole system of mutuality would be the congregation of his employees and their families into large tenement houses, called "familistères," where to some extent they were to live as one great household. The children were to be educated in common schools; a cooperative store would furnish supplies to all the members; mutual aid societies and insurance funds against accidents, sickness, and old age would develop the spirit of solidarity; and bath and wash houses, a theater, restaurants, etc., were to be erected for common use.

Second, the industry was gradually to be transformed into one conducted on a strictly cooperative basis. The employees were to own not only all the familistères where they lived, the schools, theaters, etc., but the manufacturing plant as well. Nothing, probably, shows the remarkable business capacity of M. Godin better than his recognition of the impossibility of putting his plans, in their entirety, into immediate execution. For success it was necessary first to create a stable body of workingmen devoted to the interests of his establishment and to develop in them the principles of mutuality. It was his intention to create all the institutions necessary for the life of an industrial settlement, such as he planned, while he was yet in complete control, and then, when the ground had been completely prepared and the essential parts of the machinery of government had been running satisfactorily for some time, to bring into existence that should take over its management complete mutuality, both in the ordinary life of the employees. To do years. The first step was taken in 1859, left wing of his principal familistère was laid and occupied in 1861. The following year was commenced, and in 1865 it was ready for until 1877 that this familistère was completed by right wing. The increased demand for apartments, tion of the society, led to the erection in 1883 of the of Landrecies, and, in the following year, the Cambrai. School buildings, a theater, a restaurant, and houses were in the meantime erected, until finally the factories all the buildings necessary for the self- members of the society.

## ORGANIZATION AND DESCRIPTION OF THE SOCI

The character of the Familistère Society of Guise at the time differs little from what it was in 1880, when the society constituted, except in such particulars as would necessarily re an enlargement of the business and the construction of two familistères. The description that follows is of conditions as at the present time:

The business comprises the manufacture of stoves, iron eled with porcelain, lamps, kitchen utensils, etc. The total land owned by the society for its various purposes amounts hectares (103 acres), of which 35.3813 hectares (87.4 acres) are and 6.3019 hectares (15.6 acres) at Laeken, Belgium. The plan ken is operated in every way as a branch of the central establ at Guise. A familistère has been built there for the housin employees, and the industry and life of the employees are under the constitution and regulations that prevail at further special mention will be made of Laeken. The figures, given in all the statistical tables, excepting those on pages 593 which pertain only to Guise, relate to the entire industry as at both places.

The land owned at Guise is situated on the river Oise jus outskirts of the village. Of the total of 35.3813 hectares (87 the factories and their dependencies occupy 10.6103 hectares the familistères, schools, etc., and the open place and gar rounding them, 2.6514 hectares (6.5 acres), and the remaini 22.1196 hectares (54.7 acres), is given over to parks, la gardens, etc. The river Oise divides the ground into one bank are located all the factory buildings, familistères, and the bath and

factory understanding of the basis upon which the association is organized and its affairs conducted will be discussed. But even to do this will require considerable space, as the system here put into operation provides not only for the organization of an industry along special lines, but the regulation of all the details of life of its members grouped together in a unique community.

The association is named Familistère Society of Guise: Cooperative Association of Capital and Labor (*Société du Familistère de Guise: Association Coopérative du Capital et du Travail*). Its objects are declared to be the organization of a solidarity of interests among its members, by means of the participation of capital and labor in profits, and the maintenance of common institutions for their mutual welfare. Its membership consists of persons of both sexes, who, having signed the constitution, are the possessors of one or more shares of the society's certificates of stock and cooperate in the work of the society.

The scheme of membership provides for a regular hierarchy through the division of members into classes, according to the extent to which they are interested in the affairs of the society and the length of time they have been connected with it. The members proper are divided into three classes, viz, associés, sociétaires, and participants. To become an associé a person must be 25 years of age, a resident of one of the familistères and an employee of the society for at least five years, able to read and write, an owner of certificates of stock of the society to the value of at least 500 francs ($96.50), and he must be admitted by the general assembly as an associé. A sociétaire is required to be 21 years of age and free from military duty, a resident of a familistère, an employee of the society three years, and he must be admitted by the managing council as a sociétaire. A participant must be 21 years of age, free from military duty, he must be admitted by the addition there are two other and auxiliaires, who are not The first consists of outsiders who tificates of the society, but take no composed of workingmen who are employed for extra help is felt, but who have not membership.

This division into classes is of prime basis of the entire organization of the society, ileges of each class differ widely. This difference principally to the division of profits, the right of to live in the familistères, and participation in the society. In the division of profits, as will be participates on the basis of double the amount of his the sociétaire on the basis of one and one-half times his ea the participant on the basis of the exact amount of his ear regards retention in case of lack of employment, the preferei first to the associé, secondly to the sociétaire, and thirdly to pant. The same order of preference is followed in the dis apartments in the familistères.

The administration of the affairs of the society is exclusi hands of the associés. They constitute the general asse elects all the officers and committees. The intéressés and not being members of the association, do not, of course, either in profits or in the management of affairs. The ri first are limited to the receipt of interest on and the pr profits apportioned to the certificates owned by them; tl second to their wages and a certain participation in the vari aid societies organized within the society.

To the society thus constituted, Godin turned over the enti connected with his industrial establishment. But it was b his intention to give this property to the society. In foundin he had a double aim in view—to show, by a practical e advantages of conducting a business on strictly cooperativ and to demonstrate to manufacturers that the transfer of th ment from their hands to those of their employees could plished without the former making any sacrifice, either in loss of capital invested or in a legitimate interest. In bo aims Godin achieved a complete success.

The means by which this transfer was accomplished cient to mark the society as one of the most undertakings, from a social point of view, A careful inventory of the value of the total value was estimated at the

The criticism has never been made that this was an excessive valuation. In return for the property thus handed over to the society, the latter issued founder's certificates (*titres d'apports*) to a like amount. These founder's certificates constituted the provisional capital stock of the society. They bore interest at the rate of 5 per cent, a very fair return on a commercial investment in France, and were also entitled to share in the division of the profits, as hereafter described. The payment of this interest constituted a first charge on the receipts of the society after operating expenses had been paid. The society, therefore, started with all its capital stock in the possession of M. Godin himself. The constitution, however, expressly stipulated that these certificates could be purchased at any time by the society at their face value. This was done in the following way: After all expenses of production, including interest on certificates and certain statutory charges, have been met, the net profits remaining, instead of being divided in cash among the members of the society, are applied to the purchase of founder's certificates. As fast as the latter are purchased they are canceled, and in their place there are issued to like amount association certificates or savings certificates (*titres d'épargnes*), as they are called, which latter are distributed among the members of the society as profits. The total of founder's certificates and association certificates thus always equals the sum of 4,600,000 francs ($887,800). There is absolutely no difference between the two kinds of certificates regarding the right to interest or participation in profits. When all the founder's certificates have been replaced by association certificates the society acquires the absolute ownership of its property.

But it was not sufficient to make provisions by which the society could replace founder's certificates by those of its own. As members owning certificates died or resigned from the society, certificates would be constantly passing into the hands of persons not members of the society. It was therefore necessary to devise means by which the certificates could be kept in the possession of members. This was accomplished by the provision of the constitution that association certificates are at all times redeemable, and that as soon as all the founder's certificates have been canceled, the profits should be applied to the purchase of those association certificates bearing the earliest date, those held by outsiders being purchased whenever possible. These are then canceled and an equal amount of new certificates are issued, which, being distributed as profits, go to members of the society. In this way there goes on a constant process of cancellation of old certificates and the issue of new certificates in their place, the total amount outstanding always remaining the same.

The adjustment of financial relations, such as the determination of the remuneration of labor and capital and the division of profits, was arranged with no less ingenuity than that regarding the acquisition and retention of the stock of the society. The essential parts

performed by labor and capital in the [illegible] tinctly recognized. The wages system is retained [illegible] the right of capital both to interest and to a [illegible] affirmed with no less emphasis. The determination of [illegible] of profits going to each is made in accordance with [illegible] ory: The capital of capital, if the expression may be [illegible] sists of its potentiality to purchase, i. e., to give to its [illegible] ities of value. The capital of labor consists of its power to [illegible] to give to its owner commodities of value. The wages of capital consists of the amount it can earn during a certain period, i. e., its interest. The wages of labor consists of the amount it can earn, i. e., wages strictly so called. An equitable division of the products of industry, according to this theory, therefore, accords to each: First, its wages, i. e., to capital its interest and to laborers their earnings, these items entering into the cost of production before any profits can be realized; second, a participation in net profits according to the amount of work done by each, i. e., in proportion to the interest earned by capital and the wages earned by labor. A clear distinction is thus made between what capital earns in the way of interest and what as profits. This is but the broad theory. The actual apportionment of profits is a considerably more complicated matter. The exact method of division is fixed by the constitution. The sum remaining after the payment of all operating expenses, in which latter sum is included the wages of all employees, constitutes gross profits. From these gross profits the following four fixed charges must be met: (1) The payment of 5 per cent of the value of the buildings, 10 per cent of the value of the tools and machinery, and 15 per cent of the value of models into a fund to meet depreciation in the value of the plant; (2) the payment of a sum equal to 2 per cent of the total amount paid out in wages to the fund for the insurance of workingmen; (3) the payment of not less than 25,000 francs ($4,825) into a fund for the education of children of members; (4) the payment of 5 per cent interest on 4,600,000 francs ($887,800) of capital stock, whether represented by founder's or association certificates. The sum remaining after making these four payments constitutes net profits for distribution on a cooperative basis. Three classes of services are deemed to have a right to participate in net profits, viz, labor, capital, and directors, or those officials who manage the affairs of the society. The apportionment is as follows:

First. Twenty-five per cent is applied to a reserve fund until such fund has amounted to 10 per cent of the capital stock, or 460,000 francs ($88,780). This amount was reached in 1881, and since then this portion of profit is added to that allotted to labor and capital.

Second. Fifty per cent (since the constitution of the reserve fund [illegible] per cent) is apportioned to labor and capital to be divided according to the principles described above.

Third. Twenty-five per cent to those officials who manage the affairs of the society.

It was realized that it was indispensable to the success of the undertaking to secure men of ability for these services, and it was an example of great wisdom to thus make them so largely and directly interested in the financial results of the society. The details of the division of profits within the two classes of directors and labor and capital are as follows:

Of the 25 per cent allotted to directors 4 per cent is given to the general manager, 16 per cent for division among the members of the managing council, 2 per cent to the council of audit and control (*a*), 2 per cent to the managing council for distribution to workingmen who have performed exceptional services, and 1 per cent for the preparation and maintenance of one or more scholars in the advanced schools of the State.

The 75 per cent allotted to labor and capital is distributed in the following manner: The wages of capital represented by the interest of 5 per cent on the capital stock of 4,600,000 francs ($887,800) is always a fixed sum, 230,000 francs ($44,390). The wages of labor of course varies, but the total amount for the purpose of division of profits is calculated at a sum considerably greater than that actually paid. The members of the society, as has been described, are divided into a hierarchy of classes, the higher of which enjoys privileges greatly superior to those of the lower. The chief privilege is the relative extent to which they participate in profits. According to the formula of Godin, incorporated into the constitution, associés share in profits on the basis of twice the amount of their earnings, the sociétaires, one and one-half times, and the participants and auxiliaires, the exact amount of their earnings. The sociétaires and participants, however, living in the familistères and having been twenty years in the employ of the society, are included for this purpose among the associés, and participants not inhabiting the familistères, but having been twenty years in the employ of the society are included among the sociétaires. The amount apportioned to auxiliaires, morever, is not paid to them, as they are not members, but is turned over to the insurance fund. The total of these various items, then, represents the total claims of labor. This seemingly complicated but really simple method of profit sharing will be much more easily understood by the use of the following hypothetical example:

| | Francs. |
|---|---|
| Net annual profits | 300,000 |
| Apportioned to directors, 25 per cent | 75,000 |
| Apportioned to capital and labor, 75 per cent (reserve already constituted) | 225,000 |

*a* The duties of these officers are described on pages 578 and 579.

For the division of the latter amount the [illegible] required:

| | |
|---|---|
| Profit-sharing wages of capital (5 per cent interest on [illegible] | [illegible] |
| Profit-sharing value of wages of labor— | |
| Associés (240,000 francs × 2) | [illegible] |
| Sociétaires (430,000 francs × 1½) | [illegible] |
| Participants (actual amount of wages) | [illegible] |
| Auxiliaires (actual amount of wages) | [illegible] |
| Total profit-sharing value of wages of capital and labor | [illegible] |

The per cent of profits is obtained from the formula—

$$\frac{\text{Profits multiplied by } 100}{\text{Profit-sharing value of wages of capital and labor}} = \frac{225,000 \times 100}{2,500,000} = 9 \text{ per cent.}$$

To each class, therefore, are apportioned the following profits:

| | Francs. |
|---|---|
| Capital, 230,000 francs at 9 per cent | [illegible] |
| Associés, 480,000 francs at 9 per cent | [illegible] |
| Sociétaires, 645,000 francs at 9 per cent | [illegible] |
| Participants, 675,000 francs at 9 per cent | [illegible] |
| Auxiliaires, 470,000 francs at 9 per cent | [illegible] |
| Total | [illegible] |

It is unnecessary to follow the distribution to the individual members of each class. An inspection of this scheme shows that the rewards of labor, in addition to wages, are by no means represented by the sums set down in the annual reports as the proportion of profits allotted to labor. In the first place, two items of expenditure, those of a sum equal to 2 per cent of the amount paid out in wages given to the insurance fund and the sum of not less than 25,000 francs ($4,825) to the fund for the education of children of members, are carried to operating expenses before net profits are obtained. Again, in the actual division of profits 25 per cent of the total amount is given to directors (high-class labor) before capital is allowed to participate; and here, again, labor enters proportionately to an amount considerably in excess of the sum actually paid in wages, and finally, capital profits not at all in the amount apportioned to the earnings of auxiliaires which is carried to the insurance fund. These conditions, however, are not so unfair toward capital as at first appears. The insurance of workingmen and the aid in the education of children of employees is by a great many of the more important industrial concerns of France considered as a legitimate item of the cost of production. The proportion of profits allotted to directors can be considered as a portion of their remuneration, and therefore an item in the cost of operation. The main departure from the theory of a division of profits according to the wages of labor and capital is in the multiplication of the earnings of the associés and sociétaires by two and one and one-half, respectively, before an apportionment of profits is made. This [illegible], however, has not the importance that it would have were the capital [illegible] by others than members of the society.

Fixing the rate of wages in a cooperative enterprise is in itself, frequently, a matter of considerable delicacy. This difficulty is entirely avoided at Guise by the retention of the wages system in its entirety. As the society is always under the necessity of employing workingmen who are not members, its policy has necessarily been to fix the rate of wages according to the rules that pertain elsewhere in similar establishments. The general manager, the directors of the different services, and the office employees are paid by the month. The wages of the workingmen proper are, as far as possible, fixed by the piece, in order that each one can earn according to his capacity. When this can not be done wages are paid by the hour, and every precaution is taken to have the rate so determined that justice will be done between the piece and time workers. The rate is determined by the general manager, with the advice of the managing council. In case of dissatisfaction, complaint can be made to a workingmen's committee (*syndicat du travail*), which, together with the managing council, regulates the difficulty.

Thus far attention has been given only to the industrial branch of the society. The work of the society, however, comprehends two distinct spheres of activity, viz, the conduct of an industrial enterprise on a cooperative basis and the creation and management of various social institutions for the common benefit of its members in order that certain ideals concerning mutuality might be realized. To do this four main classes of institutions were organized: (1) Familistères and annexes (bath and wash houses, etc.), in order that the members may, to a great extent, live in common; (2) the provision of schools and schooling facilities for the care of the children of members, and their education in common from their earliest years until they are able to commence work; (3) the maintenance of funds for the mutual aid of members in need, and their insurance against accidents and sickness, and their pensioning in old age; (4) the organization of a cooperative system for the purchase and supply to members of articles of necessary consumption. It is the existence of these last features that gives to the work of the society its unique character and makes it of special interest as a social study.

The machinery of government is so constructed that its control must always fall into the hands of the older members, and therefore presumably into the hands of those most interested in the welfare of the society and most competent to be at its head. This is accomplished by taking advantage of the division of members into grades of classes. Sovereignty, if it is permissible to use the word in this connection, resides in a general assembly composed of all members of the society with the rank of associé. The general assembly meets once a year in regular session to hear annual reports, to elect officers, etc. Extraordinary meetings may be called if necessity arises for them. The duties of the general assembly include the approval of annual reports, the sanctioning of all important matters of policy, the [illegible] manager, the admission of members to the [illegible] cation of constitution and by-laws, and, in general, [illegible] the affairs of the society.

The actual business interests of the society are [illegible] general manager (*administrateur-gérant*), assisted by a [illegible] cil (*conseil de gérance*), and three councils—council of [illegible] (*conseil de l'industrie*), council of management of the [illegible] *seil du familistère*), and council of audit and control ([illegible] *lance*). The general manager is the executive head of [illegible] unites in himself not only the duties of the ordinary [illegible] name, but many of those that in the case of a private [illegible] to the proprietor of the establishment. He is responsible to [illegible] eral assembly alone. He appoints and dismisses working[illegible] certain limits, however, and in general manages all the [illegible] society. He is elected by the general assembly for life, and [illegible] removed only on account of the failure of the society to pay [illegible] on the capital invested or because of misconduct or infraction [illegible] provisions of the constitution. His salary is 15,000 francs ([illegible]) and 4 per cent of the net profits in addition to participating in profits in proportion to the amount of his salary and the capital of the society that he may possess.

The general manager is aided in all his work by a managing council, the consultation of which by him is in many cases obligatory. This council consists of the general manager himself, who is its president, a maximum of ten ex officio members, three associés elected annually by the general assembly, and such other persons the importance of whose positions renders it desirable that they should have a place on the council. The ex officio members are the chiefs of the different services into which the establishment has been divided, viz, commerce, manufactures proper, setting up, foundry, materials, supplies, chief accountant, accounts and control, models, and familistère. It would be impossible to enumerate all the duties of the board. It oversees the interests of the society, it controls expenditures, it authorizes repairs, it admits members as sociétaires and participants, and, in general, its advice must be taken on all important questions when submission to the general assembly itself is not required. Members of the council receive no additional salary other than the supplemental participation in net profits, as described under the division of profits.

But a few words will be necessary to describe the rôle of the three other councils that have been mentioned among the administrative agencies of the society. The council on industrial matters is composed of the same members as the managing council. It meets once a [illegible] to decide all questions relating to the practical operations of the [illegible] duct of the industry. The council of audit and control is [illegible] three members, elected by the general assembly at its [illegible] and has as its special duty the auditing of all accounts, the verification of reports, etc. The council on management of the familistères is composed of all associés who are members of the managing council, and is presided over by the general manager. Its duties consist of the management of all affairs relating to the familistères. It makes regulations concerning the care of the buildings; it gives its advice concerning the purchase of supplies for the cooperative stores, and concerning the renting and vacating of apartments. Members receive no extra compensation for their services on these councils.

## OPERATIONS OF THE SOCIETY FROM 1880 TO THE PRESENT TIME.

The history of the practical operations of the society during the fifteen years that it has now been in existence offers no fact in the way of a change of policy of sufficient importance to be recorded. The society has continued steadfastly in the way mapped out for it by its constitution. The greatest strain that could occur to it was the death in 1888 of M. Godin, who had until then remained its general manager, and was of course the moving spirit of the whole undertaking. So firmly, however, had M. Godin founded his work that his death had practically no effect on the prosperity of the society. A new general manager, M. Dequenne, was elected, as provided for by the constitution, and the work of the society went on without break or modification of policy. As regards its growth in importance and numbers, the society has held its own, increasing, if at all, but slowly. The following table shows for each year since 1880 the total number of employees, the total amount paid out in wages, and the total value of the product:

EMPLOYEES, WAGES OF EMPLOYEES, AND VALUE OF PRODUCT OF THE FAMILISTÈRE SOCIETY OF GUISE, 1879-80 TO 1894-95.

| Year. | Employees. | Wages. | Value of product. | Year. | Employees. | Wages. | Value of product. |
|---|---|---|---|---|---|---|---|
| 1879-80 | 1,663 | $294,967.81 | $764,581.97 | 1887-88 | 1,691 | $374,008.01 | $848,178.10 |
| 1880-81 | 1,577 | 361,553.02 | 963,450.13 | 1888-89 | 1,881 | 421,498.29 | 742,229.30 |
| 1881-82 | 1,493 | 338,181.14 | 881,756.68 | 1889-90 | 1,768 | 405,704.51 | 783,536.51 |
| 1882-83 | 1,401 | 364,354.99 | 787,503.23 | 1890-91 | 1,713 | 416,962.31 | 794,859.21 |
| 1883-84 | 1,448 | 350,145.90 | 679,465.88 | 1891-92 | 1,706 | 429,502.49 | 791,993.08 |
| 1884-85 | 1,347 | 351,999.27 | 668,851.82 | 1892-93 | 1,724 | 408,934.65 | 766,324.69 |
| 1885-86 | 1,540 | 343,896.29 | 680,326.04 | 1893-94 | 1,676 | 410,454.11 | 774,816.48 |
| 1886-87 | 1,526 | 338,428.75 | 669,018.97 | 1894-95 | 1,720 | 421,217.87 | 800,927.79 |

The growth of membership, as distinct from the growth of the total number of employees, is a feature of prime importance in the history of the enterprise. The democratic principles underlying the scheme of the society look to the gradual development of a system where practically all employees will be members, and, as far as possible, all members on the same footing. Though grades of membership were created, this was done only to insure that the management of affairs should be in the older and more competent hands. There has not been manifested the slightest desire on the part of the associés to restrict their number, but, on the other hand, those in the [illegible] encouraged to fulfill the conditions [illegible] higher class. The kind of organization [illegible] members of the society would be associés, [illegible] on an equality in the management of affairs and [illegible] following tables showing the mutation of membership [illegible] est interest, indicating as they do the steady [illegible] for which the society stands pledged:

MEMBERS ENTERING AND LEAVING THE FAMILISTÈRE SOCIETY [illegible] TO 1894-95

| Year. | Associés. | | | Sociétaires. | | | Participants. | | | Total members. |
|---|---|---|---|---|---|---|---|---|---|---|
| | Admitted. | Leaving. | Number. | Admitted. | Leaving. | Number. | Admitted. | Leaving. | Number. | |
| 1879-80 | 46 | ...... | 46 | 62 | ...... | 62 | 443 | ...... | 443 | [illegible] |
| 1880-81 | 11 | 2 | 55 | 11 | 8 | 65 | 120 | 16 | 548 | [illegible] |
| 1881-82 | 8 | 2 | 61 | 14 | 9 | a 71 | 28 | ...... | a 561 | [illegible] |
| 1882-83 | 12 | 5 | 68 | 47 | 13 | a 104 | 112 | 96 | 577 | [illegible] |
| 1883-84 | 5 | ...... | 73 | 57 | 13 | 148 | 69 | 75 | 571 | [illegible] |
| 1884-85 | 13 | 3 | 83 | 60 | 20 | 188 | 64 | 92 | 543 | [illegible] |
| 1885-86 | 11 | 2 | 92 | 37 | 21 | a 207 | 41 | 73 | a 514 | [illegible] |
| 1886-87 | 3 | 2 | 93 | 16 | 18 | a 209 | 23 | 43 | a 491 | b [illegible] |
| 1887-88 | 13 | 4 | 102 | 67 | 19 | a 250 | 52 | 79 | 464 | [illegible] |
| 1888-89 | 31 | 2 | 131 | 53 | 40 | 263 | 66 | 67 | a 535 | [illegible] |
| 1889-90 | 42 | 1 | 172 | 44 | 58 | 249 | 138 | 96 | 565 | [illegible] |
| 1890-91 | 30 | 1 | 201 | 22 | 38 | 233 | 58 | 59 | 564 | [illegible] |
| 1891-92 | 21 | 6 | 216 | 25 | 30 | 228 | 47 | 67 | 544 | [illegible] |
| 1892-93 | 28 | 8 | 236 | 20 | 38 | 210 | 68 | 43 | 570 | 1,016 |
| 1893-94 | 31 | 3 | 264 | 51 | 47 | 214 | 68 | 102 | 536 | 1,014 |
| 1894-95 | 21 | 8 | 277 | 29 | 34 | 209 | 71 | 52 | 555 | 1,041 |

a These figures do not balance. They are given, however, as published by the [illegible]
b This total does not equal the sum of the items. The figures are given, howe[illegible] the society.

PER CENT OF EACH CLASS OF MEMBERS OF TOTAL MEMBERS [illegible] EMPLOYEES OF THE FAMILISTÈRE SOCIETY OF GUISE, 1879-8[illegible]

| Year. | Associés. | | | [illegible] |
|---|---|---|---|---|
| | Number. | Per cent of total members. | Per cent of total employees. | Number. |
| 1879-80 | 46 | 8.37 | 2.73 | 62 |
| 1880-81 | 55 | 8.26 | 3.49 | 65 |
| 1881-82 | 61 | 8.80 | 4.09 | 71 |
| 1882-83 | 68 | 9.06 | 4.85 | 104 |
| 1883-84 | 73 | 9.22 | 5.04 | 148 |
| 1884-85 | 83 | 10.20 | a 6.16 | 188 |
| 1885-86 | 92 | 11.22 | 5.97 | 207 |
| 1886-87 | 93 | a 11.83 | a 6.09 | 209 |
| 1887-88 | 102 | 12.50 | 6.08 | 250 |
| 1888-89 | 131 | 14.26 | 6.97 | 263 |
| 1889-90 | 172 | 17.45 | 9.73 | 249 |
| 1890-91 | 201 | 20.14 | 11.73 | 233 |
| 1891-92 | 216 | 21.96 | 12.66 | 228 |
| 1892-93 | 236 | 23.23 | 13.69 | 210 |
| 1893-94 | 264 | 26.04 | 15.75 | 214 |
| 1894-95 | 277 | 26.61 | 16.19 | 209 |

a Based on totals given; see note b.

[illegible] on totals given; see note b.

b This total does not equal the sum of the items. The figures are given, however, as published by the society.

From these tables it can be seen that, while the total number of employees has remained nearly constant, the number of members has constantly increased, and this increase has been almost wholly in the upper classes. The participants have gained on the auxiliaires, the sociétaires on the participants, and the associés on the sociétaires. Another important consideration in this connection is that of the stability of the corps of employees. Statistics on this point are almost the only ones available for determining the extent to which the advantages of a particular establishment are appreciated. In the accompanying table, showing the number and percentage of employees by the number of years employed by the society, the figures for the factories of Guise and Laeken and the familistères are left independent. Of the three the figures for the factory of Guise are of the greatest significance, as the factory of Laeken has not been in operation for so long a time, and the employees about the familistères are mostly women, among whom a great stability is not to be expected.

Of the total employees at Guise, as shown in the table, only about one-third have been employed less than ten years. Ten years may be taken as a fair standard by which to judge of stability. If men continue in the same employ that length of time they are not apt to make a voluntary change. A statement, therefore, in which 67 per cent of all employees have been employed ten years or over shows great stability; for it should be remembered that included among those employed a shorter length of time are a great many youths whose ages render it impossible for them to have been employed ten years.

NUMBER AND PER CENT OF EMPLOYEES JULY 1, [illegible]

| Years of service. | Factory at Guise. | | Factory at Laeken. | | [illegible] | | [illegible] | |
|---|---|---|---|---|---|---|---|---|
| | Number. | Per cent. | Number. | Per cent. | Number. | Per cent. | [illegible] | [illegible] |
| Less than 5 | 227 | 19.29 | 128 | 48.67 | [illegible] | [illegible] | [illegible] | [illegible] |
| 5 and less than 10 | 156 | 13.25 | 64 | 24.34 | [illegible] | [illegible] | [illegible] | [illegible] |
| 10 and less than 15 | 245 | 20.82 | 26 | 9.89 | [illegible] | [illegible] | [illegible] | [illegible] |
| 15 and less than 20 | 154 | 13.08 | 20 | 7.60 | 7 | [illegible] | [illegible] | [illegible] |
| 20 and less than 25 | 194 | 16.48 | 11 | 4.18 | 6 | [illegible] | [illegible] | [illegible] |
| 25 and less than 30 | 96 | 8.16 | 10 | 3.80 | 7 | [illegible] | [illegible] | [illegible] |
| 30 and less than 35 | 66 | 5.61 | 3 | 1.14 | 3 | [illegible] | [illegible] | [illegible] |
| 35 and less than 40 | 36 | 3.06 | 1 | .38 | 1 | .76 | [illegible] | [illegible] |
| 40 and over | 3 | .25 | ...... | ...... | ...... | ...... | [illegible] | [illegible] |
| Total | 1,177 | 100.00 | 263 | 100.00 | 132 | 100.00 | a1,[illegible] | [illegible] |

a This total does not agree with the total employees shown for 1890-91, pages [illegible] to [illegible]. [illegible] nation is not known. The figures in both cases are as published by the society.

The financial results of the workings of the society will naturally excite the greatest interest. The industry carried on by the society is one offering no especial advantages for its conduct on a cooperative basis, and its successful application here furnishes a valuable demonstration of the practicability of cooperation where a firm basis for its introduction has been previously laid. In this case the first important class of facts is that concerning the success achieved by the society in the replacing of founder's certificates by those of its own. The accompanying table shows the condition of the account at the end of each fiscal year. The present time is a peculiarly opportune one in which to make this study, for at the close of the year, June 30, 1894, the last founder's certificate was purchased and canceled, and the society thus accomplished the work of completely paying for its plant. This fact marks an important epoch in the history of the society. Prior to that date all the profits were devoted to the purchase of founder's certificates, and the actual members of the society received only association certificates, which merely entitled the holder to the interest they earned and the prospect of their redemption in the future. It was thus not until the year 1894–95 that the members began to have their profits paid in cash. As has been explained before, this payment is made in the way of the redemption of the earliest issues of certificates. It will be seen from the table showing the financial operations of [illegible] society that there were realized during the year ending June 30, [illegible] gross profits to the amount of 880,382.38 francs ($169,913.80). A[illegible] the payment of the statutory charges for insurance, education, [illegible] the sum of 287,602.01 francs ($55,507.19) net profits remained for [illegible] tribution. Of this amount 46,201.01 francs ($8,916.80) were [illegible] various uses as required by the statutes, and 241,401 francs ([illegible]) were devoted to the redemption of the association certificat[illegible] issued. In addition to this there was available from the [illegible] former years the sum of 8,550 francs ($1,650.15) which could be [illegible] to the same purpose, making a total of 249,951 francs ([illegible]) were used for the redemption of old association certificates during the year 1894-95. The general manager in his annual report thus comments upon this date in the society's history: "We have now entered upon the regular disbursement in cash of the net profits realized by the association. We have occasion to be satisfied with this result. After fifteen years of existence under the direction of our lamented founder and his successor we have repaid the 4,600,000 francs ($887,800) of founder's certificates, which are now entirely transformed into association certificates held by the workingmen themselves. We have, in addition, constituted a reserve fund of 460,000 francs ($88,780), and with the profits realized during the past year are about to redeem early issues of association certificates to the amount of 249,951 francs ($48,240.54)."

VALUE OF FOUNDER'S AND ASSOCIATION CERTIFICATES OF THE FAMILISTÈRE SOCIETY OF GUISE, 1880-81 TO 1894-95.

| Year. | Founder's certificates canceled and replaced by association certificates. | Certificates outstanding at end of year. | | |
|---|---|---|---|---|
| | | Association certificates. | Founder's certificates. | Total. |
| 1880-81 | $86,534.64 | $86,534.64 | $801,265.36 | $887,800.00 |
| 1881-82 | 115,784.09 | 202,318.73 | 685,481.27 | 887,800.00 |
| 1882-83 | 89,030.21 | 291,348.94 | 596,451.06 | 887,800.00 |
| 1883-84 | 88,717.47 | 380,066.41 | 507,733.59 | 887,800.00 |
| 1884-85 | 43,094.58 | 423,160.99 | 464,639.01 | 887,800.00 |
| 1885-86 | 47,399.06 | 470,560.05 | 417,239.95 | 887,800.00 |
| 1886-87 | 20,023.37 | 490,583.42 | 397,216.58 | 887,800.00 |
| 1887-88 | 40,102.70 | 530,686.12 | 357,113.88 | 887,800.00 |
| 1888-89 | 78,954.37 | 609,640.49 | 278,159.51 | 887,800.00 |
| 1889-90 | 86,307.86 | 695,948.35 | 191,851.65 | 887,800.00 |
| 1890-91 | 57,183.00 | 753,131.35 | 134,668.65 | 887,800.00 |
| 1891-92 | 23,706.58 | 776,837.93 | 110,962.07 | 887,800.00 |
| 1892-93 | 27,640.50 | 804,478.43 | 83,321.57 | 887,800.00 |
| 1893-94 | 41,098.19 | 845,576.62 | 42,223.38 | 887,800.00 |
| 1894-95 | 42,223.38 | 887,800.00 | .............. | 887,800.00 |

In the following series of tables the attempt has been made to make an elaborate analysis of the financial operations of the society since its foundation. They have been compiled directly from the financial reports of the society, the effort being made to so present the facts as to indicate first the receipts and then to trace their application to the various items of expenses and profits. There is given, therefore, first a statement of gross receipts, and then the expenses divided into the two classes of wages and other expenditures. The amount remaining after deducting total expenses from total receipts constitutes gross profits of operation. From this are deducted the fixed charges for depreciation, social work, education, and interest on certificates. The remainder constitutes net profits. The apportionment of these net profits is then shown in detail. A complete financial history of the society is given in these tables. From them it is possible to know exactly how much has been earned and the disposition of receipts to the minutest details.

RECEIPTS OF THE FAMILISTÈRE SOCIETY [illegible]

| Year. | From industry proper. | From social institutions, rents, etc. | Total. | Year. | [illegible] |
|---|---|---|---|---|---|
| 1879-80... | $764,581.97 | $88,463.61 | $[illegible],045.[illegible] | [illegible] | [illegible] |
| 1880-81... | 963,450.13 | 95,792.26 | 1,05[illegible],242.[illegible] | [illegible] | [illegible] |
| 1881-82... | 861,756.68 | 98,425.26 | 960,181.[illegible] | [illegible] | [illegible] |
| 1882-83... | 787,508.23 | 97,[illegible].78 | [illegible],[illegible].01 | [illegible] | [illegible] |
| 1883-84... | 679,465.88 | 91,090.04 | 770,555.[illegible] | 1891-92.. | [illegible] |
| 1884-85... | 668,851.82 | 101,503.78 | 770,[illegible] | [illegible] | [illegible] |
| 1885-86... | 680,326.04 | 106,684.62 | 787,010.[illegible] | [illegible] | [illegible] |
| 1886-87... | 669,018.97 | 107,818.33 | 776,837.30 | [illegible] | [illegible] |

EXPENDITURES AND GROSS PROFITS OF THE FAMILISTÈ[RE] [illegible]
1879-80 TO 1894-95.

| Year. | Wages. Employees of industrial establishments. | Wages. Employees of social institutions. | Wages. Total. | Other expenditures. | Total expenditures. | [illegible] |
|---|---|---|---|---|---|---|
| 1879-80..... | $282,516.11 | $12,451.70 | $294,967.81 | $[illegible] | $[illegible] | [illegible] |
| 1880-81..... | 348,702.74 | 12,850.28 | 361,553.02 | 457,[illegible] | [illegible] | [illegible] |
| 1881-82..... | 375,425.38 | 12,755.76 | 388,181.14 | [illegible] | 781,[illegible] | [illegible] |
| 1882-83..... | 350,111.08 | 14,243.81 | 364,354.89 | 339,297.14 | 703,[illegible] | [illegible] |
| 1883-84..... | 334,228.28 | 15,917.62 | 350,145.90 | 285,372.97 | [illegible],518.[illegible] | [illegible] |
| 1884-85..... | 333,544.48 | 18,454.79 | 351,999.27 | 275,041.56 | 627,040.[illegible] | [illegible] |
| 1885-86..... | 324,737.60 | 19,156.60 | 343,896.20 | 318,000.56 | 661,[illegible] | [illegible] |
| 1886-87..... | 318,790.50 | 19,638.25 | 338,428.75 | 287,719.97 | [illegible],147.[illegible] | [illegible] |
| 1887-88..... | 355,071.41 | 18,931.60 | 374,003.01 | 416,163.[illegible] | 790,166.70 | [illegible] |
| 1888-89..... | 402,623.17 | 18,875.12 | 421,498.29 | [illegible],726.52 | 690,224.[illegible] | [illegible] |
| 1889-90..... | 387,366.75 | 18,337.76 | 405,704.51 | [illegible],743.58 | 789,448.[illegible] | [illegible] |
| 1890-91..... | 396,624.89 | 20,363.42 | 416,988.31 | 432,862.61 | 849,[illegible] | [illegible] |
| 1891-92..... | 408,041.20 | 21,461.29 | 429,502.49 | 423,284.16 | [illegible] | [illegible] |
| 1892-93..... | 387,691.26 | 21,243.39 | 408,934.65 | 398,490.17 | 807,[illegible] | [illegible] |
| 1893-94..... | 389,402.12 | 20,951.99 | 410,354.11 | 394,431.75 | 804,[illegible] | [illegible] |
| 1894-95..... | 399,449.73 | 21,768.14 | 421,217.87 | 406,077.14 | 827,[illegible].01 | [illegible] |

STATUTORY CHARGES, ETC., PAID BY THE FAMILISTÈRE [illegible]
NET PROFITS REMAINING, 1879-80 TO 1[894-95.]

| Year. | Depreciation. | Social institutions. | Distributed to purchasers at cooperative stores. | Education. | [illegible] |
|---|---|---|---|---|---|
| 1879-80 | $38,543.72 | ........ | ........ | $3,[illegible] | $44,[illegible] |
| 1880-81 | 27,309.42 | ........ | ........ | 3,778.12 | 44,[illegible] |
| 1881-82 | 27,401.53 | $960.75 | ........ | 4,[illegible] | [illegible] |
| 1882-83 | 29,829.20 | 830.48 | $2,232.34 | [illegible] | 44,[illegible] |
| 1883-84 | 38,142.90 | 795.16 | 2,719.[illegible] | [illegible] | 44,[illegible] |
| 1884-85 | 37,567.27 | 562.98 | 3,[illegible] | [illegible] | [illegible] |
| 1885-86 | 47,971.40 | 545.08 | 3,[illegible] | [illegible] | [illegible] |
| 1886-87 | 48,840.14 | 806.93 | 4,[illegible] | [illegible] | [illegible] |
| 1887-88 | 20,716.45 | 901.50 | 4,030.40 | [illegible] | [illegible] |
| 1888-89 | 22,434.90 | [illegible] | [illegible] | [illegible] | [illegible] |
| 1889-90 | 24,237.90 | 814.90 | 13,347.[illegible] | [illegible] | [illegible] |
| 1890-91 | 31,562.72 | 3,337.14 | 16,[illegible] | [illegible] | [illegible] |
| 1891-92 | 34,190.36 | 4,[illegible] | 15,[illegible] | [illegible] | [illegible] |
| 1892-93 | 32,063.07 | 6,[illegible] | 17,[illegible] | [illegible] | [illegible] |
| 1893-94 | 31,890.66 | 8,[illegible] | 19,[illegible] | [illegible] | [illegible] |
| 1894-95 | 32,266.36 | 12,188.72 | 15,[illegible] | [illegible] | [illegible] |

| | | | | | | | |
|---|---|---|---|---|---|---|---|
| | 12,150.50 | 17,413.62 | 17,973.32 | 2,754.49 | 13,140.60 | 63,441.61 | 6,118.94 |
| | 20,[illegible].16 | 15,633.52 | 11,426.18 | 2,330.67 | 16,643.55 | 69,897.08 | 5,611.20 |
| | 15,[illegible].95 | 12,779.69 | 8,718.77 | 2,[illegible].84 | 5,988.79 | 45,063.64 | 2,[illegible].16 |
| | 5,[illegible].97 | 3,687.84 | 2,914.69 | 900.15 | 2,052.75 | 14,822.40 | 1,168.68 |
| | 7,[illegible].65 | 5,184.37 | 4,110.13 | 1,032.16 | 3,214.03 | 21,077.24 | 1,611.26 |
| | 12,[illegible].30 | 7,215.30 | 6,473.61 | 1,711.91 | 4,038.33 | 31,725.53 | 2,515.76 |
| | 13,[illegible].36 | 5,452.05 | 6,772.18 | 2,272.60 | 3,806.92 | 36,[illegible].09 | 2,756.62 |
| | 15,251.61 | 8,960.01 | 7,427.02 | 2,634.45 | 4,266.46 | 38,678.55 | 2,951.94 |

| ear. | Reserve fund. | Directors. | | | | | Total net profits. |
|---|---|---|---|---|---|---|---|
| | | General managers. | Boards of managers and overseers. | Education fund. | Pupils in State schools. | Working-men for special services. | |
| | $27,633.67 | $13,284.12 | $12,158.51 | | | $2,210.70 | $110,534.79 |
| | 41,687.83 | 19,707.72 | 18,065.41 | | | 3,284.63 | a164,231.12 |
| | 20,088.50 | 14,547.66 | 13,335.53 | | | 2,434.72 | a121,240.50 |
| | | 11,654.69 | 10,683.51 | | | 1,942.39 | 97,122.07 |
| | | 5,100.41 | 4,675.23 | | | 850.12 | 42,562.57 |
| | | 6,111.92 | 6,602.60 | | | 1,018.74 | 50,932.98 |
| | | 887.03 | 3,326.36 | $665.46 | $222.00 | 443.51 | 22,177.29 |
| | | 1,878.28 | 7,043.53 | 1,408.71 | 469.56 | 939.14 | 46,964.72 |
| | | 3,709.84 | 13,911.83 | 2,782.29 | 927.36 | 1,854.98 | 92,744.80 |
| | | 4,026.94 | 15,101.86 | 3,020.26 | 1,006.69 | 2,013.61 | 100,677.72 |
| | | 2,599.71 | 9,749.59 | 1,949.30 | 650.02 | 1,300.04 | 64,994.48 |
| | | 852.67 | 3,197.82 | 639.60 | 213.27 | 426.31 | 21,318.75 |
| | | 1,210.11 | 4,537.82 | 907.49 | 302.43 | 604.99 | 30,251.53 |
| | | 1,826.17 | 6,848.22 | 1,369.73 | 456.65 | 913.06 | 45,655.13 |
| | | 2,027.08 | 7,602.27 | 1,520.07 | 506.47 | 1,013.63 | 50,678.17 |
| | | 2,220.27 | 8,326.02 | 1,665.20 | 555.07 | 1,110.14 | 55,507.19 |

is total does not equal the sum of the items. The explanation is not known. The figures are as furnished by the society.

is apparent, from an inspection of these tables, that the item of ; profits" is somewhat misleading, if this is meant to indicate the ings of the enterprise over and above operating expenses. Gross its plus the amount paid in wages constitute the real earnings. s, if the last year (1894–95) be taken, it will be seen that the total ipts amounted to 5,166,885.04 francs ($997,208.81). Taking from the item of 2,104,026.61 francs ($406,077.14), the amount of operg expenses other than wages, and 230,000 francs ($44,390) interest apital, there remains the sum of 2,832,858.43 francs ($546,741.67) e distributed either in the way of wages or social expenditures ash profits. There was paid to auxiliaires as wages the sum of 306.45 francs ($64,231.64). If this be subtracted there remain ,951,98 francs ($482,510.03) for distribution among the members ar. The total number of members in that year was 1,041. If had been an equal division, each would have earned 2,401.59

francs ($463.51). The scheme, however, as has been [illegible] means contemplates an equal division.

The members of the society, as previously stated, are divided into associés, sociétaires, and participants, each class participating [illegible] different extent in net profits. Furthermore, it should be noted that the higher positions are, in general, held by the associés and those next in importance by the sociétaires, so that these classes not only share in profits to a greater extent than the participants, but receive the higher wages. In other words, the system does not include any scheme of equal division of profits, but provides for a systematic attempt to divide benefits in accordance with the length of service of each member, his skill, and the economy which he has displayed in the conservation of his capital.

Were it possible to do so, it would be exceedingly interesting to show the average amounts received each year by members of each class from each particular source of wages, profits, interest, or other participation and the total value of these benefits. Unfortunately the complication of elements involved, such for instance as the fact that members are constantly passing from one class to another, renders it impossible to make such a reduction. Nevertheless it has been possible to construct the following table, which, though partly hypothetical, is yet for the most part based on the actual work of the society, and gives information on this point of almost as great value and accuracy as if the actual average could be obtained.

The three cases are taken of a member entering the society as an associé, as did a number of the original members, of a member entering as a sociétaire, and after the constitutional limit of five years becoming an associé, and of a member entering as a participant, and after the constitutional limit of three and five years, respectively, becoming a sociétaire and then an associé. It is presumed that he earns during the whole time an average salary of 5.50 francs ($1.06) per day during 300 working days, or 1,650 francs ($318.45) per year. Though the rate of 5.50 francs ($1.06) per day is higher than that earned by an average of all employees, it is only a fair average of wages of members who earn much higher wages than auxiliaires. The basis of 300 working days to a year is also eminently fair, as during the period covered the actual number of days worked during the year has always exceeded this number. It is further assumed that none of the certificates received in the way of profits are disposed of, a legitimate assumption, for it is undoubtedly true that certificates are thus retained in the majority of cases; their transference to an outside party is rarely permitted by the society, and if transferred to a fellow member the table would still represent the average case that it purports to do. Starting, then, with this case of a member receiving an average rate of [illegible] working an average number of days, and passing through each [illegible] three possible phases of membership, it is an easy [illegible] rate of profits declared, to trace his economic condition [illegible]

described at length, a few words of farther explanation in connection with the actual results obtained will be of use. The receipts of a member from the operations of the society consist of: (1) His wages; (2) the amount of profits apportioned to his wages; (3) the amount of profits apportioned to the certificates of stock owned by him; (4) 5 per cent interest on such certificates; (5) participation in the social institutions maintained by the society, chiefly insurance and old-age pensions and schooling; (6) profits realized by the cooperative store, if patronized by him. In the table, however, only those benefits derived in the way of wages, profits, and interest are considered, as it would be difficult to accurately estimate the value of the other benefits, though they amount to a no insignificant sum. As the receipts of members consist partly of cash and partly of certificates of association stock, the economic condition of a member is represented by both the amount of actual money received each year and the increase in the amount of his certificates. The two columns giving the total cash income during the year and the amount of capital possessed at the end of the year represent, therefore, the summary of the direct pecuniary advantages enjoyed by members.

ESTIMATED PROFITS, CASH RECEIPTS, AND CAPITAL OF A MEMBER OF THE FAMILISTÈRE SOCIETY OF GUISE, 1879-80 TO 1894-95.

ENTERING AS AN ASSOCIÉ.

| Year. | Yearly wages. | Profit-sharing value of wages. | Rate per cent of profits. | Profits earned by wages. | Capital stock at end of year. | Profits on capital stock. | | Interest on capital (5 per cent). | Total cash income during year. |
|---|---|---|---|---|---|---|---|---|---|
| | | | | | | Per cent. | Amount. | | |
| 1879-80 | $318.45 | $636.90 | 15.10 | $96.17 | $96.17 | ........ | ........ | ........ | $318.45 |
| 1880-81 | 318.45 | 636.90 | 18.60 | 118.46 | 214.63 | .93 | $0.89 | $4.81 | 324.15 |
| 1881-82 | 318.45 | 636.90 | 15.00 | 95.54 | 310.17 | .75 | 1.61 | 10.73 | 330.79 |
| 1882-83 | 318.45 | 636.90 | 15.95 | 101.59 | 411.76 | .80 | 2.48 | 15.51 | 336.44 |
| 1883-84 | 318.45 | 636.90 | 7.07 | 45.03 | 456.79 | .35 | 1.44 | 20.59 | 340.48 |
| 1884-85 | 318.45 | 636.90 | 8.21 | 52.29 | 509.08 | .41 | 1.87 | 22.84 | 343.16 |
| 1885-86 | 318.45 | 636.90 | 3.60 | 22.93 | 532.01 | .18 | .92 | 25.45 | 344.82 |
| 1886-87 | 318.45 | 636.90 | 7.66 | 48.79 | 580.80 | .38 | 2.02 | 26.60 | 347.07 |
| 1887-88 | 318.45 | 636.90 | 13.68 | 87.13 | 667.93 | .69 | 4.01 | 29.04 | 351.50 |
| 1888-89 | 318.45 | 636.90 | 12.64 | 80.50 | 748.43 | .63 | 4.21 | 33.40 | 356.06 |
| 1889-90 | 318.45 | 636.90 | 8.25 | 52.54 | 800.97 | .41 | 3.07 | 37.42 | 358.94 |
| 1890-91 | 318.45 | 636.90 | 2.62 | 16.69 | 817.66 | .13 | 1.04 | 40.06 | 359.54 |
| 1891-92 | 318.45 | 636.90 | 3.63 | 23.12 | 840.78 | .18 | 1.47 | 40.88 | 360.80 |
| 1892-93 | 318.45 | 636.90 | 5.67 | 36.11 | 876.89 | .28 | 2.35 | 42.04 | 362.84 |
| 1893-94 | 318.45 | 636.90 | 6.21 | 39.55 | 916.44 | .31 | 2.72 | 43.84 | 365.01 |
| 1894-95 | 318.45 | 636.90 | 6.65 | 42.35 | 916.44 | .33 | 3.03 | 45.82 | 409.65 |

ENTERING AS A SOCIÉTAIRE.

| Year. | Yearly wages. | Profit-sharing value of wages. | Rate per cent of profits. | Profits earned by wages. | Capital stock at end of year. | Per cent. | Amount. | Interest on capital (5 per cent). | Total cash income during year. |
|---|---|---|---|---|---|---|---|---|---|
| 1879-80 | 318.45 | 477.68 | 15.10 | 72.13 | 72.13 | ........ | ........ | ........ | 318.45 |
| 1880-81 | 318.45 | 477.68 | 18.60 | 88.85 | 160.98 | .93 | .67 | 3.61 | 322.73 |
| 1881-82 | 318.45 | 477.68 | 15.00 | 71.65 | 232.63 | .75 | 1.21 | 8.05 | 327.71 |
| 1882-83 | 318.45 | 477.68 | 15.95 | 76.19 | 308.82 | .80 | 1.86 | 11.63 | 331.94 |
| 1883-84 | 318.45 | 477.68 | 7.07 | 33.77 | 342.59 | .35 | 1.06 | 15.44 | 334.97 |
| 1884-85 | 318.45 | 636.90 | 8.21 | 52.29 | 394.88 | .41 | 1.40 | 17.13 | 336.98 |
| 1885-86 | 318.45 | 636.90 | 3.60 | 22.93 | 417.81 | .18 | .71 | 19.74 | 338.90 |
| 1886-87 | 318.45 | 636.90 | 7.66 | 48.79 | 466.60 | .38 | 1.59 | 20.89 | 340.93 |
| 1887-88 | 318.45 | 636.90 | 13.68 | 87.13 | 553.73 | .69 | 3.22 | 23.33 | 345.00 |
| 1888-89 | 318.45 | 636.90 | 12.64 | 80.50 | 634.23 | .63 | 3.49 | 27.69 | 349.63 |
| 1889-90 | 318.45 | 636.90 | 8.25 | 52.54 | 686.77 | .41 | 2.60 | 31.71 | 352.76 |
| 1890-91 | 318.45 | 636.90 | 2.62 | 16.69 | 703.46 | .13 | .89 | 34.34 | 353.66 |
| 1891-92 | 318.45 | 636.90 | 3.63 | 23.12 | 726.58 | .18 | 1.27 | 35.17 | 354.99 |
| 1892-93 | 318.45 | 636.90 | 5.67 | 36.11 | 762.69 | .28 | 2.03 | 36.33 | 356.81 |
| 1893-94 | 318.45 | 636.90 | 6.21 | 39.55 | 802.24 | .31 | 2.36 | 38.12 | 358.94 |
| 1894-95 | 318.45 | 636.90 | 6.65 | 42.35 | 802.24 | .33 | 2.65 | 40.11 | 403.56 |

[illegible] paid in cash for the surrender of an equal amount of his early issues of certificates, which latter were then reissued to him as certificates of 1894-95, or a total of 2,122.54 francs ($409.65). While cash receipts are now much greater than during the first period, the amount of capital owned by the members must remain the same, or 4,600,000 francs ($887,800). Should the operations of the society increase in importance the actual value of these certificates will of course be greatly increased, or the society is permitted by its constitution to further increase the amount of its capital stock.

Turning now to the general results as shown in these tables, the showing can not but be considered as evidencing a remarkable success from the standpoint of an effort to carry on an important industrial establishment on a cooperative plan and entirely for the benefit of the workingmen there employed. It shows that, according to the actual results achieved, a workingman entering in 1879 at the lowest grade of membership, or as a participant, and then becoming successively a sociétaire and an associé after the statutory length of connection with the society, and receiving during this period the average wages of 1,650 francs ($318.45), has had his cash income increased each year from the original sum of 1,650 francs ($318.45) to 1,829.97 francs ($353.18) in 1893-94, during which time he has also acquired certificates of stock to the value of 3,594.28 francs ($693.70), and that he starts upon the new régime with this amount of capital and receives 2,061.03 francs ($397.78) in cash. If he has entered the society as a sociétaire or an associé, his receipts and the amount of his capital would be still greater. This increase can be due only to a slight extent to the greater prosperity of the industrial work of the society, as gross profits have by no means increased in an equal ratio. The increase in the total receipts of individual members as shown in this table has been, therefore, entirely due to the fact that more and more they have become the owners of the works in which they are employed; and their receipts as owners of capital figure to a constantly increasing extent. The same average wages have, in order to simplify the calculation, been maintained throughout the table. The chances are that the member would have earned considerably greater wages than this during the more recent years, in which case, of course, his receipts would be correspondingly greater. The important point should also be noticed in this connection that these results do not apply only to a selected few. The table giving the increase in membership shows that not only the total but the higher classes of membership are constantly increasing—a greater and greater number thus becoming capital owners and participants in profits.

In addition to this it should be remembered that all members, including even auxiliaires, are at the same time acquiring a right to an old-age pension that ultimately will represent a decided addition to the benefits derived from their connection with the society. The following table shows the minimum income a representative [illegible] class would enjoy after his retirement on account of old age:

MINIMUM ANNUAL INCOME OF MEMBERS OF THE FAMILISTÈRE SOCIETY OF GUISE, RETIRED ON ACCOUNT OF OLD AGE.

| Kind of member. | Years of service. | Annual income. | | | Kind of member. | Years of service. | Annual income. | | |
|---|---|---|---|---|---|---|---|---|---|
| | | Minimum pension. | Dividends and interest on capital stock. | Total. | | | Minimum pension. | Dividends and interest on capital stock. | Total. |
| Associé | 15 | $176.02 | $56. | $232.64 | Participant | 15 | [illegible] | [illegible] | [illegible] |
| | 20 | 176.02 | | 251.50 | | 20 | 105.67 | [illegible] | [illegible] |
| | 25 | 176.02 | | 270.36 | | 25 | [illegible] | [illegible] | [illegible] |
| | 30 | 176.02 | | 289.21 | | 30 | 176.02 | [illegible] | [illegible] |
| | 35 | 176.02 | | 308.08 | | 35 | 176.02 | 119.71 | [illegible] |
| Sociétaire | 15 | 140.89 | | 191.18 | Auxiliaire | 15 | 70.45 | | [illegible] |
| | 20 | 140.89 | | 210.04 | | 20 | 105.67 | | [illegible] |
| | 25 | 140.89 | | 228.90 | | 25 | 140.89 | | [illegible] |
| | 30 | 176.02 | 75.62 | 282.88 | | 30 | 176.02 | | [illegible] |
| | 35 | 176.02 | [illegible] | 301.74 | | 35 | 176.02 | | [illegible] |

The fixing of wages in a cooperative concern is a matter of peculiar importance; for their determination being made by the members themselves, they can elect to receive their awards either in the way of increased wages or larger profits. The retention of the wages system, and the necessity on the part of the society to employ workingmen who are not members, checks, at Guise, any tendency to fix wages unduly high. Nevertheless, since the creation of the society the rate of wages has shown a constant tendency to rise. This is shown in the table that follows. In order to show that this rise is not due to the fact that a larger proportion of skilled workmen (and therefore more highly paid) are employed than formerly, the average daily wages earned each year by the ten workingmen in each of the two principal occupations of molder and setter-up receiving the highest earnings during the year are also given.

AVERAGE DAILY WAGES OF WORKINGMEN OF THE FAMILISTÈRE SOCIETY OF GUISE, 1879-80 TO 1894-95.

| Year. | Average daily wages. | | | Year. | Average daily wages. | | |
|---|---|---|---|---|---|---|---|
| | All workingmen. | Ten workingmen earning highest annual wages. | | | All workingmen. | Ten workingmen earning highest annual wages. | |
| | | Molders. | Setters-up. | | | Molders. | Setters-up. |
| 1879-80 | $0.84 | $1.19 | $0.96 | 1887-88 | [illegible] | [illegible] | [illegible] |
| 1880-81 | .85 | 1.21 | 1.01 | 1888-89 | 1.04 | [illegible] | [illegible] |
| 1881-82 | .87 | 1.22 | 1.03 | 1889-90 | 1.02 | 1.75 | [illegible] |
| 1882-83 | .89 | 1.26 | 1.05 | 1890-91 | 1.05 | [illegible] | [illegible] |
| 1883-84 | .91 | 1.47 | 1.08 | 1891-92 | 1.00 | a[illegible] | [illegible] |
| 1884-85 | .92 | 1.39 | 1.16 | 1892-93 | 1.01 | a[illegible] | [illegible] |
| 1885-86 | .95 | 1.44 | 1.24 | 1893-94 | [illegible] | a[illegible] | [illegible] |
| 1886-87 | 1.01 | 1.50 | 1.31 | 1894-95 | [illegible] | a[illegible] | [illegible] |

a Average of the twenty workingmen earning highest annual wages.

The following table is introduced to show in greater detail [illegible] ation in the rates of wages of [illegible] the latest [illegible]

An important fact is brought out by the grouping according to classes of members. The members of the higher classes—associés and sociétaires—receive much higher wages than the participants and auxiliaires, thus adding to the many other advantages they enjoy that of earning greater wages. The distinction between those paid by the month and those paid by the week is maintained because the former represents the office force, the directors of services, etc., and the latter the general run of the shop employees.

WAGES OF MEMBERS OF THE FAMILISTÈRE SOCIETY OF GUISE, BY CLASSES, JULY 1, 1891.

| Wages per month. | Members paid by the month. | | | | | Wages per week. | Members paid by the week (workingmen proper). | | | | |
|---|---|---|---|---|---|---|---|---|---|---|---|
| | Associés. | Sociétaires. | Participants. | Auxiliaires. | All classes. | | Associés. | Sociétaires. | Participants. | Auxiliaires. | All classes |
| $3.86 to $4.83 | ...... | ...... | ...... | 6 | 6 | $1.93 to $2.90 | ...... | ...... | 1 | 24 | 25 |
| $5.02 to $6.76 | ...... | 1 | ...... | 8 | 9 | $3.09 to $3.86 | ...... | 2 | 2 | 17 | 21 |
| $6.95 to $9.65 | ...... | 2 | 3 | 6 | 11 | $4.05 to $4.83 | ...... | 3 | 5 | 19 | 27 |
| $9.84 to $14.48 | 2 | 9 | 1 | 13 | 25 | $5.02 to $5.79 | 1 | ...... | 4 | 27 | 32 |
| $14.67 to $19.30 | 3 | ...... | 3 | 15 | 21 | $5.98 to $7.72 | ...... | 5 | 42 | 67 | 114 |
| $19.49 to $24.13 | 2 | 5 | 3 | 5 | 15 | $7.91 to $9.65 | ...... | 6 | 65 | 55 | 126 |
| $24.32 to $28.95 | 8 | 8 | 8 | 2 | 26 | $9.84 to $11.58 | 5 | 9 | 132 | 131 | 277 |
| $29.14 to $33.78 | 21 | 6 | 1 | 1 | 29 | $11.77 to $13.51 | 16 | 28 | 97 | 59 | 200 |
| $33.97 to $38.60 | 12 | 6 | 2 | 1 | 21 | $13.70 to $15.44 | 40 | 53 | 96 | 59 | 248 |
| $38.79 to $48.25 | 8 | 6 | 2 | ...... | 16 | $15.63 to $17.37 | 29 | 35 | 44 | 25 | 133 |
| $48.44 to $57.90 | 5 | 1 | ...... | 1 | 7 | $17.56 to $19.30 | 22 | 27 | 35 | 23 | 107 |
| $58.09 to $67.55 | ...... | 1 | ...... | ...... | 1 | $19.49 to $21.23 | 12 | 13 | 11 | 10 | 46 |
| $67.74 to $77.20 | 2 | ...... | ...... | ...... | 2 | $21.42 or over | 8 | 6 | 6 | ...... | 20 |
| $77.39 to $96.50 | 3 | 1 | ...... | ...... | 4 | | | | | | |
| $96.69 or over | 2 | ...... | 1 | ...... | 3 | | | | | | |

## SOCIAL INSTITUTIONS.

It is more than probable that M. Godin himself never realized what were the most important features of his work. Combined with great business shrewdness, Godin had a distinct leaning toward speculative philosophy. As a result, he had worked out a complete theory of the rights of man; and it was in putting into practical operation these theories that he believed he was most entitled to credit as a social reformer. These rights or principles related to the communization of all the interests of members of a society; to the grouping together of the members in large apartment or tenement houses; to the guarantee to all members, under all conditions, of a minimum amount necessary for their subsistence; to their insurance against accidents and sickness, and their pensioning in old age; to the education of the children in common—not so much because existing educational facilities were lacking as because in that way the children would be taught from their earliest years the beauties and advantages of a common habitation and life—and to the development of mutuality in every possible way. Though not at all necessary incidents to his scheme of profit sharing and cooperation, he has made them integral parts of his whole organization, and seems to have regarded them, especially the familistère idea, as the most important part of his work.

Though referred to repeatedly in the account of the workings of the society, a description of the character and operations of these social institutions has been left to this place, as it was feared that their consideration at an earlier point would unnecessarily complicate the description of the general plan of the society. They can be grouped under four heads, each of which will be considered in turn:

1. The familistères.
2. Insurance and mutual relief.
3. Education.
4. Cooperative distribution.

## THE FAMILISTÈRES.

The familistère is the one feature of the society that has received the most notice from the outside world, and it can without hesitation be said to be the feature the least deserving of attention. The buildings of the society at Guise now comprehend, besides the factories and warehouses, seven groups of constructions, three of which are tenement houses or familistères. In the remaining groups of buildings are comprised a nursery, a theater, two schools, and a public laundry, and baths. The first and main familistère erected consists of a central building and two wings, each of which is quadrangular in form, four stories in height, and incloses a large interior court, paved and roofed over at the height of the building with glass. The second familistère is similar to one wing of the main familistère. Its court is not covered with glass, but is open to the air. The third, a much smaller building, contains no court.

The main familistère, or "social palace" as it is usually called, contains in all 299 tenements, of which 12 are one-room, 201 two-room, 75 three-room, 1 four-room, 6 five-room, 3 seven-room, and 1 eight-room tenements. The second group contains 147 tenements, of which 11 are one-room, 79 two-room, 56 three-room, and 1 five-room tenements. The third contains 19 tenements, chiefly of two and three rooms. The three familistères together contain 465 tenements, with a total of 1,091 rooms. Fifteen rooms of the main building are occupied by the grocery, clothing, drug and other stores, the library, etc. All of the tenements of the first two buildings have balconies on the court side, to which the doors open, this being the only means by which all but the corner ones are reached. The stairways are built in the interior and lead to the exterior balconies. They are placed in the corners. Each story contains separate sets of water-closets for men and women. Artesian wells furnish the water, which rises nearly to the height of the roof, and is distributed to reservoirs from which it descends in pipes having a spigot on [illegible] floor. The reservoirs hold from 8,000 to 8,700 liters (2,113 to [illegible] gallons). Good provision has been made for sewerage.

The total cost of construction of the three groups of [illegible] and the annexes (theater, schools, and laundry) was 2,10[illegible] francs ($405,342.19), or for the [illegible] alone 1,9[illegible]

A change was therefore made, and apartments are now apportioned in the order of the demand made for them, after the applications have been carefully examined by the council on the management of the familistère and approved by the managing council. The rents are by the month and are fixed according to the desirableness of the tenements. The following is a schedule of rents in force in the familistères:

SCHEDULE OF MONTHLY RENTS IN THE FAMILISTÈRES AT GUISE.

| Location of tenement. | One room. | Two rooms. | Three rooms. | Four rooms. | Five rooms. | Six rooms. | Seven rooms. | Eight rooms. |
|---|---|---|---|---|---|---|---|---|
| LEFT WING. | | | | | | | | |
| First story | $0.66 | $1.76 | $3.01 | $3.89 | | | | |
| Second story | 1.32 | 2.64 | 3.23 | 4.55 | $5.88 | $7.20 | $8.52 | |
| Third story | 1.13 | 2.25 | 3.06 | 4.18 | 5.32 | | | |
| Fourth story | 1.06 | 2.09 | 2.76 | 3.81 | | | | |
| CENTRAL PORTION. | | | | | | | | |
| First story | 1.02 | 2.05 | | | | | | |
| Second story | 1.28 | 2.57 | 3.47 | 4.82 | 6.10 | 7.38 | 8.67 | |
| Third story | 1.21 | 2.41 | 3.23 | 4.44 | | | | |
| Fourth story | 1.10 | 2.21 | 2.83 | 3.93 | | | | |
| RIGHT WING. | | | | | | | | |
| First story | 1.38 | 2.76 | 3.71 | 5.09 | | | | |
| Second story | 1.55 | 3.11 | 4.05 | 5.63 | 7.18 | 8.73 | 10.29 | 11.84 |
| Third story | 1.39 | 2.79 | 3.50 | 4.90 | 6.31 | | | |
| Fourth story | 1.27 | 2.55 | 3.40 | 4.67 | | | | |
| LANDRECIES. | | | | | | | | |
| First story | 1.25 | 2.51 | 3.59 | | | | | |
| Second story | 1.59 | 3.18 | 4.05 | 5.65 | | | | |
| Third story | 1.50 | 2.99 | 3.76 | | | | | |
| CAMBRAI. | | | | | | | | |
| First story | 1.35 | 2.70 | 3.96 | | | | | |
| Second story | 1.59 | 3.18 | 4.54 | 5.89 | 7.48 | 9.07 | 10.66 | |
| Third story | 1.45 | 2.90 | 4.10 | | | | | |
| Fourth story | 1.31 | 2.62 | 3.40 | | | | | |

The society experiences no trouble in renting all the tenements, there being always a considerable number of applicants on the waiting list. As has been seen, it is obligatory for associés and sociétaires to reside in the familistères. This provision alone of the constitution goes a long way toward insuring that there will never be a lack of tenants. The following tables show for each year the population of the familistères and the total amount received in rents:

POPULATION OF THE FAMILISTÈRES AT GUISE, 1879 TO 1894.

| Year. | Inhabitants of familistères. | Year. | Inhabitants of familistères. | Year. | Inhabitants of familistères. | Year. | Inhabitants of familistères. |
|---|---|---|---|---|---|---|---|
| [illegible] | 718 | 1883 | 1,241 | 1887 | 1,760 | 1891 | a 1,820 |
| [illegible] | 1,170 | 1884 | 1,250 | 1888 | 1,798 | 1892 | a 1,820 |
| [illegible] | 1,204 | 1885 | 1,748 | 1889 | 1,750 | 1893 | a 1,820 |
| [illegible] | 1,365 | 1886 | 1,765 | 1890 | 1,780 | 1894 | a 1,820 |

a Average.

TOTAL RENT PAID IN THE FAMILISTÈRES [illegible]

| Year. | Total rent. | Year. | Total rent. | Year. | [illegible] |
|---|---|---|---|---|---|
| 1879-80 ..... | $10,551.96 | 1883-84..... | $24,419.07 | 1887-88..... | [illegible] |
| 1880-81..... | 12,693.27 | 1884-85..... | 18,300.34 | 1888-89..... | [illegible] |
| 1881-82..... | 13,633.87 | 1885-86..... | 20,230.24 | 1889-90..... | [illegible] |
| 1882-83..... | 14,512.23 | 1886-87..... | 20,563.23 | 1890-91..... | [illegible] |

It is an interesting, but at the same time a difficult, [illegible] mine the effect of this grouping of a large number of [illegible] the same roof. From the standpoint of the material [illegible] inhabitants, conditions in the familistères compare [illegible] those of the village of Guise. A comparison of the vital s[illegible] the familistères and of Guise is made in the following table. age for the period covered shows not only a higher birth decidedly lower death rate among both the children and adu[illegible] of the population.

POPULATION, BIRTHS, AND DEATHS IN THE VILLAGE OF GUISE A[ND] FAMILISTÈRES, 1879 TO 1891.

VILLAGE OF GUISE.

| Year. | Population. | Births. | Deaths by age periods. | | | | | |
|---|---|---|---|---|---|---|---|---|
| | | | Less than one year. | | One year and less than ten. | | Ten years and over. | |
| | | | Number. | Per 1,000 of population. | Number. | Per 1,000 of population. | Number. | Per 1,000 of population. |
| 1879.............. | 5,550 | 128 | 34 | 6.13 | 24 | 4.32 | 100 | 18.02 |
| 1880.............. | 5,600 | 150 | 33 | 5.89 | 18 | 3.22 | 92 | 16.43 |
| 1881.............. | 5,610 | 164 | 23 | 4.10 | 14 | 2.50 | 86 | 15.33 |
| 1882.............. | 5,642 | 167 | 26 | 4.61 | 22 | 3.90 | 97 | 17.19 |
| 1883.............. | 5,675 | 160 | 26 | 4.58 | 29 | 5.11 | 96 | 16.92 |
| 1884.............. | 5,700 | 154 | 23 | 4.03 | 11 | 1.93 | 112 | 19.65 |
| 1885.............. | 5,729 | 164 | 40 | 6.98 | 31 | 5.41 | 97 | 16.93 |
| 1886.............. | 5,740 | 147 | 32 | 5.57 | 32 | 5.57 | 106 | 18.47 |
| 1887.............. | 5,765 | 131 | 28 | 4.86 | 23 | 3.99 | 97 | 16.83 |
| 1888.............. | 5,782 | 148 | 28 | 4.84 | 26 | 4.50 | 120 | 20.75 |
| 1889.............. | 6,250 | 170 | 55 | 8.80 | 18 | 2.88 | 21 | 3.36 |
| 1890.............. | 6,373 | 185 | 43 | 6.75 | 15 | 2.35 | 26 | 4.08 |
| 1891.............. | 6,415 | 177 | 45 | 7.02 | 15 | 2.34 | 60 | 9.35 |
| Average.... | 5,833 | 157 | 34 | 5.76 | 21 | 3.66 | 85 | 14.64 |

FAMILISTÈRES.

| Year. | Population. | Births. | Less than one year: Number. | Per 1,000 of population. | One year and less than ten: Number. | Per 1,000 of population. | Ten years and over: Number. | Per 1,000 of population. |
|---|---|---|---|---|---|---|---|---|
| 1879.............. | 718 | 44 | 7 | 9.75 | 6 | 8.36 | 14 | 19.50 |
| 1880.............. | 1,170 | 42 | 10 | 8.55 | 6 | 5.13 | 9 | 7.69 |
| 1881.............. | 1,204 | 40 | 11 | 9.14 | 4 | 3.32 | 12 | 10.00 |
| 1882.............. | 1,265 | 43 | 6 | 4.74 | 2 | 1.58 | 19 | [illegible] |
| 1883.............. | 1,241 | 48 | 8 | 6.45 | 10 | 8.06 | 16 | [illegible] |
| 1884.............. | 1,250 | 41 | 1 | .80 | 4 | 3.20 | 24 | [illegible] |
| 1885.............. | 1,748 | 76 | 8 | 4.58 | 14 | 8.01 | 20 | [illegible] |
| 1886.............. | 1,765 | 67 | 4 | 2.27 | 12 | 6.80 | 12 | [illegible] |
| 1887.............. | 1,760 | 74 | 5 | 2.84 | 6 | 3.41 | 17 | [illegible] |
| 1888.............. | 1,798 | 62 | 14 | 7.78 | 7 | 3.89 | 20 | [illegible] |
| 1889.............. | 1,756 | 65 | 11 | 6.26 | 5 | 2.85 | [illegible] | [illegible] |
| 1890.............. | 1,786 | 53 | (a) | [illegible] | [illegible] | [illegible] | [illegible] | [illegible] |
| 1891.............. | 1,[illegible] | 56 | (a) | [illegible] | [illegible] | [illegible] | [illegible] | [illegible] |
| Average.... | 1,[illegible] | 55 | [illegible] | [illegible] | [illegible] | [illegible] | [illegible] | [illegible] |

a Not reported. b Not [illegible]

Statistics, however, can throw but a one-sided light upon the considerations involved in the organization of domestic life as existing in the familistères. It is the moral and psychological effects upon the inhabitants themselves that are of the greatest importance. It may be pardonable to dwell at some length upon this feature of the society, as it is on this account that the society as a whole has been more often condemned than praised.

The description of his impressions, given by a member of the Société Belge d'Économie Sociale who in 1891 made a visit to the familistère of Laeken, the conditions of which are in every respect similar to those at Guise, represents the type of the most hostile criticism that has been directed against the familistère. "Attempt," he says, "to fix the impression that this collection of institutions engenders. In spite of the prosperity that is visible everywhere among the workingmen who live in the familistère, in spite of the comfortable and ingenious arrangement that can be remarked even in the smallest details, that which the visitor carries away is an impression of sadness, of oppression. The familistère of Laeken is a superb barrack, but yet a barrack. It is an ideal Fourier's convent, but everyone does not care to be an inmate of a convent. An exasperating uniformity, a lack of freedom, of independence, the sensation of an inflexible rule bringing everything to a level, impresses you gradually with sadness. The walls seem impregnated with ennui."

Whatever the opinion of one may be concerning the justness of this criticism, and it is a case where each one must form an opinion for himself, the result should not be allowed to influence the judgment concerning the merits of the whole organization. One can quite easily condemn the familistère and yet admire the ingenuity of the scheme through which the workingmen have become the owners of the plant in which they work and by which the principle of cooperation has been introduced throughout the workings of the society. To one studying in detail the organization of the society it is evident that the familistère idea is but an incidental feature, and that it is immaterial to the workings of the system whether the employees are housed under one roof or in individual cottages. But to Godin the familistère was considered as the most important part of his work. It was his hobby. Though his works were situated in the outskirts of a small country village, where land was abundant and there was no necessity whatever for the erection of huge tenements, Godin deliberately chose this type, believing that under all conditions the community of life thus created by bringing all or the greater portion of the employees of an establishment under the same roof represented the ideal type of the constitution of society. It is unnecessary to say that in so doing he acted contrary to the whole trend of modern thought in regard to the best methods of housing the working classes. The importance that he attached to this idea is seen in the great attention given to it in his

Solutions Sociales and in all the details of the [illegible] society. The associés and sociétaires are thus [illegible] familistères; and the policy pursued in the instruction of [illegible] was to have this idea constantly in view.

The familistère, therefore, should not be regarded as constituting the work of the society. This has too often been done. It requires a great deal of study to understand the workings of an elaborate scheme such as this is; but, if made, it is seen that the familistère idea is but an unessential part as far as the practical working of the whole society is concerned.

Two important features in connection with the familistère are the laundry and the allotment of land to the members for gardening purposes. It has been the policy of the society to group around the familistères institutions ministering to all the needs of the inhabitants. The conditions of tenement-house life made it extremely undesirable that clothes should be washed inside the buildings. Contemporaneous with the erection of the familistère, therefore, there was constructed a building called the buanderie, containing a swimming pool, baths, and rooms for the washing, drying, and ironing of clothes. Here members can do their own washing or hire others to do it for them.

The society possesses considerable land immediately adjoining the works. This it has divided into lots of from 1 to 3 ares (1,076.4 to 3,229.2 square feet) each, which are rented to members for vegetable gardens. From 500 to 600 members are in this way able to raise their own vegetables. The annual rent is from 1.25 to 5 francs (24 to 97 cents) per are (1,076.4 square feet). This privilege seems to be one highly prized. A very pleasant park on high land adjoining the grounds of the familistère is maintained at the expense of the society. A library of nearly 3,000 volumes is at the disposal of members free. The members are also encouraged to form societies for mutual amusement and cultivation, and there have been formed a musical, a gymnastic, a shooting, a fencing, an archery, and other societies. A commodious theater has been erected that is utilized for assemblies of all kinds, and during the winter months is occupied twice a month by traveling theatrical troupes.

## INSURANCE AND MUTUAL RELIEF.

As one would expect, insurance and mutual aid institutions play a prominent part in the social work of the society. The idea of the mutual duties of members to each other is emphasized in every possible way. "One for all and all for one" is the motto of the society and the basis of the whole theory upon which it has been organized, [illegible] certainly there is no better opportunity for the exercise of mutual[illegible] than that afforded by the care of old workingmen and the aid of [illegible] sick and injured. The pensioning of old workingmen, and the [illegible]anty of a minimum subsistence to all members, [illegible] not deemed by the society as a privilege accorded to its members, but as a natural right to which labor is entitled in a degree scarcely second to that of receiving wages. The expense of accomplishing these objects, therefore, is counted as a legitimate item of expense in the cost of operation to be met before even the payment of interest on founder's certificates. Insurance against sickness and accidents, however, is put upon an entirely different basis. For this purpose all members and employees are organized into various mutual aid societies, in which membership is obligatory and which are supported, for the most part, by the dues of members. The society stands pledged by its constitution to insure the operation of these societies, and in actual practice it has frequently been necessary for it to supplement their receipts by grants from its common fund. Four distinct societies have been organized to cover this field of insurance and aid in case of sickness:

1. Society for the guaranty of a minimum subsistence to members and for the pensioning of old or incapacitated employees.
2. Mutual aid society for the insurance of employees against sickness.
3. Mutual aid society for the insurance of women inhabiting the familistères against sickness.
4. Mutual aid society for the supply of medicines and medical supplies to inhabitants of the familistères in case of sickness.

As the constitutions and funds of these societies are entirely distinct, one from the other, their separate consideration will be necessary. The results of their practical operation, however, can be presented statistically in the same tables.

### SOCIETY FOR THE GUARANTY OF A MINIMUM SUBSISTENCE TO MEMBERS AND FOR THE PENSIONING OF OLD OR INCAPACITATED EMPLOYEES.

The first society, that for the guaranty of a minimum subsistence and for the pensioning of old or incapacitated employees, is of much the greatest interest. In describing its organization, as well as that of the other societies, the greatest precision can be obtained by translating freely the essential portions of their constitutions. The following are the important portions of the constitution of the society under discussion:

The resources of the society are derived from: (1) A payment from the gross profits of the familistère society of a yearly sum equal to 2 per cent of the total amount paid in wages to members of the society. (2) The profits apportioned to the wages of auxiliaires.

Every person employed in the establishments of the familistère society is placed under its protection, and in case of incapacity incurred during his connection with the society is accorded a pension. This pension is granted to associés, sociétaires, and participants, and, in certain cases, to auxiliaires.

After fifteen years of service the right to a pension is regulated as follows: For associés, men and women, the pension is fixed at two-fifths of their wages, and for sociétaires at one-third of [illegible] vided, however, that the monthly pension of associés shall [illegible] than 75 francs ($14.48) for the men and 45 francs ($8.69) for the [illegible], and of the sociétaires not less than 60 francs ($11.58) for the [illegible] 35 francs ($6.76) for the women. The pensions for participants [illegible] auxiliaires are fixed by the following schedule:

DAILY PENSIONS PAID TO PARTICIPANTS AND AUXILIAIRES BY THE MINIMUM SUBSISTENCE AND PENSION FUND OF THE FAMILISTÈRE SOCIETY OF GUISE.

| Years of continuous service. | Men. | Women. |
|---|---|---|
| After 15 years | $0.19 | $0.14 |
| After 20 years | .29 | .19 |
| After 25 years | .39 | .24 |
| After 30 years | .48 | .29 |

In the intermediate periods the amount of the pension is proportional to the length of service. The right of a member to a pension is suspended if, without an authorization of the board of directors, he accepts a salaried position outside of the society.

In case the resources of the fund permit, the board of directors can, on the authorization of the general assembly, increase the amount of the pension paid in cases where especially meritorious services have been rendered to the society.

If the years of service entitling a member to a pension have been accomplished at various times, each voluntary departure and each year of absence diminishes the pension paid by 2 centimes ($\frac{4}{10}$ cent) per day. The years of service performed before the age of 20 years count as half. The time passed in military service is counted as an involuntary absence and no reduction is made in consequence.

A member able to work, who retires voluntarily from the society or who is expelled or severs his connection with the society for any reason other than the lack of work, loses all right to a pension.

If before fifteen years of service an employee of the society, who is without means of support, is incapacitated for work through an accident received while at work, he is entitled to the pension of one having served twenty years. If the accident occurs after fifteen years of service, he is entitled to the pension of one having served thirty years. The indemnities accorded to those incapacitated before fifteen years of service, and not due to an accident received while at work, are left to the discretion of the insurance committee and the board of directors.

To associés and sociétaires who have been sick more than three months there is granted during one year a sum equal to their wages.

The society guarantees a minimum subsistence to its members. In cases, therefore, where the wages of a household does not equal this sum the deficiency is paid from the insurance fund. The minimum daily subsistence is fixed by the following schedule:

| | |
|---|---|
| For a man or woman | $0.45 |
| For a widower or widow, head of a family | .29 |
| For a widow without a family | .19 |
| For a man sick in a family | .29 |
| For a woman sick in a family | [illegible] |
| For children over 16 years of age (each) | [illegible] |
| For children from 14 to 16 years of age (each) | [illegible] |
| For children from 2 to 14 years of age (each) | [illegible] |
| For children under 2 years of age (each) | [illegible] |

These last have, in addition, a right to the nursery. In calculating the amount of earnings of a family, in order to determine by what amount the total family income falls below the minimum of subsistence, the gains of all the members of the family, whether from wages or from the various insurance funds, are taken into account. These daily gains that can not be absolutely determined are valued according to the following schedule:

| | |
|---|---|
| Males from 14 to 16 years of age | $0.14 |
| Males from 16 to 18 years of age | .19 |
| Males from 18 to 20 years of age | .29 |
| Males over 20 years of age | .43 |
| Females from 14 to 17 years of age | .10 |
| Females from 17 to 21 years of age | .14 |
| Females over 21 years of age | .19 |

It is considered that a mother with five children unable to work can gain nothing; one with four children, 25 centimes (5 cents); one with three children, 50 centimes (10 cents); one with two children, 75 centimes (14 cents); and one with one child unable to work can earn 1 franc (19 cents) per day. If, however, it is shown that she does earn something, it is taken into account. The minimum of subsistence is accorded to widows and orphans whose husbands and fathers have been in the service of the society at least fifteen years. For those whose husbands and fathers have died before having served that length of time, the right to aid and of living in the familistères is limited to one year. In case of sickness the rate of subsistence is only paid to those whose payment from the insurance fund against sickness does not amount to that sum and only for one month after they have stopped work.

The orphans of associés and sociétaires are taken care of by the insurance committee, which, after the approval of the board of directors, places them in families where they receive all necessary care and education.

The protection of insurance extends also to families of participants and auxiliaires in cases of exceptional misfortune and when the board of directors think such action to be necessary.

Every individual, after he has been pensioned, ceases to pay dues to the insurance fund against sickness, and will likewise cease to participate in its benefits, the pension being considered as taking its p .
In special cases, however, permission can be granted for a continued participation.

MUTUAL AID SOCIETY FOR THE INSURANCE OF EMPLOYEES AGAINST SICKNESS.

The organization of the society for the insurance of employees against sickness is as follows:

The resources of the society are derived from: (1) The payment as dues by each employee of the familistère society inhabiting the familistères of one-half of 1 per cent, and by each employee not inhabiting the familistères of 1 per cent of the amount he or she receives in wages. If the average monthly earnings of a member do not equal 100 francs ($19.30) for the head of a family, his payment is calculated as if it were that amount. If his average monthly earnings exceed 150 francs ($28.95), the member has the option of paying his percentage on either the full or on two-thirds of the amount of his wages. This rate can be raised or lowered according to the financial requirements of the fund. (2) The product of fines levied by the familistère society on workingmen for infringement of rules or breakage of material. (3) A grant by the familistère society of an amount within the discretion of the managing council.

Every employee of the familistère society, male or female, on commencing work, either in the shops, the offices, or the stores of the society, is enrolled as a member of the organization for insurance against sickness, and the payment of dues by him to the sick fund is obligatory. An exception, however, is made in the case of women inhabiting the familistères, inasmuch as they have an organization of their own.

After six months, during which regular payments have been made, every member, on becoming incapacitated for work through sickness or accident, is entitled to receive during a period not to exceed one year: (1) The professional attendance of physicians appointed and paid by the familistère society; (2) the payment of a daily sick benefit, the minimum amount of which is fixed as follows: For members less than 45 years of age on entering the employ of the society, during the first three months, two times, during the second three months, one and one half times, and during the remaining six months, one time the amount of the monthly dues they have been paying; for members over 45 years of age on entering the employ of the society, during the first three months, one and one-fourth times, during the second three months, one time, and during the remaining six months, three-fourths of the amount of the monthly dues that they have been paying.

In the case of sickness or injury occurring to an employee before he has made payments during six months, the patient receives a daily sick benefit equal to the amount of his monthly dues. If the incapacity for work continues more than one year the payment of a sick benefit ceases, and if the patient is declared to be permanently incapacitated he is pensioned as provided for in the scheme of old-age and invalidity pensions.

## MUTUAL AID SOCIETY FOR THE INSURANCE OF WOMEN INHABITING THE FAMILISTÈRES AGAINST SICKNESS.

The organization of the society for the insurance of women inhabiting the familistères against sickness is as follows:

The resources of the society are derived from: (1) The payment as dues by each woman 14 years of age or over inhabiting the familistères of 50 centimes (10 cents) per month, or 2 per cent of her earnings when the amount thus determined is superior to 50 centimes (10 cents); in no case, however, can the dues exceed 3 francs (58 cents); (2) the product of fines for infraction of rules of the familistères; (3) a grant of a lump sum by the familistère society, which can in no case exceed the total amount paid in as dues.

Every woman over 14 years of age and inhabiting the familistères is enrolled as a member of the organization for insurance against sickness, and the payment of dues by her is obligatory. After six months, during which regular payments have been made, each member on becoming incapacitated for work through sickness is entitled to receive: (1) the professional attendance of a physician and of a midwife of her own selection; (2) the payment of a daily sick benefit, the amount of *which is* fixed as follows: During the time she is confined to her [illegible] to one and one-half times, and during the period of her convalescence, or of any sickness not entirely incapacitating her for light work, to three fourths of the amount of the monthly dues that she has been paying. In case of childbirth, the higher rate is paid during the first nine days and the lower rate during the remaining time of her illness.

### MUTUAL AID SOCIETY FOR THE SUPPLY OF MEDICINES AND MEDICAL SUPPLIES TO INHABITANTS OF THE FAMILISTÈRES IN CASE OF SICKNESS.

It would seem that the supplying of medicines was a legitimate duty of the other societies, and it is difficult to see what necessity there was for the creation of a societiy for that purpose. However, it was determined to keep such service separate. Its organization is as follows:

The resources of the society are derived from: (1) A payment as dues by every inhabitant of the familistère over 14 years of age of 50 centimes (10 cents) per month. (2) A grant of a lump sum by the familistère society which can in no case exceed the total amount paid in as dues.

After six months, during which regular payments have been made, members are entitled to receive from the fund, both for themselves and their families in case of sickness, medicines, medical supplies, and instruments that have been ordered by the physician or midwife. Such provision, however, is made only in cases where the sickness is of sufficient severity as to incapacitate the patient for work. The fund also defrays the civil expenses of burials of members or members of member's families.

The administration of these various societies is in the hands of a committee for each society, elected by its members for a term of one year. The managing council exercises a general control over all their operations. It is not necessary to go into the details of the organization of these committees nor of the methods of management of their affairs. The practical workings of these societies are shown in the series of tables that follow. There are a number of points of great importance to be noticed in connection with this showing. The first of these is that there has been a failure to correctly estimate the relationship between receipts and probable expenses. It is a difficult thing to correctly calculate the mathematical basis upon which any insurance or beneficial scheme must rest. The difficulty lies in the fact that, while receipts remain nearly stationary, expenditures must necessarily increase as the scheme and its members become older. The situation of affairs at Guise is thus set forth by the general manager in his annual report for 1893–94:

If the financial and industrial situation is good, by no means the same thing can be said regarding our insurance funds. In effect, all of our relief funds at Guise have an excess of expenditures over receipts.

The expenditures of the fund for pensions and necessary subsistence have exceeded receipts by 7,031.78 francs ($1,357.13).

The insurance fund against sickness has had an excess of expenditures amounting to 5,118.37 francs ($987.85). We had in bank June

30, 1893, a surplus of 1,333.76 francs ($257.42), which had been transformed into a deficit of 3,784.61 francs ($730.43) on June 30 last.

The fund for the insurance of women against sickness has also an excess of expenditures. There remained in bank on June 30, 1894, a balance of 12,069.69 francs ($2,329.45), a reduction of 1,563.80 francs ($301.81) as against the preceding year.

The fund for the supply of medicines has still a deficit, which during this year reached 2,343.20 francs ($452.24), and which the society had to cover, as it did the deficits for previous years.

* * * * * * *

The fund for pensions and minimum subsistence, which, at the date of creation of the society, June 30, 1880, had a cash balance of 108,846.94 francs ($21,007.46) has had this amount successively increased until June 30, 1893, when it amounted to 290,552.40 francs ($56,076.61), an increase of 181,705.46 francs ($35,069.15), or an average annual increase of about 14,000 francs ($2,702). The annual increase was greater in the earlier and smaller in the later years, until during the year 1892–93, it amounted to only 2,032.51 francs ($392.27), and in the last year, 1893–94, there was a deficit of 7,031.78 francs ($1,357.13).

The permanent fund of certificates which last year amounted to 855,178 francs ($165,049.35) reached this year about 878,033 francs ($169,460.37), an increase of 22,855 francs ($4,411.02), * * * which makes the total assets of the fund 1,161,553.62 francs ($224,179.85); it was 1,145,725.60 francs ($221,125.04) last year, or a gain of 15,828.02 francs ($3,054.81), the difference resulting from the increase in the amount of certificates held and the diminution in the cash balance on hand.

* * * * * * *

It is necessary to be prudent and to seek to establish an equilibrium between our budgets of insurance and benevolence.

* * * * * * *

I should also call attention to the other insurance funds * * * of which the deficits are permanent. There, also, it is necessary to economize—it is necessary to exercise efficacious supervision and suppress the abuses. * * *

I shall count on the insurance committees to appreciate the importance of their mission, and not to permit a renewal of certain violations and certain abuses which are prejudicial to our funds and also injurious to the health of their associates and to the welfare of their families.

The abuses here spoken of were the too great leniency and liberality displayed in the allotting of benefits.

The experience of Guise has thus been commented upon, not because it represents an aggravated case of miscalculation, for it is quite otherwise, but because it may serve to emphasize the necessity of an ample provision for the future increase of disbursements. There has been a failure to do this in the case of a great many insurance schemes organized in France and elsewhere. At Guise, however, great credit is due to the policy that has been pursued from the start, in the case of the pension fund, of building up a permanent fund of certificates. Thus, in spite of the note of warning sounded by the general manager, and the fact that disbursements exceed ordinary receipts, the fund has continued to increase and the condition of the society to show improvement.

RECEIPTS AND EXPENDITURES OF THE MINIMUM SUBSISTENCE AND PENSION FUND OF THE FAMILISTÈRE SOCIETY OF GUISE, 1879-80 TO 1893-94.

| Year. | Total receipts. | Expenditures. | | | | | Balance at end of year. |
|---|---|---|---|---|---|---|---|
| | | Pensions. | Minimum subsistence. | Temporary aid. | Medical attendance and medicines. | Total. | |
| 1879-80 | a $41,[illegible] | $1,254.50 | $946.70 | $[illegible].50 | $44.70 | $3,113.49 | $[illegible] |
| 1880-81 | 35,907.62 | 1,244.66 | [illegible].09 | 951.49 | 109.47 | 3,384.65 | 71,[illegible].64 |
| 1881-82 | 32,939.43 | 2,342.47 | 1,140.87 | 3,067.55 | 273.90 | 6,725.88 | [illegible] |
| 1882-83 | 30,775.17 | 4,512.96 | 2,369.43 | 1,042.80 | 356.76 | 8,282.04 | 129,748.73 |
| 1883-84 | 31,887.38 | 5,449.73 | 2,472.74 | 1,029.70 | 611.10 | 9,564.33 | 132,[illegible].73 |
| 1884-85 | 29,157.06 | 5,512.48 | 1,505.79 | 1,657.91 | 1,121.62 | 9,797.80 | 142,427.96 |
| 1885-86 | 16,288.65 | 5,[illegible].86 | 2,[illegible].96 | 1,668.16 | 1,204.63 | 10,718.93 | 149,[illegible].70 |
| 1886-87 | 17,476.51 | 5,882.81 | 2,525.97 | 2,654.90 | 1,225.14 | 11,[illegible].62 | 154,612.09 |
| 1887-88 | 27,317.20 | 6,280.07 | 2,[illegible].01 | 2,572.90 | 1,180.36 | 12,330.43 | 169,791.36 |
| 1888-89 | 31,648.06 | 7,673.73 | 2,388.98 | 1,206.50 | 1,171.47 | 12,435.68 | 189,[illegible].74 |
| 1889-90 | 22,224.65 | 8,459.50 | 2,588.86 | 1,125.22 | 1,317.28 | 13,490.86 | 197,747.85 |
| 1890-91 | 34,004.56 | 9,726.84 | 2,266.47 | 1,239.62 | 1,817.52 | 15,050.45 | 206,761.66 |
| 1891-92 | 20,943.01 | 10,671.95 | 2,072.42 | 1,616.33 | 1,549.09 | 15,909.79 | 211,724.48 |
| 1892-93 | 17,712.00 | 12,071.09 | 1,985.11 | 1,714.65 | 1,548.88 | 17,319.73 | 212,127.15 |
| 1893-94 | 18,264.34 | 13,444.74 | 2,377.50 | 2,191.68 | 1,607.47 | 19,621.48 | 210,770.01 |

a Including $25,600.23 in fund at time of organization of society.

RECEIPTS AND EXPENDITURES PER EMPLOYEE OF THE MINIMUM SUBSISTENCE AND PENSION FUND OF THE FAMILISTÈRE SOCIETY OF GUISE, 1879-80 TO 1893-94.

| Year. | Total employees. | Total receipts per employee. | Average expenditures for— | | | | | | Total expenditures per employee. |
|---|---|---|---|---|---|---|---|---|---|
| | | | Pensions. | | Minimum subsistence. | | Temporary aid. | | |
| | | | Total pensioners. | Average pension. | Families aided. | Average per family. | Families aided. | Average per family. | |
| 1879-80 | 1,190 | a $34.96 | 9 | $129.39 | 17 | $55.63 | 22 | $30.48 | $2.62 |
| 1880-81 | 1,440 | 24.94 | 12 | 112.06 | 19 | 52.00 | 25 | 38.06 | 2.35 |
| 1881-82 | 1,500 | 21.96 | 21 | 106.83 | 24 | 47.54 | 55 | 37.50 | 5.82 |
| 1882-83 | 1,387 | 22.19 | 39 | 115.72 | 46 | 51.51 | 31 | 33.64 | 5.97 |
| 1883-84 | 1,293 | 16.99 | 46 | 118.47 | 28 | 88.35 | 46 | 22.38 | 7.40 |
| 1884-85 | 1,275 | 15.81 | 49 | 112.50 | 37 | 40.70 | 50 | 33.16 | 7.68 |
| 1885-86 | 1,217 | 13.38 | 45 | 119.76 | 45 | 53.09 | 38 | 43.90 | 8.80 |
| 1886-87 | 1,130 | 15.47 | 52 | 112.74 | 51 | 49.53 | 50 | 41.10 | 10.33 |
| 1887-88 | 1,237 | 22.08 | 53 | 116.98 | 41 | 58.30 | 42 | 61.26 | 9.96 |
| 1888-89 | 1,495 | 21.17 | 61 | 125.80 | 42 | 56.76 | 40 | 30.16 | 8.32 |
| 1889-90 | 1,471 | 15.12 | 67 | 126.26 | 43 | 60.21 | 42 | 26.79 | 9.17 |
| 1890-91 | 1,502 | 15.96 | 77 | 126.32 | 40 | 56.66 | 47 | 26.37 | 10.02 |
| 1891-92 | 1,486 | 13.98 | 86 | 124.09 | 35 | 59.21 | 32 | 50.51 | 10.62 |
| 1892-93 | 1,364 | 12.99 | 92 | 131.21 | 38 | 52.34 | 49 | 34.99 | 12.70 |
| 1893-94 | 1,364 | 13.39 | 104 | 129.28 | 39 | 60.96 | 61 | 35.93 | 14.39 |

a Including amount in fund at time of organization of society.

OPERATIONS OF THE SICK INSURANCE FUND FOR EMPLOYEES OF THE FAMILISTÈRE SOCIETY OF GUISE, 1879-80 TO 1893-94.

| Year. | Receipts. | Expenditures. | Members sick. | Total days sick. | Average days sick per member sick. | Average expenditure per day sick. |
|---|---|---|---|---|---|---|
| 1879-80 | a $8,261.09 | $4,877.35 | 442 | 9,678 | 21.90 | $0.50 |
| 1880-81 | 7,094.35 | 8,[illegible].99 | 750 | 16,456 | 21.94 | .50 |
| 1881-82 | 9,218.37 | 10,261.19 | 931 | 20,360 | 21.87 | .50 |
| 1882-83 | 8,396.75 | 8,350.15 | 708 | 17,035 | 24.06 | .49 |
| 1883-84 | 7,407.41 | 6,859.45 | 642 | 14,496 | 22.58 | .47 |
| 1884-85 | 7,113.68 | 8,064.89 | 670 | 15,236 | 22.74 | .53 |
| 1885-86 | 6,771.62 | 7,143.51 | 602 | 12,904 | 21.44 | .55 |
| 1886-87 | 6,853.24 | 5,540.45 | 523 | 10,830 | 20.71 | .51 |
| 1887-88 | 7,194.14 | 5,139.50 | 577 | 11,252 | 19.50 | .46 |
| 1888-89 | 6,806.54 | 6,133.47 | 764 | 12,965 | 16.97 | .47 |
| 1889-90 | 6,680.94 | 8,201.76 | 1,001 | 17,150 | 17.13 | .48 |
| 1890-91 | 7,146.49 | 8,328.77 | 896 | 17,033 | 19.01 | .49 |
| 1891-92 | 7,673.54 | 8,496.90 | 896 | 17,281 | 19.29 | .49 |
| 1892-93 | 7,372.45 | 8,080.39 | 696 | 16,507 | 23.65 | .49 |
| 1893-94 | 7,641.50 | 8,029.44 | 697 | 18,617 | 26.71 | .48 |

a Including $[illegible].44 in fund at time of organization of society.

OPERATIONS OF THE SICK INSURANCE FUND FOR [illegible] FAMILISTÈRES OF THE FAMILISTÈRE SOCIETY OF [illegible]

| Year. | Receipts. | Expenditures. | Members sick. | Total days sick. | [illegible] |
|---|---|---|---|---|---|
| 1879-80 | a$2,169.12 | $641.51 | 111 | [illegible] | [illegible] |
| 1880-81 | 613.95 | 955.60 | 160 | [illegible] | [illegible] |
| 1881-82 | 758.66 | 1,336.59 | 215 | [illegible] | [illegible] |
| 1882-83 | 1,028.33 | 1,012.98 | 146 | [illegible] | [illegible] |
| 1883-84 | 1,062.94 | 987.91 | 139 | [illegible] | [illegible] |
| 1884-85 | 1,311.92 | 1,064.66 | 190 | 5,117 | [illegible] |
| 1885-86 | 1,412.68 | 1,030.69 | 211 | [illegible] | [illegible] |
| 1886-87 | 1,394.07 | 954.29 | 185 | [illegible] | [illegible] |
| 1887-88 | 1,410.46 | 1,036.64 | 194 | [illegible] | [illegible] |
| 1888-89 | 907.18 | 1,279.94 | 262 | [illegible] | [illegible] |
| 1889-90 | 1,458.72 | 1,450.80 | 278 | [illegible] | [illegible] |
| 1890-91 | 1,339.61 | 1,186.58 | 184 | [illegible] | [illegible] |
| 1891-92 | 1,262.92 | 1,406.64 | 225 | [illegible] | [illegible] |
| 1892-93 | 1,257.96 | 1,206.28 | 184 | [illegible] | [illegible] |
| 1893-94 | 1,302.93 | 1,604.75 | 210 | [illegible] | [illegible] |

a Including $1,367.48 in fund at time of organization of society.

OPERATIONS OF THE MEDICAL SUPPLIES FUND OF THE FAMILISTÈRE SOCIETY OF GUISE, 1879-80 TO 1893-94.

| Year. | Receipts. | Expenditures. | Sick members receiving supplies. | Average expenditure per member sick. | Year. | Receipts. | Expenditures. | Sick members receiving supplies. | Average expenditure per member sick. |
|---|---|---|---|---|---|---|---|---|---|
| 1879-80 | a$2,877.61 | $641.18 | 247 | $2.60 | 1887-88 | $1,230.17 | $1,340.99 | [illegible] | [illegible] |
| 1880-81 | 1,026.27 | 739.78 | 469 | 1.58 | 1888-89 | 1,540.46 | [illegible] | 741 | [illegible] |
| 1881-82 | 830.29 | 1,422.01 | 583 | 2.44 | 1889-90 | 1,913.65 | [illegible] | [illegible] | [illegible] |
| 1882-83 | 834.04 | 1,121.01 | 465 | 2.41 | 1890-91 | [illegible] | [illegible] | [illegible] | [illegible] |
| 1883-84 | 899.38 | 1,260.76 | 473 | 2.66 | 1891-92 | 1,957.53 | 2,015.68 | [illegible] | [illegible] |
| 1884-85 | 1,106.29 | 1,822.84 | 611 | 2.98 | 1892-93 | 1,520.11 | 1,049.83 | [illegible] | [illegible] |
| 1885-86 | 1,217.78 | 1,567.66 | 610 | 2.57 | 1893-94 | 1,008.83 | 2,061.07 | [illegible] | [illegible] |
| 1886-87 | 1,216.59 | 1,459.35 | 606 | 2.41 | | | | | |

a Including $1,393.55 in fund at time of organization of society.

RECEIPTS AND EXPENDITURES OF THE SICK INSURANCE AND MEDICAL SUPPLIES FUNDS OF THE FAMILISTÈRE SOCIETY OF GUISE, 1879-80 TO 1893-94.

| Year. | Receipts. | Expenditures. | Cash on hand at end of year. | Year. | Receipts. | Expenditures. | Cash on hand at end of year. |
|---|---|---|---|---|---|---|---|
| 1879-80 | a$13,327.82 | $6,159.54 | $7,168.28 | 1887-88 | $9,834.77 | [illegible] | [illegible] |
| 1880-81 | 8,734.57 | 9,989.46 | 5,913.39 | 1888-89 | 9,344.18 | [illegible] | [illegible] |
| 1881-82 | 10,822.31 | 12,921.79 | 3,813.91 | 1889-90 | 10,053.31 | [illegible] | [illegible] |
| 1882-83 | 10,259.12 | 10,484.09 | 3,588.94 | 1890-91 | 10,479.14 | [illegible] | [illegible] |
| 1883-84 | 9,389.73 | 9,067.12 | 3,911.55 | 1891-92 | [illegible] | [illegible] | [illegible] |
| 1884-85 | 9,531.99 | 10,952.39 | 2,491.15 | 1892-93 | 10,150.54 | [illegible] | [illegible] |
| 1885-86 | 9,401.43 | 9,741.86 | 2,150.72 | 1893-94 | 10,553.35 | [illegible] | b[illegible] |
| 1886-87 | 9,463.90 | 7,954.09 | 3,650.53 | | | | |

a Including $6,153.47 in fund at time of organization of society.

b Deficit.

## EDUCATION.

The care and education of the children of members is [illegible] the society as one of the most important features of its [illegible] to the familistère, education was the object of the [illegible] Godin. As early as 1860 he [illegible]

Like the expenditures required for the pensioning of old workingmen and the guaranty of a minimum subsistence, the entire expense of education is made a charge on the receipts of the society before any interest or profits are paid. The great attention to education receives its explanation in the fact that the society intends, through the instruction of the children under its supervision, to make of its schools a means of perpetuating in future generations the same ideas concerning mutuality and the association that actuated its founders. The following extracts from the constitution and by-laws indicate clearly this policy: "The administration," says the constitution, "ought above all to make sure that the children receive good moral instruction with the special object of developing in them the sentiment of the duties of solidarity that exists between members. It should seek to make them understand the grandeur and benefits of the association in order that in the future they may become worthy successors in the work of their predecessors." Again, in the by-laws, a paragraph reads: "The fact should never be lost sight of that the object of the instruction ought to be to form for the society men and women valuable to it both on account of their moral character and intellectual attainments. The children, therefore, should be accustomed from their very earliest years to contribute by their good conduct to the charms of a common habitation and to the general welfare. * * * The effort should be made to instill in the minds of the children the love of the society, the love of the principles which have given birth to the present constitution and by-laws, and a desire to aid in their practical application."

To carry out these ideas the constitution made the following provisions concerning the organization of instruction. Institutions for the instruction of children are organized into five divisions or grades:

1. The nourricerie, or nursery, the object of which is to aid the mothers in the care of their infants until they have reached the age of 2 years.

2. The pouponnat, or the first infants' garden, an institution designed to give all the necessary care and amusement to children who are between the ages of 2 and 4 years.

3. The bambinat, or second infants' garden, for children between the ages of 4 and 6 years, where the first instructive exercises will be commenced.

4. Schools which will insure to all children of members of the society good primary instruction until they have reached the age of 14 years.

5. Finally, the administration makes provision for the further education of those students who have shown special aptitudes and faculties.

The education and instruction provided by the society is wholly gratuitous. The wages of instructors and all others expenses are paid by the society. The amount thus expended for each year can not be less than 25,000 francs ($4.825). This amount may be increased by the society, but in case the whole amount appropriated is not expended the balance is not carried to the next year, but to a special educational fund reserved for emergencies. The system thus created means, practically, the removal of the children, at least during their earlier years, from their parents during a greater part of the day and their elevating in common. It is unnecessary to call attention to the tremendous power thus given to the society to educate believers in the principles practiced at Guise. Handsome and well-arranged school buildings have been erected, and, without going into details, competent teachers and suitable courses of study have, apparently, been provided. The nourricerie seems to be a particularly well-conducted institution. It answers all the requirements of a well-conducted crêche, an institution existing in many manufacturing centers of France and Germany, but almost unknown in this country. Mothers, on going to work in the morning, leave their babies here, where they are fed, bathed, put to sleep or amused until the evening, when the mothers return to take them away. The following table shows the expenditure of the society each year for the care and education of children of members:

EXPENDITURES OF THE FAMILISTÈRE SOCIETY FOR THE CARE AND EDUCATION OF CHILDREN AT GUISE, 1879-80 TO 1894-95.

| Year. | Nursery. | Infant schools. | Primary schools. | Total. | Year. | Nursery. | Infant schools. | Primary schools. | Total. |
|---|---|---|---|---|---|---|---|---|---|
| 1879-80 .. | $1,351.18 | $516.40 | $1,588.51 | a $3,680.86 | 1887-88 . | $1,334.12 | $1,306.48 | $3,240.18 | $5,880.78 |
| 1880-81 .. | 1,334.25 | 473.85 | 1,970.02 | a 4,103.73 | 1888-89 . | 1,375.38 | 1,357.49 | 2,974.94 | 5,707.81 |
| 1881-82 .. | 1,351.89 | 630.13 | 2,834.83 | a 4,822.89 | 1889-90 . | 1,263.89 | 1,384.23 | 3,214.67 | 5,862.79 |
| 1882-83 .. | 1,393.50 | 767.05 | 4,475.84 | 6,636.00 | 1890-91 . | 1,067.43 | 1,586.95 | 3,317.29 | 5,971.67 |
| 1883-84 .. | 1,268.82 | 567.21 | 4,648.47 | 6,484.50 | 1891-92 . | 1,136.94 | 1,513.75 | 3,355.59 | 6,006.28 |
| 1884-85 .. | 1,513.20 | 709.99 | 3,938.33 | 6,161.52 | 1892-93 . | 1,280.72 | 1,444.99 | [illegible] | [illegible] |
| 1885-86 .. | 1,641.67 | 842.61 | 3,580.01 | 6,064.29 | 1893-94 . | 1,090.74 | 1,390.30 | 3,321.96 | 5,812.00 |
| 1886-87 .. | 1,308.45 | 863.49 | 3,530.31 | 5,702.25 | 1894-95 . | 1,179.78 | 1,175.59 | 3,494.97 | 5,850.34 |

a This total is greater than the sum of the items. The explanation is not known. The figures are given as published by the society.

## COOPERATIVE DISTRIBUTION.

The creation of the first store for the sale of necessary articles of consumption to employees of the establishment dates from the foundation of the first familistère in 1859. Previous to 1880 the store remained but a branch of the general enterprise of M. Godin, and whatever profits were realized were retained by him. On the organization of the society the principle of cooperation was introduced in this department as well as in the conduct of the industry proper. Though an independent cooperative society was not created, the service of distribution was made a distinct branch of the general undertaking, and, as its accounts have been kept separate from those relating to the industrial branch, its operations can be studied apart. Since the foundation of the society this service of distribution has not ceased to prosper. New departments have from time to time been added until, at the present time, there can be purchased through it every article necessary for the ordinary life of the members of the society.

As at present organized, the service of distribution includes two departments: (1) That of production and (2) that of sales. The first is divided into two sections, viz, a bakery and a charcuterie (an establishment for the preparation of hog products). The charcuterie was in existence prior to the organization of the society. The bakery was established in 1886. It is equipped with two furnaces or ovens and other necessary appliances, affording a productive capacity of from 700 to 800 kilograms (1,543 to 1,764 pounds) of bread per day. The flour used is first and second class, mixed in equal proportions. The price of bread is regulated by the minimum price charged by independent bakers of the village of Guise. The average annual profit realized is about 3.81 per cent. This is exclusive of the profits that may be realized by the sale of the bread through the sales department, as the bread is not sold at the bakery, but is transferred to the sales department. Previous to 1887 a dairy had been maintained by the society for supplying milk to members. It was, however, abandoned in that year, and milk is now purchased at wholesale rates from a neighboring dairyman and retailed through the sales department.

The sales department is divided into the following sections:

1. Dry goods and miscellaneous, including such categories as dry goods proper, ready-made articles of wearing apparel, hats, shoes, furniture, stationery, watches and jewelry, etc.
2. Groceries, including groceries proper, kitchen utensils, beer, wine, and other liquors.
3. Food products, such as meats, fruits, vegetables, etc.
4. Fuel.
5. Administration of the bath and wash rooms.
6. Administration of the tavern, casino, etc.
7. Auxiliary service for the sale of liquors wherever permitted.

The sale of furniture and of watches and jewelry was commenced in 1889.

The following tables, giving the total value of products and of articles of each class sold during each year since the foundation of the society, will show the progress in importance of the Cooperative Distributive Association:

VALUE OF PRODUCTS OF THE COOPERATIVE DISTRIBUTIVE ASSOCIATION OF THE FAMILISTÈRE SOCIETY OF GUISE, 1879-80 TO 1894-95.

| Year. | Charcuterie. | Dairy. | Bakery. | Total. | Year. | Charcuterie. | Dairy. | Bakery. | Total |
|---|---|---|---|---|---|---|---|---|---|
| 1879-80 .. | $5,010.58 | $799.21 | .......... | $5,809.79 | 1887-88 . | $6,033.32 | .......... | $7,280.55 | $13,313.87 |
| 1880-81 .. | 4,940.50 | 749.69 | .......... | 5,690.19 | 1888-89 . | 7,223.69 | .......... | 8,270.83 | 15,494.52 |
| 1881-82 .. | 5,424.08 | 749.04 | .......... | 6,173.12 | 1889-90 . | 9,347.95 | .......... | 9,607.84 | 18,955.79 |
| 1882-83 .. | 6,376.25 | 574.37 | .......... | 6,950.62 | 1890-91 . | 10,624.82 | .......... | 14,114.05 | 24,738.87 |
| 1883-84 .. | 5,822.02 | 520.58 | .......... | 6,342.60 | 1891-92 . | 10,945.76 | .......... | 17,525.37 | 28,471.13 |
| 1884-85 .. | 6,033.84 | 616.86 | .......... | 6,650.70 | 1892-93 . | 11,421.78 | .......... | 15,637.68 | 27,059.46 |
| 1885-86 .. | 6,532.13 | 644.13 | .......... | 7,176.26 | 1893-94 | 11,527.08 | .......... | 15,120.68 | 26,647.76 |
| 1886-87 .. | 6,278.73 | 547.82 | $6,991.78 | 13,818.33 | 1894-95 . | 12,425.34 | .......... | 14,670.90 | 27,096.24 |

INCOME FROM SALES, ETC., OF THE COOPERATIVE DISTRIBUTIVE ASSOCIATION OF THE FAMILISTÈRE SOCIETY OF GUISE, 1879–80 TO 1894–95.

| Year. | Wearing apparel. | Groceries. | Green groceries, meats, fruits, etc. | Restaurant and casino. | Beer consumed at home. | Fuel. | Baths and wash houses. | Total. |
|---|---|---|---|---|---|---|---|---|
| 1879–80... | $9,282.56 | $38,147.29 | $14,752.14 | $8,641.80 | .......... | $6,330.65 | $757.21 | [illegible] |
| 1880–81... | 10,001.06 | 41,475.28 | 15,036.94 | 9,579.91 | .......... | 6,194.51 | 812.29 | [illegible] |
| 1881–82... | 11,420.68 | 40,890.85 | 15,515.70 | 9,830.14 | .......... | 6,263.48 | 875.05 | [illegible] |
| 1882–83... | 12,339.70 | 38,087.44 | 17,060.01 | 7,810.56 | .......... | 6,377.82 | 912.52 | [illegible] |
| 1883–84... | 13,080.86 | 34,569.16 | 15,859.94 | 6,714.29 | .......... | 5,472.01 | 954.71 | [illegible] |
| 1884–85... | 15,467.90 | 39,213.80 | 15,651.79 | [illegible].410.39 | .......... | 5,270.13 | 1,084.43 | [illegible] |
| 1885–86... | 15,846.31 | 39,636.40 | 16,927.03 | 6,872.09 | .......... | 5,940.51 | 1,132.96 | [illegible] |
| 1886–87... | 15,213.30 | 40,002.62 | 16,448.22 | 7,079.66 | $196.24 | 6,456.48 | 1,236.60 | [illegible] |
| 1887–88... | 15,969.71 | 41,901.70 | 16,549.27 | 5,936.38 | 637.24 | 7,345.72 | 1,576.06 | [illegible] |
| 1888–89... | 24,004.03 | 49,834.90 | 21,061.41 | 6,962.91 | 1,002.77 | 7,675.84 | 1,600.91 | [illegible] |
| 1889–90... | 28,769.57 | 57,077.13 | 26,155.10 | 6,303.77 | 1,929.12 | 8,098.40 | 1,630.72 | 131,072.91 |
| 1890–91... | 33,214.59 | 70,068.18 | 29,317.94 | 7,639.10 | 2,366.53 | 12,809.90 | 1,505.64 | a[illegible] |
| 1891–92... | 34,404.34 | 79,450.50 | 29,289.79 | 7,905.02 | 2,896.90 | 13,331.44 | 1,588.24 | [illegible] |
| 1892–93... | 33,796.51 | 76,941.57 | 28,755.10 | 6,831.20 | 2,986.68 | 12,066.94 | 1,676.50 | [illegible] |
| 1893–94... | 33,307.58 | 76,430.66 | 29,515.56 | 6,841.39 | 3,213.23 | 11,799.55 | 1,484.59 | [illegible] |
| 1894–95... | 34,348.40 | 77,629.28 | 31,509.46 | 6,237.07 | 3,170.48 | 12,423.86 | 1,377.80 | [illegible] |

a This total is greater than the sum of the items. The explanation is not known. The figures are given as published by the society.

The administration of the service of distribution is under the general control of the board of directors and the familistère committee. A general director of the service is appointed, who is the executive head of this branch of the industry. The personnel required for the service numbers between fifty and sixty persons, the greater part of whom are women. With rare exceptions, purchases are made directly from producers, without the intervention of middlemen. For the supply of certain articles the society is affiliated with the Federation of the Cooperative Societies of France.

The stores are for the most part located in rooms on the ground floor of the main familistère. The stores for the green groceries, meats, and fruits, the tavern, etc., are, however, installed in separate buildings. Every member of the society, every employee of the establishment, or even persons utter strangers to the society, have free access to the stores and can participate equally in the profits. After all expenses have been paid the net profits remaining are divided among the purchasers in proportion to the value of their purchases during the year. Prior to 1881–82 these profits were distributed in cash. In that year the system was changed so that 50 per cent of the profits were distributed in cash and the remainder as a credit for the future purchase of articles. This was done in order to encourage members to make their purchases at the stores of the society. In 1888–89 the amount distributed as a credit for future purchases was raised to 85 per cent and only the remaining 15 per cent distributed in cash.

In order to participate in profits purchasers must secure a pass book, in which all purchases are entered. Purchases must be paid for in cash. The financial results of the operations of this branch of the general industry are shown in the following table. In explanation it should be said that "capital invested" represents only the value of the stock carried, and that the rent of the rooms occupied is included in expenses.

FINANCIAL OPERATIONS OF THE COOPERATIVE DISTRIBUTIVE ASSOCIATION OF THE FAMILISTÈRE SOCIETY OF GUISE, 1879-80 TO 1894-95.

| Year. | Capital invested. | Total sales. | Profits. | | | Per cent of profits of— | |
|---|---|---|---|---|---|---|---|
| | | | Distributed in cash. | Distributed in purchases. | Total. | Capital invested. | Total sales. |
| 1879-80 | $17,325.88 | $77,911.65 | $7,074.61 | .......... | $7,074.61 | 40.83 | 9.08 |
| 1880-81 | 13,071.77 | 83,100.09 | 7,531.62 | .......... | 7,531.62 | 57.62 | 9.06 |
| 1881-82 | 12,409.63 | 84,791.90 | 7,903.00 | $396.00 | 8,299.00 | 66.86 | 9.79 |
| 1882-83 | 12,698.61 | 82,587.55 | 5,708.05 | 1,836.26 | 7,544.31 | 59.41 | 9.13 |
| 1883-84 | 12,410.32 | 76,070.97 | 5,385.42 | 2,719.92 | 8,105.34 | 65.31 | 10.57 |
| 1884-85 | 12,112.88 | 83,118.52 | 5,901.81 | 3,699.98 | 9,601.79 | 79.22 | 11.55 |
| 1885-86 | 13,750.09 | 86,384.30 | 6,291.76 | 3,965.80 | 10,297.56 | 74.89 | 11.92 |
| 1886-87 | 16,530.72 | 87,236.12 | 6,418.93 | 4,003.44 | 10,422.37 | 63.05 | 11.95 |
| 1887-88 | 15,275.02 | 89,986.08 | 6,211.72 | 4,030.40 | 10,242.12 | 67.05 | 11.39 |
| 1888-89 | 16,973.19 | 112,041.77 | 3,972.09 | 9,752.06 | 13,725.15 | 80.86 | 12.25 |
| 1889-90 | 16,285.03 | 131,073.81 | 4,650.21 | 13,247.35 | 17,807.66 | 109.30 | 13.65 |
| 1890-91 | 18,734.47 | 157,065.82 | 5,396.43 | 16,330.66 | 21,727.09 | 115.97 | 13.83 |
| 1891-92 | 22,825.34 | 168,465.29 | 5,723.97 | 18,355.75 | 24,079.72 | 105.50 | 14.28 |
| 1892-93 | 26,142.06 | 163,068.56 | 5,017.00 | 16,967.80 | 21,964.80 | 84.10 | 13.48 |
| 1893-94 | 25,140.99 | 162,612.55 | 5,327.51 | 18,324.06 | 23,651.57 | 94.08 | 14.54 |
| 1894-95 | 25,171.02 | 166,761.85 | 5,255.28 | 18,681.37 | 23,936.65 | 95.10 | 14.35 |

As has been stated above, only holders of pass books in which purchases are entered can participate in profits. Other sales are treated as are purchases in private stores. The following table gives interesting details concerning the extent to which the stores are patronized, the amount purchased on pass books, the total number of pass books in use each year, the average value of sales, and the average amount of profits realized per pass book:

SALES AND PROFITS TO HOLDERS OF PASS BOOKS OF THE COOPERATIVE DISTRIBUTIVE ASSOCIATION OF THE FAMILISTÈRE SOCIETY OF GUISE, 1879-80 TO 1894-95.

| Year. | Number of pass books. | Sales on pass books. | Net profits. | Per cent of profits of sales on pass books. | Average yearly sales per— | | Average profits distributed per pass book. |
|---|---|---|---|---|---|---|---|
| | | | | | Pass book. | Family. | |
| 1879-80 | 60 | $4,023.59 | $7,074.61 | 175.83 | $67.06 | $96.34 | $117.91 |
| 1880-81 | 88 | 5,945.01 | 7,531.62 | 126.67 | 67.57 | 74.57 | 85.59 |
| 1881-82 | 110 | 7,643.08 | 8,299.00 | 108.58 | 69.48 | 72.77 | 75.45 |
| 1882-83 | 594 | 39,516.48 | 7,544.31 | 19.09 | 66.53 | 78.04 | 12.70 |
| 1883-84 | 699 | 54,363.38 | 8,105.34 | 14.91 | 77.77 | 83.30 | 11.60 |
| 1884-85 | 765 | 62,790.11 | 9,601.79 | 15.29 | 82.08 | 79.95 | 12.55 |
| 1885-86 | 819 | 66,018.60 | 10,297.56 | 15.60 | 80.61 | 88.54 | 12.57 |
| 1886-87 | 808 | 65,744.32 | 10,422.37 | 15.85 | 81.37 | 100.18 | 12.90 |
| 1887-88 | 864 | 68,324.88 | 10,242.12 | 14.99 | 79.08 | 95.42 | 11.85 |
| 1888-89 | 1,004 | 92,681.24 | 13,725.15 | 14.81 | 92.31 | 108.67 | 13.67 |
| 1889-90 | 1,128 | 114,201.34 | 17,807.66 | 15.67 | 101.24 | 121.42 | 15.87 |
| 1890-91 | 1,335 | 138,719.58 | 21,727.09 | 15.66 | 103.91 | 123.66 | 16.27 |
| 1891-92 | 1,400 | 151,700.41 | 24,079.72 | 15.87 | 108.36 | 130.31 | 17.20 |
| 1892-93 | 1,605 | 147,785.28 | 21,964.80 | 14.88 | 92.08 | .......... | 13.70 |
| 1893-94 | 1,630 | 147,937.71 | 23,651.57 | 15.99 | 90.76 | .......... | 14.51 |
| 1894-95 | 1,691 | 153,172.18 | 23,936.65 | 15.63 | 90.58 | .......... | 14.16 |

The steady progression, not only in the number of pass books that are in use but in the average amount purchased on each book and by each family, shows, in a striking way, the constantly increasing extent to which the benefits of cooperation in the purchasing of supplies have been utilized. The increase is not due to the increased number of members of the society, for that has been but slight during recent years. The steady increase in the amount purchased by each family, though undoubtedly due in large part to the fact that fewer purchases are made elsewhere, would tend strongly to indicate the increasing material welfare of the members.

## COOPERATIVE DISTRIBUTION.

BY EDWARD W. BEMIS, PH. D.

Many forms of cooperation are very little developed in the United States. We have none of the credit associations so common in Germany, nor the raw-material societies for the cooperative purchase of the raw materials used in small manufacture, also common in Germany. We do not have the cooperative labor gangs or societies for the collective undertaking of contracts for public and private work which are common in Russia and Italy, while our cooperative manufacturing is insignificant compared with the beginnings in Great Britain and France. The cooperative cooper shops in Minneapolis are only moderately successful. Of the eight shops existing in 1886 only the following four survive. Sharing in the general depression, none of these have been able to pay any dividends during the past three years other than 5 per cent interest on the capital in the case of the two older companies and 6 per cent in the case of the two others.

The Cooperative Barrel Manufacturing Company, started in 1874 with a small paid-up capital and a membership of less than a score, had a membership of 120 in 1885 and a capital of $50,000 in 1888. It now does a business of about $150,000 a year.

The North Star Barrel Company, started in 1877, is now doing a business of about $195,000 a year on a capital of $43,350. The membership was 77 in 1885, is 51 now, and will soon be reduced to 45. The capital, however, in 1885, was only $30,800.

The Northwestern Barrel Company, started in 1881, does a business of about $100,000 on a capital of $18,500. The capital has increased somewhat since 1886, but the membership has declined from 45 to 27.

The Hennepin Barrel Company, started in 1880, now does a business of about $275,000 a year on a capital of $47,200, and has a membership of 59. The capital was $38,000 in 1886, and the membership 52.

The introduction of machinery has led to less need of skilled coopers, and, therefore, many have withdrawn. All the members work in [illegible] own shops, and each of the latter employs from twenty to thirty [illegible] tenders and boys. The cooperative shops could do all the work [illegible] Minneapolis mills, but the latter refuse to give it all to them [illegible] lower prices. There may be a fear of combination if the [illegible] shops once secured all the trade. In the interesting [illegible] movement in the first report of [illegible] shops can not do so. When wages fall, the cooperative workers can stand it better than the others. Attempts to establish pooling arrangements between all the shops have always failed, sometimes through the refusal of a cooperative shop to unite, and sometimes through the breaking of the agreement by some private shop. The following very significant statement, taken from that report, deserves insertion, as it applies equally to many illustrations of cooperative distribution: "It may be worth while to remark that cooperation is not a religion with these coopers. They are not experimenting for the benefit of humanity. One of them might withdraw with his savings and set himself up as the proprietor of a boss shop without the slightest twinge of conscience or the remotest chance of being charged with the sin of apostacy. The president of one of the smaller shops had formerly been a member of one or two of the older and larger establishments, and withdrew to found a shop of his own in another town. He failed in the business for some reason, and came back to cooperation in Minneapolis."

Mr. William Angus, a graduate student of the University of Minnesota, who has supplied many of the facts in this statement, writes: "The cooperatives do not pay themselves quite as high wages during the busy season as the private companies pay their men; but the wages amount to more by the year, as the private companies generally rush business for a time and then close up entirely. The wages of nonstockowning employees are in every case exactly the same as those of stockowning employees for doing the same work; but in very few cases are the nonstockowning employed at the same work. They are generally busied with running the machinery, engines, etc., rough out-of-door work, and, of course, here do not receive as much pay."

Outside of these associations scarcely anything of the kind exists in America. The few small cooperative coal mines in Illinois are said by the State mine inspectors to have a bad effect on wages in their neighborhood by their readiness to sell coal at any price when trade is dull. The so-called cooperative furniture factories of Rockford, Ill., are really joint stock companies, with small shares, widely scattered among employees. Disaster has recently overtaken many of those at Rockford. Where every stockholder has an equal vote there is some tendency to keep up wages, even ruinously, at the expense of profits.

On the other hand, our farmers have made as great strides as those of Holland, Denmark, France, or Germany in the matter of cooperative creameries, and fire and tornado insurance companies. Our many very large fraternal life insurance companies are also cooperative. Our cooperative banks, or building and loan associations, have already been investigated by the Department. (a)

The present study, however, deals with another great branch of the cooperative movement which, beginning in England over fifty years

---

a Ninth Annual Report of the Commissioner of Labor, 1894.

ago, has still its greatest development there, but is now able to boast of a large growth in all the countries of Western Europe and in Italy. (a)

Cooperative distribution, the term commonly applied to the work of consumers' societies, is almost entirely confined, in many States, to the simple form of securing trade discounts for the members of farmers' organizations who concentrate their purchases of farm machinery and supplies and household goods in selected stores and factories. Such methods, which are most largely developed in France, are common with the Patrons of Husbandry, sometimes called the Grange. This organization, starting in 1866, now numbers about 250,000 members, pretty generally scattered through the northern and western States. The secretaries of the State granges of such widely separated States as Oregon, Nebraska, Ohio, Connecticut, and Rhode Island report extensive buying of this kind.

In Ohio, from 25 to 33 per cent is saved by buying for cash and concentrating trade with 45 business houses of manufacturers and jobbers. The Rhode Island Grange publishes a list of 14 Providence and 10 Newport houses that give discounts ranging from 5 per cent on gloves, hosiery, millinery, and harnesses to 15 per cent on watches and jewelry and 20 per cent at a restaurant. Through purchasing agents, also, implements, seeds, and fertilizers have been obtained in large and small quantities at a great saving and grain has been purchased by several granges in car-load lots with equally satisfactory results. In Connecticut, the State purchasing agent ships grain, coal, etc., to the granges, where they are divided among the members. The Watertown grange has its own storehouse and a spur track running to it. The Wallingford grange does a business of $18,000 a year.

The supreme secretary of the Patrons of Industry, another farmers' organization in the northern States, west of New England, writes that their local associations, in many cases, establish cooperative stores. "In other cases they contract with the local dealer for their supplies, while in other cases they buy direct from the manufacturer. The amount of goods handled by the organization as such has been enormous. The saving has been, at a fair estimate, at least 10 per cent on all purchases, the expense of handling also being very light." Another organization, the Farmers' Alliance, which has been strongest in the South and Southwest, has pursued similar methods. The Alliance papers of North Carolina and other States publish a whole page list of articles that can be obtained at designated [illegible] through the State business agency. In these and other ways [illegible]nations to charge unreasonably high prices for binding twine, [illegible]zers, barbed wire, and other farm supplies have been broken or [illegible] to be more moderate in their demands.

## THE COOPERATIVE STORE.

That form of cooperative distribution or consumers' societies, known as the cooperative store, had its American beginnings with the so called union stores in New England from 1847 to 1859. Limiting dividends and selling a little above cost, these stores either failed, or were transformed into private enterprises. None survive; yet 769 of these stores were started, and 350 of them, mostly in New England, reported in 1857 a capital of $291,000 and an annual trade of $2,000,000.

The next important effort was made by the Patrons of Husbandry, organized in 1866. All their early grange stores seem to have followed the methods of the union stores and to have met with a similar fate.

In 1864 the Rochdale methods of cooperative storekeeping were introduced in a Philadelphia store by twenty-three members who had secured from Rochdale, England, the constitution and other documents of the famous Rochdale Pioneers. In the second quarter of 1866 the sales were $7,751.34, and three branches were too hastily established. The undue ratio of expense to trade, and especially the lack of interest in the movement, led to its speedy failure in November. The oldest cooperative store in this country, at the time of its failure in 1896, was the Danvers Cooperative Union Society, with a capital of $5,500 in the shoe manufacturing town of Danvers, Mass. At first, however, from 1865 to 1869, it sold goods exclusively to its members and at cost, after the methods of the other union stores already referred to. It did not adopt the Rochdale plan until 1869. The failure is ascribed to incompetent agents, and was so disastrous as to leave the stockholders only 25 per cent of their investment.

The longest successful American trial of the Rochdale plan seems to have been by the Cooperative Store Company at Silver Lake, in the town of Kingston, Mass., which began June 14, 1875. A small store in a small place of nearly stationary population and with a trade of only $9,517.92 and a capital of $1,850 in 1895–96, or nearly the same as in 1886, its continued success under one manager seems to prove the presence there of what has been found far more important than even the Rochdale methods—a cooperative spirit, which is thus defined in the copies of the by-laws as printed in every edition for many years: "A true coopcrator has three qualities—good sense, good temper, and good will—good sense to dispose him to make the most of his means; good temper to enable him to associate with others, and good will to incline him to serve them and be at trouble to serve them, and go on serving them, whether they are grateful or not in return, caring only that he does good, and finding it a sufficient reward to see that others are benefited through his unthanked exertions."

This last enterprise, like eight or nine others still in successful operation, owed its origin to the third wave of cooperative enthusiasm which swept over New England and a few other sections in the seventies, and which was chiefly fostered at that time by the Sovereigns of Industry during its brief history from 1874 to 1880; but the Rochdale methods, then popularized, were at once taken up by a few grange stores, such as the Johnson County Cooperative Association at Olathe, Kans. During two years the Sovereigns of Industry kept two paid lecturers in the field, who devoted much of their time to instructing the people in cooperation.

In 1877, reports from 94 councils, mostly in New England, New York, New Jersey, Pennsylvania, and Ohio, showed an average membership of 77, an average capital of $884, and a total trade of $1,089,372, on which the consumers were estimated to have saved 14 per cent through cooperation. Perhaps one-half of these stores sold at market prices and returned the profits to the consumers as dividends on their trade, as in the Rochdale system.

The next epoch in the cooperative movement came with the rapid rise of the Knights of Labor, during the years 1884 to 1888. Scores of cooperative workshops, coal mines, and factories were started all over the country, without any connection with cooperative stores or knowledge of cooperative methods elsewhere. Most of these experiments failed. The few successful ones were transformed into joint-stock or private enterprises.

The same fate has befallen the numerous cooperative stores started by the farmers' associations known as the Wheel and the Alliance in the southern States from 1886 to 1892.

In 1886 a fairly exhaustive investigation of cooperation was made by five graduates of the Johns Hopkins University.(*a*) Not including the partially cooperative enterprises among the Mormons and at Allegan, Mich., to be considered later, these investigators found 30 stores outside of New England. Of these, 17, for the most part probably small ones, made no report of their business then, and seem to be extinct now or transformed into private enterprises. Seven others that reported a trade of $357,673.78 in 1886 have also disappeared. Of these seven the only ones that did a business of over $32,000 were the Philadelphia Industrial Cooperative Society, which had a trade of $168,816.54, and the Trenton Cooperative Business Association, with a trade of about $72,000.

Of 27 cooperative associations that have started in New Jersey since 1873, only 8 are now running, and in Texas cooperation has very greatly declined.(*b*)

---

*a* History of Cooperation in the United States. Volume VI of the Johns Hopkins University Studies in Historical and Political Science. By Edward W. Bemis, Albert Shaw, Amos G. Warner, Charles Howard Shinn, and Daniel R. Randall. Baltimore, 1888.

*b* For information on the New Jersey societies, prior to 1895, see the [illegible] of the New Jersey Bureau of Labor Statistics for 1895, Part V. For the [illegible], and Philadelphia societies, see How to Cooperate, by Herbert [illegible] *Company*, New York, 1892, pp. 78–83, 97–105, 121–126.

On the other hand, there are some favorable reports. The Trenton (N. J.) Cooperative Society, the Johnson County Cooperative Association, at Olathe, Kans., and the Hammonton (N. J.) Fruit Growers' Union had a trade of $49,958.20, $210,588.79, and $45,940.45, respectively, in 1886, and of $51,300, $231,141.63, and $61,427.43 in 1895–96. Sixteen associations which have started since 1886 in 15 places, viz, Pittsburg, Pa.; Dover, Raritan, and Phillipsburg, N. J.; Jamestown, N. Y.; Ishpeming, Mich.; Zumbrota, Minn.; Eureka, Emporia, Overbrook, Wakefield, and Green, Kans., and Los Angeles, Santa Paula, and Poplar, Cal., had a trade in 1895–96 of about $450,000.

The total cooperative trade outside of New England, so far as reported, was about $900,000 in 1895, as contrasted with about $1,000,000 in the associations making even partial returns in 1886. Although it is believed that no large societies have been overlooked, it is quite possible that as complete a survey of the field as was made in 1886 might reveal a small growth in the cooperative trade outside of New England during the past ten years. Nearly all the associations that have been reached in this inquiry give dividends on trade, but in Kansas these dividends are very small at present, owing to the agricultural depression there which is causing the failure of many private stores.

In New England the outlook is more encouraging. While 6 of the stores that had a trade of $134,000 in 1886 are now closed, the trade of the remaining 13 of those in existence in the former period has grown from $479,900 to $978,951.48, and 9 new stores report a trade of $251,409.49. The total cooperative trade in New England, almost entirely on the Rochdale plan, is thus over twice as great as ten years ago. To the figures for the western and southern States should be added 134 labor exchanges, with 6,000 members, pursuing methods to be described later.

The stores following the Rochdale plan, so thoroughly tested by experience, manage business in substantially the following way:

1. Small shares of stock are issued, usually limited to $5. One can, in most stores, become a member by paying only $2 of this, in addition to an initiation fee or entrance dues of 50 cents, but in that case the new member must let his profits remain in the business until 1 and sometimes 5 or more shares are paid for. Most associations will buy back the shares of withdrawing members at par or permit their transfer to another. The former method is preferred by some associations, in order to prevent possible sacrifice of stock below par by needy and ignorant members.

2. The number of shares one can hold is usually limited to 100 and sometimes to 40.

3. Each stockholder has but one vote.

4. Goods are sold at the market price. The old method of selling at cost has been found weak in two respects. It often leads to sales below cost, because of miscalculation as to expenses and depreci[illegible]

Then the attempt to sell at cost arouses the antagonism of private traders, who will sell staple articles below cost as an advertisement, and thus draw away the more ignorant and undeveloped cooperators. Even though the attempt be made, however, to sell at the market price, that price is often somewhat reduced in a place where there is a strong cooperative association, because of the efforts of private traders to retain their business.

5. Stockholders receive only a fixed rate of interest. In the East the average rate is 6 per cent and in the rest of the country it is 7 per cent.

6. Surplus earnings are divided among the customers according to their purchases. Usually stockholders receive twice as large a percentage as other customers.

7. Advertising, expensive rent, and some expenses of clerk hire and delivery of light goods are supposed to be saved by the presence of the cooperative spirit. In many American cooperative stores, however, orders are taken and goods delivered as in private stores.

The two tables following show the rate per cent of dividends returned to members on their trade for the year 1895–96—the first in 20 New England societies, the second in 13 societies outside of New England.

In 1895 the New England societies paid 6 per cent on their share capital, with the following exceptions in Massachusetts: The Arlington, at Lawrence, which paid 5 per cent; the Industrial, at New Bedford, which paid 5; the Hampden County, at Springfield, which changed to 6 in January, 1896; the Lawrence Equitable, which paid 5; the Lowell, which paid 4; the German Association, at Lawrence, which paid 7; the Plymouth Rock Cooperative Company, at Plymouth, which paid 8, and the Harvard, at Cambridge, and the Beverly associations, which, as usual, did not attempt to pay anything. Of the societies outside of New England those at Santa Paula, Cal., and Olathe and Wakefield, Kans., paid 8, 9, and 10 per cent interest, respectively, on their capital. Eight per cent was also paid at Ishpeming, Mich., and Green, Kans., 7 at Overbrook, Kans., and 6 at Cadmus, Kans., Trenton, N. J., and Jamestown, N. Y. Five was paid elsewhere, save at Los Angeles, Cal., and Emporia and Eureka, Kans., where nothing was paid in 1895, although 10 per cent has usually been paid at Eureka and 6 to 8 at Emporia.

DIVIDENDS ON MEMBERS' TRADE IN TWENTY NEW ENGLAND ROCHDALE SOCIETIES, 1895-96.

| Name of society. | Location. | Dividends (per cent). |
| --- | --- | --- |
| Lewiston Cooperative Society (a) | Lewiston, Me | 5.0 |
| Lisbon Falls Cooperative Association | Lisbon Falls, Me | 8.0 |
| Sabattus Cooperative Association | Sabattus, Me | 12.0 |
| Farmers' and Mechanics' Exchange | Brattleboro, Vt | 5.0 |
| Arlington Cooperative Association | Lawrence, Mass | 7.0 |
| Beverly Cooperative Association | Beverly, Mass | 5.0 |
| Cambridge Cooperative Society | Cambridge, Mass | 7.5 |
| Cooperative Store Company | Silver Lake, Kingston, Mass | 5.5 |
| First Swedish Cooperative Store Company | Quinsigamond, Worcester, Mass | 10.0 |
| German Cooperative Association | Lawrence, Mass | 7.0 |
| Hampden County Cooperative Association | Springfield, Mass | 3.0 |
| Harvard Cooperative Society (a) | Cambridge, Mass | 7.0 |
| Industrial Cooperative Association | New Bedford, Mass | 2.0 |
| Knights of Labor Cooperative Boot and Shoe Association. | Worcester, Mass | 10.0 |
| Lawrence Equitable Cooperative Society | Lawrence, Mass | 7.0 |
| Lowell Cooperative Association | Lowell, Mass | 10.5 |
| Plymouth Rock Cooperative Company | Plymouth Mass | 6.0 |
| Riverside Cooperative Association | Maynard, Mass | 5.5 |
| West Warren Cooperative Association | Wes. Warren, Mass | 7.5 |
| Woodlawn Cooperative Association | Pawtucket, R. I | 2.8 |
| Average dividend | | 6.8 |

a This society differs from the Rochdale plan in that it pays no interest on stock.

DIVIDENDS ON MEMBERS' TRADE IN THIRTEEN ROCHDALE SOCIETIES OUTSIDE OF NEW ENGLAND, 1895-96.

| Name of society. | Location. | Dividends (per cent). |
| --- | --- | --- |
| Trenton Cooperative Society | Trenton, N. J | 8.0 |
| Jamestown Cooperative Supply Company | Jamestown, N. Y | 6.5 |
| Ishpeming Cooperative Society | Ishpeming, Mich | 6.0 |
| Alliance Cooperative Association | Green, Kans | |
| Greenwood County Cooperative Association | Eureka, Kans | |
| Johnson County Cooperative Association | Olathe, Kans | 4.0 |
| Lyon County Alliance Exchange Company | Emporia, Kans | 2.0 |
| Osage County Cooperative Association | Overbrook, Kans | |
| Patrons' Cooperative Association | Cadmus, Kans | 7.0 |
| Wakefield Alliance Cooperative Association | Wakefield, Kans | 2.0 |
| Poplar Cooperative Association | Poplar, Cal | |
| Santa Paula Cooperative Association | Santa Paula, Cal | 3.0 |
| Socialists' Cooperative Store and Productive Association. | Los Angeles, Cal | 7.0 |
| Average dividend | | 3.5 |

The average rate of dividend on the trade of the members in 20 New England Rochdale stores in 1895 was 6.8 per cent, and in 13 outside of New England 3.5 per cent. According to the returns from 1,036,992 English cooperators, as published in the English Labor Gazette for June, 1896, only 1.3 per cent of the members received 5 per cent or less. while 14.2 per cent received from 5 to 10 per cent, 54 per cent of the members received from 10 to 15 per cent, and the remaining 30.5 per cent received over 15 per cent. In the 33 American associations just referred to, only 8 received over 7 per cent, and 4 of these over 9, the highest dividend of 12 per cent being given by an association with $31,000 trade, at Sabattus, Me.

The larger dividends in the English associations may be partly to a larger excess of retail over wholesale prices [illegible] in part, at least, due to the larger expenses here. The [illegible] ble and the Oldham Industrial cooperative societies do [illegible] for goods, and deliver comparatively few. Hence, they did [illegible] 1895 with an expense of only 4.4 per cent and 6.6 per [illegible] trade, respectively, and declared dividends on their [illegible] cent. The Manchester Equitable Society, on the other [illegible] ing orders and delivering goods, had an expense of 10 per [illegible] trade, and so could pay only 10 per cent dividends on the [illegible] members. In all of the 1,486 cooperative societies of Great [illegible] reported in the proceedings of the Twenty-eighth Annual [illegible] Congress of that country, the expenses in 1895, aside from [illegible] and depreciation, were 6.2 per cent of the total business of [illegible] ($166,555,062.20).

The expenses of the 15 New England societies from which returns on this head were obtained averaged 10.9 per cent of their trade of $1,008,977.24. If the Harvard society, which does very little delivering to its college members, and the Sovereigns' Trading Company of New Britain, Conn., be omitted from the list, the other 13 will show an average expense of 13.3 per cent of their trade of $723,825.83. Similarly the 13 societies from which returns were obtained, outside of New England, had an expense in 1895 of 9.6 per cent on their trade of $577,368.16.

The two tables which follow give for 1895 the amount of sales, expenses, the per cent of expenses of sales, the number of employees, and the trade per employee for the New England societies and those outside of New England, respectively.

SALES, EXPENSES, AND NUMBER OF EMPLOYEES OF FIFTEEN ROCHDALE SOCIETIES IN NEW ENGLAND, 1895–96.

| Name of society. | Expenses. | Sales. | Per cent of expenses of sales. | Employees. | Trade per employee. |
|---|---|---|---|---|---|
| Sovereigns' Trading Company (a) | $6,000.00 | $150,000.00 | 4.0 | [illegible] | [illegible] |
| Lisbon Falls Cooperative Association | 4,417.61 | 53,967.19 | 8.2 | 6 | [illegible] |
| Sabattus Cooperative Association | 2,500.00 | 31,000.00 | 8.1 | 3 | [illegible] |
| Farmers' and Mechanics' Exchange | 10,045.50 | 67,634.87 | 14.9 | [illegible] | [illegible] |
| Arlington Cooperative Association | [illegible] | [illegible] | [illegible] | [illegible] | [illegible] |
| Beverly Cooperative Association | 6,545.42 | [illegible] | 10.3 | [illegible] | [illegible] |
| Cooperative Store Company | 953.87 | 9,517.98 | 10.2 | [illegible] | [illegible] |
| German Cooperative Association | 2,124.00 | [illegible] | 5.9 | [illegible] | [illegible] |
| Hampden County Cooperative Association | [illegible] | 30,000.00 | 15.7 | [illegible] | [illegible] |
| Harvard Cooperative Society (b) | [illegible] | 136,151.41 | 6.2 | [illegible] | [illegible] |
| Knights of Labor Cooperative Boot and Shoe Association | [illegible] | 15,417.85 | [illegible] | [illegible] | [illegible] |
| Lawrence Equitable Cooperative Society | [illegible] | [illegible] | [illegible] | [illegible] | [illegible] |
| Lowell Cooperative Association | [illegible] | [illegible] | [illegible] | [illegible] | [illegible] |
| Riverside Cooperative Association | [illegible] | [illegible] | [illegible] | [illegible] | [illegible] |
| Woodlawn Cooperative Association | [illegible] | 17,[illegible] | [illegible] | [illegible] | [illegible] |
| Totals and averages | [illegible] | [illegible] | [illegible] | [illegible] | [illegible] |

a This society differs from the Rochdale plan [illegible]

b This society differs from the Rochdale plan [illegible]

SALES, EXPENSES, AND NUMBER OF EMPLOYEES OF THIRTEEN ROCHDALE SOCIETIES OUTSIDE OF NEW ENGLAND, 1895-96.

| Name of society. | Expenses. | Sales. | Per cent of expenses of sales. | Employees. | Trade per employee. |
|---|---|---|---|---|---|
| Trenton Cooperative Society | $6,889.58 | $51,300.00 | 13.4 | 9 | $5,700.00 |
| Jamestown Cooperative Supply Company | 4,957.26 | 31,229.94 | 15.9 | 5 | 6,245.79 |
| Ishpeming Cooperative Society | 9,228.78 | 109,021.36 | 8.5 | 14 | 7,787.24 |
| Alliance Cooperative Association | 680.00 | 7,775.00 | 8.7 | 1¼ | 6,220.00 |
| Greenwood County Cooperative Association | 1,650.00 | 11,500.00 | 14.3 | 2½ | 4,600.00 |
| Johnson County Cooperative Association | 18,406.29 | 231,141.63 | 8.0 | 30 | 7,704.72 |
| Lyon County Alliance Exchange Company | 2,482.00 | 14,500.00 | 17.1 | 3 | 4,833.33 |
| Osage County Cooperative Association | 3,494.43 | 42,869.15 | 8.2 | 4 | 10,717.29 |
| Patrons' Cooperative Association | 2,200.00 | 28,000.00 | 7.9 | 4 | 7,000.00 |
| Wakefield Alliance Cooperative Association | 2,886.54 | 31,174.97 | 9.3 | 4 | 7,793.77 |
| Poplar Cooperative Association | 375.64 | 2,594.99 | 14.5 | 1 | 2,594.99 |
| Santa Paula Cooperative Association | 600.00 | 7,000.00 | 8.6 | 2 | 3,500.00 |
| Socialists' Cooperative Store and Productive Association | 1,586.52 | 9,262.13 | 17.1 | 7 | 1,323.16 |
| Totals and averages | 55,438.24 | 577,368.16 | 9.6 | 86¾ | 6,655.54 |

From these two tables it appears that the societies that have an expense of over 14 per cent of the sales have also less than the average sales per employee. This average for the New England societies is $7,364.80, and for the others $6,655.54. There seems no connection between the size of a society and its expenses.

The test of expense, however, can not be fairly applied to a society until it is known to what extent rent of real estate is included in expense and how much should be added for interest on the capital, whether borrowed or owned by the society. In these tables rent is included where paid, but not interest.

In the following tables the capital of the societies hitherto considered is given, the business per $100 of capital, the surplus and amount borrowed, together with the interest on the capital and on the amount borrowed, and the total expenses including interest.

CAPITAL, SURPLUS, AMOUNT BORROWED, INTEREST AT 6 PER CENT, AND TOTAL EXPENSES OF FIFTEEN ROCHDALE SOCIETIES IN NEW ENGLAND, 1895-96.

| Name of society. | Capital. | Business per $100 capital. | Surplus. | Amount borrowed. | Interest on capital and amount borrowed. | Total expenses. | Per cent of total expenses of trade. |
|---|---|---|---|---|---|---|---|
| Sovereigns' Trading Company (a) | $21,000.00 | $714 | ........ | ........ | $1,260.00 | $7,260.00 | 4.8 |
| Lisbon Falls Cooperative Association | 8,155.00 | 661 | $2,210.32 | ........ | 489.30 | 4,906.91 | 9.1 |
| Sabattus Cooperative Association | 4,630.00 | 670 | 425.00 | ........ | 277.80 | 2,777.80 | 9.0 |
| Farmers' and Mechanics' Exchange | 45,000.00 | 150 | 1,026.00 | ........ | 2,700.00 | 12,745.50 | 18.8 |
| Arlington Cooperative Association | 120,664.99 | 222 | 19,000.00 | ........ | 7,239.89 | 44,065.85 | 16.5 |
| Beverly Cooperative Association | 8,000.00 | 795 | 1,971.00 | ........ | 480.00 | 7,028.42 | 11.1 |
| Cooperative Store Company | 1,850.00 | 514 | 1,447.00 | ........ | 111.00 | 1,064.87 | 11.2 |
| German Cooperative Association | 1,690.00 | 2,132 | 1,000.00 | ........ | 101.40 | 2,226.40 | 6.2 |
| Hampden County Cooperative Association | 5,500.00 | 545 | 275.00 | ........ | 330.00 | 5,930.00 | 19.9 |
| Harvard Cooperative Society (b) | 16,368.96 | 829 | ........ | ........ | 982.14 | 9,418.41 | 7.0 |
| Knights of Labor Cooperative Boot and Shoe Association | 3,405.00 | 453 | 593.17 | $1,237.50 | 284.55 | 3,993.87 | 35.9 |
| Lawrence Equitable Cooperative Society | 4,121.50 | 575 | 425.00 | ........ | 247.20 | 3,737.42 | 15.8 |
| Lowell Cooperative Association | 15,525.00 | 383 | 4,657.90 | ........ | 931.50 | 12,952.80 | 21.8 |
| Riverside Cooperative Association | 11,895.00 | 405 | 4,714.41 | 9,850.06 | 1,304.71 | 5,969.21 | 12.4 |
| Woodlawn Cooperative Association | 2,068.06 | 868 | 344.11 | 800.00 | 172.08 | 3,198.92 | 17.9 |
| Totals and averages | 269,873.50 | 374 | 38,098.90 | 11,887.56 | 16,911.67 | 127,275.88 | 13.6 |

a This society differs from the Rochdale plan in that it pays no interest but gives trade dividends.
b This society differs from the Rochdale plan in that it pays no interest on stock.

CAPITAL, SURPLUS, AMOUNT BORROWED, INTEREST AT [illegible] EXPENSES OF THIRTEEN ROCHDALE SOCIETIES OUTSIDE OF [illegible] 1895-96.

| Name of society. | Capital. | Business per $100 capital. | Surplus. | Amount borrowed. | [illegible] on capital and amount borrowed. | [illegible] | [illegible] |
|---|---|---|---|---|---|---|---|
| Trenton Cooperative Society | $10,214.00 | $502 | $12,700.00 | [illegible] | [illegible] | [illegible] | [illegible] |
| Jamestown Cooperative Supply Company | 2,240.00 | 1,294 | 844.96 | [illegible] | [illegible] | [illegible] | [illegible] |
| Ishpeming Cooperative Society | 49,635.00 | 220 | 3,191.56 | ........ | [illegible] | [illegible] | [illegible] |
| Alliance Cooperative Association | 1,345.00 | 578 | 2,000.00 | [illegible] | [illegible] | [illegible] | [illegible] |
| Greenwood County Cooperative Association | 2,440.00 | 471 | ........ | [illegible] | [illegible] | [illegible] | [illegible] |
| Johnson County Cooperative Association | 100,000.00 | 231 | 28,000.00 | ........ | [illegible] | [illegible] | [illegible] |
| Lyon County Alliance Exchange Company | 2,024.00 | 496 | [illegible] | ........ | 175.44 | [illegible] | [illegible] |
| Osage County Cooperative Association | 2,075.00 | 1,441 | 6,822.92 | [illegible] | [illegible] | [illegible] | [illegible] |
| Patrons' Cooperative Association | 3,500.00 | 800 | 6,500.00 | 4,000.00 | 450.00 | [illegible] | [illegible] |
| Wakefield Alliance Cooperative Association | 5,348.00 | [illegible] | ........ | 2,818.00 | [illegible] | [illegible] | [illegible] |
| Poplar Cooperative Association | 1,103.35 | 235 | ........ | 400.00 | [illegible] | [illegible] | [illegible] |
| Santa Paula Cooperative Association | 2,000.00 | 350 | 57.00 | ........ | [illegible] | [illegible] | [illegible] |
| Socialists' Cooperative Store and Productive Association | 578.30 | [illegible] | [illegible] | ........ | [illegible] | [illegible] | [illegible] |
| Totals and averages | 184,302.65 | 313 | 62,075.09 | 16,944.20 | 12,074.81 | [illegible] | [illegible] |

It appears that 15 New England societies, with a total capital of $269,873.50, a surplus of $38,098.60, and an amount borrowed of $11,987.58, had an average business in 1895-96 of $374 per $100 of capital and had a total expense of $127,275.88, if interest at 6 per cent on the capital and amount borrowed but not on the surplus be included. This means an average expense of 12.6 per cent on the trade of $1,008,977.24. Thirteen societies outside of New England, with a capital of $184,302.65, a surplus of $62,075.09, and with $16,944.20 borrowed, did a business of $313 per $100 of capital and had a total expense of 11.7 per cent of the trade of $577,368.16. There seem to be greater inequalities of expense in proportion to trade than can be accounted for by differences of location and of progressiveness in pushing forward, which, in the absence of a strong cooperative spirit, entails expense.

Outside of New England, and even in a measure there, as a [illegible] New York firm has written, any business house that could [illegible] ends meet the past year is to be congratulated. It is a very [illegible] therefore, that these cooperative societies present.

In most of them a depreciation of 10 per cent is yearly [illegible] the fixtures. The Arlington and some other societies have [illegible] lent habit of never allowing in their inventory of merchandise [illegible] valuation than the original cost, and putting a lower [illegible] the market price has fallen.

Of 1,570 cooperative societies in Great Britain [illegible] with respect to trusting, 59 per [illegible] is contrary to the principles [illegible] returns in this country, only 9 give no credit whatever. Many of the others, however, give but little credit, or limit it, as in England, to the value or to four-fifths of the value of the stock owned by the debtor. Others have a large amount of "bills receivable" among their assets. Trusting has been a prolific source of failure of cooperative societies in America, and many severe condemnations of it have been sent the writer by the secretaries or managers of some of these societies that have failed, as also of some of the existing societies. The New England societies do not trust as much as those farther west, although two Massachusetts societies, with a capital of about $8,000 and $15,000, respectively, report bills due them of nearly half of this. The second oldest cooperative society in the country a year ago (the Sovereigns' Cooperative Association of Webster, Mass., started in 1874), had in its by-laws for many years the words, "Never depart from the principle of buying and selling for ready money." But just before its failure in 1895 the manager, replying to the question to what extent the association trusted, wrote, "No limit, if a man is good for it!"

Ten societies report a considerable trade by their members at special discounts at private stores dealing in goods not kept by the societies. In some cases the members obtain this discount directly. In others, the discount is paid to the society and added to the other profits, an account being kept of the purchases of each member so that he may share in any dividend on these purchases. For example, the Lawrence Equitable Cooperative Society has a trade in its own store of $23,707, and receives a 10 per cent discount on an outside trade of $17,338.77. The Arlington Cooperative Association, with a trade of $267,508.21, secures 10 per cent discount on an outside trade of $67,541. The percentage in most at least of the few other cases is from 3 to 8 per cent.

A few cooperative societies secure discounts elsewhere, not only on boots and shoes, dry goods, coal and wood, hardware, oil, meat, bread, clothing, and furniture, but on bicycles, jewelry, watches, milk, musical instruments, laundry, photographs, athletic goods, and the services of the tailor, dentist, and physician.

The business directly done by the societies consists in almost all cases of the sale of groceries and the other goods usually sold in connection therewith in the ordinary private grocery. Dry goods, boots and shoes, and coal are sometimes added, more rarely meat.

The average wages of the 96 employees in 10 of the larger societies is $609.64 per year. The average of the 102 employees in 24 societies of about the same character in the Manchester, Oldham, and Rochdale districts of England in 1895 was $377.50. The average wages seem to be much higher in England when only one or two are employed than when there are many, but there seems to be no such difference here. There is hardly enough data to show whether this equality of wages in small and large American societies is due to our cooperative societies paying nearly as high wages to all employees as to the manager, or to the location of the smaller societies in small places where all wages are low.

Appropriations for education along cooperative and other sociological lines do not equal in America even the three-fourths of 1 per cent of the net profits which is thus set aside in England. A very few societies, especially some of the new ones, are beginning to do something in this direction, notably those affiliated with the Cooperative Union to be described later.

Lawrence, Mass., has a larger proportion of members of cooperative societies than any other American city. In three societies there—the Arlington, the Equitable, and the German cooperative societies—there are 3,751 members, embracing about 19,000 persons, if each member is considered to represent a family of 5 persons. This is 36 per cent of the population of the city.

The oldest cooperative society is the Sovereigns' Trading Company of New Britain, Conn., which gives credit only for 30 days and not above the value of the share, worth $94, which every one of the two hundred members owns. All the profits are absorbed by a 10 per cent discount to members on their trade. Even interest is not paid. This society, with a business of $150,000, is one of the very few successful societies not on the Rochdale plan.

Another interesting society is the Harvard Cooperative Society, Cambridge, Mass., which places its shares at $1, allows only one share to a member, pays no interest, but adds to the capital, now $16,368.96, 10 per cent of the yearly profits and gives the rest as a dividend on trade. This dividend amounted in 1895 to 7 per cent of $135,151.41.

Very few societies have branches, as in England, save the Arlington, with two branches and a large warehouse for storage in Lawrence, and the Johnson County Cooperative Association at Olathe, Kans., with branches at Gardner, Edgerton, Prairie Center, and Stanley, Kans. The latter society, although not having as rapid a growth as the Arlington, and only earning a little over 2 per cent on its total trade the past three years, after paying 10 per cent interest on its $100,000 capital, has prospered fully as much as the community about it.

The Allegan Cooperative Association, started at Allegan, Mich., in 1874, does not give dividends on trade, but sells so near cost as to leave only enough to pay 5 per cent interest on its $15,000 capital. The society is not considered cooperative by many because entirely controlled by its manager, who has held that cffiee from the start, and who buys on his own personal credit instead of on that of the association. The number of members, once 840, has declined to 150, but the old members can still buy at the low prices, and the sales in [illegible] amounted to $140,000. Farmers' organizations as far west as [illegible] buy through the society because of its low prices.

The Texas Cooperative Association, started in [illegible] did a business of over half a million dollars a [illegible] direct sales of merchandise to members in the city and at wholesale to small cooperative societies in other parts of the State. Cotton was also marketed for the farmers and formed part of the large total trade mentioned above. The small local societies have now to a large extent died out. The trade of the central wholesale society was $65,000 in 1894–95 and the commission business amounted to $222,661.91. In 1895 the auditors wrote off $70,139 of bad debts, and there was still left as due the association $33,241.28. The other assets amounted to $20,586.38. The surplus was thus made to disappear and the capital was virtually reduced from $87,930.49 to $17,490.49. Of the original capital, however, all but $2,000 or $3,000 had been withdrawn by those who had paid it in, so that the present capital has been formed out of the surplus profits. From this source has also come $150,000 of cash dividends in the past.

The Zion's Cooperative Mercantile Institution, managed by the Mormons at Salt Lake City, Utah, since 1868, gives all its profits as dividends on its capital of $1,077,000, owned by 600 stockholders, and so is not technically cooperative in any full sense, but is generally acknowledged to have in practice many cooperative elements. Little data on the subject have been obtained. It appears that the profits have averaged 9½ per cent during the past twenty-seven years and amounted to $124,914.60 in 1895, although only 8 per cent dividend was paid that year. It not only does a retail business, but also sells largely at wholesale to the many smaller cooperative societies in the State. The employees consist of 207 men and 77 women.

At Lehi City, Utah, is located the Utah Sugar Company, with 700 members, 300 employees, $320,000 stock (on which 10 per cent dividends are usually paid), and $400,000 of 6 per cent bonds. Beet sugar is made there. All the profits appear to go to the stockholders, but there are some cooperative features.

Few societies in any part of the country limit membership save that the approval of new members by the directors is usually required. In Kansas, however, most of the societies admit only members of the Farmers' Alliance, the Patrons of Husbandry, or other farmers' or laborers' organizations. The society at Los Angeles, Cal., requires that the officers and employees shall be "Socialists in good standing in the sections where the business is done."

A few other societies are not mentioned in the tables. The Sovereigns' Cooperative Association at Dover, N. J., was started by the Sovereigns of Industry in 1874, and has continuously prospered on the Rochdale plan ever since. The 253 members receive 7 per cent interest on their $28,000 capital. Only $1,300 is in "debts receivable." From 5 to 9 per cent dividends are always divided on the trade, which in 1894 amounted to about $62,000. There are at least four other Rochdale societies in New Jersey besides the one at Trenton. The Raritan

Cooperative Association, starting in 1886, has [illegible] its capital, now amounting to $6,715, and 7 per cent [illegible] which has averaged $75,000 a year. There were, [illegible]

The Cooperative Association No. 1, at Phillipsburg, [illegible] 1879, has had an average membership since 1890 of [illegible] business of $28,000, a net profit of $4,320, and a capital of [illegible] Phillipsburg is another society known as The People's [illegible] Association, with a trade of $9,886 in 1895, on a capital of [illegible] a membership of 35 at the end of the year.

As a part of the Hammonton (N. J.) Fruit Growers' Union [illegible] markets fruit cooperatively for its members, there is a [illegible] dale store. With total expenses, including interest at 6 per c[illegible] $11,000 of capital, of only 10.5 per cent of the $61,427.43 trade, [illegible] profits, besides 5 per cent of the capital for the reserve fund, w[illegible] per cent on the trade in 1895–96. This, however, and similar [illegible] in the shipping department had to be devoted to paying a cl[illegible] $3,067.76 damages for accidental injury to an employee in the sh[illegible] department. The society has a membership of 646, and for ever[illegible] of capital in the store did a business of $528 in 1895–96.

There are two cooperative societies managed by the employ[illegible] N. O. Nelson—the St. Louis Cooperative Store, started in 1894, an[illegible] having about 150 members and $15,000 trade on the Rochdal[illegible] and, conducted in the same way, the Leclaire Cooperative Store [illegible] Nelson's works at Edwardsville, Ill., started in 1893, and having [illegible] 50 members and $3,500 trade.

Partial returns have been received from a few other societies. [illegible] is a Rochdale society in Pittsburg, Pa., the Integral Coope[illegible] Association, started July 1, 1894. During its sixth quarter it [illegible] business of $8,498.33, and during its eighth quarter, $9,494.84. [illegible] trade dividends averaged over 8 per cent during 1895–96 on a bu[illegible] of about $30,000. There should also be mentioned the Centra[illegible] York Cooperative Company at Oneida, open only to the Patr[illegible] Husbandry, and other small but very promising Rochdale socie[illegible] South Worcester, Rockland, Haverhill, Lynn, Dorchester, and [illegible] ton, Mass. Most of these latter societies, as well as several othe[illegible] yet fully organized in Massachusetts, began in 1895 as purc[illegible] clubs. Saving the profits and following the lead of the ene[illegible] Cambridge Cooperative Society, regular cooperative societies ha[illegible] been organized.

In the following tables is given, according to the latest returns available, the membership of most of the cooperative societies in New England and outside of New England:

MEMBERSHIP OF NEW ENGLAND COOPERATIVE SOCIETIES.

| Name of society. | Location. | Membership. |
|---|---|---|
| Sovereigns' Trading Company | New Britain, Conn | 200 |
| Lewiston Cooperative Society | Lewiston, Me | 135 |
| Lisbon Falls Cooperative Association | Lisbon Falls, Me | 275 |
| Sabattus Cooperative Association | Sabattus, Me | 125 |
| Farmers' and Mechanics' Exchange | Brattleboro, Vt | 800 |
| Woodlawn Cooperative Association | Pawtucket, R. I | 174 |
| First Swedish Cooperative Store Company | Quinsigamond, Worcester, Mass. | 265 |
| Harvard Cooperative Society | Cambridge, Mass | 1,000 |
| Cambridge Cooperative Society | do | 135 |
| Riverside Cooperative Association | Maynard, Mass | 268 |
| Hampden County Cooperative Association | Springfield, Mass | 134 |
| Knights of Labor Cooperative Boot and Shoe Association | Worcester, Mass | 201 |
| Cooperative Store Company | Silver Lake, Kingston, Mass. | 138 |
| Plymouth Rock Cooperative Company | Plymouth, Mass | 83 |
| Arlington Cooperative Association | Lawrence, Mass | 2,350 |
| Lawrence Equitable Cooperative Society | do | 732 |
| German Cooperative Association | do | 160 |
| Industrial Cooperative Association | New Bedford, Mass | 410 |
| Beverly Cooperative Association | Beverly, Mass | 185 |
| Dorchester Cooperative Association | Dorchester, Mass | 20 |
| Lowell Cooperative Association | Lowell, Mass | 1,130 |
| West Warren Cooperative Association | West Warren, Mass | 196 |
| Haverhill Cooperative Society | Haverhill, Mass | 45 |
| Lynn Cooperative Society | Lynn, Mass | 60 |
| Rockland Cooperative Society | Rockland, Mass | 60 |
| The Hub Cooperative Emporium | Boston, Mass | 80 |
| Total membership | | 10,692 |

MEMBERSHIP OF SOCIETIES OUTSIDE OF NEW ENGLAND.

| Name of society. | Location. | Membership. |
|---|---|---|
| Trenton Cooperative Society | Trenton, N. J | 465 |
| Hammonton Fruit Growers' Union and Cooperative Society | Hammonton, N. J | 648 |
| Vineland Fruit Growers' Union and Cooperative Society | Vineland, N. J | 54 |
| Sovereigns' Cooperative Association | Dover, N. J | 253 |
| Raritan Cooperative Association | Raritan, N. J | 175 |
| Cooperative Association No. 1 | Phillipsburg, N. J | 114 |
| Jamestown Cooperative Supply Company | Jamestown, N. Y | 217 |
| Integral Cooperative Association | Pittsburg, Pa | 300 |
| Allegan Cooperative Association | Allegan, Mich | 150 |
| Ishpeming Cooperative Society | Ishpeming, Mich | 350 |
| Zumbrota Mercantile and Elevator Company | Zumbrota, Minn | 394 |
| Texas Cooperative Association | Galveston, Texas | 300 |
| Greenwood County Cooperative Association | Eureka, Kans | 165 |
| Lyon County Alliance Exchange Company | Emporia, Kans | 378 |
| Johnson County Cooperative Association | Olathe, Kans | 900 |
| Osage County Cooperative Association | Overbrook, Kans | 108 |
| The Alliance Cooperative Association | Green, Kans | 97 |
| Wakefield Cooperative Association | Wakefield, Kans | 154 |
| Patrons' Cooperative Association | Cadmus, Kans | 217 |
| Zion's Cooperative Mercantile Institution | Salt Lake City, Utah | 600 |
| Socialists' Cooperative Store and Productive Association | Los Angeles, Cal | 102 |
| Santa Paula Cooperative Association | Santa Paula, Cal | 50 |
| Poplar Cooperative Association | Poplar, Cal | 36 |
| Total membership | | 6,115 |

The New England membership early in 1896 was 10,692, and the membership elsewhere, 6,115. These, together with 6,000 in the labor exchanges, constitute a total of 22,807. There are known to be several, and perhaps over 20 cooperative societies outside of and even in New

England, from which no reports have been made. These [illegible] the total membership to 25,000 at least. The trade of the [illegible] cooperative societies in New England with a membership [illegible] amounts to $1,174,000, or $114.63 per member. The trade of 21 others with 5,465 members was $1,198,000 in 1895, or $219.21 per member. The Zion's Cooperative Mercantile Institution is not included. The sales of 1,711 British cooperative societies with 1,414,158 members in 1895 averaged £37 2s. 8d. ($180.71) worth of goods per member, or a total of £52,512,126 ($255,550,261.18).

## COOPERATION AMONG TRADE UNIONS.

Cooperation among the trade unions has almost no permanent success, although it has sometimes proved a temporary resource while men were on a strike. Mr. Henry White, general secretary of the United Garment Workers of America, writes that the many cooperative efforts in his trade have resulted in failure because of disagreement respecting the management and the selection of officials and because the attempt was made to pay higher wages and exact less work than in the other shops. Mr. Henry Weissmann, head of the bakers and confectioners, writes: "We have had cooperation in Brooklyn, Boston, Baltimore, and Philadelphia; all unsuccessful. It is lack of education and business qualifications, and more especially the latter, that produced these failures."

## UNION OF COOPERATIVE SOCIETIES.

Apart from Utah, there are five centers of the cooperative movement in America—California, Kansas, Texas, New Jersey, and Massachusetts. An attempt is now being made to federate the various Kansas societies engaged in different forms of cooperative effort, and the first congress for this purpose was held at Topeka in April. An attempt was made to secure a similar union of all parts of the country in a convention at St. Louis the last of July, but it was largely swallowed up by the Populist convention there the same week. Most of those attending came not as delegates of cooperative societies, but as individuals. Yet a national organization was effected, called the American Cooperative Union, with Alonzo Wardall, of Topeka, Kans., as president, and Imogene C. Fales, of Bensonhurst, N. Y., secretary.

Another attempt at union was made in September, 1895, when the Cooperative Union of America was founded in the eastern part of Massachusetts. At the first conference under the auspices of this union, June 6, 1896, representatives of twelve societies and of nearly 1,[illegible] members were present. Rev. Robert E. Ely and Mr. James [illegible] of the Cambridge Cooperative Society, were made president [illegible] tary, respectively, and after an enthusiastic meeting [illegible] made for a larger conference in October. After [illegible] of the union as given above, and declaring that the object is to promote cooperation on the Rochdale plan in America, the rules proceed:

ARTICLE III.—*Membership.*

1. Persons in sympathy with the Rochdale plan of cooperation may, upon the approval of the office committee, become members of the union by paying a fee of $1 per year. This fee includes a year's subscription to the organ of the union, the Cambridge Magazine.(a)

2. A cooperative society, a trade union, and any other organization may, upon approval of the office committee, become a member by paying an annual fee equal to not less than 2 cents per annum for each of its members.

ARTICLE IV.—*Government.*

1. Until a sufficient number of societies shall have become members to enable them to elect representatives after the manner of the cooperative union of Great Britain, the union shall be managed by a general board, called the central board. The members of the central board shall be appointed by the office committee, subject to confirmation by the union at the annual meeting in each year. Twelve members of the central board shall constitute a quorum.

2. The central board after its election at the annual meeting shall choose from its membership an office committee for the ensuing year of nine persons. This committee shall have direct charge of the work of the union.

3. The office committee shall elect from its own number a president, vice-president, treasurer, recording secretary, and general secretary of the union.

ARTICLE V.—*Meetings.*

1. The annual meeting of the union shall be held sometime about October in each year. Fifteen members shall constitute a quorum. Special meetings may be held at any time at the call of the office committee.

2. The central board shall meet just prior to the annual meeting. Special meetings may be held at any time, at the call of the office committee.

3. The office committee shall meet at least once a month.

ARTICLE VI.—*Amendments.*

These rules may be amended by a two-thirds vote of the members present at any meeting of the central board.

In cooperation two things are necessary for any large or enduring benefit. First is needed a moral enthusiasm, a true cooperative spirit, joined to a careful study, by all the cooperators, of the underlying principles of the movement. Cooperation is in spirit a sense of brotherhood—a willingness to subordinate the individual to the general good of at least all who can be induced to affiliate. As Thomas Hughes once stated, where financial success, by the accident, it may be, of an excellent manager, comes to a society that is devoid of this spirit, the very success gives birth "to a greedy desire for gain rather than to those higher and more elevating feelings which we have all supposed to be the legitimate result of a true and earnest cooperation." Referring to the ease with which cooperative conventions dissolve without accomplishing anything, an active worker in the movement writes of

a Later known as the American Cooperative News, published monthly.

how much easier it is for the
"co-work." On the other hand, however,
have failed for lack of a good manager and of
among the members. This is the second
that the Arlington Cooperative Society is
bership by 99 out of nearly 1,500 English
only 149 has undoubtedly stimulated the
While a truly cooperative society is not likely to
or work its employees long hours, it is apparently
proposition, that cooperative distribution is not a
business, but a new method of dividing profits,
cratic system of control on the part of members
and ready to accept the leadership of those best

## LABOR EXCHANGES.

A very novel development has lately appeared in
movement. It is known as the Labor Exchange, and
owed its origin to Mr. G. B. De Bernardi, of Independen
organized it under the laws of Missouri in 1889 and is still
the national organization.(a)

Those desiring to form a labor exchange in a place us
apiece as entrance fee. The local exchange thus formed
$2.50 to the central office at Independence for labor excha
and invites any member to deposit with it any product
receive therefor an exchange check of the same denomin
officers think would be the local wholesale price in money.
are then marked up to the usual retail price. A depositor
check to buy from the exchange anything he finds there th
By virtue of this redeemability in goods, though never in
private merchants and others are to some extent induce
these checks as money, writes Mr. De Bernardi, but somet
a discount. For the present all profits are devoted to the
the movement. In theory, however, the profits belong to
itors, and if ever withdrawn would be apportioned in prop
length of time of the individual deposits. The principal
interest is thus acknowledged, but no direct interest pay
made. Goods deposited may be sold to outsiders for the
for legal money, and in order to enable the societies to
needed supplies elsewhere members are encouraged
money as well as other property. A depositor
check or a certificate of deposit. The latter has
lines for the name of the depositor, the date,
deposit. On the reverse side

a See The Trials and Triumphs of
ive Handbook of the Labor Exchange

deposit is not redeemable in legal tender, but receivable by the Labor Exchange Association in payment for merchandise, for services, and for all debts and dues to the same, and it is based upon and secured by the real and personal property in the keeping of the association, at the branch of issue. The property held for the redemption of this certificate can not, as per charter, be mortgaged nor pledged for debts, nor can it be withdrawn, but may be exchanged by the association for other property of equal value. Pay to ——— ———, depositor." Labor checks are issued of all denominations from $\frac{1}{100}$, which is really of the same value, or meant to be, as 1 cent, to 20, which is the equivalent of $20. On the face of the checks appear the value and the words: "Balance due bearer in labor or the products of labor by Labor Exchange, Branch No. —," followed by the location of the exchange and the year of issue. Since these obligations are not redeemable in money, the United States Treasury Department has decided that they are not subject to the 10 per cent tax on circulation.

The Labor Exchange of Topeka, Kans., did a business in 1895 of $10,000 at an expense of $1,500 and with a net profit of $1,000. The goods on hand at the end of the year belonging to the 40 members was $1,200. The capital is returned as $1,000.

Although the oldest branches were started among the wage workers of large cities, more success has come with branches in small towns and rural districts.

Mr. De Bernardi writes:

The distinctive characteristic of the labor exchange from other forms of cooperation is, that we regard true cooperation impossible within one industry. We do not believe that farmers, for instance, can organize cooperation among themselves. They can only combine against other classes. What is true of farmers is true of manufacturers, of railroad men, of carpenters, tailors, cigar makers, etc. Each separate industry can only combine to raise the price of their labor or lower the price of others' labor to their own personal benefit. True equitable cooperation is only operative between two or more branches of industry.

Farmers can cooperate with other industries, and interchange products and keep each other employed. So in every other industry, we are endeavoring to form a universal cooperation of this character, and have now (June 26, 1896) 135 branches scattered in 32 States, and a membership of 6,000, comprising almost every trade, profession, and calling.

Many of these branches are engaged in profitable enterprises. Especially will I name Branch No. 11, of Pfafftown, N. C. This branch, situated in a village of 100 souls, all poor, with no money or employment visible, went to work first at making brick, then building storehouses, then putting up a tannery and tanning hides on shares, then a sawmill and gristmill, etc. Thus the labor exchange infused life and energy in a financially dead country, and what has been done, and is being accomplished at Pfafftown we believe can be done all over the United States, and thus bring prosperity to the country without binding ourselves to foreign capitalists or mortgaging our property for the use of useless money.

* * * * * * *

Our leading object is to employ idle labor [illegible] tomary wages, and interchange labor's products [illegible] with customary prices. As we issue certificates [illegible] have no fear of losing our customers, for we hold [illegible] for the outstanding certificates at par. Even if [illegible] should honor our checks (as in many places they [illegible] checks would finally come home for final redemption. [illegible] no stimulus in low prices or rebates to attract [illegible]

We are not a profit-sharing, but a benevolent, [illegible] our increase, if any, is used to extend the field of [illegible] additional enterprises, and thus employ more labor—all [illegible] the benefit of the associates. But should a member or [illegible] in need of assistance, appropriation would be made for [illegible] benefit. Beyond this we let the future take care of the [illegible]

We aim to capitalize the deposits of work and [illegible] benefit of depositors; or, to express it more clearly, we [illegible] the members insure to themselves employment and [illegible] revenues which, under the present monetary system, would [illegible] by the capitalists.

Following is also appended the distinctive parts of the by-la[illegible] one of the twenty-two southern California labor exchanges, [illegible] though only about a year old, has over 200 members.

## BY-LAWS.

### ARTICLE I.—*Name.*

This association shall be known as the "Los Angeles Branch of the Labor Exc[illegible] No 39," acting under a charter granted by the General Labor Exchange wit[illegible] quarters at Independence, Mo., and in conformity with the laws, rules, regul[illegible] and charter granted to G. B. De Bernardi and others, under the laws of the S[illegible] Missouri, as a benevolent, industrial, and educational association.

### ARTICLE II.—*Location.*

The business headquarters of this branch shall be in the city of Los A[illegible] County of Los Angeles, State of California.

### ARTICLE III.—*Purposes.*

SECTION 1. The objects of this association are to provide employment for [illegible] who are idle by facilitating the interchange of commodities and services amo[illegible] members and the public.

SEC. 2. To alleviate the suffering incident to, and avert the social dangers [illegible] may arise from a constantly increasing class of unemployed by furnishing [illegible] occupation and saving the wealth thus produced for the use and benefit of [illegible] producers and their dependents.

SEC. 3. To lighten the burden of charitable institutions by establishing on[illegible] employs the idle on a basis of justice.

### ARTICLE IV.—*Membership.*

Any member of the General Labor Exchange may become a member of this [illegible] by signing the constitution and by-laws, and there shall be no initiation [illegible] or assessments.

### ARTICLE V.—*Duties of members.*

It shall be the duty of members to interest themselves [illegible] in the welfare of the branch, to promote [illegible] by [illegible] give employment to its members in [illegible] to trade at the depository so long as it offers equal advantages in quality and price of goods. All such members shall be considered active.

ARTICLE VI.—*Withdrawals.*

A member may withdraw from this branch at any meeting by giving notice of intention at any previous regular meeting and directing the secretary to strike his name from the roll. Such action shall not affect his standing in the General Labor Exchange. Any property conveyed to the branch can not be recovered; the title to which and the possession thereof must remain in and with the Labor Exchange.

ARTICLE VII.—*Officers.*

SECTION 1. The officers of this branch shall be a president, vice president, secretary, accountant, statistician, and seven directors.

SEC. 2. Excepting the first term, all officers shall be elected for one year.

ARTICLE VIII.—*Election of officers.*

The annual election of officers shall occur on the first Tuesday in September of each year. All officers shall hold office until their successors are elected and installed.

ARTICLE IX.—*Removal of officers.*

SECTION 1. Any officer, director, or committeeman may be removed at any time by a majority vote of all the members present at a regular meeting. Such officer, director, or committeeman must first be notified in writing in a regular meeting to appear before a subsequent special or regular meeting, the date of which to be then determined in open meeting, to show cause why he shall not be removed.

ARTICLE X.—*Duties of officers.*

SECTION 1. It shall be the duty of the president to preside over all meetings, to sign labor checks and other checks, contracts and papers requiring an official signature, and attend to any matter directed by the General Labor Exchange, and to all duties devolving upon and accustomed to be performed by that officer.

SEC. 4. The accountant shall keep the books and accounts of the association, have charge of all bills, vouchers, etc. He shall countersign and issue all labor checks after they have been signed by the president and perform all the clerical duties belonging to his office and such as may be required by the board of directors. He shall give bond in the sum of $5,000, which sum may be changed at any time by the branch and raised at the discretion of the board. He shall also act as treasurer of the branch.

SEC. 7. The statistician shall have charge of the collection and distribution of statistical reports, presenting as full exhibits as can be obtained of the condition and wants of the association.

ARTICLE XI.—*Neglect of official duty.*

The absence of any officer—unless from some unavoidable cause, the same to be determined by the board of directors—from three consecutive meetings, or neglect to perform official duties, shall be deemed equivalent to a resignation of office, and the directors shall proceed to fill such vacancy until the next meeting of the branch.

ARTICLE XII.—*Committees.*

SECTION 7. It shall be the duty of the relief committee to visit sick members and to provide for their necessities so far as the funds set apart for that purpose by the branch may admit.

SEC. 8. The library committee shall devise ways and means for a free library. They shall select the books, papers, magazines, etc., and, with the advice and consent of the board of directors, provide and maintain a room to which the public may be invited.

SEC. 9. The programme committee shall [illegible] may be given by the branch.

SEC. 10. The auditing committee shall audit the [illegible] once in three months and make a written report [illegible] meeting.

ARTICLE XVI.—Discipline.

Charges may be preferred against any member at any [illegible] to be signed by at least two members and written in duplicate [illegible] to the accused, the other filed with the secretary. At the [illegible] the date of presenting charges a jury of nine, chosen by lot [illegible] ent, shall be selected to try the case as soon as practicable, the [illegible] trial to be determined by the jury. Their decision shall be [illegible] verdict is expulsion, in which case appeal may be had to the [illegible] meeting, date to be announced at the first subsequent regular [illegible] is concluded. If the defendant does not signify his readiness [illegible] limit of time above stated, the case shall, at the expiration of [illegible] though he had.

The accused may be assisted by any member as counsel.

ARTICLE XVII.—Employment bureau.

There shall be a free employment bureau in charge of the secretary [illegible] the exchange, which shall be open to all members. All who so desire [illegible] for help or situations wanted. The secretary shall make effort to fill all [illegible]

## COOPERATIVE SHIPPING ASSOCIATIONS.

Although it is not the purpose of this paper to describe the c[illegible] tive shipping of fruit by the farmers, a brief reference may be [illegible] the existence of such associations in California, northern Oh[illegible] Jersey, Georgia, and some other States. In California there a[illegible] four fruit exchanges that both dry fruit and market it in the gr[illegible] kets of the country. Eight of the southern California fruit ex[illegible] have united for some purposes and selected an executive boa[illegible] headquarters at Los Angeles, but the attempts that have bee[illegible] at different times to form an efficient State association have [illegible] failed.

About one-fourth of the grapes grown in northern Ohio in 18[illegible] shipped out by the Northern Ohio Grape Company, a coo[illegible] marketing society. Since nonmembers, however, receive man[illegible] benefits resulting from the steadiness of the market price eff[illegible] the distribution of fruit with this end in view by the cooperati[illegible] ties, many do not join these associations. Other difficulties [illegible] been met. Little success seems to have been attained in [illegible] but 90 per cent of the fruit growers of Georgia have joined [illegible] ciation of that State, and among other things have [illegible] in reducing freight rates on their shipments.

Success has also attended the Fruit Growers' Union [illegible] Society at Hammonton, N. J., whose [illegible] on a previous page. This [illegible] of berries, or about the [illegible] shipped 24,487 baskets of peaches, 2,884 barrels of pears, 621 barrels of apples, 29,370 pounds of grapes, and 10,680 quarts of plums. Although the shippers received a good price through the society, the usual profits of both this and the grocery department were absorbed in the payment of a claim for damages to an employee. In 1895 a by-law was adopted that "members shipping to commission houses not entered on our books, except in cities where we do no business, shall forfeit the sum of 3 per cent on the gross sales of all such shipments, this amount to be taken from such offending member's dividend. They shall also forfeit all privileges of the society for that year."

As an example of another type of shipping association, mention may be made of the Springfield (Mass.) Cooperative Milk Association, which collects milk from its 159 members, sells from its own carts at retail about one-fifth of the 3,000,000 quarts thus yearly gathered, wholesales about one-half of it to milk peddlers, and makes the rest into butter. The members of the association thus have a steadier market and, if living many miles from the city, they also get a higher price than would be secured without such an association. The average price obtained by the farmers usually is between 2.8 and 3 cents a quart, the retail price being 6 cents. This association is steadily growing and claims to supply one-half of the milk used in Springfield. The capital now amounts to $23,640 and members receive 6 per cent interest on their investment. Many not desiring to join this association sell their milk to it at 95 per cent of the price paid the members. No member is allowed to sell milk save to the association, either directly or indirectly, under penalty of forfeiture of his stock, or such other punishment as the directors see fit to impose, while there are very minute and rigid regulations for the proper care of the milk.

## LAWS RELATING TO COOPERATION.

In Massachusetts the law allows any seven persons to form a cooperative society, but requires that the capital must be at least $1,000, and that paid in before business is begun. This is said by some to stand in the way of cooperative societies that otherwise might be able to start in a small way and then grow, as did the famous Rochdale society. In practice, some feeble societies are able to start by beginning as purchasing clubs, dividing a large order among themselves at the common retail price and saving the profits, until the $1,000 is accumulated. When the capital stock is increased, it can not be sold below par. Such a company can not alter its business from that specified in its articles of incorporation without the unanimous consent of its stockholders. This is thought to prevent the formation of a wholesale society by the retail cooperative societies, as in England and Scotland; but if that be the correct interpretation of the law, it will doubtless be amended as soon as there is any desire for such a society. In common with [illegible] other corporations, a cooperative society in Massachusetts is obliged to make a yearly report to the secretary of state as to its capital stock, the amount paid up, the name and holdings of each shareholder, "and the assets and liabilities of the corporation, in such form and with such detail as the commissioner of corporations shall require or approve." This section is superior to that of most States, but in the case of cooperative societies should be supplemented, as in Great Britain, with the requirement that in the publication of the returns there should be included the number of members, the annual trade, the rate of dividends, the expenses by items, and a few other items. There is, of course, no secret about these items in the town where the society is located, but their wider publicity would help along the movement.

In Massachusetts no one can hold over $1,000 of stock in a cooperative association. Dividends, if there are any to divide, must be declared at least once a year, but before any distribution can be made at least 10 per cent of the net profits must be added to a sinking fund until it equals 30 per cent of the capital. Cooperative societies are also exempted from the weekly payment law, unless the employees request it, and $20 worth of stock in such a society in the hands of a member is exempted from attachment for debt.

In contrast with the many excellences of the Massachusetts law is that of Illinois, which prohibits shares under $50 and allows them to be $2,000, while no one can hold more than one share. Dividends must be in proportion to the work or product of each shareholder. "If in any kind of industry it should be impossible to assign all shareholders to equally advantageous positions or locations in work, the association may provide that shareholders shall periodically change places, or provide any other method of equalizing such matters in accordance with justice and equity." No association can employ anyone except a stockholder, but the latter can do so when sick, and in case of his death his legal representative may appoint a worker and so keep the dividends. The principle of unlimited liability by the stockholders for all debts prevails. It is evident that the framers of this law either knew nothing of cooperation or intended to have none in Illinois.

California enacted a law of some value on this subject in 1895. Connecticut, Minnesota, New York, Michigan, Wisconsin, Ohio, and a few other States have legislated, but usually in a very defective way. Outside of Massachusetts the only two States that have any at all adequate cooperative legislation are Pennsylvania and New Jersey.

The distinctive features of the Pennsylvania law are, (1) prohibition of all giving of credit (as in Wisconsin in cooperative distribution); (2) requirement that at least 25 per cent of the net profits be set aside for propaganda and as a social fund; (3) dividends must be given on the wages and salaries of employees at the same rate as on the trade of members, while nonmembers must receive one-half as high a rate on their trade, but the directors may credit profit dividends toward the payment of stock, until it reaches the limit allowed; (4) before profits are divided, as they must be quarterly according to these general Rochdale principles, the fixtures and machinery must be depreciated at the rate of 10 per cent yearly and the buildings 25 per cent; (5) besides certain other purposes for which deductions must be made, the law requires the formation of a reserve fund, to which all fines and forfeitures must be carried (the Massachusetts law, which requires the setting aside of 10 per cent of the profits until a surplus is formed equal to 30 per cent of the capital, is better on this point); (6) division of the capital into (*a*) "permanent," nonwithdrawable stock, transferable as the by-laws may determine, and paying 6 per cent interest, of which stock every member must have at least one share, and (*b*) "ordinary" stock, paying 5 per cent, and which may be repaid, transferred, or withdrawn, in accordance with the by-laws; (7) shares of either kind may be from $5 to $25 value, and no one can hold over $1,000 without consent by a vote of the members.

The New Jersey law, while lacking a few excellent features, and perhaps not on the whole much superior to that of Massachusetts or Pennsylvania, is yet sufficiently complete to merit insertion in full. The chief feature of the law, passed in 1884, is its requirement of full details in the by-laws, that there may be no misunderstanding afterwards. Reports to the State, though not sufficiently complete, are required. Proxy voting is forbidden. Every stockholder is liable for the debts of the society, up to the full par value of his share, if all has not been paid in. A cooperative society may invest one-third of its capital in another society, thus rendering legal a wholesale society on the English basis. The following is the law of New Jersey relating to cooperative societies, approved March 10, 1884:

AN ACT TO PROVIDE FOR THE FORMATION AND REGULATION OF COOPERATIVE SOCIETIES OF WORKINGMEN.

1. That it shall be lawful for any number of persons, not less than seven, residents in this State, to associate themselves into a society for the purpose of carrying on any lawful mechanical, mining, manufacturing, or trading business, or for the purpose of trading and dealing in goods, wares, and merchandise or chattels, or for the purpose of buying, selling, settling, owning, leasing, and improving real estate and erecting buildings thereon, within this State, upon making and filing a certificate of association, in writing, in manner hereinafter mentioned, and as such shall be deemed to be a corporation and to possess all the powers incident thereto.

2. That such certificate of association shall set forth:

I. The name assumed to designate such society and to be used in its business and dealings, which name shall have the word "cooperative" as a distinguishing part thereof, but shall in no respect be similar to that of any other society organized under this act.

II. The place or places in this State where the business of such society is to be conducted and the location of the principal office of the same.

III. The objects for which the society shall be formed.

IV The total amount of capital stock of such society, the number of shares into which the same is divided, the par value of each share, the manner in which the installments on the shares shall be paid, ... amount actually paid in cash on account of ...

V. The terms of admission of the members.

VI. Mode of application of profits.

VII. The mode of altering and amending the ... by-laws of the society.

3. That the said certificate of association shall be ... associating themselves together, and shall be proved ... seven of them, before an officer qualified to take acknowledgment ... estate, and after being approved by the chief of the bureau ... industries shall be recorded in the office of the clerk of the ... cipal office or place of business of such society shall be ... such certificate shall be filed in the office of the chief of the ... labor and industries.

4. That the business of every such society shall be managed ... board of not less than five directors, who shall respectively ... society, and shall be annually elected at such time and place ... in the by-laws of the society, and one of such directors shall ... and one of them shall be chosen treasurer, and such directors ... their respective offices until their successors are duly qualified; ... shall also have a secretary and such other officers, agents, and factors ... necessary to carry on its business, and shall choose them in the manner ... in the by-laws thereof.

5. That the first meeting of such society shall be called by a notice si... majority of the persons named in the certificate of association, and desig... time, place, and purpose of the meeting, and shall be personally serv... the persons signing said certificate, or by advertisement in a newspaper ... in the county where such society shall have been incorporated, if such ... service can not be made; and at such meeting so called, or at any adjour... ing thereof, a majority of the persons so signing shall constitute a quoru... transaction of business, and shall have power to elect the directors and oth... provided for in section 4 of this act, who shall serve until their succes... qualify, and to adopt by-laws, rules, and regulations for the governme... society.

6. That the by-laws of such society shall provide:

I. For an annual meeting of the members thereof, and such other re... special meetings as may be deemed desirable, the number of members ne... constitute a quorum for the transaction of business and the right of vot... same.

II. For the election of directors and other officers, agents, and factors, ... respective powers and duties.

III. For the limitation of the amount of such real and personal estate a... poses of the society shall require.

IV. Whether the shares, or any number of them, shall be transferable, a... it be determined that the same shall be transferable, provision for thei... and registration, and the consent of the board of directors to the same; a... it shall be determined that the shares shall not be transferable, provisio... ing to members the balance due to them on withdrawal, or of paying no... cases hereinafter mentioned.

V. How members may withdraw from the society.

VI. Whether and by what authority any part of the capital may ... on security of another society through which its products ... supplies secured.

VII. Whether and to what extent credit ... or taken.

VIII. In what sum and with what ...

officers or agents shall give bonds for the faithful performance of their respective duties.

IX. For the audit of accounts.

X. For the distribution of the net profits.

XI. For the custody, use, and device of the seal, which shall bear the corporate name of the society.

7. That every society incorporated under this act shall paint or affix, and shall keep painted or affixed, its name on the outside of every office or place in which the business of the association is carried on, in a conspicuous position, in letters easily legible.

8. That every society incorporated under this act shall have a registered office to which all communications and notices may be addressed, and notices in writing of the location of such office, and of any change therein, shall be filed with the chief of the bureau of statistics of labor and industries, and in the office of the clerk of the county where the office of such society is located.

9. That the capital stock of such society shall be divided into shares the par value of which shall not be more than $50, and no share shall be issued for less than its par value; and that no certificate of shares shall be issued to any member until the shares are fully paid up.

10. That no member of such society shall be entitled to more than one vote upon any subject, which vote must be cast in person; and that the board of directors shall have power, unless otherwise provided in the by-laws of the society, to fix and regulate the number of shares to be held by any one member.

11. That any society incorporated under this act may hold in its corporate name any amount of interest in any other society through which its products are disposed of or its supplies secured: *Provided*, That such interest so held shall not exceed one-third in value of the paid-up capital of the society holding said interest.

12. That the board of directors of every society incorporated under this act shall annually make a statement in writing of the condition of such society, setting forth the amount of capital stock, the number of shares issued and the par value thereof, the number of stockholders and the number of shares held by each, the amount and character of the property of the society and of its debts and liabilities; and said statement shall be signed and sworn to by a majority of directors, including the treasurer, and filed in the office of the clerk of the county where the principal office of such society is located, and that immediately thereafter a copy of such statement shall be forwarded to the chief of the bureau of statistics of labor and industries, who, if he shall have reason to doubt the correctness of such statement or upon the written request of five members of such society, shall cause an examination of the books and affairs of such society to be made and render a correct statement to the members thereof; and every member or creditor thereof shall be entitled to receive from the secretary a copy of such annual statement; and every director or other officer refusing to comply with the requirements of this section, or making and signing a false annual statement of the condition of the society, shall forfeit for each offense the sum of $100, to be recovered in an action of debt in any court of competent jurisdiction in this State by any member or creditor of the society who shall sue for the same.

13. That any member or other person having an interest in the fund of any such society may inspect the books thereof, at all reasonable hours, at the office thereof.

14. That there shall be such distribution of the profits of such society, among the workmen, purchasers, and members, as shall be prescribed in the certificate of association, at such times as therein prescribed, as often at least as once in twelve months: *Provided*, That no such distribution shall be made until a sum equal to 5 per cent of the net profits shall have been appropriated for a contingent or sinking fund, and that such appropriation shall continue to be made until there shall be accumulated a sum equal to 30 per cent of the capital stock of such society.

15. That any member of such society, by [illegible] office of the society, may nominate any person, [illegible] mother, child, brother, sister, nephew or niece, or [illegible] whom his or her share of the capital stock of the society [illegible] or her decease, and from time to time may revoke [illegible] writing similarly delivered; and such society shall keep a [illegible] of all persons so nominated and the number of shares [illegible] recorded: *Provided, nevertheless*, That in lieu of making such [illegible] may provide for payment to all such nominees of the full [illegible] to be transferred: *Provided also*, That if by the by-laws of the [illegible] are transferable, this section shall not be construed to forbid [illegible] shares by sale or will, or otherwise, subject to the consent of the [illegible]

16. That any such society may be dissolved in the manner in [illegible] corporation may be dissolved under existing laws.

17. That where the whole capital of such society shall not have [illegible] the assets of such society shall be insufficient for the payment of [illegible] ties, and obligations, each stockholder shall be bound to pay, on [illegible] him, the sum necessary to complete the amount of such share, as [illegible] tificate of association, or such proportion as shall be required to [illegible] liabilities, and obligations: *Provided, however*, That no such [illegible] required from any person after the expiration of one year from the time he [illegible] to be a member, or for any debt, liability, or obligation contracted [illegible] ceased to be a member of such society.

The following are copies of two typical sets of by-laws. [illegible] by-laws are sent to new societies, as a model for them, by the [illegible] tive Union of America, Cambridge, Mass.; the second are the [illegible] of the Arlington Cooperative Association, Lawrence, Mass., [illegible] because the society is the largest in the country on the Rochdale [illegible]

BY-LAWS OF THE ——.

ARTICLE I.—*Name.*

This society shall be known as the ——.

ARTICLE II.—*Objects.*

The object of this society shall be cooperation on what is known as the Ro[illegible] plan. All sales to be for cash.

ARTICLE III.—*Capital stock.*

The capital stock of the society shall be in shares of $2 each.

Shares may be withdrawn at thirty days' notice or earlier at the option [illegible] committee of management.

ARTICLE IV.—*Membership.*

Any person may upon approval of the committee become a member of the s[illegible] by paying an entrance fee of 25 cents each and giving name and address.

ARTICLE V.—*Meetings.*

SECTION 1. The regular quarterly meetings of the society shall be held [illegible] last —— in January, April, July, and October, the one held in January [illegible] sidered the annual meeting.

SEC. 2. Monthly general meetings shall also be held on the last [illegible] for the discussions of some question connected with the welfare [illegible]

Nominations for the coming vacancies on the committee shall be made at the monthly meeting next previous to the quarterly meeting.

ARTICLE VI.—*Officers of the society.*

SECTION 1. The management of this society shall consist of a president, treasurer, recording clerk, and nine committeemen, all of whom shall have a vote and be elected by ballot.

They shall have the general management of all business carried on by or on account of the society, supervise the buying, and appoint the secretary, salesmen, and other persons necessary for conducting such business, and may assign such persons such remuneration and on such terms as they shall deem fit.

SEC. 2. Two auditors shall be chosen, and after the first term one shall retire at the meeting in January and one at the meeting in July in each year, the first to retire being the one receiving the lowest number of votes at his election. They shall be eligible for reelection.

SEC. 3. The members at a general meeting may remunerate the committee and auditors at such a rate as they deem fit.

SEC. 4. All officers of the society shall hold their offices till others have been chosen and qualified to take their place, excepting those who, having been elected for a period of —— years continuously, may not be reelected until the expiration of one year. Any member is eligible for office.

SEC. 5. After the first election three committeemen shall retire at each of the April, July, and October quarterly meetings, the order of retirement being determined by the votes, those receiving the lowest number of votes at their election retiring first, and at the annual meeting in January the president, treasurer, and recording clerk shall retire, but, with the committeemen, except in the limitation named in section 4, shall be eligible for reelection, or others may be qualified in their stead. After the first term the order of retirement will be in the order of election.

ARTICLE VII.—*Duties of officers.*

SECTION 1. The president shall be chosen by the committee and shall preside at all meetings of the society and committee, countersign all checks, and be ex officio chairman of all committees; and in case of a tie, shall have a casting vote. In his absence a chairman may be chosen for the occasion.

SEC. 2. The treasurer shall be chosen by the committee and shall sign all checks and account for all moneys paid him on behalf of the society.

SEC. 3. The recording clerk shall be chosen by the committee and attend all meetings of the society and committee and record the minutes of the proceedings and all votes in a book kept for that purpose. He may not hold any other remunerated office under the society.

ARTICLE VIII.—*Profits.*

The profits of the society shall be disposed of as follows:

SECTION 1. Interest shall be declared on capital stock at the rate of 6 per cent per annum on $10 or more, interest to commence on the 1st of the month.

SEC. 2. After paying the expenses of the society not less than 5 per cent per annum of the net profits shall be set aside for a sinking or reserve fund, and 2½ per cent for an educational fund.

SEC. 3. The net profits after providing for interest, sinking or reserve fund, and all other claims and expenses, shall be divided among purchasers and employees in such proportions and disposed of in such manner as a majority of members present at a meeting called for that purpose shall deem fit, non-members to receive a share of the profits to the extent of one-half that given to members.

SEC. 4. Members must leave one-half their dividend until owning ten shares.

ARTICLE IX.—...

Should the society find that they have ... employ, the committee may, with the ... invest such surplus or portion thereof in such ...

ARTICLE X.—*Special meetings.*

Special general meetings may be convened by a ... members, being sent to the committee, or by three ...

ARTICLE XI.—*Quorum.*

Ten members at a general meeting and five at a ... constitute a quorum.

ARTICLE XII.—*Vacancies.*

A vacancy of an office occurring during a term, the place ... the members at a regular meeting or at a meeting called for ... committee until the next regular quarterly meeting.

ARTICLE XIII.—*Bonds.*

The treasurer, secretary, salesmen, or assistants and all other ... the members or committee shall give such bonds or securities as ... deem sufficient.

ARTICLE XIV.—*Alteration of rules.*

Any alteration or amendment to these by-laws must be proposed ... meeting, at least, previous to action being taken thereon, and shall not ... unless by a majority of members present and voting.

ARTICLE XV.—*Expulsion of members.*

The committee shall have power to expel any member who persists in ... detrimental to the society's welfare.

ARTICLE XVI.—*Inspection of accounts.*

Any member may, during reasonable office hours, inspect all and ... society's books and accounts, except the private share account of any other ... which he may not inspect except by written authority of said member.

---

BY-LAWS OF THE ARLINGTON COOPERATIVE ASSOCIATION.

ARTICLE I.—*Name.*

SECTION 1. This association shall be known as the Arlington Cooperative ... ciation.

SEC. 2. The place or places of business shall be at such location or locations ... the association may from time to time determine. All books of account, ... and documents of the association, other than such as are required for ... on of business on account of the same elsewhere, shall be kept at ... office at Lawrence, Mass.

ARTICLE II.

The object of this association is to carry on the trade of ... hold supplies, on the cooperative plan.

ARTICLE III.

SECTION 1. Any person, upon ... member of this association after ...

fee of 50 cents, and signing a declaration of his readiness to take at least one share of stock, and willingness to conform to the by-laws of this association. Such proposal shall give the person's name, trade, and address, and shall be signed by the member making the proposal, which shall be sent to the clerk with the entrance fee, and entered by him in a book kept for that purpose. If approved by the directors, he shall be considered a member upon payment of at least $1 on account of his subscription, as otherwise provided.

SEC. 2. Candidates for membership rejected by the directors shall have the power of appeal through any member to the general meeting. Any person rejected shall have his entrance fee returned on application.

SEC. 3. At the general office a list of members' names, trades, and residences shall be kept, and no person shall be deemed a member unless his name appears on this list. The names of all persons who under these by-laws cease to be members shall be erased therefrom.

ARTICLE IV.—*Funds and revenues.*

SECTION 1. The permanent capital of this association shall be in shares of $5 each. Each member of this association can hold from one to two hundred shares. The capital of the association shall be invested by the board of directors in purchase of stock and fixtures agreeable to the articles of agreement.

SEC. 2. From the absolute profits, after paying the expenses of the association, not less than 10 per cent per annum shall be set aside for a sinking fund, which shall be allowed to accumulate until it amounts to a sum equal to 30 per cent in excess of the capital stock.

SEC. 3. Interest shall be declared on the capital stock at the rate of 5 per cent per annum, payable quarterly. The directors shall at any time have power, with the sanction of the quarterly meeting, to reduce the rate of interest.

SEC. 4. After deducting from the profits the amounts provided in sections 2 and 3 of this article, and the amount assessed as State and city tax, the remainder shall be divided quarterly in proportion to the amount expended in purchase of goods. Non-members will receive a share of the profits to the extent of one-half of that given to members.

SEC. 5. Receipts from entrance fees shall be added to the sinking fund of the association.

SEC. 6. Interest will commence on the first of each month upon all paid-up shares.

SEC. 7. No interest will be paid on shares withdrawn before the end of the quarter.

SEC. 8. Dividends and interest may remain to the credit of the shareholders at their option, and whenever such accumulation amounts to the par value of one share, interest shall be declared according to the provisions of section 6 of this article, provided that this amount, with that of the original shares invested, shall not exceed the par value of 200 shares.

SEC. 9. Should the directors find that they have more cash in the treasury than they can profitably employ in the ordinary business of the association, they may, with the sanction of the quarterly meeting, invest any such portion of the cash in such manner as may be deemed advisable.

ARTICLE V.—*Officers.*

SECTION 1. The officers of this association shall consist of a president, clerk, treasurer, and ten other persons, who, with the exception of the clerk, shall constitute the board of directors.

SEC. 2. The clerk, treasurer, and eleven directors shall be chosen annually by the stockholders by ballot, and shall hold their office for one year, and until others are chosen and qualified in their stead, provided that it shall be considered requisite, unless they may be otherwise disqualified, to reelect one-half of the old board at each election. The president shall be chosen by the board of directors from their number.

SEC. 3. Candidates for the board of directors ... tion shall be nominated at the quarterly ... meeting at which they are proposed for ... not be sufficient to fill the vacancies, the ... number not so nominated.

SEC. 4. An auditor should be chosen annually by ...

SEC. 5. No person shall be eligible for the office ... member of the association for at least six months, ... has not been a director twelve months; or if he hold ... profit under the association; if he is concerned in or ... any contract with the association; or if he carries ... association; or if he has a relative employed by the association ...

ARTICLE VI.—*Duties of officers.*

SECTION 1. It shall be the duty of the president to preside ... association and board of directors, and, in case of equal division, ... vote, besides his own vote as a member. He shall sign all documents ... issued by the association or board of directors; shall hold in his ... bonds of the treasurer, or any of the employees where bonds are ... perform all other duties usually appertaining to his office.

SEC. 2. It shall be the duty of the clerk to keep a correct record ... meetings of the association and board of directors, and, in case of ... the president, to call the meeting to order. He shall keep a correct ... names and residences of the members, and shall serve all notices ... board; he shall carry on all correspondence and shall attest all documents ... the association or board of directors, and shall deliver to his successor ... property in his possession belonging to the association.

SEC. 3. It shall be the duty of the treasurer to receive all moneys due fro... bers and others, and to disburse the same in payment of claims against the ... tion when approved by an advisory committee. He shall keep such rec... business transactions of the association not otherwise provided for by th... of directors, and shall make a full and complete report at each quarterly me... the financial condition of the association. He shall give such bonds for the ... performance of his duties as the board of directors may require, and shall ... to his successor all books, money, vouchers, and other property of the asso... in his possession.

SEC. 4. It shall be the duty of the auditor, at the close of each quarter's b... to audit the accounts of the treasurer and other officers, including stock o... demanding for this purpose any information he may see fit, and report cond... the same at the quarterly meeting.

ARTICLE VII.—*Management.*

SECTION 1. The board of directors shall have control of all business carried ... or on account of, the association; the purchase and sale of goods; the ... managers and of all other persons necessary for conducting the business; ... for places of meeting; the rates of payment for work or services done ... the association, and the regulation of salaries and securities of the ... employees, to whom it may assign such duties as it shall deem proper ...

SEC. 2. The board of directors shall, with the consent of a ... power to lease, purchase, or erect any building or buildings ... association, and to mortgage, rent, or sell such buildings or ...

SEC. 3. The board of directors shall meet ... shall provide for the detailed work of the ... mittees. It shall in all things act for ...

acts and orders under the powers delegated to it shall have like force and effect as if they were acts and orders of a majority of the members of the association at a general meeting thereof. Every question at such meeting of the board shall be decided by a majority of votes cast. Seven members of the board shall constitute a quorum.

SEC. 4. The president, at the request of three members of the board, shall call a special meeting thereof by giving one day's notice in writing to the clerk; but no business shall be taken into consideration other than that specified in the notice. The board shall convene special meetings of the stockholders at their discretion, allowing three days' notice of the same.

SEC. 5. The board shall cause the accounts of all business carried on to be regularly entered in proper books, and a quarterly report to be made out covering all business to the end of each quarter, which, together with all necessary vouchers, shall be submitted to the auditor not less than seven days previous to such meeting, and shall be printed and distributed to the members as early as possible thereafter.

SEC. 6. All meetings of the board of directors shall be open to the attendance of other members, but no such member shall take part in its discussions.

ARTICLE VIII.—*Meetings.*

SECTION 1. The general meetings of this association will be held quarterly, on the last Wednesdays of January, April, July, and October of each year.

SEC. 2. The annual meeting of this association for the election of officers shall be held on the last Wednesday of April of each year.

SEC. 3. The president shall cause a special meeting of the association to be called, upon a written request, signed by ten members of the association. At such meeting no other business shall be transacted than that named in such requisition.

SEC. 4. Notices of all meetings shall be posted by the clerk in a conspicuous place in the store or stores, three days previous to the same. In the case of special meetings, such notice shall state the object of the meeting.

SEC. 5. At all meetings of the association fifteen members shall constitute a quorum for the transaction of business.

SEC. 6. No subject foreign to the purpose of the association shall be introduced at any of its meetings for business.

ARTICLE IX.—*General Regulations.*

SECTION 1. Quarters shall commence on the first of January, April, July, and October.

SEC. 2. All purchases from the association shall be made strictly for cash.

SEC. 3. Each member of the association shall be provided with a book of account, in which shall be entered a statement of shares held, with quarterly dividends and interest.

SEC. 4. Each purchaser shall be provided with checks or a pass book at the option of the directors, which shall show the amount of purchases on which dividend will be declared. The same to be returned, as provided from time to time by the board of directors.

SEC. 5. No intoxicating liquors shall be allowed on sale by the association.

SEC. 6. Any complaint as to quality or prices of goods sold by the association, or respecting the conduct of any of its employees, should be made to the directors in writing, signed by the party making the complaint, and such complaint shall be investigated and decided by the board.

SEC. 7. All sales are to be made at the average retail market price.

SEC. 8. Each stockholder shall be entitled to a certificate of his stock, under the seal of the corporation, signed by the president and the treasurer.

SEC. 9. No stockholder shall be entitled to more than one vote in any business meeting of the association.

SEC. 10. The board of directors may ... the benefits of the association who ... until it shall submit the matter to a ... days' notice thereof, which meeting shall ... directors, or otherwise, as it may think proper.

ARTICLE X.—*Withdrawals.*

SECTION 1. Any member of this association desiring to ... ciation the whole or any part of his or her stock shall ... to the directors, and within thirty days from the date of ... shall pay or cause to be paid such applicant the amount of ... withdraw, but if the board fail to pay or cause the same to be ... days, said applicant may transfer his or her shares to any ... the member has not the full number of shares allowed in the by-laws ... shall he transfer his share to persons not members of the association ... obtain consent of the board of directors, which consent shall ... president and clerk, and entered on the records of the association. ... ber transfer his share or shares he must surrender his certificate ... directors, and the board shall cause a new certificate to be issued ... whom he makes such transfer.

SEC. 2. Any member being in distress may withdraw any shares he ... the association at the discretion of the board of directors.

ARTICLE XI.—*Amendments.*

These by-laws shall not be altered or amended unless such alteration or ... ment be proposed in writing one meeting previous to action being taken t... providing, also, that two-thirds of the members present vote in the affirmati...

ARTICLE XII.—*Vacancies.*

In case of a vacancy in the offices of clerk, treasurer, auditor, or any ... directors, by resignation or otherwise, a substitute shall be elected at t... quarterly meeting, or at a special meeting called for that purpose.

ARTICLE XIII.

That the duties of the treasurer be so modified as to make him simply cu... of the money, and to entail no other bookkeeping than a simple cash acco... signing of checks and certificates, with a true rendering of said accoun... called upon.

That the duties of the clerk be made to include all bookkeeping and mone... actions with the members required by the association, as well as all present ... and that he be made the executive officer, under the directors, of all duties a... by them from time to time. That his position be sufficiently remunerate... insure a perfectly reliable and satisfactory incumbent, whose whole attent... be given to the calls of the association, and that he be placed under equal ... with the treasurer.

# RECENT REPORTS OF STATE BUREAUS OF LABOR STATISTICS.

## KANSAS.

*Eleventh Annual Report of the Kansas Bureau of Labor and Industry.* 1895. William G. Bird, Commissioner. vi, 211 pp.

This report treats of the following subjects: The milling industry, 31 pages; manufacturing industries, 40 pages; the salt industry, 11 pages; labor, 58 pages; labor organizations, 24 pages; strikes, 7 pages; sociology, 27 pages.

THE MILLING INDUSTRY.—Reports were secured from 190 milling establishments located in different parts of the State, but of this number 21 failed to give sufficient data to justify tabulation. The information contained in the reports is presented in detail and summarized for the State. The following statement shows the principal facts for the State:

MILLING STATISTICS OF KANSAS FOR THE YEAR ENDING DECEMBER 31, 1895.

| Items. | Mills reporting. | Total. |
|---|---|---|
| Cost of plant | 160 | $4,084,851 |
| Commercial value of plant | 127 | $2,832,100 |
| Capacity, 24 hours' run, barrels | 156 | 25,762 |
| Output for year, barrels | 103 | 2,241,066 |
| Sets of rolls | 150 | 1,401 |
| Sets of buhrs | 96 | 137 |
| Materials— | | |
| Wheat, bushels | 115 | 11,982,268 |
| Cost | 103 | $7,194,106 |
| Corn, bushels | 89 | 1,241,669 |
| Cost | 77 | $292,932 |
| All other grain, bushels | 52 | 567,379 |
| Cost | 38 | $57,984 |
| Products— | | |
| Flour, wheat, pounds | 96 | 428,826,360 |
| Value | 83 | $5,600,484 |
| Corn meal, pounds | 66 | 25,871,654 |
| Value | 64 | $697,968 |
| Other flour or meal, pounds | 22 | 5,787,900 |
| Value | 20 | $61,232 |
| Offal, pounds | 86 | 371,022,796 |
| Value | 78 | $1,643,569 |
| Employees— | | |
| Greatest number | 141 | 1,314 |
| Smallest number | 128 | 672 |
| Average number | 139 | 929 |
| Total wages paid during year | 112 | $458,126 |

MANUFACTURING INDUSTRIES.—Reports were secured from a number of establishments engaged in various kinds of industry. These reports represent 42 counties from all sections of the State, and hence represent the State at large. The reports are presented in detail, by industries, for each county. Answers are given to numerous questions concerning employees, wages, strikes, principal markets for produ[illegible] freight rates, etc. The totals for the six most important que[illegible] are

are as follows: Capital invested, $111,11[illegible] [illegible] $7,768,095; repairs, $2,661,376.04; products, $[illegible] [illegible] $50,985,283.06; wages, $4,508,078.18.

The Salt Industry.—A historical and descriptive [illegible] of the salt industry of the State, and statistics are [illegible] industry by county totals. The seven salt plants from [illegible] were secured reported capital at $875,000, cost of plant [illegible] value of product at $261,051.80, cost of materials at $[illegible] at $57,760.77.

Labor.—A detailed tabular account is given of reports [illegible] 519 workmen engaged in various industries throughout the [illegible] following statement, presenting the totals for five of the [illegible] treated, indicates the character of the information solicited:

REPORTS OF LABORERS, 1896.

| Items. | Carpenters. | Miners. | Barbers. | Compositors. | [illegible] |
|---|---|---|---|---|---|
| Number reporting | 20 | 58 | 36 | 42 | [illegible] |
| Number reporting earnings | 20 | 56 | 35 | 41 | [illegible] |
| Earnings for year | $12,818.00 | $22,512.80 | $19,299.65 | $[illegible] | [illegible] |
| Average | $640.90 | $402.01 | $551.41 | $[illegible] | [illegible] |
| Number reporting cost of living | 15 | 46 | 31 | [illegible] | [illegible] |
| Cost of living for year | $5,286.00 | $13,896.00 | $11,544.00 | $[illegible] | [illegible] |
| Average | $352.40 | $302.06 | *a* $372.35 | $670.84 | [illegible] |
| Number reporting savings | 4 | 3 | 3 | 9 | [illegible] |
| Savings for year | $632.75 | $350.00 | $500.00 | $1,512.[illegible] | [illegible] |
| Average | *b* $158.20 | $116.66 | $166.66 | $168.75 | [illegible] |
| Number reporting debt | 6 | 28 | 6 | 3 | [illegible] |
| Debt for year | $265.00 | $1,666.00 | $362.00 | $135.00 | [illegible] |
| Average | $44.16 | $59.50 | $60.33 | $45.00 | [illegible] |

*a* Figures here apparently should be $372.38; those given are, however, according to the [illegible]

*b* Figures here apparently should be $158.18; those given are, however, according to the [illegible]

Labor Organizations.—Of 142 schedules sent to labor orga[illegible] zations throughout the State, 92 replies, sufficiently complete for t[illegible] ulation, were received. Some of the facts presented for th[illegible] organizations are summarized in the following statement:

LABOR ORGANIZATIONS, 1896.

| Items. | Number of organizations reporting. | Items. | Number of organizations reporting. |
|---|---|---|---|
| Membership— | | Immigration— | |
| Increased | 41 | Has affected craft | [illegible] |
| Decreased | 22 | Has not affected craft | [illegible] |
| No change | 25 | | |
| Total answering question | 88 | Total answering question | [illegible] |
| | | Strikes— | |
| Wages— | | Involved in | [illegible] |
| Increased | 3 | Not involved in | [illegible] |
| Decreased | 6 | | |
| No change | 78 | Total answering question | [illegible] |
| | | Boycotts— | |
| Total answering question | 87 | Involved in | [illegible] |
| | | Not involved in | [illegible] |
| Hours of work— | | | |
| Increased | 1 | Total answering question | [illegible] |
| Decreased | 13 | [illegible] | [illegible] |
| No change | 70 | [illegible] | [illegible] |
| Total answering question | 84 | [illegible] | [illegible] |

STRIKES.—An account is given of the important strikes and labor troubles that occurred in the State during the year.

SOCIOLOGY.—Under this title is presented a number of specially prepared articles on the labor problem from college professors, economists, and labor leaders.

## MONTANA.

*Third Annual Report of the Bureau of Agriculture, Labor and Industry of Montana for the year ended November 30, 1895.* James H. Mills, Commissioner. 205 pp.

The subject-matter of this report is grouped as follows: Introductory, 21 pages; economic statistics of counties, 18 pages; fruit growing, 19 pages; official directory of United States, State, and county officials, 24 pages; labor organizations, railroad traffic, wages, production, and investments, 34 pages; precious and semiprecious metals, 12 pages; agriculture and stock growing, 32 pages; miscellaneous statistics, 11 pages; laws of special interest to wage earners, 16 pages.

ECONOMIC STATISTICS OF COUNTIES.—The following statement summarizes for the entire State some of the facts presented under this title for the different counties:

REVENUE, EXPENSES, ETC., OF COUNTIES FOR CALENDAR YEAR 1894.

| | |
|---|---|
| Revenue from licenses | $369,674.77 |
| Total expenses (a) | $1,909,169.80 |
| Amount collected for public school fund (a) | $618,210.87 |
| Amount expended for public schools (a) | $541,867.14 |
| Number of criminal cases begun by finding informations and indictments | 465 |
| Number of civil cases begun | 2,706 |
| Number of letters of administration granted | 307 |
| Number of executions issued during the year | 515 |
| Number of real estate transfers | 8,151 |
| Amount of real estate transfers | $13,200,072.77 |
| Number of real estate and chattel mortgages recorded | 6,965 |
| Amount of real estate and chattel mortgages recorded | $10,470,249.62 |
| Number of mechanics' liens filed | 479 |
| Amount of mechanics' liens filed | $118,914.04 |
| Number of divorces granted | 228 |
| Number of marriage licenses granted | 1,598 |
| Number of persons to whom final naturalization papers were issued | 9,934 |
| Indebtedness of counties March 1, 1894 | $2,584,910.11 |
| Indebtedness of counties March 1, 1895 | $2,856,053.71 |

LABOR ORGANIZATIONS, RAILROAD TRAFFIC, WAGES, PRODUCTION, AND INVESTMENTS.—The name, location, date of organization, and amount of benefits paid are given for 82 labor organizations in the State. These organizations had a membership of 9,186, of whom 36 were females. Excluding persons engaged in farming and stock

a For fiscal year ended February 28, 1895.

raising, it was estimated that 81.95 per cent [illegible] State were in employment on July 1, 1895. [illegible] in farming and stock raising, 89.37 per cent [illegible] wages, employees, and traffic are shown for the [illegible] the State, and the wages, employees, capital, and [illegible] for selected industries.

PRECIOUS AND SEMIPRECIOUS METALS.—The follow[illegible] summarizes the statistics given for the State under this [illegible]

ESTIMATED VALUE OF METALS PRODUCED.

| Kind of metal. | 1895. |
|---|---|
| Gold | [illegible] |
| Silver | *a* [illegible] |
| Copper | 17,[illegible] |
| Lead | [illegible] |
| Total | *b* [illegible] |

*a* At coining value. *b* Silver included at coining value.

AGRICULTURE AND STOCK GROWING.—The average produ[illegible] acre and the average prices are shown for farm products; [illegible] proportional amount of the sales in Montana of the produ[illegible] State as compared with the sales of imported products. The [illegible] living is indicated by tables showing the average prices of d[illegible] commodities.

The average wages per month of employees on farms an[illegible] ranges is shown for each county. The following statement p[illegible] the averages for the entire State:

AVERAGE WAGES PER MONTH, INCLUDING BOARD AND LODGING, OF EM[illegible] ON FARMS AND STOCK RANGES.

| Class of employees. | Year [illegible] Jun[illegible] 1894. |
|---|---|
| Foremen | [illegible] |
| Herders | [illegible] |
| Range riders | [illegible] |
| Farm hands | [illegible] |
| Cooks | [illegible] |

## NEW JERSEY.

*Eighteenth Annual Report of the Bureau of Statistics of Labor an[illegible] tries of New Jersey.* For the year ending October 31, 1895. [illegible] H. Simmerman, Chief. viii, 251 pp.

The following subjects are treated in this report: [illegible] 1894–95, 61 pages; effect of occupation on the health [illegible] the trade-life of workmen, 99 pages; the distrib[illegible] pages; free public employment offices, [illegible] in New Jersey, 17 pages; cooperative building and loan associations of New Jersey, 53 pages.

PANIC INQUIRY, 1894–95.—This inquiry is a continuation of that of 1893–94, the results of which were published in the report of the bureau for 1894 in the chapter relating to the industrial depression of 1893–94. Statistics covering the year ending June, 1895, were obtained from 196 manufacturers in 40 general industries, showing, month by month, the changes in the number of persons employed and in the amount of wages paid. Returns were also secured from 93 manufacturers in 39 general industries, showing the value of the goods produced by them during the year ending June, 1895. The figures are brought into comparison with the value of the output of the same establishments during the preceding year.

All the establishments from which returns were secured giving the number of employees and the amount of wages paid were in operation in June, 1893, and a comparative showing of the reported facts is made for June, 1893, June, 1894, and each succeeding month to and including May, 1895. Of the establishments reporting these facts, 86 were engaged in the manufacture of textiles and textile products and 57 were engaged in the manufacture of metals and metallic products. The establishments engaged in these two lines of production constituted more than two-thirds of the entire number reporting.

The following statement shows the number of employees and the amount of wages paid in the establishments reporting in June, 1893, June, 1894, and in each succeeding month to and including May, 1895, with the percentages of increase or decrease.

EMPLOYEES AND WAGES, JUNE, 1893, AND JUNE, 1894, TO MAY, 1895.

[Each percentage of decrease or increase after June, 1894, indicates the decrease or increase since June, 1894.]

| Month. | Employees. | | | Wages. | | |
|---|---|---|---|---|---|---|
| | Average number. | Per cent of increase. | Per cent of decrease. | Amount. | Per cent of increase. | Per cent of decrease. |
| June, 1893 | 35,457 | .......... | .......... | $1,086,537 | .......... | .......... |
| June, 1894 | 31,857 | .......... | 10.2 | 907,558 | .......... | 16.5 |
| July, 1894 | 31,843 | .......... | .......... | 868,319 | .......... | 4.3 |
| August, 1894 | 32,218 | 1.1 | .......... | 922,960 | 1.7 | .......... |
| September, 1894 | 33,140 | 4.0 | .......... | 938,154 | 3.4 | .......... |
| October, 1894 | 33,933 | 6.5 | .......... | 1,004,121 | 10.6 | .......... |
| November, 1894 | 34,853 | 9.4 | .......... | 971,822 | 7.1 | .......... |
| December, 1894 | 34,186 | 7.3 | .......... | 1,003,131 | 10.5 | .......... |
| January, 1895 | 34,506 | 8.3 | .......... | 941,373 | 3.7 | .......... |
| February, 1895 | 34,684 | 8.9 | .......... | 955,646 | 5.3 | .......... |
| March, 1895 | 35,138 | 10.3 | .......... | 1,068,703 | 17.8 | .......... |
| April, 1895 | 36,594 | 14.9 | .......... | 1,081,226 | 19.1 | .......... |
| May, 1895 | 37,057 | 16.3 | .......... | 1,139,174 | 25.5 | .......... |

The figures secured from the 93 establishments in 39 general industries, showing the value of their products for the year ending June, 1895, when compared with similar figures from the same establishments for the preceding year, show a substantial [illegible] business. For the year ending June, 1894, the [illegible] factured products of the establishments [illegible] for the succeeding year the total value of such [illegible] 730; an increase of $2,870,847, or 20.5 per cent.

EFFECT OF OCCUPATION ON THE HEALTH AND DU[illegible] TRADE-LIFE OF WORKMEN.—The inquiry in regard [illegible] first undertaken by the bureau in 1889. The information [illegible] was obtained from individual journeymen at work at [illegible] trades, and the data thus gathered are presented in tables, [illegible] occupations, showing the age at which the workman began [illegible] his trade, his present age, the age at which he first began [illegible] (i. e., lose his activity as a workman), the number of years act[illegible] work, and whether American or foreign born. The inquiry h[illegible] carried on for a period of six years, and returns have been secu[illegible] 19,947 workmen, employed in nearly 80 occupations. The [illegible] secured in 1895 from 1,167 individuals are tabulated in connect[illegible] those previously secured.

The following statement shows the principal results of the [illegible] gation as to certain occupations in which returns were secur[illegible] comparatively large numbers of workmen:

AGE, YEARS AT WORK, ETC., OF WORKMEN IN CERTAIN OCCUPATI[illegible]

| Occupations. | Number of individuals reported. | Average. | | Per cen[illegible] beginning to decline. |
|---|---|---|---|---|
| | | Present age. | Years at work. | |
| Bakers | 933 | 33.4 | 16.9 | [illegible] |
| Bricklayers and masons | 1,022 | 36.4 | 18.5 | [illegible] |
| Carpenters | 2,732 | 36.5 | 18.7 | [illegible] |
| Cigar makers | 1,061 | 31.3 | 14.4 | [illegible] |
| Hat finishers | 1,257 | 32.8 | 15.0 | 12.[illegible] |
| Hat makers | 1,247 | 34.5 | 15.5 | 17.[illegible] |
| Miners, iron ore | 1,269 | 34.1 | 16.0 | 7.[illegible] |
| Potters, pressers | 455 | 30.5 | 15.7 | [illegible] |
| Potters, kiln men | 297 | 32.6 | 14.5 | 10.[illegible] |
| Painters | 1,235 | 34.5 | 15.0 | 12.[illegible] |
| Plumbers | 661 | 32.0 | 16.0 | 10.[illegible] |
| Printers | 462 | 31.2 | 14.5 | [illegible] |
| Railroad, locomotive engineers | 449 | 43.1 | 13.5 | [illegible] |
| Railroad, locomotive firemen | 411 | 31.4 | 6.6 | [illegible] |
| Railroad, brakemen | 1,445 | 30.2 | 8.2 | [illegible] |
| Stonecutters | 701 | 33.5 | 19.0 | ........ |

THE DISTRIBUTION OF WEALTH.—The distribution of visi[illegible] able wealth among those liable to taxation in four of the princi[illegible] of the State, namely, Camden, Newark, Jersey City, and Pater[illegible] been ascertained by means of transcripts made of the tax du[illegible] for 1895 of the cities named. The data thus secured show [illegible] assessed valuations of $1,000 and upward covered over 90 [illegible] of the total assessed valuation of taxable property in each of [illegible] cities.

The following statement shows the total [illegible] erty and the number of persons, [illegible]

**assessed for taxation in 1895 in each of the cities named, in amounts under $1,000, from $1,000 to $5,000, and $5,000 and over:**

ASSESSED VALUATION OF PROPERTY AND NUMBER OF PERSONS ASSESSED IN FOUR PRINCIPAL CITIES OF NEW JERSEY, 1895.

| Items. | Camden. | Newark. | Jersey City. | Paterson. | Total. |
|---|---|---|---|---|---|
| Valuations under $1,000— | | | | | |
| Amount | $2,756,644 | $6,676,423 | $7,460,417 | $3,281,942 | a $20,182,652 |
| Per cent of total valuation | 8.3 | 5.2 | 8.4 | 8.3 | 7.0 |
| Persons assessed (b) | 7,102 | 21,482 | 21,681 | 10,916 | 61,181 |
| Valuations from $1,000 to $5,000— | | | | | |
| Amount | $12,214,874 | $40,392,941 | $30,154,810 | $15,750,917 | c $98,506,316 |
| Per cent of total valuation | 36.9 | 31.6 | 33.7 | 40.0 | 34.0 |
| Persons assessed (b) | 5,818 | 16,933 | 11,958 | 5,780 | 40,489 |
| Valuations of $5,000 and over— | | | | | |
| Amount | $18,128,480 | $80,805,770 | $51,784,395 | $20,354,607 | $171,072,252 |
| Per cent of total valuation | 54.8 | 63.2 | 57.9 | 51.7 | 59.0 |
| Persons assessed (b) | 962 | 4,746 | 3,275 | 730 | 9,713 |
| Total valuation— | | | | | |
| Amount | $33,099,998 | $127,875,134 | $89,399,622 | $39,387,466 | $289,762,220 |
| Persons assessed (b) | 13,882 | 43,161 | 36,914 | 17,426 | 111,383 |

a Figures here apparently should be $20,175,426; those given are, however, according to the original.
b Including firms and corporations.
c Figures here apparently should be $98,513,542; those given are, however, according to the original.

COOPERATIVE MOVEMENT IN NEW JERSEY.—A historical and descriptive statement is made of the formation and development of cooperative societies in the State, and separate accounts are given of 24 such organizations, showing the nature of the business carried on by them, the extent of their operations, and the success or failure attending their efforts.

The following extract is made from the text on this subject:

In the early seventies, under the patronage of the Sovereigns of Industry, the Patrons of Husbandry, and the Knights of Labor, various attempts were made in this State to organize cooperative associations for the purchase of supplies, farming implements, and articles of household necessity. These were, however, mostly voluntary associations without any legal status as corporate bodies. A few of them incorporated under the general corporation law of the State, but most of them simply adopted a constitution and by-laws and a form of certificate of shares, etc. Under this plan of association the members were simply joint partners and individually liable for all debts contracted. This for a time gave them almost unlimited credit in the purchase of such articles as they chose to handle, but it also operated to discourage members who possessed property, who on the first reverse would become alarmed and withdraw. Consequently most of these enterprises were short lived. But in 1881 the legislature passed an act entitled "An act to encourage the formation of cooperative associations among workingmen," since which time certificates of association of 40 cooperative societies have been filed in the office of the bureau of statistics of labor and industries and approved by the chief as the law requires. Seven of these were organized for manufacturing or productive cooperation, and 33 for distribution or exchange. Not one of the productive associations ever began business. Of the 33 distributive societies organized 25 began business and 8 did not. Of those that began business 10 are still in operation.

COOPERATIVE BUILDING AND LOAN [illegible]
SEY.—The new legislative requirement [illegible]
made as of October 31, and compiled within [illegible]
vented the complete tabulation of the statistical [illegible]
by the building and loan associations of the State [illegible]
rogatories issued by the bureau in August, 1895, and [illegible]
as were made could not be analyzed and summarized [illegible]
manner.

In the following statement the general statistics [illegible]
shareholders, and capital and profits are summarized [illegible]
tions from which returns were received, the returns [illegible]
in operation less than one year not being included:

| | |
|---|---|
| Total associations reporting | [illegible] |
| Shares in force | [illegible] |
| Shares borrowed on | [illegible] |
| Shareholders | [illegible] |
| Borrowers | [illegible] |
| Installment dues on shares in force | [illegible] |
| Net profits on shares in force | [illegible] |
| Net assets | [illegible] |

The total outstanding indebtedness reported by 234 asso[illegible]
amounted to $1,486,081, including overpayments and amounts
canceled shares, but not including net worth, unearned premiu[illegible]
undelivered loans. One hundred and fifty-seven associations bo[illegible]
$1,842,202, and 164 associations repaid borrowed money to the [illegible]
of $1,858,319 during the year, the latter amount including paym[illegible]
borrowings of previous years. Two hundred and ninety-four [illegible]
tions reported receipts and disbursements during the year am[illegible]
to $17,272,430 and $16,206,831, respectively.

*La Petite Industrie: Salaires et Durée du Travail; Tome II, Le Vêtement à Paris.* Office du Travail, Ministère du Commerce, de l'Industrie, des Postes et des Télégraphes. 1896. 721 pages.

Since its organization the French labor bureau has been carrying on a series of investigations concerning the wages and hours of labor of workingmen in France. Several volumes giving the partial results of the investigation have been issued, but the final volume, in which the information that has been obtained will be summarized and analyzed, has not yet appeared. As a branch of this general investigation, a special investigation has at the same time been carried on concerning the conditions of labor in the minor industries at Paris. By minor industries is meant such work as is carried on either in the homes of the workingmen in connection with stores and shops, or in such small industrial establishments that they scarcely merit the designation of independent factories. Such, for example, are the industries of tailoring, dressmaking, bread baking, slaughtering, and preparation of meats, etc.

The first volume related to industries in connection with food products, such as bakeries, butcher shops, supplies of milk, etc., and was issued in 1893. (*a*) The second volume, giving the results of this investigation as far as the clothing trades at Paris are concerned, has just been published by the bureau.

The nature of the industries treated in these two volumes has necessitated the adoption of a special method of investigation. Instead of conducting the inquiry along identical lines for the different industries, so that results could be assembled in general tables from which general statements concerning the condition of labor could be deduced, it has been found necessary to take up each individual industry, or in cases each individual occupation separately, and to write a monographic treatise concerning the conditions pertaining to it, to which are added, somewhat in the nature of exhibits, a series of studies of the conditions of labor in a number of particular establishments.

The general portion of each monograph thus first gives an account of the estimated number of such establishments in Paris, the conditions under which the industry is carried on, that is, the mechanical

---

*a* La Petite Industrie: Salaires et Durée du Travail; Tome I, L'Alimentation à Paris. Office du Travail, Ministère du Commerce, de l'Industrie et des Colonies. 1893.

appliances necessary, the number of [illegible] the occupations of these assistants, their wages [illegible] chances that they have for themselves becoming [illegible] ments, etc. The study of conditions in particular [illegible] upon schedules filled in by M. Maroussem, to whom [illegible] was intrusted by the bureau, then follows.

In other words, the object of the investigation has [illegible] either a complete or even partial statistical census of the [illegible] considered, but rather to present the information in such [illegible] show the conditions of labor in each and the general [illegible] rounding its conduct.

*Répartition des Salaires du Personnel Ouvrier dans les Man[illegible] de l'État et les Compagnies de Chemins de Fer.* Office du [illegible] Ministère du Commerce, de l'Industrie, des Postes et d[illegible] graphes. 1896. 154 pp.

This report relating to the wages of employees of the Gov[illegible] manufactories and the railway companies of France in 1895 [illegible] pared by the French labor bureau in pursuance of a reques[illegible] committee on labor of the Chamber of Deputies. The Gov[illegible] manufactories referred to comprise the 20 establishments for [illegible] ufacture of tobacco and the 7 establishments for the manuf[illegible] matches, both of which industries are in France state mo[illegible] The railway companies include all of the 7 great companies in [illegible] the railway system of France is divided, as well as 4 compan[illegible] ating what are called secondary lines.

The information is given for each of these three branches se[illegible] and relate to the amount of wages earned during a single w[illegible] ment period. In other words, the information is not based on [illegible] age for the year. In no instance are wages given either by [illegible] even by the most general grouping of occupations, the only [illegible] being those according to sex, the age periods 12 to 18, 19 to [illegible] 45, 46 to 60, and over 60 years, and the amount of the earnin[illegible] employees. As the report says, therefore, "the present inve[illegible] ought not to be regarded as furnishing indications rigorou[illegible] concerning the absolute amount of wages, but rather as givin[illegible] a sufficiently accurate idea of the relative situation of the [illegible] workingmen from the standpoint of their earnings."

The inquiry covers a total of 174,864 employees, of whic[illegible] were employed in the manufacture of tobacco, 2,120 in the man[illegible] of matches, and 157,624 in railway transportation.

Of the 15,120 employees of the tobacco factories, 1,44[illegible] cent, were males, and 13,674, or 90 per cent, were females. T[illegible] sion according to age periods as well as sex is shown [illegible] table, in which the amount of the earnings per [illegible] is shown according to sex and age periods. It [illegible] wages are here paid almost exclusively by the piece and have been reduced to the hour basis. The number of hours worked per day is ten.

AVERAGE EARNINGS PER HOUR OF EMPLOYEES OF GOVERNMENT TOBACCO FACTORIES.

| Age. | Number of employees receiving— | | | | | | | | | | All employees. | | | |
|---|---|---|---|---|---|---|---|---|---|---|---|---|---|---|
| | $0.0048 to $0.0676 per hour. | | $0.0677 to $0.0801 per hour. | | $0.0801 to $0.1063 per hour. | | $0.1063 to $0.1255 per hour. | | Over $0.1255 per hour. | | Male. | | Female. | |
| | Male. | Female. | Male. | Female. | Male. | Female. | Male. | Female. | Male. | Female. | Number. | Average wages. | Number. | Average wages. |
| 12 to 18 years | 7 | 34 | 1 | 10 | ...... | 6 | ...... | ...... | ...... | ...... | 8 | $0.0[illegible] | 50 | $0.0[illegible] |
| 19 to 25 years | 2 | 377 | 5 | 66 | 15 | 107 | 4 | 9 | 2 | 1 | 28 | .085 | 560 | .059 |
| 26 to 45 years | 19 | 6,262 | 113 | 1,431 | 562 | 1,255 | 225 | 149 | 88 | 58 | 947 | .102 | 9,145 | .062 |
| 46 to 60 years | 13 | 2,771 | 48 | 431 | 186 | 301 | 84 | 58 | 42 | 11 | 298 | .104 | 3,007 | .060 |
| Over 60 years | 6 | 281 | 9 | 28 | 43 | 34 | 11 | 9 | 8 | ...... | 77 | .100 | 352 | .056 |
| Total | 47 | 9,715 | 176 | 1,966 | 758 | 1,703 | 234 | 230 | 141 | 70 | 1,446 | .102 | 12,674 | .062 |

The above table shows a marked difference in the wages of male and female employees. While but 3 per cent of the males earn $0.0676 or less per hour, 71 per cent of the females receive not more than that amount of pay. The average wages of all male employees is stated at $0.102 per hour and of female employees at $0.062 per hour.

In the Government match factories the proportion of employees who are males is somewhat larger, 676, or 32 per cent, being of that sex, and 1,444, or 68 per cent, being females. A table similar to the one just given for tobacco factory employees shows their average earnings according to sex and age periods. Here also ten hours constitute a day's work.

AVERAGE EARNINGS PER HOUR OF EMPLOYEES OF GOVERNMENT MATCH FACTORIES.

| Age. | Number of employees receiving— | | | | | | | | | | All employees. | | | |
|---|---|---|---|---|---|---|---|---|---|---|---|---|---|---|
| | $0.0048 to $0.0676 per hour. | | $0.0677 to $0.0801 per hour. | | $0.0801 to $0.1063 per hour. | | $0.1063 to $0.1255 per hour. | | Over $0.1255 per hour. | | Male. | | Female. | |
| | Male. | Female. | Male. | Female. | Male. | Female. | Male. | Female. | Male. | Female. | Number. | Average wages. | Number. | Average wages. |
| 12 to 18 years | 12 | 63 | 4 | 17 | 8 | 9 | 1 | 1 | ...... | 1 | 25 | $0.073 | 91 | $0.066 |
| 19 to 25 years | 39 | 349 | 16 | 133 | 23 | 114 | 20 | 5 | 11 | 3 | 109 | .087 | 604 | .062 |
| 26 to 45 years | 102 | 294 | 69 | 164 | 122 | 184 | 88 | 17 | 70 | 2 | 451 | .100 | 661 | .073 |
| 46 to 60 years | 14 | 40 | 14 | 22 | 31 | 16 | 6 | ...... | 8 | ...... | 73 | .098 | 78 | .069 |
| Over 60 years | 3 | 6 | 7 | 3 | 8 | 1 | ...... | ...... | ...... | ...... | 18 | .083 | 10 | .067 |
| Total | 170 | 752 | 110 | 339 | 192 | 324 | 115 | 23 | 89 | 6 | 676 | .097 | 1,444 | .068 |

Though still considerable, the difference between the earnings of male and female employees as here shown is not so great as in the case of the tobacco workers. Twenty-five per cent of the men and 52 per cent of the women received $0.0676 per hour or under. The average earnings for all the employees were, for the men $0.097, and for the women $0.068 per hour.

The total number of employees [illegible] information was obtained was 157,[illegible] cent, were male, and 19,687, or 12 per [illegible] almost entirely gate keepers or watchmen [illegible] generally provided with houses and [illegible] well. The hours of labor of the men vary [illegible] those of the women from 10 to 15. In [illegible] work of the women, however, a statement of [illegible] little significance. The following tables show [illegible] way employees according to sex and age periods [illegible] sary to make a showing separately for those employ[illegible] by the month and who constitute the higher grades [illegible] those paid by the day, hour, or piece, or the working[illegible]

AVERAGE EARNINGS PER HOUR OF WORKING PEOPLE ON [illegible] RAILWAYS.

| Age. | Number of employees receiving— | | | | | | | | | |
|---|---|---|---|---|---|---|---|---|---|---|
| | $0.0048 to $0.0676 per hour. | | $0.0677 to $0.0801 per hour. | | $0.0801 to $0.1062 per hour. | | $0.10[illegible] to $0.12[illegible] per hour. | | Over [illegible] per [illegible] | |
| | Male. | Female. | Male. | Female. | Male. | Female. | Male. | Female. | Male. | Female. |
| 12 to 18 years | 1,325 | 364 | 82 | ...... | 46 | ...... | 3 | ...... | ...... | [illegible] |
| 19 to 25 years | 5,311 | 1,005 | 1,202 | ...... | 1,071 | ...... | 223 | ...... | [illegible] | [illegible] |
| 26 to 45 years | 22,808 | 1,827 | 9,115 | 4 | 9,026 | ...... | 3,117 | ...... | [illegible] | [illegible] |
| 46 to 60 years | 2,732 | 457 | 2,339 | ...... | 3,008 | 1 | 1,214 | ...... | [illegible] | [illegible] |
| Over 60 years | 199 | 59 | 191 | 1 | 369 | ...... | 172 | ...... | [illegible] | [illegible] |
| Total | 32,375 | 3,712 | 12,920 | 5 | 13,520 | 1 | 4,728 | ...... | [illegible] | [illegible] |

AVERAGE EARNINGS PER MONTH OF HIGHER EMPLOYEES OF GOVE[illegible] RAILWAYS.

| Age. | Employees receiving per month— | | | | | | | | | |
|---|---|---|---|---|---|---|---|---|---|---|
| | Less than $13.70. | | From $13.70 to $21.23. | | From $21.42 to $31.85. | | From $32.04 to $43.43. | | $43.62 or over. | |
| | Male. | Female. | Male. | Female. | Male. | Female. | Male. | Female. | Male. | Female. |
| 12 to 18 years | 8 | 13 | 5 | ...... | 1 | ...... | ...... | ...... | ...... | ...... |
| 19 to 25 years | 28 | 877 | 500 | 5 | 285 | 2 | 77 | ...... | 21 | ...... |
| 26 to 45 years | 802 | 11,549 | 24,201 | 16 | 20,085 | 1 | 4,099 | ...... | 2,904 | ...... |
| 46 to 60 years | 52 | 2,744 | 6,342 | 1 | 8,118 | 5 | 1,864 | ...... | 1,407 | ...... |
| Over 60 years | 5 | 105 | 176 | ...... | 141 | 1 | 47 | ...... | 5 | ...... |
| Total | 895 | 15,288 | 31,224 | 22 | 28,630 | 9 | 6,087 | ...... | 4,337 | ...... |

In considering this table account should be taken of the [illegible] the railway companies contribute materially in a great man[illegible] the support of their employees. About 15 per cent of all [illegible] are housed by the companies. In addition, most of them pro[illegible] medical attendance and supplies in case of sickness, and make [illegible] for pensions for their employees after they have become inca[illegible] for work through accident or old age. The average value [illegible] these benefits is estimated to be equivalent to a 15 per cent ad[illegible] the wages. The average wages of employees according to age periods is not given, but the average wages of the men is stated to be $0.0869 per hour.

*Statistica degli Scioperi avvenuti nell' Industria e nell' Agricoltura durante l' anno 1894.* Ministero di Agricoltura, Industria e Commercio, Direzione Generale della Statistica. 1896. 52 pp.

This report emanates from the general bureau of statistics of the Italian department of agriculture, industry, and commerce. In presenting it here the same general plan is followed with the data for 1894 as in "Strikes in Italy in recent years," published in Bulletin No. 1, which covered the period from 1879 to 1893, inclusive.

The report shows that in 1894 there were 109 strikes, of which 104 reported the number of strikers as 27,595, an average of 265 strikers to each strike. It appears that about one-half of the strikes took place in northern Italy, the main seat of the great industries; but Sicily and Lazio report also a great number. The total number of strikes in 1894 was less than during any one of the five years preceding (less strikers in 1889—23,322).

Of the 109 strikes, 12 were accompanied by acts of violence and 11 by minor disturbances. The rest were conducted in a perfectly orderly manner. It is an established fact that in 11 strikes the strikers received assistance from associations of resistance, but this number is certainly below the truth, as in many instances the assistance is not tendered openly but secretly.

A table, the details in which are given by provinces, shows that 19,766 of the 27,595 strikers were adult males, 3,890 adult females, and 3,939 children (15 years of age and under of both sexes), from which it appears that the females and children formed 28 per cent of the total number of strikers.

The most important strikes, so far as the number of persons involved is concerned, took place in the sulphur mines of Sicily and in the Government cigar factory at Lucca. None of these strikes terminated successfully.

The following table shows the average number of strikers to each strike for the years 1878 to 1894:

AVERAGE NUMBER OF STRIKERS PER STRIKE, 1878 to 1894.

| Year. | Average number of strikers per strike. | Year. | Average number of strikers per strike. | Year. | Average number of strikers per strike. |
|---|---|---|---|---|---|
| 1878 | 165 | 1884 | 296 | 1890 | 289 |
| 1879 | 144 | 1885 | 398 | 1891 | 272 |
| 1880 | 227 | 1886 | 177 | 1892 | 263 |
| 1881 | 212 | 1887 | [illegible] | 1893 | 253 |
| 1882 | 130 | 1888 | 293 | 1894 | 265 |
| 1883 | 198 | 1889 | 187 | | |

This exhibit is followed by a [illegible] strikers and per cent of [illegible] for which strikes were undertaken:

CAUSES [illegible]

| Cause or object. | [illegible] |
|---|---|
| For increase of wages | [illegible] |
| For reduction of hours | [illegible] |
| Against reduction of wages | [illegible] |
| Against increase of hours | [illegible] |
| Other causes | [illegible] |
| Total classified | [illegible] |
| Not classified | [illegible] |
| Grand total | [illegible] |

The following table shows the results of [illegible]

RESULTS OF STRIKES, 1894.

| Cause or object. | Succeeded. | | | | Succeeded partly. | | | | [illegible] |
|---|---|---|---|---|---|---|---|---|---|
| | Strikes. | | Strikers. | | Strikes. | | Strikers. | | [illegible] |
| | Number. | Per cent. | Number. | Per cent. | Number. | Per cent. | Number. | Per cent. | [illegible] |
| For increase of wages | 18 | 39 | 2,720 | 15 | 18 | [illegible] | 4,511 | [illegible] | [illegible] |
| For reduction of hours | 2 | 17 | 155 | 6 | 4 | [illegible] | 1,[illegible] | 54 | [illegible] |
| Against reduction of wages | 4 | 33 | 1,007 | 67 | 2 | 17 | 67 | 5 | [illegible] |
| Against increase of hours | ..... | ..... | ..... | ..... | ..... | ..... | ..... | ..... | [illegible] |
| Other causes | a 11 | 36 | a1,472 | 28 | 5 | 16 | [illegible] | 11 | [illegible] |
| All causes | a 35 | 34 | a5,354 | 19 | [illegible] | 28 | 6,[illegible] | [illegible] | [illegible] |

a The figures as given are according to the summary table as printed in the [illegible] from the detail table, however, that one strike of 250 strikers has been omitted in [illegible]mary.

In the following table are shown the per cents of [illegible] ceeded, succeeded partly, and failed, together with the [illegible] strikers engaged therein, for the years 1878–1891 to 1894.

RESULTS OF STRIKES, 1878–1891 to 1894.

| Year. | Per cent of strikes. | | | Per cent [illegible] | |
|---|---|---|---|---|---|
| | Successful. | Partly successful. | Failed. | Successful. | [illegible] |
| 1878–1891 | 16 | [illegible] | 41 | [illegible] | [illegible] |
| 1892 | [illegible] | [illegible] | [illegible] | [illegible] | [illegible] |
| 1893 | [illegible] | [illegible] | [illegible] | [illegible] | [illegible] |
| 1894 | [illegible] | [illegible] | [illegible] | [illegible] | [illegible] |

From this table it will be [illegible] steadily from 16 per cent in 187[illegible] ing tendency, however, [illegible] benefited is considered, which is shown to have been less in 1894 than in preceding years.

The following table shows, for 1894, a classification of strikes according to the industries in which the strikers were employed:

STRIKES, BY INDUSTRIES, 1894.

| Industry. | Strikes. | | Strikers. |
|---|---|---|---|
| | Total. | Reporting number of strikers. | |
| Weavers, spinners, and carders | 15 | 15 | 2,514 |
| Miners and ore diggers | 21 | 16 | 12,318 |
| Mechanics | 3 | 3 | 182 |
| Founders | 2 | 2 | 35 |
| Railroad employees | 3 | 3 | 1,597 |
| Day laborers | 12 | 12 | 3,044 |
| Masons and stonecutters | 8 | 8 | 1,130 |
| Kiln and furnace tenders | 3 | 3 | 482 |
| Hat makers | 2 | 2 | 258 |
| Tanners | 11 | 11 | 603 |
| Dyers | 5 | 5 | 412 |
| Bakers and pastry cooks | 3 | 3 | 525 |
| Joiners | 1 | 1 | 50 |
| Glass workers | 1 | 1 | 25 |
| Omnibus drivers and conductors | 3 | 3 | 298 |
| Boatmen | 1 | 1 | 32 |
| Cart drivers | 3 | 3 | 540 |
| Porters and coal carriers | 2 | 2 | 105 |
| Other industries | 10 | 10 | 2,53 |
| Total | 109 | 104 | 27,595 |

In 1894 there were 323,261 days of work lost in 103 strikes reported, which shows an average loss of 3,138 days of work per strike.

By the law of June 15, 1893, boards of arbitration and conciliation were authorized to be organized in the various industrial centers of the Kingdom, but up to the end of 1894 none had been created.

The report closes with a fully annotated table showing, by provinces and for each strike, the locality, date of beginning, occupation, sex, and number of strikers, cause or object of strike, result, whether accompanied by threats or acts of violence, duration of strike in days, and total number of days of work lost.

## DECISIONS OF COURTS AFF...

[This subject, begun in Bulletin No. 2, will be continued ... ing with the decisions as they occur. All material ... duced in the words of the courts, indicated, when short, ... when long, by being printed solid. In order to save space ... simply by way of explanation, is given in the words of the ...

### DECISIONS UNDER STATUTORY L...

APPROPRIATION OF MONEY IN AID OF TEXTILE ... STITUTIONALITY OF STATUTE—*Hanscom et al. v. City ... Northeastern Reporter, page 196.*—This case was brought ... judicial court of Massachusetts, on a petition of ten taxpayers ... of Lowell, against the defendant city, to restrain the payment ... appropriated by the city council, to be paid to the trustees of ... Textile School. The money was voted in accordance with ... of the acts of 1895, which provides that citizens of cities in ... tain conditions exist may associate themselves together, by ... in writing, for the purpose of establishing and maintaining ... school, and that any city in which such a corporation is organ... appropriate and pay to such corporation a sum of money, no... $25,000.

The court rendered its decision February 29, 1896, and dism... petition. The opinion of said court was delivered by Judge ... is as follows:

The establishment of a textile school in a large manufactur... may be of such special and direct benefit to the city as to war... appropriation by it of a sum of money in aid of the school, ... persons from elsewhere may be members or trustees of the corp... or may be admitted to be taught therein. It is in aid of manufa... which the constitution (part 2, c. 5, sec. 2) enjoins the legisla... encourage, and the statute (St. 1895, c. 475) falls within the ... of Merrick v. Inhabitants of Amherst, 12 Allen, 500, and is co... tional. See also Jenkins v. Inhabitants of Andover, 103 Mass., ... Petition dismissed.

---

ASSIGNMENT FOR THE BENEFIT OF CREDITORS—P... CLAIMS—WAGES—*In re Scott et al. 42 Northeastern Reporter, ... 1079.*—In the final accounting by William E. Scott and A... Brown, assignees of William A. Drake, the supreme court ... general term, rendered a judgment ... court of Orange County, denying a ... ance of claim for wages, ...

1886. From this judgment the National Bank of Port Jervis, a creditor, appealed to the court of appeals, which rendered its decision February 25, 1896, reversing the judgment of the supreme court. The following is quoted from the opinion of the court of appeals as delivered by Judge Haight:

The question raised upon this appeal calls for a construction of chapter 283 of the laws of 1886, which is as follows: "In all distributions of assets under all assignments made in pursuance of this act the wages or salaries actually owing to the employees of the assignor or assignors, at the time of the execution of the assignment, shall be preferred before any other debt, and should the assets of the assignor or assignors not be sufficient to pay in full all the claims preferred, pursuant to this section, they shall be applied to the payment of the same pro rata to the amount of each such claim."

It is contended on behalf of the appellants that the wages or salaries preferred under the provisions of this act are those only of employees who are actually in the employ of the assignor at the time of his executing the assignment. We are unable to adopt this view. It would practically nullify the provisions of the act. The assignor by discharging his employees the day before the execution of the assignment could evade its provisions. The language of the act does not require such a construction. It is the "wages or salaries actually owing to the employees of the assignor or assignors at the time of the execution of the assignment" that are preferred, and the preference is not limited to the wages and salaries of those in the employ of the assignor at the time of the assignment. This, we think, is the fair reading and meaning of the provision.

There is, however, one question which we think the general term has overlooked, and that is that the act was not intended to be retroactive, and to create a preference in favor of employees for wages earned prior to its passage. This question was considered in the case of People v. Remington, supra [16 Northeastern Reporter, p. 680], with reference to the act creating a preference of the wages of employees, etc., in corporations where a receiver had been appointed. It was held that that act was not retroactive, and, as we have seen, the views there expressed have been approved by this court. That act and the one we now have under review, so far as this question is concerned, are similar, and the same construction should be given to each.

---

ATTORNEYS' FEES IN SUITS FOR WAGES—LIEN FOR WAGES—*Ackley v. Black Hawk Gravel Mining Co. et al. 44 Pacific Reporter, page 330.*—This action was brought in the superior court of Sierra County, Cal., by one David Ackley to establish and enforce a lien for wages. A judgment was rendered for the plaintiff and the defendants carried the case on appeal to the supreme court of the State. Said court rendered its decision March 24, 1896, and its opinion, delivered by Commissioner Searls, and containing a statement of the facts in the case, is given below:

This is an action to recover from the corporation defendant for work, labor, and services performed by plaintiff for said defendant, and to

have the value thereof declared a lien
ant under and pursuant to an act of the
California approved March 31, 1891, entitled
payment of the wages of mechanics and
tions." The other defendants are made
or claim to have, some interest in the property
lien thereon, etc. Plaintiff had judgment
$397 for wages, a counsel fee of $100, and costs
a total of $524.35, which was declared to be a valid
erty of the corporation defendant described therein,
paramount to any lien or interest of the other defendants
the property so subject to said lien to be sold in
Defendants appealed from the judgment, and the
the judgment roll without a statement or bill of exceptions
ute in question is brief, and we quote it in full. It is

"SECTION 1. Every corporation doing business in this
pay the mechanics and laborers employed by it the wages
and due them weekly or monthly, on such day in each week
as shall be selected by said corporation.

"SEC. 2. A violation of the provisions of section one of
shall entitle each of the said mechanics and laborers to a l
the property of said corporation for the amount of their wa
lien shall take preference over all other liens, except duly
mortgages or deeds of trust; and in any action to recover th
of such wages, or to enforce said lien, the plaintiff shall be
a reasonable attorney's fee, to be fixed by the court, and w
form part of the judgment in said action, and shall also be
an attachment against said property."

Various points are made by appellant in favor of reversal,
of which need be noticed. The complaint is for work, labor,
ices performed by plaintiff for the corporation defendant, an
as essential to the point in hand, is as follows: "That betwee
day of May, 1892, and the 1st day of May, 1894, plaintiff p
twenty-four months' service for defendant, Black Hawk Grave
Company, as watchman of its property, known as the 'Jenki
property,' hereinafter described, which said services were perf
plaintiff for said defendant corporation at its request; that s
ices were reasonably worth $2,190; that no part thereof has be

The findings are in almost the exact language of the complai
that the court finds the value of the services were $480, and
has been paid on account thereof. There is no allegation o
that plaintiff was employed by the week or month, or that
were "due weekly or monthly."

In Keener v. Irrigation Company (decided by department
court, December 31, 1895), 43 Pac., 14, this same statute w
review, and the court, speaking through Harrison, J., said:
terms of the first section of this act, it does not apply to all
tions, but only to those who, while doing business in this Stat
laborers and mechanics by the week or month, whose wag
the terms of their employment, are payable weekly or mo
does not purport to impose upon those corporations any
bility toward all the mechanics or laborers whom it may
create a right in favor of those of its employees wh
earned or payable by the week or by the month. A
to be enforced herein exists only by virtue of the
bent upon the plaintiff to bring

and to show that the wages earned by him were 'due weekly or monthly.' His complaint is, however, defective in this respect, and contains no allegation concerning the times at which the wages were payable, or that he was employed at weekly or monthly wages; and, from the allegations in reference thereto, it would seem that there was no agreement upon this point." Following the case above quoted, we add: As the plaintiff is not entitled to avail himself of the provisions of the act of 1891, that provision of the judgment allowing him a counsel fee was unauthorized. We recommend that the judgment in favor of the plaintiff for the sum of $397, as wages and costs, be affirmed, and that that portion of the judgment awarding counsel fees, and declaring that plaintiff is entitled to a lien upon the property of the corporation defendant, and directing a sale of such property, be reversed, and that appellant recover its costs on this appeal.

---

CHINESE LABORERS—DEPORTATION—*United States v. Sing Lee. 71 Federal Reporter, page 680.*—This case arose under chapter 60 of the acts of Congress of 1891-92, as amended by chapter 14 of the acts of Congress of 1893-94, providing for the restriction of Chinese immigration. The action was brought in the United States district court for the district of Oregon, and the decision of said court was rendered January 7, 1896.

The facts in the case are fully set forth in the opinion delivered by District Judge Bellinger, which appears below in part:

This is a proceeding for the deportation of the defendant on the ground that he is a Chinese laborer, and that he has failed to register as required by law. The defendant is a resident of the State of California, where he has lived many years. He swears, and proves by the testimony of a number of Chinese witnesses, that he is a merchant doing business in San Francisco. It appears that he is the lessee of some fruit land in California for a term of years, upon which he employs laborers; and the fact is stipulated in the case that he performed manual labor in aiding the laborers employed by him to work said farm in caring for the fruit during its growth and picking. It is * * * established that a Chinese person who works for others for short periods of time, or who regularly works in the manufacture of fabrics for sale by himself, is a laborer, within the meaning of the Chinese exclusion acts. The lease referred to, as appears from the copy in evidence, was entered into August 6, 1894, and is for a certain ranch adjoining the town of Penryn, in Placer County, Cal. The term of the lease is from October 1, 1894, to the last day of September, 1897. The labor performed by the defendant was performed, it must be presumed, within the term of the lease, and after October 1, 1894; and this presents the question whether a person shown to have been a merchant at the date of the act in question [chapter 60, acts of 1891-92], and who continued to be such merchant until after the time within which laborers were permitted to register, is liable to deportation if, after such time, he becomes a laborer. I am of the opinion that such a person is not liable to deportation. * * *

The act in question made it the duty of all Chinese laborers within the United States at the time of the passage of the act to procure a certificate of residence. A merchant ... certificate. Being, therefore, lawfully ... the requirements of the acts of Congress ... policy of the law to make an act of labor ... a crime punishable with heavy penalties.

In addition to these considerations, the fact that ... had for many years a fixed residence in California ... arrested here while en route to Montana, to which ... through ticket, is against this proceeding. A practice ... to result in oppression and injustice. If the defendant ... is claimed, and has been so during the lease in question ... to deportation in that jurisdiction, where the facts ... easily obtainable, and where the law can be vindicated ... venience or danger of injustice. The findings will be ... and the order of the court is that he be discharged from ...

---

COAL MINING—WEIGHING PRODUCT BEFORE SCREENING ... *v. State.* *42 Northeastern Reporter, page 911.*—The appellant ... victed and fined $100 on information and affidavit filed in th... court of Parke County, Ind., for a violation of sections 5 an... act approved March 2, 1891 (p. 57, acts of 1891), concerning ... ing, and appealed to the supreme court of the State. Section ... act reads as follows:

All coal mined in this State under contract for payment b... or other quantity, shall be weighed before being screened, an... weight thereof shall be credited to the miner of such coal, a... pounds of such coal as mined shall constitute a bushel, and ... sand pounds of coal as mined shall constitute a ton: *Provided*, T... ing in this act shall be so construed as to compel payment for ... rock, slate, black-jack, or other impurities, including dirt, w... be loaded with or amongst the coal.

The affidavit and information charged that appellant, Marti... ruary 13, 1894, was superintendent of mine No. 8 of the Park... Coal Company, and that one William Cherry was employed ... company to mine coal at the rate and price of 70 cents per ... said Cherry mined a quantity of coal, exceeding $3\frac{1}{2}$ tons, for ... pany, and that said Martin unlawfully failed and neglected ... to said Cherry the full weight of such coal before it was scre... screened it before it was weighed. The supreme court, Ja... 1896, reversed the judgment of the trial court and instruc... sustain appellant's motion for a new trial. The opinion, deli... Judge McCabe, contains the following:

Among other things, it is said by the learned counsel for ... lant that it is undisputed that the coal mined by Cherry ... sulphur, slate, black-jack, and other impurities. It further ... appears that section 5 of the act of 1891 is impossible of ... from the following evidence of Cherry: "Q. Is there any ... ticable way in the business of bituminous coal mining ... sulphur, slate, black-jack, dirt, ...

or the slack?—A. No; the only thing that can be done is to separate the lump coal from the slack. The part that does not pass through the screen is called 'lump coal,' and is the coal of commerce. The part that passes through the screen is called 'slack,' and is composed of the fine coal mixed with sulphur, slate, black-jack, dirt, and other impurities. There is no way known to the business of mining by which the fine coal of the slack can be separated from the sulphur, black-jack, dirt, and other impurities contained in the slack. Q. How about the lump coal?—A. Most of the impurities are contained in the slack; but the lump coal also contains some sulphur, slate, black-jack, and other impurities. It is impossible, in any practicable way, to separate the coal of the lump coal from the sulphur, slate, black-jack, and the like contained in the lump coal. Q. Is there any practicable way in the business of coal mining, either before or after screening, by which the pure coal may be completely and fully separated from the sulphur, slate, black-jack, and other impurities?—A. No; all that can be done is to separate the lump coal, with its impurities, from the slack, with its impurities, by screening. Q. State whether or not, in your opinion, it was practical to mine the coal you were working in that day and weigh and pay for it before screening such dirt and other small refuse matter?—A. It would have been impossible to have weighed the coal that I mined that day before screening it, and given me credit with the weight thereof, without also including in such weights the weight of such dirt and other refuse matter." While section 5 imperatively requires the coal to be weighed before it is screened, and the full weight thereof to be credited to the miner of the same, yet the proviso qualified such requirement by providing "that nothing in this act shall be so construed as to compel payment for sulphur, rock, slate, black-jack, or other impurities," etc. This was the State's evidence, and stands uncontradicted; and from it the conclusion is irresistible that the only way possible to avoid paying for mining the impurities excepted out of the statute was by screening before weighing the coal, and even then some impurities would be paid for, but small in quantity compared with that resulting from weighing before screening the particular coal mined by Cherry involved in this prosecution. The provision as imperatively requires the statute to be so construed as not to compel payment for such impurities as it requires any coal mined to be weighed before it is screened; and, as the evidence makes it clear that the only way to avoid payment for such impurities in the case of the particular coal here involved was to screen it before it was weighed, therefore the failure to weigh before it was screened was not a violation of the fifth section, and hence the appellant, under the evidence, was not liable to the penalty provided in the seventh section. Therefore the finding of the court was contrary to the law and the evidence, which was assigned as a cause in the motion for a new trial.

---

EMPLOYERS' LIABILITY—ASSUMPTION OF RISKS BY EMPLOYEE—CONSTRUCTION OF THE "FACTORY ACT"—*Knisley v. Pratt et al. 42 Northeastern Reporter, page 986.*—This action was brought by Sarah Knisley against Pascal P. Pratt and others to recover damages for the loss of her left arm. The plaintiff, who was operating a "punching machine" in a hardware factory, was engaged in cleaning it while it was in motion, by rubbing the dirt and [illegible] with a piece of waste held in her left hand, [illegible] between the cogwheels, causing such injuries [illegible] necessitate amputation of the latter near the [illegible] of the trial court was given for the defendants and [illegible] to the supreme court of the State (New York) which rendered a judgment granting the plaintiff a new [illegible] judgment the defendants appealed to the court of app[eals] which, on February 18, 1896, reversed the order of [illegible] and ordered a judgment for the defendants against the [illegible] suit with costs. The opinion of the court of appea[ls] Judge Bartlett, shows the facts in the case, and the fo[llowing] therefrom:

Plaintiff, being of full age, entered the employ of de[fendants] or June, 1890, and the accident happened Septembe[r] testified that she worked, off and on, about three [illegible] punching machines prior to the accident, sometimes times half a day at a time, as required; that up to accident she had worked twelve different times, on dif[illegible] one of the punching machines, and had cleaned it abo[ut] times. It is insisted, on behalf of the plaintiff, that e[illegible] admitted facts, she would by the rules of the commo[n] to have assumed, not only the ordinary but the obv[ious] business, yet the provisions of the statute commonl[y] "Factory act" would enable her to recover. Laws 18[illegible] amending laws 1886, c. 409, entitled "An act to regu[illegible] ment of women and children in manufacturing estab[illegible] provide for inspectors to enforce the same."

The statute [section 12 of above act] provides tha[t] duty of the owner of any manufacturing establishme[nt] the discretion of the factory inspector, belt shifters, of throwing on or off belts or pulleys. It also provide[s] cogs shall be properly guarded. It is admitted tha[t] were unguarded on this machine. The plaintiff was i[illegible] of that condition of the cogwheels, and it appears tha[t] iron, put in between the upright and the gear wheels, tected the hand of the operative while cleaning the [illegible] motion. The defendants are chargeable, therefore, w[illegible] only under the statute, viz, a failure to properly guard the punching machine.

In order to sustain the judgment in favor of plainti[ff] to hold that, where the statute imposes a duty upon [illegible] performance of which will afford greater protection [illegible] it is not possible for the latter to waive the protectio[n] under the common-law doctrine of obvious risks. We new and startling doctrine, calculated to establish a me[illegible] unknown to the common law, and which is contrary t[o] Massachusetts and England under similar statutes.

It should be remarked at the outset that the factory does not, in terms, give a cause of action to one suff[ering] reason of the failure of the employer to discharge hi[s] An action for such injury is the ordinary common-la[w] gence, and subject to the rules of the common law. [illegible]

The principle contended for seems to rest, if it can be maintained at all, upon a question of public policy. The factory act, it is said, is passed to regulate the employment of women and children, and imposes upon the employer certain duties, and subjects him to specified penalties in case of default; that a sound u i policy requires the rigid enforcement of this act, and it would contravene that policy to permit an employee, by implied contract or promise, to waive the protection of the statute. We think this proposition is essentially unsound, and proceeds upon theories that can not be maintained. It is difficult to perceive any difference in the quality and character of a cause of action, whether it has its origin in the ancient principles of the common law, in the formulated rules of modern decisions, or in the declared will of the legislature. Public policy in each case requires its rigid enforcement, and it was never urged in the common-law action for negligence that the rule requiring the employee to assume the obvious risks of the business was in contravention of that policy.

We are of the opinion that there is no reason in principle or authority why an employee should not be allowed to assume the obvious risks of the business, as well under the factory act as otherwise. There is no rule of public policy which prevents an employee from deciding whether, in view of increased wages, the difficulties of obtaining employment, or other sufficient reasons, it may not be wise and prudent to accept employment subject to the rule of obvious risks. The statute does, indeed, contemplate the protection of a certain class of laborers, but it does not deprive them of their free agency and the right to manage their own affairs.

The facts in the case at bar, whether it be considered as an action for negligence at common law or under the statute, show conclusively that the plaintiff assumed the obvious risk of working on the machine in operating which she was injured. It is impossible not to feel great sympathy for this unfortunate plaintiff, who has been maimed for life, but her recovery is barred by legal principles that are salutary and proper in the general administration of justice.

---

EMPLOYERS' LIABILITY—DEFECTIVE WAYS, WORKS, OR MACHINERY—*Geloneck v. Dean Steam Pump Co. 43 Northeastern Reporter, page 85.*—Action was brought in the superior court of Hampden County, Mass., by Otto G. Geloneck against the Dean Steam Pump Company to recover for personal injuries in moving machinery on trucks furnished by the defendants. The plaintiff relied principally upon clause 1 of section 1 of chapter 270 of the acts of 1887, which reads as follows:

Where, after the passage of this act, personal injury is caused to an employee, who is himself in the exercise of due care and diligence at the time:

(1) By reason of any defect in the condition of the ways, works, or machinery connected with or used in the business of the employer, which arose from or had not been discovered or remedied owing to the negligence of the employer or of any person in the service of the employer and intrusted by him with the duty of seeing that the ways, works, or machinery were in proper condition; * * * the employee, or in case

the injury results in death, the legal representative shall have the same right of compensation and employer as if the employee had not been an service of the employer, nor engaged in its work.

The facts in the case were substantially as follows: loaded onto a truck and the foreman in charge undertook a room, but the pole handle, not being properly secured, and caused the pump to topple over and fall upon the him severely.

A verdict was given for the plaintiff in the superior court ment was rendered in his favor. The defendants carried exceptions to the charge of the judge, to the supreme court which rendered its decision February 25, 1896, sustaining the of the lower court. In the opinion of said court, delivered Barker, the following language was used:

The jury were instructed, in substance, that, to constitute a the condition of the ways, works, or machinery, it was not that any particular instrument should be defective in itself; instance, the plaintiff need not show that there was a fault in that it had a cracked wheel or a broken axletree, or somethin kind, which gave way; that, in the sense of the law, a thing found to be not reasonably safe and suitable, if it is insuffic unsuitable for the purposes to which it is applied, and is inten applied, and under the conditions in which it is used, and is to be used; that the question is not limited to whether there thing which has a weak spot or crack, or is decayed, but it the inquiry whether the appliances, as they are put toge used, and intended to be used, are reasonably safe and suit connection with this instruction, the jury was also told that th ant was not obliged to have a faultless arrangement, or which nobody could find any fault, but only to use reasonab have things reasonably safe and suitable. These instructi correct. An unsuitableness of ways, works, or machinery, intended to be done, and actually done, by means of them, is within the meaning of St. 1887, c. 270, sec. 1, cl. 1, alth ways, works, or machinery are perfect of their kind, in goo and suitable for some work done in the employer's busine than the work in doing which their unsuitableness causes the workmen.

In such a case the employer is wrong in furnishing applian use for which they are unsuitable, and, in effect, in so orde carrying on his work that, without fault of the ordinary work natural consequence will be that the appliances will be used poses for which they are unsuitable. The circumstance that ployer intends that his work shall be done in the manner an means in use when the accident occurs distinguishes the those in which he furnishes a stock of appliances from which man is to select such as are fit for the particular work in han an unsuitableness is neither accidental nor temporary, nor negligence of a workman who is not charged with the duty to the fitness of the ways, works, and machinery. ruled.

Master and Servant—Term of Employment—*Rosenberger v. Pacific Coast Railway Company.* · *43 Pacific Reporter, page 963.*—Action was brought by E. B. Rosenberger against the railroad company in the superior court of San Luis Obispo County, Cal. Judgment was given for Rosenberger and the railroad company appealed to the supreme court of the State, which rendered its decision February 21, 1896, affirming the judgment of the lower court. The facts of the case are stated fully in the opinion of the court, which is given in part below:

This cause was tried by the court and a jury, the plaintiff had judgment, and the defendant appeals therefrom, and from an order denying a new trial. The plaintiff claimed that he was employed by the defendant as an accountant for the period of one year, commencing November 24, 1893, at a salary of $1,800 per annum, payable in monthly installments of $150, and that he was discharged without cause March 24, 1894. This action was commenced, July 17, 1894, to recover three months' salary, from March 24 to June 24, 1893, amounting to $450, and for that sum he recovered judgment. The defense was that the employment was not for any definite time, and that his discharge was justifiable, because of his failure to give bond for the faithful discharge of his duties.

Prior to the employment in question, plaintiff was employed in a similar capacity in the offices of the Union Pacific Railway at Omaha, and had made an application for employment to the auditor of the Oregon Improvement Company, with which company defendant is connected. C. O. Johnson, defendant's superintendent, wrote plaintiff, November 4, 1893, to know whether he was still an applicant, and for information as to experience, salary required, etc. To this letter plaintiff replied, saying, as to salary, that it should be worth $1,800 or $2,000 a year. To this Johnson replied by telegraph, saying: "Yours ninth. Salary eighteen hundred. Change immediate if possible. Could probably arrange transportation. Wire earliest date."

Plaintiff arrived at San Luis Obispo the night of November 24, and on the morning of the 25th talked with Mr. Johnson an hour and a half, but plaintiff was unable to state with any particularity what was said, but, as he expressed it: "As near as I can give it, we arrived at an understanding to furnish me one year's employment upon satisfactory performance of my duties which I had undertaken. * * * I say that as near as I can recollect, we arrived at that understanding." Mr. Johnson testified that in this conversation nothing was said as to salary, or the term for which he was employed, but that it related wholly to the duties he was to perform. Though plaintiff was paid monthly, up to the time of his discharge, at the rate $150 per month, the correspondence constituted a hiring for a year. Section 2010 of the Civil Code provides: "A servant is presumed to have been hired for such length of time as the parties adopt for the estimation of wages. A hiring at a yearly rate is presumed to be for one year; a hiring at a daily rate for one day; a hiring by piecework for no specified time." The payment by the month, at the monthly proportion of the yearly rate, would not of itself be sufficient to change the contract to a hiring by the month; nor would the custom of the defendant, or of all railroad companies, to hire by the month convert the contract, created by the correspondence between the parties, which under the code constituted a contract for a year, into a contract by the month. If the correspondence had not fixed the term for which the plaintiff was hired, section 2011 of the Civil Code, as well as the custom, would have made a hiring ... this point the instruction to the jury was right, ... by the evidence.

While it is the duty of an employee who has been ... charged to seek other employment, and thus diminish ... tained by him, he is not required, as a condition of ... that he has made such endeavor and failed. The burden ... defendant to show that he could, by diligence, have obtained ... ment elsewhere. Whatever compensation may have been ... such employment is also to be shown by the defendant in ... of damages; otherwise the damages will be measured by the ... wages agreed to be paid.

---

SUNDAY LABOR—BARBERS—CONSTITUTIONALITY OF ST... *Ex parte Jentzsch. 44 Pacific Reporter, page 803.*—This case wa... before the supreme court of California on a petition by Leo Je... a writ of habeas corpus to secure his discharge from custod... conviction for violation of section 310½ of the Penal Code, pr... barbers from keeping open their places of business or workin... trade on Sundays after 12 o'clock noon. The court rendered it... April 17, 1896, declaring the above section unconstitutional, ... its opinion, delivered by Judge Henshaw, the following is quo...

Petitioner was convicted under section 310½ of the Penal Co... is a new section, enacted in 1895, and which provides as ... "Every person who as proprietor, manager, lessee, employee, ... keeps open or conducts, or causes to be kept open or condu... barber shop, bath house and barber shop, barber shop of a ... establishment, or hair-dressing establishment, or any place fo... or hair dressing used and conducted in connection with any ot... of business or resort, or who engages at work or labor as a ... any such shop or establishment on Sunday, or on a legal holi... the hour of 12 o'clock m. of said day, is guilty of a misdemea...

It is contended that the section is a violation of the follo... stitutional provisions:

Article 1, § 1: "All men are by nature free and indepen... have certain inalienable rights, among which are those of enjo... defending life and liberty; acquiring, possessing and protect... erty; and pursuing and obtaining safety and happiness."

Article 1, § 21: "No special privileges or immunities sha... granted which may not be altered, revoked or repealed by th... ture; nor shall any citizen, or class of citizens, be granted ... or immunities which, upon the same terms, shall not be gran... citizens."

Article 4, § 25, subsecs. 2, 33: "The legislature shall not ... or special laws in any of the following enumerated cases, that ... Second. For the punishment of crimes and misdemeanors." ... cases where a general law can be made applicable."

In construing so-called "Sunday laws" courts have variously ... them, some from a religious view, others from a secular, ... from an anomalous commingling of both. In this State the... been upheld from a religious standpoint. Under ...

guaranties to all equal liberty of religion and conscience, any law which forbids an act not in itself contra bonos mores, because that act is repugnant to the beliefs of one religious sect, of necessity interferes with the liberty of those who hold to other beliefs or to none at all. Liberty of conscience and belief is preserved alike to the followers of Christ, to Buddhist, and Mohammedan, to all who think that their tenets alone are illumined by the light of divine truth; but it is equally preserved to the skeptic, agnostic, atheist, and infidel, who says in his heart, "There is no God." So it has come to be the established rule in this State to view and construe such laws as civil and secular enactments. Thus construed, these laws, when decreed valid, are upheld as a proper exercise of the police power, as an exercise of the legislative prerogative to regulate the relations, contracts, intercourse, and business of society at large, and its particular members with respect to each other. Herein it is held that the legislature may pass laws for the preservation of health and the promotion of good morals—propositions which are indisputable

Says Mr. Justice Field, in Ex parte Newman, 9 Cal., 518: "Labor is a necessity, imposed by the condition of our race, and to protect labor is the highest office of our laws." It is this language which respondent quotes and relies upon in support of the validity of this penal statute. Upon the question thus presented of the proper limits of the police power much might be written, and much, indeed, will have to be written, ere just bounds are set to its exercise. But in this case neither time permits nor necessity demands the consideration. Still it may be suggested in passing that our Government was not designed to be paternal in form. We are a self-governing people, and our just pride is that our laws are made by us as well as for us. Every individual citizen is to be allowed so much liberty as may exist without impairment of the equal rights of his fellows. Our institutions are founded upon the conviction that we are not only capable of self-government as a community, but, what is the logical necessity, that we are capable, to a great extent, of individual self-government. If this conviction shall prove ill founded, we have built our house upon sand. The spirit of a system such as ours is therefore at total variance with that which, more or less veiled, still shows in the paternalism of other nations.

It may be injurious to health to eat bread before it is twenty-four hours old, yet it would strike us with surprise to see the legislature making a crime of the sale of fresh bread. We look with disfavor upon such legislation as we do upon the enactment of sumptuary laws. We do not even punish a man for his vices, unless they be practiced openly, so as to lead to the spread of corruption or to breaches of the peace or to public scandal. In brief, we give to the individual the utmost possible amount of personal liberty, and, with that guaranteed him, he is treated as a person of responsible judgment, not as a child in his nonage, and is left free to work out his destiny as impulse, education, training, heredity, and environment direct him. So, while the police power is one whose proper use makes most potently for good, in its undefined scope and inordinate exercise lurks no small danger to the Republic, for the difficulty which is experienced in defining its just limits and bounds affords a temptation to the legislature to encroach upon the rights of citizens with experimental laws none the less dangerous because well meant. We think the act under consideration gives plain evidence of such encroachment. It is sought to be upheld by the argument that it is a police regulation; that it seeks to protect labor against the oppression of capital. The people have passed the law; let not the courts interfere with it. If [illegible] they may amend or repeal it.

It is not easy to see where or how this law [illegible] unjust exactions of capital. A man's constitutional [illegible] than his personal freedom. It means, with many other [illegible] freely to labor, and to own the fruits of his toil. It is a [illegible] the protection of labor which punishes the laborer for [illegible] that is precisely what this law does. The laboring barber, [illegible] a most respectable, useful, and cleanly pursuit, is singled out [illegible] thousands of his fellows in other employments and told that [illegible] he shall not work upon holidays and Sundays after 12 o'[illegible] His wishes, tastes, or necessities are not consulted. If he lab[illegible] a criminal. Such protection to labor, carried a little furth[illegible] send him from the jail to the poorhouse. How comes it that [illegible] lative eye was so keen to discern the needs of the oppressed ba[illegible] yet was blind to his toiling brethren in other vocations? S[illegible] and street-car operatives toil through long and weary Sunda[illegible] so do mill and factory hands. There is no Sunday period of [illegible] no protection for the overworked employees of our daily pap[illegible] these not need rest and protection? The bare suggestion of t[illegible] siderations shows the injustice and inequality of this law. [illegible] whether or not a general law to promote rest from labor in all [illegible] vocations may be upheld as within the due exercise of the poli[illegible] as imposing for its welfare a needed period of repose upon t[illegible] community, a law such as this certainly can not. A law is no[illegible] general because it operates upon all within a class. There [illegible] back of that a substantial reason why it is made to operate o[illegible] a class, and not generally upon all. As was said in Passadena [illegible] son, 91 Cal., 238, 27 Pac., 604: "The conclusion is that, althou[illegible] is general and constitutional when it applies equally to all [illegible] embraced in a class founded upon some natural or intrinsic [illegible] tutional distinction, it is not general or constitutional if i[illegible] particular privileges or imposes peculiar disabilities or burdens [illegible] ditions in the exercise of a common right upon a class of pers[illegible] trarily selected from the general body of those who stand in [illegible] the same relation to the subject of the law."

And in Darcy v. City of San Jose, 104 Cal., 642, 38 Pac., 50[illegible] classification, however, must be founded upon differences w[illegible] either defined by the constitution or natural, and which will s[illegible] reason which might rationally be held to justify the diversity [illegible] lation." In the case of our cities the constitution itself decre[illegible] sification by population, and the differing exigencies of m[illegible] government require that laws operating upon any class shoul[illegible] general; otherwise the constitutional scheme itself is ove[illegible] But in a law such as this no reason has been or can be shown [illegible] followers of one useful and unobjectionable employment sh[illegible] debarred from the right to labor upon certain days, and othe[illegible] classes of employment be not so debarred. If it be constitu[illegible] single out one such class, and debar its members from the right [illegible] on one day in the week, it would be constitutional to prohibit th[illegible] following their vocation upon six days of the week. When [illegible] such class is singled out and put under the criminal ban of a [illegible] as this, the law not only is special, unjust, and unreasonable [illegible] operation, but it works an invasion of individual liberty [illegible] free labor, which it pretends to protect. Here [illegible] nated class is based upon no distinction, and [illegible]

Cooley: "Everyone has a right to demand that he be governed by general rules; and a special statute that singles his case out as one to be regulated by a different law from that which is applied in all similar cases would not be legitimate legislation, but an arbitrary mandate, unrecognized by the law." The prisoner is discharged.

---

SUNDAY LABOR—BARBERS—CONSTITUTIONALITY OF STATUTE—*State v. Granneman. 33 Southwestern Reporter, page 784.*—William Granneman was convicted in the St. Louis court of criminal corrections of plying his trade as a barber on Sunday in violation of the act approved March 18, 1895, and from the judgment of said court he appealed to the supreme court of Missouri.

The act under which the conviction was had reads as follows:

SECTION 1. It shall be a misdemeanor for any person to carry on the business of barbering on Sunday.

SEC. 2. Anyone found guilty of violating the first section of this act shall be fined not less than twenty-five dollars, nor more than fifty dollars, or imprisoned in the county jail not less than fifteen nor more than thirty days, or both, in the discretion of the court.

January 21, 1896, the supreme court rendered its decision in this case, declaring the above act unconstitutional and reversing the judgment of the lower court. From the opinion of said court, delivered by Judge Burgess, the following is quoted:

Defendant's first contention is that the act of the general assembly, under which the conviction was had, is in conflict with section 53, article 4, of the constitution of this State, and therefore void.

The section of the constitution with which it is claimed the law is in conflict provides that "the general assembly shall not pass any local or special law. * * * And where a general law can be made applicable, no local or special law shall be enacted; and whether a general law could have been made applicable in any case is hereby declared a judicial question, and as such shall be judicially determined, without regard to any legislative assertion on that subject."

Barbering is laboring, and the object of the act is to enforce an observance of the Sabbath, and to prohibit that kind of labor on that day. The policy of our laws is to compel the observance of Sunday as a day of rest, and if this may be done by a general law, applicable alike to all classes and kinds of labor, then the act falls within the inhibition of the paragraph of the constitution quoted, which prohibits the legislature from passing any local or special law, where a general law can be made applicable. That a general law prohibiting all kinds of labor on Sunday may not only be passed, but that we have such a law now upon our statute book, is indisputable. 1 Rev. St. 1889, § 3852. The fact that laboring on Sunday may be prohibited by proper legislation, as a police regulation, does not place the act beyond or without the inhibition of the constitution.

If the act is valid, then why may not the legislature by one act prohibit the farmer from laboring on Sunday, by another a blacksmith, and so on until all kinds of labor on that day are prohibited? Clearly, this may be done by a general law embracing all kinds of labor. The object of the constitution is manifest. It was to ... lation, and to substitute general law in ... eral law the same ends could be accomplished. ... wholesome rule that the invalidity of an act of ... conformity with the mandates of the constitution ... a reasonable doubt before we assume to pronounce ... another rule, alike obligatory on us, which requires ... a law invalid when it clearly appears to be so by ... conflict with the constitution. Our conclusion is that ... because in conflict with the constitution. The judgment ... and defendant discharged.

---

SUNDAY LABOR—BARBERS—CONSTITUTIONALITY O... *People v. Havnor. 43 Northeastern Reporter, page 541.*—He was convicted of carrying on the business of a barber ... New-York after 1 o'clock p. m. on Sunday in violation of ... of chapter 823 of the acts of 1895. The judgment of ... was affirmed by the appellate division of the supreme ... York and the defendant appealed to the court of appeals ... which rendered its decision April 14, 1896, and affirmed t... of the lower courts. The following is quoted from the o... court, which was delivered by Judge Vann:

The main ground upon which the defendant asks us ... judgment against him is that the statute under which he ... is in conflict with that provision of the constitution w... that "no person shall be deprived of life, liberty or pro... due process of law." Const., art. 1, § 6. The statute in ... tled "An act to regulate barbering on Sunday," provid... person who carries on or engages in the business of shav... ting or other work of a barber on the first day of the ... deemed guilty of a misdemeanor * * * provided that ... New York and the village of Saratoga Springs barber s... may be kept open and the work of a barber performed the... o'clock of the afternoon of the first day of the week." ... 823. The defendant claims that this statute deprives hi... extent of his "liberty," by preventing him from carryin... calling as he wishes, and also of his "property," by preve... use of his premises, tools, and labor, and thus renderi... productive. It is not claimed that his occupation is of a ... or that he so carried on his business as to disturb the p... good order of the neighborhood, or that the act for whi... victed, if done on any day of the week other than the fi... hour of that day prior to 1 o'clock in the afternoon, wou... a violation of law. Nor is it claimed that the convictio... ized by the common law, or that it was based upon any ... the one above cited, and, indeed, the judgment of the co... sessions expressly refers to that act, and adjudges the def... of a misdemeanor because he violated its command.

The phrase "due process of law" is not satisfied, ... pronounced, after an opportunity to be heard, by ... jurisdiction, in accordance with the provisions of ... statute accords with the provisions of the funda...

sense, whatever prevents a man from following a useful calling is an invasion of his "liberty," and whatever prevents him from freely using his lands or chattels is a deprivation of his "property." Yet, during the history of our State, many laws have been passed which, to some extent, have interfered with the right to liberty and property; but their accord with the constitution has seldom been questioned, and, when questioned, has been generally sustained. The power of taxation, the right to preserve the public health, to protect the public morals, and to provide for the public safety, may interfere somewhat with both liberty and property, yet proper statutes to effect these ends have never been held to invade the guaranties of the constitution. While the confinement of the insane or of those afflicted with contagious diseases infringes upon personal liberty, and the destruction of buildings to prevent the spread of fire, the exercise of the power of eminent domain, and the prevention of cruelty to animals encroach upon the right to property, still the proper exercise of these powers, under the authority of the legislature, although constant and known of all men, gives rise to no question of moment under the constitution. The sanction for these apparent trespasses upon private rights is found in the principle that every man's liberty and property is, to some extent, subject to the general welfare, as each person's interest is presumed to be promoted by that which promotes the interest of all. Dependent upon this principle is the great police power, so universally recognized, but so difficult to define, which guards the health, the welfare, and the safety of the public. While this power may not be employed ostensibly for the common good, but really for an ulterior purpose, when its object and effect are manifestly in the public interest, as was said in the Jacobs Case, "it is very broad and comprehensive, and * * * under it the conduct of an individual and the use of property may be regulated so as to interfere, to some extent, with the freedom of the one and the enjoyment of the other." In the exercise of this power the legislature has the right, generally, to determine what laws are needed to preserve the public health and protect the public safety, yet its discretion in this respect is not wholly without limit, for our courts have been steadfast in holding that the statute must have some relation to the general welfare; that the purpose to be reached must be a public purpose, and that "the law must in fact be a police law." Thus it has been held that "an act to improve the public health by prohibiting the manufacture of cigars and preparation of tobacco in any form in tenement houses in certain cases" (Laws 1884, c. 272) was unconstitutional, because it did not tend to promote the public health, and that this was not the end actually aimed at. For the same reason "an act to prevent deception in sales of dairy products" (Laws 1884, c. 202) was declared to conflict with the constitution, as it absolutely prohibited an innocent industry that was not fraudulently conducted, solely for the reason that it competed with another, and might reduce the price of an article of food. In a recent case an act prohibiting the sale of any article of food, upon the inducement that something would be given to the purchaser as a premium or reward (Laws 1887, c. 691), was held to be an unauthorized invasion of the rights of property and an improper exercise of the police power of the State. It was expressly declared in that case that the courts must be able to see, upon a perusal of the enactment, that there is some fair, just, and reasonable connection between it and the common good, and that, unless such relation exists, the statute can not be upheld as an exercise of the police power. Subject, however, to the limitation that the real object of the statute must appear, upon inspection, to have a reasonable connection [illegible] public, the exercise of the police power by the [illegible] lished as not in conflict with the constitution. [illegible] even if the effect is to interfere to some extent with [illegible] or the prosecution of a lawful pursuit, it is not regarded [illegible] tion of property or an encroachment upon liberty, because [illegible] vation of order and the promotion of the general welfare, [illegible] organized society, of necessity involve some sacrifice of [illegible]

The vital question, therefore, is whether the real purpose [illegible] statute under consideration has a reasonable connection with [illegible] lic health, welfare, or safety. The object of the act, as gathered [illegible] its title and text, was to regulate the prosecution of a particular [illegible] on Sunday, by prohibiting it from being carried on as a busin[illegible] that day, except in two localities, to which the prohibition appli[illegible] after a certain hour. It does not require the observance of the [illegible] bath as a holy day, or in any sense as a religious institution, as [illegible] dent from the fact that the entire day is left open to all s[illegible] employments but one, and a part of the day, in certain places, t[illegible] There is nothing in the act to prevent the defendant from carryi[illegible] his trade "in an manner or in any place that he pleases. [illegible] simply prohibited from carrying on that trade upon Sunday." [illegible] peculiar character of the first day of the week, not simply on a[illegible] of the obligations of religion, but as a day of rest and recreatio[illegible] been recognized for time out of mind both by the legislature a[illegible] courts. Statutes passed upon the subject while we were a col[illegible] Great Britain, as well as under the various constitutions in force [illegible] our organization as a State, have, so far as appears, been uni[illegible] enforced by the courts. Similar laws in other States, and esp[illegible] those which require the closing of places of business on Sunday [illegible] generally been sustained.

While questions have been raised as to noiseless and inoffensiv[illegible] pations that can be carried on by one individual without requiri[illegible] services of others, as well as to persons who observe the seventh i[illegible] of the first day of the week, still the rule is believed to be g[illegible] throughout the Union, although not generally enforced, that th[illegible] nary business of life shall be suspended on Sunday, in order that t[illegible] the physical and moral well-being of the people may be advanced. [illegible] inconvenience to some is not regarded as an argument against th[illegible] stitutionality of the statute, as that is an incident to all general [illegible] Sunday statutes have been sustained as constitutional almost w[illegible] exception; the most notable instance to the contrary (Ex parte [illegible] man, 9 Cal., 502), decided by a divided court in an early day in C[illegible] nia, having been subsequently overruled by the courts of that [illegible] (Ex parte Andrews, 18 Cal., 685; Ex parte Koser, 60 Cal., 202). [illegible] leading case in our own State upon the subject is that of Lindenm[illegible] v. People (33 Barb., 548), in which Judge Allen discussed the co[illegible] law, as well as legislation affecting the Sabbath, with great for[illegible] clearness. He held, in substance, that the body of the consti[illegible] recognizes Sunday as a day of rest, and an institution to be resp[illegible] by not counting it as a part of the time allowed to the govern[illegible] examining bills submitted for his appoval; that the Sabbath ex[illegible] a day of rest by the common law, without the necessity of legis[illegible] action to establish it; and that the legislature has the right to [illegible] late its observance as a civil and political institution. That ca[illegible] expressly approved in Neuendorff v. Duryea (69 N. Y., 557, [illegible] was referred to as one "which has never been question[illegible] higher or equal authority," and "as declaring the law of this State." It was cited with approval in People *v.* Moses (140 N. Y., 214, 215; 35 N. E., 499), where Judge Earl, speaking for a majority of the court, said: "The Christian Sabbath is one of the civil institutions of the State, and that the legislature, for the purpose of promoting the moral and physical well-being of the people, and the peace, quiet, and good order of society, has authority to regulate its observance and prevent its desecration by any appropriate legislation, is unquestioned." While works of charity and necessity have usually been excepted from the effect of laws relating to the Sabbath, and sometimes, also, those persons who keep another day of the week, still quiet pursuits have not, even when they can be carried on without the labor of others, because general respect and observance of the day, so far as practicable, have been deemed essential to the interest of the public, including as a part thereof, those who prefer not to keep the day, as their health and morals are entitled to protection, even against their will, the same as those of any other class in the community. According to the common judgment of civilized men, public economy requires, for sanitary reasons, a day of general rest from labor, and the day naturally selected is that regarded as sacred by the greatest number of citizens, as this causes the least inconvenience through interference with business. (Lindenmuller *v.* People, supra.)

It is to the interest of the State to have strong, robust, healthy citizens, capable of self-support, of bearing arms, and of adding to the resources of the country. Laws to effect this purpose, by protecting the citizen from overwork, and requiring a general day of rest to restore his strength and preserve his health, have an obvious connection with the public welfare. Independent of any question relating to morals or religion, the physical welfare of the citizen is a subject of such primary importance to the State, and has such a direct relation to the general good, as to make laws tending to promote that object proper under the police power, and hence valid under the constitution, which "presupposes its existence, and is to be construed with reference to that fact." (Village of Carthage *v.* Frederick, 122 N. Y., 268, 273; 25 N. E., 480.) The statute under discussion tends to effect this result, because it requires persons, engaged in a kind of business that takes many hours each day, to refrain from carrying it on during one day in seven. This affords an opportunity, recurring at regular intervals, for rest, needed both by the employer and the employed, and the latter, at least, may not have the power to observe a day of rest without the aid of legislation. As Mr. Tiedeman says, in his work on Police Powers: "If the law did not interfere, the feverish, intense desire to acquire wealth, * * * inciting a relentless rivalry and competition, would ultimately prevent not only the wage earners, but likewise the capitalists and employers themselves, from yielding to the warnings of nature, and obeying the instinct of self-preservation, by resting periodically from labor." (Tied. Lim., 181.) As barbers generally work more hours each day than most men, the legislature may well have concluded that legislation was necessary for the protection of their health. We think that this statute was intended and is adapted to promote the public health, and thereby to serve a public purpose of the utmost importance, by promoting the observance of Sunday as a day of rest. It follows, therefore, that it does not go beyond the limits of legislative power by depriving anyone of liberty or property within the meaning of the constitution.

The learned counsel for the defendant, however, criticises the act in question as class legislation, and claims th[illegible] teenth amendment to the Constitution of the [illegible] denies to barbers who do not reside in New [illegible] protection of the laws. That amendment d[illegible] arrangements made for different portions of a State [illegible] which, in carrying out a public purpose, is li[illegible] but, within the sphere of its operation, affects [illegible] larly situated. It was not designed to interfere wi[illegible] the police power by the State for the protection of [illegible] preservation of morals. The statute treats all barbers [illegible] same localities, for none can work on Sunday outside of [illegible] Saratoga, but all may work in those places until a cert[illegible] are, therefore, treated alike under like circumstances [illegible] both in the privileges conferred and in the liabilities [illegible] v. Missouri, 120 U. S., 68; 7 Sup. Ct., 350.) As was said [illegible] appellate division in deciding this case: "If the legislature [illegible] to regulate the observance and prevent the desecration of [illegible] it has the power to say what acts in the different localities of [illegible] it is necessary to prohibit to accomplish this purpose. It is [illegible] ceivable that an act in one locality, thickly settled, should be [illegible] which in sparsely settled districts of the State could be allow[illegible] for this reason an act might be objectionable in one district [illegible] another. All of these regulations have in view the proper [illegible] of the day, and are within the discretion of the legislature."

We think that the statute violates no provision of either the [illegible] or State constitution, and that the judgment appealed from [illegible] therefore, be affirmed.

This case was decided by a divided court, and strong dissentin[illegible] ions were delivered by Judges Gray and Bartlett.

---

VALIDITY OF CONTRACTS MAKING THE ACCEPTANCE OF BE[illegible] FROM RAILWAY RELIEF FUND BY EMPLOYEES A WAIVER OF C[illegible] FOR PERSONAL INJURIES—CONSTITUTIONALITY OF STATUTE— [illegible] *v. Pennsylvania Co.* *71 Federal Reporter, page 931.*—This case wa[illegible] in the United States circuit court for the northern district of Oh[illegible] the decision of said court was given January 28, 1896. The o[illegible] was delivered by District Judge Ricks and contains a full staten[illegible] the facts in the case. The following language is quoted therefr[illegible]

This suit was originally instituted in the court of common pl[illegible] Lucas County, and, by due proceedings had, was removed [illegible] defendant, which is a nonresident corporation, to this court[illegible] plaintiff sues to recover for damages, because of certain neglig[illegible] the defendant and its agents in failing to have properly filled th[illegible] between the ties at a certain junction or cross-over in said [illegible] by reason of which plaintiff's foot was caught while undertak[illegible] uncouple cars. There are certain other acts of negligence [illegible] the petition, which it is not necessary here to consider. [illegible] permanent injury, and damages in the sum of $25,000 [illegible]

To this petition the defendant filed an answer, which [illegible] the negligent acts charged in the petition [illegible] that said plaintiff, at the time he received the injuries complained of, was a member of the Voluntary Relief Department of the Pennsylvania Lines West of Pittsburg; that said Voluntary Relief Department is an organization formed for the purpose of establishing and managing a fund, known as a "relief fund," for the payment of definite amounts to employees contributing to the fund who, under the regulations, are entitled thereto when they are disabled by accident or sickness, and, in the event of their death, to their relatives or other beneficiaries specified in the application for insurance; that said relief fund is formed from voluntary contributions of the employees of the road, from contributions given by said defendant, the Pennsylvania Company, when necessary to make up any deficit, from income or profits derived from investments, or profits of the moneys of the fund, and such gifts and legacies as may be made for the use of the fund. The regulations governing said Voluntary Relief Department require that those who participate in the benefits of the relief fund must be employees in the service of the Pennsylvania Company, and be known as members of the relief fund. Defendant, further answering, says that no employee of the company is required to become a member of said relief fund; that the same is purely voluntary; that anyone who has become a member may withdraw at any time, upon proper notice; that contributions from such members cease by so withdrawing. The defendant further says that participation in the benefits of such relief fund is based upon the application of the beneficiaries; that on the 3d day of January, 1894, the plaintiff in this case, being in the employ of the defendant company, applied for membership, and in said application agreed to be bound by the regulations of the said fund. Defendant further says that the application for membership was approved and accepted at the office of the superintendent of the relief department, and that thereupon said plaintiff became a member of said relief fund. Defendant further says that, when said plaintiff received the injuries complained of, he thereupon became entitled to the benefits growing out of his membership in said relief fund, by reason of the injury so received while in said service; that said plaintiff thereupon immediately applied to said department for such benefits, and received monthly payments therefrom, amounting in all to the sum of $399, until the commencement of this action, on the 25th day of May, 1895. Defendant says that the plaintiff, in his application for membership, expressly agreed that, should he bring suit against either of the companies now associated in the administration of the relief department for damages on account of injury or death, payment of benefits from the relief fund on account of the same shall not be made until such suit is discontinued, or, if prosecuted to judgment or compromise, any payment of judgment or amount of compromise shall preclude any claim upon the relief fund for such injury or death. Defendant says that, the plaintiff having commenced suit against the defendant, payments to the plaintiff for the benefits accruing under said contract were suspended; and defendant says that by virtue of the agreement aforesaid, and the acceptance by the plaintiff of the benefits from said relief fund on account of said injuries, the said defendant thereupon became discharged from any and all liability to the plaintiff on account of said injuries. The plaintiff has demurred to this answer. He contends, first, that the contract set up in the answer is invalid; and, next, that it is in violation of an act of the legislature of Ohio passed in 1890, in 87 Ohio Laws, page 149.

There are two questions to be determined upon the demurrer thus interposed. The first question is whether this contract between the plaintiff and the defendant is a valid one. the court, rests entirely upon the pleadings. and the allegations of the defendant's answer by the plaintiff having demurred thereto. It tant to emphasize the facts thus admitted. voluntarily became a member of this relief knowledge of its rules and regulations. The avers, and the demurrer admits it, that, by his to become a member of such relief department, that the acceptance by him of benefits from the relief or death, should operate as a release of all claims said defendant arising from such injury or death. It that it is the acceptance of benefits from this relief fund agreement, releases the railroad company from a claim If the employee injured does not accept such benefits, sue for damages, his right of action is unimpaired, and waived. This is the case as presented by the pleadings facts. It is not the question of whether a railroad tract with its employees, can exempt itself from suits for ries caused by its negligence. That, as a general rule, can not This case does not present that question, neither does it pre issue of fact as to whether this contract for insurance is a vo one or not. If the pleadings and evidence in a case should sh an employee entered into such a contract, ignorant of its terms, under restraint or duress or compulsion, the court would then be ized, and it would be its duty, to inquire into that fact, and against any wrong of that nature. But, as before stated, question is now in any way presented. The pleadings do not ev gest such an issue. The sole question is whether, under the a facts already stated, this contract is valid. There are decision supreme courts of the States of Iowa, Maryland, Pennsylvania, State courts in Ohio, and of circuit courts of the United States and Maryland, holding such contracts legal and binding. Un plan, employees of railroads are afforded protection by a spe insurance. This sort of protection is not available to them in o insurance companies, except at such high cost as to make it subst unobtainable. Members sick or injured are entitled to benefits, less of what causes their temporary disabilities. They will thus benefits in cases where no claim against the railroad company c made. They could receive benefits, also, in cases where the inj the result of their own contributory negligence, or of that of servants in the same department of service, in both of which c a rule, no right of action would arise against their employer. employees desire to enjoy the benefits of such contracts, they have the right to make them. They are capable of deciding fo selves whether they want to contract for such protection. I within the powers of a legislature to assume that this class of m paternal legislation, and that, therefore, they will protect t depriving them of the power to contract as other men may.

At this point in its opinion the court cites and quotes from lowing cases: Johnson *v.* Railroad Co., 163 Pa. St. Reports, p. 29 Atlantic Reporter, p. 854; Donald *v.* Railroad Co., 61 Reporter, p. 971; Fuller *v.* Association, 67 Md. Atlantic Reporter, p. 237; Owens *v.* Railway Co.

p. 715, and Martin v. Railroad Co., 41 Federal Reporter, p. 125. The opinion then continues as follows:

In view of these authorities and of the reasons given in support of the conclusions reached, I feel justified in holding the contract in this case valid and binding upon the plaintiff.

The next question to be determined is whether the act of 1890 of Ohio [87 Ohio Laws, p. 149] is constitutional. The latter part of the first section of said act reads as follows:

"And no railroad company, insurance company, or association of other persons shall demand, expect, require or enter into any contract, agreement or stipulation with any other person about to enter, or in the employment of any railroad company whereby such person stipulates or agrees to surrender, or waive any right to damages against any railroad company thereafter arising from personal injury or death, or whereby he agrees to surrender or waive, in case he asserts the same, any other right whatsoever, and all such stipulations or agreements shall be void."

This act has been declared unconstitutional in the case of Cox v. Railway Company (33 Ohio Law J.), April 22, 1895, in a well-considered opinion by Judge Dilatush, of the Warren County court of common pleas; and the court reaches that conclusion because said act violates section 1, article 1, of the bill of rights, as interfering with the rights of private contract. That provision of the bill of rights is as follows:

"All men are by nature free and independent, and have certain inalienable rights, among which are those of enjoying and defending life and liberty, acquiring, possessing and defending property, and seeking and obtaining happiness and safety."

Article 2, section 26, of the constitution of Ohio, provides:

"All laws of a general nature shall have uniform operation throughout the State."

The act under consideration, while it is general in its nature, applies only to railroad companies and their employees, and is not, therefore, general in its application, and does not operate uniformly on all classes of citizens. Under this statute railroad companies are prohibited from making contracts which other corporations in the State are allowed to ma e.

Article 1, section 10, of the Constitution of the United States, prohibits any State from passing any law impairing the obligation of contracts.

Article 8, section 16, of the bill of rights of the State of Ohio, prohibits making any law impairing the obligation of contracts.

Article 2 of the Northwestern Territorial Government (1787) provides as follows:

"In the just preservation of rights and property, it is understood and declared that no law ought ever to be made or have force in said Territory that shall in any manner interfere with or affect private contracts or engagements bona fide without fraud, previously made."

This extract from the ordinance of 1787 shows how jealously the right of personal liberty in the making of private contracts was regarded, and how carefully any restriction of said right was restrained.

The act under consideration is certainly one which impairs the rights of a large number of the citizens of Ohio to exercise a privilege which is dear to all persons, namely, that of making contracts concerning their own labor and the fruits thereof, and, so far as it relates to such contracts already made, impairs their validity. The act seems to assume that a large class of [illegible] employed by railroad corporations [illegible] for their own labor.

As hereinbefore stated, this contract [illegible] no unfair advantage is taken of these [illegible] in its broadest and fullest sense, is a [illegible] protection and assistance. If in [illegible] the employee insured, that is in itself [illegible] of an unconscionable nature, or unfair in the [illegible] sufficient pretext to assume that all such contracts [illegible] of the legislative body, or that so large a class [illegible] restricted in their right of personal liberty.

The Ohio statute, in denying to the employees [illegible] tion the right to make their own contracts concerning [illegible] depriving them of "liberty," and of the right to [illegible] of manhood, "without due process of law." Being [illegible] employees of railroads, it is class legislation of the [illegible] acter. Laws must be not only uniform in their application [illegible] the territory over which the legislative jurisdiction [illegible] must apply to all classes of citizens alike. There [illegible] for railroad employees, another law for employees [illegible] another law for employees on a farm or the highways. [illegible] is dangerous. Statutes intended to favor one class often [illegible] sive, tyrannical, and prescriptive to other classes never [illegible] affected thereby; so that the framers of our constitution, [illegible] experience, wisely provided that laws should be general [illegible] and uniform throughout the State.

For the reasons stated, I am of the opinion, first, that [illegible] set out in the defendant's answer is a valid contract; and [illegible] the act of the legislature of Ohio, which declares it [illegible] invalid, is unconstitutional. The demurrer to the answer [illegible] overruled.

---

WAGES EXEMPT FROM GARNISHMENT—*Chapman et al.* [illegible] *Southern Reporter, page 918.*—J. P. Berry recovered a judgment [illegible] Henry Chapman in the circuit court of Hinds County, Miss. [illegible] of garnishment was issued April 8, 1895, and served on the [illegible] and Vicksburg Railroad Company on the same day. The [illegible] answered, setting up that at the time the garnishment was [illegible] owed Chapman $81, and during the month of April Chapman's [illegible] amounted to $81 and Chapman drew from its supply car [illegible] amounting to $56.65; that during the month of May C[illegible] wages amounted to $81 and he drew supplies amounting [illegible] that Chapman's wages for the month of June were $81 and [illegible] supplies amounting to $71; that, adding the amount of its [illegible] to Chapman at the time the writ was served to the balances [illegible] from month to month, it owed him $121 at the time [illegible] filed. Chapman claimed the amount due him as exempt [illegible] 1963 subsection 10*a* of the Code of 18[illegible], which [illegible]

And the following property shall be exempt [illegible] other legal process, to wit: [illegible]

(a) The wages of every laborer [illegible]

the head of a family, to the amount of one hundred dollars, and every other person to the amount of twenty dollars.

From a judgment of the circuit court in favor of the plaintiff, Berry, for the full amount of all the wages due the defendant, Chapman, at the time the writ of garnishment was served, and what became due before the answer was filed, less $100, defendant and the railroad company appealed to the supreme court of the State. The decision of said court was rendered October 28, 1895, and reversed the decision of the lower court.

The opinion of the court, delivered by Judge Woods, contained the following language:

In the case at bar the judgment creditor sought to tie up and have applied to his judgment the entire monthly wages of his debtor for three or four months under one writ of garnishment, executed in April and returnable in July, less $100, exempt as wages. This view was adopted by the court below, and judgment was entered accordingly. The action was erroneous and rests upon a rigid adherence to the letter of the law of garnishment as contained in chapter 55, Code 1892, and in failing to give vital efficacy to that provision, subdivision *a* of the tenth head of the first section of our law of exempt property, found in chapter 45 of the Code, by which "the wages of every laborer or person working for wages, being the head of a family, to the amount of one hundred dollars," are made exempt from seizure under legal process.

The humane and wise purpose of this exemption law was to secure, not only to the laborer, but the family of which he was head, and for which, by every obligation, legal as well as moral, it was his duty to provide, the necessaries and comforts of life. His wages, by law, are set apart and dedicated to that righteous end. They are made absolutely exempt from seizure under legal process, and we must give our exemption law such construction as will carry its beneficent designs into effect. The chapter on garnishment and the independent chapter on exemption must be so construed as to give harmonious effect to both. The view which we decline to follow would practically render nugatory this salutary provision exempting the wages of laborers.

If, by successive service of writs of garnishment, as was attempted in Chandler *v.* White [71 Miss. Reports, p. 16], or by a single writ returnable to a term of court long subsequent to its execution, as was done in the present case, the debtor [creditor] can aggregate the small monthly wages to an amount in excess of the exemption, and seize this excess, no matter how great his judgment may be, it will be readily seen that both the laborer and his family may come to actual want, and the statute for their protection would be rendered nugatory.

Under the mistaken view which prevailed below, the laborer would be deprived of his small wages, payable at short intervals, until his creditor's debt had been satisfied, if the laborer continued in the service of the same employer, or he would be driven to seek employment, and if successful in finding it, he would again be forced to leave the second employment, after a writ of garnishment had been served, and once more engage in his wandering search for a third engagement, and so on indefinitely, his family meanwhile subjected to all the hardship and want incident to such vicissitudes of evil fortune.

The true view is that on the 1st of each month, or whenever, by the contract of employment, the wages, not exceeding $100, are due and payable, the laborer has the right [illegible] withstanding his employer may have [illegible] garnishment writ was served in this case [illegible] railway company's answer showed that [illegible] wages, the judgment creditor took nothing [illegible] then due were not garnishable, and the [illegible] demand payment of his employer, notwith[illegible] writ; and, in like manner, the wages for April, [illegible] respectively fell due (provided, always, the [illegible] more than $100 of the laborer's wages in his [illegible] exempt from seizure under legal process, no matter [illegible] writ or a half dozen successive writs, and the railway [illegible] and should have paid, according to the tenor of [illegible] monthly wages to the laborer. Our statute was design[illegible] laborers and their families the small fruits of their [illegible] bound to give it such proper and liberal interpretation [illegible] and force to that wise and humane design.

It is contended, however, by counsel for appellee, that, [illegible] contention is thought by us to be not sound, yet, under [illegible] answer of the garnishee, judgment should be rendered for [illegible] answer does not admit an indebtedness of $120.10, but [illegible] facts by which the various balances, after deducting from [illegible] wages of $100 the amounts paid to or accounted for with [illegible] monthly, if added to the $81 due when the garnishment [illegible] in April, amount to $121. But we have already seen [illegible] wages for March was exempt from seizure under legal [illegible] not garnishable, and that the judgment creditor acquired [illegible] the service of his writ at that time. Let the garnishee be di[illegible] on its answer.' Judgment accordingly.

## DECISIONS UNDER COMMON LAW.

Breach of Contract of Employment—Damages—*Tarr Co. v. Kimbrough. 34 Southwestern Reporter, page 528.*-[illegible] was brought in the court of common pleas of Fayette County, [illegible] J. Mc. Kimbrough against the William Tarr Company to reco[illegible] contract of employment. Judgment was rendered for the plai[illegible] the defendant appealed to the court of appeals of the Stat[illegible] court rendered its decision February 26, 1896, and rever[illegible] judgment of the lower court. It appears from the evide[illegible] Kimbrough made a contract with the appellant by which [illegible] employed to do certain work for one year, commencing June [illegible] at a salary of $1,200 per annum, payable in monthly install[illegible] $100. He was paid $100 for the month of June; $68.02 on [illegible] salary, and was discharged without cause on the first day of [illegible] The opinion of the court of appeals, delivered by Judge [illegible] contains the following language:

The court erred in giving instructions to the jury. [illegible] that if the plaintiff was employed by the defendant [illegible] August at the contract price of $100, and was disch[illegible] then he was entitled to recover the $100 in dam[illegible] the contract, and so told the jury. If the contract existed as claimed by plaintiff, he was only entitled to recover such damages as he had actually sustained, and which the jury should have fixed on proof. A refusal to allow one to perform a stipulated service is not equivalent to performance, so as to entitle the party to recover the agreed price for full performance.

EMPLOYERS' LIABILITY—ASSUMPTION OF RISKS—*Greene v. Western Union Telegraph Co. 72 Federal Reporter, page 250.*—This was an action brought by John C. Greene against the Western Union Telegraph Company, in the United States circuit court for the district of Minnesota, to recover damages for personal injuries received while in the employ of said company.

The plaintiff was in the employ of the telegraph company as a lineman, and was one of a crew engaged in repairing its telegraph lines, under the immediate charge of a foreman, who had authority to hire and discharge men, and direct them where and how to work. He was ordered by the foreman to climb a certain "gin pole," and while climbing it, pursuant to said order, the pole fell, apparently because not sufficiently guyed, and he was injured. The plaintiff claimed that the foreman was a vice-principal, for whose negligence in the erection of said pole the defendant was liable.

Upon the evidence given the defendant moved the court to direct a verdict in its favor, and the court, March 11, 1892, granted the motion. The opinion was delivered by District Judge Nelson, and was as follows:

I think this was a risk that the plaintiff assumed when he was hired. It was a part of his duty. He was not only to help erect and climb poles and string wires, but to help to put up those gin poles. While the business may have been a hazardous one, he assumed the ordinary risks incident thereto, and among them that of a pole not being properly guyed, owing to negligence on the part of his fellow-workmen. Conceding that the foreman was a representative of the company, I do not think it has anything to do with the case. Plaintiff was not taken from any particular duty and put into one that he was not hired to do, which was extra hazardous, and unnecessarily exposed him to a danger which he did not contemplate by virtue of his employment; but he was hired to do just what he was ordered to do, and in so doing the accident happened. I think this man was injured by a risk which he assumed by virtue of his employment, and I instruct you that the defendant is entitled to your verdict.

---

EMPLOYERS' LIABILITY—RAILROAD COMPANIES—*Union Pacific Ry. Co. v. O'Brien. 16 Supreme Court Reporter, page 618.*—This was an action brought by Nora O'Brien against the Union Pacific Railway Company in the circuit court for the district of Colorado to recover

## RECENT GOVERNMENT CONTRACTS

[The Secretaries of the Treasury, War, and Navy Departments have conse furnish statements of all contracts for constructions and repairs entered into b These, as received, will appear from time to time in the Bulletin.]

The following contracts have been made by the office of the vising Architect of the Treasury:

DETROIT, MICH.—July 17, 1896. Contract with Forster & l Minneapolis, Minn., for joinery work, marble work, iron stairs, p ing, etc., in courthouse, post office, etc., $216,740. Work to be pleted within seventeen months.

PUEBLO, COLO.—July 18, 1896. Contract with L. L. Leach & Chicago, Ill., for erection and completion of post office, except h apparatus, plumbing, elevator car and machinery, electric wir conduits, $187,774. Work to be completed within eighteen mon

SIOUX CITY, IOWA.—August 7, 1896. Contract with Optenb Sonneman, Sheboygan, Wis., for low-pressure, steam-heating mechanical ventilating apparatus, power boilers, etc., for courtl post office, and customhouse, $10,699. Work to be completed ninety working days.

WASHINGTON, D. C.—August 10, 1896. Contract with Geor Howard, Philadelphia, Pa., for extending and remodeling one f elevator at the Bureau of Engraving and Printing, $1,500. W be completed within three months.

SOUTH BEND, IND.—August 17, 1896. Contract with Hend Bros. & O'Neill, Valparaiso, Ind., for erection and completion o office, except heating apparatus, $35,036. Work to be completed eleven months.

MADISON, IND.—August 24, 1896. Contract with Bailey, K & Co., Louisville, Ky., for erection and completion of post office, e heating apparatus, using Portage red sandstone and Indiana lime $22,460. Work to be completed within six months.

NEWARK, N. J.—August 29, 1896. Contract with Morse, Wi & Co., Philadelphia, Pa., for hydraulic elevator in the custom and post office, $3,995. Work to be completed within seventy from approval of bond.

MERIDIAN, MISS.—September 1, 1896. Contract with F. B. S & Co., Chicago, Ill., for erection and completion of post office, e heating apparatus, $51,000. Work to be completed within twenty

BEAVER FALLS, PA.—September 2, 1896. Contract with William Miller & Sons, Pittsburg, Pa., for erection and completion of post office, except heating apparatus, $30,746.

RACINE, WIS.—September 9, 1896. Contract with Adam H. Harcus, Racine, Wis., for erection and completion of customhouse and post office, except heating apparatus, $44,347. Work to be completed within ten months.

TAUNTON, MASS.—September 10, 1896. Contract with A. H. Kleinecke, Chicago, Ill., for interior finish, plumbing, and gas piping for post office, $16,060. Work to be completed within six months.

KANSAS CITY, MO.—September 25, 1896. Contract with J. J. Hanighen, Omaha, Nebr., for horizontal drain pipes, clean outs, and clean out manholes for post office and courthouse, $3,733. Work to be completed within fifty working days.

YOUNGSTOWN, OHIO.—September 25, 1896. Contract with Bailey, Koerner & Co., Louisville, Ky., for erection and completion of post office, except heating apparatus, $31,849. Work to be completed within nine months.

DETROIT, MICH.—September 26, 1896. Contract with Henry Carew & Co. for isolated ward building for marine hospital, $1,090. Work to be completed within sixty days.

BUFFALO, N. Y.—September 26, 1896. Contract with John Peirce, New York City, for superstructure, etc., of Jonesboro granite for post office, $719,900. Work to be completed within twenty months.

KANSAS CITY, MO.—September 30, 1896. Contract with the Huagh-Noelke Iron Works, Indianapolis, Ind., for steel and iron work of roof, etc., of post office and courthouse, $7,005. Work to be completed within sixty days.

# BULLETIN

OF THE

# DEPARTMENT OF LABOR.

No. 7—NOVEMBER, 1896.

ISSUED EVERY OTHER MONTH.

EDITED BY

CARROLL D. WRIGHT,
COMMISSIONER.

OREN W. WEAVER,
CHIEF CLERK.

WASHINGTON:
GOVERNMENT PRINTING OFFICE.
1896.

# CONTENTS.

BULLETIN

OF THE

DEPARTMENT OF LABOR.

No. 7. WASHINGTON. NOVEMBER, 1896.

## INDUSTRIAL COMMUNITIES. (a)

BY W. F. WILLOUGHBY.

### CHAPTER VI.

### OTHER INDUSTRIAL COMMUNITIES.

The places that have been described in detail by no means constitute all the industrial centers in Europe coming under the designation of industrial communities. There is no hard and fast distinction that marks off a place as an industrial community. Under varying conditions are formed the practically self-contained industrial community, as Guise, the community dominated by and almost wholly given up to the interests of a single industrial establishment, as Essen or Anzin, and the ordinary manufacturing town, all of which may be styled industrial communities. In the middle class may be numbered a good many other places only slightly less important than those of which a detailed account has been given. While it is impracticable, if indeed it were desirable, to give to each of these the same full treatment that has been accorded to the others, it will nevertheless be of value to make mention of their existence and to give some account of their conditions or institutions most worthy of investigation. The following is believed to be a practically complete list of important places with sufficient of the character of industrial communities to cause the grouping around them of systems of special institutions, thus giving to them a special life and character of their own:

In France there are, first, the other more important mining communities, notably Douchy, Courrières, Liévin, and Bessèges, and, second, Le Creuzot, the seat of the important iron and steel works of MM. Schneider & Co., the "Krupp" of France; Noisiel, a unique community of its kind, the seat of the chocolate works of M. Menier, and Thaon, a village

a See footnote to the beginning of this series of articles in Bulletin No. 3.

almost wholly devoted to the enterprise of the Thaon [illegible] Cloth Works. In Belgium those most worthy of [illegible] ing community of Mariemont and Bascoup, the Vieille-[illegible] Mining and Smelting Company, and Willebroek, the seat of the [illegible] mills of De Naeyer & Co. In Holland there is the village of [illegible] Park, the seat of the Netherland Yeast and Alcohol Factory, [illegible] Delft. In Switzerland the most prominent example is that of the Suchard Chocolate Works at Neuchâtel; in England, the not very fortunate enterprise of Saltaire, and in Italy, the works of Senator Rossi, at Schio. In Germany there is no industrial community at all approaching Essen in importance. Mention might be made of Bochum, where are located the iron works of the Bochumer Verein, and the works of Gebrüder Stumm at Neunkirchen.

A more or less full account of the more important of these places is given in this chapter.

## THE MARIEMONT AND BASCOUP COAL MINING COMPANY, BELGIUM.

A number of considerations mark the Mariemont and Bascoup Coal Mining Company as the most worthy of study, from the standpoint of the present report, of all the industrial concerns in Belgium. The company is one of the most important in the country. The seat of its operations, the two adjoining villages of Mariemont and Bascoup, is so largely given up to this one concern as to constitute it a distinct industrial community, and, more than all, its workingmen's institutions are among the most remarkable in Europe.

The present company was formed through the consolidation of the two separate companies of Mariemont and Bascoup, whose properties were contiguous. The mining of coal in this region dates back several centuries, but the first regular operations may be said to date from 1794, while the first concession to the companies was made in 1802. The two companies have constantly grown in importance through the acquisition of additional property. Legally they are still distinct companies, though they are under the same direction and for all practical purposes are one undertaking.

In 1894 the company employed in round numbers 6,500 men, and its annual production was between 1,000,000 and 1,200,000 tons.

In the organization of labor by the company the principle of piecework has probably been carried further than in any other mine in Belgium, if not on the Continent. Wherever possible the work is let out by contract to the workingmen, and in addition to the wages thus provided for certain premiums or bonuses are added for good work, participation in any savings that may be made in the use of [illegible] and supplies. This is done by making an estimate of the [illegible] supplies and tools under the system of day labor in [illegible] and any saving made on this sum is shared equally by the men and by the company. M. Weiler, the engineer of the company, is very enthusiastic over the success of this system. He says that marked economy, which has profited the men and the company alike, is manifest in the use of both tools and materials. Every effort is made to interest the workingmen in their work and in the general prosperity of the company. One of the unique methods employed is to make a large graphic representation of the organization of the company into departments, branches, etc., a copy of which is given to each workingman who deserves it, so that he can get an intelligent idea of the whole work of the company in which he is employed and the exact position that he occupies in this system. This is but a sample of the efforts made to give to the men an intelligent conception of what they are doing. They are encouraged to find an education in their work.

In 1888 the company prepared a statement of the average wages paid since 1869, as well as for the year 1888, in greater detail. These statistics of wages are the only ones available and are given in the tables that follow:

AVERAGE DAILY WAGES OF EMPLOYEES OF THE MARIEMONT AND BASCOUP COAL MINING COMPANY, 1869 TO 1888.

| Year. | Average daily wages. | Year. | Average daily wages. | Year. | Average daily wages. | Year. | Average daily wages. |
|---|---|---|---|---|---|---|---|
| 1869 | $0.56 | 1874 | $0.85 | 1879 | $0.61½ | 1884 | $0.67½ |
| 1870 | .60 | 1875 | .81½ | 1880 | .70 | 1885 | .62 |
| 1871 | .62 | 1876 | .79½ | 1881 | .68½ | 1886 | .62½ |
| 1872 | .69½ | 1877 | .65½ | 1882 | .69 | 1887 | .62½ |
| 1873 | .89 | 1878 | .62 | 1883 | .72 | 1888 | .63 |

AVERAGE DAILY WAGES OF EMPLOYEES OF THE MARIEMONT AND BASCOUP COAL MINING COMPANY, BY CLASSES, 1888.

MARIEMONT.

| Class of employees. | Employees above ground. | | Employees below ground. | | Total employees. | |
|---|---|---|---|---|---|---|
| | Number. | Average daily wages. | Number. | Average daily wages. | Number. | Average daily wages. |
| Men | 700 | $0.67 | 1,701 | $0.73½ | 2,401 | $0.71½ |
| Women | 127 | .23½ | | | 127 | .23½ |
| Boys under 16 years of age | 57 | .22 | 260 | .28 | 317 | .27 |
| Girls under 16 years of age | 80 | .19½ | | | 80 | .19½ |
| Total | 964 | .54½ | 1,961 | .67½ | 2,925 | .63 |

BASCOUP.

| Class of employees. | Employees above ground: Number. | Employees above ground: Average daily wages. | Employees below ground: Number. | Employees below ground: Average daily wages. | Total employees: Number. | Total employees: Average daily wages. |
|---|---|---|---|---|---|---|
| Men | 569 | $0.62 | 1,864 | $0.76 | 2,433 | $0.73 |
| Women | 132 | .27½ | | | 132 | .27½ |
| Boys under 16 years of age | 38 | .19½ | 319 | .34 | 357 | .32½ |
| Girls under 16 years of age | 92 | .19 | | | 92 | .19 |
| Total | 831 | .50½ | 2,183 | .70 | 3,014 | .64½ |

## WORKINGMEN'S INSTITUTIONS.

The workingmen's institutions of Mariemont and Bascoup include a complete scheme of institutions similar to those already described as existing at Anzin and Blanzy, such as the housing of employees, medical service, mutual-aid societies, cooperative distributive stores, schools, etc. To describe these in detail would result in unnecessary repetition. There are two institutions, however, which are unlike any yet studied and which form the central feature of the Mariemont and Bascoup Company's social work. These two, both of which are of great importance and interest, will therefore be described in detail, while a general description of the other work of the company will be sufficient. The first of these is the scheme of chambers of explanations (*chambres d'explications*) and councils of conciliation and arbitration (*conseils de conciliation et d'arbitrage*), organized for the prevention and adjustment of difficulties between the company and its employees, an effort which is without doubt the most important in Europe for the settlement of labor disputes without invoking the aid of the Government. The second is the system of workingmen's insurance against accidents, sickness, and old age. The latter is important, not because it is unique, but because it affords an example of the general system for the insurance of coal miners in Belgium.

Before commencing their consideration a few words should be said regarding the spirit in which these institutions have been created and are now administered, for it should never be forgotten that this spirit usually determines the real benefits that result. The following extract is quoted from an address delivered by M. Julien Weiler before the Société Belge d'Économie Sociale, entitled the Spirit of the Workingmen's Institutions of Mariemont. M. Weiler is the engineer in chief in charge of the practical operations of mining at Mariemont, and is the official to whom has been given the especial charge of all the company's work for its employees, and his remarks are therefore in the nature of an authoritative statement of the motives actuating the company in its social work. He says:

> The question can be raised why we do not allow the pension and aid funds to be entirely managed by the workingmen. The reply to this is that we are desirous of diminishing our intervention each time that the workingmen manifest the desire to augment theirs. But the workingmen have not yet arrived at the stage where they can be disfranchised from all guidance. The proof of this is that as soon as they are released from control by the company they have recourse to some other guidance.
>
> We think, however, and this is a point on which I wish to lay great stress, for it indicates the whole spirit of our institution, that the employer ought to continue his intervention provided that his intervention is limited to that which the workingmen themselves can [illegible] In other words, the employer, as regards the matter of tutelage, [illegible] to prepare the way for his own abdication.

You do not find among us, therefore, patriarchal institutions. * * * On the contrary all our efforts tend to emancipate our workingmen intellectually and morally. Self-help is the sentiment we desire to inculcate, and I hope to show you that we have in great measure succeeded in our efforts.

## COUNCILS OF CONCILIATION AND ARBITRATION.

Without doubt the most worthy of study of all the institutions of the Mariemont and Bascoup Coal Mining Company is the system adopted by it for the prevention and settlement of labor disputes. Its efforts in this direction are the most remarkable made by any private company in Europe. "The entire history of arbitration in Belgium," says the French Labor Bureau in its exhaustive report on Arbitration and Conciliation, "is contained in that of the councils of conciliation of Mariemont and Bascoup and in that of the councils of industry and labor."

The first application of the idea of conciliation in labor disputes by the Mariemont and Bascoup Company, and indeed the first application of that idea in Belgium, was made in the year 1877. As the result of a serious strike of his employees in 1875-76, the manager of the company requested his engineer, M. Julien Weiler, to make a thorough study of the question of conciliation and arbitration with the view of the introduction of some such system in the organization of labor in his company.

The first action, as the result of this investigation, took the form of the creation of chambers of explanation (*chambres d'explications*) in the workshops of the company. These were meetings of clerks or foremen and workingmen, with the object of furnishing a medium of communication between the employees and their employer, in order that misunderstandings might be avoided. In this way an opportunity was afforded for the settlement of demands that in themselves would often be readily granted were they only brought to the attention of the other side, but which if not explained might cause great trouble and friction. Each of the 9 trades in the workshops was represented by a committee of 6 workingmen and 6 clerks or foremen, with a central committee composed of a delegate from each group for the discussion of general questions.

At first, owing to lack of previous organization among the workingmen, and to the fact that the establishment of the chambers coincided with a serious reduction in wages, the results were not satisfactory. However, after the chambers had been in existence several years, and after several modifications had been made in their organization, they began to render valuable services. Fines were entirely abolished, while the officials were unanimously of the opinion that their regulations had never been more respected; moreover, the regulations themselves were revised by the committees. The system of task work which had previously been considered impracticable was successfully organized, and the cost of production diminished by at least 20 per cent, while wages were advanced to an equal extent.

In 1880 the system was extended to the employees [illegible] transportation of the coal, and in 1889 further extended so as to [illegible] all the different branches of the company's work and all its [illegible]

The creation of these chambers of explanation necessarily [illegible] the creation of an institution which, while leaving to these chambers the oversight of daily details, should itself have a higher character and take cognizance of more serious disputes. On January 1, 1888, therefore, there was created both at Mariemont and at Bascoup a council of conciliation and arbitration (*conseil de conciliation et d'arbitrage*). These councils constitute the supplement to the chambers of explanations. The constitutions of the two councils are practically identical. Each is organized in such a way as to secure to both the workingmen and the company an absolute equality. The council is composed of 6 representatives of the company and 6 representatives of the workingmen. The company chooses its representatives from among its higher employees, including the general manager. The 6 workingmen representatives are elected by a species of electoral college. Thirty-six delegates are elected by the workingmen, 6 from each branch of the company's service. Only those workingmen can vote who are 21 years of age and have been in the service of the company at least six months. This body, all the members of which have the right to attend the meetings of the council, elects 6 representatives, who alone have the right of voting in the deliberations. These representatives must be over 30 years of age and they must have been in the service of the company five or more years.

The council meets monthly. A president and vice-president are appointed annually, one of whom must be chosen from among the workingmen. The president convokes the meetings, at which two-thirds of the representatives of each side must be present in order to constitute a quorum.

All disputes of general interest must be discussed before the whole council. Prior to the examination of a disputed point, and during its discussion, work must be continued under the conditions which existed when the difficulty arose. Both parties agree to accept the decisions of the council for at least three months and the same question can not be raised again within that period. All the expenses of the council are defrayed by the company. Each representative and delegate is allowed 2 francs (39 cents) for each meeting that he attends, and is indemnified for all time lost.

The proof of the value of any institution is in the practical results of its operations. A survey of the operations of these councils during the first six years of their existence shows how many important questions have been brought to their attention and decided by them.

During the first year, 1888, there were 18 meetings of the [illegible] which 39 questions were considered. Nineteen of these were [illegible] questions, that is, such as concerned more than one group [illegible] men; 10 were questions relating to one group or branch of workingmen, and 10 related to individuals.

Among the more important general questions was a demand made in April for increased pay based on newspaper reports of an alleged revival of trade. The administration made a formal agreement to increase wages as soon as better times should set in and the other party was satisfied. On November 19 a second demand was made, resulting in a recommendation of a 5 per cent increase, which the administration granted.

Another question related to a demand for shorter hours and certain regulations regarding time of ascending and descending the mines. This question evoked heated discussions during three meetings of the council, when a committee of 2 from each side was appointed to thoroughly examine the question. The committee's report was in the form of a compromise and was adopted, to the satisfaction of both parties.

As frequent disputes had arisen regarding the time of arrival of employees, the council succeeded in having clocks placed at convenient points about the mines.

Several questions relating to health, security, and convenience were satisfactorily settled.

Among the minor questions were such as related to complaints on account of fines, promotions, discharges, errors in wage payments, etc. All these were disposed of after careful examination either by the council alone or with the assistance of special committees.

In 1889 there were 12 meetings of the council, at which 30 questions were considered. Of these 17 were general questions, 6 related to one group of workingmen, and 7 to individuals. On September 23 the council considered a demand for a further increase of wages in proportion to the higher price of coal. The company offered its books in evidence, and the meeting adjourned pending their investigation. At the next meeting, October 22, the company agreed to increase the wages 5 per cent, to take effect November 1. Another increase of 5 per cent was made December 1, and still another of 10 per cent on January 1, 1890. Other general questions considered related to hours of work, the fine system, indemnities to workingmen, etc.

In 1890 there were 16 meetings of the council, at which 57 questions were considered. Of these 39 were general questions, 15 related to one group of workingmen, and 3 related to individuals. The general questions related to the health service, wages, the sale of coal to workingmen, hours of labor, aid and provident funds, fines, holidays, handling of explosives, etc.

In 1891 there were 13 meetings, at which 47 questions were considered. Of these 33 were general, 12 special, and 2 individual questions. The general questions related to the health service, the provident fund, wages, fines, fuel, tools, etc.

In 1892 there were 12 meetings of the council, at which 27 general questions, 7 special, and 6 individual [illegible] considered.

In 1893 there were 13 meetings of the [illegible] 8 special, and 2 individual questions, or a total of [illegible] The nature of the questions was about the same as [illegible]

## WORKINGMEN'S SICK FUND AND OLD-AGE [illegible]

The investigation of the insurance funds of the [illegible] coup Company offers the opportunity for a study of the [illegible] particular case, of the general system of the [illegible] gium, one of the most interesting insurance systems on the [illegible]

In Belgium the insurance of coal miners by their [illegible] tory. For this purpose there was created during the [illegible] in each of the 6 great mining districts into which Belgium is divided a central insurance fund, in which all mining companies were required to participate. They have all been in active operation since then, and have therefore had an existence of over 50 years.

The scheme comprehended, in addition to the central funds [illegible] more regular insurance of workingmen, the creation of special [illegible] funds by each mining company for purposes of relief in minor cases of sickness. Each mining company, therefore, participates in two funds, a central fund and a special individual one.

The central fund is maintained by (1) the payment of three-fourths of 1 per cent of their wages by the workingmen, (2) the payment of an equal amount by the company, (3) a subsidy from the State, (4) a smaller subsidy from the province, and (5) gifts and donations.

This fund is invested in approved securities. The schedule of monthly indemnities is as follows:

1. To an employee injured so as to be permanently and totally unable to work—
   - (a) Workingman, married, 19 years of age or over.......... [illegible]
   - (b) Workingman, unmarried, 19 years of age or over.......... [illegible]
   - (c) Workingman, unmarried, less than 19 years of age.......... [illegible]
   - (d) Workingwoman, 16 years of age or over.......... [illegible]
   - (e) Workingwoman, less than 16 years of age.......... [illegible]
2. To an employee injured so as to be able to earn not more than 50 per cent of his usual wages—
   - (a) Workingman, married, 19 years of age or over.......... [illegible]
   - (b) Workingman, unmarried, 19 years of age or over.......... [illegible]
   - (c) Workingman, unmarried, less than 19 years of age.......... [illegible]
   - (d) Workingwoman, 16 years of age or over.......... [illegible]
   - (e) Workingwoman, less than 16 years of age.......... [illegible]
3. To an employee injured so as to be able to earn from 51 to 70 per cent of his usual wages—
   - (a) Workingman, married, 19 years of age or over.......... [illegible]
   - (b) Workingman, unmarried, 19 years of age or over.......... [illegible]
   - (c) Workingman, unmarried, less than 19 years of age.......... [illegible]
   - (d) Workingwoman, 16 years of age or over.......... [illegible]
   - (e) Workingwoman, less than 16 years of age.......... [illegible]

| | |
|---|---|
| 4. To an employee injured so as to be able to earn from 71 to 85 per cent of his usual wages— | |
| (a) Workingman, married, 19 years of age or over | $1.54 |
| (b) Workingman, unmarried, 19 years of age or over | 1.16 |
| (c) Workingman, unmarried, less than 19 years of age | .77 |
| (d) Workingwoman, 16 years of age or over | .69 |
| (e) Workingwoman, less than 16 years of age | .46 |
| 5. To an employee injured so as to be able to earn over 85 per cent of his usual wages, no allowance. | |
| 6. To a widow of a workingman killed by an accident— | |
| (a) Personal allowance | 2.90 |
| (b) For each male child under 12 or female child under 15 years of age | .39 |
| 7. To a widow of a workingman pensioned on account of injuries not fatal— | |
| (a) Personal allowance, 50 per cent of husband's pension. | |
| (b) For each male child under 12 or female child under 15 years of age | .39 |
| 8. To a father and mother or a grandfather and grandmother of a workingman killed by an accident when unable to support themselves—maximum for division among all | 1.74 |
| 9. To an orphan of a workingman killed by an accident— | |
| (a) Male, until 12 years of age | 1.16 |
| (b) Female, until 15 years of age | 1.16 |
| 10. To a workingman unable to work and at least 55 years of age, and having been employed at least 30 years, or of any age and having been employed 35 years | 2.90 |
| 11. To a workingman, whether able to work or not, 60 years of age, or 65 if employed more than half the time above ground, and having been employed 35 years | 3.86 |
| 12. To the widow of a workingman pensioned on account of old age, if she had been his wife 30 years | 1.54 |

The operations of this fund may be seen from the following figures, taken from the latest annual report obtainable, showing receipts and expenditures during that fiscal year and during the entire period of its existence:

RECEIPTS AND EXPENDITURES OF THE MINERS' INSURANCE FUND OF THE CENTRAL DISTRICT OF BELGIUM, 1893, AND AGGREGATE, 1841-1893.

| Items. | 1893. | 1841-1893. |
|---|---|---|
| RECEIPTS. | | |
| Contributions of employees, retained from wages | $36,638.08 | $713,397.48 |
| Contributions of mine owners | 36,638.08 | 713,397.48 |
| Subsidy of central government | 1,361.28 | 46,650.53 |
| Subsidy of province | 283.55 | 9,289.19 |
| Interest on funds invested | 8,352.93 | 217,547.97 |
| Total | 83,272.92 | 1,700,282.65 |
| EXPENDITURES. | | |
| Running expenses other than cash benefits: | | |
| Salary of secretary | 386.00 | (a) |
| Salaries of commission of physicians | 115.80 | (a) |
| Attendance fees of workingmen's delegates | 247.04 | (a) |
| Rent of room | 38.60 | (a) |
| Printing annual report | 38.60 | (a) |
| Stamps and miscellaneous | 27.03 | (a) |
| Total running expenses | 853.07 | 17,748.39 |
| Pensions and aid | 74,444.00 | 1,386,287.16 |
| Indemnities to widows remarrying | 593.59 | 18,903.14 |
| Total | 75,890.66 | 1,422,938.69 |
| Excess of receipts over expenditures | 7,382.26 | 277,[illegible].96 |

a Not reported.

During the year 1893 the number of pensioners increased from 2,241 to 2,366 and the value of the average yearly pension from 166.69 francs ($32.17) to 167.12 francs ($32.25). Among this number of 2,366 pensioners there were 1,162 pensioned for old age (old workingmen or widows of old workingmen). During the life of the fund 6,219 pensions have been recorded. Of these 3,853 have become extinguished, leaving on December 31, 1893, as has been stated, 2,366 still in existence. There are nine corporations participating in the fund, of which that of Mariemont and Bascoup is easily the most important. Of the total amount paid for all the companies in 1893, it contributed 153,980.66 francs ($29,718.27), or 40.56 per cent, and among the total number of pensioners it was credited with 923, or 36.31 per cent.

As has been said, each mining company is required also to maintain a special fund for relief purposes rather than insurance operations, properly speaking. In compliance with this obligation, two special relief funds have been organized by the Mariemont and Bascoup Company, one in Mariémont and the other in Bascoup. Their constitutions are practically identical.

The special relief funds are constituted by (1) the payment by the workingmen of a sum equal to three-eighths of 1 per cent of their wages, (2) the payment by the company of an equal amount, and (3) the payment by the company of all sums collected by it through fines levied upon the workingmen.

From this fund benefits are paid to workingmen as follows: In case of accidents an indemnity equal to 30 per cent of their wages during three months if they are incapacitated for work that length of time. After that patients are cared for by the general insurance fund. In case of sickness an indemnity is paid during the first six months equal to 22 per cent of the patient's wages, during the second six months one equal to 15 per cent, and during the second year one equal to 7½ per cent of his average wages. In case the sickness continues longer than two years special provision is made in each case.

An important point to be noted in connection with this as well as the general fund is the large extent to which the workingmen themselves participate in its management.

The following statement of the receipts and expenditures of the funds of both Mariemont and Bascoup for the year 1893 will show the extent of their operations:

| | |
|---|---|
| Cash on hand January 1, 1893 | $633.67 |
| Receipts | 10,992.85 |
| Total | 11,626.52 |
| Expenditures | 11,871.71 |
| Deficit borne by the company December 31, 1893 | 245.19 |

In addition to the work done by these two funds a great deal is done by the company along the same lines through its old-age pension fund for its office employees, engineers, etc., and through its pension fund for boss miners, foremen, chief machinists, and laborers of a similar grade, and by the workingmen themselves through their mutual aid societies.

In 1868 the company organized a pension fund for its higher employees which provides for a maximum pension equal to 60 per cent of the recipient's usual wages after he is 60 years of age and has been in the employ of the company 35 years, or has become incapacitated for work, whatever his age or length of service. In case the incapacity is due to an accident the pension can equal 70 per cent of his wages. Provisions are also made for pensioning the widows of members. The funds are raised through the retention of a certain percentage of the wages of members and the payment of a varying amount by the company.

The company also, entirely at its own expense, grants pensions to its boss miners, foremen, chief machinists, etc., (1) in the case of boss miners when they are 55 years of age, and of the others when they are 60 years of age; (2) when incapacitated for work, whatever their age, as the result of old age or sickness; (3) when incapacitated for work as the result of an accident. In the first and second cases the pension is equal to 2 per cent of the recipient's wages for each of the first 15 years that he has been in the service of the company and 1 per cent for each succeeding year. In the third case the pension is equal to 5 per cent of the wages for each of the first 5 years and three-fourths of 1 per cent for each succeeding year of his service. In no case, however, can the pension exceed 40 per cent of the wages nor be less than 35 francs ($6.76) per month. Provision is also made for the pensioning of the widows of these workingmen. If these pensioners are in receipt of pensions from the two regularly constituted funds, only the excess of the pension which they are entitled to under the present regulations over those received from the regular funds is paid to them.

It will be unnecessary to describe the character and operations of the mutual-aid societies organized by the workingmen, as they offer little variation from the aid societies described in the cases of Anzin and Blanzy. It is sufficient to say that they are entirely independent of the management of the company, their funds are obtained through membership dues and interest on capital invested, and their objects are the granting of aid in cases of sickness, the defrayal of funeral expenses of deceased members, etc. Membership in these societies is almost universal on the part of the workingmen.

## OTHER SOCIAL INSTITUTIONS.

The remaining institutions need only a brief mention. They include savings banks, technical and primary schools, cooperative societies, workingmen's houses, etc. The company has itself erected 562 houses for its employees. They are arranged in groups of two, four, and six. Each house possesses a small garden. Their average cost was about

3,500 francs ($675.50) and the average rent is 7.50 francs ([illegible]) per month, which gives a return on the money invested of 25 per cent. The company has also advanced money to its employees with which to build houses of their own. About 40 houses are annually built by this means, and it is estimated that one-quarter of the workingmen have become house owners. Several cooperative societies for the sale of commodities, organized by the employees themselves, have been in operation since 1869. Though moderately successful, they have far from driven private dealers from the field. The great benefit derived from them is that they have compelled the other dealers to lower their prices to their own standard of minimum prices.

## VIEILLE-MONTAGNE ZINC MINING AND SMELTING COMPANY, BELGIUM.

There are few more interesting aggregations of workingmen's institutions than that organized under the auspices of the Zinc Mining and Smelting Company of Vieille-Montagne, of Belgium, with its headquarters at Chénée. At the same time there is not the same opportunity for studying the conditions of a particular locality as in the case of Anzin, Essen, Guise, or Blanzy, for the company has constantly extended its operations, acquiring other establishments in other countries as well as in Belgium, until to-day it almost controls the zinc trade of Europe. As in each place, however, it has pursued the policy of organizing its workingmen into special communities with their special institutions, their study enters into the scope of the report.

The history of the zinc industry is practically that of the Vieille-Montagne Company. The industry may be said to have been created by it through the discovery of an easy process for the reduction of zinc ore by Abbé Dony, the real founder of the Vieille-Montagne Company. The existence of ore rich in zinc at Vieille-Montagne was known as early as the middle of the fifteenth century, but the efforts that culminated in the present company only date from 1806, when Abbé Dony was granted his concession. Dony, through his discovery of the means of extracting the pure zinc from the ore, laid the foundation for the industry. In 1837 the present Vieille-Montagne Company was created in order to concentrate various interests that had become scattered among different heirs. Even at this time the industry was in a feeble condition. The company, however, early inaugurated the policy of bringing in and absorbing other mines and establishments for the extraction and production of zinc. Thus at the present time, though the seat of the company and the most important center of production is at Vieille-Montagne, it has in its possession twenty-one large establishments situated at different points in Belgium and other countries.

The company has now reached a position where it is much the largest zinc mining and manufacturing company in the world. In [illegible] the total production of zinc in Europe was 267,494 tons. Of this [illegible] the Vieille-Montagne Company produced 53,710 tons, or between one-fifth and one-sixth of the total amount, or 30,000 tons more than that mined by the next largest company.

## WORKINGMEN'S INSTITUTIONS.

The fact that all of the company's establishments are organized in one central system and that the same policy regarding social institutions has been followed throughout makes it possible to study the social work of the company as a whole in much the same manner as if there were one instead of a number of industrial centers.

In 1889 the company, in a pamphlet published in connection with its exhibit at the Paris Exposition, made the following concise statement of the principles by which it has been guided in the treatment of its employees. A translation of it is given:

1. The best mode of remunerating workingmen is that which interests them not only in the general profits of the enterprise, but also in the industrial results over which they themselves can exercise a direct and personal influence.

2. In order that wages may be sufficient, they ought to be of an amount that would permit workingmen not only to live, but also to make savings in order to provide for future as well as present wants.

3. But even if the workingmen receive such wages, they will not save nor acquire possession of homes unless the employer facilitates their so doing either through the creation of savings banks or by advances made with liberality and yet with caution.

4. Even with this aid but a minority of workingmen will be in a position to profit by these institutions. For the protection of the others there should be organized funds aiding them in cases of sickness, invalidity, and old age.

5. Two considerations which too often are neglected are absolutely necessary in order to insure the normal operation of these funds and prevent their disaster.

(*a*) The first consists in keeping an exact account of present and especially of prospective expenditures and in providing resources and the necessary reserves.

(*b*) The second consists in avoiding the systems of the management of the funds by the employer alone or by the employees alone, in favor of a mixed system.

In that way only can workingmen be interested in the management and the necessary control by the employer be at the same time exercised.

To carry out these principles, the company has organized a system for the determination of wages and for limited profit sharing, institutions to aid workingmen to become house owners, aid and insurance funds, savings banks, and other institutions intended to better the intellectual and moral condition of its employees.

THE SYSTEM OF THE [illegible]

All those who take part in any way in the [illegible] from the manager down to the ordinary [illegible] within the limits of their respective spheres of [illegible] returns of the work carried on. This result has been [illegible] the application of a system of extra wages, premiums, [illegible] regular wages. The remuneration of employees is [illegible] two parts, the one fixed, or wages, properly speaking, [illegible] able, called a premium. The first is intended to be, in a [illegible] the time consecrated by the employee to the service of [illegible] the second is a reward for personal efforts—a return for [illegible] activity and intelligence displayed by the workingman.

The basis on which the premium is determined varies accor the nature of the service. The premiums are proportional eithe quantity of ore mined or to the economy in the use of materi cially that of fuel, or to the amount of the final product produced a given time. The amount of the premium is greater or less, ac to the relative importance of the work; but whatever the bas this is a matter of prime importance—the workingman always exactly the rate, and each day, according to the results obtain himself calculate his share. The premium account is made u same time as that of his wages, but only half of the amount ea paid over. The other half is carried to his account, and is nc dated until the end of the year. In this way the workingman h an accumulation of savings at the end of each year.

The following table shows the total number of employees average daily wages per employee for each year since the orga of the company:

AVERAGE DAILY WAGES OF EMPLOYEES OF THE VIEILLE-MONTAGN MINING AND SMELTING COMPANY, 1837 TO 1888.

| Year. | Employees. | Average daily wages. | Year. | Employees. | Average daily wages. | Year. | Employees. |
|---|---|---|---|---|---|---|---|
| 1837 | 932 | $0.26 | 1855 | 6,763 | $0.37 | 1873 | 6,[illegible] |
| 1838 | 945 | .26 | 1856 | 5,417 | .42 | 1874 | 6,751 |
| 1839 | 960 | .26[illegible] | 1857 | 5,492 | .43 | 1875 | 6,477 |
| 1840 | 982 | .26 | 1858 | 6,068 | .46 | 1876 | 6,472 |
| 1841 | 1,080 | .26[illegible] | 1859 | 4,947 | .43[illegible] | 1877 | 7,121 |
| 1842 | 1,072 | .26[illegible] | 1860 | 5,034 | .43 | 1878 | 7,193 |
| 1843 | 1,124 | .27 | 1861 | 5,264 | .43 | 1879 | 6,[illegible] |
| 1844 | 1,148 | .27 | 1862 | 5,332 | .44 | 1880 | 6,362 |
| 1845 | 1,187 | .27[illegible] | 1863 | 5,559 | .45[illegible] | 1881 | 6,[illegible] |
| 1846 | 1,211 | .28[illegible] | 1864 | 5,842 | .46[illegible] | 1882 | 6,[illegible] |
| 1847 | 1,432 | .28 | 1865 | 6,232 | .46[illegible] | 1883 | 6,903 |
| 1848 | 1,247 | .31[illegible] | 1866 | 6,[illegible] | .48 | 1884 | 6,647 |
| 1849 | 1,511 | .35[illegible] | 1867 | 6,[illegible] | .49 | 1885 | 6,161 |
| 1850 | 1,953 | .35 | 1868 | 6,802 | .49 | 1886 | 5,[illegible] |
| 1851 | 2,083 | .40[illegible] | 1869 | 7,115 | .50 | 1887 | [illegible] |
| 1852 | 2,343 | .40[illegible] | 1870 | 6,[illegible] | .51 | 1888 | [illegible] |
| 1853 | 4,481 | .37 | 1871 | 7,[illegible] | .52 | | |
| 1854 | 5,960 | .36[illegible] | 1872 | 7,[illegible] | .5[illegible] | | |

These figures are of course of value only from the standpoint of a comparison of wages from year to year. But for this purpose they are of exceptional interest, covering, as they do, a period of over fifty years in the most important establishment in the world of the industry to which it relates. Average wages, it will be seen, have gradually risen from 1.35 francs ($0.26) in 1837 to 3.24 francs ($0.62½) in 1875, since when the average has been almost stationary. According to the report of the company for 1893 the average for that year was exactly that paid in 1875. Of this amount the premium represents a proportion ranging from 10 to 25 per cent.

In commenting on these figures for the year 1888 the company said:

> It should be remarked that the average wages, 3.18 francs ($0.61), relate to workingmen very differently situated. We can therefore say that, although the average wages paid amount to 3.18 francs ($0.61) per day, or 1,019 francs ($196.67) per year, in reality 50 per cent of the employees earn yearly wages of 1,200 francs ($231.60) or more, and the remainder earning less is composed in large part of youths whose wages are added to the family income.

As in the case of Anzin, it would be a matter of interest to compare the variation in wages with the variation in the prices of commodities that enter into the cost of living for the workingmen. In the present case material is at hand to do this in but the single case of wheat. The table that follows shows this relative variation by expressing the quantity of wheat that the average salary will buy.

QUANTITY OF WHEAT PURCHASABLE BY AVERAGE WAGES OF EMPLOYEES OF THE VIEILLE-MONTAGNE ZINC MINING AND SMELTING COMPANY, 1837 TO 1888.

| Years. | Average daily wages. | Price of wheat per hectoliter (a). | Bushels of wheat average daily wages would purchase. |
|---|---|---|---|
| 1837 to 1846 | $0.27 | $5.06 | 0.151 |
| 1847 to 1856 | .36¼ | 5.82 | .178 |
| 1857 to 1866 | .45 | 4.98 | .255 |
| 1867 to 1876 | .55½ | 6.06 | .239 |
| 1877 to 1886 | .60¼ | 4.87 | .354 |
| 1887 | .61 | 3.70 | .467 |
| 1888 | .61½ | 3.76 | .463 |

a Hectoliter equals 2.8377 bushels.

This table shows that though the period of the greatest increase in nominal average wages was between the years 1837 and 1876, and that wages were comparatively steady during the period 1877 to 1888, yet that if wages were expressed in wheat this latter period would be the one of the greatest increase. It thus affords an illustration of the necessity of distinguishing between nominal and actual wages.

## INSTITUTIONS TO AID WORKINGMEN TO BECOME HOUSE OWNERS.

It has been the consistent policy of the company to encourage the workingmen to become the proprietors of their houses. To do this it has, according to circumstances, either bought and subdivided land which it afterwards offered to its employees on favorable terms, or has made advances to them with which to build, repayable in installments, or has itself built houses that it sold on the installment plan. In this way over one thousand of its employees have become house owners. In addition to this the company has also encouraged the acquisition of garden plots by employees. The acquisition of gardens has gone along with that of houses. A great many of the owners of houses are reported as owning a cow, and almost all of them keep one or more pigs.

The company has also made notable efforts along the line of constructing houses to rent to those of its employees who are unable or unwilling to become house owners. In general these houses are constructed in groups of two or four, with but a single tenement to each house. A house of four rooms, with shed attached and a small garden, rents for from 80 to 100 francs ($15.44 to $19.30) per year, or about 10 per cent of the wages of the renter. The company says regarding the question of housing:

Our experience has demonstrated that isolated and independent houses form the preferable system from every point of view for workingmen.

## AID AND INSURANCE FUNDS.

Workingmen's aid societies in connection with the Vieille-Montagne Company consist of three institutions by which the European workingman usually provides for the future or the contingencies of sickness and old age: (1) Sick and acccident fund, (2) old-age pension fund, and (3) savings banks.

The sick and accident fund was organized as early as 1847. Its objects are:

1. To furnish gratuitously in cases of sickness or accident medical aid to workingmen and members of their families dependent upon them.

2. To grant to those thus temporarily incapacitated for work pecuniary aid during their invalidity. The amounts of these indemnities have varied at different periods. Under the present regulations, which went into effect in 1886, the amount paid in each case is determined by a permanent commission, but can not exceed one-third of the recipient's average wages in case of sickness or one-half of his wages in case of incapacity resulting from an accident.

3. To contribute by a grant of money toward defraying the expense consequent upon childbirth in a workingman's family.

4. To contribute to the funeral expenses of a workingman or of a member of his family by the grant of a fixed indemnity of 20 francs ($3.86) in the case of a married workingman, 15 francs ($2.90) for unmarried adults, and 5 francs ($0.97) for children, in addition to furnishing the coffin.

5. To furnish temporary aid to widows, orphans, and parents of deceased workingmen. The rate of these indemnities is as follows: 50 centimes (10 cents) per day to the widow of a workingman; 75 centimes (14 cents) per day to the widow of an overseer; 10 centimes (2 cents) per day to each child until he or she is 14 years of age, or in case the other parent is also dead 25 centimes (5 cents) per day, and 25 centimes (5 cents) to the parents of the deceased.

Previous to April 11, 1891, the resources of this fund were obtained by a contribution on the part of the workingmen of an amount equal to from 1 to 5 per cent of their wages, the company making up any balance necessary for the complete carrying out of its provisions. On that date, however, the company assumed the entire charge of the fund, the workingmen henceforth contributing in no way to the support of the fund.

The old-age pension fund likewise dates from 1847, and was established in order to grant life pensions to workingmen who had become unable to work either as the result of old age or sickness after having been in the employment of the company a certain period of years. The amount of the pension is fixed at one-fifth of the amount of the pensioner's average annual wages, obtained by taking an average of the three years in which he earned the most during the last five years of his service. This amount, however, can not be less than 50 centimes (10 cents) nor more than 1 franc (19 cents) per day. In order to be entitled to a pension the workingman must have had 15 years of continuous service with the company and be totally incapacitated for work. If, however, he is incapacitated as the result of an accident received in the service of the company through no fault of his own, he is entitled to a pension irrespective of the length of his service. The expenses of this fund have always been borne by the company. These two funds supplement each other, and together form one system. To some extent, therefore, space will be economized by showing the results of their operations in the same tables. The first table shows the total expenditures of each fund each year since 1850, the average expenditure per member, and the relation that the total expenditure bears to average wages.

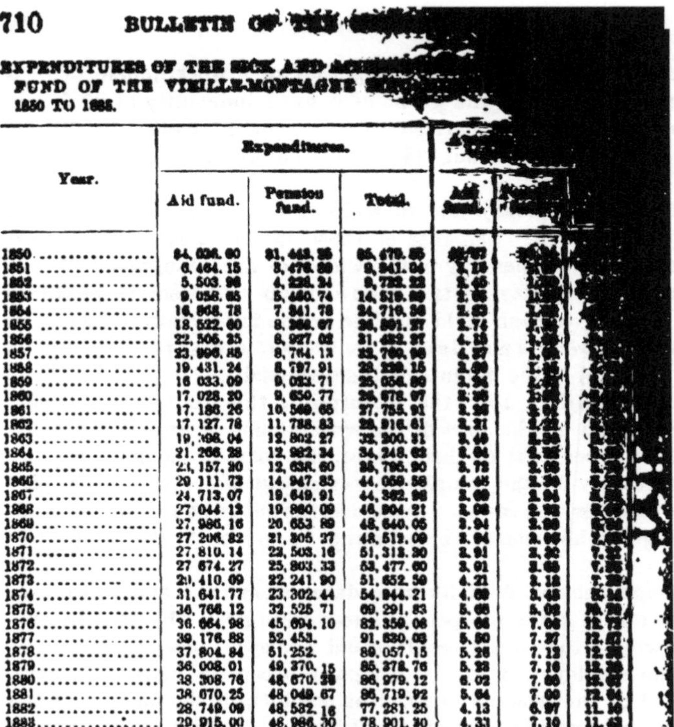

EXPENDITURES OF THE SICK AND ACCIDENT [illegible] FUND OF THE VIEILLE-MONTAGNE [illegible] 1850 TO 1888.

| Year. | Expenditures. Aid fund. | Pension fund. | Total. | [illegible] Aid fund. | [illegible] | [illegible] |
|---|---|---|---|---|---|---|
| 1850 | $4,036.60 | $1,442.95 | $5,479.55 | [illegible] | [illegible] | [illegible] |
| 1851 | 6,464.15 | 3,476.89 | 9,941.04 | 1.15 | [illegible] | [illegible] |
| 1852 | 5,503.98 | 4,228.24 | 9,732.22 | 2.45 | [illegible] | [illegible] |
| 1853 | 9,058.65 | 5,460.74 | 14,519.39 | 2.05 | [illegible] | [illegible] |
| 1854 | 16,868.78 | 7,841.78 | 24,710.56 | 2.53 | [illegible] | [illegible] |
| 1855 | 18,522.60 | 8,368.67 | 26,891.27 | 2.74 | [illegible] | [illegible] |
| 1856 | 22,505.25 | 8,927.02 | 31,432.27 | 4.15 | [illegible] | [illegible] |
| 1857 | 23,996.85 | 8,764.13 | 32,760.98 | 4.37 | [illegible] | [illegible] |
| 1858 | 19,431.24 | 8,797.91 | 28,229.15 | 2.60 | [illegible] | [illegible] |
| 1859 | 16,033.09 | 9,021.71 | 25,054.80 | 2.94 | [illegible] | [illegible] |
| 1860 | 17,028.20 | 9,650.77 | 26,678.97 | 2.36 | [illegible] | [illegible] |
| 1861 | 17,186.26 | 10,569.65 | 27,755.91 | 2.26 | [illegible] | [illegible] |
| 1862 | 17,127.78 | 11,788.83 | 28,916.61 | 2.21 | [illegible] | [illegible] |
| 1863 | 19,398.04 | 12,802.27 | 32,200.31 | 3.40 | [illegible] | [illegible] |
| 1864 | 21,266.28 | 12,982.34 | 34,248.62 | 3.04 | [illegible] | [illegible] |
| 1865 | 23,157.30 | 12,638.60 | 35,795.90 | 3.73 | [illegible] | [illegible] |
| 1866 | 29,111.73 | 14,947.85 | 44,059.58 | 4.45 | [illegible] | [illegible] |
| 1867 | 24,713.07 | 19,649.91 | 44,362.98 | 3.60 | [illegible] | [illegible] |
| 1868 | 27,044.12 | 19,860.09 | 46,904.21 | 3.08 | [illegible] | [illegible] |
| 1869 | 27,986.16 | 20,653.89 | 48,640.05 | 3.94 | 2.90 | [illegible] |
| 1870 | 27,206.82 | 21,305.27 | 48,512.09 | 3.94 | 2.96 | [illegible] |
| 1871 | 27,810.14 | 23,503.16 | 51,313.30 | 2.91 | 3.32 | [illegible] |
| 1872 | 27,674.27 | 25,803.33 | 53,477.60 | 3.91 | 2.65 | [illegible] |
| 1873 | 29,410.69 | 22,241.90 | 51,652.59 | 4.21 | 3.18 | [illegible] |
| 1874 | 31,641.77 | 23,302.44 | 54,944.21 | 4.69 | 3.45 | [illegible] |
| 1875 | 36,766.12 | 32,525.71 | 69,291.83 | 5.65 | 5.02 | [illegible] |
| 1876 | 36,664.98 | 45,694.10 | 82,359.08 | 5.66 | 7.06 | 12.72 |
| 1877 | 39,176.88 | 52,453. | 91,630.03 | 5.50 | 7.37 | 12.87 |
| 1878 | 37,804.84 | 51,252. | 89,057.15 | 5.26 | 7.12 | 12.38 |
| 1879 | 36,008.01 | 49,370. 15 | 85,378.76 | 5.23 | 7.16 | 12.39 |
| 1880 | 38,308.76 | 48,670.36 | 86,979.12 | 6.02 | 7.60 | 13.67 |
| 1881 | 38,670.25 | 48,049.67 | 86,719.92 | 5.64 | 7.00 | 12.64 |
| 1882 | 28,749.09 | 48,532. 16 | 77,281.25 | 4.13 | 6.97 | 11.10 |
| 1883 | 29,915.00 | 48,986.30 | 78,901.30 | 4.33 | 7.10 | 11.43 |
| 1884 | 35,713.11 | 49,949.36 | 85,662.47 | 5.45 | 7.03 | 12.08 |
| 1885 | 29,469.94 | 53 | 86,888.47 | 4.83 | 9.25 | 14.08 |
| 1886 | 44,988.11 | 56,418. 57 | 79,706.68 | 7.63 | 5.88 | 13.51 |
| 1887 | 40,970.81 | 35,634.36 | 76,605.17 | 6.50 | 5.73 | 12.23 |
| 1888 | 39,859.71 | 34,004.48 | 73,864.19 | 6.12 | 5.22 | 11.34 |

a Not reported.

The following table shows in greater detail the objects of the [illegible] tures of the combined funds for the years 1850 to 1888, incl[illegible] well as the proportion each item of expense bears to the total [illegible] expended:

AGGREGATE EXPENDITURES OF THE SICK AND ACCIDENT FUND AND [illegible] AGE PENSION FUND OF THE VIEILLE MONTAGNE ZINC MINING AND [illegible] COMPANY, BY OBJECTS OF EXPENDITURE, 1850 TO 1888.

| Objects of expenditure. | Expenditures, 1850 to 1888. | Average yearly expenditures. | Per ce[illegible] |
|---|---|---|---|
| Indemnities for accidents, sickness, and partial invalidity | [illegible] | [illegible] | [illegible] |
| Medicines | [illegible] | [illegible] | [illegible] |
| Medical service | [illegible] | [illegible] | [illegible] |
| Funeral expenses and other expenditures | [illegible] | [illegible] | [illegible] |
| Old-age and invalidity pensions | [illegible] | [illegible] | [illegible] |
| Payments to widows | [illegible] | [illegible] | [illegible] |
| Payments to orphans | [illegible] | [illegible] | [illegible] |
| Total | [illegible] | [illegible] | [illegible] |

These two tables serve a double purpose. In the first place, they show to how great an extent the wages of the workingmen are supplemented by their participation in these relief funds. The addition thus made to wages is seen to have steadily increased, both absolutely and relatively to wages, until 1880, after which it has varied more or less, with a slight tendency to decrease. Thus in 1850 each workingman received on an average benefits to the amount of 14.55 francs ($2.81), which represented an increase to his wages of 2.65 per cent. In 1860 the amount was 27.46 francs ($5.30), or 4.10 per cent; in 1870, 36.42 francs ($7.03), or 4.61 per cent, and in 1880, 70.83 francs ($13.67), or 7.64 per cent, while in 1888 the amount of benefits had sunk to 58.77 francs ($11.34), or 6.32 per cent. The average for the whole period represented an addition to wages of 5 per cent.

On the other hand, these tables, in common with similar tables given for the other industrial communities, have a theoretical interest to students of workingmen's insurance as showing how great an average expenditure is necessary on the part of a company if the funds are supplied by it, or how great a proportion of their wages must be sacrificed by the workingmen if the funds are maintained by them, in order to insure the steady operation of (1) a sick fund providing for the supply of medicine and medical attendance and a fixed sick benefit and (2) a pension fund for old and incapacitated workingmen and their widows and orphans.

The last table is of especial interest as showing what portion of this expenditure is devoted to each specific purpose.

In considering the significance of these figures the important facts should be noted that (1) expenses of administration are not included, (2) that the figures relate to a body of men that has been constantly increasing in numbers, (3) that necessarily the expenditures for pensions during the first few years after the inauguration of the fund would be comparatively light, but that as the fund became older the number of pensioners would increase materially. The company believe that a normal number has now been reached, varying between 450 and 500.

## SAVINGS BANK.

The company first organized a savings bank in 1842. In the beginning but slow progress was made. For a good many years the number of depositors and the amount of deposits increased scarcely at all. Commencing with 1853, however, the use of the bank by employees has increased yearly. Thus while in that year there were but 126 depositors, representing scarcely 2 per cent of the personnel, in 1888 there were 900 depositors, or nearly 14 per cent of all employees. The following table shows the number of depositors, the amount of deposits, etc.,

for each year from 1853 to 1888. [illegible]
been paid from the start:

DEPOSITORS AND DEPOSITS IN THE SAVINGS [illegible]
ZINC MINING AND SMELTING [illegible]

| Year. | Deposi-tors. | Amount deposited. | Amount withdrawn. | [illegible] | [illegible] |
|---|---|---|---|---|---|
| 1853 | 136 | ........ | ........ | ........ | [illegible] |
| 1854 | 152 | $9 777.57 | $8,[illegible] | [illegible] | [illegible] |
| 1855 | 147 | 6,767.[illegible] | 7,749.[illegible] | [illegible] | [illegible] |
| 1856 | 147 | 7 958.[illegible] | 7,[illegible] | [illegible] | [illegible] |
| 1857 | 181 | 8 001.39 | 5,[illegible] | [illegible] | [illegible] |
| 1858 | 202 | 9,674.13 | 6,[illegible] | 1,[illegible] | [illegible] |
| 1859 | 209 | 8,839.59 | 7,[illegible] | 1,[illegible] | [illegible] |
| 1860 | 218 | 11,[illegible] | 10,[illegible] | 1,[illegible] | [illegible] |
| 1861 | 192 | 7,846.41 | 11,851.16 | 1,[illegible] | [illegible] |
| 1862 | 203 | 6,5[illegible] 47 | 6,621.65 | 1,[illegible] | [illegible] |
| 1863 | 212 | 8,371.37 | 10,618 67 | 1,[illegible] | [illegible] |
| 1864 | 222 | 10,926.[illegible] | 9,992.77 | 1,[illegible] | [illegible] |
| 1865 | 260 | 15,312.81 | 10,[illegible] | 1,[illegible] | [illegible] |
| 1866 | 259 | 12,170.19 | 14,072.[illegible] | 1,[illegible] | [illegible] |
| 1867 | 288 | 12,576.85 | 10,576 79 | 1,[illegible] | [illegible] |
| 1868 | 320 | 15,921 15 | 11,461.11 | 1,[illegible] | [illegible] |
| 1869 | 462 | 19,846.38 | 10,920 [illegible] | 2,[illegible] | a [illegible] |
| 1870 | 462 | 20,984 31 | 15,[illegible] | 2,[illegible] | a [illegible] |
| 1871 | 461 | 21,114.01 | 16,476.[illegible] | 2,165.51 | [illegible] |
| 1872 | 485 | 22,195.19 | 21,[illegible] | 2,514.[illegible] | [illegible] |
| 1873 | 518 | 23,705 03 | 21,261.65 | [illegible] | [illegible] |
| 1874 | 569 | 24,276.31 | 19,091.37 | 4,213.[illegible] | [illegible] |
| 1875 | 611 | 36,551.11 | 24,257.01 | 4,[illegible].47 | [illegible] |
| 1876 | 651 | 42,170.11 | 28,766.[illegible] | 5,540.64 | [illegible] |
| 1877 | 686 | 43,609.70 | 38,379.[illegible] | 6,126.01 | a [illegible] |
| 1878 | 714 | 42,708.97 | 35,532.07 | 6,655.[illegible] | a [illegible] |
| 1879 | 728 | 55,012.53 | 42,072.46 | 7,109.[illegible] | [illegible] |
| 1880 | 744 | 51,231 85 | 44,030.05 | 8,423.[illegible] | [illegible] |
| 1881 | 757 | 63,986.25 | 52,465.31 | 9,[illegible] | [illegible] |
| 1882 | 784 | 76,655.93 | 65,498.99 | 10,454.62 | a [illegible] |
| 1883 | 815 | 72,606.60 | 62,945.98 | 10,510.97 | a [illegible] |
| 1884 | 839 | 66,009.48 | 58,091.07 | 12,[illegible] | a [illegible] |
| 1885 | 804 | 77,740.59 | 80,077.82 | 13,[illegible].16 | a [illegible] |
| 1886 | 845 | 78,362.82 | 59,835.21 | 14,500.16 | [illegible] |
| 1887 | 880 | 77,503.78 | 67,553.28 | 15,834.11 | a [illegible] |
| 1888 | 900 | 76,644.93 | 79,824.90 | 16,748.93 | 345,[illegible] |

a These figures do not balance. They are given, however, as published by the comp[illegible]

## LIFE INSURANCE SOCIETY.

The aid and pension funds are intended solely for the benefit [illegible] workingmen, strictly speaking. For the administrative force, [illegible]ing the engineer's office employees, etc., the company organize[illegible] uary 1, 1877, a life insurance society, believing that this form of [illegible] provision for the future was the most successful for the high[illegible] employees. According to the provisions of this society, each em[illegible] desiring to become a member is required to pay into the fund 3 p[illegible] of his annual wages. To this the company adds an amount equ[illegible] per cent of his wages. The amount of insurance is then cal[illegible] according to the tables in use by insurance companies. Eve[illegible] years a balance sheet of assets and liabilities is struck, and if [illegible] a surplus half of it is apportioned among the members, increa[illegible] rata their insurance, and the other half is carried to a [illegible] The constitution determines the [illegible] always made in approved securities. The society is administered by a central committee elected from among the members. The entire expense of administration and bookkeeping is borne by the company, so that there is absolutely no expense of management. Although participation in this society is purely voluntary, with scarcely an exception all of the employees of the company have become policy holders.

## CONCLUSION.

What has been the result of the social work of the company? Concerning this there are available but two kinds of evidence: The testimony of the company itself and that of the employees as shown through statistics of the stability of their employment. Concerning the first, the company speaks unequivocally:

And now, first of all from the social point of view, has the company obtained that harmony between all who cooperate by their labor in the work of production; has it realized that ideal of social peace of which all economists and politicians dream? We do not venture to assert that this result has been definitely obtained, and that the workingmen of our establishments have already reached this degree of moral superiority. All that we can state is that up to the present time, despite the diversity of countries and variety of industries in which we are engaged, we have never had to suffer a strike in any one of our establishments, and that we have succeeded in obtaining a remarkable degree of stability in our corps of employees as regards the length of their employment, since, with an average of 6,500 employees, the average length of the time they have been employed has exceeded 12 years.

These facts find their natural explanation in all the advantages accorded to the workingmen, their high wages, participation in industrial profits, facilities for making savings and becoming owners of property, and security for the future through the organization of gratuitous aid in case of sickness, and the provision of pensions for workingmen incapacitated by old age or infirmities. There are more than 900 depositors in the savings bank, and more than 1,000 workingmen are the owners of their own homes. More than a fourth of all the employees are either house owners or are in possession of savings, the interest on which increases the resources of the family. All this is conducive to good order, good work, and peace. Such are the results of the system consistently practiced by the company.

Moreover, it is important to add that, in spite of the expenditures entailed by this course, these results have been obtained without at all sacrificing the financial prosperity of the company, but on the other hand have contributed to it. If the company has been able to pay to its shareholders each year during 50 years an average dividend of 20 per cent on their original investment, * * * it can be affirmed that the honor is due to the liberality with which the administration has treated its employees.

These results appear to be of such a nature as to determine the company to persevere in its course and to encourage others who have not yet entered this work to do likewise.

The following table showing the stability of the [illegible] the employees:

YEARS OF SERVICE OF EMPLOYEES OF THE VIEILLE-MONTAGNE [illegible] AND SMELTING COMPANY, [illegible].

| Years of service. | Employees | Per cent. | Years of service | Employees | Per cent. | Years of service. | Employees | Per cent. |
|---|---|---|---|---|---|---|---|---|
| Under 2 | 1,065 | 16.95 | 19 but under 20 | 151 | 2.36 | 37 but under 38 | [illegible] | [illegible] |
| 2 but under 3 | 574 | 8.97 | 20 but under 21 | 94 | 1.47 | 38 but under 39 | [illegible] | [illegible] |
| 3 but under 4 | 349 | 5.45 | 21 but under 22 | 83 | 1.30 | 39 but under 40 | [illegible] | [illegible] |
| 4 but under 5 | 376 | 5.87 | 22 but under 23 | 104 | 1.63 | 40 but under 41 | [illegible] | [illegible] |
| 5 but under 6 | 294 | 4.59 | 23 but under 24 | 112 | 1.75 | 41 but under 42 | [illegible] | [illegible] |
| 6 but under 7 | 277 | 4.33 | 24 but under 25 | 95 | 1.48 | 42 but under 43 | [illegible] | [illegible] |
| 7 but under 8 | 214 | 3.34 | 25 but under 26 | 64 | 1.00 | 43 but under 44 | [illegible] | [illegible] |
| 8 but under 9 | 200 | 3.12 | 26 but under 27 | 66 | 1.03 | 44 but under 45 | [illegible] | [illegible] |
| 9 but under 10 | 177 | 2.77 | 27 but under 28 | 49 | .77 | 45 but under 46 | [illegible] | [illegible] |
| 10 but under 11 | 148 | 2.31 | 28 but under 29 | 66 | 1.03 | 46 but under 47 | [illegible] | [illegible] |
| 11 but under 12 | 177 | 2.77 | 29 but under 30 | 38 | .59 | 47 but under 48 | [illegible] | [illegible] |
| 12 but under 13 | 173 | 2.70 | 30 but under 31 | 57 | .89 | 48 but under 49 | [illegible] | [illegible] |
| 13 but under 14 | 204 | 3.19 | 31 but under 32 | 81 | 1.27 | 49 but under 50 | [illegible] | [illegible] |
| 14 but under 15 | 229 | 3.58 | 32 but under 33 | 25 | .39 | 50 or over | [illegible] | [illegible] |
| 15 but under 16 | 178 | 2.78 | 33 but under 34 | 24 | .38 | | | |
| 16 but under 17 | 168 | 2.63 | 34 but under 35 | 22 | .34 | Total | [illegible] | [illegible] |
| 17 but under 18 | 175 | 2.73 | 35 but under 36 | 36 | .56 | | | |
| 18 but under 19 | 112 | 1.75 | 36 but under 37 | 20 | .31 | | | |

## THE NETHERLANDS YEAST AND ALCOHOL FACTORY, AGNETA PARK, DELFT, HOLLAND.

Like the Familistère at Guise, this community represents the deliberate attempt to organize an industry on special lines so as to increase the solidity of interests of all concerned in the undertaking, to develop mutuality, and to create social institutions of all kinds. As at Guise, therefore, the introduction of a plan for participation in profits by the workingmen and the creation of a special industrial community were made integral parts of the scheme. While at Guise, however, a system of pure cooperation was adopted and a peculiar plan of tenement houses or familistères selected, at Delft a system of partial profit sharing and a village of isolated cottages were introduced. At both places, however, each of the various social institutions is related to all the others and together form one system.

This industry, founded by Mr. J. C. van Marken, is that of the manufacture of yeast, for which he has a market in a great many countries of Europe and in Great Britain. Alcohol, gin, and some other products are also manufactured as by-products. The first factory was built and operations commenced in 1869. The capital invested was 150,000 florins ($60,300). Since then a number of buildings have been added and the capital successively increased until in 1892 it was something over 1,200,000 florins ($482,400). The Netherlands Oil Manufactory was established in 1884, adjoining the yeast works, as a separate company. As Mr. van Marken, however, has a controlling interest in this company, the two enterprises, as far as social institutions are concerned, have for the most part been treated as one undertaking. The company employs between 250 and 300 men.

The key to the social system that has been created is found in the hierarchal organization of all the employees and the practice of a system of profit sharing, combined with the opportunity accorded to workingmen to become part owners of the capital stock of the company. Combined with these are a great variety of institutions, such as cooperative stores, sick funds, old-age, accident, and life insurance, institutions for promoting house ownership, etc.

## THE ORGANIZATION OF LABOR.

The personnel of the company is divided into the following classes: Higher officials, officials and superintendents, lower officials and under superintendents, clerks and workingmen, and assistant clerks and helpers. The latter class consists of persons under 20 years of age who have not yet fully learned their calling. The salaries of higher officials, officials, and superintendents are determined by the directors. All employees under these grades are paid by the hour. The wages of an under superintendent are 14 florins ($5.63) per week of sixty hours. Under that grade employees are paid in accordance with the nature of their work, and are divided into six wage classes, receiving, respectively, 16.5, 17, 17.5, 18, 18.5, and 22 Dutch cents (6⅝, 6⅞, 7, 7⅛, 7⅜, and 8⅞ cents) per hour.

Care has been taken that the wages for a full day's work with proper effort can not fall below that necessary for the maintenance of an ordinary family. The minimum is fixed at 12 florins ($4.82) per week. Helpers receive 12 Dutch cents (4⅞ cents) per hour, with a 10 per cent advance for overtime work. The office force receives the following weekly wages on the basis of 7½ hours' work per day: Assistant clerks, 7 florins ($2.81); clerks, 10 florins ($4.02); minor officials, 15 florins ($6.03), and 15, 22, and 23 Dutch cents (6, 8⅞, and 9¼ cents) per hour, respectively, for overtime work. Members of the force of a lower grade than minor officials and superintendents whose work requires special knowledge receive enhanced wages.

Besides the above classification as to the nature of the occupation, there is another division of employees into five classes, in accordance with the energy and zeal displayed while at work. The object of this classification is to form the basis for a scheme of premium payments. All employees above the first class receive premiums, amounting to 2 per cent of their wages for the second class, 5 per cent for the third class, 10 per cent for the fourth class, and 20 per cent for the fifth class. New employees are placed in the first class. Promotions are then made annually on the 1st of January according to the devotion to duty that has been shown by the employees, one step only being made at a time. Neglect of duty is punished by failure to promote or by reduction to a lower class.

In addition to this, extra premiums may be paid for special services, and finally premiums for economy in the use of working time, raw materials, fuel, etc. The classification of the personnel and the premiums awarded are kept secret by the directors.

### PARTICIPATION IN PROFITS.

The participation in profits consists of 10 per cent of the net gains of the company. Of this not more than one-fourth may be devoted to purposes of public utility benefiting all the employees. The balance is distributed in accordance with the grade and merit class of the employees. The employees of the five different classes receive their share in proportion to two, three, four, six, and eight times the amount of their salaries. Thus in 1890 the employees received 1¾, 2⅝, 3½, 5¼, and 7 per cent of the amount of their wages as shares in profits.

From 1870 to 1890, 2,554,000 florins ($1,026,708) were paid out in wages, 315,000 florins ($126,630) in premiums, 81,000 florins ($32,562) as shares in profits, and 100,000 florins ($40,200) for the insurance of employees, or a total of 3,050,000 florins ($1,226,100). During this same period 2,024,000 florins ($813,648) were credited to the capital account, of which 984,000 florins ($395,568) were paid into the amortization fund, 96,000 ($38,592) to the reserve fund, 360,000 ($144,720) were counted as net profits, 389,000 ($156,378) interest on capital, and 195,000 ($78,390) interest on bonds.

### THE HOUSING OF EMPLOYEES.

As a matter of course the provision of houses for the workingmen constitutes an essential feature of an attempt such as that of Mr. van Marken to build up a special industrial center. In doing this Mr. van Marken desired to have the employees cooperate as far as possible in the work. Instead, therefore, of directly providing houses, he organized among his employees a cooperative society with a capital of 160,000 florins ($64,320). To this society he gave 3,000 florins ($1,206) in cash for founders' shares to that amount and turned over a tract of 4 hectares (about 9⅞ acres) in return for founders' shares to the amount of 29,000 florins ($11,658). The remaining 128,000 florins ($51,456) of the capital stock was raised by a loan at 4½ per cent interest, to secure which a mortgage on the property was given. With this money the ground, which was named Agneta Park, was laid out as a park, and in it were erected 86 houses. The community thus created is one of unusual attractiveness. The streets are in all cases winding and bordered by shade trees and lawns. In the center of the grounds an artificial lake has been constructed of sufficient size to permit of boating. The houses are built on the Mulhouse plan, each having a garden adjoining. The material is of brick, and various plans were selected so as to give to each house a distinct character. The rents average 2.25 florins (90 cents) per week.

The system by which the members of the cooperative society were to become the owners of the property is somewhat similar to that by which the workingmen of Guise became the owners of the plant in which they labored. The total rent is so fixed as to represent 7½ per cent return on the cost of the land and buildings. Out of the total receipts are paid, first, the expenses of maintenance and management, then 4½ per cent interest on loans and 5 per cent dividends on founders' shares. Ten per cent of the balance goes to the reserve fund, and the remainder is devoted to liquidating the mortgage. At the same time this latter amount is credited to the individual tenants in proportion to the rents paid. As soon as a tenant has in this way 100 florins ($40.20) credited to his account he comes into possession of a share of stock bearing interest at 3 per cent. It is then only a matter of time when all the stock of the cooperative society, including the founders' shares, which will be replaced by ordinary shares, will be held by the tenants, who thus become owners of the property. Shares of stock can only be transferred with the consent and through the agency of the board of directors. Once the shares are all held by the tenants the system adopted at Guise of canceling and redeeming the old shares and issuing new ones in their place will be practiced.

In addition to providing dwelling houses, this cooperative association also runs a cooperative store, a building for which has been erected in Agneta Park. This store sells food products, clothing, and dry goods. Profits from its operations are added to rents and distributed in the same way, credit being given in proportion to the value of the purchases of each tenant.

## OTHER SOCIAL INSTITUTIONS.

In addition to these efforts relating to housing and profit sharing, there has been created a great variety of social institutions. Among these may be mentioned savings banks, funds for the insurance of employees against sickness, accidents, old age, and death and for the pensioning of widows of workingmen, educational and recreative institutions, and an assembly of workingmen and heads of departments for the purpose of discussing all matters relating to the mutual interests of the workingmen and the establishments and the settlement of difficulties that may arise. Without entering into details, the general features of these institutions can be briefly given.

The insurance of employees of course constitutes the most important class of workingmen's institutions. In case of sickness employees receive from the company their full wages for a period not exceeding eight weeks. In addition all employees are compelled to belong to a fund for securing medical aid and medicines. Their weekly contributions are 12 Dutch cents (4¾ cents) for men, 8 (3¼ cents) for women, and 3 (1¼ cents) for each child. For this the members, in case of sickness, receive free medicines and medical attendance, except that patients are required to pay 5 Dutch cents (2 cents) per visit in the case of clerks and workingmen and 10 and 15 Dutch cents (4 and 6 cents) in the case of higher officials.

In case of accidents while at work employees receive full wages until their recovery. In 1881 an accident insurance fund was inaugurated. This fund pays two full years' wages in case of death, loss of sight, or loss of two limbs; one year's wages for the loss of an eye, a foot, or the right hand; 35 per cent of the past year's wages for the loss of the left hand, and from 5 to 25 per cent for the loss of one or more fingers. This insurance is paid whether the accident occurs while at work or not. The premium required from members is 0.65 per cent of their wages.

In regard to the provision of old-age insurance, the company has adopted a peculiar but at the same time effective system. The company retains 7 per cent of the wages of each employee who has been a year or more in its service. With this money it purchases each year a paid-up old-age pension from an insurance society that will begin to run when the employee reaches the age of 60 years. By this arrangement the longer a man remains in the company the greater the value of the old-age pension to which he acquires a right. Each year's operation constitutes a separate transaction. A person thus, after ten years' service, is in possession of ten paid-up old-age pension policies. This system presents the very great advantage that should a man leave the service of the company he retains possession of his policies, and the acquisition of an old-age pension is thus in no way dependent upon employees remaining in the employment of the company.

A system of life insurance has been added to that of old-age pensions. By paying 2 per cent of their wages in addition to the 7 per cent for old-age pensions the company secures for its employees a life-insurance policy equal to 9 per cent of all the wages received by them during the time that they are insured.

As the above life insurance is insufficient to prevent distress in the case of families of persons who died after a short period of insurance, a special widows' fund has been established. For the support of this fund each director and employee contributes 1 per cent of his wages. The company adds an amount equal to one-half of the total of these contributions. Any widow who is unable to provide for her own or her children's subsistence, either by her earnings or from other sources, receives 4 florins ($1.61) per week for herself and 1 florin (40½ cents) for each child, the total, however, to in no case exceed 8 florins ($3.22). In September, 1893, 15 widows and 31 children were receiving aid from this fund.

The organization for bringing together the workingmen and the heads of departments for the purpose of the common consideration of questions relating to their interests, or "The Kernel" as it is called, though somewhat differently organized from the councils of arbitration and conciliation at Mariemont and Bascoup, yet follows so closely the general lines of action of these institutions that it would be superfluous to give the details of its organization and operations.

The remaining institutions, such as the savings banks, of which there are two, a kindergarten, manual-training, housekeeping, and other schools, and a gymnasium, reading room, and casino, though of course of importance, do not require any extended comment.

## THE CHOCOLATE FACTORY OF MENIER, NOISIEL, FRANCE.

The village of Noisiel, the seat of the chocolate factory of Menier, though interesting in itself, offers but few points for an extended study. Though it is a pure type of an industrial village, the management of affairs has been so completely retained in the hands of the proprietor that no opportunity has been afforded for the development of workingmen's institutions on the principle of mutuality. The village has been managed entirely as a branch of the general industrial enterprise, so that information concerning it can not well be distinguished from that of the manufacturing branch proper.

Though M. Menier's industry dates from 1825, the industrial village itself was only constructed in 1874. The village is absolutely removed from all other houses and forms a distinct community presenting all the necessary houses, such as schools, restaurants, stores, etc., for a self-contained existence. At the present time there are 295 tenements. They are built in groups of two and are exceptionally well constructed and comfortable. The general appearance of the group with its open spaces and shade trees is that of a charming country village. (*a*) The total number of employees of the works is about 1,500, but these houses are sufficient to accommodate practically the entire personnel, as the industry is such as to permit of the advantageous employment of women and children, and in consequence it is customary for the wives and daughters of workingmen to be employed as well.

Absolutely no opportunity is given for the development of institutions by the workingmen themselves, nor has M. Menier himself created many such. The work in this direction consists chiefly in the maintenance of a relief fund and the support of primary schools. Every care, however, seems to have been taken to make the condition of the employees as comfortable as possible. The village is so managed as to produce a moderate income on the money invested in it.

*a* For details concerning these houses see the Eighth Special Report of the Department of Labor on the Housing of the Working People.

## IRON AND STEEL WORKS OF SCHNEIDER & CO. LE CREUZOT, FRANCE.

Though Le Creuzot is not an industrial center possessing a unified and homogeneous character of such a nature as to warrant its study in detail as an industrial village, it should at least be noticed on account of the important industrial and social work of the iron and steel works of Schneider & Co. This establishment is one of the most important in France, and it employs between 12,000 and 13,000 workingmen. In connection with its works at Le Creuzot the company has organized a great many institutions for the benefit of its employees. It has erected 466 houses for renting to its workingmen; it has facilitated the acquisition of homes through the advancement of money with which to build; it has provided liberally for the support of schools; an efficient medical service for the gratuitous care of sick and injured workingmen is maintained. Since 1877 it has provided for the pensioning of its old employees through contributions to the National Old-Age and Invalidity Fund. In 1889 the company reported to the Paris Exposition that during the year 1888 the total amount expended by it for the benefit of its employees was 1,632,000 francs ($314,976).

# RATES OF WAGES PAID UNDER PUBLIC AND PRIVATE CONTRACT.

BY ETHELBERT STEWART.

The tables which immediately follow are the results of an original investigation in the cities of Baltimore, Boston, New York, and Philadelphia as to the wages paid, first, to those engaged on public work employed directly by the city or State, second, to those engaged on public work employed by contractors, and, third, to those engaged on private work employed by contractors or firms. The rates given in these three divisions are not only for the same occupations, but these occupations represent similar work so far as it was possible to obtain such data. Table I shows the number of persons in each occupation and the rate of wages per hour; also in addition to the wages per hour the hours of work per week are shown. Each city is taken up separately and in the following order: Baltimore, Boston, New York, and Philadelphia. The occupations for which wages were secured appear under each city in alphabetical order and are subdivided into the three classes previously given. For brevity's sake these classes are indicated in the tabulations by (1) public, by city or State, (2) public, by contractors, and (3) private, by contractors or firms.

Whenever it could be done wages were secured in each class for each occupation. This, however, was not always possible. Taking Baltimore in Table I, for instance, it is seen that for the first occupation, asphalt layers, no data are given for persons engaged on public work directly by the city or State, none having been found in the city; all such work, both public and private, having presumably been done through contractors or firms. The same is true for the next occupation, asphalt mixers; but for blacksmiths it is seen that data were secured in each of the three classes. In the first two occupations, asphalt layers and asphalt mixers, it is shown that the same wages per hour were paid in each of the two classes of work, and also that the same number of hours per week were worked. Blacksmiths employed on public work directly by the city or State, working 54 hours per week, were paid from 22½ to 30½ cents per hour. Those employed on public work by contractors worked 60 hours per week, but were paid lower wages—from 17½ to 26 cents per hour, the majority being paid 22½ cents per hour. Those employed on private work by contractors or firms, working 60 hours per week, were paid also from 17½ to 26 cents per hour, the majority being paid 22½ cents. In this manner each of the occupations for which data are given may be analyzed and comparisons made.

Table II summarizes the data given in Table I so far as the wages per hour for each occupation are concerned, the data relating to the number of employees and hours worked per week being omitted. In this table each occupation is taken up in order, and for each of the three classes of employment the highest, lowest, and average wages per hour are given. For example, taking up the city of Baltimore, it is seen that carpenters on public work employed directly by the city or State were paid a maximum wage of 33⅓ cents per hour, the lowest wage paid being 27⅜ cents per hour, while the average was 32¼ cents per hour. Carpenters on public work employed by contractors were paid as much as 31¼ cents per hour and as little as 22½ cents, the average wage per hour being 25⅞ cents. For those working at the same occupation employed on private work by contractors or firms a maximum of 28 cents and a minimum of 18 cents per hour were paid, the average being 26½ cents. These averages, being based on the entire number of carpenters shown in Table I, afford a fair comparison. As shown, those employed on public work directly by the city or State were paid the highest average wage, 32¼ cents per hour; those engaged on public work under contractors were paid the lowest, 25⅞ cents per hour, while those engaged on private work by contractors or firms were paid 26½ cents per hour.

It is strongly asserted in some quarters that the tendency of letting public contracts to the lowest bidder is to lower the wages of labor; that the lowest bidder is, generally speaking, the man who pays lowest wages or expects to use poorest material; that the idea that the lowest bidder is the one willing to accept least profits for himself is erroneous.

The legislature of the State of New York seems to have been convinced of the tendency of the contract system to lower the rates of wages, and in 1894 passed a law that all contractors on public works, State and municipal, must pay the prevailing rate of wages in the locality in which the work is being done. Pursuant to this law a clause like the following is inserted in all contracts:

All work to be performed and materials furnished to be according to the laws of the State and municipality applicable thereto, especially of chapter 622 of the laws of 1894, entitled "An act to amend chapter three hundred and eighty-five of the laws of eighteen hundred and seventy, entitled 'An act to regulate the hours of labor of mechanics, workingmen and laborers in the employ of the State, or otherwise engaged on public works,'" which provides that * * * "all mechanics, workingmen, and laborers now or hereafter employed by the State, or any municipal corporation therein, through its agents or officers, or in the employ of persons contracting with the State or such corporation for performance of public works, * * * shall receive not less than the prevailing rate of wages in the respective trades or callings in which such mechanics, workingmen, and laborers are employed in said locality. And in all such employment none but citizens of the United States shall be employed by the State, any municipal corporation therein, and by persons contracting with the State or any municipal corporation thereof."

An inquiry as to whether or not the inspectors on public works were expected to enforce this clause was answered by the statement that its enforcement was left to the trades unions and working people in whose interest the law was enacted. Aside, however, from all questions of the efficiency of this law, or the manner of its enforcement or nonenforcement, the fact of its existence can hardly be construed in any other light than as an admission by the legislature of the State of New York of a tendency in the contract system to lower the rates of wages and an attempt upon its part to check that tendency.

A glance at the tables will show that the highest rates of wages paid to unskilled labor are paid to those employed directly by the municipality or State. This is, generally speaking, true also of the skilled trades. Some of the exceptions will be noted further on. With the exception of Boston, all cities included in this investigation fix the rate of wages paid unskilled labor by city ordinance or, as in the case of New York City, by State law.

The city of Baltimore, by an ordinance passed May 1, 1883, provides that:

> The pay of all laborers shall be ten dollars per week, and no deduction from said amount shall be made except for such time as any of said laborers may lose by absenting themselves from the work upon which they may be at the time employed. (*a*)

The law also fixes the hours of labor, which must not exceed 9 in any one day, and provides that all city employees must be citizens and registered voters.

Philadelphia fixes the rate of pay for laborers at $1.75 per day of 9 hours, except for those employed in the public parks. In its annual appropriation ordinance it has a most elaborate scale, fixing the rate of wages for nearly every trade and occupation employed by any of the city departments. These rates are the maximum union rates in cases of organized trades, and corresponding rates for other occupations.

The same may be said of the laws of New York fixing rates of wages to be paid by New York City.

As a general statement it may be affirmed that the public, when employing directly by the day, pays the highest prevailing rate of wages for the shortest prevailing day's labor. This is especially true of the United States Government, where in the navy-yards of Boston and Brooklyn the highest outside rates for a 10-hour day are made the prevailing rates for an 8-hour day. Certain apparent exceptions to this rule have been omitted from the tables, as without an explanatory note in each case some misunderstanding might arise, and because they become, upon examination, not really exceptions. For instance, the city gas works of Philadelphia employs bricklayers at a rate considerably below the prevailing one. These men, however, work steadily or are paid in the absence of work—that is, they lose no time. The

---

*a* Ordinance No. 48, sec. 221, art. 48, p. 1020, City Code of Baltimore.

superintendent states that these men were given their choice of the maximum union rate of $4.05 and take work when they could get it, or $3 per day and steady employment. It will be readily seen that yearly earnings would be greater at the lesser daily rate. The exigencies of statistical tabulation would require that these men be brought in direct contrast with other bricklayers working on public contracts or on private work. It is only just, however, to say that these men—in fact, all employees of the city gas department—are paid higher wages than are paid either by the company furnishing gas to the city under contract or the company producing gas for private sale.

Another instance is that of 10 pavers employed by the city of Baltimore at rates much lower than the prevailing contract rate and lower than that paid to other pavers by the city. These men work for the waterworks department. They are steadily employed, are not union men, never required to do new work, and are in fact paid 25 cents more per day of 9 hours than other nonunion pavers get for 10 hours. With this explanation of the two most striking apparent exceptions, the statement may be repeated that the public when employing directly by day labor pays the highest prevailing rate for the shortest prevailing day. It would appear also, at least in a great many instances, that this can be done with the best economic results.

Probably the highest wages paid (in the occupations employed) are those paid by the trustees of the New York and Brooklyn bridge, being considerably higher than wages in like occupations by private concerns. Yet the trustees of this bridge have been enabled to reduce fares continuously, to abolish tolls for pedestrians, and magnificently improve the plant.

The city of Boston sprinkles its own streets by day labor, having practically abandoned the contract system for this work. On page 414 of the Annual Report of the Street Department of the city of Boston for 1895, the superintendent summarizes the result of the two systems as follows:

COMPARISON OF DAY WORK WITH CONTRACT WORK IN BACK BAY AND SOUTH END, BOSTON.

| | |
|---|---|
| Back Bay: | |
| Contract Work, 1894 | $6,696.02 |
| Day Work, 1895 | 4,990.00 |
| Saving in day work | 1,706.02 |
| South End: | |
| Contract work, 1894 | 5,128.50 |
| Day work, 1895 | 2,540.00 |
| Saving in day work | 2,588.50 |
| Total saving in day work over contract work | 4,294.52 |

The above comparison is one of the most satisfactory evidences of the good results accomplished by the division this year. To it attention is specially directed.

The cost for watering in the Back Bay with fresh water in 1894 under contract was $575 per mile, while this year the same service was more efficiently rendered at an expense of $424 per mile.

The expense in the South End last year for watering was $460 per mile for fresh water, $630 per mile for salt water, while this year the watering was done under day work at an expense of $277 per mile.

As will appear in the tables following, the rate of wages paid by the city to the men who actually did the street watering was considerably greater than the wages paid by the contractors.

The city of Baltimore does all its street cleaning by day labor, except the machine sweeping. This is at present let by contract at a very low figure. Nevertheless, the street commissioner states upon computations made by him that he could more efficiently do the work and save the city enough in one year to pay for the machines and horses; that is, he could pay for the contractor's plant in one year and do better work. This computation was based upon a proposition to pay the legal city rate of wages, $1.66⅔ per day, whereas the present contractor pays the machine drivers but 80 cents per day. The cost of supervision and inspection increases as the contract price decreases, until it sometimes costs almost as much to make the contractor do his work as it would to do it.

The cost of inspection is enormous for cities, still greater proportionately for the Federal Government, and this cost must be added to the contract price before it can be determined whether or not the contract figure is a low one.

It is to be regretted that a minute and satisfactory statement of the cost of sewer construction, excavation, etc., could not be obtained in Boston, where an enormous amount is being done under both systems. The work performed by day labor is, however, generally experimental, and even where the conditions were practically similar no accounts that could be used for comparison were obtainable. One very significant statement was, however, made by the secretary of the metropolitan sewer commission to the effect that one piece of sewer had been constructed entirely by day labor because it undermined some private property, and notwithstanding their inspection the commissioners were afraid to risk the contract system because of the heavy damage suits that might result from faulty work.

Three of the cities in which this investigation was conducted clean their streets by day labor. These are Boston, New York, and Baltimore (except the machine sweeping). Baltimore does not keep a record of cost of cleaning per mile. Philadelphia lets her contract for a lump sum, and keeps no record of cost of cleaning per mile. The city of Brooklyn is included as a part of the metropolitan district of New York for purposes of comparison.

The cost of cleaning the streets in Boston is shown in [illegible] statement:

COST PER MILE OF CLEANING STREETS IN EACH DISTRICT OF BOSTON, INCLUDING SUPERVISION, LABOR, YARD, AND STABLE EXPENSES.

| Old districts. | Miles of streets cleaned. | Cost of cleaning streets. | 64 per cent of the total cost of supervision. | 73 per cent of the total cost of yard and stable expenses. | Total expenses. | Total cost per mile. |
|---|---|---|---|---|---|---|
| 1 | 439.64 | $5,496.26 | $210.33 | [illegible] | [illegible] | [illegible] |
| 2 | 486.80 | 5,488.50 | 210.00 | [illegible] | [illegible] | [illegible] |
| 3 | 652.00 | 7,395.90 | 283.04 | 1,344.04 | [illegible] | [illegible] |
| 4 | 399.53 | 6,207.84 | 237.56 | 1,128.16 | [illegible] | [illegible] |
| 5 | 167.66 | 4,722.78 | 180.73 | [illegible] | [illegible] | [illegible] |
| 6 | 207.89 | 4,776.40 | 183.78 | [illegible] | [illegible] | [illegible] |
| 7 | 94.35 | 4,201.00 | 160.77 | 763.46 | [illegible] | [illegible] |
| 9 | 204.96 | 4,091.73 | 156.56 | 743.61 | [illegible] | [illegible] |
| Total | 2,652.83 | 42,380.41 | 1,621.82 | 7,701.80 | 51,7[illegible] | ........ |

Average cost per mile of cleaning streets in eight old districts, including supervision, etc., $19.49.

| New districts. | Miles of streets cleaned. | Cost of cleaning streets. | 64 per cent of the total cost of supervision. | 73 per cent of the total cost of yard and stable expenses. | Total expenses. | Total cost per mile. |
|---|---|---|---|---|---|---|
| 1 | 858.50 | $12,004.77 | $460.77 | $2,188.11 | $14,[illegible] | [illegible] |
| 2, 3 | 799.28 | 9,924.00 | 380.89 | 1,808.85 | 12,113.74 | [illegible] |
| 7 | 433.71 | 12,427.71 | 476.99 | 2,265.20 | 15,[illegible] | [illegible] |
| 8 | 3,671.47 | 35,109.57 | 1,347.55 | 6,[illegible] | 42,[illegible] | [illegible] |
| 9 | 459.13 | 6,902.07 | 264.92 | 1,258.04 | 8,[illegible] | [illegible] |
| 10 | 2,544.07 | 27,084.99 | 1,039.54 | 4,936.76 | 33,061.[illegible] | [illegible] |
| Total | 8,766.16 | 103,453.11 | 3,970.66 | 18,856.36 | 12[illegible] | ........ |

Average cost per mile of cleaning streets in six new districts, including supervision, etc., $14.40.

The average cost of cleaning the 11,418.99 miles actually swept, and comprising all the districts new and old, was $15.58 per mile. This includes much that is charged to other accounts in other cities, such, for instance, as dumping, etc. It will be seen from the tables that notwithstanding this low cost per mile, Boston paid higher rates of wages than any city except New York.

The cost of street cleaning in New York City for 1895 is shown in the following statement:

The whole cost of cleaning the streets (including every expense incurred in the administration of the department) was, per mile of street swept per day... $22.94

The whole cost of cleaning the streets (including every expense incurred in the administration of the department) was—

Per cart load of material collected (including permits)................ 1.[illegible]

Per cart load of material collected (excluding permits)................ [illegible]

The cost of sweeping, not including supervision or any other expense, was, per mile of street swept per day................ [illegible]

The cost of carting ashes, garbage, and street sweepings, not including supervision, was, per cart load of material collected................ [illegible]

The cost of collecting and removing snow and ice, not including supervision, was, per cart load of snow and ice removed................ [illegible]

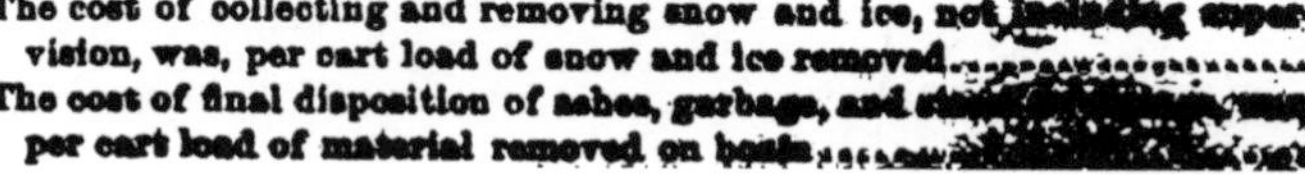

The cost of final disposition of ashes, garbage, and [illegible] was, per cart load of material removed on boats................ [illegible]

At the same time the contract price in Brooklyn was $23 per mile for street cleaning, and $215,000 as a lump sum for removal of ashes, etc. To this contract price of $23 per mile must be added the cost of administration, which was 60 cents per mile, as estimated for this report. At that time the Brooklyn contractor was paying $1.50 per day of twelve hours. The contract price for 1896 is $17 per mile and $210,000 lump sum for removal of ashes. With the reduction in contract price, the wages of street sweepers are reduced to $1.25 per day.

Perhaps the most striking instance of reduction in wages as a result of contract employment is that shown for New York City under the occupation of snow shovelers. Formerly the city did this work by day labor, employing extra men as occasion required at $2 per day of eight hours. Last winter the work was let by contract, and the wages paid, as shown in the first table, was $1.25 per day of ten hours.

One feature of the investigation as originally designed, but which could not appear in tabular form, was the effect of the contract system or rates of wages paid under it upon wages paid on private work. When times are good the contractors who bid lowest get the public contracts and the smaller private ones, thus absorbing both this class of contractors and the cheaper labor which they employ, leaving the larger private contracts to the contractors styling themselves "legitimate," and who employ only union or high-priced labor. In times of business depression, however, such as the country has been experiencing, there is little else but public contracts to be had, and the employer of the higher-priced labor must bid low enough to get it, or have nothing to do. To enable their employers to meet these conditions, the Bricklayers' Union of Baltimore during last year reduced its scale of wages from $4 for eight hours to $3.60 for eight hours, and then again to $3 for nine hours, with eight hours on Saturday, this being the rate generally paid nonunion bricklayers. The carpenters of Baltimore and the painters of New York City are now having a similar struggle. Notwithstanding the State law and indictments under it, the painters of New York find it impossible to maintain their scale on public works. As a rule, however, it must be said that the effect of the contract system in reducing wages is largely confined to unskilled and unorganized labor, and that the trades are but slowly and slightly affected.

TABLE I.—HOURS OF WORK AND WAGES OI

1. ON PUBLIC WORK, EMPLOYED DIRECTLY BY
2. ON PUBLIC WORK, EMPLOYED BY CONTRA
3. ON PRIVATE WORK, EMPLOYED BY CONTRAC

BALTIMORE.

| Number of employees. | Hours of work per week. | Wages per hour. |
|---|---|---|
| **Asphalt layers:** | | |
| Public, by contractors. | | |
| 25 | 60 | $0.20 |
| Private, by contractors or firms. | | |
| 13 | 60 | .20 |
| **Asphalt mixers:** | | |
| Public, by contractors. | | |
| 8 | 60 | .15 |
| Private, by contractors or firms. | | |
| 6 | 60 | .15 |
| **Blacksmiths:** | | |
| Public, by city or State. | | |
| 2 | 54 | .22½ |
| 1 | 54 | .25 |
| 1 | 54 | .30½ |
| Public, by contractors. | | |
| 1 | 60 | .17½ |
| 20 | 60 | .22½ |
| 4 | 60 | .25 |
| 2 | 60 | .26 |
| Private, by contractors or firms. | | |
| 1 | 60 | .17½ |
| 3 | 60 | .18 |
| 7 | 60 | .20 |
| 20 | 60 | .22½ |
| 2 | 60 | .26 |
| **Blacksmiths' helpers:** | | |
| Public, by city or State. | | |
| 1 | 54 | 22½ |
| Public, by contractors. | | |
| 15 | 60 | .13 |
| 2 | 60 | .15 |
| 10 | 60 | .16⅔ |
| 2 | 60 | .20 |
| Private, by contractors or firms. | | |
| 15 | 60 | .13½ |
| 5 | 60 | .15 |
| 10 | 60 | .16⅔ |
| **Boiler makers:** | | |
| Public, by contractors | | |
| 50 | 60 | .22½ |
| Private, by contractors or firms. | | |
| 50 | 60 | .22½ |
| **Bricklayers:** | | |
| Public, by city or State. | | |
| 1 | 54 | .22½ |
| 28 | 53 | .39⅛ |
| Public, by contractors. | | |
| 120 | 54 | .33⅓ |
| 63 | 53 | .34 |
| 28 | 48 | .37½ |
| Private, by contractors or firms. | | |
| 13 | 54 | .33⅓ |
| 35 | 53 | .34 |
| 10 | 55 | .39⅛ |
| **Carpenters:** | | |
| Public, by city or State. | | |
| 4 | 54 | $0.27¾ |
| 18 | 54 | .33⅓ |
| Public, by contractors. | | |
| 70 | 54 | .22½ |
| 15 | 60 | .22½ |
| 140 | 54 | .25 |
| 2 | 54 | .27¾ |
| 12 | 48 | .31¼ |
| 65 | 54 | .31¼ |
| Private, by contractors or firms. | | |
| 20 | 60 | .18 |
| 4 | 60 | .20 |
| 15 | 60 | .22½ |
| 134 | 54 | .27¾ |
| 70 | 54 | .28 |
| **Cement finishers:** | | |
| Public, by contractors. | | |
| 1 | 60 | .25 |
| 2 | 60 | .35 |
| Private, by contractors or firms. | | |
| 4 | 60 | .35 |
| **Cement mixers:** | | |
| Public, by contractors. | | |
| 25 | 60 | .12½ |
| Private, by contractors or firms. | | |
| 10 | 60 | .12½ |
| **Chandelier fitters:** | | |
| Public, by contractors. | | |
| 10 | 60 | .30 |
| Private, by contractors or firms. | | |
| 10 | 60 | .30 |
| **Chandelier makers:** | | |
| Public, by contractors. | | |
| 15 | 60 | .25 |
| Private, by contractors or firms. | | |
| 20 | 60 | .25 |
| **Coal handlers:** | | |
| Public, by contractors. | | |
| 5 | a 60 | a .15 |
| Private, by contractors or firms. | | |
| 5 | a 60 | a .15 |
| **Cornice workers:** | | |
| Public, by contractors. | | |
| 3 | 54 | .33⅓ |
| 3 | 54 | .35 |
| 3 | 54 | .37½ |
| Private, by contractors or firms. | | |
| 6 | [illegible] | [illegible] |
| 2 | [illegible] | [illegible] |
| 6 | [illegible] | [illegible] |
| **Curbstone setters:** | | |
| Public, by [illegible] | | |
| 3 | [illegible] | [illegible] |

TABLE 2.—HOURS OF WORK AND WAGES OF EMPLOYEES—

1. ON PUBLIC WORK, EMPLOYED DIRECTLY BY THE CITY OR STATE,
2. ON PUBLIC WORK, EMPLOYED BY CONTRACTORS,
3. ON PRIVATE WORK, EMPLOYED BY CONTRACTORS OR FIRMS—Cont'd.

BALTIMORE—Continued.

| Number of employees. | Hours of work per week. | Wages per hour. |
|---|---|---|
| **Engineers, chief:** | | |
| Public, by contractors. | | |
| 1 | a 70 | a $0.28½ |
| Private, by contractors or firms. | | |
| 1 | a 70 | a .28½ |
| **Engineers, hoisting:** | | |
| Public, by city or State. | | |
| 6 | 54 | .30½ |
| Public, by contractors. | | |
| 2 | 60 | .20 |
| 7 | 54 | .22½ |
| 6 | 60 | .22½ |
| 2 | 53 | .22½ |
| Private, by contractors or firms. | | |
| 8 | 53 | .28½ |
| **Engineers, stationary:** | | |
| Public, by contractors. | | |
| 6 | a 70 | a .20 |
| 2 | 60 | .22½ |
| 4 | 60 | .25 |
| Private, by contractors or firms. | | |
| 6 | 60 | .20 |
| 6 | a 70 | a .20 |
| 4 | 60 | .22½ |
| 3 | 60 | .25 |
| **Engineers, stationary, assistant:** | | |
| Public, by contractors. | | |
| 3 | a 70 | a .18½ |
| Private, by contractors or firms. | | |
| 3 | a 70 | a .18½ |
| **Engineers, steam roller:** | | |
| Public, by contractors. | | |
| 2 | 60 | .40 |
| Private, by contractors or firms. | | |
| 1 | 60 | .40 |
| **Firemen:** | | |
| Public, by contractors. | | |
| 1 | a 60 | a .15 |
| 4 | a 60 | a .17½ |
| 2 | a 77 | a .18½ |
| 10 | a 70 | a .20 |
| Private, by contractors or firms. | | |
| 1 | a 60 | a .15 |
| 1 | 60 | .16⅔ |
| 4 | a 60 | a .17½ |
| 2 | a 77 | a .18½ |
| 6 | 60 | .20 |
| 10 | a 70 | a .20 |
| **Foremen, carpenters:** | | |
| Public, by contractors. | | |
| 10 | 54 | .27⅞ |
| 1 | 54 | .33⅓ |
| Private, by contractors or firms. | | |
| 5 | 54 | .30½ |
| 2 | 54 | .33⅓ |
| 1 | 54 | .39 |
| 2 | 54 | .41⅔ |
| **Foremen, cornice workers:** | | |
| Public, by contractors. | | |
| 1 | 51 | $0.33⅓ |
| Private, by contractors or firms. | | |
| 3 | 54 | .33⅓ |
| **Foremen, electric lighting:** | | |
| Public, by contractors. | | |
| 1 | a 70 | a .33 |
| Private, by contractors or firms. | | |
| 1 | a 70 | a .33 |
| **Foremen, electric lighting, assistant:** | | |
| Public, by contractors. | | |
| 1 | a 70 | a .21½ |
| Private, by contractors or firms. | | |
| 1 | a 70 | a .21½ |
| **Foremen, laborers:** | | |
| Public, by city or State. | | |
| 10 | 54 | .22½ |
| 6 | 54 | .25 |
| 1 | 54 | .27½ |
| 11 | 54 | .29½ |
| Public, by contractors. | | |
| 4 | 60 | .20 |
| 6 | 60 | .25 |
| 8 | 60 | .30 |
| 4 | 60 | .35 |
| Private, by contractors or firms. | | |
| 4 | 60 | .17½ |
| 6 | 60 | .20 |
| 2 | 60 | .25 |
| **Foremen, linemen:** | | |
| Public, by contractors. | | |
| 1 | a 70 | a .28½ |
| Private, by contractors or firms. | | |
| 1 | a 70 | a .28½ |
| **Galvanized and sheet iron workers:** | | |
| Public, by contractors. | | |
| 24 | 54 | .22½ |
| 2 | 54 | .25 |
| Private, by contractors or firms. | | |
| 16 | 54 | .22½ |
| 4 | 54 | .25 |
| **Gardeners:** | | |
| Public, by city or State. | | |
| 3 | 60 | .18½ |
| 1 | 60 | .21½ |
| Private, by contractors or firms. | | |
| 4 | 60 | .15 |
| 1 | 60 | .20 |
| **Gas-holder riveters:** | | |
| Public, by contractors. | | |
| 150 | 60 | .27½ |
| Private, by contractors or firms. | | |
| 150 | 60 | .27½ |
| **Granite cutters:** | | |
| Public, by contractors. | | |
| 43 | 48 | $0.46⅞ |
| Private, by contractors or firms. | | |
| 51 | 48 | .46⅞ |
| **Ground men, electric lighting:** | | |
| Public, by contractors. | | |
| 6 | a 70 | a .12⅞ |
| 6 | 60 | .15 |
| Private, by contractors or firms. | | |
| 6 | a 70 | a .12⅞ |
| 6 | 60 | .15 |
| **Hod carriers:** | | |
| Public, by contractors. | | |
| 3 | 53 | .22⅝ |
| 6 | 48 | .25 |
| 10 | 54 | .27⅞ |
| Private, by contractors or firms. | | |
| 6 | 53 | .23⅝ |
| 7 | 53 | .28¼ |
| **Hostlers:** | | |
| Public, by contractors. | | |
| 2 | a 84 | a .07½ |
| 2 | a 70 | a .10 |
| 1 | a 70 | a .12⅞ |
| Private, by contractors or firms. | | |
| 1 | a 70 | a .12⅞ |
| **Inspectors, electric lighting:** | | |
| Public, by contractors. | | |
| 8 | a 70 | a .17½ |
| 2 | a 77 | a .18½ |
| 2 | a 77 | a .19½ |
| Private, by contractors or firms. | | |
| 8 | a 70 | a .17½ |
| 2 | a 77 | a .18½ |
| 3 | a 77 | a .19½ |
| **Iron workers, ornamental:** | | |
| Public, by contractors. | | |
| 12 | 60 | .24 |
| Private, by contractors or firms. | | |
| 12 | 60 | .24 |
| **Iron workers', ornamental, helpers:** | | |
| Public, by contractors. | | |
| 6 | 60 | .15 |
| Private, by contractors or firms. | | |
| 6 | 60 | .15 |
| **Iron workers, structural:** | | |
| Public, by contractors. | | |
| 85 | 60 | .17½ |
| 30 | 60 | .20 |
| 12 | 54 | .22½ |
| 20 | 60 | .22½ |
| 4 | 60 | .25 |
| Private, by contractors or firms. | | |
| 86 | 60 | .17½ |
| 15 | 60 | .20 |
| 30 | 60 | .22½ |
| 18 | 54 | .25 |

a Work 7 days per week.

TABLE I.—HOURS OF WORK AND WAGES OF EMPLOYEES—

1. ON PUBLIC WORK, EMPLOYED DIRECTLY BY THE CITY OR STATE.
2. ON PUBLIC WORK, EMPLOYED BY CONTRACTORS.
3. ON PRIVATE WORK, EMPLOYED BY CONTRACTORS OR FIRMS—Cont'd.

BALTIMORE—Continued.

| Number of employees. | Hours of work per week. | Wages per hour. |
|---|---|---|
| **Iron workers', structural, helpers:** | | |
| Public, by contractors. | | |
| 33 | 60 | $0.15 |
| Private, by contractors or firms. | | |
| 30 | 60 | .15 |
| **Laborers:** | | |
| Public, by city or State. | | |
| 661 | 54 | .18⅓ |
| 1 | 54 | .22⅔ |
| Public, by contractors. | | |
| 30 | 60 | .11⅓ |
| 1,750 | 60 | .12½ |
| 112 | 54 | .14 |
| 83 | 60 | .15 |
| 10 | 48 | .15⅝ |
| Private, by contractors or firms. | | |
| 100 | 60 | .11⅓ |
| 691 | 60 | .12½ |
| 46 | 54 | .14 |
| 177 | 60 | .15 |
| 20 | 54 | .16⅔ |
| **Laborers, building:** | | |
| Public, by contractors. | | |
| 26 | 60 | .15 |
| Private, by contractors or firms. | | |
| 14 | 54 | .14 |
| 15 | 54 | .16⅔ |
| **Linemen:** | | |
| Public, by contractors. | | |
| 3 | a 70 | a .21⅓ |
| 3 | a 70 | a .24⅔ |
| 6 | 60 | .25 |
| Private, by contractors or firms. | | |
| 8 | a 84 | a .20 |
| 3 | a 70 | a .21⅓ |
| 3 | a 70 | a .24⅔ |
| 6 | 60 | .25 |
| **Machinists:** | | |
| Public, by city or State. | | |
| 1 | 54 | .27⅞ |
| 1 | 54 | .33⅓ |
| 1 | 54 | .36 |
| Public, by contractors. | | |
| 4 | 60 | .15 |
| 1 | 60 | .16⅔ |
| 56 | 60 | .22½ |
| 4 | 60 | .25 |
| Private, by contractors or firms. | | |
| 1 | 60 | .16⅔ |
| 8 | 60 | .17 |
| 10 | 60 | .18 |
| 2 | 60 | .19½ |
| 6 | 60 | .20 |
| 36 | 60 | .22½ |
| **Machinists' helpers:** | | |
| Public, by contractors. | | |
| 8 | 60 | [illegible] |
| **Machinists' helpers—Conc'd.** | | |
| Private, by contractors or firms. | | |
| 1 | 60 | $0.12 |
| 1 | 60 | .12½ |
| 2 | 60 | .13⅓ |
| 11 | 60 | .15 |
| **Masons, stone:** | | |
| Public, by city or State. | | |
| 19 | 53 | .40⅝ |
| Public, by contractors. | | |
| 32 | 60 | .30 |
| 6 | 60 | .32½ |
| 8 | 60 | .35 |
| 6 | 54 | .36 |
| 5 | 54 | .39 |
| Private, by contractors or firms. | | |
| 18 | 54 | .33⅓ |
| 2 | 54 | .36 |
| 25 | 53 | .39 |
| **Molders:** | | |
| Public, by contractors. | | |
| 46 | 60 | .22½ |
| 80 | 60 | .27⅞ |
| Private, by contractors or firms. | | |
| 46 | 60 | .22½ |
| 80 | 60 | .27⅞ |
| **Molders' helpers:** | | |
| Public, by contractors. | | |
| 40 | 60 | .12½ |
| Private, by contractors or firms. | | |
| 40 | 60 | .12½ |
| **Oilers:** | | |
| Public, by contractors. | | |
| 2 | a 84 | a .14⅜ |
| Private, by contractors or firms. | | |
| 2 | a 84 | a .14⅜ |
| **Painters:** | | |
| Public, by city or State. | | |
| 2 | 54 | .22½ |
| Public, by contractors. | | |
| 2 | 54 | .16⅔ |
| 30 | 54 | .22½ |
| Private, by contractors or firms. | | |
| 6 | 54 | .16⅔ |
| 4 | 60 | .17½ |
| 12 | 60 | .18 |
| 26 | 54 | .25 |
| 23 | 53 | [illegible] |
| **Pattern makers:** | | |
| Public, by city or State. | | |
| 1 | 54 | [illegible] |
| Public, by contractors. | | |
| 16 | [illegible] | [illegible] |
| 26 | [illegible] | [illegible] |
| Private, by contractors or firms. | | |
| [illegible] | [illegible] | [illegible] |
| [illegible] | [illegible] | [illegible] |
| **Pavers:** | | |
| Public, by city or State. | | |
| 11 | 54 | $0.30⅓ |
| Public, by contractors. | | |
| 20 | 60 | .25 |
| **Pavers, Belgian block:** | | |
| Public, by city or State. | | |
| 33 | 53 | .51 |
| Public, by contractors. | | |
| 19 | 54 | .39 |
| 17 | 54 | .34 |
| 10 | 53 | .31 |
| Private, by contractors or firms. | | |
| 20 | [illegible] | [illegible] |
| 2 | 60 | [illegible] |
| 16 | 54 | [illegible] |
| **Pavers, cobblestone:** | | |
| Public, by city or State. | | |
| 24 | 53 | .41⅔ |
| Public, by contractors. | | |
| 4 | 60 | [illegible] |
| 42 | 54 | [illegible] |
| 11 | 54 | [illegible] |
| 16 | 53 | [illegible] |
| Private, by contractors or firms. | | |
| 29 | 60 | [illegible] |
| 10 | 60 | [illegible] |
| 10 | 54 | [illegible] |
| **Plumbers:** | | |
| Public, by contractors. | | |
| 10 | 54 | [illegible] |
| 20 | 54 | [illegible] |
| 4 | 54 | [illegible] |
| Private, by contractors or firms. | | |
| 40 | 54 | [illegible] |
| 4 | 54 | [illegible] |
| **Rammers, Belgian block:** | | |
| Public, by city or State. | | |
| 17 | 53 | .36 |
| Public, by contractors. | | |
| 4 | 54 | [illegible] |
| 9 | 54 | [illegible] |
| 6 | 53 | [illegible] |
| Private, by contractors or firms. | | |
| 20 | 60 | [illegible] |
| 6 | 54 | [illegible] |
| **Rammers, cobblestone:** | | |
| Public, by city or State. | | |
| 17 | [illegible] | [illegible] |
| Public, by contractors. | | |
| [illegible] | [illegible] | [illegible] |

## 3. ON PRIVATE WORK, EMPLOYED BY CONTRACTORS OR FIRMS—Cont'd.

### BALTIMORE—Concluded.

| Number of employees. | Hours of work per week. | Wages per hour. |
|---|---|---|
| **Rammers, paving:** | | |
| Public, by contractors. | | |
| 4 | 60 | $0.17½ |
| Private, by contractors or firms. | | |
| 6 | 60 | .17½ |
| **Repairmen, electric lighting:** | | |
| Public, by contractors. | | |
| 1 | 60 | .08⅓ |
| 1 | a 70 | a .12⅞ |
| Private, by contractors or firms. | | |
| 1 | 60 | .08⅓ |
| 1 | a 70 | a .12⅞ |
| **Roofers, slate:** | | |
| Public, by contractors. | | |
| 12 | 58 | .24 |
| Private, by contractors or firms. | | |
| 16 | 58 | .24 |
| **Steam and gas fitters:** | | |
| Public, by contractors. | | |
| 20 | 54 | .27¾ |
| Private, by contractors or firms. | | |
| 26 | 54 | .27¾ |
| **Stonecutters:** | | |
| Public, by city or State. | | |
| 10 | 48 | .43¾ |
| **Stonecutters—Conc'd.** | | |
| Public, by contractors. | | |
| 20 | 60 | $0.22½ |
| 6 | 54 | .26 |
| 1 | 54 | .30 |
| Private, by contractors or firms. | | |
| 18 | 48 | .40⅝ |
| 27 | 48 | .43 |
| **Stone setters:** | | |
| Public, by contractors. | | |
| 7 | 48 | .50 |
| Private, by contractors or firms. | | |
| 12 | 48 | .40⅝ |
| **Tappers, waterworks:** | | |
| Public, by city or State. | | |
| 3 | 54 | .22½ |
| Private, by contractors or firms. | | |
| 6 | 60 | .16⅔ |
| **Tinsmiths:** | | |
| Public, by contractors. | | |
| 20 | 60 | .20 |
| Private, by contractors or firms. | | |
| 20 | 60 | .20 |
| **Tool sharpeners:** | | |
| Public, by contractors. | | |
| 3 | 48 | .40⅝ |
| **Tool sharpeners—Conc'd.** | | |
| Private, by contractors or firms. | | |
| 8 | 48 | $0.40⅝ |
| **Trimmers, electric lighting:** | | |
| Public, by contractors. | | |
| 17 | a 70 | a .12⅞ |
| 14 | a 70 | a .14⅜ |
| Private, by contractors or firms. | | |
| 14 | a 70 | a .14⅜ |
| **Watchmen:** | | |
| Public, by city or State. | | |
| 1 | a 84 | a .14 |
| 22 | (b) | .22⅛ |
| Public, by contractors. | | |
| 4 | a 80 | a .11⅛ |
| Private, by contractors or firms. | | |
| 4 | a 80 | a .11⅛ |
| **Wire men:** | | |
| Public, by contractors. | | |
| 10 | 54 | .27¾ |
| Private, by contractors or firms. | | |
| 2 | 55 | .16⅓ |
| 4 | 55 | .22 |
| 1 | 55 | .24¼ |
| 5 | 55 | .27¾ |
| 10 | 54 | .27¾ |

### BOSTON.

| Number of employees. | Hours of work per week. | Wages per hour. |
|---|---|---|
| **Blacksmiths:** | | |
| Public, by city or State. | | |
| 1 | a 63 | a .25 |
| 1 | 54 | .25⅜ |
| 1 | 54 | .27¾ |
| 3 | 50 | .28 |
| 1 | 48 | .29 |
| 1 | 48 | .32 |
| 1 | 54 | .32 |
| 14 | 54 | .33⅓ |
| 1 | 48 | .35 |
| 1 | 54 | .36 |
| 3 | 48 | .38 |
| 3 | 54 | .39 |
| 4 | 48 | .41 |
| Public, by contractors. | | |
| 3 | 54 | .22½ |
| 2 | 59 | .23 |
| 2 | 54 | .25 |
| 5 | 60 | .25 |
| 3 | a 77 | a .26 |
| 5 | 54 | .27¾ |
| 7 | 60 | .30 |
| Private, by contractors or firms. | | |
| 6 | 56 | .24 |
| 3 | 60 | .26 |
| 15 | 56 | .26⅔ |
| 1 | 56 | .27 |
| 5 | 56 | .29 |
| **Blacksmiths' helpers:** | | |
| Public, by city or State. | | |
| 1 | 54 | .22½ |
| 4 | 48 | .23 |
| 1 | 54 | .25 |
| 6 | 54 | .27¾ |
| 10 | 48 | .32½ |
| Public, by contractors. | | |
| 5 | 60 | .18⅓ |
| Private, by contractors or firms. | | |
| 15 | 56 | .16 |
| 1 | 56 | .18 |
| 5 | 60 | .18⅓ |
| 5 | 56 | .19 |
| **Boiler makers:** | | |
| Public, by city or State. | | |
| 3 | 54 | .33⅓ |
| Private, by contractors or firms. | | |
| 2 | 56 | .20 |
| 4 | 56 | .21 |
| **Boiler makers' helpers:** | | |
| Public, by city or State. | | |
| 1 | 54 | .27¾ |
| Private, by contractors or firms. | | |
| 4 | 56 | .14 |
| 2 | 56 | .18 |
| **Bracers:** | | |
| Public, by city or State. | | |
| 1 | 54 | .22½ |
| 11 | 50 | .25 |
| 3 | 54 | .26 |
| 5 | 60 | .25 |
| 3 | 54 | .27¾ |
| 1 | 54 | .30⅝ |
| 3 | 54 | .33⅓ |
| Public, by contractors. | | |
| 2 | 60 | .20 |
| 1 | 60 | .22½ |
| 78 | 60 | .22½ |
| 6 | 59 | .23 |
| 16 | a 84 | a .23 |
| 35 | 54 | .26 |
| 6 | 60 | .26 |
| 14 | a 77 | a .26 |
| **Bricklayers:** | | |
| Public, by city or State. | | |
| 1 | 48 | .44 |
| 2 | 48 | .44 |
| Public, by contractors. | | |
| 208 | 48 | .42 |
| 19 | 54 | .42 |
| 3 | 60 | .42 |

a Work 7 days per week.

b Not reported.

TABLE I.—HOURS OF WORK AND WAGES

1. ON PUBLIC WORK, EMPLOYED DIRECTLY BY
2. ON PUBLIC WORK, EMPLOYED BY CONTRACTORS
3. ON PRIVATE WORK, EMPLOYED BY CONTRAC

BOSTON—Continued.

| Number of employees. | Hours of work per week. | Wages per hour. |
|---|---|---|
| **Bricklayers—Conc'd.** | | |
| **Private, by contractors or firms.** | | |
| 25 | 54 | $0.35 |
| 133 | 48 | .42 |
| **Bricklayers, sewer:** | | |
| **Public, by city or State.** | | |
| 24 | 48 | .56 |
| 3 | 50 | .60 |
| 6 | 50 | .70 |
| **Public, by contractors.** | | |
| 3 | 54 | .44½ |
| 12 | a 84 | a .58⅓ |
| 3 | 48 | .60 |
| 11 | 54 | .60 |
| 8 | 48 | .65 |
| 50 | 48 | .70 |
| 7 | a 77 | a .70 |
| 8 | 48 | .75 |
| **Private, by contractors or firms.** | | |
| 10 | 48 | .70 |
| **Calkers, iron:** | | |
| **Public, by city or State.** | | |
| 34 | 54 | .24½ |
| 2 | 54 | .25 |
| 1 | 54 | .26⅔ |
| 2 | 54 | .30⅝ |
| **Public, by contractors.** | | |
| 37 | 60 | .22½ |
| 12 | 60 | .23 |
| 18 | 60 | .23 |
| 6 | 59 | .25½ |
| **Calkers, wood:** | | |
| **Public, by city or State.** | | |
| 4 | 54 | .33⅓ |
| **Public, by contractors.** | | |
| 2 | 59 | .30½ |
| 9 | 48 | .40⅝ |
| **Private, by contractors or firms.** | | |
| 9 | 48 | .40⅝ |
| **Carpenters:** | | |
| **Public, by city or State.** | | |
| 1 | 54 | .21⅓ |
| 15 | 54 | .27⅞ |
| 6 | 50 | .28 |
| 8 | 66 | .28 |
| 3 | 50 | .30½ |
| 2 | 54 | .30⅝ |
| 1 | 48 | .32 |
| 20 | 54 | .33⅓ |
| 2 | 48 | .35 |
| 3 | 48 | .38 |
| 3 | 48 | .40 |
| 3 | 54 | .41⅔ |
| 1 | 48 | .45 |
| **Public, by contractors.** | | |
| 4 | 60 | .20 |
| 3 | 60 | .22½ |
| 4 | 60 | .25 |
| 9 | a 77 | a .25 |
| 4 | 59 | .25½ |
| 14 | a 84 | a .27¾ |

| Number of employees. | Hours of work per week. | Wages per hour. |
|---|---|---|
| **Carpenters—Conc'd.** | | |
| **Public, by contractors—Conc'd.** | | |
| 9 | 54 | $0.27½ |
| 32 | 53 | .28 |
| 37 | 54 | .28 |
| 8 | 58 | .30 |
| 25 | 54 | .30 |
| **Private, by contractors or firms.** | | |
| 119 | 56 | .24 |
| 15 | 56 | .26¼ |
| 5 | 56 | .27 |
| 9 | 54 | .27½ |
| 18 | 53 | .28 |
| 34 | 54 | .28 |
| 20 | 53 | .30 |
| 38 | 54 | .30 |
| **Carpenters' helpers:** | | |
| **Public, by city or State.** | | |
| 6 | 54 | .22½ |
| 3 | 50 | .24½ |
| 3 | 54 | .25 |
| **Public, by contractors.** | | |
| 5 | a 77 | a .20 |
| **Private, by contractors or firms.** | | |
| 2 | 56 | .18 |
| 4 | 56 | .19 |
| 24 | 56 | .20 |
| **Cement finishers:** | | |
| **Public, by contractors.** | | |
| 3 | 54 | .30 |
| 16 | 54 | .44½ |
| **Private, by contractors or firms.** | | |
| 11 | 54 | .44½ |
| **Cement finishers' helpers:** | | |
| **Public, by contractors.** | | |
| 16 | 54 | .22½ |
| 3 | 54 | .28 |
| **Private, by contractors or firms.** | | |
| 11 | 54 | .22½ |
| **Coal handlers:** | | |
| **Public, by city or State.** | | |
| 3 | a 63 | a .22½ |
| 2 | a 63 | a .22½ |
| 3 | a 63 | a .23½ |
| **Private, by contractors or firms.** | | |
| 6 | a 70 | a .17½ |
| 3 | a 70 | a .17½ |
| 4 | a 70 | a .18 |
| **Cooks:** | | |
| **Public, by city or State.** | | |
| 2 | a 63 | a .18½ |
| **Public, by contractors.** | | |
| 2 | a 75 | a .21½ |
| **Private, by contractors or firms.** | | |
| 1 | a 75 | a .18½ |
| **Draftsmen:** | | |
| **Public, by city or State.** | | |
| 1 | 54 | [illegible] |
| 2 | 54 | [illegible] |
| 3 | 54 | [illegible] |
| 1 | [illegible] | [illegible] |
| 1 | [illegible] | [illegible] |
| 1 | [illegible] | [illegible] |
| 3 | [illegible] | [illegible] |

| Number of employees. | Hours of work per week. | Wages per hour. |
|---|---|---|
| **Drivers with double team:** | | |
| Public, by city or State. | | |
| 1 | 54 | $0.50 |
| Public, by contractors. | | |
| 10 | 60 | .40 |
| 50 | 60 | .45 |
| 51 | 54 | .55½ |
| **Drivers with single team:** | | |
| Public, by city or State. | | |
| 1 | 54 | .33⅓ |
| Public, by contractors. | | |
| 1 | 60 | .30 |
| 28 | 54 | .33⅓ |
| **Dump men:** | | |
| Public, by city or State. | | |
| 9 | 54 | .27¾ |
| Public, by contractors. | | |
| 4 | a 77 | a .30 |
| **Engineers, hoisting:** | | |
| Public, by city or State. | | |
| 2 | 50 | .34½ |
| 3 | 48 | .38 |
| Public, by contractors. | | |
| 17 | 54 | .22¼ |
| 14 | 60 | .25 |
| 4 | a 77 | a .25 |
| 3 | 60 | .27½ |
| 2 | a 77 | a .30 |
| 5 | 54 | .30½ |
| 4 | 48 | .34⅜ |
| Private, by contractors or firms. | | |
| 6 | 48 | .34½ |
| **Engineers, stationary:** | | |
| Public, by city or State. | | |
| 2 | a 70 | a .21½ |
| 1 | a 70 | a .22 |
| 1 | a 63 | a .22⅔ |
| 6 | a 70 | a .24⅝ |
| 1 | a 63 | a .27⅞ |
| 3 | 54 | .27⅞ |
| 8 | a 63 | a .27⅞ |
| 6 | a 70 | a .28⅝ |
| 16 | a 63 | a .30⅝ |
| 17 | 54 | .33⅓ |
| 1 | a 63 | a .33⅓ |
| 1 | a 63 | a .33⅓ |
| 4 | 48 | .35 |
| 2 | 50 | .36 |
| 4 | a 63 | a .36½ |
| 2 | 48 | .38 |
| 1 | 54 | .42⅝ |
| 2 | a 63 | a .45⅝ |
| 1 | 63 | .76 |
| Public, by contractors. | | |
| 1 | 60 | .17½ |
| 2 | a 77 | a .22 |
| 7 | 60 | .25 |
| 4 | a 77 | a .25 |
| 8 | a 84 | a .27½ |
| Private, by contractors or firms. | | |
| 1 | a 70 | a .20 |
| 14 | a 70 | a .22⅞ |
| 1 | a 70 | a .25 |
| **Engineers, stationary—Conc'd.** | | |
| Private, by contractors or firms—Conc'd. | | |
| 12 | a 70 | a $0.28⅝ |
| 1 | a 70 | a .28½ |
| 7 | a 70 | a .30 |
| 4 | a 70 | a .35⅝ |
| 1 | a 70 | a .38½ |
| 1 | a 70 | a .40 |
| **Firemen:** | | |
| Public, by city or State. | | |
| 8 | a 70 | a .16½ |
| 1 | a 63 | a .22⅔ |
| 8 | a 63 | a .22⅔ |
| 2 | 48 | .25 |
| 17 | a 63 | a .25 |
| 1 | a 63 | a .27⅞ |
| 8 | 48 | 29 |
| Public, by contractors. | | |
| 3 | 60 | .18⅓ |
| 1 | a 70 | a .21½ |
| Private, by contractors or firms. | | |
| 1 | a 70 | a .18½ |
| 16 | a 70 | a .20 |
| 38 | a 70 | a .22⅞ |
| **Foremen, blacksmiths:** | | |
| Public, by contractors. | | |
| 1 | 60 | .32½ |
| Private, by contractors or firms. | | |
| 2 | 56 | .32 |
| 1 | 60 | .33⅓ |
| 1 | 56 | .35⅝ |
| 1 | 56 | .40 |
| **Foremen, bricklayers:** | | |
| Public, by contractors. | | |
| 1 | 60 | .50 |
| 15 | 48 | .52½ |
| 5 | 48 | .62½ |
| Private, by contractors or firms. | | |
| 3 | 48 | .45 |
| 5 | 48 | .53⅛ |
| 1 | 48 | .55 |
| **Foremen, carpenters:** | | |
| Public, by city or State. | | |
| 1 | 54 | 44½ |
| Public, by contractors. | | |
| 1 | a 77 | a .30 |
| 4 | 54 | .32⅛ |
| 2 | 53 | .39½ |
| Private, by contractors or firms. | | |
| 1 | 56 | .30 |
| 2 | 56 | .30⅛ |
| 6 | 54 | .23⅛ |
| 1 | 56 | .34 |
| 3 | 53 | .39½ |
| **Foremen, dredge runners:** | | |
| Public, by contractors. | | |
| 1 | 60 | .38½ |
| 1 | 60 | .50 |
| Private, by contractors or firms. | | |
| 1 | 60 | 58⅓ |
| **Foremen, laborers:** | | |
| Public, by city or State. | | |
| 3 | 48 | $0.25 |
| 4 | 54 | .25 |
| 4 | 54 | .27⅞ |
| 2 | 54 | .33⅓ |
| 2 | 54 | .36 |
| 1 | 66 | .40 |
| 3 | 50 | .42 |
| 1 | 66 | .45 |
| 2 | 50 | .60 |
| Public, by contractors. | | |
| 2 | 60 | .20 |
| 2 | 60 | .22½ |
| 2 | 60 | .22⅞ |
| 6 | 60 | .25 |
| 20 | 60 | .30 |
| 2 | 54 | .33⅓ |
| 1 | 60 | .33⅓ |
| 3 | a 77 | a .36⅛ |
| 6 | a 84 | a .38⅞ |
| 17 | 60 | .40 |
| 5 | 50 | .40⅝ |
| 10 | a 84 | a .41⅞ |
| 3 | 60 | .50 |
| 2 | a 77 | a .54½ |
| Private, by contractors or firms. | | |
| 1 | a 70 | a .16½ |
| 1 | a 70 | a .22⅞ |
| 1 | 54 | .27 |
| 4 | 60 | .30 |
| **Foremen, machinists:** | | |
| Public, by city or State. | | |
| 1 | 48 | .41 |
| Public, by contractors. | | |
| 1 | 60 | .32½ |
| Private, by contractors or firms. | | |
| 4 | 50 | .33 |
| 1 | 60 | .33⅓ |
| **Foremen, masons:** | | |
| Public, by city or State. | | |
| 3 | 48 | .62½ |
| Private, by contractors or firms. | | |
| 1 | 56 | .42⅞ |
| **Foremen, painters:** | | |
| Public, by contractors. | | |
| 12 | 48 | .32 |
| 3 | 48 | .35 |
| 4 | 48 | .37½ |
| Private, by contractors or firms. | | |
| 1 | 56 | .30½ |
| 25 | 48 | .32 |
| 6 | 45 | .35 |
| 10 | 48 | .35 |
| 8 | 48 | .37½ |
| **Foremen, pavers:** | | |
| Public, by contractors. | | |
| 3 | 54 | .44½ |
| 6 | 54 | .55⅝ |
| Private, by contractors or firms. | | |
| 1 | 54 | .44½ |

a Work 7 days per week.

TABLE I.—HOURS OF WORK AND WAGES O

1. ON PUBLIC WORK, EMPLOYED DIRECTLY BY T
2. ON PUBLIC WORK, EMPLOYED BY CONTRACTO
3. ON PRIVATE WORK, EMPLOYED BY CONTRACT

BOSTON—Continued.

| Number of employees. | Hours of work per week. | Wages per hour. |
|---|---|---|
| **Foremen, track men:** | | |
| Public, by contractors. | | |
| 1 | 54 | $0.50 |
| 1 | 54 | .55½ |
| 1 | 54 | .61 |
| Private, by contractors or firms. | | |
| 2 | 54 | .50 |
| 1 | 54 | .55½ |
| 1 | 54 | .61 |
| **Gatemen:** | | |
| Public, by city or State. | | |
| 5 | a 63 | a .22¾ |
| 4 | 54 | .23½ |
| Private, by contractors or firms. | | |
| 1 | a 70 | a .17⅛ |
| 2 | a 70 | a .20 |
| 2 | a 70 | a .21⅜ |
| **Glaziers:** | | |
| Public, by contractors. | | |
| 3 | 54 | .27¾ |
| Private, by contractors or firms. | | |
| 7 | 54 | .27¾ |
| **Harness makers:** | | |
| Public, by city or State. | | |
| 1 | 54 | .27¾ |
| 1 | 54 | .30⅝ |
| 1 | 54 | .33⅓ |
| 1 | 54 | .44⅜ |
| Public, by contractors. | | |
| 1 | 54 | .25 |
| Private, by contractors or firms. | | |
| 1 | 56 | .24 |
| 4 | 56 | .26¾ |
| **Hod carriers:** | | |
| Public, by contractors. | | |
| 72 | 48 | .25 |
| Private, by contractors or firms. | | |
| 7 | 60 | .23½ |
| 43 | 48 | .25 |
| **Horse feeders:** | | |
| Public, by city or State. | | |
| 2 | a 63 | a .25 |
| Private, by contractors or firms. | | |
| 6 | a 70 | a .15 |
| **Horseshoers:** | | |
| Public, by city or State. | | |
| 1 | 54 | .27¾ |
| 1 | 54 | .33⅓ |
| 2 | 54 | .30⅝ |
| 1 | 54 | .36 |
| Public, by contractors. | | |
| 1 | 58 | .[illegible] |
| 1 | 58 | .31[illegible] |
| **Hostlers:** | | |
| Public, by city or State. | | |
| 4 | a 63 | a .[illegible] |
| 4 | a 63 | a .[illegible] |
| **Hostlers—Conc'd.** | | |
| Public, by city or State—Conc'd. | | |
| 1 | 48 | $0.25 |
| 10 | a 63 | a .25 |
| 3 | a 63 | a .27¾ |
| Public, by contractors. | | |
| 1 | a 70 | a .14⅜ |
| Private, by contractors or firms. | | |
| 51 | a 63 | a .16⅝ |
| 1 | a 63 | a .18⅜ |
| **Laborers:** | | |
| Public, by city or State. | | |
| 62 | 48 | .19 |
| 75 | 54 | .19⅛ |
| 19 | 48 | .20 |
| 12 | 50 | .20 |
| 68 | 54 | .20 |
| 105 | 66 | .20 |
| 3 | 48 | .22 |
| 1,145 | 54 | .22½ |
| 1 | a 63 | a .22½ |
| 239 | 54 | .22⅝ |
| 4 | 50 | .22 |
| 58 | 54 | .23⅛ |
| 81 | 50 | .24 |
| 36 | 48 | .25 |
| 3 | 54 | .25 |
| 4 | 54 | .27¾ |
| 2 | 48 | .29 |
| 1 | 48 | .32 |
| 2 | 48 | .35 |
| 2 | 48 | .39 |
| Public, by contractors. | | |
| 1 | 60 | .12½ |
| 485 | 60 | .12⅝ |
| 597 | 60 | .14 |
| 2,500 | 60 | .15 |
| 217 | a 77 | a .15 |
| 120 | 59 | .15⅛ |
| 20 | 60 | .16 |
| 114 | 54 | .16⅝ |
| 106 | 60 | .16⅝ |
| 96 | a 84 | a .16⅝ |
| 698 | 60 | .17⅛ |
| 30 | a 77 | a .17⅛ |
| 50 | 59 | .17¾ |
| 60 | 50 | .18 |
| 331 | 54 | .19⅛ |
| 30 | 60 | .20 |
| 8 | a 77 | a .20 |
| 88 | a 84 | a .20 |
| 1 | 60 | .21⅛ |
| 10 | 54 | .22⅛ |
| 1 | 60 | .24 |
| 2 | 60 | .25 |
| Private, by contractors or firms. | | |
| 43 | a 70 | a .15 |
| 61 | 56 | .16 |
| 90 | 54 | .16[illegible] |
| 702 | 60 | .17[illegible] |
| 16 | 58 | .1[illegible] |
| 115 | 54 | .[illegible] |
| 12 | 56 | .[illegible] |
| **Laborers, building:** | | |
| Public, by contractors. | | |
| 10 | [illegible] | [illegible] |
| 6 | [illegible] | [illegible] |
| 154 | [illegible] | [illegible] |

| Number of employees. | Hours of work per week. | Wages per hour. |
|---|---|---|
| **Masons' helpers:** | | |
| Public, by city or State. | | |
| 9 | 48 | $0.26¼ |
| 1 | 54 | .30¼ |
| 4 | 60 | .33⅓ |
| 1 | 54 | .36 |
| Public, by contractors. | | |
| 43 | 48 | .25 |
| 10 | 48 | .30 |
| 8 | a77 | a.31 |
| Private, by contractors or firms. | | |
| 136 | 48 | .25 |
| **Miners:** | | |
| Public, by city or State. | | |
| 6 | 60 | .35 |
| 3 | 60 | .25 |
| Public, by contractors. | | |
| 4 | 60 | .22½ |
| 20 | a84 | a.23 |
| **Oilers:** | | |
| Public, by city or State. | | |
| 1 | a63 | a.22¼ |
| 7 | a63 | a.22¾ |
| 3 | a63 | a.25 |
| 2 | a63 | a.27½ |
| Private, by contractors or firms. | | |
| 19 | a70 | a.20 |
| 4 | a70 | a.21¼ |
| 7 | a70 | a.22¾ |
| **Painters:** | | |
| Public, by city or State. | | |
| 1 | 54 | .21¼ |
| 1 | 54 | .22¾ |
| 5 | 54 | .26 |
| 8 | 54 | .27½ |
| 1 | 48 | .29 |
| 2 | 48 | .32 |
| 2 | 54 | .33⅓ |
| 3 | 48 | .35 |
| 1 | 54 | .41⅔ |
| Public, by contractors. | | |
| 121 | 48 | .30 |
| 12 | 48 | .31¼ |
| Private, by contractors or firms. | | |
| 9 | 56 | .21 |
| 19 | 56 | .22 |
| 1 | 56 | .23 |
| 30 | 56 | .24 |
| 2 | 56 | .25 |
| 224 | 48 | .30 |
| 18 | 48 | .31¼ |
| **Painters, decorative:** | | |
| Public, by contractors. | | |
| 6 | 48 | .35 |
| Private, by contractors or firms. | | |
| 2 | 48 | .35 |
| **Pattern makers:** | | |
| Public, by city or State. | | |
| 1 | 48 | .41 |
| Private, by contractors or firms. | | |
| 1 | 56 | .30 |
| 11 | 56 | .33 |
| **Pavers:** | | |
| Public, by contractors. | | |
| 282 | 54 | $0.44½ |
| 10 | 54 | .50 |
| Private, by contractors or firms. | | |
| 56 | 60 | .37½ |
| 20 | 54 | .44½ |
| 5 | 54 | .50 |
| **Pipe layers:** | | |
| Public, by city or State. | | |
| 2 | 54 | .24½ |
| 4 | 54 | .25 |
| 2 | 54 | .26¾ |
| 7 | 54 | .27¾ |
| 2 | 54 | .28¾ |
| Public, by contractors. | | |
| 32 | 60 | .20 |
| 4 | 54 | .22¼ |
| 90 | 60 | .22¾ |
| 56 | 59 | .23 |
| Private, by contractors or firms. | | |
| 4 | 54 | 27½ |
| **Plasterers:** | | |
| Public, by contractors. | | |
| 63 | 47 | .43 |
| 17 | 48 | .50 |
| Private, by contractors or firms. | | |
| 90 | 47 | .43 |
| 10 | 48 | .50 |
| **Plasterers' helpers:** | | |
| Public, by contractors. | | |
| 46 | 47 | .32 |
| Private, by contractors or firms. | | |
| 52 | 47 | .32 |
| **Plumbers:** | | |
| Public, by city or State. | | |
| 12 | 54 | .33⅓ |
| 1 | 54 | .36 |
| 1 | 48 | .41 |
| 1 | 48 | .44 |
| Public, by contractors. | | |
| 156 | 48 | .47 |
| 17 | 48 | .56¼ |
| Private, by contractors or firms. | | |
| 1 | 56 | .28⅛ |
| 3 | 56 | .34 |
| 102 | 48 | .47 |
| 4 | 48 | .56¼ |
| **Plumbers' helpers:** | | |
| Public, by contractors. | | |
| 135 | 48 | .12½ |
| Private, by contractors or firms. | | |
| 90 | 48 | .12½ |
| 7 | 48 | .15⅝ |
| **Rammers, paving:** | | |
| Public, by contractors. | | |
| 191 | 54 | .28¾ |
| **Rammers, paving—Cont'd.** | | |
| Private, by contractors or firms. | | |
| 55 | 60 | $0.20 |
| 8 | 54 | .22¼ |
| 8 | 54 | .25 |
| **Riggers:** | | |
| Public, by city or State. | | |
| 4 | 54 | .22¼ |
| 4 | 54 | .25 |
| 3 | 54 | .27½ |
| 2 | 48 | .29 |
| 5 | 48 | .32 |
| 1 | 48 | .35 |
| 2 | 48 | .36 |
| Public, by contractors. | | |
| 2 | a84 | a.31¼ |
| Private, by contractors or firms. | | |
| 1 | 60 | .25 |
| **Rock men:** | | |
| Public, by city or State. | | |
| 74 | 54 | .22½ |
| Public, by contractors. | | |
| 9 | 60 | .16⅔ |
| 55 | 60 | .17½ |
| 25 | 54 | .19⅛ |
| 44 | 60 | .20 |
| 167 | 54 | .22¼ |
| Private, by contractors or firms. | | |
| 5 | 60 | .17½ |
| 30 | 60 | .20 |
| **Spikers:** | | |
| Public, by contractors. | | |
| 19 | 54 | .25 |
| Private, by contractors or firms. | | |
| 14 | 60 | .22½ |
| 25 | 54 | .25 |
| **Sprinklers, street:** | | |
| Public, by city or State. | | |
| 7 | 54 | .22¼ |
| 20 | 54 | .23⅛ |
| Public, by contractors. | | |
| 11 | 60 | .16⅔ |
| 7 | 54 | .18⅛ |
| Private, by contractors or firms. | | |
| 15 | 54 | .19⅛ |
| **Stonecutters:** | | |
| Public, by city or State. | | |
| 2 | 54 | .27¾ |
| 17 | 54 | .33⅓ |
| 2 | 48 | .38 |
| 4 | 50 | .38 |
| Public, by contractors. | | |
| 1 | 60 | .35 |
| 1 | 60 | .44 |
| Private, by contractors or firms. | | |
| 3 | 60 | .27½ |

a Work 7 days per week.

TABLE I.—HOURS OF WORK AND [illegible]

1. ON PUBLIC WORK, EMPLOYED [illegible]
2. ON PUBLIC WORK, EMPLOYED BY [illegible]
3. ON PRIVATE WORK, EMPLOYED BY [illegible]

BOSTON—Concluded.

| Number of employees. | Hours of work per week. | Wages per hour. |
|---|---|---|
| **Sweepers, street:** | | |
| Public, by city or State. | | |
| 101 | 54 | $0.22½ |
| 51 | 54 | .22½ |
| 107 | 54 | .22½ |
| Private, by contractors or firms. | | |
| 45 | 60 | .17½ |
| **Timekeepers:** | | |
| Public, by city or State. | | |
| 2 | 54 | .27¾ |
| 1 | 54 | .33⅓ |
| Public, by contractors. | | |
| 2 | 60 | .15 |
| 3 | 54 | .27¾ |
| 1 | 60 | .33⅓ |
| Private, by contractors or firms. | | |
| 4 | 54 | .27¾ |
| **Tinsmiths:** | | |
| Public, by city or State. | | |
| 1 | 54 | .33⅓ |
| Private, by contractors or firms. | | |
| 1 | 60 | .20 |
| 1 | 60 | .22 |
| 2 | 56 | .24 |
| 2 | 60 | .24 |
| 1 | 56 | .29 |
| **Track men:** | | |
| Public, by contractors. | | |
| 19 | 54 | .27¾ |
| **Track men—Conc'd.** | | |
| Private, by contractors or firms. | | |
| 66 | 60 | $0.22½ |
| 25 | 54 | .27¾ |
| **Watchmen:** | | |
| Public, by city or State. | | |
| 1 | a 70 | a .20 |
| 5 | a 70 | a .21 |
| 1 | a 70 | a .22½ |
| 3 | a 63 | a .22½ |
| 4 | 48 | .28 |
| 2 | 54 | .25 |
| 3 | a 63 | a .25 |
| 4 | a 70 | a .25 |
| Public, by contractors. | | |
| 2 | a 84 | a .14½ |
| 3 | a 77 | a .15½ |
| 2 | a 84 | a .16½ |
| Private, by contractors or firms. | | |
| 1 | a 70 | a .14½ |
| 2 | a 70 | a .15 |
| 1 | a 70 | a .17½ |
| 6 | a 70 | a .20 |
| 8 | a 70 | a .22½ |
| **Water boys:** | | |
| Public, by city or State. | | |
| 3 | 54 | .08⅓ |
| 3 | 66 | .12 |
| **Water bo[illegible]** | | |
| Public, [illegible] | | |
| 6 | [illegible] | [illegible] |
| 6 | [illegible] | [illegible] |
| 6 | [illegible] | [illegible] |
| Private, [illegible] | | |
| 4 | [illegible] | [illegible] |
| 2 | [illegible] | [illegible] |
| **Wheelwright[illegible]** | | |
| Public, by [illegible] | | |
| 2 | [illegible] | [illegible] |
| 2 | [illegible] | [illegible] |
| 1 | [illegible] | [illegible] |
| Public, by con[illegible] | | |
| 2 | [illegible] | [illegible] |
| 1 | [illegible] | [illegible] |
| 1 | [illegible] | [illegible] |
| **Wire men:** | | |
| Public, by cit[illegible] | | |
| 2 | [illegible] | [illegible] |
| 1 | [illegible] | [illegible] |
| 4 | [illegible] | [illegible] |
| Private, by con[illegible] | | |
| 1 | [illegible] | [illegible] |
| 2 | [illegible] | [illegible] |
| 2 | [illegible] | [illegible] |
| 1 | [illegible] | [illegible] |
| 1 | [illegible] | [illegible] |
| 4 | [illegible] | [illegible] |

NEW YORK CITY.

| Number of employees. | Hours of work per week. | Wages per hour. |
|---|---|---|
| **Ash and garbage lifters:** | | |
| Public, by city or State | | |
| 430 | 48 | .28⅛ |
| Public, by contractors. | | |
| 250 | 66½ | .15⅝ |
| **Ash tenders:** | | |
| Public, by city or State. | | |
| 10 | 48 | .33 |
| Private, by contractors or firms. | | |
| 23 | a 84 | a .14½ |
| 8 | a 70 | a .15 |
| 1 | a 70 | a .16 |
| 1 | a 70 | a .17½ |
| 1 | a 70 | a .19 |
| **Asphalt layers:** | | |
| Public, by contractors. | | |
| 31 | 60 | .20 |
| 5 | 60 | .22½ |
| 8 | 60 | .25 |
| Private, by contractors or firms. | | |
| 5 | 60 | .20 |
| 13 | 60 | .22½ |
| 28 | 60 | .25 |
| 15 | 48 | .28⅛ |
| **Asphalt mixers:** | | |
| Public, by contractors. | | |
| 1 | 60 | .22½ |
| Private, by contractors or firms. | | |
| 6 | 60 | .22½ |
| 3 | 60 | .24 |
| 5 | 48 | .25 |
| 20 | 60 | .25 |
| **Axmen:** | | |
| Public, by city or State. | | |
| 2 | 48 | .24 |
| 25 | 48 | .28⅛ |
| 56 | 48 | .31¼ |
| 3 | 48 | .36 |
| 1 | 48 | .37½ |
| 2 | 48 | .43¾ |
| Private, by contractors or firms. | | |
| 4 | 60 | .26 |
| 10 | 54 | .27¾ |
| **Blacksmiths:** | | |
| Public, by city or State. | | |
| 1 | 48 | .31 |
| 5 | (b) | .35 |
| 11 | 48 | .36 |
| 4 | [illegible] | .37 |
| 16 | [illegible] | [illegible] |
| 1 | (b) | [illegible] |
| 3 | [illegible] | [illegible] |
| 4 | [illegible] | [illegible] |
| 4 | [illegible] | [illegible] |
| **Blacksmiths—Co[illegible]** | | |
| Public, by con[illegible] | | |
| 1 | [illegible] | [illegible] |
| 11 | [illegible] | [illegible] |
| 1 | [illegible] | [illegible] |
| 15 | [illegible] | [illegible] |
| 6 | [illegible] | [illegible] |
| Private, by con[illegible] | | |
| 28 | [illegible] | [illegible] |
| 2 | [illegible] | [illegible] |
| 1 | [illegible] | [illegible] |
| 26 | [illegible] | [illegible] |
| 19 | [illegible] | [illegible] |
| 6 | [illegible] | [illegible] |
| 15 | [illegible] | [illegible] |
| 11 | [illegible] | [illegible] |
| 1 | [illegible] | [illegible] |
| 6 | [illegible] | [illegible] |
| 6 | [illegible] | [illegible] |
| 2 | [illegible] | [illegible] |
| 2 | [illegible] | [illegible] |
| 15 | [illegible] | [illegible] |
| 12 | [illegible] | [illegible] |
| 4 | [illegible] | [illegible] |
| **Blacksmiths' hel[illegible]** | | |
| Public, by city [illegible] | | |
| [illegible] | [illegible] | [illegible] |

## I. ON PRIVATE WORK, EMPLOYED BY CONTRACTORS OR FIRMS—Cont'd.

### NEW YORK CITY—Continued.

| Number of employees. | Hours of work per week. | Wages per hour. |
|---|---|---|
| **Blacksmiths' helpers—Conc'd.** | | |
| Public, by contractors. | | |
| 1 | 60 | $0.32½ |
| 14 | 48 | .25 |
| 3 | 48 | .28½ |
| Private, by contractors or firms. | | |
| 21 | 55 | .17½ |
| 10 | 60 | .17½ |
| 2 | 55 | .19 |
| 4 | 54 | .19½ |
| 8 | 60 | .20 |
| 1 | 55 | .21½ |
| 4 | 54 | .25 |
| 15 | 48 | .28½ |
| 10 | 48 | .31¼ |
| **Boiler makers:** | | |
| Public, by city or State. | | |
| 5 | 48 | .33 |
| 3 | 48 | .37½ |
| 14 | 48 | .38 |
| 15 | 48 | .41 |
| Private, by contractors or firms. | | |
| 3 | 55 | .24½ |
| 1 | 60 | .25 |
| 11 | 60 | .27½ |
| 9 | 55 | .28½ |
| 5 | 55 | .30 |
| 12 | 54 | .31 |
| 1 | 55 | .42 |
| **Boiler makers' helpers:** | | |
| Public, by city or State. | | |
| 1 | 48 | .24 |
| 15 | 48 | .27 |
| Private, by contractors or firms. | | |
| 10 | 55 | .17½ |
| 1 | 60 | .17½ |
| 3 | 55 | .19 |
| 10 | 60 | .20 |
| **Boiler setters:** | | |
| Public, by contractors. | | |
| 7 | 48 | .50 |
| Private, by contractors or firms. | | |
| 10 | 48 | .50 |
| **Boiler setters' helpers:** | | |
| Public, by contractors. | | |
| 10 | 48 | .28½ |
| Private, by contractors or firms. | | |
| 40 | 48 | .25 |
| **Bracers:** | | |
| Public, by contractors. | | |
| 16 | 59 | .20 |
| 25 | 60 | .20 |
| Private, by contractors or firms. | | |
| 20 | 29 | .20½ |
| **Brass finishers:** | | |
| Public, by contractors. | | |
| 6 | 60 | .30 |
| Private, by contractors or firms. | | |
| 6 | 60 | .30 |
| **Bricklayers:** | | |
| Public, by city or State. | | |
| 52 | 48 | .50 |
| **Bricklayers—Conc'd.** | | |
| Public, by contractors. | | |
| 12 | 48 | $0.40 |
| 2 | 48 | .43¾ |
| 127 | 48 | .50 |
| Private, by contractors or firms. | | |
| 5 | 60 | .40 |
| 126 | 48 | .50 |
| **Calkers, iron:** | | |
| Public, by city or State. | | |
| 11 | 48 | .31 |
| 19 | 48 | .31¼ |
| 10 | 48 | 35 |
| Public, by contractors. | | |
| 12 | 59 | .20½ |
| Private, by contractors or firms. | | |
| 35 | 60 | .20 |
| 25 | 59 | .20½ |
| **Calkers, wood:** | | |
| Public, by city or State. | | |
| 1 | 48 | .36 |
| 1 | (a) | .39 |
| 22 | 48 | .41 |
| Public, by contractors. | | |
| 20 | 60 | .22½ |
| Private, by contractors or firms. | | |
| 237 | 54 | .36 |
| **Car cleaners:** | | |
| Public, by city or State. | | |
| 45 | 48 | .26½ |
| Private, by contractors or firms. | | |
| 12 | 60 | .15 |
| 289 | b 70 | b .15 |
| 6 | b 70 | b .17½ |
| 2 | b 70 | b .19 |
| **Car couplers:** | | |
| Public, by city or State. | | |
| 14 | 48 | .26½ |
| Private, by contractors or firms. | | |
| 63 | b 84 | b .12½ |
| 7 | b 70 | b .15 |
| **Carpenters:** | | |
| Public, by city or State. | | |
| 5 | 48 | .37½ |
| 2 | 48 | .38 |
| 4 | 48 | .41 |
| 47 | 48 | .43¾ |
| 28 | 48 | .47 |
| Public, by contractors. | | |
| 17 | 59 | 25 |
| 8 | 48 | .30 |
| 24 | 48 | .37½ |
| 154 | 48 | .43¾ |
| Private, by contractors or firms. | | |
| 7 | 60 | .17½ |
| 3 | b 70 | b .22½ |
| 1 | b 70 | b .24 |
| 8 | 55 | .24½ |
| 6 | 60 | .25 |
| 1 | b 70 | b .25 |
| 6 | [illegible] | .25½ |
| 9 | [illegible] | .26½ |
| **Carpenters—Conc'd.** | | |
| Private, by contractors or firms—Conc'd. | | |
| 34 | 55 | $0.27½ |
| 12 | b 70 | b .27½ |
| 7 | 55 | [illegible] |
| 1 | b 70 | b .[illegible] |
| 1 | 55 | .30 |
| 14 | 60 | .30 |
| 16 | b 70 | b .30 |
| 1 | b 70 | b .34½ |
| 1 | 55 | [illegible] |
| 1 | b 70 | b .[illegible] |
| 197 | 48 | .43¾ |
| 1 | b 70 | b .[illegible] |
| 1 | b 70 | b .48½ |
| **Carpenters' helpers:** | | |
| Public, by city or State. | | |
| 1 | 48 | .25 |
| 1 | 48 | .28½ |
| 3 | 48 | .31¼ |
| 3 | 48 | .33 |
| Public, by contractors. | | |
| 8 | 48 | .25 |
| 20 | 60 | .20 |
| Private, by contractors or firms. | | |
| 1 | 60 | .15 |
| 1 | 55 | .19 |
| 2 | b 70 | b .20 |
| 1 | b 70 | b .22½ |
| 25 | 48 | .30 |
| **Cement mixers:** | | |
| Public, by city or State. | | |
| 1 | 48 | .28½ |
| Public, by contractors. | | |
| 3 | 60 | .16⅔ |
| 30 | 60 | .17½ |
| Private, by contractors or firms. | | |
| 50 | 60 | .17½ |
| 4 | 60 | .20 |
| 10 | 48 | .25 |
| **Cleaners:** | | |
| Public, by city or State. | | |
| 49 | 48 | .25 |
| 3 | 48 | .31¼ |
| Private, by contractors or firms. | | |
| 42 | b 84 | b .20¼ |
| **Conductors:** | | |
| Public, by city or State. | | |
| 140 | 48 | .34½ |
| Private, by contractors or firms. | | |
| 22 | b 70 | b .17½ |
| 20 | b 70 | b .19 |
| 1,075 | b 70 | b .20 |
| 27 | b 70 | b .22 |
| 310 | b 70 | b .23 |
| **Coppersmiths:** | | |
| Public, by city or State. | | |
| 2 | 48 | .37½ |
| 8 | 48 | .38 |
| 16 | 48 | .41 |
| Private, by contractors or firms. | | |
| 6 | 60 | .25 |
| 6 | 60 | .27½ |
| 2 | 54 | .30¼ |

a Not reported.

b Work 7 days per week.

TABLE I.—HOURS OF WORK AND W[illegible]

1. ON PUBLIC WORK, EMPLOYED DIRECTLY BY T[illegible]
2. ON PUBLIC WORK, EMPLOYED BY CONTRACTO[illegible]
3. ON PRIVATE WORK, EMPLOYED BY CONTRACT[illegible]

NEW YORK CITY—Continue[illegible]

| Number of employees. | Hours of work per week. | Wages per hour. |
|---|---|---|
| **Coppersmiths' helpers:** | | |
| Public, by city or State. | | |
| 3 | 48 | $0.25 |
| 16 | 48 | .25 |
| Private, by contractors or firms. | | |
| 3 | 54 | .27½ |
| **Curbstone setters:** | | |
| Public, by contractors. | | |
| 29 | 60 | .30 |
| 4 | 54 | .50 |
| Private, by contractors or firms. | | |
| 7 | 54 | .38½ |
| **Curbstone setters' helpers:** | | |
| Public, by contractors. | | |
| 19 | 60 | .15 |
| Private, by contractors or firms. | | |
| 3 | 54 | .19[illegible] |
| **Derrick men:** | | |
| Public, by contractors. | | |
| 6 | 48 | .34½ |
| 7 | 48 | .37½ |
| Private, by contractors or firms. | | |
| 10 | 48 | .34½ |
| **Derrick men's helpers:** | | |
| Public, by contractors. | | |
| 16 | 48 | .31¼ |
| Private, by contractors or firms. | | |
| 20 | 48 | .31¼ |
| **Divers:** | | |
| Public, by city or State. | | |
| 1 | 48 | 1.20 |
| 6 | 24 | 1.25 |
| Public, by contractors. | | |
| 10 | 60 | .50 |
| 25 | (a) | 1.00 |
| Private, by contractors or firms. | | |
| 8 | (a) | 1.00 |
| **Divers' tenders:** | | |
| Public, by city or State. | | |
| 2 | (a) | .35 |
| Public, by contractors. | | |
| 30 | 60 | .25 |
| Private, by contractors or firms. | | |
| 3 | 50 | .25 |
| **Dock builders:** | | |
| Public, by city or State. | | |
| 147 | (a) | .30 |
| 12 | (a) | .30 |
| Public, by contractors. | | |
| 86 | 60 | .20 |
| 745 | 60 | .22½ |
| 186 | 60 | .25 |
| 11 | 48 | .30 |
| 22 | 60 | .30 |
| **Dock builders—Conc'd.** | | |
| Private, by contractors or firms. | | |
| 360 | 60 | $0.22½ |
| 780 | 60 | .25 |
| 20 | 56 | .28½ |
| **Draftsmen:** | | |
| Public, by city or State. | | |
| 1 | 48 | .31¼ |
| 1 | 48 | .37½ |
| 5 | 48 | .38 |
| 4 | 48 | .41 |
| 10 | 48 | .50 |
| 1 | 48 | .56½ |
| 2 | 48 | .62½ |
| 7 | 48 | .63 |
| Public, by contractors. | | |
| 2 | 54 | .27½ |
| 3 | 54 | .33½ |
| Private, by contractors or firms. | | |
| 3 | 60 | .30 |
| **Drillers:** | | |
| Public, by city or State. | | |
| 2 | 48 | .24 |
| 49 | 48 | .28 |
| Public, by contractors. | | |
| 28 | 50 | .26½ |
| Private, by contractors or firms. | | |
| 39 | 50 | .25½ |
| **Drivers:** | | |
| Public, by city or State. | | |
| 222 | 48 | .24 |
| 28 | 48 | .25 |
| 867 | 48 | .28½ |
| 14 | 48 | .31 |
| 1 | 48 | .31¼ |
| 3 | b 56 | b .31¼ |
| 1 | 48 | .36 |
| Public, by contractors. | | |
| [illegible] | 66½ | .15½ |
| 28 | 60 | .17½ |
| 877 | 60 | .20 |
| 3 | 54 | .22½ |
| Private, by contractors or firms. | | |
| 28 | b 70 | b .17½ |
| 6 | 54 | .19½ |
| 65 | 60 | .20 |
| 725 | b 70 | b .20 |
| 20 | 54 | .22½ |
| 30 | 60 | .25 |
| 1 | 60 | .30 |
| **Drivers with double team:** | | |
| Public, by city or State. | | |
| 2 | 48 | [illegible] |
| 1 | 48 | [illegible] |
| 8 | 56 | [illegible] |
| Public, by contractors. | | |
| 28 | 60 | .[illegible] |
| **Drivers with [illegible]:** | | |
| Public, [illegible] | | |
| 6 | [illegible] | [illegible] |
| 1 | [illegible] | [illegible] |
| 1 | [illegible] | [illegible] |
| 20 | [illegible] | [illegible] |
| 20 | [illegible] | [illegible] |

a Not reported.

| | | |
|---|---|---|
| **[illegible]—Cont'd.** | | |
| Public, by contractors. | | |
| 20 | 59 | $0.25 |
| 22 | 60 | .25 |
| 25 | 60 | .27½ |
| 1 | 53 | .34 |
| 2 | 60 | .35 |
| Private, by contractors or firms. | | |
| 2 | a 84 | a .16⅔ |
| 3 | a 84 | a .18¾ |
| 1 | a 84 | a .20 |
| 7 | a 84 | a .[illegible] |
| 31 | a 84 | a .22½ |
| 6 | a 70 | a .25 |
| 2 | a 84 | a .25 |
| 26 | 59 | .25½ |
| 12 | a 84 | a .27 |
| 52 | 60 | .27½ |
| 1 | 55 | .29½ |
| 1 | a 70 | a .30 |
| 1 | 60 | .33⅓ |
| 2 | a 70 | a .32 |
| 2 | a 56 | a .37½ |
| 1 | 54 | .39 |
| 2 | a 70 | a .41 |
| 1 | a 70 | a .49½ |
| **Engineers, steam roller:** | | |
| Public, by city or State. | | |
| 3 | 48 | 37½ |
| Public, by contractors. | | |
| 12 | 60 | .30 |
| 2 | 60 | .35 |
| **Engine hostlers:** | | |
| Public, by city or State. | | |
| 5 | 48 | .50 |
| Private, by contractors or firms. | | |
| 5 | a 70 | a .16⅔ |
| 2 | a 84 | a .19⅝ |
| 2 | a 70 | a .20 |
| 50 | a 84 | a .20⅝ |
| 1 | a 84 | a .22 |
| 8 | a 84 | a .25 |
| 9 | a 84 | a .29½ |
| **Engine wipers:** | | |
| Public, by city or State. | | |
| 10 | 48 | .26½ |
| Private, by contractors or firms. | | |
| 14 | a 84 | a .12½ |
| 42 | a 70 | a .15 |
| 5 | 56 | .16⅔ |
| 1 | a 70 | a .17⅞ |
| **Firemen:** | | |
| Public, by city or State. | | |
| 1 | 48 | .25 |
| 11 | 48 | .28 |
| 16 | 50 | .30 |
| 12 | 48 | .31¼ |
| 8 | 48 | .33⅓ |
| Public, by contractors. | | |
| 4 | 60 | 17½ |
| **Firemen—Cont'd.** | | |
| Private, by contractors or firms. | | |
| 28 | a 84 | a $0.16⅔ |
| 8 | a 70 | a .17⅞ |
| 11 | a 63 | a .17½ |
| 7 | a 84 | a .18¾ |
| 2 | a 63 | a .19⅝ |
| 4 | a 70 | a .20 |
| 12 | a 70 | a .22½ |
| **Firemen, locomotive:** | | |
| Public, by city or State. | | |
| 8 | 48 | .28 |
| 18 | 48 | .29 |
| Private, by contractors or firms. | | |
| 52 | a 70 | a .17½ |
| 476 | a 63 | a .22⅝ |
| **Foremen, blacksmiths:** | | |
| Public, by city or State. | | |
| 1 | 48 | .43¾ |
| Private, by contractors or firms. | | |
| 7 | 60 | .30 |
| 2 | 60 | .32 |
| 2 | 55 | .37½ |
| **Foremen, elevator constructors:** | | |
| Public, by contractors. | | |
| 15 | 48 | .43¾ |
| 1 | 48 | .47 |
| 1 | 48 | .50 |
| Private, by contractors or firms. | | |
| 11 | 48 | .43¾ |
| **Foremen, iron workers, assistant:** | | |
| Public, by contractors. | | |
| 6 | 48 | .37½ |
| Private, by contractors or firms. | | |
| 6 | 48 | .37½ |
| **Foremen, laborers:** | | |
| Public, by city or State. | | |
| 5 | (b) | .35 |
| Public, by contractors. | | |
| 41 | 60 | .30 |
| **Foremen, linemen:** | | |
| Public, by city or State. | | |
| 1 | 48 | .40 |
| Public, by contractors. | | |
| 2 | 48 | 43¾ |
| **Foremen, machinists:** | | |
| Public, by city or State. | | |
| 1 | 48 | 62½ |
| Public, by contractors. | | |
| 1 | 53 | .30¼ |
| Private, by contractors or firms. | | |
| 9 | 60 | .35 |
| 1 | 60 | .40⅝ |
| **Foremen, steam fitters:** | | |
| Public, by contractors. | | |
| 4 | 48 | .47 |
| 4 | 48 | .50 |
| **Foremen, steam fitters—Cont'd.** | | |
| Private, by contractors or firms. | | |
| 4 | 48 | $0.47 |
| 6 | 48 | .50 |
| **Foremen, sweepers, street:** | | |
| Public, by city or State. | | |
| 3 | 48 | .28 |
| 2 | 48 | .31¼ |
| 56 | 48 | .40 |
| Public, by contractors. | | |
| 60 | 72 | .20⅝ |
| **Galvanized and sheet iron workers:** | | |
| Public, by city or State. | | |
| 2 | 48 | .31 |
| 4 | 48 | .41 |
| 21 | 48 | .44 |
| Public, by contractors. | | |
| 56 | 48 | .43¾ |
| Private, by contractors or firms. | | |
| 8 | 60 | .25 |
| 8 | 55 | .32⅔ |
| 1 | 55 | .37½ |
| 55 | 48 | .43¾ |
| **Galvanized and sheet iron workers' helpers:** | | |
| Public, by contractors. | | |
| 1 | 48 | .18¾ |
| 1 | 48 | .22 |
| 2 | 48 | .25 |
| Private, by contractors or firms. | | |
| 3 | 60 | .15 |
| 7 | 55 | .24½ |
| 5 | 48 | .25 |
| **Gardeners:** | | |
| Public, by city or State. | | |
| 50 | 48 | .25 |
| Private, by contractors or firms. | | |
| 225 | 60 | .16⅔ |
| **Gatemen, street railway:** | | |
| Public, by city or State. | | |
| 34 | 48 | .31¼ |
| Private, by contractors or firms. | | |
| 132 | a 84 | a .10⅛ |
| 67 | a 70 | a .12⅞ |
| 409 | a 84 | a .12½ |
| 71 | a 84 | a .13⅞ |
| 6 | a 84 | a .14⅞ |
| **Harness makers:** | | |
| Public, by city or State. | | |
| 1 | 48 | .31¼ |
| 2 | 48 | .33 |
| 13 | 48 | .37½ |
| Private, by contractors or firms. | | |
| 12 | 60 | .22½ |
| **Hod carriers:** | | |
| Public, by city or State. | | |
| 1 | 48 | .34 |

a Work 7 days per week.

b Not reported.

TABLE I.—HOURS OF WORK AND

1. ON PUBLIC WORK, EMPLOYED DIRE
2. ON PUBLIC WORK, EMPLOYED BY CO
3. ON PRIVATE WORK, EMPLOYED BY C

NEW YORK CITY

| Number of employees. | Hours of work per week. | Wages per hour. |
|---|---|---|
| **Hod carriers—Conc'd.** | | |
| Public, by contractors. | | |
| 5 | 48 | $0.30 |
| 4 | 46 | .25 |
| 150 | 48 | .30 |
| 45 | 48 | .31¼ |
| Private, by contractors or firms. | | |
| 111 | 48 | .30 |
| **Holders-on:** | | |
| Public, by city or State. | | |
| 7 | 48 | .28 |
| Public, by contractors. | | |
| 25 | 48 | .39⅛ |
| Private, by contractors or firms. | | |
| 12 | 54 | .24⅜ |
| 65 | 48 | .28⅛ |
| 35 | 48 | .34⅜ |
| **Hostlers:** | | |
| Public, by city or State. | | |
| 1 | a 70 | a .19⅝ |
| 19 | 48 | .24 |
| 86 | a 56 | a .24⅜ |
| 18 | a 56 | a .25 |
| Public, by contractors. | | |
| 12 | a 84 | a .13⅜ |
| 12 | a 84 | a .14⅜ |
| 3 | a 70 | a .20 |
| Private, by contractors or firms. | | |
| 491 | a 70 | a .17⅛ |
| 20 | 60 | .22½ |
| **Iron workers, structural:** | | |
| Public, by contractors. | | |
| 46 | 48 | .28⅛ |
| 25 | 48 | .30 |
| 7 | 60 | .30 |
| 25 | 48 | .31¼ |
| 10 | 48 | .34⅜ |
| 47 | 48 | .37½ |
| 24 | 48 | .43¾ |
| Private, by contractors or firms. | | |
| 43 | 48 | .28⅛ |
| 40 | 48 | .30 |
| 10 | 60 | .30 |
| 45 | 48 | .31¼ |
| 25 | 48 | .34⅜ |
| 45 | 48 | .37½ |
| 35 | 48 | .43¾ |
| 6 | 48 | .47 |
| **Iron workers', structural, helpers:** | | |
| Public, by contractors. | | |
| 55 | 48 | .25 |
| 25 | 60 | .25 |
| Private, by contractors or firms. | | |
| 56 | 48 | .20 |
| 5 | 48 | .25 |
| 25 | 60 | .25 |
| **Joiners:** | | |
| Public, by city or State. | | |
| 3 | 48 | .[illegible] |
| 13 | [illegible] | [illegible] |

| Number of employees. | Hours of work per week. |
|---|---|
| **Joiners—Conc'd.** | |
| Public, by contractors | |
| 3 | 60 |
| 25 | 60 |
| 10 | 60 |
| Private, by contractors | |
| 2 | 54 |
| 10 | 54 |
| **Laborers:** | |
| Public, by city or State | |
| 1 | 48 |
| 6 | 48 |
| 261 | 48 |
| 256 | (b) |
| 1,175 | 48 |
| 114 | 48 |
| 19 | 48 |
| 4 | 48 |
| 78 | 48 |
| 11 | 48 |
| Public, by contractors | |
| 54 | 60 |
| 1 | 53 |
| 95 | 48 |
| 2,394 | 60 |
| 588 | 59 |
| 21 | 48 |
| 1 | 53 |
| 1,956 | 60 |
| 7 | 48 |
| 35 | 54 |
| 56 | 48 |
| 22 | 48 |
| Private, by contractors | |
| 10 | a 84 |
| 280 | a 84 |
| 74 | a 84 |
| 199 | 60 |
| 497 | a 70 |
| 828 | 59 |
| 75 | 60 |
| 25 | 55 |
| 25 | 60 |
| 43 | a 84 |
| 107 | a 70 |
| 656 | 60 |
| 372 | a 70 |
| 24 | 55 |
| 309 | 54 |
| 48 | 48 |
| 15 | 54 |
| 1 | a 70 |
| 165 | 48 |
| 5 | a 70 |
| 1 | a 70 |
| **Laborers, building:** | |
| Public, by city or State | |
| 1 | 48 |
| 6 | 48 |
| Public, by contractors | |
| 12 | [illegible] |
| 35 | [illegible] |
| [illegible] | [illegible] |
| [illegible] | [illegible] |

| | | |
|---|---|---|
| Private, by contractors or firms. | | |
| 7 | 60 | $0.17½ |
| 2 | a 70 | a .17½ |
| [illegible] | 55 | .19 |
| [illegible] | 60 | .20 |
| Masons: | | |
| Public, by city or State. | | |
| 4 | 48 | .40½ |
| 45 | 48 | .50 |
| 8 | (b) | .50 |
| Public, by contractors. | | |
| 19 | 48 | .40 |
| 120 | 48 | .50 |
| Private, by contractors or firms. | | |
| 5 | 60 | .25 |
| 20 | 60 | .35 |
| 130 | 48 | .50 |
| Masons, stone: | | |
| Public, by city or State. | | |
| 1 | 48 | .50 |
| Public, by contractors. | | |
| 4 | 48 | .40 |
| 126 | 48 | .50 |
| Private, by contractors or firms. | | |
| 12 | 48 | .50 |
| Masons' helpers: | | |
| Public, by city or State. | | |
| 4 | 48 | .29½ |
| Private, by contractors or firms. | | |
| 7 | 66 | .15 |
| 14 | 60 | .20 |
| 10 | 60 | .25 |
| Molders: | | |
| Public, by city or State. | | |
| 9 | 48 | .38 |
| 11 | 48 | .41 |
| 6 | 48 | .44 |
| Private, by contractors or firms. | | |
| 1 | 55 | .33⅓ |
| Oakum spinners: | | |
| Public, by city or State. | | |
| 3 | 48 | .25 |
| Private, by contractors or firms. | | |
| 1 | 54 | .22½ |
| Painters: | | |
| Public, by city or State. | | |
| 6 | 48 | .36 |
| 11 | 48 | .41 |
| 22 | 48 | .42[illegible] |
| 48 | 48 | .44 |
| Public, by contractors. | | |
| 36 | 54 | .37½ |
| 12 | 48 | .30 |
| 4 | 48 | .31¼ |
| 20 | 55 | [illegible] |
| 25 | 55 | .37½ |
| 8 | 55 | .42[illegible] |
| Private, by contractors or firms. | | |
| 11 | 60 | $0.20 |
| 9 | 60 | .30 |
| 4 | 54 | .40 |
| 244 | 48 | .43[illegible] |
| Painters, sign: | | |
| Public, by city or State. | | |
| 3 | 48 | .31¼ |
| 3 | 48 | .35 |
| 4 | 48 | .37½ |
| 3 | 48 | .41 |
| Private, by contractors or firms. | | |
| 1 | 60 | .17½ |
| 1 | 60 | .20 |
| 9 | 55 | .21[illegible] |
| 38 | 60 | .22[illegible] |
| 40 | 55 | .24[illegible] |
| 10 | 55 | .27[illegible] |
| 1 | 55 | [illegible] |
| 1 | 55 | .30 |
| Painters' helpers: | | |
| Public, by city or State. | | |
| 2 | 48 | .25 |
| 4 | 48 | .25½ |
| 3 | 48 | .30 |
| 3 | 48 | .34⅜ |
| Public, by contractors. | | |
| 6 | 48 | .25 |
| Private, by contractors or firms. | | |
| 1 | 55 | .08⅓ |
| 7 | 60 | .15 |
| 2 | 60 | .16 |
| 51 | 55 | .16⅔ |
| Pavers, Belgian block: | | |
| Public by city or State. | | |
| 47 | 48 | .50 |
| Public, by contractors. | | |
| 9 | 54 | .44½ |
| 143 | 54 | .50 |
| Private, by contractors or firms. | | |
| 14 | 60 | .22½ |
| 9 | 60 | .25 |
| 53 | 54 | .50 |
| Pipe fitters: | | |
| Public, by city or State. | | |
| 1 | 48 | .25 |
| 3 | 48 | .44 |
| Private, by contractors or firms. | | |
| 1 | 55 | .21⅞ |
| 1 | 60 | .24 |
| 2 | 55 | .24½ |
| 1 | 60 | .25 |
| 1 | 60 | .26 |
| 1 | 55 | .26¼ |
| 4 | 55 | .27½ |
| 1 | 60 | .27½ |
| 1 | 55 | .30 |
| 2 | 55 | .32½ |
| Public, by contractors. | | |
| 268 | 44 | $0.34⅜ |
| Private, by contractors or firms. | | |
| 207 | 44 | .34⅜ |
| Plasterers' helpers: | | |
| Public, by contractors. | | |
| 141 | 44 | .20 |
| 20 | 48 | .31¼ |
| Private, by contractors or firms. | | |
| 156 | 44 | .20 |
| Plumbers: | | |
| Public, by city or State. | | |
| 1 | 48 | .37½ |
| 3 | 48 | .41 |
| 1 | 48 | .43¾ |
| 14 | 48 | .44 |
| 7 | 48 | .50 |
| Public, by contractors. | | |
| 6 | 48 | .40 |
| 141 | 48 | .47 |
| Private, by contractors or firms. | | |
| 1 | 56 | .24 |
| 4 | 60 | .27½ |
| 2 | 58 | .29[illegible] |
| 6 | 60 | .30 |
| 1 | 55 | .31¼ |
| 1 | 55 | .33[illegible] |
| 5 | 58 | .35[illegible] |
| 1 | 55 | .42 |
| 120 | 48 | .47 |
| Plumbers' helpers: | | |
| Public, by city or State. | | |
| 2 | 48 | .16 |
| 1 | 48 | .22 |
| 3 | 48 | .31¼ |
| Public, by contractors. | | |
| 1 | 48 | .09⅜ |
| 1 | 48 | .10⅜ |
| 13 | 48 | .12½ |
| 61 | 48 | .15⅝ |
| Private, by contractors or firms. | | |
| 2 | 60 | .10 |
| 8 | 60 | .15 |
| 50 | 48 | .15⅝ |
| 3 | 56 | .18[illegible] |
| 8 | 56 | .24 |
| Rammers, Belgian block: | | |
| Public, by city or State. | | |
| 16 | 48 | .27½ |
| 3 | 48 | .35 |
| Public, by contractors. | | |
| 64 | 54 | .30 |
| Private, by contractors or firms. | | |
| 15 | 60 | .17½ |
| 22 | 54 | .30 |

a Work 7 days per week.

b Not reported.

TABLE I.—HOURS OF WORK AND [WAGES …]

1. ON PUBLIC WORK, EMPLOYED DIRE[CTLY …]
2. ON PUBLIC WORK, EMPLOYED BY C[ONTRACTORS …]
3. ON PRIVATE WORK, EMPLOYED BY [CONTRACTORS OR FIRMS …]

NEW YORK CIT[Y …]

| Number of employees. | Hours of work per week. | Wages per hour. |
|---|---|---|
| **Riggers:** | | |
| Public, by city or State. | | |
| 12 | 48 | $0.31 |
| 4 | 48 | .31¼ |
| 2 | 48 | .35 |
| 10 | 48 | .37½ |
| 2 | 48 | .38 |
| 2 | 48 | .43¾ |
| 3 | 48 | .44 |
| Public, by contractors. | | |
| 10 | 60 | .17½ |
| 15 | 59 | .20 |
| 39 | 60 | .20 |
| 20 | 48 | .25 |
| 29 | 48 | .31¼ |
| Private, by contractors or firms. | | |
| 24 | 60 | .20 |
| 15 | 59 | .20⅚ |
| 27 | 60 | .22½ |
| 20 | 60 | .25 |
| 66 | 48 | .31¼ |
| 18 | 48 | .37½ |
| 15 | 54 | .39 |
| **Riggers' helpers:** | | |
| Public, by contractors. | | |
| 6 | 60 | .17½ |
| 35 | 48 | .22 |
| Private, by contractors or firms. | | |
| 10 | 60 | .17½ |
| 26 | 48 | .22 |
| **Riveters:** | | |
| Public, by city or State. | | |
| 6 | 48 | .32 |
| 18 | 48 | .35 |
| 4 | 48 | .37½ |
| Public, by contractors. | | |
| 15 | 60 | .25 |
| 10 | 48 | .31¼ |
| 40 | 48 | .34⅜ |
| Private, by contractors or firms. | | |
| 1 | 55 | .13⅗ |
| 1 | 55 | .16⅓ |
| 7 | 55 | .19 |
| 30 | 55 | .21⅚ |
| 2 | 55 | .24½ |
| 25 | 60 | .25 |
| 2 | 55 | .27¼ |
| 2 | 55 | .28⅜ |
| 60 | 48 | .31¼ |
| 52 | 48 | .34⅜ |
| **Rivet heaters:** | | |
| Public, by contractors. | | |
| 5 | 60 | .15 |
| Private, by contractors or firms. | | |
| 4 | 54 | .16⅔ |
| **Rock men:** | | |
| Public, by city or State. | | |
| 6 | 48 | .28⅛ |
| Public, by contractors. | | |
| 60 | [illegible] | .17½ |
| 127 | [illegible] | [illegible] |
| [illegible] | [illegible] | [illegible] |

| Number of employees. | Hours of work per week. |
|---|---|
| **Rock men—Conc'd.** | |
| Private, by contract[ors or firms.] | |
| 190 | 59 |
| 10 | 60 |
| 20 | 59 |
| **Roofers, slate:** | |
| Public, by city or S[tate.] | |
| 1 | 48 |
| Private, by contract[ors or firms.] | |
| 8 | 48 |
| 2 | 48 |
| **Sawmill men:** | |
| Public, by city or S[tate.] | |
| 5 | 48 |
| Private, by contract[ors or firms.] | |
| 3 | 54 |
| **Shipsmiths:** | |
| Public, by city or S[tate.] | |
| 3 | 48 |
| 11 | 48 |
| Private, by contract[ors or firms.] | |
| 21 | 54 |
| **Shipsmiths' helpers:** | |
| Public, by city or S[tate.] | |
| 22 | 48 |
| Private, by contract[ors or firms.] | |
| 10 | 54 |
| **Shipwrights:** | |
| Public, by city or S[tate.] | |
| 1 | 48 |
| 30 | 48 |
| Private, by contract[ors or firms.] | |
| 186 | 54 |
| **Snow shovelers:** | |
| Public, by city or S[tate.] | |
| 2,300 | 48 |
| Public, by contract[ors.] | |
| 3,000 | 60 |
| **Spar makers:** | |
| Public, by city or S[tate.] | |
| 2 | 48 |
| Private, by contract[ors or firms.] | |
| 26 | 54 |
| **Steam and gas fitters:** | |
| Public, by city or S[tate.] | |
| 3 | 48 |
| 1 | 48 |
| 8 | 48 |
| Public, by contract[ors.] | |
| 12 | [illegible] |
| 54 | [illegible] |
| Private, [illegible] | |
| 6 | [illegible] |
| [illegible] | [illegible] |
| [illegible] | [illegible] |

| Occupation / class | No. | Hours | Rate |
|---|---|---|---|
| [illegible]: | | | |
| Public, by city or State. | 1 | 48 | $0.43[illegible] |
| Public, by contractors. | 10 | 60 | .30 |
| | 2 | 54 | .22½ |
| | 5 | 60 | .22½ |
| | 4 | 60 | .25 |
| Private, by contractors or firms. | 2 | 60 | .22½ |
| | 2 | a 70 | a .25 |
| | 7 | a 70 | a .30 |
| [illegible]: | | | |
| Public, by city or State. | 1 | 48 | .37½ |
| | 1 | 48 | .38 |
| Private, by contractors or firms. | 1 | 55 | .21⅞ |
| | 1 | 60 | .25 |
| | 2 | 55 | .26¼ |
| | 2 | 55 | .28⅛ |
| Track men: | | | |
| Public, by city or State. | 6 | 48 | .50 |
| Private, by contractors or firms. | 1 | a 70 | a .15 |
| | 27 | a 70 | a .17½ |
| | 194 | a 70 | a .20 |
| Track men—Cont'd. | | | |
| Private, by contractors or firms—Cont'd. | 14 | a 70 | a $0.22½ |
| | 2 | a 70 | a .23 |
| | 5 | a 70 | a .24⅜ |
| | 1 | a 70 | a .26 |
| | 1 | a 70 | a .28⅛ |
| Train dispatchers: | | | |
| Public, by city or State. | 6 | 48 | .42⅞ |
| Private, by contractors or firms. | 2 | a 84 | a .20½ |
| | 2 | a 84 | a .22 |
| | 23 | a 84 | a .24⅜ |
| Watchmen: | | | |
| Public, by city or State. | 3 | (b) | .16⅔ |
| | 5 | (b) | .18⅛ |
| | 5 | (b) | .21⅞ |
| | 5 | 48 | .26 |
| | 2 | 48 | .29 |
| | 1 | 48 | .30 |
| | 3 | 48 | .31 |
| Private, by contractors or firms. | 5 | a 84 | a .12½ |
| | 6 | a 84 | a .14⅜ |
| | 1 | a 70 | a .15 |
| Watchmen—Cont'd. | | | |
| Private, by contractors or firms—Cont'd. | 2 | a 84 | a $0.16⅔ |
| | 1 | a 70 | a .20 |
| | 1 | 56 | .21⅞ |
| Wire men: | | | |
| Public, by contractors. | 6 | a 56 | a .31¼ |
| | 49 | 48 | .37½ |
| Private, by contractors or firms. | 6 | a 56 | a .31¼ |
| | 20 | 48 | .37½ |
| Wire men's helpers: | | | |
| Public, by contractors. | 15 | 48 | .22 |
| | 15 | 48 | .25 |
| Private, by contractors or firms. | 10 | 48 | .26 |
| Woodworkers: | | | |
| Public, by city or State. | 1 | 48 | .37½ |
| Private, by contractors or firms. | 30 | 60 | .22½ |

## PHILADELPHIA.

| Occupation / class | No. | Hours | Rate |
|---|---|---|---|
| Asphalt layers: | | | |
| Public, by contractors. | 46 | 60 | .17½ |
| Private, by contractors or firms. | 14 | 60 | .17½ |
| Asphalt mixers: | | | |
| Public, by contractors. | 16 | 60 | .15 |
| Private, by contractors or firms. | 3 | 60 | .15 |
| Blacksmiths: | | | |
| Public, by city or State. | 2 | 54 | .27½ |
| | 24 | 60 | .30 |
| | 16 | 54 | .22½ |
| Public, by contractors. | 3 | 60 | .22½ |
| | 141 | 60 | .27⅞ |
| | 4 | 60 | .30 |
| | 2 | 54 | .30½ |
| Private, by contractors or firms. | 9 | 60 | .20 |
| | 12 | 60 | .20½ |
| | 1 | 60 | .21⅞ |
| | 37 | 60 | .22½ |
| | 4 | 60 | .25 |
| | [illegible] | 60 | .27½ |
| | 6 | 60 | .30 |
| | 2 | 54 | .30½ |
| Blacksmiths' helpers: | | | |
| Public, by city or State. | 2 | 54 | .18½ |
| | 16 | 60 | .22 |
| | 7 | 54 | .22½ |
| | 2 | 54 | .25 |
| Public, by contractors. | 2 | 60 | .15 |
| | 3 | 60 | .17½ |
| | 6 | 54 | .18½ |
| Private, by contractors or firms. | 12 | 60 | .14 |
| | 29 | 60 | .15 |
| | 4 | 60 | .17 |
| | 6 | 54 | .18½ |
| Boiler makers: | | | |
| Public, by contractors. | 147 | 60 | .25 |
| Private, by contractors or firms. | 147 | 60 | .25 |
| Bricklayers: | | | |
| Public, by city or State. | 18 | 50 | .46¼ |
| Public, by contractors. | 4 | 60 | .26 |
| | 15 | 50 | .43¾ |
| | 28 | 54 | .45 |
| | 120 | 50 | .46¼ |
| Bricklayers—Cont'd. | | | |
| Private, by contractors or firms. | 59 | 50 | .45 |
| | 145 | 50 | .46¼ |
| Bricklayers, sewer: | | | |
| Public, by contractors. | 91 | 54 | .50 |
| | 15 | 60 | .50 |
| | 84 | 54 | .55 |
| | 30 | 54 | .55½ |
| | 45 | 54 | .61 |
| Private, by contractors or firms. | 25 | 54 | .50 |
| Bricklayers' helpers: | | | |
| Public, by city or State. | 6 | 54 | .24¼ |
| Public, by contractors. | 9 | 50 | .18⅜ |
| Calkers, iron: | | | |
| Public, by city or State. | 30 | 54 | .26 |
| | 140 | 54 | .27⅞ |
| Public, by contractors. | 5 | 54 | .22½ |
| | 19 | 60 | .22½ |
| | 9 | 54 | .25 |
| Private, by contractors or firms. | 5 | 60 | .22½ |
| | 2 | 60 | .24 |
| | 3 | 60 | .26 |

a Work 7 days per week.

b Not reported.

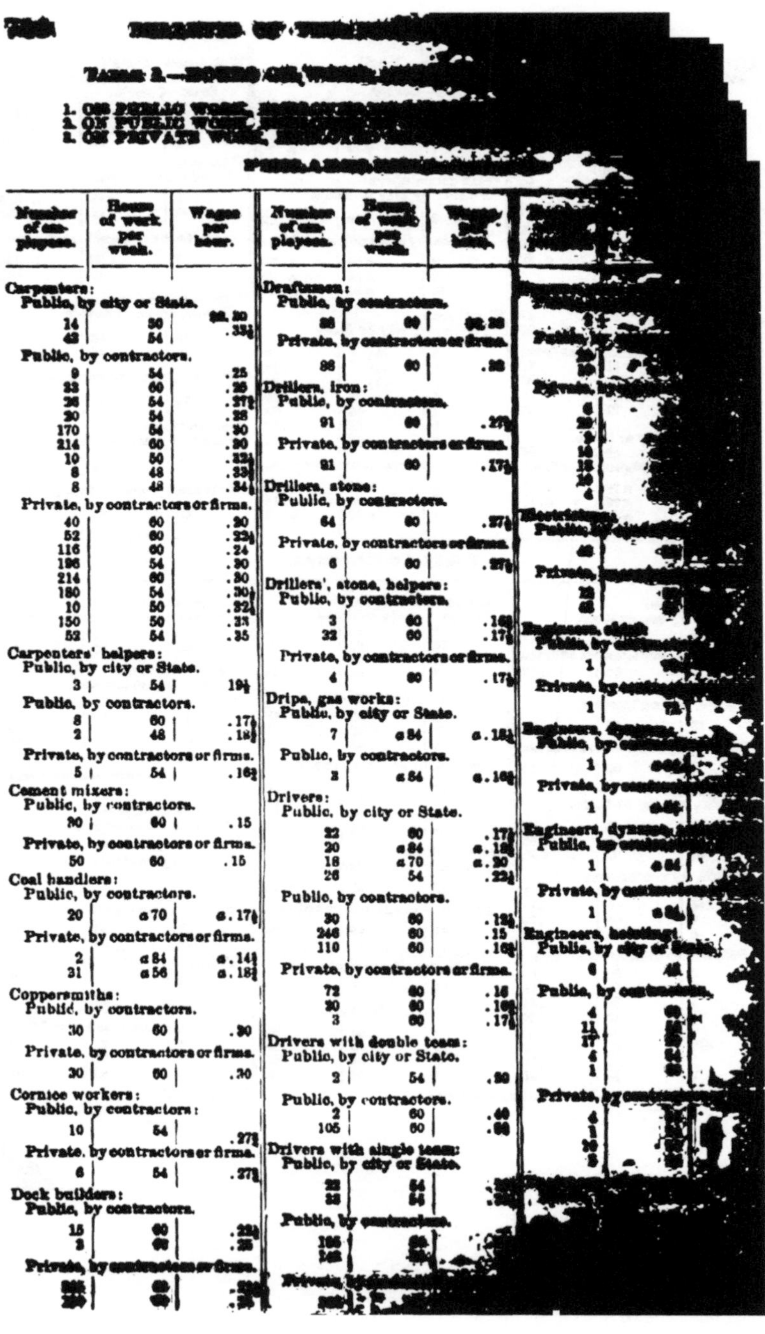



TABLE 2.—[illegible]

1. ON PUBLIC WORK, [illegible]
2. ON PUBLIC WORK, [illegible]
3. ON PRIVATE WORK, [illegible]

| Number of employees. | Hours of work per week. | Wages per hour. |
|---|---|---|
| **Carpenters:** | | |
| Public, by city or State. | | |
| 14 | 30 | $0.30 |
| 42 | 54 | .33⅓ |
| Public, by contractors. | | |
| 9 | 54 | .25 |
| 22 | 60 | .26 |
| 26 | 54 | .27⅞ |
| 20 | 54 | .28 |
| 170 | 54 | .30 |
| 214 | 60 | .30 |
| 10 | 50 | .32½ |
| 8 | 48 | .33⅓ |
| 8 | 48 | .34⅜ |
| Private, by contractors or firms. | | |
| 40 | 60 | .20 |
| 52 | 60 | .22½ |
| 116 | 60 | .24 |
| 196 | 54 | .30 |
| 214 | 60 | .30 |
| 180 | 54 | .30½ |
| 10 | 50 | .32½ |
| 150 | 50 | .33 |
| 52 | 54 | .35 |
| **Carpenters' helpers:** | | |
| Public, by city or State. | | |
| 3 | 54 | 19⅘ |
| Public, by contractors. | | |
| 8 | 60 | .17½ |
| 2 | 48 | .18⅜ |
| Private, by contractors or firms. | | |
| 5 | 54 | .16⅔ |
| **Cement mixers:** | | |
| Public, by contractors. | | |
| 20 | 60 | .15 |
| Private, by contractors or firms. | | |
| 50 | 60 | .15 |
| **Coal handlers:** | | |
| Public, by contractors. | | |
| 20 | a 70 | a .17⅛ |
| Private, by contractors or firms. | | |
| 2 | a 84 | a .14⅓ |
| 31 | a 56 | a .18⅜ |
| **Coppersmiths:** | | |
| Public, by contractors. | | |
| 30 | 60 | .30 |
| Private, by contractors or firms. | | |
| 30 | 60 | .30 |
| **Cornice workers:** | | |
| Public, by contractors. | | |
| 10 | 54 | .27⅞ |
| Private, by contractors or firms. | | |
| 6 | 54 | .27⅞ |
| **Dock builders:** | | |
| Public, by contractors. | | |
| 15 | 60 | .22½ |
| 3 | 60 | .25 |
| Private, by contractors or firms. | | |
| [illegible] | 60 | [illegible] |
| [illegible] | 60 | [illegible] |
| **Draftsmen:** | | |
| Public, by contractors. | | |
| 26 | 60 | $0.33 |
| Private, by contractors or firms. | | |
| 86 | 60 | .33 |
| **Drillers, iron:** | | |
| Public, by contractors. | | |
| 91 | 60 | .27½ |
| Private, by contractors or firms. | | |
| 81 | 60 | .17½ |
| **Drillers, stone:** | | |
| Public, by contractors. | | |
| 64 | 60 | .27½ |
| Private, by contractors or firms. | | |
| 6 | 60 | .27½ |
| **Drillers', stone, helpers:** | | |
| Public, by contractors. | | |
| 3 | 60 | .16⅔ |
| 32 | 60 | .17⅜ |
| Private, by contractors or firms. | | |
| 4 | 60 | .17½ |
| **Drips, gas works:** | | |
| Public, by city or State. | | |
| 7 | a 84 | a .18⅓ |
| Public, by contractors. | | |
| 2 | a 84 | a .16⅔ |
| **Drivers:** | | |
| Public, by city or State. | | |
| 22 | 60 | .17⅜ |
| 20 | a 84 | a .18⅜ |
| 18 | a 70 | a .20 |
| 26 | 54 | .22⅓ |
| Public, by contractors. | | |
| 30 | 60 | .12½ |
| 246 | 60 | .15 |
| 110 | 60 | .16⅔ |
| Private, by contractors or firms. | | |
| 72 | 60 | .16 |
| 20 | 60 | .16⅔ |
| 3 | 60 | .17⅜ |
| **Drivers with double team:** | | |
| Public, by city or State. | | |
| 2 | 54 | .30 |
| Public, by contractors. | | |
| 2 | 60 | .40 |
| 105 | 60 | .60 |
| **Drivers with single team:** | | |
| Public, by city or State. | | |
| 22 | 54 | [illegible] |
| 33 | 54 | [illegible] |
| Public, by contractors. | | |
| 166 | [illegible] | [illegible] |
| 142 | [illegible] | [illegible] |
| Private, by contractors or firms. | | |
| [illegible] | [illegible] | [illegible] |
| **[illegible]** | | |
| Public, [illegible] | | |
| 2 | [illegible] | [illegible] |
| Public, by [illegible] | | |
| 20 | [illegible] | [illegible] |
| 10 | [illegible] | [illegible] |
| Private, by [illegible] | | |
| 6 | [illegible] | [illegible] |
| 22 | [illegible] | [illegible] |
| 9 | [illegible] | [illegible] |
| 13 | [illegible] | [illegible] |
| 12 | [illegible] | [illegible] |
| 16 | [illegible] | [illegible] |
| 4 | [illegible] | [illegible] |
| **Electricians:** | | |
| Public, by [illegible] | | |
| 48 | [illegible] | [illegible] |
| Private, [illegible] | | |
| 22 | [illegible] | [illegible] |
| 48 | [illegible] | [illegible] |
| **Engineers, [illegible]:** | | |
| Public, by [illegible] | | |
| 1 | [illegible] | [illegible] |
| Private, by [illegible] | | |
| 1 | 72 | [illegible] |
| **Engineers, [illegible]:** | | |
| Public, by [illegible] | | |
| 1 | a 84 | [illegible] |
| Private, by [illegible] | | |
| 1 | a 84 | [illegible] |
| **Engineers, [illegible]:** | | |
| Public, by [illegible] | | |
| 1 | a 84 | [illegible] |
| Private, by [illegible] | | |
| 1 | a 84 | [illegible] |
| **Engineers, hoisting:** | | |
| Public, by city or State. | | |
| 6 | 48 | [illegible] |
| Public, by contractors. | | |
| 4 | 60 | [illegible] |
| 11 | [illegible] | [illegible] |
| 17 | [illegible] | [illegible] |
| 4 | [illegible] | [illegible] |
| 1 | [illegible] | [illegible] |
| Private, by [illegible] | | |
| 4 | [illegible] | [illegible] |
| 1 | [illegible] | [illegible] |
| 20 | [illegible] | [illegible] |
| 2 | [illegible] | [illegible] |

[illegible]—Cont'd.
[illegible]

| | |
|---|---|
| 60 | $0.25 |
| a70 | a.25 |
| a70 | a.30 |

[illegible] contractors or firms.

| | |
|---|---|
| 60 | .21[illegible] |
| 60 | .2[illegible] |
| 60 | .25 |
| a56 | a.27[illegible] |
| a56 | a.2[illegible] |
| a56 | a.31 |
| a56 | a.24[illegible] |
| a56 | a.35 |

[illegible]stationary, assistants:
[illegible]y contractors.

| | |
|---|---|
| a84 | a.16[illegible] |

[illegible]y contractors or firms.

| | |
|---|---|
| a84 | a.16[illegible] |

[illegible]steam roller:
[illegible]y contractors.

| | |
|---|---|
| 60 | .25 |
| 60 | .2[illegible] |

[illegible]y contractors or firms.

| | |
|---|---|
| 60 | .25 |

[illegible]shipbuilding:
[illegible]y contractors.

| | |
|---|---|
| 60 | .17½ |

[illegible]y contractors or firms.

| | |
|---|---|
| 60 | .17½ |

[illegible]y city or State.

| | |
|---|---|
| a84 | a.19½ |
| a84 | a.23 |
| 48 | .28½ |

[illegible]y contractors.

| | |
|---|---|
| a84 | a.16[illegible] |
| a70 | a.22½ |

[illegible]y contractors or firms.

| | |
|---|---|
| 60 | .15[illegible] |
| a84 | a.16[illegible] |
| a56 | a.25 |

[illegible]n setters), shipbuild-

[illegible]y contractors.

| | |
|---|---|
| 60 | .26 |

[illegible]y contractors or firms.

| | |
|---|---|
| 60 | .25 |

[illegible]lacksmiths:
[illegible]y city or State.

| | |
|---|---|
| 60 | .36 |
| 60 | .37½ |

[illegible]y contractors.

| | |
|---|---|
| 60 | .30 |

[illegible]y contractors or firms.

| | |
|---|---|
| 60 | .27½ |
| 60 | .30 |

Foremen, bricklayers:
Public, by city or State.

| | | |
|---|---|---|
| 4 | 54 | $0.20 |
| 1 | 50 | .60 |

Public, by contractors.

| | | |
|---|---|---|
| 1 | 60 | .45 |

Private, by contractors or firms.

| | | |
|---|---|---|
| 1 | 60 | .45 |
| 1 | 50 | .54 |
| 3 | 50 | .57 |
| 1 | 50 | .6[illegible] |
| 7 | 50 | .6[illegible] |

Foremen, carpenters:
Public, by city or State.

| | | |
|---|---|---|
| 1 | 54 | .35½ |
| 2 | 54 | .39 |
| 1 | 50 | .42 |

Public, by contractors.

| | | |
|---|---|---|
| 1 | 48 | .27½ |
| 3 | 54 | .30 |

Private, by contractors or firms.

| | | |
|---|---|---|
| 12 | 54 | .30 |

Foremen, dock builders:
Public, by contractors.

| | | |
|---|---|---|
| 4 | 60 | .30 |

Private, by contractors or firms.

| | | |
|---|---|---|
| 20 | 60 | .36 |

Foremen, iron workers, structural:
Public, by contractors.

| | | |
|---|---|---|
| 2 | 54 | .33½ |

Private, by contractors or firms.

| | | |
|---|---|---|
| 1 | 54 | .33½ |

Foremen, laborers:
Public, by city or State.

| | | |
|---|---|---|
| 1 | 54 | .19½ |
| 1 | 54 | .2[illegible] |
| 1 | 50 | .39 |

Public, by contractors.

| | | |
|---|---|---|
| 16 | 60 | .20 |
| 85 | 60 | .25 |
| 3 | 60 | .27½ |
| 25 | 60 | .30 |
| 5 | 60 | .35 |

Private, by contractors or firms.

| | | |
|---|---|---|
| 7 | 60 | .20 |
| 20 | 60 | .25 |
| 8 | 60 | .30 |

Foremen, linemen:
Public, by contractors.

| | | |
|---|---|---|
| 1 | 60 | .2[illegible] |

Private, by contractors or firms.

| | | |
|---|---|---|
| 1 | 60 | .2[illegible] |

Foremen, machinists:
Public, by city or State.

| | | |
|---|---|---|
| 1 | 54 | .42½ |
| 1 | 54 | .44[illegible] |
| 1 | 54 | .52[illegible] |

Foremen, machinists—Cont'd:
Private, by contractors or firms.

| | | |
|---|---|---|
| 14 | 60 | $0.30 |
| 3 | 60 | .35 |

Foremen, painters:
Public, by city or State.

| | | |
|---|---|---|
| 1 | 54 | .30 |
| 1 | 54 | .3[illegible] |
| 1 | 54 | .[illegible] |
| 1 | 50 | .[illegible] |

Public, by contractors.

| | | |
|---|---|---|
| 9 | 54 | .30 |

Private, by contractors or firms.

| | | |
|---|---|---|
| 12 | 54 | .30 |
| 3 | 54 | .44½ |

Foremen, plasterers:
Public, by contractors.

| | | |
|---|---|---|
| 6 | 48 | .50 |

Private, by contractors or firms.

| | | |
|---|---|---|
| 9 | 48 | .50 |

Foremen, purifiers, gas works:
Public, by city or State.

| | | |
|---|---|---|
| 5 | a70 | a.27½ |
| 1 | a70 | a.3[illegible] |

Public, by contractors.

| | | |
|---|---|---|
| 2 | 60 | .30 |

Foremen, stokers:
Public, by city or State.

| | | |
|---|---|---|
| 1 | a84 | a.27 |

Private, by contractors or firms.

| | | |
|---|---|---|
| 2 | a84 | a.[illegible] |

Galvanized and sheet iron workers:
Public, by contractors.

| | | |
|---|---|---|
| 150 | 60 | .22½ |
| 18 | 54 | .27[illegible] |
| 10 | 54 | .30 |

Private, by contractors or firms.

| | | |
|---|---|---|
| 150 | 60 | .22½ |
| 7 | 54 | .27[illegible] |
| 12 | 54 | .30 |
| 10 | 54 | .30½ |

Gardeners:
Public, by city or State.

| | | |
|---|---|---|
| 1 | 54 | .2[illegible] |
| 6 | 54 | .2[illegible] |
| 8 | 54 | .27[illegible] |
| 1 | 54 | .2[illegible] |
| 1 | 54 | .32 |

Private, by contractors or firms.

| | | |
|---|---|---|
| 5 | 60 | .15 |
| 4 | 60 | .1[illegible] |
| 3 | 60 | .20 |
| 1 | 60 | .22½ |
| 1 | 60 | .25 |

Gardeners' helpers:
Public, by city or State.

| | | |
|---|---|---|
| 1 | 54 | .2[illegible] |

a Work 7 days per week.

## Table I.—HOURS OF WORK AND WAGES OF EMPLOYEES—

1. ON PUBLIC WORK, EMPLOYED DIRECTLY BY THE CITY OR STATE.
2. ON PUBLIC WORK, EMPLOYED BY CONTRACTORS.
3. ON PRIVATE WORK, EMPLOYED BY CONTRACTORS OR FIRMS—Cont'd.

### PHILADELPHIA—Continued.

| Number of employees. | Hours of work per week. | Wages per hour. |
|---|---|---|
| **Gardeners' helpers—Conc'd.** | | |
| Private, by contractors or firms. | | |
| 6 | 60 | $0.12½ |
| 5 | 60 | .15 |
| **Granite cutters:** | | |
| Public, by contractors. | | |
| 30 | 54 | .30 |
| Private, by contractors or firms. | | |
| 36 | 54 | .30 |
| **Hod carriers:** | | |
| Public, by city or State. | | |
| 14 | 54 | .24 |
| 8 | 50 | .30 |
| Public by contractors. | | |
| 20 | 60 | .17½ |
| 12 | 50 | .27 |
| 7 | 54 | .30 |
| 40 | 48 | .31¼ |
| 41 | 50 | .32½ |
| Private, by contractors or firms. | | |
| 20 | 60 | .17½ |
| 6 | 50 | .24 |
| 16 | 50 | .27 |
| 40 | 50 | .30 |
| 45 | 45 | .31¼ |
| 40 | 50 | .32½ |
| **Inspectors, electric lighting:** | | |
| Public, by city or State. | | |
| 8 | 60 | .37½ |
| Public, by contractors. | | |
| 10 | a 70 | a .20 |
| 2 | a 70 | a .22½ |
| Private, by contractors or firms. | | |
| 10 | a 70 | a .20 |
| 2 | a 70 | a .22½ |
| **Iron workers:** | | |
| Public, by contractors | | |
| 50 | 60 | .20 |
| Private, by contractors or firms. | | |
| 50 | 60 | .20 |
| **Iron workers, structural:** | | |
| Public, by contractors. | | |
| 20 | 54 | .22½ |
| Private, by contractors or firms. | | |
| 7 | 54 | .22½ |
| **Joiners:** | | |
| Public, by contractors. | | |
| 68 | 60 | .27½ |
| Private, by contractors or firms. | | |
| 68 | 60 | .27½ |
| **Laborers:** | | |
| Public, by city or State. | | |
| 142 | 54 | .16⅔ |
| 166 | 60 | .17½ |
| 108 | a 70 | a .17½ |
| 763 | 54 | .19¼ |
| 2 | a 70 | a .20 |
| **Laborers—Conc'd.** | | |
| Public, by contractors. | | |
| 3,540 | 60 | $0.12½ |
| 65 | 60 | .13 |
| 75 | 60 | .13½ |
| 15 | 60 | .14 |
| 885 | 60 | .15 |
| 7 | 54 | .16⅔ |
| 30 | 60 | .17½ |
| 3 | 60 | .20 |
| 2 | 48 | .25 |
| Private, by contractors or firms. | | |
| 46 | 60 | .12 |
| 64 | 60 | .12½ |
| 312 | 60 | .13 |
| 487 | 60 | .13½ |
| 506 | 60 | .15 |
| 2 | 54 | .16⅔ |
| 4 | 54 | .19¼ |
| 7 | 60 | .20 |
| 9 | 54 | .22½ |
| **Laborers, building:** | | |
| Public, by city or State. | | |
| 111 | 50 | .21 |
| 9 | 50 | .24 |
| Public, by contractors. | | |
| 58 | 60 | .15 |
| 2 | 48 | .17½ |
| 20 | 50 | .18 |
| 25 | 54 | .19¼ |
| Private, by contractors or firms. | | |
| 35 | 60 | .15 |
| 85 | 54 | .16⅔ |
| 12 | 60 | .17½ |
| 13 | 50 | .18 |
| 8 | 54 | .19¼ |
| 22 | 50 | .21 |
| **Linemen:** | | |
| Public, by city or State. | | |
| 8 | 60 | .25 |
| 8 | 60 | .27½ |
| Public, by contractors. | | |
| 8 | 60 | .25 |
| Private, by contractors or firms. | | |
| 50 | 60 | .25 |
| **Machinists:** | | |
| Public, by city or State. | | |
| 2 | 54 | .27½ |
| 1 | 60 | .30 |
| 5 | 54 | .30½ |
| 1 | 60 | .32½ |
| 34 | 54 | .33⅓ |
| Public, by contractors. | | |
| 565 | 60 | .25 |
| 20 | 54 | .27½ |
| 4 | 60 | .30 |
| Private, by contractors or firms. | | |
| 301 | 60 | .20 |
| 5 | 60 | .22 |
| 41 | 60 | .22½ |
| 567 | 60 | .25 |
| 20 | 54 | .27½ |
| **Machinists' helpers:** | | |
| Public, by city or State. | | |
| 6 | 54 | $0.22 |
| 5 | 54 | .25 |
| 1 | 60 | .25 |
| Private, by contractors or firms. | | |
| 11 | 60 | .15 |
| 23 | 60 | .17½ |
| **Masons, stone:** | | |
| Public, by contractors. | | |
| 25 | 54 | .30 |
| 15 | 54 | .39 |
| 10 | 50 | .42 |
| Private, by contractors or firms. | | |
| 40 | 54 | .33⅓ |
| 35 | 50 | .45 |
| **Meter inspectors, gas works:** | | |
| Public, by city or State. | | |
| 46 | 48 | .43 |
| Private, by contractors or firms. | | |
| 4 | 60 | .22½ |
| **Millwrights:** | | |
| Public, by contractors. | | |
| 18 | 54 | .28 |
| Private, by contractors or firms. | | |
| 18 | 54 | .28 |
| **Molders:** | | |
| Public, by contractors. | | |
| 203 | 60 | .20 |
| Private, by contractors or firms. | | |
| 203 | 60 | .20 |
| 50 | 60 | .25 |
| 10 | 60 | .27 |
| **Oilers:** | | |
| Public, by city or State. | | |
| 30 | a 84 | a .18 |
| Private, by contractors or firms. | | |
| 8 | 48 | .22½ |
| **Painters:** | | |
| Public, by city or State. | | |
| 2 | 54 | .25 |
| 25 | 54 | .27 |
| 21 | 50 | .30 |
| 6 | 54 | .33 |
| Public, by contractors. | | |
| 5 | 54 | .28 |
| 35 | 48 | .30 |
| 77 | 54 | .30 |
| 100 | 60 | .30 |
| 20 | 54 | .30 |
| Private, by contractors or firms. | | |
| 419 | 54 | .30 |
| 100 | 60 | .30 |
| 11 | 54 | .30 |
| **Pattern makers:** | | |
| Public, by city or State. | | |
| 3 | 54 | .33 |

a Work 7 days per week.

| | | |
|---|---|---|
| contractors or firms. | | |
| | 60 | .34½ |
| | 60 | .35 |
| | 60 | .37 |
| | 60 | .39 |
| | 54 | .30½ |
| m block: | | |
| contractors. | | |
| | 60 | .35 |
| | 54 | .44½ |
| contractors or firms. | | |
| | 60 | .25 |
| | 54 | .50 |
| lestone: | | |
| contractors. | | |
| | 60 | .35 |
| | 54 | .44½ |
| contractors or firms. | | |
| | 60 | .35 |
| | 54 | .50 |
| contractors. | | |
| | 60 | .22½ |
| contractors or firms. | | |
| | 60 | .22½ |
| | 60 | .25 |
| | 60 | .27½ |
| contractors. | | |
| | 54 | .30½ |
| | 48 | .40 |
| contractors or firms. | | |
| | 54 | .39 |
| | 48 | .40 |
| city or State. | | |
| | 54 | .30½ |
| | 54 | .33⅓ |
| | 54 | .43 |
| contractors. | | |
| | 54 | .33⅓ |
| | 48 | .37½ |
| | 48 | .43¾ |
| contractors or firms. | | |
| | 54 | .27¾ |
| | 54 | .33⅓ |
| | 54 | .39 |
| lpers: | | |
| city or State. | | |
| | 54 | .25 |
| | 54 | .27¾ |
| contractors. | | |
| | 54 | .16⅔ |
| city or State. | | |

| | | | |
|---|---|---|---|
| Public, by city or State. | | | |
| | 69 | a 70 | a .25 |
| Public, by contractors. | | | |
| | 22 | 60 | .17½ |
| Private, by contractors or firms. | | | |
| | 1 | a 70 | a .19 |
| Rammers, Belgian block and cobblestone: | | | |
| Public, by contractors. | | | |
| | 10 | 60 | .15 |
| | 10 | 54 | .30½ |
| Private, by contractors or firms. | | | |
| | 17 | 60 | .20 |
| | 30 | 54 | .33⅓ |
| Riggers: | | | |
| Public, by contractors. | | | |
| | 67 | 60 | .18½ |
| | 4 | 50 | .24 |
| | 2 | 50 | .30 |
| | 3 | 54 | .33⅓ |
| Private, by contractors or firms. | | | |
| | 67 | 60 | .18½ |
| | 10 | 54 | .27¾ |
| | 25 | 50 | .30 |
| | 3 | 54 | .33⅓ |
| | 4 | 50 | .48 |
| Riveters: | | | |
| Public, by contractors. | | | |
| | 190 | 60 | .20 |
| Private, by contractors or firms. | | | |
| | 190 | 60 | .20 |
| Steam and gas fitters: | | | |
| Public, by city or State. | | | |
| | 23 | 54 | .27¾ |
| | 4 | 50 | .36 |
| Public, by contractors. | | | |
| | 274 | 60 | .25 |
| | 5 | 60 | .30 |
| | 2 | 48 | .31¼ |
| | 29 | 54 | .33⅓ |
| | 4 | 48 | .37½ |
| Private, by contractors or firms | | | |
| | 8 | 60 | .20 |
| | 1 | 60 | .22½ |
| | 274 | 60 | .25 |
| | 3 | 60 | .26⅔ |
| | 12 | 60 | .27½ |
| | 5 | 54 | .27¾ |
| | 3 | 54 | .30½ |
| | 80 | 54 | .33⅓ |
| Steam and gas fitters' helpers: | | | |
| Public, by city or State. | | | |
| | 14 | 54 | .22¼ |
| | 15 | 54 | .24 |
| | 1 | 54 | .24½ |

| | | | |
|---|---|---|---|
| | 6 | 54 | .16⅔ |
| | 1 | 60 | .19½ |
| Stokers: | | | |
| Public, by city or State. | | | |
| | 212 | a 84 | a .22 |
| Private, by contractors or firms. | | | |
| | 8 | a 84 | a .19 |
| Stonecutters: | | | |
| Public, by city or State. | | | |
| | 5 | 54 | .33⅓ |
| | 3 | 50 | .42 |
| Public, by contractors. | | | |
| | 25 | 54 | .30 |
| | 50 | 50 | .42 |
| | 14 | 50 | .43 |
| Private, by contractors or firms. | | | |
| | 10 | 54 | .36 |
| | 44 | 50 | .43 |
| Stonecutters' helpers: | | | |
| Public, by city or State. | | | |
| | 3 | 50 | .24 |
| Public, by contractors. | | | |
| | 50 | 50 | .22 |
| Stone setters: | | | |
| Public, by contractors. | | | |
| | 6 | 50 | .48⅔ |
| Private, by contractors or firms. | | | |
| | 25 | 50 | .46 |
| | 2 | 50 | .48⅔ |
| Trimmers, electric lighting: | | | |
| Public, by city or State. | | | |
| | 10 | 60 | .27½ |
| Public, by contractors. | | | |
| | 35 | a 70 | a .20 |
| Private, by contractors or firms. | | | |
| | 35 | a 70 | a .20 |
| Watchmen: | | | |
| Public, by city or State. | | | |
| | 12 | a 84 | a .16⅔ |
| | 2 | a 84 | a .18 |
| | 2 | a 84 | a .20⅔ |
| Public, by contractor | | | |
| | 3 | a 84 | a .16⅔ |
| | 4 | a 70 | a .17½ |
| | 32 | 60 | .16 |
| Private, by contractors or firms. | | | |
| | 15 | 60 | .12 |
| | 32 | 60 | .18 |
| Weighers, gas works: | | | |
| Public, by city or State. | | | |
| | 4 | a 70 | a .22 |

TABLE II.—SUMMARY OF WAGES PE[illegible]

1. ON PUBLIC WORK, [illegible]
2. ON PUBLIC WORK, [illegible]
3. ON PRIVATE WORK, [illegible]

[illegible]

| Occupations. | Public, by city or State. | | | Pu[illegible] |
|---|---|---|---|---|
| | Highest. | Lowest. | Average. | Hi[illegible] |
| Asphalt layers | | | | $0[illegible] |
| Asphalt mixers | | | | |
| Blacksmiths | $0.[illegible] | $0.[illegible] | $0.25 | |
| Blacksmiths' helpers | .22½ | [illegible] | [illegible] | |
| Boiler makers | | | | |
| Bricklayers | [illegible] | [illegible] | [illegible] | |
| Carpenters | [illegible] | .27⅞ | [illegible] | |
| Cement finishers | | | | |
| Cement mixers | | | | |
| Chandelier fitters | | | | |
| Chandelier makers | | | | |
| Coal handlers | | | | |
| Cornice workers | | | | |
| Curbstone setters | | | | |
| Curbstone setters' helpers | | | | |
| Draftsmen | | | | |
| Drivers | | | | |
| Drivers with single team | .33⅓ | .25 | .27½ | |
| Dynamo men | | | | |
| Dynamo men's helpers | | | | |
| Engine cleaners | | | | |
| Engineers, chief | | | | |
| Engineers, hoisting | .30½ | .30½ | .30½ | |
| Engineers, stationary | | | | |
| Engineers, stationary, assistant | | | | |
| Engineers, steam roller | | | | |
| Firemen | | | | |
| Foremen, carpenters | | | | |
| Foremen, cornice workers | | | | |
| Foremen, electric lighting | | | | |
| Foremen, electric lighting, assistant | | | | |
| Foremen, laborers | .29½ | .22½ | .26 | |
| Foremen, linemen | | | | |
| Galvanized and sheet iron workers | | | | |
| Gardeners | .21½ | .18¾ | .19 | |
| Gas-holder riveters | | | | |
| Granite cutters | | | | |
| Ground men, electric lighting | | | | |
| Hod carriers | | | | |
| Hostlers | | | | |
| Inspectors, electric lighting | | | | |
| Iron workers, ornamental | | | | |
| Iron workers', ornamental, helpers | | | | |
| Iron workers, structural | | | | |
| Iron workers', structural, helpers | | | | |
| Laborers | .22½ | .18¾ | .18¾ | |
| Laborers, building | | | | |
| Linemen | | | | |
| Machinists | .36 | .27⅞ | .33⅓ | |
| Machinists' helpers | | | | |
| Masons, stone | .40⅝ | .40⅝ | .40⅝ | |
| Molders | | | | |
| Molders' helpers | | | | |
| Oilers | | | | |
| Painters | .22½ | [illegible] | [illegible] | |
| Pattern makers | .27⅞ | .27⅞ | [illegible] | |
| Pavers | [illegible] | [illegible] | [illegible] | |
| Pavers, Belgian block | .52 | .54 | .61 | |
| Pavers, cobblestone | .41¼ | .41¼ | [illegible] | |
| Plumbers | | | | |
| Rammers, Belgian block | .34 | .34 | [illegible] | |
| Rammers, cobblestone | [illegible] | [illegible] | [illegible] | |
| Rammers, paving | | | | |
| Repairmen, electric lighting | | | | |
| [illegible] | | | | |

| | | | | | | | | | |
|---|---|---|---|---|---|---|---|---|---|
| Blacksmiths | .41 | .25 | .24 | .30 | [illegible] | [illegible] | .20 | .24 | [illegible] |
| Blacksmiths' helpers | [illegible] | [illegible] | .20 | .18¾ | .18¾ | .18¾ | .19 | .16 | .17 |
| Boiler makers | [illegible] | [illegible] | [illegible] | ....... | ....... | ....... | .21 | .20 | [illegible] |
| Boiler makers' helpers | .27½ | .27½ | .27½ | ....... | ....... | ....... | .13 | .14 | .15½ |
| [illegible] | [illegible] | [illegible] | [illegible] | .25 | .30 | .25¼ | ....... | ....... | ....... |
| Bricklayers | .44 | .41 | .43 | .42 | .42 | .42 | .42 | .35 | .41 |
| Bricklayers, sewer | .70 | .55 | .58½ | .75 | .44½ | [illegible] | .70 | .70 | .70 |
| Calkers, iron | .30½ | .34½ | .35 | .25½ | .22½ | .25½ | ....... | ....... | ....... |
| Calkers, wood | .32½ | .32½ | .32½ | .46½ | .30½ | .30 | .46½ | .46½ | .46½ |
| Carpenters | .45 | .31 | .31½ | .30 | .30 | .27½ | .30 | .24 | [illegible] |
| Carpenters' helpers | .25 | .22½ | .22½ | .20 | .20 | .20 | .20 | .18 | [illegible] |
| Cement finishers | ....... | ....... | ....... | .44½ | .28 | .42½ | .44½ | .44½ | .64½ |
| Cement finishers' helpers | ....... | ....... | ....... | .25 | .22½ | .22½ | [illegible] | [illegible] | [illegible] |
| [illegible] | .22½ | .22½ | .22½ | ....... | ....... | ....... | .18 | .17½ | .17½ |
| Cooks | .18¾ | .18¾ | .18¾ | .21¼ | .21¼ | .21¼ | .18¾ | [illegible] | [illegible] |
| Draftsmen | .75 | .33½ | .46½ | ....... | ....... | ....... | .42½ | .24 | [illegible] |
| Dredge crane men | ....... | ....... | ....... | .30½ | [illegible] | .30½ | .30½ | [illegible] | [illegible] |
| Dredge runners | ....... | ....... | ....... | .34½ | [illegible] | .37½ | .34½ | .34½ | [illegible] |
| Drillers, stone | .25 | .25 | .25 | .30 | .17½ | .25 | .30 | .30 | .30 |
| Drivers | .28 | .19½ | .22½ | .31½ | .14½ | .17½ | .31½ | .17½ | .31½ |
| Drivers with double team | .50 | .50 | .50 | .55½ | .40 | .48½ | ....... | ....... | ....... |
| Drivers with single team | .33½ | .33½ | .33½ | .33½ | .30 | .33 | ....... | ....... | ....... |
| Dump men | .27½ | .27½ | .27½ | .30 | .30 | .30 | ....... | ....... | ....... |
| Engineers, hoisting | .38 | .34½ | .36½ | .34½ | .22½ | .35½ | .34½ | .34½ | .34½ |
| Engineers, stationary | .76 | .21½ | .32 | .27½ | .17½ | .30½ | .40 | .30 | .37 |
| Firemen | .29 | .16½ | .24 | .21½ | .19½ | .30 | .22½ | .18¼ | .21½ |
| Foremen, blacksmiths | ....... | ....... | ....... | .32½ | .28½ | .33½ | .40 | .32 | .34½ |
| Foremen, bricklayers | ....... | ....... | ....... | .62½ | .50 | .56½ | .55 | .45 | .51½ |
| Foremen, carpenters | .44½ | .44½ | .44½ | .39½ | .30 | .34½ | .39½ | .30 | .34 |
| Foremen, dredge runners | ....... | ....... | ....... | .50 | .38½ | .44½ | .55½ | .55½ | .55½ |
| Foremen, laborers | .60 | .25 | .34½ | .54½ | .20 | .35½ | .30 | .16½ | .36½ |
| Foremen, machinists | .41 | .41 | .41 | .32½ | .32½ | .32½ | .32½ | .32 | .32 |
| Foremen, masons | .62½ | .62½ | .62½ | ....... | ....... | ....... | .42½ | .42½ | .42½ |
| Foremen, painters | ....... | ....... | ....... | .37½ | .32 | .33½ | .37½ | .30½ | .33½ |
| Foremen, pavers | ....... | ....... | ....... | .55½ | .44½ | .51½ | .44½ | .44½ | .44½ |
| Foremen, track men | ....... | ....... | ....... | .61 | .50 | .55½ | .61 | .50 | .54½ |
| Gatemen | .23½ | .22½ | .23 | ....... | ....... | ....... | .21½ | .17½ | .20 |
| Glaziers | ....... | ....... | ....... | .27½ | .27½ | .27½ | .27½ | .27½ | .27½ |
| Harness makers | .44½ | .27½ | .34 | .25 | .25 | .25 | .26½ | .24 | .26½ |
| Hod carriers | ....... | ....... | ....... | .25 | .25 | .25 | .25 | .22½ | .24½ |
| Horse feeders | .25 | .26 | .26 | ....... | ....... | ....... | .15 | .15 | .15 |
| Horseshoers | .36 | .27½ | .33½ | .31½ | .22½ | .27 | ....... | ....... | ....... |
| Hostlers | .27½ | .22½ | .24½ | .14½ | .14½ | .14½ | .18½ | .16½ | .16½ |
| Laborers | .30 | .19 | .22 | .26 | .12½ | .15½ | .22½ | .15 | .17½ |
| Laborers, building | ....... | ....... | ....... | .25 | .20 | .24½ | .25 | .20 | .24½ |
| Lathers | ....... | ....... | ....... | .37½ | .35½ | .37 | .37½ | .37½ | .37½ |
| Machinists | .41 | .27½ | .34½ | .23 | .22½ | .22½ | .29 | .23½ | .26 |
| Machinists' helpers | .27½ | .24½ | .25½ | ....... | ....... | ....... | .22 | .16 | .19½ |
| Masons | .50 | .47 | .49 | .42 | .42 | .42 | .42 | .37½ | .41½ |
| Masons, stone | .47½ | .42 | .42½ | .42 | .30 | .36½ | ....... | ....... | ....... |
| Masons' helpers | .26 | .26½ | .30½ | .31 | .26 | .26½ | .25 | .25 | .25 |
| Miners | .25 | .25 | .25 | .23 | .23½ | .23 | ....... | ....... | ....... |
| Oilers | .27½ | .22½ | .24 | ....... | ....... | ....... | .22½ | .20 | .20½ |
| Painters | .41½ | .31½ | .30 | .31½ | .30 | .30 | .31½ | .21 | [illegible] |
| Painters, decorative | ....... | ....... | ....... | .35 | .35 | .35 | .36 | .35 | .35 |
| Pattern makers | .41 | .41 | .41 | ....... | ....... | ....... | .32 | .30 | .31½ |
| Pavers | ....... | ....... | ....... | .50 | .44½ | .44½ | .50 | [illegible] | .40 |
| Pipe layers | [illegible] | .24½ | .27½ | .23 | .30 | .32 | .27½ | .27½ | .27½ |
| Plumbers | ....... | ....... | ....... | .50 | .43 | .44½ | .50 | .43 | [illegible] |
| Plumbers' helpers | ....... | ....... | ....... | [illegible] | .33 | .33 | .33 | .33 | [illegible] |
| [illegible] | .30 | [illegible] | [illegible] | [illegible] | .37 | [illegible] | [illegible] | [illegible] | [illegible] |
| [illegible] helpers | ....... | ....... | ....... | .12½ | .12½ | .12½ | [illegible] | [illegible] | [illegible] |

| Occupations. | Public, by city or State. | | | Public, by [illegible] | | |
|---|---|---|---|---|---|---|
| | Highest. | Lowest. | Average. | Highest. | Lowest. | [illegible] |
| Rammers, paving | ........ | ........ | ........ | $0.2[illegible] | $0.2[illegible] | [illegible] |
| Riggers | $0.36 | $0.22½ | $0.2[illegible] | .31½ | .2[illegible] | [illegible] |
| Rock men | .23[illegible] | .22[illegible] | .2[illegible] | .2[illegible] | .1[illegible] | [illegible] |
| Spikers | ........ | ........ | ........ | .25 | .25 | |
| Sprinklers, street | .26½ | .22[illegible] | .29 | .18½ | .1[illegible] | |
| Stonecutters | .36 | .27[illegible] | .34 | .44 | .30 | |
| Sweepers, street | .23[illegible] | .22[illegible] | .22[illegible] | ........ | [illegible] | |
| Timekeepers | .33[illegible] | .27[illegible] | .2[illegible] | .26[illegible] | .15 | |
| Tinsmiths | .33[illegible] | .33[illegible] | .33[illegible] | ........ | ........ | |
| Track men | ........ | ........ | ........ | .27[illegible] | .27[illegible] | |
| Watchmen | .25 | .20 | .23½ | .1[illegible] | .14[illegible] | |
| Water boys | .12 | .08½ | .10[illegible] | .10[illegible] | .08[illegible] | |
| Wheelwrights | .38 | .33[illegible] | .36[illegible] | .37 | .27[illegible] | |
| Wire men | .33½ | .27[illegible] | .31[illegible] | ........ | ........ | |

## NEW YORK CITY.

| Occupations. | Highest. | Lowest. | Average. | Highest. | Lowest. |
|---|---|---|---|---|---|
| Ash and garbage lifters | .28[illegible] | .28[illegible] | .28[illegible] | .15[illegible] | .15[illegible] |
| Ash tenders | .33 | .33 | .33 | ........ | ........ |
| Asphalt layers | ........ | ........ | ........ | .35 | .30 |
| Asphalt mixers | ........ | ........ | ........ | .22½ | .22½ |
| Axmen | .43½ | .24 | .31 | ........ | ........ |
| Blacksmiths | .44 | .31 | .38 | .40[illegible] | .20 |
| Blacksmiths' helpers | .32[illegible] | .26 | .28[illegible] | .22½ | .22[illegible] |
| Boiler makers | .41 | .35 | .38[illegible] | ........ | ........ |
| Boiler makers' helpers | .27 | .24 | .26[illegible] | ........ | ........ |
| Boiler setters | ........ | ........ | ........ | .50 | .50 |
| Boiler setters' helpers | ........ | ........ | ........ | .28½ | .28½ |
| Bracers | ........ | ........ | ........ | .30 | .20 |
| Brass finishers | ........ | ........ | ........ | .30 | .30 |
| Bricklayers | .50 | .50 | .50 | .50 | .40 |
| Calkers, iron | .35 | .31 | .32 | .20½ | .20½ |
| Calkers, wood | .41 | .38 | .40[illegible] | .22[illegible] | .22[illegible] |
| Car cleaners | .26½ | .26½ | .26½ | ........ | ........ |
| Car couplers | .26[illegible] | .26[illegible] | .26[illegible] | ........ | ........ |
| Carpenters | .47 | .37[illegible] | .44[illegible] | .43[illegible] | .25 |
| Carpenters' helpers | .33 | .25 | .30[illegible] | .30 | .25 |
| Cement mixers | .28½ | .28½ | .28½ | .17[illegible] | .16[illegible] |
| Cleaners | .31[illegible] | .25 | .25[illegible] | ........ | ........ |
| Conductors | .34[illegible] | .34[illegible] | .34[illegible] | ........ | ........ |
| Coppersmiths | .41 | .37[illegible] | .39[illegible] | ........ | ........ |
| Coppersmiths' helpers | .28 | .25 | .27[illegible] | ........ | ........ |
| Curbstone setters | ........ | ........ | ........ | .50 | .30 |
| Curbstone setters' helpers | ........ | ........ | ........ | .15 | .15 |
| Derrick men | ........ | ........ | ........ | .37[illegible] | .34[illegible] |
| Derrick men's helpers | ........ | ........ | ........ | .31[illegible] | .31[illegible] |
| Divers | 1.25 | 1.20 | 1.24[illegible] | .56 | 1.00 |
| Divers' tenders | .35 | .35 | .35 | .25 | .25 |
| Dock builders | .39 | .30 | .30[illegible] | .30 | .20 |
| Draftsmen | .63 | .31½ | .49[illegible] | .33½ | .27[illegible] |
| Drillers | .28 | .24 | .27[illegible] | .25[illegible] | .25[illegible] |
| Drivers | .36 | .24 | .27[illegible] | .23[illegible] | .15[illegible] |
| Drivers with double team | .56½ | .43[illegible] | .53[illegible] | .35 | a.35 |
| Drivers with single team | .43[illegible] | .28[illegible] | .37[illegible] | .35 | .27½ |
| Drivers with team and truck | .75 | .75 | .75 | .45 | .45 |
| Elevator constructors | ........ | ........ | ........ | .47 | .31½ |
| Elevator constructors' | | | | | |

| | | | | | | | | | |
|---|---|---|---|---|---|---|---|---|---|
| [illegible] | [illegible] | [illegible] | [illegible] | [illegible] | [illegible] | [illegible] | | | |
| ...st iron | | | | | | | | | |
| .......... | .44 | .31 | .40[illegible] | .40[illegible] | .40[illegible] | .40[illegible] | .40[illegible] | .25 | .40[illegible] |
| ...st iron | | | | | | | | | |
| ........ | | | | .25 | [illegible] | [illegible] | .25 | .15 | [illegible] |
| ........ | .25 | .25 | .25 | | | | .16[illegible] | [illegible] | [illegible] |
| ...ilway... | .31½ | .31½ | .31½ | | | | .16[illegible] | [illegible] | [illegible] |
| ........ | .37[illegible] | .31½ | .3[illegible] | | | | [illegible] | [illegible] | [illegible] |
| ........ | .34 | .34 | .34 | .31½ | .30 | .30 | .30 | [illegible] | [illegible] |
| ........ | .28 | .28 | .28 | .2[illegible] | .2[illegible] | .2[illegible] | [illegible] | [illegible] | [illegible] |
| ........ | .25 | .18[illegible] | .24[illegible] | .30 | .1[illegible] | .14[illegible] | [illegible] | .17[illegible] | .17[illegible] |
| ...ctural | | | | .43[illegible] | .2[illegible] | .2[illegible] | .47 | [illegible] | .24 |
| ...ctural, | | | | | | | | | |
| ........ | | | | .25 | .25 | .25 | .25 | .20 | .22 |
| ........ | .41 | .25 | .40[illegible] | .25 | .30 | .30 | .30 | .25 | [illegible] |
| ........ | .38 | .16 | .34[illegible] | .25 | .12[illegible] | .16 | .2[illegible] | .1[illegible] | [illegible] |
| ........ | .31½ | .38½ | .30[illegible] | .31½ | .18[illegible] | .27[illegible] | .21[illegible] | .2[illegible] | [illegible] |
| ........ | .44 | .36 | .40[illegible] | .34[illegible] | .34[illegible] | .34[illegible] | .20 | .22 | [illegible] |
| ........ | .50 | .35 | .40[illegible] | .34[illegible] | .30 | .2[illegible] | .43[illegible] | .20 | [illegible] |
| ...s ........ | .31½ | .34 | .37[illegible] | | | | .30 | .17[illegible] | [illegible] |
| ........ | .50 | .4[illegible] | .49[illegible] | .50 | .40 | .40[illegible] | .30 | [illegible] | [illegible] |
| ........ | .50 | .50 | .50 | .50 | .40 | .40[illegible] | .30 | .30 | [illegible] |
| ........ | .28½ | .28½ | .28½ | | | | .30 | .30 | [illegible] |
| ........ | .44 | .38 | .40[illegible] | | | | .25 | .15 | [illegible] |
| ........ | .25 | .25 | .25 | | | | [illegible] | [illegible] | [illegible] |
| ........ | .44 | .38 | .43½ | .40[illegible] | .27[illegible] | .34½ | [illegible] | [illegible] | [illegible] |
| ........ | .41 | .31½ | .36½ | | | | .30 | .17[illegible] | [illegible] |
| ........ | .34½ | .25 | .30½ | .25 | .25 | .25 | .16[illegible] | [illegible] | [illegible] |
| ...ock.... | .50 | .50 | .50 | .50 | .44½ | .40[illegible] | .50 | [illegible] | [illegible] |
| ........ | .44 | .25 | .39½ | | | | .2[illegible] | [illegible] | [illegible] |
| ........ | | | | .54½ | .54½ | .54½ | [illegible] | [illegible] | [illegible] |
| ........ | | | | .31½ | .30 | .3[illegible] | .30 | .30 | [illegible] |
| ........ | .50 | .37½ | .45½ | .47 | .40 | .4[illegible] | .47 | [illegible] | [illegible] |
| ........ | .31½ | .16 | .24[illegible] | .15[illegible] | .08½ | .15 | .34 | .10 | [illegible] |
| block .. | .38 | .37½ | .37½ | .30 | .30 | .30 | .30 | .17[illegible] | [illegible] |
| ........ | .44 | .31 | .35½ | .31½ | .17½ | .2[illegible] | .30 | .30 | [illegible] |
| ........ | | | | .23 | .17[illegible] | .21[illegible] | .22 | .17[illegible] | [illegible] |
| ........ | .37½ | .32 | .34½ | .34½ | .25 | .31[illegible] | .34[illegible] | [illegible] | [illegible] |
| ........ | | | | .15 | .15 | .15 | .16[illegible] | [illegible] | [illegible] |
| ........ | .28½ | .28½ | .28½ | .20 | .17½ | .18 | [illegible] | [illegible] | [illegible] |
| ........ | .44 | .44 | .44 | | | | [illegible] | [illegible] | [illegible] |
| ........ | .28 | .28 | .28 | | | | [illegible] | [illegible] | [illegible] |
| ........ | .44 | .41 | .43½ | | | | [illegible] | [illegible] | [illegible] |
| ...rs ........ | .28 | .28 | .28 | | | | [illegible] | [illegible] | [illegible] |
| ........ | .41 | .37½ | .41 | | | | .36 | .36 | .25 |
| ........ | .25 | .25 | .25 | .12½ | .12½ | .12½ | | | |
| ........ | .44 | .44 | .44 | | | | .30 | .30 | .30 |
| ...ers........ | .44 | .37½ | .40½ | .43[illegible] | .40 | .41 | .43[illegible] | .27[illegible] | .42[illegible] |
| fitters' | | | | | | | | | |
| ........ | | | | .35 | .18[illegible] | .23[illegible] | .25 | .25 | .25 |
| ........ | .31½ | .31½ | .31½ | | | | .25 | .25 | .25 |
| ........ | .50 | .50 | .50 | .43[illegible] | .43[illegible] | .43[illegible] | | | |
| ........ | .28[illegible] | .33 | .37½ | .10[illegible] | .10[illegible] | .10[illegible] | .16[illegible] | .15 | .25 |
| ........ | .37½ | .37½ | .37½ | | | | .24[illegible] | .17[illegible] | .20[illegible] |
| ...rs........ | .51[illegible] | .51[illegible] | .51[illegible] | | | | .30 | .17[illegible] | .1[illegible] |
| ........ | .37[illegible] | .37[illegible] | .37[illegible] | | | | .18[illegible] | .12[illegible] | .17 |
| ........ | .43[illegible] | .43[illegible] | .43[illegible] | .25 | .20 | .21½ | .30 | .2[illegible] | .27 |
| ........ | .38 | .37[illegible] | .37[illegible] | | | | .2[illegible] | .21 | .2[illegible] |
| ........ | .50 | .50 | .50 | | | | .2[illegible] | .15 | .2[illegible] |
| ........ | .43[illegible] | .43[illegible] | .43½ | | | | .3[illegible] | .2[illegible] | .2[illegible] |
| ........ | .31 | .16[illegible] | .28[illegible] | | | | .21[illegible] | .17[illegible] | .15 |
| ........ | | | | .37½ | .21½ | .30[illegible] | .2[illegible] | .21[illegible] | .25 |
| ...s........ | | | | .25 | .22 | .22[illegible] | .25 | .25 | .35 |
| ........ | .37½ | .37½ | .37½ | | | | .22[illegible] | .22½ | .22[illegible] |

PHILADELPHIA.

TABLE II.—SUMMARY OF WAGES PER HOUR OF EMPL

PHILADELPHIA—Continued.

| Occupations. | Public, by city or State. | | | Public, by contractor | | |
|---|---|---|---|---|---|---|
| | Highest. | Lowest. | Average. | Highest. | Lowest. | Average. |
| Boiler makers | | | | $0.25 | $0.25 | $0.1 |
| Bricklayers | $0.48½ | $0.48½ | $0.48½ | .48½ | .35 | .4 |
| Bricklayers, sewer | | | | .61 | .50 | .5 |
| Bricklayers' helpers | .24½ | .24½ | .24½ | .18¾ | .18¾ | .1 |
| Calkers, iron | .27½ | .25 | .27½ | .25 | .22½ | .2 |
| Carpenters | .32½ | .30 | .32½ | .34½ | .25 | .2 |
| Carpenters' helpers | .19¼ | .19¼ | .19¼ | .18¾ | .17½ | .1 |
| Cement mixers | | | | .15 | .15 | .1 |
| Coal handlers | | | | .17½ | .17½ | .1 |
| Coppersmiths | | | | .30 | .30 | .3 |
| Cornice workers | | | | .27½ | .27½ | .2 |
| Dock builders | | | | .25 | .22½ | .2 |
| Draftsmen | | | | .38 | .38 | .3 |
| Drillers, iron | | | | .17½ | .17½ | .1 |
| Drillers, stone | | | | .27½ | .27½ | .2 |
| Drillers', stone, helpers | | | | .17½ | .16¾ | .1 |
| Drips, gas works | .18¾ | .18¾ | .18¾ | .16¾ | .16¾ | .1 |
| Drivers | .22½ | .17½ | .19¾ | .16¾ | .12½ | .1 |
| Drivers with double team | .50 | .50 | .50 | .50 | .40 | .4 |
| Drivers with single team | .33¼ | .27½ | .31¼ | .30 | .27½ | .2 |
| Dynamo men | .31¼ | .31¼ | .31¼ | .25 | .22½ | .2 |
| Electricians | | | | .35 | .35 | .3 |
| Engineers, chief | | | | .29½ | .29½ | .2 |
| Engineers, dynamo | | | | .22½ | .22½ | .2 |
| Engineers, dynamo, assistant | | | | .16¾ | .16¾ | .1 |
| Engineers, hoisting | .37½ | .37½ | .37½ | .30 | .22½ | .2 |
| Engineers, stationary | .33½ | .21 | .23¾ | .30 | 25 | .2 |
| Engineers, stationary, assistant | | | | .16¾ | .16¾ | .1 |
| Engineers, steam roller | | | | .35 | .25 | .2 |
| Fasteners, shipbuilding | | | | .17½ | .17½ | .1 |
| Firemen | .28¼ | .19¼ | .22¾ | .22½ | .16¾ | .2 |
| Fitters (iron setters), shipbuilding | | | | .25 | .25 | .2 |
| Foremen, blacksmiths | .37½ | .35 | .36¼ | .30 | .30 | .3 |
| Foremen, bricklayers | 60 | .39 | .43¾ | .45 | .45 | .4 |
| Foremen, carpenters | .42 | .35½ | .39 | .39 | .37½ | .3 |
| Foremen, dock builders | | | | .30 | .30 | .3 |
| Foremen, iron workers, structural | | | | .33½ | .33½ | .3 |
| Foremen, laborers | .39 | .19¼ | .29¼ | .25 | .20 | .2 |
| Foremen, linemen | | | | .29½ | .29½ | .2 |
| Foremen, machinists | .53½ | .42½ | .46¾ | | | |
| Foremen, painters | .42 | .32 | .38 | .39 | .39 | .3 |
| Foremen, plasterers | | | | .50 | .50 | .5 |
| Foremen, purifiers, gas works | .32½ | 27½ | 28¼ | .20 | .20 | .2 |
| Foremen, stokers | .27 | .27 | .27 | | | |
| Galvanized and sheet iron workers | | | | .30 | .22½ | .2 |
| Gardeners | .32 | .22½ | .27 | | | |
| Gardeners' helpers | .22½ | .22½ | .22½ | | | |
| Granite cutters | | | | .30 | .30 | .3 |
| Hod carriers | .30 | .24 | .25 | .22½ | .17½ | .2 |
| Inspectors, electric lighting | .37½ | .37½ | .37½ | .22½ | .20 | .2 |
| Iron workers | | | | .30 | .30 | .3 |
| Iron workers, structural | | | | .22½ | .22½ | .2 |
| Joiners | | | | .27½ | .27½ | .2 |
| Laborers | .20 | .16¼ | .16¾ | .25 | .12½ | .1 |
| Laborers, building | .24 | .21 | .21¼ | .19½ | .15 | .1 |
| Linemen | .27½ | .25 | .25¾ | .25 | .25 | .2 |
| Machinists | .38½ | .27½ | .32¾ | .35 | .35 | .3 |
| Machinists' helpers | .26 | .22½ | .23¾ | | | |
| Masons, stone | | | | .45 | .35 | .4 |
| Meter inspectors, gas works | .33½ | .33½ | .33½ | | | |

TABLE II.—SUMMARY OF WAGES PER HOUR OF EMPLOYEES, ETC.—Conc'd.

**PHILADELPHIA**—Concluded.

| Occupations. | Public, by city or State. | | | Public, by contractors. | | | Private, by contractors or firms. | | |
|---|---|---|---|---|---|---|---|---|---|
| | Highest. | Lowest. | Average. | Highest. | Lowest. | Average. | Highest. | Lowest. | Average. |
| Plasterers | ...... | ...... | ...... | $0.40 | $0.30½ | $0.38¾ | $0.40 | $0.39 | $0.40 |
| Plumbers | $0.42 | $0.30½ | $0.34¾ | .43¾ | .33¼ | .35⅛ | .39 | .27¾ | .34 |
| Plumbers' helpers | .27¾ | .25 | .27 | .16¾ | .16¾ | .16¾ | ...... | ...... | ...... |
| Pump men | .34⅛ | .22 | .25¾ | .22⅛ | .22⅛ | .22⅛ | ...... | ...... | ...... |
| Purifiers, gas works | .25 | .25 | .25 | .17½ | .17½ | .17½ | .19 | .19 | .19 |
| Rammers, Belgian block and cobblestone | ...... | ...... | ...... | .30⅛ | .15 | .22¾ | .33¼ | .20 | .28⅛ |
| Riggers | ...... | ...... | ...... | .33¾ | .18⅛ | .19¾ | .48 | .18⅛ | .23⅜ |
| Riveters | ...... | ...... | ...... | .20 | .20 | .20 | .20 | .20 | .20 |
| Steam and gas fitters | .36 | .27¾ | .29 | .37½ | .25 | .26 | .33¼ | .20 | .26⅜ |
| Steam and gas fitters' helpers | .24⅛ | .22½ | .23¼ | .18¾ | .16¾ | .17½ | .19¼ | .15 | .16⅜ |
| Stokers | .23 | .23 | .23 | ...... | ...... | ...... | .18 | .18 | .18 |
| Stonecutters | .42 | .33¼ | .36⅛ | .43 | .30 | .38¾ | .43 | .36 | .41⅜ |
| Stonecutters' helpers | .24 | .24 | .24 | .22 | .22 | .22 | ...... | ...... | ...... |
| Stone setters | ...... | ...... | ...... | .48⅛ | .48⅛ | .48⅛ | .48⅛ | .45 | .45⅜ |
| Trimmers, electric lighting | .37½ | .37½ | .37½ | .20 | .20 | .20 | .20 | .20 | .20 |
| Watchmen | .20¾ | .16¾ | .17⅛ | .18 | .10¾ | .17¼ | .18 | .12 | .16 |
| Weighers, gas works | .22 | .22 | .22 | .20 | .20 | .20 | ...... | ...... | ...... |

## PENNSYLVANIA.

*Annual Report of the Secretary of Internal Affairs* [illegible] *of Pennsylvania.* Vol. XXIII, 1895. Part III, [illegible] James M. Clark, Chief of Bureau. 256 pp.

This report treats of the following subjects: Black [illegible] pages; statistics of manufactures, 150 pages; silk [illegible] pages; strikes and lockouts, 22 pages.

BLACK AND TIN PLATE.—This part of the report consists of [illegible] of the tin-plate industry in general, an account of its develo[illegible] recent years in the United States, a description, with illustratio[illegible] largest tin-plate establishment in Pennsylvania, and statistics [illegible] the extent of tin-plate manufacture in the State.

There were in operation during 1895, in Pennsylvania, 10 bl[illegible] works and 17 dipping works for which returns are shown. [illegible] dipping works 557 persons were employed. The average nu[illegible] days in operation was 244 and the total amount of wages [illegible] $188,224.32. This makes an average of $337.92 per person fo[illegible] days, or $1.38 per day. The black-plate manufacturers employ[illegible] persons, who were paid $1,063,695.23 in wages during 229 days [illegible] tion. This makes an average per person for the 229 days of $4[illegible] $1.87 per day. For the entire tin-plate industry in the State th[illegible] therefore, a total of $1,251,919.55 paid out in earnings to 3,031 for an average time of 231⅔ days, a per capita for skilled and [illegible] laborers of $413.03 for the year, or $1.78 per day.

The following statement shows the number of black-plate w[illegible] dipping works in the various States in 1895:

NUMBER OF BLACK-PLATE AND DIPPING WORKS IN THE UNITED STAT[illegible] BY STATES.

| State. | Black-plate works. | | Dip[illegible] |
|---|---|---|---|
| | Plants. | Hot mills. | Pla[illegible] |
| Pennsylvania | 11 | 64 | |
| Ohio | 9 | [illegible] | |
| Indiana | 6 | 44 | |
| Maryland | [illegible] | 10 | |
| Illinois | [illegible] | [illegible] | |
| Missouri | 1 | 7 | [illegible] |
| Michigan | [illegible] | [illegible] | |
| New York | [illegible] | [illegible] | |
| Virginia | [illegible] | [illegible] | |
| West Virginia | [illegible] | [illegible] | |

| Year. | Establishments considered. | Average persons employed. | Aggregate wages paid. | Average yearly earnings. | Value of product. |
|---|---|---|---|---|---|
| [illegible] | 381 | 140,850 | $[illegible],136,287 | $492.88 | $572,985,668 |
| [illegible] | 381 | 125,611 | 57,402,682 | 456.96 | 227,491,310 |
| [illegible] | 381 | 112,271 | 46,186,794 | 411.28 | 188,462,976 |
| [illegible] | 381 | 130,925 | 57,381,100 | 438.39 | 225,730,418 |

In 1895, 36 of the 51 industries, representing 294 establishments, had increased days of operation over 1894. Eleven industries, representing 83 establishments, had a decrease, while 4 industries, representing 4 establishments, worked the same number of days. The average increase over 1894 for all persons employed was 16 days, and the average decrease, as compared with 1892, 9 days.

In 1895 there was a general increase in employment, the excess of the persons employed in the 381 establishments over 1894 being 18,654. Notwithstanding this increase, there were still 9,925 fewer persons employed in 1895 than in 1892. An analysis of the detail tables shows that 31 industries, representing 304 establishments, employed 123,436 persons in 1892 and 112,307 persons in 1895, a decrease of 11,129 in the number employed, and the remaining 20 industries, representing 77 establishments, employed 17,414 persons in 1892 and 18,618 in 1895, an increase of 1,204, leaving a net decrease in 1895 over 1892, as above stated, of 9,925 persons, or 7.04 per cent.

In 24 industries, representing 255 establishments and employing 102,233 persons, the average daily wages were lower in 1895 than in 1892; in the remaining 27 industries, representing 126 establishments and employing 28,692 persons, the wages had increased over those of 1892. The net average decrease per day, 1895 over 1892, was 11 cents.

As satisfactory returns of product were received from only 162 of the 381 establishments, a separate series of comparative tables is presented giving the product of these 162 establishments, the corresponding value, aggregate of wages, yearly earnings, and persons employed.

General statistics of iron and steel manufactures for the year 1895 are presented in a series of 67 tables, or one for each branch of the industry, and a final table summarizing the data. The various branches considered comprise the production and manufacture of iron and steel, including pig metal. The tables show the average number of persons employed, number of days in operation, aggregate amount of wages paid, total product, and value of product for each of 1,528 establishments.

These establishments employed, in 1895, a total of 185,496 persons, receiving an aggregate of $88,586,570 in wages, or a per capita of $477.56. The average number of days in operation was 286, which makes an average daily wage of $1.67. The total value of the manufactured product was $336,415,068.

SILK MANUFACTURE.—This consists of a statistical presentation of the number of spindles and hand-power and Jacquard looms, and of persons employed, amount of wages paid, and value of product for each of the 65 silk manufacturing establishments in the State.

These establishments employed, in 1895, a total of 13,815 skilled and unskilled laborers, receiving a total of $4,082,292.08 in wages, or a per capita of $295.50 for 48 weeks of operation. Salaries of clerks, salesmen, officials, and members of the firm are not included in these figures. The gross value of the product during the year was $24,184,583.84.

STRIKES AND LOCKOUTS.—During the year 1895, 151 strikes occurred in Pennsylvania that came to the knowledge of the bureau. In these strikes 17,113 persons were actually engaged.

The following statement shows, by industries, the number of strikes and the number of persons engaged:

STRIKES AND PERSONS ENGAGED, BY INDUSTRIES.

[Each strike in each establishment is considered separately, whether part of a general strike or not. So also when one establishment has several strikes it is regarded as so many separate establishments. Thus the number of strikes is made to equal the number of establishments.]

| Industry. | Strikes. | Persons engaged. | Industry. | Strikes. | Persons engaged. |
|---|---|---|---|---|---|
| Coal mining | 70 | 10,805 | Cotton, worsted, and jeans | 1 | [illegible] |
| Carpet weaving | 36 | 2,524 | Cotton and silk goods | 1 | [illegible] |
| Boiler making | 6 | 571 | Cabinetmaking | 1 | [illegible] |
| Blast furnace | 4 | 440 | Worsted, woolen, and cotton goods | 1 | [illegible] |
| Pig-iron manufacturing | 2 | 230 | Cloaks and suits | 1 | [illegible] |
| Hosiery making | 3 | 91 | Cloth weaving | 1 | [illegible] |
| Woolen and worsted goods | 4 | 191 | Chenille weaving | 1 | [illegible] |
| Cigar making | 3 | 447 | Worsted goods | 1 | [illegible] |
| Woolen goods | 2 | 123 | Chenille and upholstery manufacturing | 1 | [illegible] |
| Cotton and woolen goods | 2 | 275 | Cooking utensils | 1 | 19 |
| Upholstery manufacturing | 2 | 140 | Brickmaking | 1 | 175 |
| Iron and steel works | 1 | 64 | Shoe manufacturing | 1 | 15 |
| Foundry and machine works | 1 | 36 | Drivers of milk wagons | 1 | [illegible] |
| Electric street railway | 1 | 80 | | | |
| Iron works | 1 | 150 | | | |

Of the total number of strikes, 109 were inaugurated for an increase of wages, 19 against a reduction of wages, and the remaining 23 for various other causes. Ninety-eight strikes, or 64.9 per cent, were ordered by labor organizations; 33, or 21.8 per cent, were not so ordered, and of the remaining 20 strikes the returns were incomplete. Forty-three, or 28.4 per cent, of the strikes succeeded; 70, or [illegible] failed; 32, or 21.2 per cent, succeeded partly, and in the cases of the others the results were doubtful.

In 114 strikes the returns indicated that 9,017 persons were involved who were not actually engaged as strikers, but who were prevented from working on account of strikes. The average duration in 143 strikes, where the time was reported, was 31.8 days. In the 151 establishments where strikes occurred, 134, or 88.7 per cent, of the establishments were closed for longer or shorter periods of time; 13, or 8.6 per cent, were not closed, and for the others this fact was not reported upon.

Estimates of actual loss to strikers were received from 132, or 87.4 per cent, of all the strikes. In these there were 15,109 persons engaged, whose loss in wages, etc., amounted to $913,495, or an average of $60.46 per person. In one strike no loss to strikers was reported. As regards loss to operators, estimates were received from 89 establishments. These showed a total loss of $235,883. Sixteen establishments reported no loss.

*Bulletin de l'Institut International de Statistique.* Tome VIII, 2e son. Luigi Bodio, Secrétaire Général. xi, 346 pp.

This volume is almost entirely devoted to a reproduction report presented by E. Levasseur on behalf of the commi statistics of education, concerning the statistics of primary ed in all the civilized countries of the world. This committ appointed in 1889 and made a first report in 1891 at the sessio Institute held at Vienna. The present is a second report made ing to the same plan in which the effort is made to complete tl mation thus given and to include countries concerning which info could not be obtained for the former report. The report con two parts. In the first, primary education is historically and s ally treated of in each country separately. The second is an to make a general comparative study of the subject of primary tion in all of the countries, based upon the information give first part.

*Bulletin de l'Institut International de Statistique.* Tome IX, 2e Li Luigi Bodio, Secrétaire Général. cx, 314 pp.

This volume gives a report of the fifth session of the Inter Statistical Institute, which was held at Bern, Switzerland, 26–31, 1895. In addition to the report of proceedings, the list bers, committees, etc., it embraces the following papers whi read at this meeting:

1. Die Einkommensverteilung in alter und neuer Zeit [The of incomes in ancient and modern times], by G. Schmoller
2. L'état démographique de la Roumanie d'après le mouveme population [The demographic condition of Roumania acco the movement of population], by C. Crupenski.
3. Taxation in the United States, with suggestions for esta a form for comparing the taxation of different count Edward Atkinson.
4. L'échange international des bulletins de recensement concer étrangers: Rapport fait au nom de la première secti international exchange of census bulletins concerning the

bulletins concerning foreigners: Motion presented to the Bureau of the International Statistical Institute], by K. T. von Inama-Sternegg, F. von Juraschek, and H. Rauchberg.

6. Statistique internationale des valeurs mobilières [International statistics of personal property], by A. Neymarck.

7. La longévité dans les familles [Longevity in families], by L. Vacher.

8. Les causes des régularités statistiques [The causes of statistical regularities], by W. Lexis.

9. Sur un certain nombre de professions qu'il est particulièrement désirable de voir figurer sur les nomenclatures professionnelles des différents pays [Concerning a certain number of occupations which it is particularly desirable should be included in the nomenclature of occupations in the various countries], by J. Bertillon.

10. Trois projets de nomenclature des infirmités [Three plans for a nomenclature of infirmities], by J. Bertillon.

11. Nomenclature des accidents [The nomenclature of accidents], by J. Bertillon.

12. Zur Frage einer internationalen Berufssparkassenstatistik [The question of international statistics of the occupations of savings-bank depositors], by K. Rasp.

13. Statistique de la production, du mouvement international et de la consommation des métaux précieux [Statistics of the production, international movement, and consumption of the precious metals], by C. F. Ferraris.

14. International military medical statistics, by J. S. Billings.

15. La statistique agricole: Rapport fait à l'assemblée générale au nom de la troisième section [Agricultural statistics: Report made in behalf of the third section to the general assembly], by Th. Pilat.

16. La statistique agricole: Rapport fait à la troisième section de la session de Berne [Agricultural statistics: Report made to the third section of the session at Bern], by Th. Pilat.

17. La statistique du tonnage des transports à l'intérieur: Rapport fait au nom du comité des transports à l'intérieur [Statistics of the tonnage of internal commerce: Report made in behalf of the committee of internal commerce], by E. Cheysson.

18. La statistique des transports à l'intérieur: Supplément au rapport de Vienne en 1891 [Statistics of internal commerce: Supplement to the report of Vienna in 1891], by J. Borkowsky.

19. Les registres de population: Rapport fait au nom de la première section [Registers of population: Report made in behalf of the first section], by E. Nicolaï.

20. La statistique des divorces: Rapport fait au nom de la première section [Statistics of divorce: Report made in behalf of the first section], by E. J. Yvernès.
21. Le premier recensement de la population de l'Empire russe qui doit être fait en 1897 [The first census of the population of the Russian Empire to be taken in 1897], by N. Troïnitsky.
22. L'échange des publications statistiques [The exchange of statistical publications], by E. Nicolaï.
23. Le dénombrement général de la population en 1900: Rapport présenté au nom de la première section [The general census of population in 1900: Report presented in behalf of the first section], by L. Guillaume.
24. Die Erhebungsperiode der Handelsstatistik [The periods for the collection of trade statistics], by T. Geering.
25. Comparabilité des statistiques du commerce [Comparability of statistics of commerce], by W. A. Verkerk Pistorius.
26. La statistique commerciale internationale: Rapport fait au nom de la deuxième section [International statistics of commerce: Report made in behalf of the second section], by A. E. Bateman.
27. La statistique des salaires des ouvriers industriels: Rapport fait au nom de la quatrième section [Wage statistics of industrial workmen: Report made in behalf of the fourth section], by K. T. von Inama-Sternegg.
28. Essai de statistique internationale des produits agricoles [Essay concerning international statistics of agricultural products], by P. G. Craigie.
29. Observations et expériences concernant les dénombrements représentatifs [Observations and experiences concerning representative censuses], by A. N. Kiaer.
30. La statistique internationale de la consommation de l'alcool: Rapport fait au nom de la quatrième section [International statistics of the consumption of alcohol: Report made in behalf of the fourth section], by G. E. Milliet.
31. De la protection de l'enfance et plus spécialement de l'assistance aux enfants-trouvés en Italie et dans quelques autres États [The protection of children and especially the assistance of foundlings in Italy and some other states], by E. Raseri.
32. Statistique des forêts: Question rédigée sur la proposition de la Société Nationale d'Agriculture [Statistics of forests: A schedule of inquiry prepared in accordance with a proposition of the National Agricultural Society], by E. Levasseur.
33. Des méthodes à suivre pour l'étude des différentes classes sociales [Methods to be followed in studying the different social classes], by J. Bertillon.
34. Les progrès de la statistique en Roumanie et la création du service d'anthropométrie [The progress of statistics in Roumania and [illegible]

annual international report on the statistics of the movement of population: Report presented in behalf of the first section], by G. von Mayr.

36. Internationale Jahresberichte über die Bevölkerungsbewegung [International annual reports concerning the movement of population], by G. von Mayr.

37. L'organisation d'une statistique internationale du chômage [The organization of international statistics of nonemployment], by O. Moron.

38. L'organisation internationale de la statistique du travail [The international organization of statistics of labor], by H. Denis.

39. L'organisation internationale de la statistique du travail: Rapport sur la proposition de M. Denis, fait au nom de la quatrième section [The international organization of statistics of labor: Report concerning the proposition of M. Denis, made in behalf of the fourth section], by E. Cheysson.

40. La machine électrique à recensement: Expériences et améliorations [The electrical tabulating machine: Experiences and improvements], by H. Rauchberg.

41. La comparaison entre les statistiques du commerce des différents pays [The comparison of statistics of commerce of different countries], by A. E. Bateman.

42. Organisation et développement de l'office du travail en Angleterre [Organization and development of the labor bureau of England], by A. E. Bateman.

43. La crise du revenu [The income crisis], by E. Cheysson.

44. La monographie de commune [Monographs of communes], by E. Cheysson.

45. Conférence sur les lois statistiques [Address on statistical laws], by G. von Mayr.

46. Conférence sur l'histoire de la démographie [Address on the history of demography], by E. Levasseur.

*Minimum de Salaire. Enquête—Mai 1896. Rapport présenté au Conseil communal au nom de la Commission d'enquête, par M. le Bourgmestre, Président.* Ville de Bruxelles, Travaux Publics. 180 pp.

In May, 1896, the communal council of Brussels appointed a commission, of which M. Buls, the mayor, was made president, to make an investigation of the subject of the advisability of adopting a minimum wage scale for all workingmen employed by the municipality, and as a part of such inquiry to determine to what extent this principle had been accepted by other public bodies in Belgium.

In accordance with these instructions, the commission sent out schedules of inquiry to the provincial [illegible] trations of communes having a population of over [illegible] tions of employers and workingmen, and to certain [illegible] who had done work for the city of Brussels. The [illegible] obtained was supplemented by statements showing the actual wages paid to employees in the different services of the city and a calculation of what would be the financial results of the application of a system of minimum wages.

The replies of each body or individual making response to the schedules sent are given separately.

As regards the provinces, a résumé of the information given shows that 8 provinces make some sort of provision concerning the minimum wages to be paid workingmen employed either directly by them or by contractors working on their account. Of these, 6 stipulate expressly the minimum wages to be paid, and 2 require the contractors to state in their bids the minimum wages that they will pay. Four provinces require the contractors to submit regular accounts of wages paid by them, and 3 require them to insure their employees against accidents. All of the 8 report that they have received no complaint concerning the working of the system, either from the contractors or the workingmen; and but 2 have experienced any abnormal financial results, a slight increase in the cost of certain work being in each case noticed.

The returns from the communes show that 47, with a total population of 1,427,515, have provisions concerning the minimum wages that must be paid workingmen engaged on public work, and that 39, with a total population of 586,919, have no such provisions. Five of the latter, however, announce that such a measure will probably be introduced within a short time. Of the 47 with minimum wage provisions, 38 indicate the rate of wages that must be paid either by themselves or by contractors working for them. The other 9 require the contractors to indicate in their bids the lowest wages that they will pay to their employees. If this rate is deemed too low their bids are not considered. Twenty-four communes require the contractor to keep a special record of wages paid and to furnish other information to the administration. Seventeen communes have fixed the maximum duration of the working-day; the others have no provisions on this subject. Eighteen communes require the contractors to insure their workingmen against accidents. All of the communes except two from which information was obtained declare that they have received no complaint concerning the workings of the system; and all save one report that no abnormal financial effects have resulted. Seventeen of the communes, however, [illegible] as yet executed no work under the régime of minimum wages.

In making reply to the circulars sent to them, 2 provinces [illegible] communes gave the schedules of minimum wages adopted by [illegible] From these statements it has been possible to compile the [illegible] table, showing the minimum wages [illegible]

**principal occupations in the building trades. In constructing this table unusual occupations and those for which but a single quotation could be secured were omitted. Omissions on this account, however, were very few in number.**

MINIMUM WAGES PER HOUR IN THE PRINCIPAL OCCUPATIONS IN THE BUILDING TRADES PAID BY CERTAIN PROVINCES AND COMMUNES OF BELGIUM

| Occupation. | Province Hainaut. | Commune Jumet. | Commune Braine-le-Comte. | Commune Tournai. | Commune Roulers. | Commune Menin. | Commune Gendbrugge. | Commune Nivelles. | Commune Verviers. |
|---|---|---|---|---|---|---|---|---|---|
| Excavator | $0.048 | $0.048 | $0.048 | $0.058 | $0.041 | $0.058 | $0.054 | $0.058 | $0.058 |
| Paver | .068 | 068 | 068 | .087 | .077 | | 062 | .077 | .087 |
| Paver's helper | .044 | 044 | 044 | | 058 | | .042 | | |
| Mason | 068 | 068 | 068 | .068 | 058 | .058 | 058 | .068 | .077 |
| Mason's helper | 044 | 044 | 044 | .039 | .041 | .039 | 042 | .044 | |
| Stonecutter | 068 | 068 | .068 | .077 | .068 | .068 | 068 | .077 | .077 |
| Plasterer | 069 | 069 | .069 | .068 | 058 | .058 | .062 | .068 | .068 |
| Plasterer's helper | 050 | 050 | 050 | .034 | 041 | .039 | 042 | .044 | |
| Painter | .064 | 064 | 064 | .068 | .058 | .058 | | .073 | |
| Glazier | 068 | 068 | .068 | .077 | .058 | .058 | | | |
| Glazier's helper | 044 | 044 | .044 | | | | | | |
| Carpenter and joiner | .068 | .068 | .068 | .068 | .058 | .058 | .062 | | .068 |
| Carpenter and joiner's helper | .044 | .044 | .044 | | | .039 | | | |
| Locksmith | .069 | 069 | .069 | .073 | | | | | |
| Locksmith's helper | .044 | .044 | .044 | | | | | | |
| Driver | .054 | .054 | .054 | | | | | | .058 |
| Plumber | | | | .087 | .058 | | 062 | | .068 |
| Plumber's helper | | | | .031 | | | .042 | | |
| Lead and zinc worker | .077 | .077 | .077 | .073 | | .058 | | | |
| Lead and zinc worker's helper | .044 | 044 | .044 | | | | | | |
| Zinc worker | | | | | .058 | | | | .068 |
| Slater | | | | | .058 | | .062 | | |
| Slater's helper | | | | | 041 | | .042 | | |
| Slate roofer | .077 | 077 | .077 | | | .058 | | .068 | |
| Slate roofer's helper | .052 | .052 | .052 | | | | | .044 | |
| Roofer | | | | .077 | | | | | |
| Roofer's helper | | | | .029 | | .039 | | | |
| Blacksmith | | | | | .058 | .058 | | .077 | |
| Marble worker | .069 | .069 | .069 | | | | | | |
| Paper and tapestry hanger | .068 | .068 | .068 | | | | | | |
| Whitewasher | .058 | .058 | .058 | | | | | | |

| Occupation. | Province Brabant. | Commune Anderlecht. | Commune Alost. | Commune Grammont. | Commune Berchem-Anvers. | Commune Charleroi | Commune Herstal | Commune Schaerbeek | Commune Dison. |
|---|---|---|---|---|---|---|---|---|---|
| Excavator | $0.048 | $0.068 | $0.039 | | $0.058 | $0.058 | | $0.068 | $0.064 |
| Paver | | 077 | .058 | | 077 | .073 | $0.073 | .077 | |
| Paver's helper | .042 | 048 | 039 | | 054 | | | .048 | |
| Mason | | 068 | 052 | $0.048 | 071 | 073 | .077 | .077 | |
| Mason's helper | .042 | .048 | 039 | .039 | 054 | | | .048 | |
| Stonecutter | | | .058 | | | | | | |
| Painter | | | | .048 | | | | | |
| Glazier | | | | .068 | | | | | |
| Carpenter and joiner | | | | .048 | | | | | |
| Driver | | | | | | | | .058 | |
| Blacksmith | | | | .048 | | | | | |
| Marble worker | | | | .058 | | | | | |
| Filler-in, masonry | | | | | .068 | | .068 | | .068 |

The replies received from the associations of employers and employees are much less interesting in character and need not here be considered.

The information concerning workingmen employed directly by the city of Brussels showed a total of 1,570 persons employed, of which 800 were credited to the city gas works, 344 to the street-cleaning department, 299 to public works proper, and the remainder to other services. Of these, 7 received in wages less than 2 francs ($0.39) per day, 17 from 2 to 2.50 francs ($0.39 to $0.48), 240 from 2.50 to 3 francs ($0.48 to $0.58), 528 from 3 to 3.50 francs ($0.58 to $0.68), 348 from 3.50 to 4 francs ($0.68 to $0.77), 258 from 4 to 4.50 francs ($0.77 to $0.87), 129 from 4.50 to 5 francs ($0.87 to $0.97), 31 from 5 to 5.50 francs ($0.97 to $1.06), and 12 over 5.50 francs ($1.06). The most prevalent wage rate was that from 3 to 3.50 francs ($0.58 to $0.68), 528, or slightly over one-third of the total number of employees, being found in that class, while 606, or nearly two-fifths, received from 3.50 to 4.50 francs ($0.68 to $0.87). Eleven hundred and thirty-four workingmen, or 72 per cent of all employed, therefore, received from 3 to 4.50 francs ($0.58 to $0.87) per day.

On the basis of this information two calculations were made in order to show the extra expense that would be caused to the city under two different systems of minimum wages. The second calculation, showing the result if all employees worked eight hours a day only and all receiving less than 4 francs ($0.77) per day had their wages increased to that amount, indicated that the extra expense to the city consequent upon its introduction would be 751,740.90 francs ($145,085.99).

*Travail du Dimanche: Établissements Industriels (non compris les mines minières et carrières).* Volume I, Belgique. Office du Travail, Ministère de l'Industrie et du Travail. 1896. lxiii, 503 pp.

The present report is the first of a series of volumes under preparation by the Belgian labor bureau, the object of which is to determine the extent and character of work performed on Sunday in all industries. The investigation when complete will consist of five parts giving (1) the results of an investigation concerning Sunday work in industrial establishments other than mines and quarries, (2) the same concerning mines and quarries, (3) the opinions and advice of the councils of industry and commercial and industrial associations, (4) Sunday work in large stores, and (5) Sunday work in foreign countries (Germany, Austria, Switzerland, and England). Only the first of these, the present report, has as yet been completed.

In this report information was obtained from 1,459 establishments or divisions of establishments, representing 268 different industries and employing 119,477 working men and women.

Of the 1,459 establishments, 12 employed 1,000 or more working people, 34 from 500 to 999, 75 from 250 to 499, 326 from 50 to 249, 470 from 10 to 49, and 542 less than 10 employees.

Of the 119,477 employees, 93,275, or 78.07 per cent, were males, of which 84,762 were 16 years of age or over and 8,513 under that age; and 26,202, or 21.93 per cent, were females, of which 13,163 were 21 years of age or over, 9,127 between the ages of 16 and 21, and 3,912 less that 16 years of age.

[illegible]ysis of the material collected, the Belgian bureau has first [illegible] division of the establishments into two classes—those carrying on regular and those carrying on irregular Sunday work. Within each of these a further threefold division has been made: (1) The work of production proper, (2) the work of repair, cleaning, and maintenance, and (3) the work of guarding, transportation, and shipment.

The report contains six general tables, one for each of these branches. These tables give for each establishment separately the character of the industry, its location, the total number of employees, and the number employed on Sunday, according to sex and age periods, and the per cent of the total number of employees thus employed. These tables, it should be remarked, are not mutually exclusive, as the same establishment may and usually does figure in all six tables.

Of the 1,459 establishments considered, 513, or 35.16 per cent, show no Sunday work, while in 946, or 64.84 per cent, such work is carried on. In these latter, 183, or 19.34 per cent, exhibit regular Sunday work, 516, or 54.55 per cent, Sunday work at irregular occasions, and 247, or 26.11 per cent, both regular and irregular Sunday work.

According to the nature of the work done, 675, or 71.35 per cent, work on Sunday in productive operations, 500, or 52.85 per cent, at repairs, cleaning, etc., and 109, or 11.52 per cent, at guarding works, transportation, and shipment. The sum of these establishments of course exceeds 946, as the same establishments engaged in more than one kind of work.

If distinction be made also as to whether work is regular or not, it is found that of the 946 establishments working Sundays 308, or 32.56 per cent, work regularly at the operations of production, 143, or 15.12 per cent, regularly at repairs, etc., and 74, or 7.82 per cent, regularly at guarding, transportation, etc.; and that 462, or 48.84 per cent, work irregularly at operations of production, 392, or 41.44 per cent, irregularly at repairs, etc., and 41, or 4.33 per cent, irregularly at guarding, etc.

If the percentages be calculated according to the total number of establishments investigated, whether pursuing Sunday work or not, it will be seen that 21.11 per cent of all work regularly on Sundays in the operations of production, 9.80 per cent at repairs, etc., and 5.07 per cent at guarding, transportation, etc.; and 31.67 per cent irregularly at production, 26.87 per cent at repairs, etc., and 2.81 per cent at guarding, transportation, etc.

The reasons for Sunday work were not always stated. It is of interest, however, to know that of the 308 establishments pursuing regular Sunday work in production 119, or 38.64 per cent, carried on such work the whole 24 hours on account of the nature of the work, which was such as to demand continuous operations. The best example of this is that of blast furnaces. In 189 other establishments work was carried on during a part of Sunday only, and in 91 of these the work was due to the fact that one of the shifts of work extended over from Saturday night into Sunday morning. There, therefore, remain 98 establishments, or 31.82 per cent, where Sunday work had no relation to maintaining the continuity of work. In 462 establishments, where work at production was carried on during certain Sundays of the year only, the desire to take advantage of certain seasons of the year was responsible for 76 establishments, or 16.45 per cent, doing Sunday work.

In regard to the frequency of Sunday work in those establishments where irregular Sunday work was carried on, in 110, or 14.42 per cent, of such cases Sunday work was practiced on 26 Sundays or ever, in 244, or 31.98 per cent, on from 10 to 25 Sundays, and in 409, or 53.60 per cent, on less than 10 Sundays.

In a great many of these cases the whole establishment would not, of course, be in operation as during week days. The following table is intended to show the extent to which the entire force was employed:

ESTABLISHMENTS WITH SPECIFIED PER CENT OF EMPLOYEES AT WORK ON SUNDAY.

| Per cent of employees at work on Sunday. | Regular work. | | | Irregular work. | | |
|---|---|---|---|---|---|---|
| | Production. | Repairs, cleaning, maintenance. | Guarding, transportation, shipment. | Production. | Repairs, cleaning, maintenance. | Guarding, transportation, shipment. |
| Less than 5 per cent | 16 | 77 | 53 | 174 | [illegible] | 27 |
| From 5 to 24 per cent | 61 | 52 | 14 | 153 | 23 | [illegible] |
| From 25 to 49 per cent | 94 | 6 | 4 | 32 | 1 | 1 |
| Not less than 50 per cent | 127 | 3 | 3 | 17 | .......... | .......... |
| Not stated | 10 | 5 | .......... | 86 | [illegible] | 10 |
| Total | 308 | 143 | 74 | 462 | [illegible] | 41 |

In this table the most interest naturally attaches to establishments with regular Sunday work. Deducting the 15 establishments not reporting the per cent of their employees at work, it will be seen that of the 510 remaining, 133, or 26.08 per cent, employed on Sunday at least 50 per cent of their employees; that 104, or 20.39 per cent, employed from 25 to 49 per cent; 127, or 24.90 per cent, from 5 to 24 per cent, and 146, or 28.63 per cent, less than 5 per cent of their force.

Turning to the question of the number of working people employed on Sunday, it is found that 77,798, or 65.12 per cent, of all the employees covered by the investigation were never employed on Sundays, while 41,679, or 34.88 per cent, were occupied all or occasional Sundays. Of this last number 13,651, or 32.75 per cent, worked every Sunday; [illegible] or 35.30 per cent, every other Sunday in order to complete the [illegible], and 13,316, or 31.95 per cent, occasional Sundays only. In [illegible] the total number of persons investigated, or 119,477, [illegible] are 11.43 working every Sunday, 12.31 every other [illegible] occasional Sundays.

In regard to the nature of the work, [illegible] 41,679 workingmen who were employed, [illegible]

[illegible] the great majority, or between 89 and 90 per cent, of those [illegible] Sundays were engaged in the operations of production.

[illegible] been seen that 13,316 persons work only occasionally on Sun-[illegible]. In the cases of 1,699 of these the frequency of Sunday labor [illegible] not determined. Of the remaining 11,617 persons 950, or 8.18 per [illegible], work 26 Sundays or more, 4,352, or 37.46 per cent, from 10 to 25 [illegible], and 6,315, or 54.36 per cent, less than 10 Sundays. With [illegible] class, therefore, Sunday work is comparatively infrequent.

To obtain precise information concerning the extent of Sunday work is necessary to carry the analysis to the number of hours worked on [illegible]nday. The most important class of Sunday workers are those who e employed regularly on that day. This class numbers 13,651. Of [illegible]ese, 12,011 are employed in the operations of production. Among [illegible]ese it is found that 3,523 work the same number of hours each Sunday, [illegible]d 8,488 a different number of hours on alternate Sundays. In the [illegible]st class 917 work at least 12 hours, 1,521 from 6 to 12 hours, 1,022 less [illegible]an 6 hours, and 63 a number of hours that could not be determined. [illegible] the second class 3,869 work alternate Sundays 6 and 18 hours, 900 [illegible]ternately 7 and 17 hours, 1,603 first 6, then 12, and then 6 hours, and [illegible]116 in some other combination. Of the 1,422 employed regularly in [illegible]pair, cleaning, and maintenance work 106 were employed at least 12 [illegible]urs, 644 from 6 to 12 hours, 652 less than 6 hours, and 20 a number [illegible] hours not determined. Of the 218 employed in guarding, trans-[illegible]rtation, and shipment service 32 were employed at least 12 hours, 94 [illegible] 6 to 12 hours, 47 less than 6 hours, and the remainder a number [illegible] hours not determined.

Among the 14,712 workingmen who are employed alternate Sundays [illegible] order to complete the shift, the great majority work 6 hours, or until [illegible] o'clock in the morning, and the remaining workingmen various hours [illegible]om 4 to 12.

In regard to the 13,316 workingmen who are employed only occasion-[illegible]ly on Sundays, after deducting 536 whose hours of labor could not be [illegible]tained, it is found that 4,398 are organized in shifts which for the [illegible]st part work first 6 and then 12 or 18 hours on alternate Sundays, [illegible]590 at least 12 hours when working, 3,320 from 6 to 12 hours, and [illegible]472 less than 6 hours.

In the foregoing no account has been taken of the sex or ages of employees. The facts regarding Sunday employment have been presented in the following table in such a way as to bring out these elements:

NUMBER AND PER CENT OF EMPLOYEES WORKING SUNDAY, BY SEX AND AGE.

| Sex and age. | Employees working every Sunday. | | Employees working alternate Sundays. | | Employees working occasional Sundays. | | Total. | |
|---|---|---|---|---|---|---|---|---|
| | Number. | Per cent. | Number. | Per cent. | Number. | Per cent. | Number. | Per cent. |
| Males 16 years or over | 11,965 | 28.75 | 12,642 | 30.33 | 12,183 | 29.19 | 36,790 | 88.27 |
| Females 21 years or over | 255 | .61 | 794 | 1.91 | 429 | 1.03 | 1,478 | 3.55 |
| Males under 16 years | 866 | 2.08 | 1,168 | 2.80 | 407 | .97 | 2,441 | 5.85 |
| Females 16 to 20 years | 307 | .88 | 76 | .18 | 253 | .61 | 636 | 1.67 |
| Females under 16 years | 178 | .43 | 32 | .08 | 64 | .15 | 274 | .66 |
| Total | 13,651 | 32.75 | 14,712 | 35.30 | 13,316 | 31.95 | 41,679 | 100.00 |

From this table it will be seen that 88.27 per cent of those working Sunday were males 16 years of age or over, and 5.85 per cent were males under that age. On the other hand, 3.55 per cent of those working Sunday were females 21 years of age or over, and 2.33 per cent were under that age. It must be remembered that the work performed on alternate Sundays is really more in the nature of work Saturday night, as it is usually terminated early Sunday morning, than Sunday work proper.

There remains one other element yet to be considered in treating of Sunday work—that of the industries in which it is prevalent. The significant results of the whole investigation, from this standpoint, are brought out in the following table, wherein is given the extent of Sunday work in the 68 most important industries.

In this tabulation there are included all industries in which over 10 establishments were investigated or those employing more than 700 workingmen. The table thus restricted relates to 1,069 establishments and 102,952 employees, or 73.27 per cent of all establishments and 86.17 per cent of all employees comprehended within the investigation. A special feature of this table is the column wherein has been shown the per cent of hours worked of possible hours of Sunday work, in order that the relative prevalence of Sunday work in the different industries can be easily discernible. This per cent was calculated in the following manner: The total number of employees in each industry was multiplied by 24 in order to give the maximum number of hours of possible work. For each establishment the number of persons working each Sunday was then multiplied by the number of hours that they worked ... sum of these latter amounts was then divided by the first ... representing the possible capacity. In the case of irregular ... of the number of Sundays worked was also taken into ...

| | | | | |
|---|---|---|---|---|
| Woolen spinning | 27 | 2,991 | 3.7 | 1.0 |
| Jute and hemp spinning and weaving | 9 | 1,283 | 11.1 | 0.3 |
| Linen spinning and weaving | 18 | 7,289 | ........ | ........ |
| Copper and bronze foundry | 12 | 212 | 41.7 | 11.3 |
| Iron foundry (small articles) | 10 | 536 | 30.0 | 0.4 |
| Iron foundry and cast-steel works | 18 | 796 | 16.7 | 3.3 |
| Manufacture of iron castings | 11 | 2,284 | 18.2 | 3.6 |
| Manufacture of gas | 14 | 547 | 7.1 | 1.8 |
| Copper, lead, and zinc working | 18 | 74 | 22.2 | 21.6 |
| Ice making | 2 | 986 | 50.0 | 10.2 |
| Printing and lithographing | 31 | 1,090 | 64.5 | 12.5 |
| Manufacture of laces, ribbons, and braids | 3 | 713 | ........ | ........ |
| Iron and steel plate rolling | 5 | 543 | 60.0 | 12.3 |
| Copper and zinc rolling | 7 | 668 | 14.3 | 7.5 |
| Wool washing and singeing | 11 | 657 | 27.0 | 16.3 |
| Laundry work | 14 | 169 | 14.0 | 12.4 |
| Manufacture of machinery, metallic constructions, railway cars, and supplies | 30 | 6,196 | 70.0 | 7.8 |
| Manufacture of small machinery, accessories, and tools | 14 | 235 | ........ | ........ |
| Masonry and ceiling work | 14 | 246 | 7.1 | 1.2 |
| Marble sawing and cutting | 14 | 113 | 21.4 | 24.8 |
| Horseshoeing | 10 | 24 | 60.0 | 54.2 |
| Flour milling | 34 | 738 | 17.6 | 6.0 |
| Paper and pasteboard making | 16 | 2,839 | 31.2 | 4.4 |
| Preparation of phosphate of lime | 8 | 143 | ........ | ........ |
| Preparation of hair for hats | 4 | 750 | ........ | ........ |
| Manufacture of china and porcelain | 1 | 879 | 100.0 | 4.3 |
| Wool preparation | 10 | 338 | ........ | ........ |
| Ironing linens | 10 | 89 | 40.0 | 42.7 |
| Cotton and linen twisting | 3 | 994 | ........ | ........ |
| Flax retting and breaking | 10 | 398 | 60.0 | 44.7 |
| Soap making | 12 | 99 | 33.3 | 16.2 |
| Wood sawing | 17 | 232 | 17.6 | 20.0 |
| Locksmithing and stove manufacture | 16 | 223 | 18.7 | 18.2 |
| Sugar making and refining | 20 | 2,957 | 100.0 | 82.0 |
| Manufacture of tobacco and cigars | 18 | 727 | 5.5 | 0.3 |
| Leather tanning and currying | 16 | 478 | 31.2 | 2.5 |
| Tapestry and decorating | 8 | 83 | 25.0 | 4.8 |
| Dyeing of threads and woven goods | 25 | 741 | ........ | ........ |
| Garment dyeing and cleaning | 10 | 165 | 40.0 | 40.0 |
| Cotton weaving | 17 | 2,922 | 5.9 | 0.03 |
| Woolen weaving | 11 | 1,613 | ........ | ........ |
| Manufacture of mixed woven goods | 20 | 3,610 | 5.0 | 0.1 |
| Cooperage | 11 | 248 | 36.3 | 28.2 |
| Manufacture of window glass | 8 | 4,092 | 88.9 | 11.2 |
| Manufacture of men's and women's clothing | 23 | 99 | 52.2 | 40.5 |
| Manufacture of zinc | 10 | 4,278 | ........ | ........ |

PER CENT OF REGULAR SUNDAY WORK, BY INDUSTRIES—Concluded.

[In calculating the per cent of hours worked of possible hours of work in the last column of this table 24 has been taken as the possible hours of work for each employee.]

| Industry. | Establishments considered. | Employees considered. | Production. | | Repair, cleaning, maintenance. | | Guarding, transportation, shipment. | | Per cent of hours worked of possible hours of work. |
|---|---|---|---|---|---|---|---|---|---|
| | | | Per cent of establishments working Sundays. | Per cent of employees working Sundays. | Per cent of establishments working Sundays. | Per cent of employees working Sundays. | Per cent of establishments working Sundays. | Per cent of employees working Sundays. | |
| Paper and pasteboard making | 15 | 2,000 | 60.0 | 15.2 | 18.7 | 3.9 | | | 5.63 |
| Preparation of phosphate of lime | 11 | 312 | 54.5 | 26.6 | 36.3 | 5.1 | 45.4 | 1.5 | 12.43 |
| Preparation of hair for hats | 4 | 750 | | | | | | | |
| Manufacture of china and porcelain | 3 | 960 | 66.6 | 2.3 | 33.3 | 0.8 | 33.3 | 0.1 | .85 |
| Wool preparation | 11 | 1,047 | 18.2 | 19.8 | 18.2 | 2.9 | | | 4.74 |
| Ironing linens | 11 | 95 | | | | | | | |
| Cotton and linen twisting | 3 | 994 | 33.3 | 5.2 | | | | | 1.13 |
| Flax retting and breaking | 10 | 398 | | | | | | | |
| Soap making | 13 | 102 | 7.7 | 1.9 | | | | | .65 |
| Wood sawing | 17 | 232 | | | 11.8 | 3.0 | 5.9 | 0.4 | .61 |
| Locksmithing and stove manufacture | 17 | 225 | | | | | 5.9 | 0.9 | .18 |
| Sugar making and refining | 20 | 2,957 | | | | | | | |
| Manufacture of tobacco and cigars | 18 | 727 | 5.5 | 0.3 | | | | | .01 |
| Leather tanning and currying | 19 | 538 | 21.1 | 2.2 | | | 5.3 | 0.4 | .33 |
| Tapestry and decorating | 11 | 97 | | | | | | | |
| Dyeing of threads and woven goods | 26 | 773 | 15.4 | 6.3 | 3.8 | 0.6 | | | 1.62 |
| Garment dyeing and cleaning | 10 | 165 | | | | | | | |
| Cotton weaving | 19 | 3,405 | | | 21.1 | 0.3 | 15.8 | 0.1 | .10 |
| Woolen weaving | 12 | 1,760 | | | 16.6 | 0.6 | | | .09 |
| Manufacture of mixed woven goods | 20 | 3,610 | | | 20.0 | 0.3 | 5.0 | 0.03 | .04 |
| Cooperage | 11 | 248 | | | | | | | |
| Manufacture of window glass | 9 | 4,944 | 100.0 | 73.4 | | | | | 24.65 |
| Manufacture of men's and women's clothing | 27 | 125 | 3.7 | 1.6 | | | | | .33 |
| Manufacture of zinc | 7 | 1,898 | 100.0 | 65.5 | | | | | 22.28 |

PER CENT OF IRREGULAR SUNDAY WORK, BY INDUSTRIES.

| Industry. | Establishments considered. | Employees considered. | Production: Per cent of establishments working Sundays. | Production: Per cent of employees working Sundays. | Repair, cleaning, maintenance: Per cent of establishments working Sundays. | Repair, cleaning, maintenance: Per cent of employees working Sundays. | Guarding, transportation, shipment: Per cent of establishments working Sundays. | Guarding, transportation, shipment: Per cent of employees working Sundays. | Per cent of hours worked of possible hours of work. |
|---|---|---|---|---|---|---|---|---|---|
| Manufacture of sulphuric acid and related industries | 11 | 432 | 9.1 | 0.5 | | | | | .13 |
| Manufacture of steel | 4 | 1,336 | 50.0 | 25.9 | 42.9 | 4.5 | | | 1.45 |
| Manufacture of matches | 6 | 786 | | | 50.0 | 1.2 | | | .02 |
| Sizing of woven goods | 11 | 553 | 9.0 | 1.1 | 45.4 | 2.1 | | | .26 |
| Manufacture of arms | 13 | 2,530 | 30.8 | 2.4 | 15.4 | 0.4 | | | .07 |
| Manufacture of candles | 3 | 740 | 33.3 | 18.0 | 66.6 | 0.8 | | | .70 |
| Bread and pastry baking | 31 | 321 | 16.1 | 3.7 | 6.4 | 1.5 | | | .16 |
| Manufacture of bolts and screws | 8 | 817 | | | 75.0 | 1.8 | | | .13 |
| Brewing and malting | 63 | 666 | 68.0 | 21.0 | 4.8 | 1.5 | 3.2 | 0.3 | 4.59 |
| Manufacture of bricks | 9 | 432 | 88.8 | 4.1 | 20.0 | 1.3 | | | 1.41 |
| Manufacture of brushes | 10 | 529 | | | 22.2 | 1.0 | | | .02 |
| Manufacture of carriages and wagons | 16 | 139 | 43.8 | 8.6 | | | | | .37 |
| Manufacture of hats | 6 | 969 | 16.6 | 0.1 | 33.3 | 1.7 | | | .13 |
| Carpentry and joinery | 21 | 219 | 42.9 | 16.9 | | | | | .53 |
| Boiler making | 16 | 1,211 | 68.7 | 9.8 | 25.0 | 1.5 | | | .66 |
| Manufacture of lime | 18 | 734 | 27.8 | 19.6 | | | | | .25 |
| Manufacture of cement | 11 | 571 | 18.2 | 0.4 | 27.2 | 3.9 | 9.1 | 1.0 | .67 |
| Manufacture of shoes, etc | 20 | 466 | 20.0 | 5.4 | 10.0 | 1.5 | 10.0 | 0.8 | .09 |
| Manufacture of glass and glassware | 8 | 2,609 | 25.0 | 0.1 | 37.5 | 0.2 | | | .06 |
| Cabinetmaking | 16 | 113 | 18.7 | 6.2 | | | | | .23 |
| Iron puddling and rolling | 24 | 8,222 | 58.3 | 82.9 | 58.3 | 4.1 | 3.7 | 0.2 | 3.11 |
| Cotton spinning | 14 | 1,893 | | | 78.5 | 4.2 | | | .18 |

($981,043.60). This was used in the prosecution of public works, such as grading, road construction, forest thinning, etc., the payment of railway and steamer transportation, the furnishing of rations, special grants to municipalities for necessary works, and to benefit local destitute unemployed.

*Fifth Annual Report of the Department of Labor of New Zealand for the year ending March 31, 1896.* Hon. R. J. Seddon, Minister of Labor. xxxiv, 46 pp.

This report, like that for the preceding year, deals with the condition of the labor market; assistance rendered by the department in finding employment for persons out of work, and wages and employees in railway workshops and factories; the effect of certain features of the factory inspection and shops acts, and reports of factory inspectors. In addition to these topics the present report touches briefly upon the effects of the truck act; industrial conciliation and arbitration; the servants' registry offices; foreign immigration; cooperative works, and publishes the minutes of the proceedings of the board of conciliation.

Much space is devoted to the operations of the employment bureau of this department. During the year 2,871 persons obtained employment through this bureau, 1,880 of whom were married. Of the total number, 708 were sent to private employment, and 2,163 to Government works.

Since the organization of the department, June, 1891, 15,739 men have been assisted, and these with their dependents make a total of 53,579 persons who have been directly benefited by this branch of the labor department during the five years of its existence.

The number of factories registered under the factories act increased during the year from 4,109 to 4,647, and the number of factory hands from 29,879 to 32,387. This increase is partly due to a more complete registration of small establishments.

The other topics treated in the report are of local interest.

# DECISIONS OF COURTS AFFECTING LABOR.

[This subject, begun in Bulletin No. 2, will be continued in successive issues, dealing with the decisions as they occur. All material parts of the decisions are reproduced in the words of the courts, indicated when short by quotation marks and when long by being printed solid. In order to save space immaterial matter, needed simply by way of explanation, is given in the words of the editorial reviser.]

## DECISIONS UNDER STATUTORY LAW.

**Employers' Liability—Railroad Companies**—*Texas Central Ry. Co. v. Frazier. 34 Southwestern Reporter, page 664.*—Suit was brought in the district court of Hamilton County, Tex., by Etta Frazier, widow of J. W. Frazier, for herself and minor child, Freddie Frazier, against the Texas Central Railway Company to recover damages for the death of her husband. From a judgment in her favor the railway company appealed the case to the court of civil appeals of the State, which rendered its decision March 4, 1896, and affirmed the judgment of the lower court. The opinion of said court was delivered by Judge Key, and the following, containing a statement of the facts in the case, is quoted therefrom:

On the 15th of April, 1893, a freight train was wrecked on appellant's road near the town of Aquilla, in Hill County, Tex., one result of which was the death of appellee's husband, J. W. Frazier, who was employed and serving as a brakeman on said train.

That appellee was the wife of J. W. Frazier; that the minor, Freddie Frazier, was their only child; that the wreck occurred at the time and place alleged; and that J. W. Frazier was a brakeman on the train, and received injuries in the wreck, which caused his death in a few hours thereafter, were clearly shown, and these facts are not disputed. But appellant's contention is that the testimony fails to show the alleged negligence of the engineer, and fails to show that said engineer, if negligent, was other than a fellow-servant of J. W. Frazier, for whose negligence appellant would not be responsible. It is also contended that the death of Frazier resulted from one of the ordinary risks of the service in which he was engaged, and, therefore, that appellant is not liable.

As to the question of negligence on the part of the engineer, it [illegible] be that, if we were trying the case as jurors, we should reach a [illegible] ent conclusion, and return a different verdict; but, after a [illegible] sideration of the statement of facts, we can not say that the [illegible] without evidence to support it. By the verdict under [illegible] twelve men, presumably disinterested and honest, have [illegible] on the occasion in question the engineer did not [illegible] that a person of ordinary prudence would have [illegible] decision is not so clearly unsupported by [illegible] setting it aside.

The act approved March 10, 1891, defining who are and who are not fellow-servants, declares "that all persons engaged in the service of any railway corporations, foreign or domestic, doing business in this State, who are intrusted by such corporation with the authority of superintendence, control, or command of other persons in the employ or service of such corporation, or with the authority to direct any other employee, are vice principals of such corporation, and not fellow-servants with such employee." (Laws 22d Leg., p. 25.) The evidence in this case shows that Neal, the engineer, had authority from appellant to direct the deceased, who was head brakeman, to put on the brake, and that it was the duty of the deceased to obey such direction. This made the engineer a vice principal, under the statute above cited, and the doctrine of fellow-servants does not apply.

As to the question of Frazier's assumption of risk, it is sufficient to say that, while it is true that he assumed the risks ordinarily incident to his employment as brakeman, such assumption would not shield appellant from injuries resulting from its negligence; and, under the court's charge, the jury were not authorized to find for the plaintiff unless they found that the engineer was guilty of negligence in the respect charged, and that he was appellant's vice principal.

---

EMPLOYERS' LIABILITY—RAILROAD COMPANIES—ANNULMENT OF STATUTE BY ADOPTION OF CONSTITUTION—*Crisswell v. Montana Cent. Ry. Co. 44 Pacific Reporter, page 525.*—This case was originally brought in the district court of Cascade County, Mont., by Charles G. Crisswell against the railroad company to recover damages for injuries received while in the company's employ. A verdict was rendered for the plaintiff and the defendant appealed the case to the supreme court of the State, which rendered its decision November 25, 1895, and affirmed the judgment of the lower court. Said decision was reported in 42 Pacific Reporter, page 767, and was published in part on page 433 of Bulletin No. 4 of the Department of Labor, issued in May, 1896. Subsequently the supreme court granted a rehearing in the case upon the question as to what effect section 11 of article 15 of the State constitution had upon the statute (section 697 of the Compiled Statutes of 1887) on which the former decision in the case hinged. Section 697 of the Compiled Statutes of 1887 reads as follows:

That in every case the liability of the corporation to a servant or employee acting under the orders of his superior shall be the same in case of injury sustained by default or wrongful act of his superior, or to an employer not appointed or controlled by him, as if such servant or employee were a passenger.

The material part of section 11 of article 15 of the constitution of the State is as follows:

And no company or corporation formed under the laws of any other country, State or Territory, shall have, or be allowed to exercise, or enjoy within this State any greater rights or privileges than those possessed or enjoyed by corporations of the same or similar character created under the laws of the State.

Upon this rehearing the supreme court rendered its decision April 13, 1896, reversing its former decision and declaring that section 697 was annulled by section 11 of article 15 of the constitution.

The opinion of said court was delivered by Judge Hunt, and in the course of the same he states, in effect, that section 697 of the Compiled Statues is to be found first as section 20 of "An act to provide for the formation of railroad corporations in the Territory of Montana," passed over the governor's veto on May 7, 1873 (Laws Mont., 1873, ex. sess., p. 93 et seq.), and that an examination of the various sections of the act, taken in connection with its title above quoted, showed that the act applied to domestic railroad corporations only. The judge then continues, and the following is quoted therefrom:

Holding, therefore, that section 697 applied to domestic railroad corporations only, what effect did the adoption of the constitution have upon that section? No comment is necessary to demonstrate that a rule of liability by which a domestic railroad company may suffer heavily for negligence of an employee, where another, but foreign, railroad corporation can not be made liable at all for like negligence, is the imposition of a burden upon the former, and not upon the latter. Whether the legislature of the State may impose such different burdens is immaterial to the question under consideration. Without deciding that question, it may be here assumed they can. Still, our examination will not go beyond the point of ascertaining whether the constitution by section 11, article 15, supra, has annulled section 697, or whether it has extended it so that it has become applicable to all railroad companies, foreign and domestic.

The learned counsel for the respondent argues that section 11 is self-executing. We agree with him in that contention, but not to the extent he would apply the doctrine of self execution. The prohibition lays down a principle of protection to domestic corporations that at once, upon the adoption of the constitution and the admission of the State, became a sufficient rule by means of which the rights and privileges possessed, by domestic companies were and are protected against legislative or other discriminations extending the possession or enjoyment of rights or privileges to foreign corporations greater than those already possessed or those that may be attempted to be granted by any future action. To this extent the provision was completely self-executing, and no legislation was required to give the prohibition full force and operation. Cooley, Const. Lim., p. 99.

But we can not assent to respondent's position that the object of the constitutional provision was to establish uniformity with respect to the two classes of corporations by making laws that were applicable only to the domestic class at the time of the adoption of the constitution extend to the foreign class, in order to make an equal liability for all, or that the clause does establish uniformity by so operating upon such Territorial laws. As said, the inhibition at once, by itself, prevented the discriminations; but there is no affirmative language, and no intent, by the words used, to extend to foreign companies the burdens, rights, and privileges imposed or granted by law to domestic corporations. In this respect legislation must be had to affect [illegible] corporations by force of law. By section 1 of the schedule of the constitution all laws enacted by the legislative assembly of the Territory [illegible] and in force at the time the State [illegible]

[illegible] constitution, should be and remain in full force as [illegible] until altered or repealed, or until expired by their [illegible]. This provision is likewise self-executing. By it, rights [illegible]. It operated of itself to keep in force a system of laws [illegible] of the State, unless such laws were inconsistent with [illegible]. But, as to any such repugnant statutes, it operated as [illegible] repeal, for, when the constitution became the fundamental law, [illegible] in conflict with it yielded, and when the question of a conflict [illegible] to the court, and the conflict clearly appears, the statute [illegible] declared to be inoperative and void. Cooley, Const. Lim., p. 58. As the supreme court of Illinois has very recently said, by way of repetition of one of its earlier decisions:

"The understanding with all persons is that a law passed, either before or after the adoption of the constitution, which is repugnant to its provisions, must be held to be of no valid force, and precisely as if it had been repealed before the performance of the act." Washington Home of Chicago *v.* City of Chicago, 157 Ill. 414, 41 N. E., 893.

From these views it follows that the prohibition clause against any discrimination against a domestic corporation is self-executing as a prohibition but not as an affirmative imposition upon or securement to foreign companies of the rights or privileges expressly only accorded by the State laws to domestic companies. It also follows that by section 697 a greater burden was put upon appellant than was placed upon a foreign company of a similar character. The statute therefore, being inconsistent with the constitution, was annulled by the adoption of the constitution.

---

EMPLOYERS' LIABILITY—RAILROAD COMPANIES—FELLOW-SERVANTS—*Gulf, C. and S. F. Ry. Co. v. Warner. 35 Southwestern Reporter, page 364.*—This action was brought by Charles C. Warner against the Gulf, Colorado and Santa Fe Railway Company to recover damages for injuries received by the plaintiff while in the employ of said company. Judgment was given for the plaintiff in the lower court, and the defendant appealed to the court of civil appeals of Texas, and said court certified the case to the supreme court of the State, which rendered its decision April 27, 1896.

The opinion of the supreme court was delivered by Judge Denman, and contains a statement of the facts in this case, and a clear and definite interpretation of the fellow-servants act of 1893 (chap. 91, acts of 1893), which repealed the fellow servants act of 1891 (chap. 24, acts of 1891), and upon which the result of this action hinged.

Said opinion, practically in full, reads as follows:

The court of civil appeals have certified to this court a question and explanatory statement, as follows:

"On the 7th day of October, 1893, appellee, an employee of appellant, at that time, while engaged with his duties as switchman in the railroad yards of appellant, in Cleburne was injured by a car passing over and crushing his leg. The car that inflicted the injury was being pushed by a locomotive in charge of a switch engineer, who was an employee of appellant, and while switching was being done by a switch crew of which both appellee and the switch engineer were members.

The switch crew consisted of a foreman, the engineer, the fireman, and switchmen. The foreman directed the switching, as it was his duty to do. The engineer had no authority or control over the switchmen. The switchmen were in the transportation department, and the switch engineer in the mechanical department. The yard master employed and discharged the switchmen, and the master mechanic employed and discharged the engineers. The duties of an engineer require skilled labor, and the duties of a switchman do not.

"On motion: Was the switch engineer a fellow-servant of the switchman who was injured, under the provisions of the fellow-servants act of 1893?"

The act referred to, as far as it affects the question certified, is as follows:

"An act to define who are fellow-servants, and who are not fellow-servants, and to prohibit contracts between employer and employees, based upon contingency of the injury or death of the employees, limiting the liability of the employer for damages.

"SECTION 1. *Be it enacted by the legislature of the State of Texas*, That all persons engaged in the service of any railway corporation, foreign or domestic, doing business in this State, or in the service of a receiver, manager, or of any person controlling or operating such corporation, who are intrusted by such corporation, receiver, or person in control thereof, with the authority of superintendence, control, or command of other persons in the employment of such corporation, or receiver, manager, or person in control of such corporation, or with the authority to direct any other employee in the performance of the duty of such employee, are vice principals of such corporation, receiver, manager, or person controlling the same, and are not fellow-servants of such employee.

"SEC. 2. That all persons who are engaged in the common service of such railway corporation, receiver, manager, or person in control thereof, and who, while so employed, are in the same grade of employment, and are working together at the same time and place, and to a common purpose, neither of such persons being intrusted by such corporation, receiver, manager, or person in control thereof, with any superintendence or control over their fellow-employees, or with the authority to direct any other employee in the performance of any duty of such employee, are fellow-servants with each other: *Provided*, That nothing herein contained shall be so construed as to make employees of such corporation, receiver, manager, or person in control thereof, fellow-servants with other employees engaged in any other department or service of such corporation, receiver, manager, or person in control thereof. Employees who do not come within the provisions of this section shall not be considered fellow-servants." Gen. Laws, 1893, p. 120.

It will be observed that the caption of the act declares its purpose to be "to define who are fellow-servants and who are not fellow-servants," and that section 2 completely accomplishes such purpose by first defining who are fellow-servants, and then declaring that "employees who do not come within the provisions of this section shall not be considered fellow-servants." This section divides all employees into fellow-servants, and nonfellow-servants, and gives the distinctive characteristics of the former, but not of the latter. The purpose of the statute was accomplished by limiting and definitely determining the employees who should thereafter be classed as fellow-servants, for whose negligence the employer should not be responsible to another fellow-servant; [illegible]

[illegible] statutory definition of fellow-servants, for the employer [illegible] responsible for their negligence, whether they be termed [illegible] principals, or otherwise.

[illegible] characteristics prescribed by the statute as essential [illegible] concurring and common to two or more employees in order to [illegible] them fellow-servants are: First. They must be "engaged in the common service." As here used, "service" means the thing or work being performed for the employer at the time of the accident, and out of which it grew, and "common" means that which pertains equally to the employees sought to be held fellow-servants; and, therefore, "common service" means the particular thing or work being performed for the employer, at the time of the accident, and out of which it grew, jointly, by the employees sought to be held fellow-servants.

The members of a crew running a train, though each be in the performance of different acts in reference thereto, are all "engaged in the common service," for they are jointly performing the thing or work of managing the train for the employer; but they would not be "engaged in the common service" with the members of a crew running another train for the employer over the same road, for one crew would be jointly performing the thing or work of managing one train, while the other would be jointly performing the thing or work of managing the other train. We therefore conclude that the engineer and switchman were "engaged in the common service."

Second. They must be "in the same grade of employment." "Grade" means the rank or relative positions occupied by the employees while "engaged in the common service." This definition, however, gives us no certain means of determining whether given employees are in the same or different grades, for it furnishes no test by which their respective ranks or relative positions "in the common service" can be ascertained. In the absence of a statutory test, the grade would have depended upon the test which might have been adopted by the courts, such as authority one over the other, order of promotion, skill in the service, compensation received, etc. We are of the opinion that the legislature anticipated and settled this difficulty in the construction of the word "grade" by the use of the clause, "neither of such persons being intrusted * * * with any superintendence or control over their fellow-employees," etc., as explanatory of what was meant by the clause "in the same grade;" thus adopting the most natural test of grade in the construction of the statute, authority one over the other while "engaged in the common service." Probably the most serious difficulty in arriving at the conclusion that one clause was intended as merely explanatory of the other is the fact that the explanatory clause does not immediately follow the one it explains; but this objection is removed when we consider that, in the original section, as enacted in 1891, the qualifying clause immediately follows the words "same grade," and was evidently intended to explain their meaning. Since the engineer had no authority or control over the switchman, and vice versa, while "engaged in the common service," we conclude that they were "in the same grade of employment."

Third. They must be "working together at the same time and place." While "at" indicates nearness in time and place, it does not demand an

Fourth. They must be working "to a common purpose." By this is meant that the acts required of each in the performance of his duties at the time of the accident must be in furtherance of "the common service." We are of the opinion that the engineer in managing the engine, and the switchman in performing his duties, both having in view the switching of the cars, were working to a "common purpose." When these four distinguishing characteristics are found concurring and common to two or more employees, they must be held fellow-servants under the statute; otherwise, not.

It is urged that the proviso adds, as another distinguishing characteristic, that they must be in the same department. A proviso may be inserted for the purpose either of adding something to, or of insuring a certain construction of, the preceding language of the statute. This proviso bears upon its face unmistakable evidence of having been inserted for the latter purpose. It says: "Provided nothing herein contained shall be so construed as to make employees * * * fellow-servants with other employees engaged in any other department or service," and to complete the idea we may add the words "than the common service" above specified. The words "department or service," as here used, merely means a subdivision of business, as running a train, clearing away a wreck, repairing a track, etc., and, if employees are, at the time of the accident, engaged in the same subdivision of business, they are also "engaged in the common service," as we have hereinbefore construed that term. In other words, the proviso was merely intended to insure the strict construction above given by us to the words "engaged in the common service." In so far as section 1 of the act bears upon the question of "who are fellow-servants and who are not fellow-servants," we can not see that it adds anything to section 2. It merely selects a certain class of employees, who are non-fellow-servants under the terms of section 2, and declares that they are vice principals. It results that we must answer the question certified in the affirmative.

---

WEIGHING COAL AT MINES—CONSTITUTIONALITY OF STATUTE—*Harding et al. v. People. 43 Northeastern Reporter, page 624.*—William Harding and another were convicted of a crime in the circuit court of Vermilion County, Ill., and brought their case before the supreme court of the State on a writ of error. Said court rendered its decision March 30, 1896, and reversed the judgment of the lower court. The facts in the case are given in the opinion of the supreme court, delivered by Judge Cartwright, which reads as follows:

Plaintiffs in error were indicted and convicted for a violation of the act requiring the weighing of coal at mines, in force July 1, 1887, as amended by act in force July 1, 1891. Some of the counts upon which they were found guilty charged them with a failure to weigh all the coal delivered from the mine, and others charged them with not keeping a correct record of the weight of each miner's car. The portion of the act under which the prosecution was had, material to the same, is as follows:

"SECTION 1. That the owner, agent, or operator of every coal mine in this State at which miners are paid by weight, shall provide at such mines suitable and accurate scales of standard manufacture for the [illegible] delivered from such mines shall be carefully [illegible] as above provided, and a correct record shall be [illegible] each miner's car, which record shall be kept open [illegible] for the inspection of all miners or others pecuniarily [illegible] the product of such mine. The person designated and [illegible] weigh the coal and keep such record shall, before entering [illegible] make and subscribe to an oath before some magistrate [illegible] authorized to administer oaths, that he will accurately [illegible] carefully keep a true record of all coal delivered from such [illegible] such oath shall be kept conspicuously posted at the place of [illegible]

[illegible] Any person, owner or agent, operating a coal mine in this State who shall fail to comply with all the provisions of this act, or who shall obstruct or hinder the carrying out of its requirements, shall be fined for the first offense not less than fifty dollars ($50) nor more than two hundred dollars ($200); for the second offense not less than two hundred dollars ($200) nor more than five hundred dollars ($500); and for a third offense not less than five hundred dollars ($500), or be imprisoned in the county jail not less than six months nor more than one year: *Provided*, That the provisions of this act shall apply only to coal mines whose product shall be shipped by rail or water."

The constitutionality of this act is challenged by plaintiffs in error, and this is the only question that will be considered, although the application of the statute to this case is disputed, and questions of variance and of error in giving and refusing of instructions are also raised. It is objected that the act is in violation of section 2 of article 2 of our constitution, which provides that no person shall be deprived of life, liberty, or property without due process of law, because it singles out operators of one class of coal mines and imposes restrictions upon them not required to be borne by operators of other mines, or by persons engaged in other business, and also by interfering with the right of employer and laborer to contract with each other. The Consolidated Coal Company had owned and operated the mine where plaintiffs in error were employed for six or seven years. The greater part of i product was shipped from the mine by rail, on the Wabash Railroad, and sold in other markets. All the coal so shipped was correctly weighed on scales of standard manufacture by the company, at the mine, before being dumped into the railroad cars, and a correct record was made of the weight of each miner's car, and that record was posted and kept open at all reasonable hours for the inspection of the miners or any person interested. During this time the company had also furnished the Wabash Railroad Company with coal for its locomotives, which was delivered at the mine, into tenders of the locomotives, as they stopped there for coal. There were about 250 miners employed, and the average output of the mine was from 700 to 950 tons of screened coal per day. The miners were paid 55 cents per ton for screened coal. About the last 100 miners' cars that came up in the evening of each day would be placed on the storage tracks, for the purpose of coaling the locomotives during the night and the next day. This last coal was not weighed, but each miner was given the average weight of the cars sent up by him and weighed during the day as the weight of his last car, crediting him with the average weight of the cars mined by him that day that had been actually weighed. By the act under consideration its provisions are applied only to coal mines whose produce is shipped by rail or water, and the learned attorney-general and counsel for the people construe the provision as making the law applicable to each mine where the major portion of its product is so shipped. However that may be, it is plain that the act not only singles out the operator of a mine, and imposes restrictions and burdens upon him as to the use and enjoyment of his property that are not imposed upon other branches of business similarly situated and conducted, but it divides the operators of mines and only applies its provisions to those whose product is shipped in a certain manner. In the various constitutions the phrases "due process of law" and "the law of the land" are used interchangeably, sometimes one being employed and sometimes the other; but they are synonymous, and the meaning is the same in every case. Cooley, Const. Lim., 353. In Millett *v.* People, 117 Ill., 294, 7 N. E., 631, it was said of this phrase, "And this means general public law, binding upon the members of the community, under all circumstances, and not partial or private laws, affecting the rights of private individual or classes of individuals," citing James *v.* Reynolds, 2 Tex., 251; Wynehemer *v.* People, 13 N. Y., 376; Vanzant *v.* Waddel, 2 Yerg., 269. And the same declaration was made in Frorer *v.* People, 141 Ill., 171, 31 N. E., 395, where the statute prohibiting engaging in keeping a truck store was held unconstitutional, and in Braceville Coal Co. *v.* People, 147 Ill., 66, 35 N. E., 62, where the same conclusion was reached as to an act to provide for the weekly payment of wages by a corporation.

The right to enact such a statute does not arise out of the police power, where much latitude is allowed in determining what may tend to insure the comfort, safety, or welfare of society; and it is not authorized by section 29 of article 4 of the constitution, providing for laws to secure safety to coal miners. Millett *v.* People, supra. Each person subject to the laws has a right that he shall be governed by general public rules. Laws and regulations entirely arbitrary in their character, singling out particular persons not distinguished from others in the community by any reason applicable to such persons, are not of that class. Distinctions in rights and privileges must be based upon some distinction or reason not applicable to others. In Braceville Coal Co. *v.* People, supra, it is said: "And it is only when such distinctions exist that differentiate in important particulars, persons or classes of persons from the body of the people, that laws having operation only on such particular persons or classes of persons have been held to be valid enactments." No possible reason or distinction affecting any interest, justifying the division of mines made by the act, has been suggested, except that it might be intended to reach mines in which the larger number of miners were employed. But this is not the division or distinction made, and does not in any manner follow from such division. It is not the language or purport of the act, and, if such had been the intention of the legislature, it would certainly have been made manifest by basing the division or distinction upon the number of miners employed. The act applies equally to the owner of a small mine, where the product may not exceed a carload per day, and the owner of a mine such as that of the Consolidated Coal Company. The distinction is based solely upon the fact of the product being shipped by rail or water, and counsel have been able to suggest no reason why the legislature should require the product of such a mine to be weighed in the manner specified, and not that of another mine where the product is sold on the spot. The distinction between operators who sell their product at the mine to some shipper, who ships it away to the market, and those who themselves ship their coal by rail or water, is [illegible] special burdens and restrictions upon one class, [illegible] other, constitute an arbitrary deprivation of [illegible] makes that an offense, if committed by a person [illegible] branch of mining, which, if done by persons in another [illegible] business, is lawful, without any reason for distinction [illegible] the two, we must regard it as unconstitutional.

[illegible] v. People, supra, and Ramsey v. People, 142 Ill., 380, 32 [illegible] provisions similar to those of the act now under consideration were held to be unconstitutional and void, the general right of the laborer and employer to contract in regard to the price of labor, and the method of ascertaining the price, was asserted; and the rule was laid down that any restriction upon that right is a deprivation of both liberty and property within the meaning of the constitutional provision. In view of the discussion of the principles involved in these cases, no extended statement of them will be necessary here. This act makes it an offense against the law for the employer and laborer, at any coal mine at which the miners are paid by weight, to determine upon the weight of any miner's car, or any lot of coal, by any other method than that pointed out by the statute. A failure to weigh a car, and to keep a correct record of the weight, renders the operator liable to the penalties prescribed by the act, although he and the laborer may have agreed upon the weight of the car, or contracted for other methods of determining the weight. This is well illustrated by the facts of this case. The last cars that came up in the evening of each day, designed for coaling the locomotives, would not be weighed until the coal was dumped into the tenders, so that the miners could not obtain the weights until the next day, and they wanted them the same evening. For this reason, and at the instance of the miners, it has been the custom for six years to give each miner the weight according to the system above stated. When the company attempted to change this system and weigh the coal, the miners objected, and insisted upon the custom of averaging weights. No objection was ever made by any miner to this manner of arriving at the weight, instead of weighing the cars on the scales. Here was an arrangement, amounting to a contract between the parties, with which all the contracting parties were satisfied, and the testimony upon which plaintiffs in error were convicted came from miners who left during the miners' strike of 1894, and were not again employed by the company. It seems that a law which deprives men engaged in the business of mining from contracting with each other for the purpose of ascertaining the weight of the coal mined, or the amount due them, in any manner mutually satisfactory, can not be sustained. That such is the effect of this law is the contention of counsel for the people, and it is only upon the assumption that the law does so control the power to contract that a conviction could have been had in this case; for, as already seen, the parties had contracted otherwise. This act takes away the freedom of contracting by the parties for the ascertainment of the weight of coal, except by a certain method; and, in our opinion, it is unconstitutional. The judgment will be reversed, and the cause remanded, with direction to the circuit court to discharge the defendants. Judgment reversed.

## DECISIONS UNDER COMMON LAW.

CONSPIRACY—BOYCOTTS—*Oxley Stave Co. v. Coopers' International Union of North America et al. 72 Federal Reporter, page 695.*—This case was brought in the United States circuit court for the district of Kansas by a bill in equity filed by the Oxley Stave Company against the Coopers' International Union of North America, Lodge No. 18, of Kansas City, Kans., the Trades Assembly of Kansas City, Kans., and various individuals named, who are officers and members of such organizations, and also against "all other persons who may be members of either of said organizations, their agents, attorneys, etc.," to enjoin them from inaugurating and maintaining a boycott against the use of packages, casks, barrels, etc., made by complainant by means of certain machines constituting part of its plant. The circuit court rendered its decision March 9, 1896, and allowed the injunction asked for. The opinion of said court was delivered by District Judge Foster, and the following is quoted therefrom:

This brings us to the question whether, under the allegations of the bill, which is verified, and the other evidence presented, the complainant is entitled to the relief prayed for. The material allegations of the bill are but partially controverted by the defendants. Indeed, they are substantially admitted. Much testimony was offered to show that barrels hooped by machinery were not as serviceable or as valuable as hand-hooped barrels. It also appears that there is some little difference in the price of such barrels; that a skilled workman can hoop 14 to 16 barrels per day by hand, and that the hooping machine does the work of about six or seven men; and that boys or young men, from 16 years upward, are employed, to some extent, in operating the machines. All of this cuts but little figure in the case. Whether the work of the machine is better or worse than the hand work is not material. The barrels are made and sold as machine work, and a price fixed accordingly, and the customer must decide whether or not he will buy them; and the complainant, in operating the machines in its business, is engaged in a legitimate enterprise, and defendants had no legal right to demand that it should cease operating them. There is some testimony tending to show that the reason the packing companies had not made contracts for these barrels for this year was not on account of the threatened boycott, but because they preferred hand-hooped barrels. The purchasing agent of Fowler Sons & Co., Limited (Robert McWhittaker), however, testified that a committee of the Coopers' Union and Trades Assembly notified him, if his company purchased machine-made barrels, they would boycott the contents of the barrels, and that such notice would tend to make his company very careful about purchasing machine-made barrels. The manager of Swift & Co. testified that his company was buying hand-made barrels on account of the threatened boycott. The following is a copy of the resolution of the Trades Assembly on the subject, and indicates the purpose of the defendant associations:

"To the officers and members of the Trades Assembly, greeting: Whereas, the cooperage firms of J. R. Kelley and the Oxley Cooperage Company have placed in their plants hooping machines operated by child labor; and whereas, said hooping machines is the direct cause of at least one hundred coopers being out of employment, of which a great many are unable to do anything else, on account of age,—at a meeting held by Coopers' Union No. 18 on the 31st of December, 1895, a committee was appointed to notify the above firms that unless they discontinued the use of said machines on and after the 15th of January, [illegible] that Coopers' Union No. 18 would cause a boycott to be placed on [illegible] packages hooped by said machines, the 15th of [illegible]; and [illegible]

[illegible] to bring the matter before the Trades Assembly [illegible] and petition the Assembly to indorse our action, and to [illegible] in the hands of their grievance committee, to act in [illegible] with a committee appointed by Coopers' Union No. 18 to [illegible] before letting their contracts for their cooperage. [illegible], be it resolved, that this Trades Assembly indorse the action of Coopers' Union No. 18, and the matter be left in the hands of the grievance committe for immediate action.

"Yours, respectfully,

"J. L. COLLINS,

"*Secretary Coopers' International Union of N. A., Lodge No. 18.*"

James Coble, president of Coopers' Union, testified as follows: Unless complainant ceased using the machines—

"That the boycott would be declared by the Coopers' Union upon the contents of the tierces and barrels hooped by machinery; meaning thereby that the members of the said Coopers' Union, and of its parent association, the Trades Assembly, would thereafter cease to purchase or use any of the commodities that were packed in machine-hooped tierces or barrels."

No one can question the right of the defendants to refuse to purchase machine-made packages, or of goods packed in them, or, by fair means, to persuade others from purchasing or using them. If that is all that is implied by a boycott, as insisted by defendants, it is difficult to see where they violate any law, although it might injure the complainant's business. It has been decided, however, that while such action would not be unlawful by an individual, a combination and conspiracy to accomplish the purpose would be an illegal act. In Arthur *v.* Oakes, 11 C. C. A., 209, 63 Fed., 321, 322, Mr. Justice Harlan says:

"It is one thing for a single individual, or for several individuals, each acting upon his own responsibility, and not in cooperation with others, to form the purpose of inflicting actual injury upon the property or rights of others. It is quite a different thing, in the eye of the law, for many persons to combine or conspire together with the intent not simply of asserting their right of accomplishing lawful ends by peaceable methods, but of employing their united energies to injure others or the public. An intent upon the part of a single person to injure the rights of others or of the public is not in itself a wrong of which the law will take cognizance, unless some injurious act be done in execution of the unlawful intent. But a combination of two or more persons with such intent, and under circumstances that give them, when so combined, a power to do an injury they would not possess as individuals acting singly, has always been recognized as in itself wrongful and illegal."

The term "boycott" has acquired a significance in our vocabulary, and in the literature of the law. The resolution of the defendant associations says, unless complainant discontinue the use of said machines on and after January 15, 1896, that Coopers' Union No. 18 would cause a boycott to be placed on all packages hooped by said machines. Just what action would be taken, the resolution does not state. It does not say the defendants would not purchase the packages, or the goods packed in them, but simply says a "boycott" would issue. That term implies that a general proscription of all articles so manufactured, and the goods packed in them, would be inaugurated and maintained by the power of these assemblies, wherever they could reach. It is fair to presume, from the resolution and other testimony, that the defendants were determined to use all means, short of violence, to make the proscription effective. That has been the history of such proceedings in the past, and such is the meaning imputed to the use of the word "boycott." It has become a word carrying with it a threat and a menace, and was evidently so intended by this resolution. In Thomas *v.* Railway Co., 62 Fed. 818–821, the court says:

"But the combination was unlawful, without respect to the contract feature. It was a boycott."

Again the court says:

"The combination under discussion was a boycott. It was so termed by Debs, Phelan, and all engaged in it. Boycotts, although unaccompanied by violence, have been pronounced unlawful in every State of the United States where the question has arisen, unless it be in Minnesota, and they are held to be unlawful in England."

The court further says:

"Boycotts have been declared illegal conspiracies in State *v.* Glidden, 55 Conn. 46, 8 Atl. 890; in State *v.* Stewart, 59 Vt. 273, 9 Atl. 559; Steamship Co *v.* McKenna, 30 Fed. 48; Casey *v.* Typographical Union, 45 Fed. 135; Toledo A. A. & N. M. Ry. Co. *v.* Pennsylvania Co. 54 Fed. 730, and in other cases."

From these authorities we reach the conclusion that complainant is entitled to the relief prayed for. The labor-saving machines which modern invention has brought into every industry in life excite our wonder and admiration, but our enthusiasm is subdued by the thought that the machines must largely drive the skilled laborer out of a field he has spent years to fit himself for, and upon which, more or less, depends the means of livelihood for himself and his family; and yet it is a hopeless task for the laborer to contend against the use of machinery, wherever it can be utilized. Labor can only adjust itself to the constant progress made in all the mechanical pursuits, and it has been well said that, despite all the inventions to save hand work, there never was a time when the laborer was paid better, or had greater advantages, than he has to-day. The injunction will be allowed as prayed for by complainant.

---

CONSPIRACY—UNLAWFUL COMBINATIONS—*Elder et al. v. Whitesides et al. 72 Federal Reporter, page 724.*—This case was brought by a bill in equity filed in the United States circuit court for the eastern district of Louisiana by Elder, Dempster & Co., of Liverpool, England, owners of certain steamboats, against William Whitesides and others, citizens of Louisiana. The bill alleged the existence of an unlawful combination and conspiracy, on the part of the defendants, to prevent the loading or unloading of complainants' steamboats at Gretna, La., except by such labor as might be acceptable to said defendants; that such combination and conspiracy absolutely prevented complainants from loading or unloading their steamers at said port of Gretna by other than the said defendants and their confederates. An injunction was asked restraining the defendants from continuing their said combination and conspiracy. The court rendered its decision [illegible]

The opinion of the court was delivered by District Judge Parlange, and the following is quoted therefrom:

The defendants have been granted all the time which they have requested to present their side of the case. The argument made by their counsel may be divided under four heads. He urged: First, that there is no allegation or proof of any overt act committed by the defendants against the particular vessel mentioned in the bill; second, that a court of equity can not enjoin crime; third, that no damages have actually been inflicted upon the vessel; and, fourth, that the proof of conspiracy is insufficient.

In a recent case decided by the United States circuit court of appeals, seventh circuit (Arthur v. Oakes, 11 C. C. A., 209; 63 Fed., 310), in which Mr. Justice Harlan was the organ of the court, all the law points made by the counsel for the defendants have been passed upon, clearly and distinctly. In speaking of combinations and conspiracies, Mr. Justice Harlan said:

"According to the principles of the common law, a conspiracy upon the part of two or more persons, with the intent, by their combined power, to wrong others, or to prejudice the rights of the public, is in itself illegal, although nothing be actually done in the execution of such conspiracy. This is fundamental in our jurisprudence. So, a combination or conspiracy to procure an employee or body of employees to quit service, in violation of the contract of service, would be unlawful, and, in a proper case, might be enjoined, if the injury threatened would be irremediable in law. It is one thing for a single individual or for several individuals, each acting upon his own responsibility, and not in cooperation with others, to form the purpose of inflicting actual injury upon the property or rights of others. It is quite a different thing, in the eye of the law, for many persons to combine or conspire together with the intent, not simply of asserting their rights or of accomplishing lawful ends by peaceable methods, but of employing their united energies to injure others or the public. An intent upon the part of a single person to injure the rights of others or of the public is not in itself a wrong of which the law will take cognizance, unless some injurious act be done in execution of the unlawful intent. But a combination of two or more persons with such an intent, and under circumstances that give them, when so combined, a power to do an injury they would not possess as individuals acting singly, has always been recognized as in itself wrongful and illegal."

The justice cites approvingly the language of another court, as follows: "There is nothing in the objection that to punish a conspiracy where the end is not accomplished would be to punish a mere unexecuted intention. It is not the bare intention that the law punishes, but the act of conspiring, which is made a substantive offense by the nature of the object to be affected." State v. Buchanan, 5 Har. and J., 317.

The justice further said:

"The authorities all agree that a court of equity should not hesitate to use this power [injunction] when the circumstances of the particular case in hand require it to be done in order to protect rights of property against irreparable damages by wrongdoers. * * * That some of the acts enjoined would have been criminal, subjecting the wrongdoers to actions for damages or to criminal prosecution, does not, therefore, in itself determine the question as to interference by injunction. If the acts stopped at crime, or involved merely crime, or if the injury threatened could, if done, be adequately compensated in damages, equity would not interfere. But as the acts threatened involve irreparable injury to and destruction of property for all the purposes for which the property was adapted, as well as continuous acts of trespass, to say nothing of the rights of the public, the remedy at law would have been inadequate. 'Formerly,' Mr. Justice Story says, 'courts of equity were extremely reluctant to interfere at all, even in regard to cases of repeated trespasses. But now there is not the slightest hesitation, if the acts done or threatened to be done to the property would be ruinous, irreparable, or would impair the just enjoyment of the property in future. If, indeed, courts of equity did not interfere in cases of this sort, there would, as has been truly said, be a great failure of justice in this country.'"

So far as the question of jurisdiction is concerned, it is clearly settled, both by Arthur *v.* Oakes, supra, and by the decision of the United States circuit court of appeals of this (the fifth) circuit. Hagan *v.* Blindell, 6 C. C. A., 86; 56 Fed., 696. In both of those cases the jurisdiction depended entirely, as in the case at bar, upon the diverse citizenship of the parties and the equitable powers of the court.

The decisions above referred to clearly dispose of all the law points raised by defendants' counsel. The proof of conspiracy is made out by the affidavits offered by complainants. The only proof offered by the defendants is their affidavit, which confines itself to a denial that they interfered with the complainants or prevented the loading of the vessel *Niagara*, or caused damages to the complainants. This seems to be in line with the argument of their counsel, and to be based upon the theory that the jurisdiction of the court depends upon unlawful overt acts having been committed against the particular vessel mentioned in the bill, and upon actual damages having been caused the complainants, prior to the application for the injunction. (There is no denial of the agreement or conspiracy to do the unlawful things charged in the bill, which conspiracy is the gravamen of the case.) The preliminary injunction must issue.

---

EMPLOYERS' LIABILITY—FELLOW-SERVANTS—*Southern Pacific Co. v. McGill. 44 Pacific Reporter, page 302.*—Action was brought in the district court of Pima County, Ariz., by William McGill against the Southern Pacific Company to recover damages for injuries sustained while in the employ of said company. Judgment was rendered for McGill and the company appealed the case to the supreme court of the Territory of Arizona, which affirmed the judgment of the lower court. The court, however, granted a rehearing, and as a result of the same rendered a decision February 10, 1896, reversing the judgment of the lower court. The facts of the case were as follows:

McGill, hereinafter referred to as "the plaintiff," was a section foreman in the employ of the defendant company. He was directed by the roadmaster to go to a point on the track 6 or 7 miles west of a section called "Pantano," and there to grade and lay a track in order to [illegible] an engine which had been derailed. He went there with his [illegible] tools and worked part of a day, when the civil engineer in [illegible]

the train started, and had not gone over three-quarters of a mile when it collided with a passenger train and the plaintiff was seriously injured about the head. The charge was made in the complaint that Barrett, the conductor of the work train, ran the train negligently, and with want of care and attention to his duty, and so caused the accident.

The opinion of the supreme court was delivered by Chief Justice Baker, and contains the following:

The following instruction was given to the jury for the plaintiff: "The court instructs the jury that the conductor of a railway train, who commands its movements, directs when it shall start, at what station it shall stop, and has the general management of it, and control over the persons employed on it, represents the railway company; and is not a fellow-servant with a section foreman in the employ of said company. If the jury believe from the evidence that John Barrett was the conductor of the train upon which plaintiff was, and had the powers just stated regarding such train, the court instructs the jury that Barrett was not a fellow-servant with the plaintiff."

This instruction was not altered, changed, or modified by any instruction subsequently given, and, being objected to and duly assigned as error, constitutes the pivotal point in the case. There is an endless diversity of opinion upon this "fellow-servant" doctrine in the decisions of the various courts in this country. The cases are too numerous to cite, and it would be an idle effort to attempt to reconcile or distinguish them. I can do no better than to deduce one or two propositions applicable to the facts at bar, which the decided weight of all the cases authorizes.

(1) A person entering upon the service of a corporation assumes all the risk naturally incident to his employment, including the dangers which may arise from the negligence of a fellow-servant.

(2) That the master's liability does not depend upon gradations in the employment, unless the superiority of the person causing the injury was such as to make him principal or vice principal.

(3) The liability of the master does not depend upon the fact that the servant injured may be doing work not identical with that of the wrongdoer. The test is, the servant must be employed in different departments, which in themselves are so distinct and separate as to preclude the probability of contact and of danger of injury by the negligent performance of the duties of the servant in the other department.

In the case at bar the plaintiff and Barrett, the conductor, were brought together at the same time and place, and closely associated in the discharge of their respective duties. The very work which the plaintiff engaged to do necessitated the constant use of a train, such as the one in use at the time of the collision, to transport laborers, tools, materials, supplies, etc., to the place of operations; and he must be held to have contemplated its use when he accepted the employment. He was at work when riding upon this train in going to and from the point where the wreck occurred, just as much as he was when he was actually engaged in raising the derailed engine. Both he and the conductor were engaged in a common purpose and object—the clearing of the track and the raising of the fallen engine.

The labors of both contributed to and were intended to effect that immediate and present result. Both had a common master. That there was some gradation—some difference in the work of the two—

within themselves as to preclude the probability of contact and of danger to one servant in one department by reason of the negligence of another servant in another department. This can not be said of the plaintiff's and Barrett's employment.

The plaintiff's labors constantly exposed him to the dangers of running and moving the work train, and he must be held to have assumed the risk of such dangers.

The giving of the instructions quoted was reversible error, since, upon the facts, the conductor of the work train and the plaintiff were fellow-servants. The judgment is reversed and a new trial is ordered.

---

EMPLOYERS' LIABILITY—FELLOW-SERVANTS—*Northern Pacific R. R. Co. v. Peterson. 16 Supreme Court Reporter, page 843.*—This action was commenced by Peterson in the United States circuit court for the district of Minnesota, fourth division, to recover damages for injuries sustained while in the employ of the railroad company. The facts in the case were as follows:

The plaintiff, a day laborer, was employed on an extra gang, amounting in numbers to thirteen men, with one Holverson as foreman, at a place called "Old Superior," a station on the line of the defendant's road. Holverson had power to employ men, and also to discharge them. The men were taken each morning on hand cars to the place where they were to work during the day, and when the work was finished were brought back.

The members of the gang themselves worked the hand cars, Holverson generally occupying a place on the front hand car and taking care of the brakes. He always went with the gang, superintended their work, even if taking no part in the actual manual labor, and came home with them at the end of the day's labor. When the accident occurred Holverson held his accustomed place on the front hand car, at the brakes, and Peterson was on the same car. While going around a curve in the track Holverson thought he saw some object in front of him and applied the brakes suddenly, in consequence of which the car was abruptly stopped. He gave no warning of his intention, and the rear car was following so closely that it could not stop before running into the car ahead, the result of which was that the first car was thrown from the track, throwing the plaintiff Peterson off the car and injuring his leg by having the rear car run over it. Upon these facts the jury returned a verdict in favor of Peterson, and the case was taken by the railroad company to the United States circuit court of appeals for the eighth circuit upon a writ of error. Said court affirmed the judgment of the court below, and the railroad company then carried the case on writ of error to the United States Supreme Court, which rendered its decision April 13, 1896, reversing the judgments of the lower courts

The opinion of said court was delivered by Mr. Justice Peckham, and the following is quoted therefrom:

The sole question for our determination is whether Holverson occupied the position of fellow-servant with the plaintiff below. If he did, then this judgment is wrong, and must be reversed.

By the verdict of the jury, under the charge of the court, we must take the fact to be that Holverson was foreman of the extra gang for the defendant company, and that he had charge of and superintended the gang in the putting in of the ties, and assisting in keeping in repair the portion of the road included within the three sections; that he had power to hire and discharge the hands in his gang, then amounting to thirteen in number, and had exclusive charge of the direction and management of the gang in all matters connected with their employment; that the plaintiff below was one of the gang of hands so hired by Holverson, and was subject to the authority of Holverson in all matters relating to his duty as laborer. Upon these facts the courts below have held that the plaintiff and Holverson were not fellow-servants in such a sense as to preclude plaintiff recovering from the railroad company damages for the injuries he sustained through the negligence of Holverson, acting in the course of his employment as such foreman.

In the course of the review of the judgment by the United States circuit court of appeals, that court held that the distinction applicable to the determination of the question of a coemployee was not "whether the person has charge of an important department of the master's service, but whether his duties are exclusively those of supervision, direction, and control over a work undertaken by the master, and over subordinate employees engaged in such work, whose duty it is to obey, and whether he has been vested by the common master with such power of supervision and management." Continuing, the court said that "the other view that has been taken is that whether a person is a vice principal is to be determined solely by the magnitude or importance of the work that may have been committed to his charge; and that view is open to the objection that it furnishes no practical or certain test by which to determine in a given case whether an employee has been vested with such departmental control or has been 'so lifted up in the grade and extent of his duties' as to constitute him the personal representative of the master. That this would frequently be a difficult and embarrassing question to decide, and that courts would differ widely in their views, if the doctrine of departmental control was adopted, is well illustrated by the case of Borgman *v.* Railway Co., 41 Fed., 667, 669. We are of the opinion, therefore, that the nature and character of the respective duties devolved upon and performed by persons in the same common employment, should, in each instance, determine whether they are, or are not, fellow-servants, and that such relation should not be deemed to exist between two employees, when the function of one is to exercise supervision and control over some work undertaken by the master which requires supervision, and over subordinate servants engaged in that work, and where the other is not vested by the master with any such power of direction or management." 4 U. S. App., 574, 578; 2 C. C. A., 157; 51 Fed., 182.

The court thereupon affirmed the judgment.

It seems quite plain that Holverson was not the "chief" or "superintendent" of a separate and distinct department or branch of the busi-

is placed upon a company for the negligence of such an officer. We also think that the ground of liability laid down by the courts below is untenable.

The general rule is that those entering into the service of a common master become thereby engaged in a common service, and are fellow-servants; and, prima facie, the common master is not liable for the negligence of one of his servants which has resulted in an injury to a fellow-servant. There are, however, some duties which a master owes, as such, to a servant entering his employment. He owes the duty to provide such servant with a reasonably safe place to work in, having reference to the character of the employment in which the servant is engaged. He also owes the duty of providing reasonably safe tools, appliances, and machinery for the accomplishment of the work necessary to be done. He must exercise proper diligence in the employment of reasonably safe and competent men to perform their respective duties, and it has been held in many States that the master owes the further duty of adopting and promulgating safe and proper rules for the conduct of his business, including the government of the machinery, and the running of trains on a railroad track.

If the master be neglectful in any of these matters, it is a neglect of a duty which he personally owes to his employee, and, if the employee suffer damage on account thereof, the master is liable.

If, instead of personally performing these obligations, the master engages another to do them for him, he is liable for the neglect of that other, which in such case is not the neglect of a fellow-servant, no matter what his position as to other matters, but is the neglect of the master to do those things which it is the duty of the master to perform as such.

In addition to the liability of the master for his neglect to perform these duties, there has been laid upon him by some course a further liability for the negligence of one of his servants in charge of a separate department or branch of business, whereby another of his employees has been injured, even though the neglect was not of that character which the master owed, in his capacity as master, to the servant who was injured. In such case it has been held that the neglect of the superior officer or agent of the master was the neglect of the master, and was not that of the coemployee, and hence that the servant, who was a subordinate in the department of the officer, could recover against the common master for the injuries sustained by him under such circumstances. It has been already said that Holverson sustained no such relation to the company, in this case, as would uphold a liability for his acts based upon the ground that he was a superintendent of a separate and distinct branch or department of the master's business.

It is proper, therefore, to inquire what is meant to be included by the use of such a phrase.

A leading case on this subject in this court is that of Railway Co. v. Ross, 112 U. S., 377, 5 Sup. Ct., 184. In that case a railroad corporation was held responsible to a locomotive engineer in the employment of the company for damages received in a collision which was caused by the negligence of the conductor of the train drawn by the engine of which the plaintiff was engineer. This court held the action was maintainable, on the ground that the conductor, upon the occasion in question, was an agent of the corporation, clothed with the control [illegible] management of a distinct department, in which his duty was [illegible] that of direction and superintendence; that he had the [illegible] and management of the train, and that he occupied a [illegible] position from the brakemen, porters, and other [illegible] on it; and that he was in fact, and should be treated as, a personal representative of the corporation, for whose negligence the corporation was responsible to subordinate servants. The engineer was permitted to recover on that theory. These facts give some indication of the meaning of the phrase.

In the above case the instruction given by the court at the trial to which exception was taken was in these words: "It is very clear, I think, that if the company sees fit to place one of its employees under the control and direction of another, that then the two are not fellow-servants engaged in the same common employment, within the meaning of the rule of law of which I am speaking." That instruction, thus broadly given, was not, however, approved by this court in the Ross case. Such ground of liability—mere superiority in position, and the power to give orders to subordinates—was denied. What was approved in that case, and the foundation upon which the approval was given, is very clearly stated by Mr. Justice Brewer in the course of his opinion delivered in the case of Railroad Co. v. Baugh, 149 U. S., 368, 13 Sup. Ct., 914, at page 380, 149 U. S., and page 914, 13 Sup. Ct., and the following pages. In the Baugh case it is also made plain that the master's responsibility for the negligence of a servant is not founded upon the fact that the servant guilty of neglect had control over, and a superior position to that occupied by, the servant who was injured by his negligence. The rule is that, in order to form an exception to the general law of nonliability, the person whose neglect caused the injury must be "one who was clothed with the control and management of a distinct department, and not a mere separate piece of work in one of the branches of service in a department." This distinction is a plain one, and not subject to any great embarrassment in determining the fact in any particular case. When the business of the master or employer is of such great and diversified extent that it naturally and necessarily separates itself into departments of service, the individuals placed by the master in charge of these separate branches and departments of service, and given entire and absolute control therein, may properly be considered, with respect to employees under them, vice principals and representatives of the master, as fully and as completely as if the entire business of the master were placed by him under one superintendent. Thus, Mr. Justice Brewer in the Baugh case, illustrates the meaning of the phrase "different branches or departments of service" by suggesting that "between the law department of a railway corporation and the operating department there is a natural and distinct separation—one which makes the two departments like two independent kinds of business, in which the one employer and master is engaged. So, oftentimes, there is, in the affairs of such corporation what may be called a manufacturing or repair department, and another strictly operating department. These two departments are, in their relations to each other, as distinct and separate as though the work of each was carried on by a separate corporation. And from this natural separation flows the rule that he who is placed in charge of such separate branch of the service—who alone superintends and has the control of it—is, as to it, in the place of the master."

The subject is further elaborated in the case of Howard v. Railroad Co., 26 Fed., 837, in an opinion by Mr. Justice Brewer, then circuit judge of the eighth circuit. The other view is stated very distinctly in the cases of Borgman v. Railroad Co., 41 Fed., 667, and Woods v. Lindvall, 1 C. C. A., 37, 48 Fed., 62. This last case is much stronger for the plaintiff than the one at bar. The foreman in this case bore no resemblance, in the importance and scope of his authority, to that possessed by Murdock in the Woods case, supra. These cases which have been cited serve to illustrate what was in the minds of the courts when the various distinctions as to departments and separate branches of service were suggested. In the Baugh case the engineer and fireman of a locomotive engine running alone on the railroad, and without any train attached, were held to be fellow-servants of the company, so as to preclude the fireman from recovering from the company for injuries caused by the negligence of the engineer.

The meaning of the expression "departmental control" was again, and very lately, discussed in Railroad Co. v. Hambly, 154 U. S., 349, 14 Sup. Ct., 983, where it was held, as stated in the headnote, that a common day laborer, in the employ of a railroad company, who, while working for the company, under the orders and direction of a section boss or foreman, on a culvert on the line of the company's road, receives an injury through the neglect of a conductor and an engineer in moving a particular passenger train upon the company's road, is a fellow-servant of such engineer and of such conductor, in such a sense as exempts the railroad company from liability for the injury so inflicted.

The subject is again treated in Railroad Co. v. Keegan, 160 U. S., 259; 16 Sup. Ct. 269 (decided at this term), when the men engaged in the service of the railroad company were employed in uncoupling from the rear of trains cars which were to be sent elsewhere and in attaching other cars in their place; and they were held to be fellow-servants, although the force, consisting of five men, was under the orders of a boss who directed the men which cars to uncouple and what cars to couple, and the neglect was alleged to have been the neglect of the boss, by which the injury resulted to one of the men. This court held that they were fellow-servants, and the mere fact that one was under the orders of the other constituted no distinction, and that the general rule of nonliability applied.

These last cases exclude, by their facts and reasoning, the case of a section foreman from the position of a superintendent of a separate and distinct department. They also prove that mere superiority of position is no ground for liability.

This boss of a small gang of 10 or 15 men, engaged in making repairs upon the road, wherever they might be necessary, over a distance of three sections, aiding and assisting the regular gang of workmen upon each section as occasion demanded, was not such a superintendent of a separate department, nor was he in control of such a distinct branch of the work of the master, as would be necessary to render the master liable to a coemployee for his neglect. He was in fact, as well as in law, a fellow-workman. He went with the gang to the place of [illegible] work in the morning, stayed there with them during the day, [illegible] tended their work, giving directions in regard to it, and returned [illegible] with them in the evening, acting as a part of the crew of the [illegible] upon which they rode. The mere fact, if it be a fact, that he [illegible] actually handle a shovel or a pick, is an unimportant matter. [illegible] more than one man is engaged in doing any particular work, [illegible] almost a necessity that one should be boss, and the other [illegible] but both are, nevertheless, fellow-workmen.

If, in approaching the line of separation between a [illegible] and a superintendent of a particular and separate [illegible] may be embarrassment in determining the question [illegible] no such difficulty. It is clearly one of fellow-[illegible] [illegible] which the plaintiff has recovered in [illegible]

Holverson in not taking proper care at the time when he applied the brake to the front car. It was not a neglect of that character which would make the master responsible therefor, because it was not a neglect of a duty which the master owes, as master, to his servant, when he enters his employment.

The charge of the court to the jury in the matter complained of was erroneous, and the judgment must therefor be reversed, and the case remanded, with directions to grant a new trial.

---

EMPLOYERS' LIABILITY—FELLOW-SERVANTS—*Northern Pacific R. R. Co. v. Charless. 16 Supreme Court Reporter, page 848.*—This was a suit brought against the Northern Pacific Railroad Company by one Charless as plaintiff, to recover damages for injuries received while in the employ of said company. The plaintiff recovered a judgment and the case was carried on appeal to the United States circuit court of appeals for the ninth circuit, which sustained the judgment of the lower court. The case was then brought on writ of error before the United States Supreme Court, which rendered its decision April 13, 1896, and reversed the judgments of the courts below. The opinion of said court, delivered by Mr. Justice Peckham, gives a full statement of the facts in the case, and the following is quoted therefrom:

The plaintiff below was an ordinary day laborer, employed, under a section boss or foreman, to keep a certain portion of the roadbed of the defendant in repair. The foreman had power to employ and discharge men, and to superintend their work, and was himself a workman. He employed the plaintiff, who, with the rest of the men employed in the gang—some four, five, or six—was carried to and from his work, daily, on a hand car worked by the men themselves.

In August, 1886, on the 28th of the month, an accident occurred as the men were on their way to their work. They were using a hand car with what is alleged to have been a defective brake. The foreman had complained of it to the yardmaster a short time before, who had promised a better one. In the meantime, and as a temporary makeshift, the foreman had provided the car with a brake which consisted of a bit of wood, 4 by 4, fastened on the side of the car with a bolt; and the long arm acted as a lever, and pressed the shorter portion of the timber against the wheel. In that way the car had been run for a day or two before the morning of the accident. On that day the plaintiff, with the rest of the men in the gang, and the foreman, started on the hand car to go over a certain portion of the section to inspect the condition of the road. They were running the car very rapidly, under the direction and supervision of the foreman, and had arrived at a narrow cut in the road, around a curve, when they were suddenly confronted with a freight train coming through the cut in the opposite direction. There had been no warning or signal of any kind given by any of the employees on the freight train of its approach, and the plaintiff below knew nothing of the fact that any freight train was expected. Efforts were made to stop the hand car, and, as the speed did not seem to be slackened in time, plaintiff became frightened, and undertook to jump from the front end of the car, when he stumbled over some tools that were on the car and fell between the rails in front of it. As the hand car approached him he put his foot up against it in order to prevent its running over him; but the impetus of the car was too great, and it ran over and doubled him up and wrenched his spine, causing him great internal injuries. The other hands jumped off the car, removed it from the track, and took the plaintiff out of danger, before the freight train passed by.

The injuries of the plaintiff were of a very serious nature, and his legs became paralyzed and he was rendered a cripple for life. He commenced this action against the defendant below to recover damages on account of the negligence of the agents and servants of the defendant. The negligence claimed consisted in—

(1) The defective brake on the car, which it is alleged was an appliance for the prosecution of the work on the defendant's road, and necessary to be used to enable the employees to perform their duties, and that, as such appliance, it was the duty of the defendant to see that it was reasonably safe and fit for the purpose intended.

(2) The negligence of the foreman in charge of the gang, who directed the speed of the hand car, and ran it at a hazardous rate of speed when he knew that a train coming toward him was expected, while the other members of the gang were ignorant of that fact.

(3) The negligence of the train hands on the approaching train, in giving no signals of their approach around the curve and through the cut, although they were near a public crossing, and some signals were necessary on that account.

Upon the trial evidence was given tending to prove the above facts, and among other things the judge charged the jury as follows:

"I think that the case, when stripped of all the side issues and the incidental questions surrounding it, resolves itself into just this question for this jury to determine: Whether the injury to the plaintiff resulted directly from the negligence of the defendant in needlessly exposing him to the danger of being hurt by a collision between the hand car and the extra freight train at the place where it occurred, or whether the injury was a mere accident, which was the result of one of the ordinary hazards of the employment in which he was engaged; whether it was an ordinary risk of his employment, or whether an extraordinary danger caused by the negligence on the part of the defendant; whether that negligence was a negligence of the foreman in running the hand car too fast up to a point which he knew to be dangerous, and which he did not warn the other men working on the hand car of, so that it was impossible for them, without extreme hazard to their lives, to avoid a collision; or whether the negligence was on the part of the officers in charge of the freight train, in approaching a curve in the cut, which obstructed the train from view, or passing a public crossing, without giving warning by sounding the whistle or engine bell. If, in any of these respects, there was actual neglect on the part of defendant which placed the plaintiff in a situation of extraordinary danger—something clear beyond the ordinary risks of his employment—and his injury was not in any degree owing to his own negligence at the time, the defendant would be liable to damages."

The defendant below excepted to each of the above propositions [illegible] laid down by the learned judge in his charge, and the jury [illegible] verdict in favor of the plaintiff, which was affirmed by the [illegible] of appeals for the ninth circuit (2 C. C. A., 380, 51 Fed., [illegible] defendant below sued out a writ of error from this court [illegible] judgment.

Many of the facts surrounding the happening of this accident are similar in their nature to those existing in the case of Railroad Co. v. Peterson (just decided), 16 Sup. Ct., 843. The employment of the plaintiff below, the nature of the work, and the powers of the section boss under whom he worked, are substantially the same as those existing in the other case. We may refer to the general principles of the law of master and servant applicable to these facts which are set forth in the opinion of this court in that case, and which we think govern the case at bar, upon those facts.

In regard to the particular allegations of negligence above set forth, it is not necessary, in the view we take of this case, to express any opinion whether the alleged defect in the brake on the hand car rendered it a defective appliance, within the meaning of the law, rendering the master liable for a failure to provide a reasonably safe and proper appliance for the work to be done by his employees.

There were two other propositions submitted to the jury by the learned judge, each of which was, as we think, of a material nature and also clearly erroneous.

1. We think it was error to submit to the jury the question of the negligence of the employees on the extra freight train in failing to give the signals of its approach. This failure, assuming that it constituted negligence, was nothing more than the negligence of coservants of the plaintiff below in performing the duty devolving upon them. The principle which covers the facts of this case was laid down in Randall v. Railroad Co., 109 U. S., 478, 3 Sup. Ct., 322, and that case has never been overruled or questioned. Among the latest expressions of opinion of this court in regard to views similar to those stated in the case in 109 U. S. and 3 Sup. Ct., supra, is the case of Railroad Co. v. Hambly, 154 U. S., 349, 14 Sup. Ct., 983. It seems to us that the Randall and Hambly cases are conclusive, and necessitate a reversal of this judgment. In the Hambly case it was held that a common day laborer in the employ of a railroad company who, while working for the company, under the orders and direction of a section boss or foreman, on a culvert on the line of the company's road, received an injury through the negligence of a conductor and of an engineer in moving a particular passenger train upon the company's road, was a fellow-servant with such engineer and with such conductor, in such a sense as exempts the railroad company from liability for the injury so inflicted. We are unable to distinguish any difference in principle arising from the facts in these two cases.

The question of the negligence of the hands upon the extra freight train should not have been submitted to the jury as constituting any right to a recovery against the corporation on the ground of such negligence.

2. We also regard it as erroneous to have submitted to the jury the general question whether Kirk, the section foreman, was negligent in running his hand car at too high a speed just prior to the accident. Kirk and the plaintiff below were coemployees of the company, and the neglect of Kirk, if it existed, in driving his hand car too fast (assuming it was in proper condition), was not such negagence as would render the company responsible to Kirk's coemployee. It was not the neglect of any duty which the company, as master, was bound itself to perform. This we have held in the Peterson case, and for the reasons there stated. While it may be assumed that the master would have been liable if a defective brake had been the cause of the accident, yet the defendant below is, under the charge of the judge, permitted to be

made liable by proof of the speed of the hand car, if the jury found that Kirk, the foreman, knew it to be dangerous, and that the accident happened because of that speed, even though it would have happened if the brake had been the regular kind, and in good order. The language of the court does not separate the question of general negligence in running a hand car which was in good order, too fast, from that which might be negligence with reference to running a hand car with a defective brake at the same rate of speed. For using in a negligent manner a defective appliance furnished by the master the latter might be liable if a coemployee were thereby and in consequence thereof injured. As the master furnished the defective appliance it would be no answer to say that it was negligently used. But, on the other hand, the master would not be responsible for the negligent use of a proper appliance. From the language used by the court, the company might have been held liable if Kirk were running the hand car at a dangerous rate of speed, although the jury found the brake actually used to have been sufficient. A dangerous rate of speed was therefore held to be negligence. That neglect, we hold, the company was not responsible for.

Upon the other question of the negligence of the employees on the freight train, the error in the charge is not rendered harmless by any explanation given by the learned judge. The difficulty remains uncured. The jury might have found from the evidence that this hand car, while going at the rate of speed stated, could have been stopped with the extemporized brake in time to prevent any danger of a collision, in case the proper signals had been given by the hands on the freight train, but that the accident resulted from their failure to give those signals, and that such failure was negligence on their part. The verdict may have been based upon such negligence. We hold the company was not liable for the negligence of the hands on the freight train in failing to give proper signals.

The judgment entered upon the verdict of the jury must be reversed and the cause remanded, with instructions to grant a new trial.

---

SEAMEN—EXTRA WAGES—*The Potomac—Niagara Falls Paper Co. v. Crouckett et al.* 72 *Federal Reporter, page* 535.—This case was a libel, brought in the district court of the United States for the southern district of New York, by James Crouckett and James Hanley, against the barge Potomac (Niagara Falls Paper Company, claimant), to recover extra wages. The district court made a decree in favor of libelants, and the claimant appealed to the United States circuit court of appeals, second circuit. Said court rendered its decision February 18, 1896, and reversed the decree of the lower court. Its opinion, delivered by Circuit Judge Shipman, and containing a statement of the facts in the case, is given below:

The libelants shipped, in September, 1894, on board the barge [illegible] *mac*, one as mate and the other as seaman, and each upon [illegible] month. The barge left Buffalo in September, bound for [illegible] in Canada. On her return trip, she was laden with [illegible] on deck, consigned to Tonawanda, N. Y., and left [illegible] the morning of September 23, in tow of the tug [illegible] morning she encountered a violent gale, and after passing Cove Island light the towline parted, the barge drifted, shipped heavy seas, became waterlogged, lost part of her deck load, dropped anchor in the night near Flower Pot Island, and stayed there till morning, when the tug came and towed her to a small harbor in Canada called "Tubmerry," between 1 and 2 miles from the larger Tubmerry port. The vessel was tied up near the lighthouse, where there was a hamlet of 8 families, containing about 75 people. In order to free the barge from water, it was necessary to remove the lumber from the deck, put on steam pumps, box them in, and afterwards reload the cargo. The captain hired men from the shore to assist in this work, but the sailors exacted extra compensation before they would touch the cargo for the purpose of unloading, and demanded and received from the captain a promise to pay extra wages of 30 cents per hour. The barge was placed in proper condition, and was towed to Tonawanda. The extra compensation of each of the libelants amounted to $10.50. The owners paid the extra amount to all the sailors except the two libelants. There was no apparent reason for this discrimination. To recover the extra wages this libel was brought.

The district judge in deciding in favor of the libelants was undoubtedly influenced by the seeming unfairness of the claimants in paying a part only of the men in accordance with the promise of the captain. He furthermore says:

"If I thought that a decree for the libelants involved a departure from the old and salutary rule that seamen must not expect extra compensation for services rendered in their capacity as seamen, no matter how arduous or meritorious they may be, I should dismiss the libel. It would lead to gross insubordination, and increase the difficulties and dangers of navigation immeasurably if the court should sanction the idea that a seaman may refuse to obey the master's order on the ground that the work he is directed to perform is 'extra' and entitled him to additional compensation."

He thought that the facts took the case out of the general rule, because the *Potomac* was in port at the time in question, and says:

"The work was partly on the vessel and partly on shore and consisted in unloading and reloading a part of her cargo."

No question is made as to the general rule which the district judge stated, or that seamen are bound, without extra compensation, to render extra labor and services to save the vessel and cargo in case of wreck or impending calamity, and that a contract for extra pay "made when the ship is in distress, or obtained by any unfair practices or advantage taken by the seamen, is wholly void." (Curt. Merch. Seam., 28.) In this case the barge had become disabled, and was taken to a harbor of refuge, so as to be enabled to prosecute her voyage. She was compelled by stress of weather to stop at Tubmerry, in order to gain ability to go to her place of destination. We think that the district judge was

reasonable means of knowledge, by purchase, contract or otherwise, directly or indirectly, cause or permit any garments, or such other articles as aforesaid, to be manufactured or made up, in whole or in part, or any work to be done thereupon within this State, and in place or under circumstances involving danger to the public health, the said individual or corporation, upon conviction in any court of competent jurisdiction, shall be fined not less than ten dollars or more than one hundred dollars for each garment manufactured, made up or worked upon.

SEC. 3. This act shall take effect from the date of its passage.

Approved April 4, 1896.

## MASSACHUSETTS.

### ACTS OF 1896.

CHAPTER 241.—*Weekly payment of wages.*

SECTION 1. No person or partnership engaged in this Commonwealth in manufacturing business and having more than twenty-five employees shall, by a special contract with persons in his or its employ or by any other means, exempt himself or itself from the provisions of chapter four hundred and thirty-eight of the acts of the year eighteen hundred and ninety-five relative to the weekly payment of wages.

SEC. 2. Whoever violates the provisions of this act shall be punished by a fine not exceeding fifty dollars and not less than ten dollars.

SEC. 3. This act shall take effect upon its passage.

Approved April 6, 1896.

CHAPTER 334.—*Weekly payment of wages.*

SECTION 1. Section one of chapter four hundred and thirty-eight of the acts of the year eighteen hundred and ninety-five is hereby amended by inserting after the word "to," in the fourth line, the words:—all contractors and to,—also by inserting after the word "such," in the eighth line, the word:—contractors,—so that the section as amended will read as follows:—*Section 1.* Sections fifty-one to fifty-four, inclusive, of chapter five hundred and eight of the acts of the year eighteen hundred and ninety-four, relative to the weekly payment of wages by corporations, shall apply to all contractors and to any person or partnership engaged in this Commonwealth in any manufacturing business and having more than twenty-five employees. And the word "corporation," as used in said sections, shall include such contractors, persons and partnerships.

SEC. 2. This act shall take effect upon its passage.

Approved April 28, 1896.

CHAPTER 343.—*Traversing machinery in cotton factories.*

SECTION 1. No traversing carriage of any self-acting mule in any cotton factory shall be allowed to travel within twelve inches of any pillar, column, pier or fixed structure: *Provided*, That this section shall only apply to factories erected after the passage of this act.

SEC. 2. If the provisions of this act are violated in any such cotton factory the owner of such factory shall be punished by a fine of not less than twenty dollars nor more than fifty dollars for each offense.

Approved April 28, 1896.

CHAPTER 444.—*Suits for wages.*

In actions of contract for the recovery of money due for manual labor two or more persons may join in one action against the same defendant or defendants when the claim of no one of such persons exceeds the sum of twenty dollars, although the claims of such persons are not joint; and each of such persons so joining may recover the sum found to be due to him personally. The claim of each person so joining shall be stated in a separate count in the declaration, and the court may make such order for the trial of issues as shall be found most convenient and may enter separate judgments and issue one or more executions, and may make such order concerning costs as in its opinion justice may require.

Approved May 28, 1896.

CHAPTER 449.—*Employment of laborers by cities.*

SECTION 1. So much of chapter three hundred and twenty of the acts of the year eighteen hundred and eighty-four and the amendments thereto as relates to the employment of laborers by cities, and that portion of the civil service rules of the Commonwealth and the cities thereof as authorized by said acts and designated

[illegible]-one of the acts of the year eighteen hundred and ninety-five, shall be construed as a continuation of that chapter and not as new enactments.

SEC. [illegible]. This act shall take effect upon its passage.

Approved June [illegible], 1896.

## MISSISSIPPI.

### ACTS OF 1896.

CHAPTER 86.—*Liability of employers for death of employees.*

SECTION 1. Section 663 of the Annotated Code of 1892 [shall] be so amended as to read as follows: Whenever the death of any person shall be caused by any real, wrongful or negligent act or omission, or by such unsafe machinery, ways or appliances, as would, if death had not ensued, have entitled the party injured or damaged thereby to maintain an action and recover damages in respect thereof, and such deceased person shall have left a widow or children, or both, or husband or father or mother or sister or brother, the person or corporation, or both, that would have been liable if death had not ensued, and the representative of such person, shall be liable for damages, notwithstanding the death, and the fact that death is instantaneous shall, in no case, affect the right of recovery. The action for such damages may be brought in the name of the widow for the death of the husband, or by the husband for the death of his wife, or by a parent for the death of a child, or in the name of a child for the death of a parent, or by a brother for the death of a sister, or by a sister for the death of a brother, or by a sister for the death of a sister, or a brother for the death of a brother, or all parties interested may join in the suit, and there shall be but one suit for the same death, which suit shall inure to the benefit of all parties concerned, but the determination of such suit shall not bar another action unless it be decided upon its merits. In such action the party or parties suing shall recover such damages as the jury may assess, taking into consideration all damages of every kind to the decedent and all damages of every kind to any and all parties interested in the suit. Executors or administrators shall not sue for damages for injuries causing death except as below provided; but every such action shall be commenced within one year after the death of such deceased person.

SEC. 2. This act shall apply to all personal injuries of servants or employees received in the service or business of the master or employer, where such injuries result in death.

SEC. 3. Damages recovered under the provisions of this act shall not be subject to the payment of the debts or liabilities of the deceased, and such damages shall be distributed as follows: Damages for the injury and death of a married man shall be equally distributed to his wife and children, and if he has no children all shall go to his wife; damages for the injury and death of a married woman shall be equally distributed to her husband and children, and if she has no children all shall go to her husband; if the deceased has no husband nor wife, the damages shall be distributed equally to the children; if the deceased has no husband nor wife nor children the damages shall be distributed equally to the father, mother, brothers and sisters or to such of them as the deceased may have living at his or her death. If the deceased leave neither husband or wife or children or father or mother or sister or brother, then the damages shall go to the legal representative subject to debts and general distribution, and the executor may sue for and recover such damages on the same terms as are prescribed for recovery by the next of kin in section 1 of this act, and the fact that deceased was instantly killed shall not affect the right of the legal representative to recover.

SEC. 4. This act shall take effect immediately.

Approved March 23, 1896.

CHAPTER 87.—*Liability of corporations for injuries of employees.*

SECTION 1. Section 3559 of the Annotated Code of 1892 [shall] be amended so that the same shall read as follows, to wit: Every employee of any corporation shall have the same rights and remedies for an injury suffered by him from the act or omission of the corporation or its employees, as are allowed by [to?] other persons not employees, where the injury results from the negligence of a superior agent or officer, or of a person having the right to control or direct the services of the party injured, and also when the injury results from the negligence of a fellow-servant engaged in another department of labor from that of the party injured, or of a fellow-servant on another train of cars, or one engaged about a different piece of work. Knowledge [illegible] employee injured of the defective or unsafe character or condition of any

SEC. 2. Each such room or apartment used for the purposes aforesaid, shall be regarder [regarded] as a workshop or factory, and shall be separate from and have no door, window or other opening into any living or sleeping room of any tenement or dwelling, and no such workshop or factory shall be used at any time for living or sleeping purposes, and shall contain no bed, bedding, cooking or other utensils, excepting what is required to carry on the work therein; and every such shop or factory shall have an entrance from the outside direct, and if above the first floor shall have a separate and distinct stairway leading thereto, and every such workshop or factory shall be well and sufficiently lighted, heated and ventilated by ordinary, or, if necessary, by mechanical appliance, and shall provide for each person employed therein, no less than 250 cubic feet of air space in day time, and 400 cubic feet at night, and shall have suitable closet arrangements for each sex employed therein, as follows: Where there are ten or more persons, and three or more to the number of twenty, are of either sex, a separate and distinct water-closet, either inside the building, with adequate plumbing connections, or on the outside at least twenty feet from the building, shall be provided for each sex; when the number employed is more than twenty-five of either sex, there shall be provided an additional water-closet for such sex up to the number of fifty persons, and above that number in the same ratio, and all such closets shall be kept strictly and exclusively for the use of the employees and employer or employers of such workshop or factory; *Provided*, That where more than one room is used under the direction of one employer, all such rooms are to be regarded as one shop, or factory, and every such workshop or factory shall be kept in a clean and wholesome condition, all stairways and the premises within a radius of thirty feet shall be kept clean, and closets shall be regularly disinfected and supplied with disinfectants, and the inspector of factories or his assistants may require all necessary changes, or any process of cleaning, painting or whitewashing which he may deem essential to assure absolute freedom from obnoxious odor, filth, vermin, decaying matter or any condition liable to impair health or breed infectious or contagious diseases; he shall prevent the operation of such shops and factories that do not conform to the provisions of this act, and cause the arrest and prosecution of the person or persons operating the same.

SEC. 3. No person, for himself or for any other person, firm or corporation, shall give out work to or contract with any other person to perform such work necessary to make such goods mentioned in section one, after having received notice from the inspector of factories or his assistants, that said latter person has not complied with the provisions of section two of this act which notice shall remain in force, until said person has complied with this law, of which notice must be given to the employer by the inspector of factories or his assistants.

SEC. 4. Every such person, firm or corporation heretofore mentioned shall obtain and keep a record of all persons to whom work is given out or contracted for, including their names and addresses which record shall be opened to inspection of the State inspector of workshops and factories, when called for.

SEC. 5. No person, firm or corporation shall receive, handle, or convey to others, or sell, hold in stock or expose for sale any goods mentioned in section one, unless made under the sanitary conditions provided for and prescribed in this act; but this act shall not include the making of garments or other goods by any person for another by personal order, and when received for wear or use direct from the maker's hands, and all violations of the provisions of this act shall be prosecuted by the inspector with the advice and consent of the chief inspector of workshops and factories.

SEC. 6. Any person, firm or corporation who shall violate any of the provisions of this act shall, upon conviction thereof, be fined in any sum not less than fifty dollars nor more than one hundred dollars for each offense, or imprisoned not less than thirty nor more than sixty days or both, at the option of the court, such fine to be collected by the court in which conviction is had and turned over to the chief inspector of workshops and factories, and by him to be paid into the State treasury to be credited to the general revenue fund; and in all prosecutions brought by or under the direction of the inspector of workshops and factories for the violation of this act, he shall not be held to give security for costs, or adjudged to pay any costs, but in all cases where the accused be acquitted or is found to be indigent, the costs shall be paid out of the county treasury of the county in which proceedings are brought, the same as the costs in all other cases of misdemeanor.

SEC. 7. This act shall take effect and be in force on and after its passage.

Passed April 27, 1896.

PAGE 324.—*State board of arbitration, etc.*

SECTION 1. Sections 4, 13, and 14 of said above entitled act [act passed March 14, 1893, page 83, acts of 1893] shall be amended so as to read as follows:

SEC. 4. Whenever any controversy or difference not involving questions which may be the subject of a suit or action in any court of the State exists between an

Code has been enforced by legal proceedings in some State other than the State of Virginia, in such manner as to deprive such person to any extent of the benefit of such exemption, shall be prima facie evidence that any resident of this State, who may at any time have been owner or holder of said claim or debt, has violated this law.

SEC. 4. This act shall be in force from the date of its passage.

Approved February 11, 1896.

CHAPTER 351.—*Protection of debts due laborers on buildings, etc.*

SECTION 1. No assignment or transfer of any debt, or any part thereof, due or to become due to a general contractor by the owner for the construction, erection or repairing of any building, structure, or railroad for such owner, shall be valid or enforceable in any court of law or equity by any legal process, or in any other manner, by the assignee of any such debt, unless and until the claims of all subcontractors, supply men and laborers against such general contractor for labor performed and materials furnished in and about the construction, erection and repairing of such building, structure or railroad shall have been satisfied; *Provided*, That if such subcontractors, supply men and laborers shall give their assent in writing to such assignment, it shall be thereby made valid as to them, but the payment or appropriation of such assignment by the owner without such assent in writing shall not protect such owner from the demands of such subcontractors, supply men and laborers to the extent of such assignment.

SEC. 2. No debt or demand, or any part thereof, due or to become due by the owner of any building, structure or railroad to a general contractor for the construction, erection or repairing of such building, structure or railroad, shall be subject to the payment of any debt or the lien of any judgment, writ of fieri facias or any garnishee proceeding obtained or sued out upon any debt due such general contractor which shall have been contracted in any other manner or for any other purpose than in the construction, erection or repairing of such building, structure or railroad for such owner unless and until the claims due by such general contractor to all subcontractors, supply men and laborers for materials furnished and labor performed in and about the construction, erection or repairing of such building, structure or railroad shall have been paid.

SEC. 3. All acts and parts of acts in conflict with this act are hereby repealed.

SEC. 4. This act shall be in force from its passage.

Approved February 17, 1896.

New York City.—November 14, 1896. Contract with D. H. Hayes, Chicago, Ill., for stone and brick work, roof covering, interior finish, etc., of appraiser's warehouse, $322,500. Work to be completed within twelve months.

Richmond, Ky.—November 25, 1896. Contract with the Pittsburg Heating Supply Company, Pittsburg, Pa., for low-pressure, return-circulation, steam heating, and ventilating apparatus for post office, $3,465. Work to be completed within sixty days.

Saginaw, Mich.—November 25, 1896. Contract with Charles W. Gindele, Chicago, Ill., for erection and completion of post office, except heating apparatus, $70,900. Work to be completed within twelve months.

Newburg, N. Y.—November 25, 1896. Contract with Sproul & McGurrin, Grand Rapids, Mich., for low-pressure, return-circulation, steam heating, and ventilating apparatus for post office, $2,344. Work to be completed within sixty working days.

Lightning Source UK Ltd.
Milton Keynes UK
UKHW011833031218
333382UK00008B/544/P